Matter & Interactions I

Modern Mechanics

Ruth W. Chabay

Bruce A. Sherwood

North Carolina State University

John Wiley & Sons, Inc.

Hoboken • Chichester • Brisbane • Toronto • Singapore

This book was set by the authors.

To order books or for customer service, please call 1(800)-CALL-WILEY (225-5945).

Printed in the United States of America.

ISBN-13 978-0-471-79382-3
ISBN-10 0-471-79382-5

10 9 8 7 6 5 4 3 2 1

Preface to Volume I

Matter & Interactions I: Modern Mechanics focuses on the atomic structure of matter and the interactions that matter undergoes. This two-volume textbook emphasizes the fact that there are only a small number of fundamental principles that underlie the behavior of matter, and that using these powerful principles it is possible to construct models that can explain and predict a wide variety of physical phenomena.

Prerequisites

This book is intended for introductory college physics courses taken by science and engineering students. You will need a basic knowledge of derivatives and integrals, which can be obtained by studying calculus concurrently. A brief review of derivatives is available in an appendix at the end of this volume.

Modeling

This course places a major emphasis on constructing and using physical "models." A central aspect of science is the modeling of complex real-world phenomena. A physical model is based on what we believe to be fundamental principles; its intent is to predict or explain the most important aspects of an actual situation. Modeling necessarily involves making approximations and simplifying assumptions in order that the model can be analyzed in detail. The principles of classical mechanics and thermal physics that are the focus of Volume I have wide applicability; we can use these principles to model systems as different as molecules and galaxies.

Computer modeling

Computer modeling is important in this course, because it makes it possible for you to analyze complex systems which would otherwise require very sophisticated mathematics or which could not be analyzed at all without a computer. Numerical calculations based on the Momentum Principle give us the opportunity to watch the dynamical evolution of the behavior of a system. Simple models frequently need to be refined and extended; this can be done straightforwardly with a computer model but is often impossible with a purely analytical model.

Computer modeling has become as important as theory and experiment in contemporary science and engineering. We introduce you to serious computer modeling right away to help you build a strong foundation in the use of this important tool.

Experiments using simple equipment

Some end-of-chapter problems involve experiments using simple equipment such as weights, string, coffee filters, a stopwatch, a weak spring, a styrofoam cup, and a toy gyroscope. You will find that by experimenting with very simple equipment you can gain insight into rather deep scientific issues.

? Stop and Think

As you read the text, you will frequently come to a paragraph that asks you to stop and think by making a prediction, carrying out a step in a derivation or analysis, or applying a principle. Usually these questions are answered in the following paragraphs, but it is important that you make a serious effort to answer the questions on your own before reading further. Be honest in comparing your answers to those in the text; paying attention to surprising or counterintuitive results can be a useful learning strategy.

Exercises

Small exercises that require you to apply new concepts are found at the end of many sections of the text. These may involve qualitative reasoning or simple calculations. You should work these exercises when you come to them, to consolidate your understanding of the material you have just read. Answers to the exercises are at the end of each chapter.

Conventions used in diagrams and equations

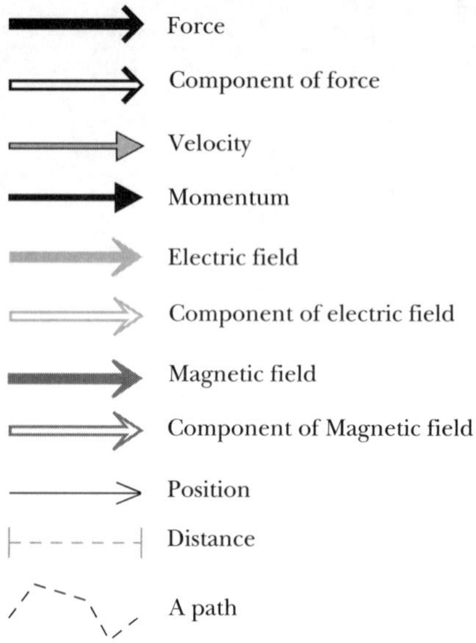

The conventions most commonly used to represent vectors and scalars in diagrams in this text are shown in the adjacent figure. In equations and text, a vector will be written with an arrow above it:

$$\vec{p}$$

The magnitude of $\vec{p}$ is written as $|\vec{p}|$ or simply as p.
The x component of $\vec{p}$ is written as p_x.
A vector may be written in terms of components or in terms of unit vectors along the x, y, and z axes:

$$\vec{p} = \langle p_x, p_y, p_z \rangle = p_x\hat{\imath} + p_y\hat{\jmath} + p_z\hat{k}$$

To the instructor

The approach to mechanics and thermal physics taken in this volume differs significantly from that in most introductory physics texts. Key emphases of the approach are these:

- Starting from fundamental principles rather than secondary formulas
- Atomic-level description and analysis
- Modeling the real world through idealizations and approximations
- Computer modeling of physical systems
- Qualitative reasoning that complements the quantitative reasoning
- Unification of mechanics and thermal physics
- Tight integration of theory and experiment

The following one-semester introductory course on electricity and magnetism is addressed by *Matter & Interactions II: Electric & Magnetic Interactions*, Ruth Chabay & Bruce Sherwood, John Wiley & Sons.

Textbook web site

For a wealth of useful information, go to the Wiley web site:

http://www.wiley.com/college/chabay

There you will find a link to the authors' public web site, which includes information of a general character, including useful educational software, and information on how to obtain access to a web site run by adopters, containing extensive resources for instructors, such as a daily log of what we do in own classrooms, sample course web sites, problem solutions, sample exams, etc. At the start of the daily log is procurement information for the simple experiments used with this volume.

Articles dealing with Matter & Interactions

Three key articles dealing with this curriculum are the following:

Chabay, R. W. & Sherwood, B. A. (1999). Bringing atoms into first-year physics. *American Journal of Physics* **67**, 1045-1050.

Chabay, R. W. & Sherwood, B. A. (2004). Modern mechanics. *American Journal of Physics* **72**, 439-445.

Chabay, R. W. & Sherwood, B. A. (2005). Restructuring Introductory E&M. *American Journal of Physics*, to appear Spring 2005.

Computer homework problems

Some critically important homework problems are designed for the student to write a computer program. We strongly recommend VPython as the best tool for student use. An adequate subset of the underlying Python programming language can be taught in an hour or two, even to students who have never written a program before. Real-time 3D animations are generated as a side effect of student computations, and these animations provide powerfully motivating and instructive visualizations of fields and motions. VPython supports true vector computations, which encourages students to begin thinking about vectors as powerful tools rather than barriers to understanding. VPython can be obtained at no cost for Windows, Macintosh, and Linux at http://vpython.org.

Homework problems

Given the emphasis on physical modeling, it is appropriate for homework to consist of a small number of rather large problems, rather than a large number of small, routine problems. In addition to the large homework problems provided at the end of a chapter, there are many small-scale problems in the form of exercises distributed throughout the chapter, and small review questions at the end of the chapter. You may wish to ask students to write out some exercises or review questions as part of the homework, or to have them work out exercises at intervals during a lecture. We have prepared an extensive and sophisticated set of questions for the WebAssign computer homework system.

Choosing topics

In our own teaching at North Carolina State University, with three 50-minute lectures and one 110-minute small-group studio lab every week, we are able to complete most but not all of this modern mechanics volume in a 15-week semester up to the middle of Chapter 10 (not including the Boltzmann distribution). You can read the organizational details in the daily log available in the instructor resources mentioned above.

What can be omitted if there is not enough time to do everything? Let us say that the one thing we feel should *not* be omitted is the introduction to entropy in terms of quantum statistical mechanics (Chapter 10). This is a climax of the theme of integrating mechanics and thermal physics. One way to decide what can be omitted is to be guided by what needs to be done to ensure dealing adequately with entropy.

Any starred section (*) can safely be omitted. Material in these sections is not referenced in later work. Next we offer chapter by chapter discussions of what other topics might be omitted, and what the ramifications are of such omissions.

Chapter 1 (Interactions and Motion): Unless your students have a particularly strong background in vectors, all of this chapter is important for later work (except for the starred section on relativity).

Chapter 2 (The Momentum Principle 1: Impulse and Momentum Change): You can omit the section on determinism, as we do. The starred sections are reference material for those interested but not essential. Note the examples that engage students right away with concepts of modeling: making approximations, simplifying assumptions, and estimates. It is important to talk about these concepts, because they usually are unfamiliar even to students with strong physics backgrounds.

Chapter 3 (The Momentum Principle 2: Instantaneous Rate of Change): We omit buoyancy and pressure (you could return to these topics during Chapter 11 on gases).

Chapter 4 (The Energy Principle): This is the foundation for the energy concept, and nothing should be omitted (other than starred sections, of course). The starred section on "a puzzle" is an amusing teaser that hints at field energy. The starred derivations at the end of the chapter are probably of interest to few students but are provided for completeness.

Chapter 5 (Energy in Macroscopic Systems): If you are pressed for time, you might choose to omit the second half of the chapter on energy dissipation, beginning with Section 5.9. We ourselves are able to mention dissipative processes only briefly.

Chapter 6 (Energy Quantization): None of the unstarred sections of this short chapter should be omitted.

Chapter 7 (Multiparticle Systems): The starred sections on modeling friction are not mathematically difficult but are conceptually challenging. Nothing else in this chapter is a candidate for omission.

Chapter 8 (Collisions: Exploring the Nucleus): A good candidate for omission is the analysis of collisions in the center-of-mass frame. Note that in this revision there is a basic introduction to collisions in Chapter 2.

Chapter 9 (Angular Momentum): The starred sections on gyroscopes are interesting but students find them difficult, presumably because the analysis includes both a nonzero torque and three dimensions. We typically show the students a bicycle wheel precessing at the end of a rope and sketch briefly the analysis, but we do not have time to go into detail.

Chapter 10 (Entropy: Limits on the Possible): We strongly recommend doing at least the first half of this chapter, on the Einstein solid. To make this work well, exposition and student computing should be interwoven. It might seem odd to ask the students simply to reproduce the results shown in the textbook, but in our experience and that of our students it is enormously educational to be required to make the ideas concrete in a computer program. You find yourself asking "What exactly is q1?" and answering very concretely by actually calculating with the initially unfamiliar quantities. It is the difference between writing an essay about entropy and calculating the actual value of the entropy.

Despite its intrinsic importance, you might omit the second half of the chapter on the Boltzmann distribution, as we do (for lack of time). However, if you do Chapter 11 you will need a few key results from this analysis.

Chapter 11 (The Kinetic Theory of Gases): One should not omit the initial sections, including the relationship of pressure to the results from Chapter 10. We should point out that the important formula for particle flow ($nA\bar{v}$) and the concept of mean free path are both used in Volume II (though they are rederived there). You might decide to omit the sections on macroscopic energy transfers (isothermal and adiabatic processes).

Chapter 12 (The Efficiency of Engines): This entire chapter is another application of the second law of thermodynamics and may be omitted. Another possibility is to omit just the sections on nonzero-power engines.

Acknowledgments for Volume I

We owe much to the unusual working environment provided by the Center for Innovation in Learning and the Department of Physics at Carnegie Mellon, which made it possible during the 1990's to carry out the research and development leading to this two-volume textbook. We are grateful for the open-minded attitude of our colleagues in the Carnegie Mellon physics department toward curriculum innovations.

We are grateful to the support of our colleagues Robert Beichner and John Risley in the Physics Education Research and Development group at North Carolina State University, and to other colleagues in the NCSU physics department.

For the writing of Chapter 10 on quantum statistical mechanics we gratefully acknowledge the inspiration provided in a splendid article by Thomas A. Moore and Daniel V. Schroeder, "A different approach to introducing statistical mechanics," *American Journal of Physics* vol. 65, pp. 26-36 (January 1997). We thank Robert Bauman, Gregg Franklin, and Curtis Meyer for helping us think deeply about energy. Vidhya Ramachandran contributed a useful take-home experiment on heat capacity.

We thank David Andersen and David Scherer for the development of tools that enabled us and our students to write associated software. We also thank Stacey Benson, Krishna Chowdary, Thomas Foster, Barry Luokkala, Matthew Kohlmyer, Robert Swendsen, Hugh Young, Chris Gould, Prabha Ramakrishnan, Joe Heafner, Aaron Titus, and Mark Haugan.

This project was supported, in part, by the National Science Foundation (grants MDR-8953367, USE-9156105, DUE-9554843, DUE-9972420, DUE-0320608, and DUE-0237132). Opinions expressed are those of the authors, and not necessarily those of the Foundation.

Ruth W. Chabay
Bruce A. Sherwood
Department of Physics
North Carolina State University
November 2005

Image Credits

Chapter 1: Randall Feenstra (Figure 1.3). Chapter 3: Randall Feenstra (Figure 3.2). Chapter 5: Stacey Benson (Figure 5.23 and Figure 5.24). Judith Harrison (Figure 5.35).

All other figures were created by the authors, using Adobe Illustrator, Adobe Photoshop, VPython, POV-Ray, and Specular Infini-D.

The text was produced by the authors as camera-ready copy, using Adobe FrameMaker. The font is 10-point New Baskerville.

Volume 1: Modern Mechanics

Chapter 1

Interactions and Motion

Chapter 1

Interactions and Motion

This course deals with the nature of matter and its interactions. The variety of phenomena that we will be able to explain and understand is very wide, including the orbit of stars around a black hole, nuclear fusion, and the speed of sound in a solid.

> The main goal of this course is to have you engage in a process central to science: the attempt to explain in detail a broad range of phenomena using a small set of powerful fundamental principles.
>
> The specific focus is on learning how to model the nature of matter and its interactions in terms of a small set of physical laws that govern all mechanical interactions, and in terms of the atomic structure of matter.

This first chapter introduces the notion of interactions and the changes they produce. The major topics are:

- The kinds of matter we will deal with
- How to detect interactions
- Precise description of position and motion in 3D space
- Momentum

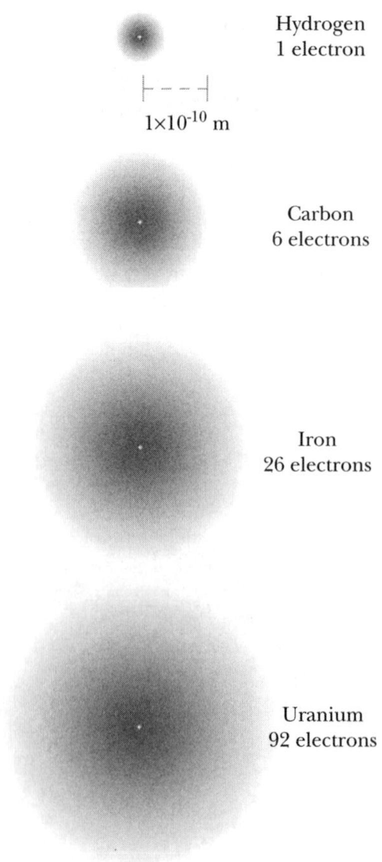

Figure 1.1 Atoms of hydrogen, carbon, iron, and uranium. The white dot shows the location of the nucleus. On this scale, however, the nucleus would be much too small to see.

1.1 Kinds of matter

In this course we will deal with material objects of many sizes, from subatomic particles to galaxies. All of these objects have certain things in common.

Atoms and nuclei

Ordinary matter is made up of tiny atoms. An atom isn't the smallest type of matter, for it is composed of even smaller objects (electrons, protons, and neutrons), but many of the ordinary everyday properties of ordinary matter can be understood in terms of atomic properties and interactions. As you probably know from studying chemistry, atoms have a very small, very dense core, called the nucleus, around which is found a cloud of electrons. The nucleus contains protons and neutrons, collectively called nucleons. Electrons are kept close to the nucleus by electric attraction to the protons (the neutrons don't interact with the electrons).

? Recall your previous studies of chemistry. How many protons and electrons are there in a hydrogen atom? In helium or carbon atoms?

Throughout this text you will encounter questions like the preceding one, which ask you to stop and think before reading further. An important part of reading and understanding a scientific text is to ask yourself questions and to try to answer them. You will learn more from reading this text if you try to answer these questions before looking at the discussion in the subsequent paragraph.

If you don't remember the properties of these atoms, see the periodic table on the inside front cover of this textbook. Hydrogen is the simplest atom, with just one proton and one electron. A helium atom has two protons and two electrons. A carbon atom has six protons and six electrons. Near the other end of the chemical periodic table, a uranium atom has 92 protons and 92 electrons. Figure 1.1 shows the approximate cloud of electrons for several elements but cannot show the nucleus to the same scale; the tiny dot marking the nucleus in the figure is much larger than the actual nucleus.

The radius of the electron cloud for a typical atom is about 1×10^{-10} meter. The reason for this size can be understood using the principles of quantum mechanics, a major development in physics in the early 20th century. The radius of a proton is about 1×10^{-15} meter, very much smaller than the radius of the electron cloud.

Nuclei contain neutrons as well as protons (Figure 1.2). The most common form or "isotope" of hydrogen has no neutrons in the nucleus. However, there exist isotopes of hydrogen with one or two neutrons in the nucleus (in addition to the proton). Hydrogen atoms containing one or two neutrons are called deuterium or tritium. The most common isotope of helium has two neutrons (and two protons) in its nucleus, but a rare isotope has only one neutron; this is called helium-3.

The most common isotope of carbon has six neutrons together with the six protons in the nucleus (carbon-12), while carbon-14 with eight neutrons is an isotope that plays an important role in dating archeological objects.

Near the other end of the periodic table, uranium-235, which can undergo a fission chain reaction, has 92 protons and 143 neutrons, while uranium-238, which does not undergo a fission chain reaction, has 92 protons and 146 neutrons.

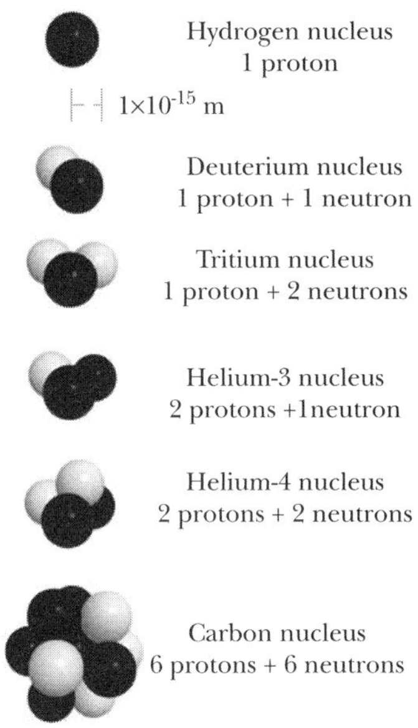

Figure 1.2 Nuclei of hydrogen, helium, and carbon. Note the *very* much smaller scale than in Figure 1.1!

Molecules and solids

When atoms come in contact with each other, they may stick to each other ("bond" to each other). Several atoms bonded together can form a molecule—a substance whose physical and chemical properties differ from those of the constituent atoms. For example, water molecules (H_2O) have properties quite different from the properties of hydrogen atoms or oxygen atoms.

An ordinary-sized rigid object made of bound-together atoms and big enough to see and handle is called a solid, such as a bar of aluminum. A new kind of microscope, the scanning tunneling microscope (STM), is able to map the locations of atoms on the surface of a solid, which has provided new techniques for investigating matter at the atomic level. Two such images appear in Figure 1.3. You can see that atoms in a crystalline solid are arranged in a regular three-dimensional array. The arrangement of atoms on the surface depends on the direction along which the crystal is cut. The irregularities in the bottom image reflect "defects," such as missing atoms, in the crystal structure.

Figure 1.3 Two different surfaces of a crystal of pure silicon. The images were made with a scanning tunneling microscope.

Liquids and gases

When a solid is heated to a higher temperature, the atoms in the solid vibrate more vigorously about their normal positions. If the temperature is raised high enough, this thermal agitation may destroy the rigid structure of the solid. The atoms may become able to slide over each other, in which case the substance is a liquid.

At even higher temperatures the thermal motion of the atoms or molecules may be so large as to break the interatomic or intermolecular bonds completely, and the liquid turns into a gas. In a gas the atoms or molecules are quite free to move around, only occasionally colliding with each other or the walls of their container.

In this course we will learn how to analyze many aspects of the behavior of solids and gases. We won't have much to say about liquids, because their properties are much harder to analyze. Solids are simpler to analyze than liquids because the atoms stay in one place (though with thermal vibration about their usual positions). Gases are simpler to analyze than liquids be-

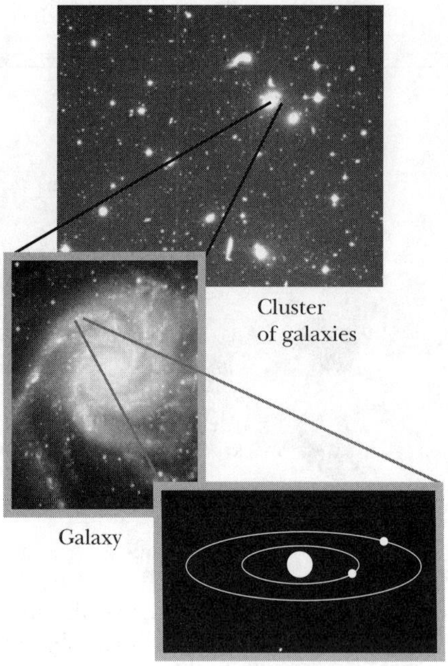

Figure 1.4 Our Solar System exists inside a galaxy, which itself is a member of a cluster of galaxies.

cause between collisions the gas molecules are approximately unaffected by the other molecules. Liquids are the awkward intermediate state, where the atoms move around rather freely, but always in contact with other atoms. This makes the analysis of liquids very complex.

Planets, stars, solar systems, and galaxies

In our brief survey of the kinds of matter that we will study, we make a giant leap in scale from atoms all the way up to planets and stars, such as our Earth and Sun. In this course we will see that many of the same principles that apply to atoms apply to planets and stars. By making this leap we bypass an important physical science, geology, whose domain of interest includes the formation of mountains and continents. We will study objects that are much bigger than mountains, and we will study objects that are much smaller than mountains, but we don't have time in one course to apply the principles of physics to every important kind of matter.

Our Sun and its accompanying planets constitute our Solar System (Figure 1.4). It is located in the Milky Way galaxy, a giant rotating disk-shaped system of stars. On a clear dark night you can see a band of light (the Milky Way) coming from the huge number of stars lying in this disk, which you are looking at from a position in the disk, about two-thirds of the way out from the center of the disk. Our galaxy is a member of a cluster of galaxies that move around each other much as the planets of our Solar System move around the Sun. The Universe contains many such clusters of galaxies.

1.2 Detecting interactions

Objects made of different kinds of matter interact with each other in various ways: gravitationally, electrically, magnetically, and through the strong and weak interactions. How can we detect that an interaction has occurred? In this section we consider various kinds of observations that indicate the presence of interactions.

> **?** Before you read further, take a moment to think about your own ideas of interactions. How can you tell that two objects are interacting with each other?

1.2.1 Change of direction

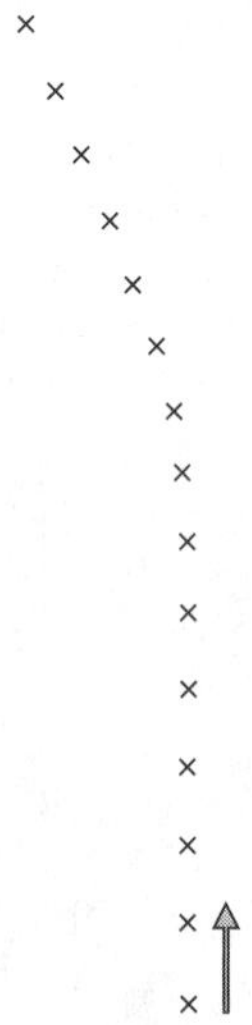

Figure 1.5 A proton moves through space, far from almost all other objects. The initial direction of the proton's motion is upward, as indicated by the arrow. The x's represent the position of the proton at equal time intervals.

Suppose you observe a proton moving through a region of outer space, far from almost all other objects. The proton moves along a path like the one shown in Figure 1.5. The arrow indicates the initial direction of the proton's motion, and the "x's" in the diagram indicate the position of the proton at equal time intervals.

> **?** Do you see evidence that the proton is interacting with another object?

Evidently a change in direction is a vivid indicator of interactions. If you observe a change in direction of the motion of a proton, you will find another object somewhere that has interacted with this proton.

> **?** Suppose that the only other object nearby was another proton. What was the approximate initial location of this second proton?

Since two protons repel each other electrically, the second proton must have been located to the right of the bend in the first proton's path.

1.2.2 Change of speed

Suppose that you observe an electron traveling in a straight line through outer space far from almost all other objects (Figure 1.6). The path of the electron is shown as though a camera had taken multiple exposures at equal time intervals.

? Where is the electron's speed largest? Where is the electron's speed smallest?

The speed is largest at the top, where the dots are farther apart, which means that the electron has moved farthest during the time interval between exposures. The speed is smallest at the bottom, where the dots are closer together, which means that the electron has moved the least distance during the time interval between exposures.

? Suppose that the only other object nearby was another electron. What was the approximate initial location of this other electron?

The other electron must have been located directly below the starting point of the path, since electrons repel each other electrically.

Evidently a change in speed is an indicator of interactions. If you observe a change in speed of an electron, you will find another object somewhere that has interacted with the electron.

Figure 1.6 An electron moves through space, far from almost all other objects. The initial direction of the electron's motion is upward, as indicated by the arrow. The x's represent the position of the electron at equal time intervals.

1.2.3 Change of velocity: change of speed or direction

In physics, the word "velocity" has a special technical meaning which is different from its meaning in everyday speech. In physics, the quantity called "velocity" indicates a combination of speed and direction. (In contrast, in everyday speech, "speed" and "velocity" are often used as synonyms. In physics, however, all words have precise meanings and there are no synonyms.)

For example, consider an airplane that is flying with a speed of 1000 kilometers/hour in a direction that is due east. We say the velocity is 1000 km/hr, east, where we specify both speed and direction. An airplane flying west with a speed of 1000 km/hr would have the same speed, but a different velocity.

We have seen that a change in an object's speed, or a change in the direction of its motion, indicates that the object has interacted with at least one other object. The two indicators of interaction, change of speed and change of direction, can be combined into one compact statement:

> A change of velocity (speed or direction or both) indicates the existence of an interaction.

Figure 1.7 Two successive positions of a particle (indicated by a dot), with arrows indicating the velocity of the particle at each location. The shorter arrow indicates that the speed of the particle at location 2 is less than its speed at location 1.

Diagrams showing changes in velocity

In physics diagrams, the velocity of an object is represented by an arrow: a line with an arrowhead. The tail of the arrow is placed at the location of the object, and the arrow points in the direction of the motion of the object. The length of the arrow is proportional to the speed of the object. Figure 1.7 shows two successive positions of a particle at two different times, with velocity arrows indicating a change in speed of the particle (it's slowing down). Figure 1.8 shows three successive positions of a different particle at three different times, with velocity arrows indicating a change in direction but no change in speed.

We will see a little later in the chapter that velocity is only one example of a physical quantity that has a "magnitude" (an amount or a size) and a direction. Other examples of such quantities are position relative to an origin in 3D space, force, and magnetic field. Quantities having magnitude and direction can be usefully described as "vectors". Vectors are mathematical

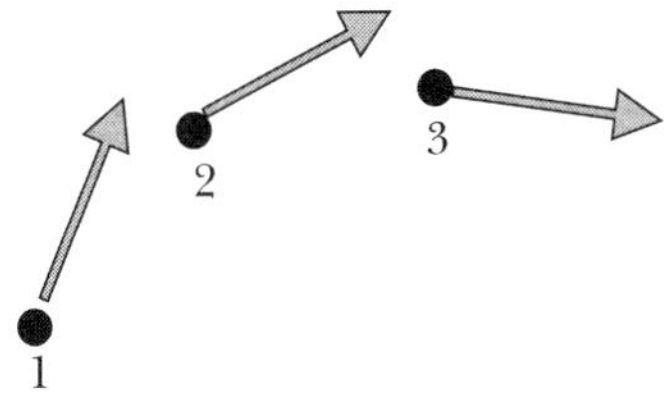

Figure 1.8 Three successive positions of a particle (indicated by a dot), with arrows indicating the velocity of the particle at each location. The arrows are the same length, indicating the same speed, but they point in different directions, indicating a change in direction and therefore a change in velocity.

quantities which have their own special rules of algebra, similar (but not identical) to the rules of ordinary algebra. Arrows are commonly used in diagrams to denote vector quantities. We will use vectors extensively in this course.

1.2.4 Uniform motion

Suppose you observe a rock moving along in outer space far from all other objects. We don't know what made it start moving in the first place; presumably a long time ago an interaction gave it some velocity and it has been coasting through the vacuum of space ever since.

It is an observational fact that such an isolated object moves at constant, unchanging speed, in a straight line. Its velocity does not change (neither its direction nor its speed changes). We call such motion with unchanging velocity "uniform motion" (Figure 1.9).

Figure 1.9 "Uniform motion"—no change in speed or direction.

An object at rest

A special case of uniform motion is the case in which an object's speed is zero and remains zero—the object remains at rest. In this case the object's speed is constant (zero) and the direction of motion, while undefined, is not changing.

Uniform motion implies no net interaction

When we observe an object in uniform motion, we conclude that since its velocity is not changing, either it is not interacting significantly with any other object, or else it is undergoing multiple interactions that cancel each other out. In either case, we can say that there is no "net" (total) interaction.

1.3 Newton's first law of motion

The basic relationship between change of velocity and interaction is summarized qualitatively by Newton's "first law of motion":

NEWTON'S FIRST LAW OF MOTION

An object moves in a straight line and at constant speed except to the extent that it interacts with other objects.

The words "to the extent" imply that the stronger the interaction, the more change there will be in direction and/or speed. The weaker the interaction, the less change. If there is no net interaction at all, the direction doesn't change and the speed doesn't change (uniform motion). This case can also be called "uniform velocity" or "constant velocity," since velocity refers to both speed and direction. It is important to remember that if an object is not moving at all, its velocity is not changing, so it too may be considered to be in uniform motion.

Newton's first law of motion is only qualitative, because it doesn't give us a way to calculate quantitatively how much change in speed or direction will be produced by a certain amount of interaction, a subject we will take up in the next chapter. Nevertheless, Newton's first law of motion is important in providing a conceptual framework for thinking about the relationship between interaction and motion.

The English physicist Isaac Newton was the first person to state this law clearly. Newton's first law of motion represented a major break with ancient tradition, which assumed that constant pushing was required to keep something moving. This law says something radically different: no interactions at all are needed to keep something moving!

Does Newton's first law apply in everyday life?

Superficially, Newton's first law of motion may at first seem not to apply to many everyday situations. To push a chair across the floor at constant speed, you have to keep pushing all the time.

> **?** Doesn't Newton's first law of motion say that the chair should keep moving at constant speed without anyone pushing it? In fact, shouldn't the speed or direction of motion of the chair change due to the interaction with your hands? Does this everyday situation violate Newton's first law of motion? Try to answer these questions before reading farther.

The complicating factor here is that your hands aren't the only objects that are interacting with the chair. The floor also interacts with the chair, in a way that we call friction. If you push just hard enough to compensate exactly for the floor friction, the sum of all the interactions is zero, and the chair moves at constant speed as predicted by Newton's first law. (If you push harder than the floor does, the chair's speed does increase.)

Motion without friction

It is difficult to observe motion without friction in everyday life, because objects almost always interact with many other objects, including air, flat surfaces, etc. This explains why it took people such a long time (Newton lived in the 1600's) to understand clearly the relationship between interaction and change.

You may be able to think of situations in which you have seen an object keep moving at constant (or nearly constant) velocity, without being pushed or pulled. One example of a nearly friction-free situation is a hockey puck sliding on ice. The puck slides a long way at nearly constant speed in a straight line (constant velocity) because there is little friction with the ice. An even better example is the uniform motion of an object in outer space, far from all other objects.

Exercises

At the end of a section you will usually find exercises such as those that follow. It is important to work through exercises as you come to them, to make sure that you can apply what you have just read. Simply reading about concepts in physics is not enough—you must be able to use the concepts in answering questions and solving problems.

Make a serious attempt to do an exercise before checking the answer at the end of the chapter. This will help you assess your own understanding.

Ex. 1.1 Which of the following objects are moving with constant velocity?

(a) A ship sailing northeast at a speed of 5 meters per second
(b) The moon orbiting the Earth
(c) A tennis ball traveling across the court after having been hit by a tennis racket
(d) A can of soda sitting on a table
(e) A person riding on a Ferris wheel which is turning at a constant rate

Ex. 1.2 Apply Newton's first law to each of the following situations. In which situations can you conclude that the object is undergoing a net interaction with one or more other objects?

(a) A book slides across the table and comes to a stop
(b) A proton in a particle accelerator moves faster and faster

(c) A car travels at constant speed around a circular race track
(d) A spacecraft travels at a constant speed toward a distant star
(e) A hydrogen atom remains at rest in outer space

Ex. 1.3 A spaceship far from all other objects uses its thrusters to attain a speed of 10^4 m/s. The crew then shuts off the power. According to Newton's first law, what will happen to the motion of the spaceship from then on?

1.4 Other indicators of interaction

Change of identity

Change of velocity (change of speed and/or direction) is not the only indicator of interactions. Another is change of identity, such as the formation of water (H_2O) from the burning of hydrogen in oxygen. A water molecule behaves very differently from the hydrogen and oxygen atoms of which it is made.

Change of shape or configuration

Another indicator of interaction is change of shape or configuration (arrangement of the parts). For example, slowly bend a pen or pencil, then hold it in the bent position. The speed hasn't changed, nor is there a change in the direction of motion (it's not moving!). The pencil has not changed identity. Evidently a change of shape can be evidence for interactions, in this case with your hand.

Other changes in configuration include "phase changes" such as the freezing or boiling of a liquid, brought about by interactions with the surroundings. In different phases (solid, liquid, gas), atoms or molecules are arranged differently. Changes in configuration at the atomic level are another indication of interactions.

Change of temperature

Another indication of interaction is change of temperature. Place a pot of cold water on a hot stove. As time goes by, a thermometer will indicate a change in the water due to interaction with the hot stove.

Other indications of interactions

Is a change of position an indicator of an interaction? That depends. If the change of position occurs simply because a particle is moving at constant speed and direction, then a mere change of position is not an indicator of an interaction, since uniform motion is an indicator of no net interaction.

? If however you observe an object at rest in one location, and later you observe it again at rest but in a different location, did an interaction take place?

Yes. You can infer that there must have been an interaction to give the object some velocity to move the object toward the new position, and another interaction to slow the object to a stop in its new position.

In later chapters we will consider interactions involving change of identity, change of shape, and change of temperature, but for now we'll concentrate on interactions that cause a change of velocity (speed and/or direction).

1.4.1 Indirect evidence for an interaction

Sometimes there is indirect evidence for an interaction. When something doesn't change although you would normally expect a change due to a known interaction, this indicates that another interaction is present. Consider a balloon that hovers motionless in the air despite the downward gravitational pull of the Earth. Evidently there is some other kind of interaction that opposes the gravitational interaction. In this case, interactions with air molecules have the net effect of pushing up on the balloon ("buoyancy"). The lack of change implies that the effect of the air molecules exactly compensates for the gravitational interaction with the Earth.

When you push a chair across the floor and it moves with constant velocity despite your pushing on it (which ought to change its speed), that means that something else must also be interacting with it (the floor).

The stability of the nucleus of an atom is another example of indirect evidence for an interaction. The nucleus contains positively charged protons that repel each other electrically, yet the nucleus remains intact. We conclude that there must be some other kind of interaction present, a nonelectric attractive interaction that overcomes the electric repulsion. This is evidence for the "strong interaction" that acts between protons and neutrons in the nucleus.

1.4.2 Summary: changes as indicators of interactions

Here then are the most common indicators of interactions:

- change of velocity (change of direction and/or change of speed)
- change of identity
- change of shape
- change of temperature
- lack of change when change is expected (indirect evidence)

The important point is this: **Interactions cause change.**

In the absence of interactions, there is no change, which is usually uninteresting. The exception is the surprise when nothing changes despite our expectations that something should change. This is indirect evidence for some interaction that we hadn't recognized was present, that more than one interaction is present and the interactions cancel each others' effects.

For the next few chapters we'll concentrate on change of velocity as evidence for an interaction (or lack of change of velocity, which can give indirect evidence for additional interactions).

1.5 Describing the 3D world: Vectors

Physical phenomena take place in the 3D world around us. In order to be able to make quantitative predictions and give detailed, quantitative explanations, we need tools for describing precisely the positions and velocities of objects in 3D, and the changes in position and velocity due to interactions. These tools are mathematical entities called 3D "vectors."

1.5.1 3D coordinates

We will use a 3D coordinate system to specify positions in space and other vector quantities. Usually we will orient the axes of the coordinate system as shown in Figure 1.10: $+x$ axis to the right, $+y$ axis upward, and $+z$ axis coming out of the page, toward you. This is a "right-handed" coordinate system: if you hold the thumb, first, and second fingers of your right hand perpendicular to each other, and align your thumb with the x axis and your first finger with the y axis, your second finger points along the z axis. (In some math and physics textbook discussions of 3D coordinate systems, the y axis points

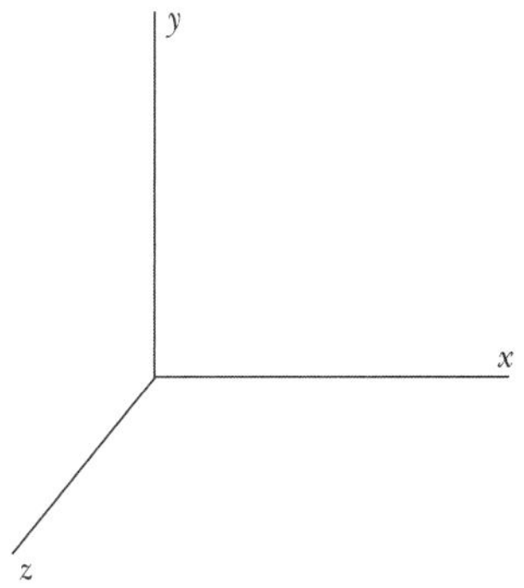

Figure 1.10 Right-handed 3D coordinate system.

out and the z axis points up, but we will also use a 2D coordinate system with y up, so it makes sense always to have the y axis point up.)

1.5.2 Basic properties of vectors: magnitude and direction

A vector is a quantity that has a magnitude and a direction. For example, the velocity of a baseball is a vector quantity. The magnitude of the baseball's velocity is the speed of the baseball, for example 20 meters/second. The direction of the baseball's velocity is the direction of its motion at a particular instant, for example "up" or "to the right" or "west" or "in the $+y$ direction." A symbol denoting a vector is written with an arrow over it:

$$\vec{v} \text{ is a vector.}$$

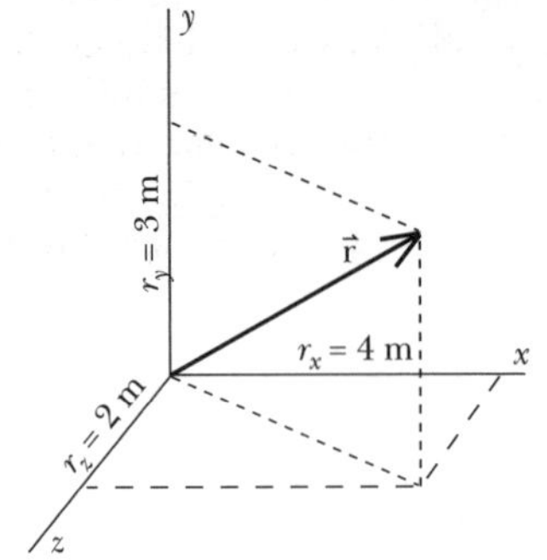

Figure 1.11 A position vector and its x, y, and z components.

A position in space can also be considered to be a vector, called a position vector, pointing from an origin to that location. Figure 1.11 shows a position vector that might represent your final position if you started at the origin and walked 4 meters along the x axis, then 2 meters parallel to the z axis, then climbed a ladder so you were 3 meters above the ground. Your new position relative to the origin is a vector that can be written like this:

$$\vec{r} = \langle 4, 3, 2 \rangle \text{ m}$$

$$x \text{ component } r_x = 4 \text{ m}$$
$$y \text{ component } r_y = 3 \text{ m}$$
$$z \text{ component } r_z = 2 \text{ m}$$

In three dimensions a vector is a triple of numbers $\langle x, y, z \rangle$. Quantities like the position of an object and the velocity of an object can be represented as vectors:

$$\vec{r} = \langle x, y, z \rangle \text{ (a position vector)}$$

$$\vec{r}_1 = \langle 3.2, -9.2, 66.3 \rangle \text{ m (a position vector)}$$

$$\vec{v} = \langle v_x, v_y, v_z \rangle \text{ (a velocity vector)}$$

$$\vec{v}_1 = \langle -22.3, 0.4, -19.5 \rangle \text{ m/s (a velocity vector)}$$

Components of a vector

Each of the numbers in the triple is referred to as a "component" of the vector. The x component of the vector $\vec{v}$ is the number v_x. The z component of the vector $\vec{v}_1$ above is -19.5 m/s. A component such as v_x is not a vector, since it is only one number.

It is important to note that the x component of a vector specifies the difference between the x coordinate of the tail of the vector and the x coordinate of the tip of the vector. It does not give any information about the location of the tail of the vector (compare Figure 1.11 and Figure 1.12).

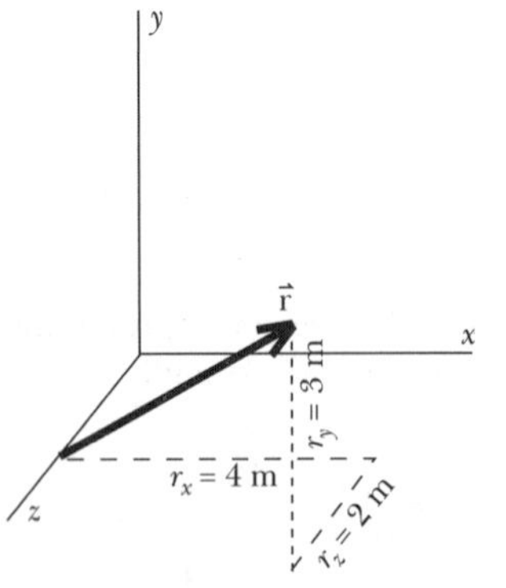

Figure 1.12 The arrow represents the vector $\vec{r} = \langle 4, 3, 2 \rangle$ m, drawn with its tail at location $\langle 0, 0, 2 \rangle$.

1.5.3 Equality of vectors

A vector is equal to another vector if and only if all the components of the vectors are equal. If $\vec{r} = \langle 4, 3, 2 \rangle$ m,

$$\vec{w} = \vec{r} \text{ means that}$$

$$w_x = r_x \text{ and } w_y = r_y \text{ and } w_z = r_z, \text{ so } \vec{w} = \langle 4, 3, 2 \rangle \text{ m}$$

If two vectors are equal, their magnitudes and directions are the same.

1.5.4 Drawing vectors

In Figure 1.11 we represented your position vector relative to the origin graphically by an arrow whose tail is at the origin and whose arrowhead is at your position. The length of the arrow represents the distance from the origin, and the direction of the arrow represents the direction of the vector, which is the direction of a direct path from the initial position to the final position (the "displacement"; by walking and climbing you "displaced" yourself from the origin to your final position).

Since it is difficult to draw a 3D diagram on paper, when working on paper you will usually be asked to draw vectors which all lie in a single plane. Figure 1.13 shows an arrow in the *xy* plane representing the vector $\langle -3, -1, 0 \rangle$.

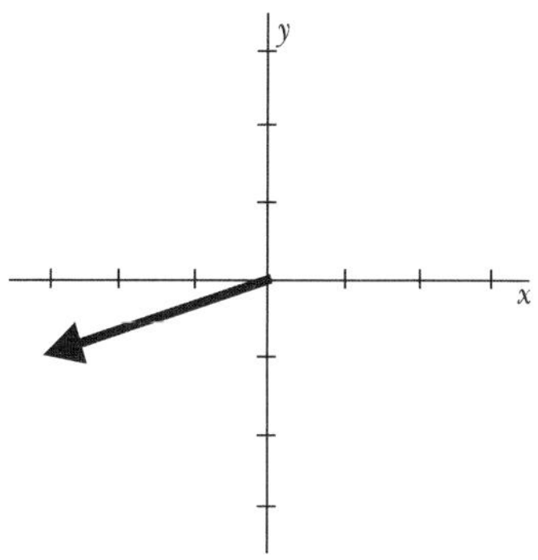

Figure 1.13 The position vector $\langle -3, -1, 0 \rangle$, drawn at the origin, in the *xy* plane. The components of the vector specify the displacement from the tail to the tip. The z axis, which is not shown, comes out of the page, toward you.

1.5.5 Vectors and scalars

A quantity which is represented by a single number is called a *scalar.* A scalar quantity does not have a direction. Examples include the mass of an object, such as 5 kg, or the temperature, such as -20 C. Vectors and scalars are very different entities; a vector can never be equal to a scalar, and a scalar cannot be added to a vector. Scalars can be positive or negative:

$$m = 5\ \text{kg}$$
$$T = -20\ \text{C}$$

Ex. 1.4 How many numbers are needed to specify a 3D position vector?

Ex. 1.5 How many numbers are needed to specify a scalar?

Ex. 1.6 Does the symbol $\vec{a}$ represent a vector or a scalar?

Ex. 1.7 Which of the following are vectors?
a) 5 m/s, b) $\langle -11, 5.4, -33 \rangle$ m, c) $\vec{r}$, d) v_z

Ex. 1.8 $\vec{a} = \langle -3, 7, 0.5 \rangle$. If $\vec{b} = \vec{a}$, what is the *y* component of $\vec{b}$?

1.5.6 Magnitude of a vector

Consider again the vector in Figure 1.14, showing your displacement from the origin. Using a 3D extension of the Pythagorean theorem for right triangles (Figure 1.15), the net distance you have moved from the starting point is

$$\sqrt{(4\ \text{m})^2 + (3\ \text{m})^2 + (2\text{m})^2} = \sqrt{29}\ \text{m} = 5.39\ \text{m}$$

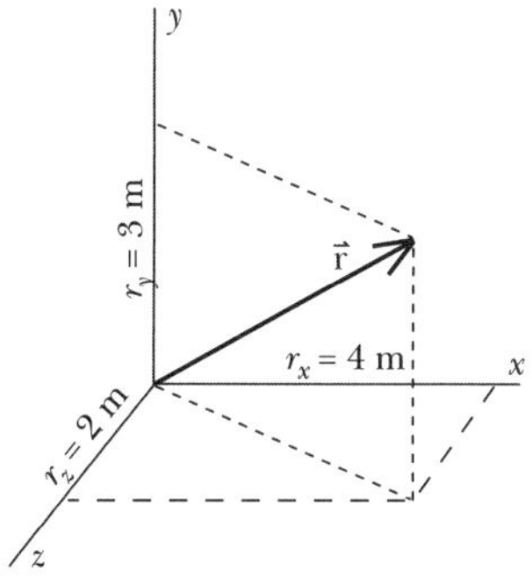

Figure 1.14 A vector representing a displacement from the origin.

We say that the *magnitude* $|\vec{r}|$ of the position vector $\vec{r}$ is

$$|\vec{r}| = 5.39\ \text{m}$$

The magnitude of a vector is written either with absolute-value bars around the vector as $|\vec{r}|$, or simply by writing the symbol for the vector without the little arrow above it, *r.*

In general, the magnitude of a vector can be calculated by taking the square root of the sum of the squares of its components (see Figure 1.15).

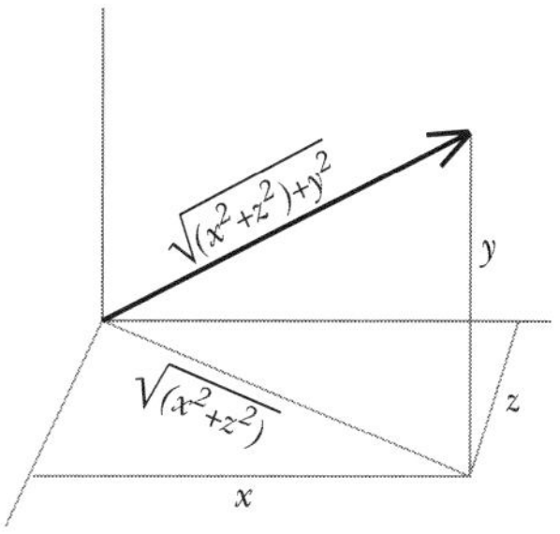

Figure 1.15 The magnitude of a vector is the square root of the sum of the squares of its components (3D version of the Pythagorean theorem).

MAGNITUDE OF A VECTOR

If the vector $\vec{r} = \langle r_x, r_y, r_z \rangle$ then $|\vec{r}| = \sqrt{r_x^2 + r_y^2 + r_z^2}$ (a scalar).

Note that the magnitude of a vector is always a positive number. Since the magnitude of a vector is a single number, and not a triple of numbers, it is a scalar, not a vector.

Can a *vector* be positive or negative?

? Consider the vector $\vec{v} = \langle 8\times10^6, 0, -2\times10^7 \rangle$ m/s. Is this vector positive? Negative? Zero?

None of these descriptions is appropriate. The x component of this vector is positive, the y component is zero, and the z component is negative. Vectors aren't positive, or negative, or zero. Their components can be positive or negative or zero, but these words just don't mean anything when used with the vector as a whole.

On the other hand, the *magnitude* of a vector such as $|\vec{v}|$ is always positive.

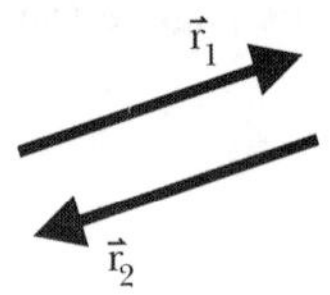

Figure 1.16 Two vectors (Exercise 1.10).

Ex. 1.9 Does the symbol $|\vec{v}|$ represent a vector or a scalar?

Ex. 1.10 Consider the vectors $\vec{r}_1$ and $\vec{r}_2$ represented by arrows in Figure 1.16. Are these two vectors equal?

Ex. 1.11 If $\vec{r} = \langle -3, -4, 1 \rangle$ m, find $|\vec{r}|$.

Ex. 1.12 Can the magnitude of a vector be a negative number?

Ex. 1.13 What is the magnitude of the vector $\vec{v}$, where $\vec{v} = \langle 8\times10^6, 0, -2\times10^7 \rangle$ m/s?

1.5.7 Mathematical operations involving vectors

Although the algebra of vectors is similar to the scalar algebra with which you are very familiar, it is not identical. There are some algebraic operations that cannot be performed on vectors.

Algebraic operations that *are* legal for vectors include the following operations, which we will discuss in this chapter:

- adding one vector to another vector: $\vec{a} + \vec{w}$
- subtracting one vector from another vector: $\vec{b} - \vec{d}$
- finding the magnitude of a vector: $|\vec{r}|$
- finding a unit vector (a vector of magnitude 1): $\hat{r}$
- multiplying (or dividing) a vector by a scalar: $3\vec{v}$
- finding the rate of change of a vector: $\dfrac{\Delta\vec{r}}{\Delta t}$ or $\dfrac{d\vec{r}}{dt}$

In later chapters we will also see that there are two more ways of combining two vectors:

the vector dot product, whose result is a scalar
the vector cross product, whose result is a vector

Operations that are *not* legal for vectors

Although vector algebra is similar to the ordinary scalar algebra you have used up to now, there are certain operations that are not legal (and not meaningful) for vectors:

A vector cannot be set equal to a scalar.

A vector cannot be added to or subtracted from a scalar.

A vector cannot occur in the denominator of an expression. (Although you can't divide by a vector, note that you can legally divide by the *magnitude* of a vector, which is a scalar.)

1.5.8 Multiplying a vector by a scalar

A vector can be multiplied (or divided) by a scalar. If a vector is multiplied by a scalar, each of the components of the vector is multiplied by the scalar:

$$\text{If } \vec{r} = \langle x, y, z\rangle \text{ then } a\vec{r} = \langle ax, ay, az\rangle$$

$$\text{If } \vec{v} = \langle v_x, v_y, v_z\rangle \text{ then } \frac{\vec{v}}{b} = \langle \frac{v_x}{b}, \frac{v_y}{b}, \frac{v_z}{b}\rangle$$

$$\left(\frac{1}{2}\right)\langle 6, -20, 9\rangle = \langle 3, -10, 4.5\rangle$$

Multiplication by a scalar "scales" a vector, keeping its direction the same but making its magnitude larger or smaller (Figure 1.17). Multiplying by a negative scalar reverses the direction of a vector.

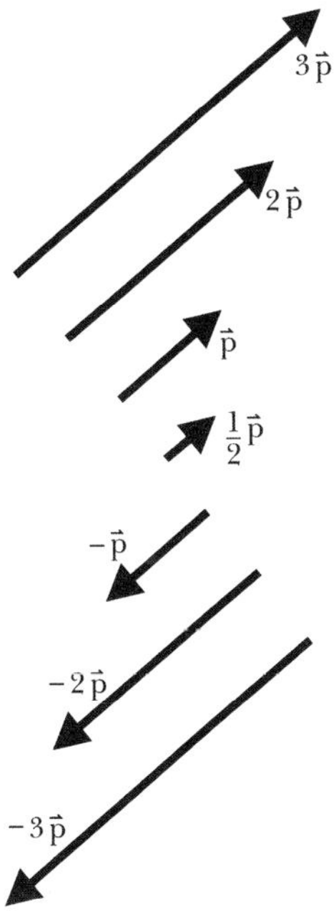

Figure 1.17 Multiplying a vector by a scalar changes the magnitude of the vector. Multiplying by a negative scalar reverses the direction of the vector.

Magnitude of a scalar

You may wonder how to find the magnitude of a quantity like $-3\vec{r}$, which involves the product of a scalar and a vector. This expression can be factored:

$$|-3\vec{r}| = |-3| \cdot |\vec{r}|$$

The magnitude of a scalar is its absolute value, so:

$$|-3\vec{r}| = |-3| \cdot |\vec{r}| = 3\sqrt{r_x^2 + r_y^2 + r_z^2}$$

Ex. 1.14 If $\vec{v} = \langle 2, -3, 5\rangle$ m/s, what is $3\vec{v}$?

Ex. 1.15 If $\vec{r} = \langle 2, -3, 5\rangle$ m, what is $\frac{\vec{r}}{2}$?

Ex. 1.16 What is the result of multiplying the vector $\vec{a}$ by the scalar f, where $\vec{a} = \langle 0.02, -1.7, 30.0\rangle$ and $f = 2.0$?

Ex. 1.17 How does the direction of the vector $-\vec{a}$ compare to the direction of the vector $\vec{a}$?

Ex. 1.18 Is $3 + \langle 2, -3, 5\rangle$ a meaningful expression? If so, what is its value?

Ex. 1.19 Is $\frac{4}{\langle 6, -7, 4\rangle}$ a meaningful expression? If so, what is its value?

1.5.9 Direction of a vector: Unit vectors

One way to describe the direction of a vector is by specifying a *unit vector*. A unit vector is a vector of magnitude 1, pointing in some direction. A unit vector is written with a "hat" (caret) over it instead of an arrow. The unit vector $\hat{a}$ is called "a-hat".

? Is the vector $\langle 1, 1, 1\rangle$ a unit vector?

The magnitude of $\langle 1, 1, 1\rangle$ is $\sqrt{1^2 + 1^2 + 1^2} = 1.73$. so this is not a unit vector.

The vector $\langle 1/\sqrt{3}, 1/\sqrt{3}, 1/\sqrt{3}\rangle$ is a unit vector, since its magnitude is 1:

$$\sqrt{\left(\frac{1}{\sqrt{3}}\right)^2 + \left(\frac{1}{\sqrt{3}}\right)^2 + \left(\frac{1}{\sqrt{3}}\right)^2} = 1$$

Note that every component of a unit vector must be less than or equal to 1.

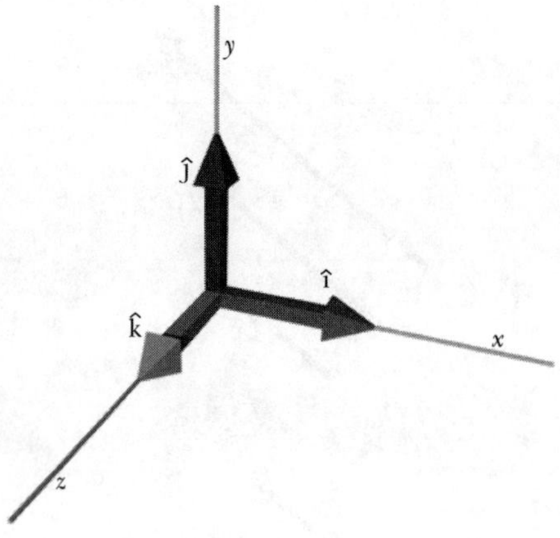

Figure 1.18 The unit vectors $\hat{\imath}$, $\hat{\jmath}$, $\hat{k}$.

In our 3D Cartesian coordinate system, there are three special unit vectors, oriented along the three axes. They are called i-hat, j-hat, and k-hat, and they point along the *x*, *y*, and *z* axes, respectively (Figure 1.18):

$$\hat{\imath} = \langle 1, 0, 0\rangle$$
$$\hat{\jmath} = \langle 0, 1, 0\rangle$$
$$\hat{k} = \langle 0, 0, 1\rangle$$

One way to express a vector is in terms of these special unit vectors:

$$\langle 0.02, -1.7, 30.0\rangle = 0.02\hat{\imath} + (-1.7)\hat{\jmath} + 30.0\hat{k}$$

We will usually use the $\langle x, y, z\rangle$ form rather than the ijk form in this book, because $\langle x, y, z\rangle$ is the familiar notation used in many calculus textbooks.

Not all unit vectors point along an axis, as shown in Figure 1.19. For example, the vectors

$$\hat{g} = \langle 0.5774, 0.5774, 0.5774\rangle \text{ and } \hat{F} = \langle 0.424, 0.566, 0.707\rangle$$

are both unit vectors, since the magnitude of each is equal to 1. Note that every component of a unit vector is less than or equal to 1.

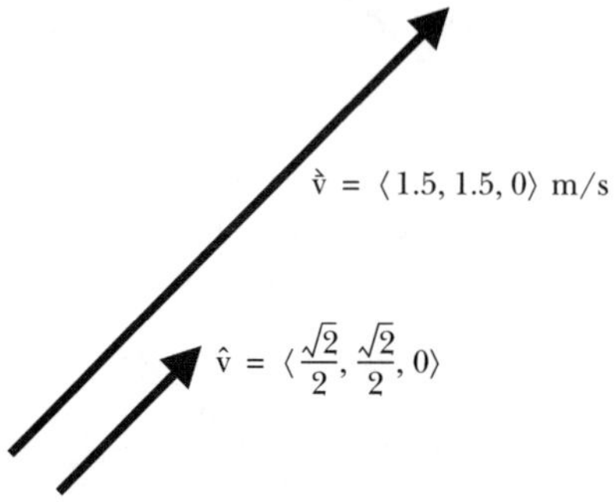

Figure 1.19 The unit vector $\hat{v}$ has the same direction as the vector $\vec{v}$, but its magnitude is 1, and it has no physical units.

Calculating unit vectors

Any vector may be factored into the product of a unit vector in the direction of the vector, multiplied by a scalar equal to the magnitude of the vector.

$$\vec{w} = |\vec{w}| \cdot \hat{w}$$

For example, a vector of magnitude 5, aligned with the y axis, could be written as:

$$\langle 0, 5, 0\rangle = 5\langle 0, 1, 0\rangle$$

Therefore, to find a unit vector in the direction of a particular vector, we just divide the vector by its magnitude:

$$\hat{r} = \frac{\vec{r}}{|\vec{r}|} = \frac{\langle x, y, z\rangle}{\sqrt{(x^2 + y^2 + z^2)}}$$

$$\hat{r} = \left\langle \frac{x}{\sqrt{(x^2+y^2+z^2)}}, \frac{y}{\sqrt{(x^2+y^2+z^2)}}, \frac{z}{\sqrt{(x^2+y^2+z^2)}} \right\rangle$$

For example, if $\vec{v} = \langle -22.3, 0.4, -19.5\rangle \text{m/s}$, then

$$\hat{v} = \frac{\vec{v}}{|\vec{v}|} = \frac{\langle -22.3, 0.4, -19.5\rangle \text{ m/s}}{\sqrt{(-22.3)^2 + (0.4)^2 + (-19.5)^2} \text{ m/s}} = \langle -0.753, 0.0135, -0.658\rangle$$

Remember that to divide a vector by a scalar, you divide each component of the vector by the scalar. The result is a new vector. Note also that a unit vector has no physical units (such as meters per second), because the units in the numerator and denominator cancel.

Ex. 1.20 What is the unit vector in the direction of $\langle 0, 6, 0\rangle$?

Ex. 1.21 What is the unit vector in the direction of $\langle -300, 0, 0\rangle$?

Ex. 1.22 What is the unit vector in the direction of $\langle 2, 2, 2\rangle$? What is the unit vector in the direction of $\langle 3, 3, 3\rangle$?

Ex. 1.23 What is the unit vector $\hat{a}$ in the direction of $\vec{a}$, where $\vec{a} = \langle 400, 200, -100\rangle \text{ m/s}^2$?

Ex. 1.24 Write the vector $\vec{a} = \langle 400, 200, -100\rangle \text{ m/s}^2$ as the product $|\vec{a}| \cdot \hat{a}$.

1.5.10 Vector addition

The sum of two vectors is another vector, obtained by adding the components of the vectors:

$$\vec{A} = \langle A_x, A_y, A_z \rangle$$

$$\vec{B} = \langle B_x, B_y, B_z \rangle$$

$$\vec{A} + \vec{B} = \langle (A_x + B_x), (A_y + B_y), (A_z + B_z) \rangle$$

For example,

$$\langle 1, 2, 3 \rangle + \langle -4, 5, 6 \rangle = \langle -3, 7, 9 \rangle$$

Don't add magnitudes!

The magnitude of a vector is *not* in general equal to the sum of the magnitudes of the two original vectors! For example, the magnitude of the vector $\langle 3, 0, 0 \rangle$ is 3, and the magnitude of the vector $\langle -2, 0, 0 \rangle$ is 2, but the magnitude of the vector $(\langle 3, 0, 0 \rangle + \langle -2, 0, 0 \rangle)$ is 1, not 5!

Adding vectors graphically: Tip to tail

The sum of two vectors has a geometric interpretation. In Figure 1.20 you first walk along displacement vector $\vec{A}$, followed by walking along displacement vector $\vec{B}$. What is your net displacement vector $\vec{C} = \vec{A} + \vec{B}$? The x component C_x of your net displacement is the sum of A_x and B_x. Similarly, the y component C_y of your net displacement is the sum of A_y and B_y.

To add two vectors $\vec{A}$ and $\vec{B}$ graphically (Figure 1.20):

- Draw the first vector $\vec{A}$
- Move the second vector $\vec{B}$ (without rotating it) so its tail is located at the *tip* of the first vector
- Draw a new vector from the tail of vector $\vec{A}$ to the tip of vector $\vec{B}$

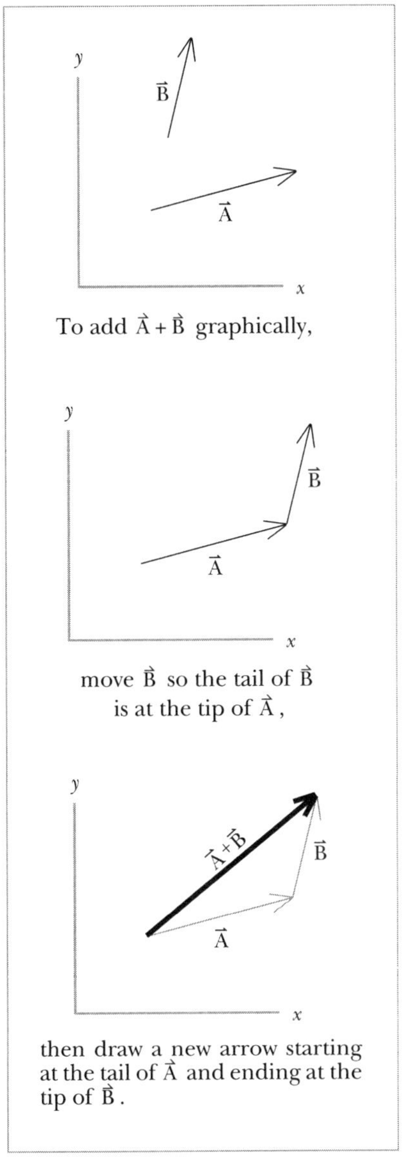

Figure 1.20 The procedure for adding two vectors graphically: draw vectors tip to tail.

1.5.11 Vector subtraction

The difference of two vectors will be very important in this and subsequent chapters. To subtract one vector from another, we subtract the components of the second from the components of the first:

$$\vec{A} - \vec{B} = \langle (A_x - B_x), (A_y - B_y), (A_z - B_z) \rangle$$

$$\langle 1, 2, 3 \rangle - \langle -4, 5, 6 \rangle = \langle 5, -3, -3 \rangle$$

Subtracting vectors graphically: Tail to tail

To subtract one vector $\vec{B}$ from another vector $\vec{A}$ graphically:

- Draw the first vector $\vec{A}$
- Move the second vector $\vec{B}$ (without rotating it) so its tail is located at the *tail* of the first vector
- Draw a new vector from the tip of vector $\vec{B}$ to the tip of vector $\vec{A}$

Note that you can check this algebraically and graphically. As shown in Figure 1.21, since the tail of $\vec{A} - \vec{B}$ is located at the tip of $\vec{B}$, then the vector $\vec{A}$ should be the sum of $\vec{B}$ and $\vec{A} - \vec{B}$, as indeed it is:

$$\vec{B} + (\vec{A} - \vec{B}) = \vec{A}$$

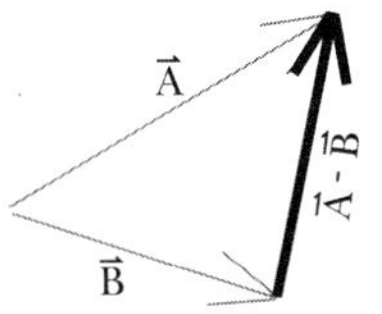

Figure 1.21 The procedure for subtracting vectors graphically: draw vectors tail to tail; draw new vector from tip of second vector to tip of first vector.

1.5.12 The zero vector

It is convenient to have a compact notation for a vector whose components are all zero. We will use the symbol $\vec{0}$ to denote a zero vector, in order to distinguish it from a scalar quantity that has the value 0.

$$\vec{0} = \langle 0, 0, 0 \rangle$$

For example, the sum of two vectors $\vec{B} + (-\vec{B}) = \vec{0}$.

1.5.13 Change in a quantity: The Greek letter Δ

Frequently we will want to calculate the change in a quantity. For example, we may want to know the change in an moving object's position or the change in its velocity during some time interval. The Greek letter Δ (capital delta) is used to denote the change in a quantity (either a scalar or a vector).

We typically use the subscript *i* to denote an *initial* value of a quantity, and the subscript *f* to denote the *final* value of a quantity. If a vector $\vec{r}_i$ denotes the initial position of an object relative to the origin (its position at the beginning of the time interval), and $\vec{r}_f$ denotes the final position of the object, then

$$\Delta\vec{r} = \vec{r}_f - \vec{r}_i$$

$\Delta\vec{r}$ means "change of $\vec{r}$" or $\vec{r}_f - \vec{r}_i$ (displacement)

Δt means "change of t" or $t_f - t_i$ (time interval)

The symbol Δ (delta) always mean "final minus initial", not "initial minus final". For example, when a child's height changes from 1.1 m to 1.2 m, the change is $\Delta y = +0.1$ m, a positive number. If your bank account dropped from \$150 to \$130, what was the change in your balance? Δ(bank account) = −20 dollars.

Ex. 1.25 If $\vec{F}_1 = \langle 300, 0, -200 \rangle$ and $\vec{F}_2 = \langle 150, -300, 0 \rangle$, what is the sum $\vec{F}_1 + \vec{F}_2$?

Ex. 1.26 What is the magnitude of $\vec{F}_1$ (see Exercise 1.25)? What is the magnitude of $\vec{F}_2$? What is the magnitude of $\vec{F}_1 + \vec{F}_2$?

Ex. 1.27 What is the magnitude of $\vec{F}_1$ (see Exercise 1.25) plus the magnitude of $\vec{F}_2$? Is $|\vec{F}_1 + \vec{F}_2| = |\vec{F}_1| + |\vec{F}_2|$?

Ex. 1.28 What is the difference $\vec{F}_1 - \vec{F}_2$? What is $\vec{F}_2 - \vec{F}_1$?

Ex. 1.29 A snail is initially at location $\vec{r}_1 = \langle 3, 0, -7 \rangle$ m. At a later time the snail has crawled to location $\vec{r}_2 = \langle 2, 0, -8 \rangle$ m. What is $\Delta\vec{r}$, the change in the snail's position?

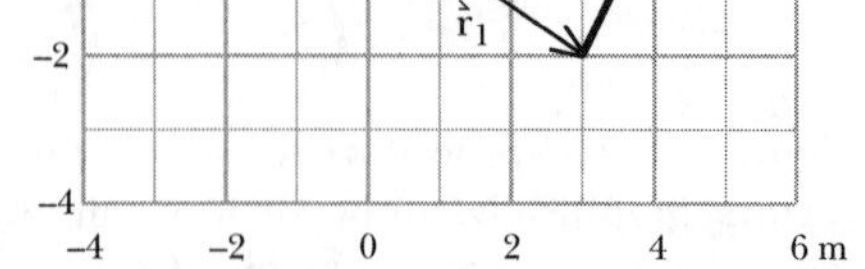

Figure 1.22 Relative position vector.

1.5.14 Relative position vectors

Vector subtraction is used to calculate relative position vectors, vectors which represent the position of an object relative to another object. In Figure 1.22 object 1 is at location $\vec{r}_1$ and object 2 is at location $\vec{r}_2$. We want the components of a vector that points from object 1 to object 2. This is the vector obtained by subtraction: $\vec{r}_{2\text{ relative to }1} = \vec{r}_2 - \vec{r}_1$. Note that the form is always "final" minus "initial" in these calculations.

Ex. 1.30 In Figure 1.22, $\vec{r}_1 = \langle 3, -2, 0 \rangle$ m and $\vec{r}_2 = \langle 5, 2, 0 \rangle$ m. Calculate the position of object 2 relative to object 1, as a relative position vector. Before checking the answer at the back of this chapter, see whether your answer is consistent with the appearance of the vector $\vec{r}_{2 \text{ relative to } 1} = \vec{r}_2 - \vec{r}_1$ shown in Figure 1.22.

Ex. 1.31 What is the position of object 1 relative to object 2, as a vector?

1.6 SI units

In this book we use the SI (Système Internationale) unit system. The SI unit of mass is the kilogram (kg), the unit of distance is the meter (m), and the unit of time is the second (s). In later chapters we will encounter other SI units, such as the newton (N), which is a unit of force. It is essential to use SI units in physics equations; this may require that you convert from some other unit system to SI units. If mass is known in grams, you need to divide by 1000 and use the mass in kilograms. If a distance is given in centimeters, you need to divide by 100 and convert the distance to meters. If the time is measured in minutes, you need to multiply by 60 to use a time in seconds. A convenient way to do such conversions is to multiply by factors which are equal to 1, such as (1 min)/(60 s) or (100 cm)/(1 m). As an example, consider converting 60 miles per hour to SI units, meters per second. Start with the 60 mi/hr and multiply by factors of 1:

$$\left(60\ \frac{\cancel{\text{mi}}}{\cancel{\text{hr}}}\right)\left(\frac{1\ \cancel{\text{hr}}}{60\ \cancel{\text{min}}}\right)\left(\frac{1\ \cancel{\text{min}}}{60\ \text{s}}\right)\left(\frac{5280\ \cancel{\text{ft}}}{1\ \cancel{\text{mi}}}\right)\left(\frac{12\ \cancel{\text{in}}}{1\ \cancel{\text{ft}}}\right)\left(\frac{2.54\ \cancel{\text{cm}}}{1\ \cancel{\text{in}}}\right)\left(\frac{1\ \text{m}}{100\ \cancel{\text{cm}}}\right) = 26.8\ \frac{\text{m}}{\text{s}}$$

Observe how most of the units cancel, leaving final units of m/s.

Ex. 1.32 A snail moved 80 cm (80 centimeters) in 5 minutes. What was its average speed in SI units? Write out the factors as was done above.

1.7 Velocity

We use vectors not only to describe the position of an object but also to describe velocity (speed and direction). If we know a object's present speed in meters per second and the object's direction of motion, we can predict where it will be a short time into the future. As we have seen, *change* of velocity is an indication of interaction. We need to be able to work with velocities of objects in 3D, so we need to learn how to use 3D vectors to represent velocities. After learning how to describe velocity in 3D, we will also learn how to describe *change* of velocity, which is related to interactions.

1.7.1 Average speed

The concept of speed is a familiar one. Speed is a single number, so it is a scalar quantity (speed is the magnitude of velocity). A world class sprinter can run 100 meters in 10 seconds. We say the sprinter's average speed is (100 m)/(10 s) = 10 m/s. In SI units speed is measured in meters per second, abbreviated "m/s".

A car that travels 100 miles in 2 hours has an average speed of (100 miles)/(2 hours) = 50 miles per hour, (or about 22 m/s). In symbols,

$$v_{avg} = \frac{d}{t}$$

where v_{avg} is the "average speed," d is the distance the car has traveled, and t is the elapsed time.

There are other useful versions of the basic relationship among average speed, distance, and time. For example,

$$d = v_{avg}t$$

expresses the fact that if you run 5 m/s for 7 seconds you go 35 meters. Or you can use

$$t = \frac{d}{v_{avg}}$$

to calculate that to go 3000 miles in an airplane that flies at 600 miles per hour will take 5 hours.

Units

While it is easy to make a mistake in one of the formulas relating speed, time interval, and change in position, it is also easy to catch such a mistake by looking at the units. If you had written $t = v_{avg}/d$, you would discover that the right hand side has units of (m/s)/m, or 1/s, not s. Always check units!

Instantaneous speed compared to average speed

If a car went 70 miles per hour for the first hour and 30 miles an hour for the second hour, it would still go 100 miles in 2 hours, with an average speed of 50 miles per hour. Note that during this two hour interval, the car was almost never actually traveling at its average speed of 50 miles per hour.

To find the "instantaneous" speed—the speed of the car at a particular instant—we should observe the short distance the car goes in a very short time, such as a hundredth of a second: If the car moves 0.3 meters in 0.01 s, its instantaneous speed is 30 meters per second.

1.7.2 Vector velocity

Earlier we calculated vector differences between two different objects. The vector difference $\vec{r}_2 - \vec{r}_1$ represented a *relative* position vector—the position of object 2 relative to object 1 at a particular time. Now we will be concerned with the change of position of *one* object during a time interval, and $\vec{r}_f - \vec{r}_i$ will represent the "displacement" of this single object during the time interval, where $\vec{r}_i$ is the initial 3D position and $\vec{r}_f$ is the final 3D position (note that as with relative position vectors, we always calculate "final minus initial"). Dividing the (vector) displacement by the (scalar) time interval $t_f - t_i$ (final time minus initial time) gives the average (vector) velocity of the object:

DEFINITION: AVERAGE VELOCITY

$$\vec{v}_{avg} = \frac{\vec{r}_f - \vec{r}_i}{t_f - t_i}$$

Another way of writing this expression, using the "Δ" symbol (Greek capital delta) to represent a change in a quantity, is:

$$\vec{v}_{avg} = \frac{\Delta\vec{r}}{\Delta t}$$

remembering that this is a compact notation for:

$$\vec{v}_{avg} = \frac{\Delta\vec{r}}{\Delta t} = \left\langle\frac{\Delta x}{\Delta t}, \frac{\Delta y}{\Delta t}, \frac{\Delta z}{\Delta t}\right\rangle$$

The magnitude of the average velocity, $|\vec{v}_{avg}|$, is called the average speed.

1.7.3 Determining average velocity from change in position

Consider a bee in flight (Figure 1.23). At time t_i = 15 s after 9:00 AM, the bee's position vector was $\vec{r}_i = \langle 2, 4, 0\rangle$ m. At time t_f = 15.1 s after 9:00 AM, the bee's position vector was $\vec{r}_f = \langle 3, 3.5, 0\rangle$ m. On the diagram, we draw and label the vectors $\vec{r}_i$ and $\vec{r}_f$.

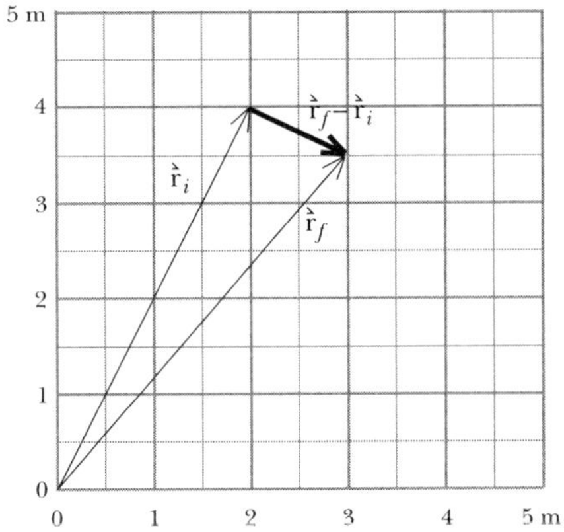

Figure 1.23 The displacement vector points from initial position to final position.

Next, on the diagram, we draw and label the vector $\vec{r}_f - \vec{r}_i$, with the tail of the vector at the bee's initial position. One useful way to think about this graphically is to ask yourself what vector needs to be added to the initial vector $\vec{r}_1$ to make the final vector $\vec{r}_f$, since $\vec{r}_i$ can be written in the form $\vec{r}_f = \vec{r}_i + (\vec{r}_f - \vec{r}_i)$.

The vector we just drew, the change in position $\vec{r}_f - \vec{r}_i$, is called the "displacement" of the bee during this time interval. This displacement vector points from the initial position to the final position, and we always calculate displacement as "final minus initial".

Note that the displacement $\vec{r}_f - \vec{r}_i$ refers to the positions of one object (the bee) at two different times, not the position of one object relative to a second object at one particular time ("relative position vector"). However, the vector subtraction is the same kind of operation for either kind of situation.

We calculate the bee's displacement vector numerically by taking the difference of the two vectors, final minus initial:

$$\vec{r}_f - \vec{r}_i = \langle 3, 3.5, 0\rangle \text{ m} - \langle 2, 4, 0\rangle \text{ m} = \langle 1, -0.5, 0\rangle \text{ m}$$

This numerical result should be consistent with our graphical construction. Look at the components of $\vec{r}_f - \vec{r}_i$ in Figure 1.23. Do you see that this vector has an x component of +1 and a y component of −0.5 m? Note that the (vector) displacement $\vec{r}_f - \vec{r}_i$ is in the direction of the bee's motion.

The average velocity of the bee, a vector quantity, is the (vector) displacement $\vec{r}_f - \vec{r}_i$ divided by the (scalar) time interval, $t_f - t_i$. Calculate the bee's average velocity:

$$\vec{v}_{avg} = \frac{\vec{r}_f - \vec{r}_i}{t_f - t_i} = \frac{\langle 1, -0.5, 0\rangle \text{ m}}{(15.1 - 15) \text{ s}} = \frac{\langle 1, -0.5, 0\rangle \text{ m}}{0.1 \text{ s}} = \langle 10, -5, 0\rangle \text{ m/s}$$

Since we divided $\vec{r}_f - \vec{r}_i$ by a scalar $(t_f - t_i)$, the average velocity $\vec{v}_{avg}$ points in the direction of the bee's motion, if the bee flew in a straight line.

What is the speed of the bee?

$$\text{speed of bee} = |\vec{v}_{avg}| = \sqrt{10^2 + (-5)^2 + 0^2} \text{ m/s} = 11.18 \text{ m/s}$$

What is the direction of the bee's motion, expressed as a unit vector?

$$\text{direction of bee: } \hat{v}_{avg} = \frac{\vec{v}_{avg}}{|\vec{v}_{avg}|} = \frac{\langle 10, -5, 0\rangle \text{ m/s}}{11.18 \text{ m/s}} = \langle 0.894, -0.447, 0\rangle$$

Note that the "m/s" units cancel; the result is dimensionless. We can check that this really is a unit vector:

$$\sqrt{0.894^2 + (-0.447)^2 + 0^2} = 0.9995$$

This is not quite 1.0 due to rounding the velocity coordinates and speed to three significant figures.

Put the pieces back together and see what we get. The original vector factors into the product of the magnitude times the unit vector:

$$|\vec{v}|\hat{v} = (11.18 \text{ m/s})\langle 0.894, -0.447, 0\rangle = \langle 10, -5, 0\rangle \text{ m/s}$$

This is the same as the original vector $\vec{v}$.

Ex. 1.33 At a time 0.2 seconds after it has been hit by a tennis racket, a tennis ball is located at $\langle 5, 7, 2\rangle$ m, relative to an origin in one corner of a tennis court. At a time 0.7 seconds after being hit, the ball is located at $\langle 9, 2, 8\rangle$ m.
(a) What is the average velocity of the tennis ball?
(b) What is the average speed of the tennis ball?
(c) What is the unit vector in the direction of the ball's velocity?

Ex. 1.34 A spacecraft is observed to be at a location $\langle 200, 300, -400\rangle$ m relative to an origin located on a nearby asteroid, and 5 seconds later is observed at location $\langle 325, 25, -550\rangle$ m.
(a) What is the average velocity of the spacecraft?
(b) What is the average speed of the spacecraft?
(c) What is the unit vector in the direction of the spacecraft's velocity?

1.7.4 Scaling a vector to fit on a graph

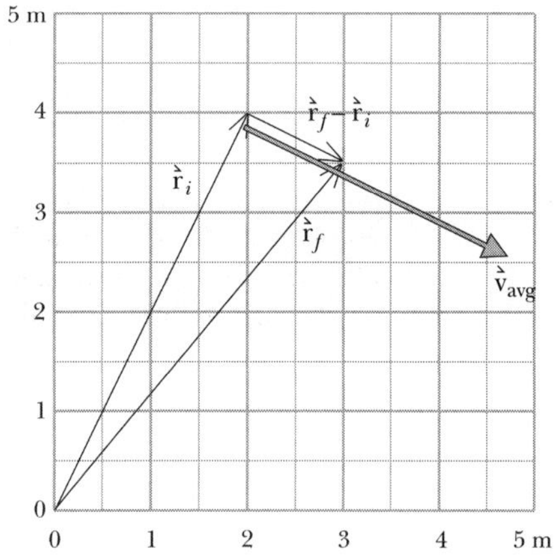

Figure 1.24 Average velocity vector: displacement divided by time interval.

We can plot the average velocity vector on the same graph that we use for showing the vector positions of the bee (Figure 1.24). However, note that velocity has units of m/s while positions have units of m, so in a way we're mixing apples and oranges.

Moreover, the magnitude of the vector, 11.18 m/s, doesn't fit on a graph that is only 5 units wide (in meters). It is standard practice in such situations to scale the arrow representing the vector down to fit on the graph, preserving the correct direction. In Figure 1.24 we've scaled the velocity vector down by about a factor of 3 to make the arrow fit on the graph. Of course if there is more than one velocity vector we use the same scale factor for all the velocity vectors. The same kind of scaling is used with other physical quantities that are vectors, such as force and momentum, which we will encounter later.

1.7.5 Predicting a new position

We can rewrite the velocity relationship in the form

$$(\vec{r}_f - \vec{r}_i) = \vec{v}_{avg}(t_f - t_i)$$

That is, the (vector) displacement of an object is its average (vector) velocity times the time interval. This is just the vector version of the simple notion that if you run at a speed of 7 m/s for 5 s you move a distance of (7 m/s) (5 s) = 35 m, or that a car going 50 miles per hour for 2 hours goes (50 mi/hr) (2 hr) = 100 miles.

Is $(\vec{r}_f - \vec{r}_i) = \vec{v}_{avg}(t_f - t_i)$ a valid vector relation? Yes, multiplying a vector $\vec{v}_{avg}$ times a scalar $t_f - t_i$ yields a vector. We make a further rearrangement to obtain a relation for updating the position when we know the velocity:

THE POSITION UPDATE FORMULA

$$\vec{r}_f = \vec{r}_i + \vec{v}_{avg}(t_f - t_i)$$

This equation says that if we know the starting position, the average velocity, and the time interval, we can predict the final position. This equation will be important throughout this course.

Using the position update formula

The position update formula $\vec{r}_f = \vec{r}_i + \vec{v}_{avg}(t_f - t_i)$ is a vector equation, so we can write out its full component form:

$$\langle x_f, y_f, z_f \rangle = \langle x_i, y_i, z_i \rangle + \langle v_{avg,x}, v_{avg,y}, v_{avg,z} \rangle (t_f - t_i)$$

Because the x component on the left of the equation must equal the x component on the right (and similarly for the y and z components), this compact vector equation represents three separate component equations:

$$x_f = x_i + v_{avg,x}(t_f - t_i)$$

$$y_f = y_i + v_{avg,y}(t_f - t_i)$$

$$z_f = z_i + v_{avg,z}(t_f - t_i)$$

Example

At time $t_i = 12.18$ s after 1:30 PM a ball's position vector is $\vec{r}_i = \langle 20, 8, -12 \rangle$ m. The ball's velocity at that moment is $\hat{v} = \langle 9, -4, 6 \rangle$ m/s. At time $t_f = 12.21$ s after 1:30 PM, where is the ball, assuming that its velocity hardly changes during this short time interval?

$$\vec{r}_f = \vec{r}_i + \vec{v}(t_f - t_i) = \langle 20, 8, -12 \rangle \text{m} + (\langle 9, -4, 6 \rangle \text{m/s})(12.21 - 12.18)\text{s}$$

$$\vec{r}_f = \langle 20, 8, -12 \rangle \text{ m} + \langle 0.27, -0.12, 0.18 \rangle \text{ m}$$

$$\vec{r}_f = \langle 20.27, 7.88, -11.82 \rangle \text{ m}$$

Note that if the velocity changes significantly during the time interval, in either magnitude or direction, our prediction for the new position may not be very accurate. In this case the velocity at the initial time could differ significantly from the average velocity during the time interval.

Ex. 1.35 A proton traveling with a velocity of $\langle 3\times10^5, 2\times10^5, -4\times10^5 \rangle$ m/s passes the origin at a time 9.0 seconds after a proton detector is turned on. Assuming the velocity of the proton does not change, what will its position be at time 9.7 seconds?

Ex. 1.36 How long does it take a baseball with velocity $\langle 30, 20, 25 \rangle$ m/s to travel from location $\vec{r}_1 = \langle 3, 7, -9 \rangle$ m to location $\vec{r}_2 = \langle 18, 17, 3.5 \rangle$ m?

Ex. 1.37 A "slow" neutron produced in a nuclear reactor travels from location $\langle 0.2, -0.05, 0.1 \rangle$ m to location $\langle -0.202, 0.054, 0.098 \rangle$ m in 2 microseconds ($1\,\mu\text{s} = 1\times10^{-6}$ s).
(a) What is the average velocity of the neutron?
(b) What is the average speed of the neutron?

1.7.6 Instantaneous velocity

The curved gray line in Figure 1.25 shows the path of a ball through the air. The gray dots mark the ball's position at time intervals of one second. While the ball is in the air, its velocity is constantly changing, due to interactions with the Earth (gravity) and with the air (air resistance).

Suppose we ask: What is the velocity of the ball at the precise instant that it reaches location *B*? This quantity would be called the "instantaneous velocity" of the ball. We can start by approximating the instantaneous velocity of the ball by finding its average velocity over some larger time interval.

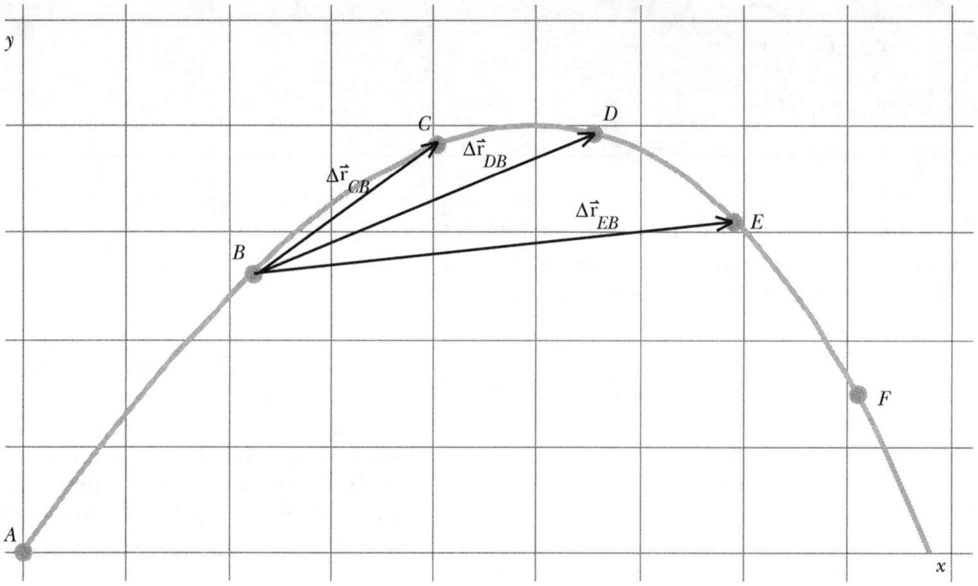

Figure 1.25 The trajectory of a ball through air. The axes represent x and y distance from the ball's initial location; each square on the grid corresponds to 10 meters. Three different displacements, corresponding to three different time intervals, are indicated by arrows on the diagram.

The table in Figure 1.26 shows the time and the position of the ball for each location marked by a gray dot in Figure 1.25. We can use these data to calculate the average velocity of the ball over three different intervals, by finding the ball's displacement during each interval, and dividing by the appropriate Δt for that interval:

loc.	t (s)	position (m)
A	0.0	$\langle 0, 0, 0\rangle$
B	1.0	$\langle 22.3, 26.1, 0\rangle$
C	2.0	$\langle 40.1, 38.1, 0\rangle$
D	3.0	$\langle 55.5, 39.2, 0\rangle$
E	4.0	$\langle 69.1, 31.0, 0\rangle$
F	5.0	$\langle 80.8, 14.8, 0\rangle$

Figure 1.26 Table showing elapsed time and position of the ball at each location marked by a dot in Figure 1.25.

$$\vec{v}_{EB} = \frac{\Delta\vec{r}_{EB}}{\Delta t} = \frac{\vec{r}_E - \vec{r}_B}{t_E - t_B} = \frac{(\langle 69.1, 31.0, 0\rangle - \langle 22.3, 26.1, 0\rangle)\text{m}}{(4.0 - 1.0)\text{s}}$$
$$= \langle 15.6, 1.6, 0\rangle\frac{\text{m}}{\text{s}}$$

$$\vec{v}_{DB} = \frac{\Delta\vec{r}_{DB}}{\Delta t} = \frac{\vec{r}_D - \vec{r}_B}{t_D - t_B} = \frac{(\langle 55.5, 39.2, 0\rangle - \langle 22.3, 26.1, 0\rangle)\text{m}}{(3.0 - 1.0)\text{s}}$$
$$= \langle 16.6, 6.55, 0\rangle\frac{\text{m}}{\text{s}}$$

$$\vec{v}_{CB} = \frac{\Delta\vec{r}_{CB}}{\Delta t} = \frac{\vec{r}_C - \vec{r}_B}{t_C - t_B} = \frac{(\langle 40.1, 38.1, 0\rangle - \langle 22.3, 26.1, 0\rangle)\text{m}}{(2.0 - 1.0)\text{s}}$$
$$= \langle 17.8, 12.0, 0\rangle\frac{\text{m}}{\text{s}}$$

Not surprisingly, the average velocities we calculate over these different time intervals are not the same, because both the direction of the ball's motion and the speed of the ball were changing continuously during its flight. The three average velocity vectors that we calculated are shown in Figure 1.27.

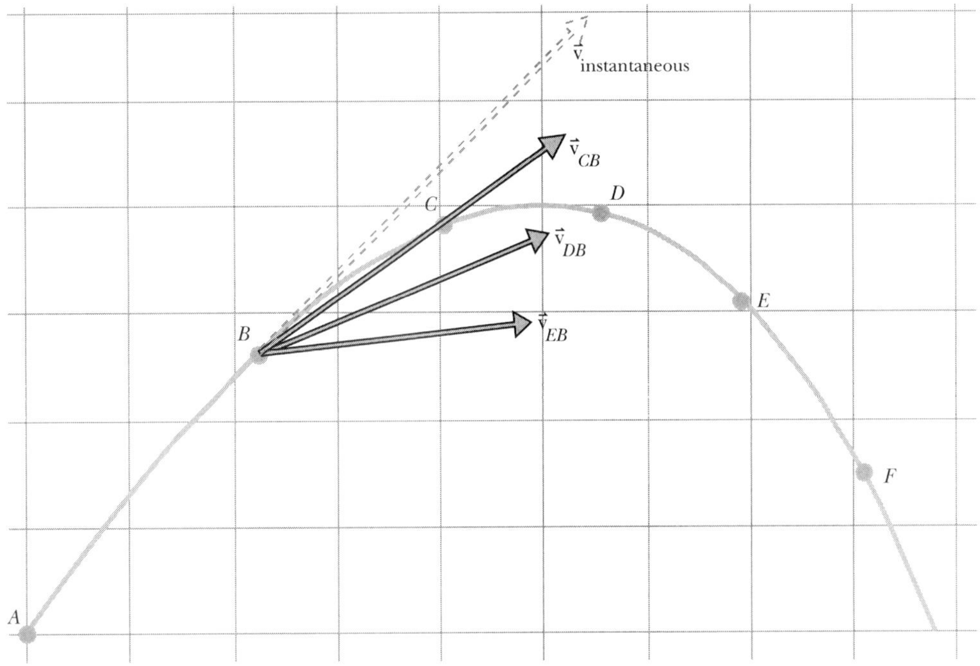

Figure 1.27 The three different average velocity vectors calculated above are shown by three arrows, each with its tail at location *B*. Note that since the units of velocity are m/s, these arrows use a different scale from the distance scale used for the path of the ball. The three arrows representing average velocities are drawn with their tails at the location of interest. The dashed arrow represents the actual instantaneous velocity of the ball at location *B*.

? Which of the three average velocity vectors depicted in Figure 1.27 best approximates the instantaneous velocity of the ball at location *B*?

Simply by looking at the diagram, we can tell that $\vec{v}_{CB}$ is closest to the actual instantaneous velocity of the ball at location *B*, because its direction is closest to the direction in which the ball is actually traveling. Because the direction of the instantaneous velocity is the direction the ball is moving at a particular instant, the instantaneous velocity is tangent to the ball's path. Of the three average velocity vectors we calculated, $\vec{v}_{CB}$ best approximates a tangent to the path of the ball. Evidently $\vec{v}_{CB}$, the velocity calculated with the shortest time interval, $t_C - t_B$, is the best approximation to the instantaneous velocity at location *B*. If we used even smaller values of Δt in our calculation of average velocity, such as 0.1 second, or 0.01 second, or 0.001 second, we would presumably have better and better estimates of the actual instantaneous velocity of the object at the instant when it passes location *B*.

Two important ideas have emerged from this discussion:

- The direction of the instantaneous velocity of an object is tangent to the path of the object's motion.
- Smaller time intervals yield more accurate estimates of instantaneous velocity.

1.7.7 Connection to calculus

You may already have learned about derivatives in calculus. The instantaneous velocity is a derivative, the limit of $\Delta\vec{r}/\Delta t$ as the time interval Δt used in the calculation gets closer and closer to zero:

$$\vec{v} = \lim_{\Delta t \to 0} \frac{\Delta \vec{r}}{\Delta t}, \text{ which is written as } \vec{v} = \frac{d\vec{r}}{dt}$$

In Figure 1.27, the process of taking the limit is illustrated graphically. As smaller values of Δt are used in the calculation, the average velocity vectors approach the limiting value: the actual instantaneous velocity.

A useful way to see the meaning of the derivative of a vector is to consider the components:

$$\vec{v} = \frac{d\vec{r}}{dt} = \frac{d}{dt}\langle x, y, z\rangle = \langle \frac{dx}{dt}, \frac{dy}{dt}, \frac{dz}{dt}\rangle = \langle v_x, v_y, v_z\rangle$$

The derivative of the position vector $\vec{r}$ gives components that are the components of the velocity, as we should expect.

Informally, you can think of $d\vec{r}$ as a very small ("infinitesimal") displacement, and dt as a very small ("infinitesimal") time interval. It is as though we had continued the process illustrated in Figure 1.25 to smaller and smaller time intervals, down to an extremely tiny time interval dt with a correspondingly tiny displacement $d\vec{r}$. The ratio of these tiny quantities is the instantaneous velocity.

The ratio of these two tiny quantities need not be small. For example, suppose an object moves in the x direction a tiny distance of 1×10^{-15} m, the radius of a proton, in a very short time interval of 1×10^{-23} s:

$$\vec{v} = \frac{\langle 1\times10^{-15}, 0, 0\rangle \text{ m}}{1\times10^{-23} \text{ s}} = \langle 1\times10^{8}, 0, 0\rangle \text{ m/s},$$

which is one-third the speed of light (3×10^{8} m/s)!

Change of magnitude and/or change in direction

Note that the time rate of change of a vector $\vec{r} = |\vec{r}|\hat{r}$ has two parts:

rate of change of the magnitude of the vector $\dfrac{d|\vec{r}|}{dt}$

rate of change of the direction of the vector $\dfrac{d\hat{r}}{dt}$

We will discuss this further in later sections.

1.7.8 Summary of velocity

DEFINITION OF AVERAGE VELOCITY $\vec{v}_{avg} = \dfrac{\Delta\vec{r}}{\Delta t} = \dfrac{\vec{r}_f - \vec{r}_i}{t_f - t_i}$

Equivalently: POSITION UPDATE FORMULA $\vec{r}_f = \vec{r}_i + \vec{v}_{avg}\Delta t$

DEFINITION OF INSTANTANEOUS VELOCITY $\vec{v} = \lim\limits_{\Delta t \to 0}\dfrac{\Delta\vec{r}}{\Delta t} = \dfrac{d\vec{r}}{dt}$

The symbol Δ (delta) means "change of": $\Delta t = t_f - t_i$, $\Delta\vec{r} = \vec{r}_f - \vec{r}_i$

The instantaneous velocity of an object is tangent to the path of the object.

To approximate the instantaneous velocity of an object, calculate its average velocity over a very short time interval.

1.8 Momentum

Newton's first law of motion:

An object moves in a straight line and at constant speed except to the extent that it interacts with other objects.

gives us a conceptual connection between interactions and their effects on the motion of objects. However, this law does not allow us to make quantitative (numerical) predictions or explanations—we could not have used this law to predict the exact trajectory of the ball shown in Figure 1.25, and we could not use this law alone to figure out how to send a rocket to the Moon. In order to make quantitative predictions or explanations of physical phenomena, we need a quantitative measure of interactions and a quantitative measure of effects of those interactions.

Newton's first law of motion does contain the important idea that if there is no interaction, a moving object will continue to move in a straight line, with no change of direction or speed, and an object that is not moving will remain at rest. A quantitative version of this law would provide a means of predicting the motion of an object, or of deducing how it must have moved in the past, if we could list all of its interactions with other objects.

1.8.1 Changes in velocity

? What factors make it difficult or easy to change an object's velocity?

You have probably noticed that if two objects have the same velocity but one is much more massive than the other, it is more difficult to change the heavy object's speed or direction. It is easier to stop a baseball traveling at a hundred miles per hour than to stop a car traveling at a hundred miles per hour! It is easier to change the direction of a canoe than to change the direction of a large, massive ship such as the Titanic (which couldn't change course quickly enough after the iceberg was spotted).

Momentum involves both mass and velocity

To take into account both an object's mass and its velocity, we can define a vector quantity called "momentum" that involves the product of mass (a scalar) and velocity (a vector). Instead of saying "the stronger the interaction, the bigger the change in the velocity," we now say "the stronger the interaction, the bigger the change in the momentum."

Momentum, a vector quantity, is usually represented by the symbol $\vec{p}$. We might expect that the mathematical expression for momentum would be simply $\vec{p} = m\vec{v}$, and indeed this is almost, but not quite, correct.

Experiments on particles moving at very high speeds, close to the speed of light $c = 3\times10^{8}$ m/s, show that changes in $m\vec{v}$ are not really proportional to the strength of the interactions. As you keep applying a force to a particle near the speed of light, the speed of the particle barely increases, and it is not possible to increase a particle's speed beyond the speed of light.

Through experiments it has been found that changes in the following quantity are proportional to the amount of interaction:

DEFINITION OF MOMENTUM

$$\vec{p} = \gamma m\vec{v}$$

where the proportionality factor γ (lower-case Greek gamma) is defined as

$$\gamma = \frac{1}{\sqrt{1 - \left(\frac{|\vec{v}|}{c}\right)^2}}$$

In these equations $\vec{p}$ represents momentum, $\vec{v}$ is the velocity of the object, m is the mass of the object, $|\vec{v}|$ is the magnitude of the object's velocity (the speed), and c is the speed of light. Momentum has units of $\text{kg}\cdot\text{m/s}$. To calculate momentum in these units, you must specify mass in kg and velocity in meters per second.

This is the "relativistic" definition of momentum. Albert Einstein in 1905 in his special theory of relativity predicted that this would be the appropriate definition for momentum at high speeds, a prediction that has been abundantly verified in a wide range of experiments.

Example

Suppose that a proton (mass 1.7×10^{-27} kg) is traveling with a velocity of $\langle 2\times10^7, 1\times10^7, -3\times10^7\rangle$ m/s. What is the momentum of the proton?

$$|\vec{v}| = \sqrt{(2\times10^7)^2 + (1\times10^7)^2 + (-3\times10^7)^2} = 3.7\times10^7\ \frac{\text{m}}{\text{s}}$$

$$\frac{|\vec{v}|}{c} = \frac{3.7\times10^7\ \frac{\text{m}}{\text{s}}}{3\times10^8\ \frac{\text{m}}{\text{s}}} = 0.12$$

$$\gamma = \frac{1}{\sqrt{1-(0.12)^2}} = 1.007$$

$$\vec{p} = (1.007)(1.7\times10^{-27}\text{kg})\langle 2\times10^7, 1\times10^7, -3\times10^7\rangle\frac{\text{m}}{\text{s}}$$

$$\vec{p} = \langle 3.4\times10^{-20}, 1.7\times10^{-20}, -5.1\times10^{-20}\rangle\frac{\text{kg}\cdot\text{m}}{\text{s}}$$

$\lvert\vec{v}\rvert$ m/s	$\lvert\vec{v}\rvert/c$	γ
0	0	1.0000
3	1×10^{-8}	1.0000
300	1×10^{-6}	1.0000
3×10^6	1×10^{-2}	1.0001
3×10^7	0.1	1.0050
1.5×10^8	0.5	1.1547
2.997×10^8	0.999	22.3663
2.9997×10^8	0.9999	70.7124
3×10^8	1	infinite! (and impossible)

Figure 1.28 Values of γ calculated for some speeds. γ is shown to four decimal places, which is more accuracy than we will usually need in this course.

1.8.2 Approximate formula for momentum

In the example above, we found that $\gamma = 1.007$. Since in that calculation we used only two significant figures, we could have used the approximation that $\gamma \approx 1.0$ without affecting our answer. Let's examine the expression for γ to see if we can come up with a guideline for when it is reasonable to use the approximate expression

$$\vec{p} \approx 1 \cdot m\vec{v} \quad \text{approximate expression for momentum}$$

Looking at the expression for γ

$$\gamma = \frac{1}{\sqrt{1-\left(\frac{|\vec{v}|}{c}\right)^2}}$$

we see that it depends only on the ratio of the speed of the object to the speed of light (the object's mass doesn't appear in this expression).

If $\frac{|\vec{v}|}{c}$ is a very small number, then $1-\left(\frac{|\vec{v}|}{c}\right)^2 \approx 1-0 \approx 1$, so $\gamma \approx 1$.

Some values of $(|\vec{v}|/c)$ and γ are displayed in Figure 1.28. From this table you can see that even at the very high speed where $|\vec{v}|/c = 0.1$, which means that $|\vec{v}| = 3\times10^7$ m/s, the relativistic factor γ is only slightly different from 1. For large-scale objects such as a space rocket, whose speed is typically only about 1×10^4 m/s, we can ignore the factor γ, and momentum is to a good approximation $\vec{p} \approx m\vec{v}$. It is only for high-speed cosmic rays or particles produced in high-speed particle accelerators that we need to use the full relativistic definition for momentum, $\vec{p} = \gamma m\vec{v}$.

From this table you can also get a sense of why it is not possible to exceed the speed of light. As you make a particle go faster and faster, approaching the speed of light, additional increases in the speed become increasingly difficult, because a tiny increase in speed means a huge increase in momentum, requiring huge amounts of interaction. In fact, for the speed to equal the speed of light, the momentum would have to increase to be infinite! There is a speed limit in the Universe, 3×10^8 m/s.

We will repeatedly emphasize the role of momentum throughout this course because of its fundamental importance not only in classical (prequantum) mechanics but also in relativity and quantum mechanics. The use of momentum clarifies the physics analysis of certain complex processes such as collisions, including collisions at speeds approaching the speed of light.

1.8.3 Direction of momentum

Like velocity, momentum is a vector quantity, so it has a magnitude and a direction.

? A leaf is blown by a gust of wind, and at a particular instant is traveling straight upward, in the $+y$ direction. What is the direction of the leaf's momentum?

The mathematical expression for momentum can be looked at as the product of a scalar part times a vector part. Since the mass m must be a positive number, and the factor gamma (γ) must be a positive number, this scalar factor cannot change the direction of the vector (Figure 1.29). Therefore the direction of the leaf's momentum is the same as the direction of its velocity: straight up (the $+y$ direction).

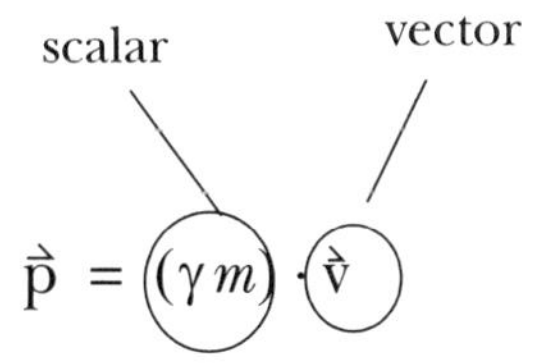

Figure 1.29 The expression for momentum is the product of a scalar times a vector. The scalar factor is positive, so the direction of an object's momentum is the same as the direction of its velocity.

Ex. 1.38 A good sprinter can run 100 meters in 10 seconds. What is the magnitude of the momentum of a sprinter whose mass is 65 kg and who is running at a speed of 10 m/s?

Ex. 1.39 A baseball has a mass of about 155 g. What is the magnitude of the momentum of a baseball thrown at a speed of 100 miles per hour? (Note that you need to convert mass to kilograms and speed to meters/second. See the inside back cover of the textbook for conversion factors.)

Ex. 1.40 What is the magnitude (in $\text{kg}\cdot\text{m/s}$) of the momentum of a 1000 kg airplane traveling at a speed of 500 miles per hour? (Note that you need to convert speed to meters per second.)

Ex. 1.41 What is the magnitude of the momentum of an electron traveling at a speed of 2×10^8 meters per second? (Masses of particles are given on the inside back cover of this textbook.)

Ex. 1.42 Show that when the speed is within one percent of the speed of light ($|\vec{v}| = 0.99c$), the ratio of the correct relativistic momentum to the approximate nonrelativistic momentum $m|\vec{v}|$ is quite large. Such speeds are attained in particle accelerators.

Ex. 1.43 If a particle has momentum $\vec{p} = \langle 4, -5, 2\rangle$ kg·m/s, what is the magnitude $|\vec{p}|$ of its momentum?

1.8.4 Change of momentum

In the next chapter we will introduce "the momentum principle" which quantitatively relates change in momentum $\Delta\vec{p}$ to the strength and duration of an interaction. In order to be able to use the momentum principle we need to know how to calculate changes in momentum.

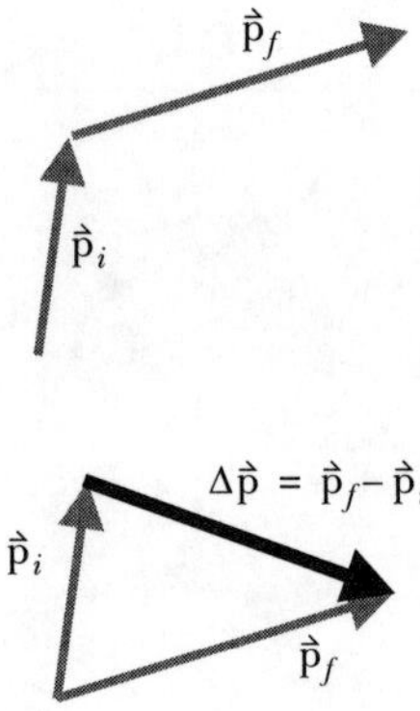

Figure 1.30 Calculation of $\Delta\vec{p}$.

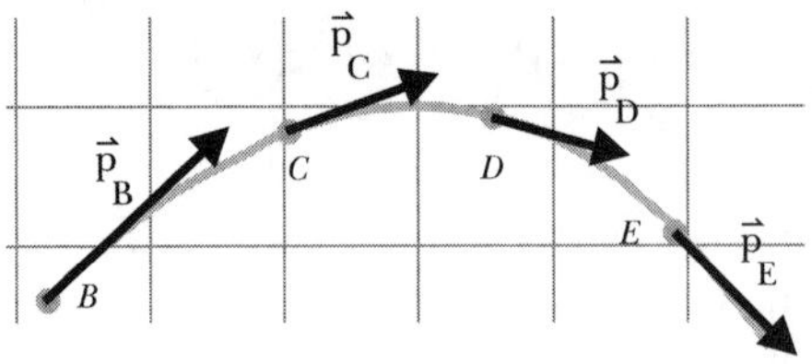

Figure 1.31 A portion of the trajectory of a ball moving through air, subject to gravity and air resistance. The arrows represent the momentum of the ball at the locations indicated by letters.

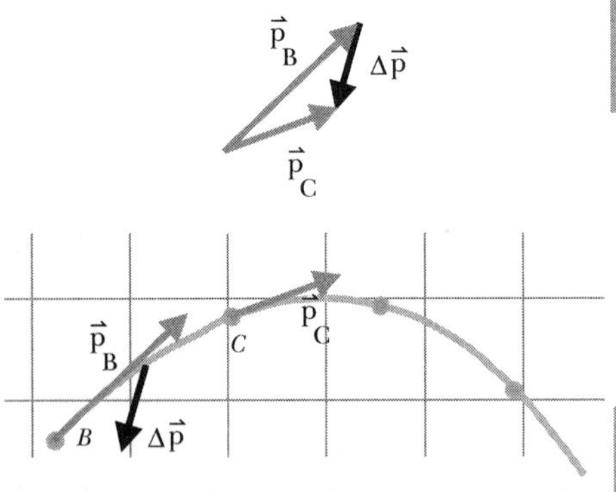

Figure 1.32 Graphical calculation of $\Delta\vec{p}$. The result is also shown superimposed on the ball's path, midway between the initial and final locations.

Momentum is a vector quantity, so just as was the case with velocity, there are two aspects of momentum that can change: magnitude and direction. A mathematical description of change of momentum must include either a change in the magnitude of the momentum, or a change in the direction of the momentum, or both.

Change of the vector momentum

The change in the momentum $\Delta\vec{p}$ during a time interval is a vector: $\Delta\vec{p} = \vec{p}_f - \vec{p}_i$. This vector expression captures both changes in magnitude and changes in direction. Figure 1.30 is a graphical illustration of a change from an initial momentum $\vec{p}_i$ to a final momentum $\vec{p}_f$. Place the initial and final momentum vectors tail to tail, then draw a vector from initial to final. This is the vector representing $\Delta\vec{p} = \vec{p}_f - \vec{p}_i$. This is the same procedure you used to calculate relative position vectors by subtraction, or displacement vectors by subtraction. The rule for subtracting vectors is always the same: Place the vectors tail to tail, then draw from the tip of the initial vector to the tip of the final vector. This resultant vector is "final minus initial".

Example

Figure 1.31 shows a portion of the trajectory of a ball in air, subject to gravity and air resistance. When the ball is at location B, its momentum is $\vec{p}_B = \langle 3.03, 2.83, 0\rangle\ \text{kg}\cdot\text{m/s}$. When it is at location C, its momentum is $\vec{p}_C = \langle 2.55, 0.97, 0\rangle\ \text{kg}\cdot\text{m/s}$. Find the change in the ball's momentum between these locations, and show it on the diagram.

$$\Delta\vec{p} = \vec{p}_C - \vec{p}_B = \langle 2.55, 0.97, 0\rangle\ \text{kg}\cdot\text{m/s} - \langle 3.03, 2.83, 0\rangle\ \text{kg}\cdot\text{m/s}$$
$$= \langle -0.48, -1.86, 0\rangle\ \text{kg}\cdot\text{m/s}$$

Both the x and y components of the ball's momentum decreased, so $\Delta\vec{p}$ has negative x and y components. This is consistent with the graphical subtraction shown in Figure 1.32.

Change in magnitude of momentum

If an object's speed changes (that is, the magnitude of its velocity changes), the magnitude of the object's momentum also changes. However, note that the change of the magnitude of momentum is not in general equal to the magnitude of change of momentum.

Example

Suppose you are driving a 1000 kilogram car at 20 m/s in the $+x$ direction. After making a 180 degree turn, you drive the car at 20 m/s in the $-x$ direction. (20 m/s is about 45 miles per hour or 72 km per hour.)

(a) What is the change of magnitude of the momentum of the car $\Delta|\vec{p}|$?

These speeds are very small compared to the speed of light, so we can use the approximate nonrelativistic formula for momentum.

$$\Delta|\vec{p}| = |\vec{p}_2| - |\vec{p}_1| \approx |m\vec{v}_2| - |m\vec{v}_1| = |m||\vec{v}_2| - |m||\vec{v}_1|$$
$$\Delta|\vec{p}| = (1000\ \text{kg})(20\ \text{m/s}) - (1000\ \text{kg})(20\ \text{m/s})$$
$$\Delta|\vec{p}| = 0\ \text{kg}\cdot\text{m/s}$$

(b) What is the magnitude of the change of momentum of the car $|\Delta\vec{p}|$?

$$\Delta\vec{p} = \vec{p}_2 - \vec{p}_1 = 1000 \text{ kg } \langle 20, 0, 0\rangle \text{ m/s} - 1000 \text{ kg } \langle -20, 0, 0\rangle \text{ m/s}$$

$$= \langle 4\times10^4, 0, 0\rangle \text{ kg}\cdot\text{m/s}$$

$$|\Delta\vec{p}| = 4\times10^4 \text{ kg}\cdot\text{m/s}$$

So $|\Delta\vec{p}| \neq \Delta|\vec{p}|$.

Change of direction of momentum

There are various ways to specify a change in the direction of motion. For example, if you use compass directions, you could say that an airplane changed its direction from 30° east of North to 45° east of North: a 15° clockwise change. One can imagine various other schemes, involving other kinds of coordinate systems. The standard way to deal with this is to use vectors, as in Figure 1.33, which shows the changing momentum of an object traveling at constant speed along a circular path.

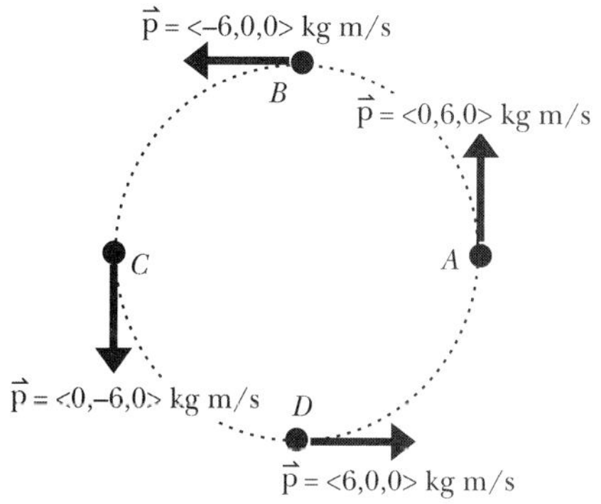

Figure 1.33 The momentum of an object traveling in a circle changes, even if the magnitude of momentum does not change.

Ex. 1.44 A tennis ball of mass 57 g travels with velocity $\langle 50, 0, 0\rangle$ m/s toward a wall. After bouncing off the wall, the tennis ball is observed to be traveling with velocity $\langle -48, 0, 0\rangle$ m/s.
(a) Draw a diagram showing the initial and final momentum of the tennis ball.
(b) What is the change in the momentum of the tennis ball?
(c) What is the change in the magnitude of the tennis ball's momentum?

Ex. 1.45 The planet Mars has a mass of 6.4×10^{23} kg, and travels in a nearly circular orbit around the Sun, as shown in Figure 1.34. When it is at location A, the velocity of Mars is $\langle 0, 0, -2.5\times10^4\rangle$ m/s. When it reaches location B, the planet's velocity is $\langle -2.5\times10^4, 0, 0\rangle$ m/s.
(a) What is $\Delta\vec{p}$, the change in the momentum of Mars between locations A and B?
(b) On a copy of the diagram in Figure 1.34, draw two arrows representing the momentum of Mars at locations *C* and D, paying attention to both the length and direction of each arrow.
(c) What is the direction of the change in the momentum of Mars between locations *C* and *D*? Draw the vector $|\Delta\vec{p}|$ on your diagram.

Ex. 1.46 A 50 kg child is riding on a carousel (merry-go-round) at a constant speed of 5 m/s. What is the magnitude of the change in the child's momentum $|\Delta\vec{p}|$ in going all the way around (360°)? In going halfway around (180°)? Draw a diagram showing the initial vector momentum and the final vector momentum, then subtract, then find the magnitude.

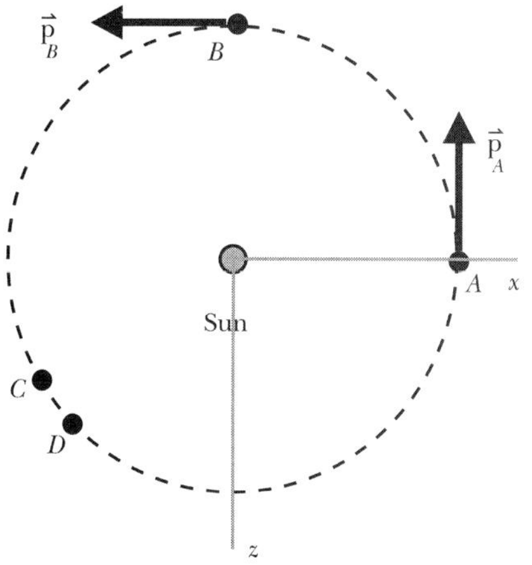

Figure 1.34 The nearly circular orbit of Mars around the Sun, viewed from above the orbital plane. Not to scale: the sizes of the Sun and Mars are exaggerated.

1.8.5 Average rate of change of momentum

The rate of change of the vector position is such an important quantity that it has a special name: "velocity". In Section 1.7 we discussed how to find both the average rate of change of position (average velocity) and the instantaneous rate of change of position (instantaneous velocity) of an object.

The average rate of change of momentum and the instantaneous rate of change of momentum are also extremely important quantities. In some situations, we will only be able to find an average rate of change of momentum:

AVERAGE RATE OF CHANGE OF MOMENTUM

$$\frac{\Delta\vec{p}}{\Delta t} = \frac{\vec{p}_f - \vec{p}_i}{t_f - t_i}$$

This quantity is a vector, and points in the direction of $\Delta\vec{p}$.

Example

If the momentum of a ball changes from $\langle 1, 2, 0\rangle$ kg · m/s to $\langle 0.5, 0, 0.5\rangle$ kg · m/s in half a second, the average rate of change of momentum of the ball is

$$\frac{(\langle 0.5, 0, 0.5\rangle - \langle 1, 2, 0\rangle)\ \text{kg}\cdot\text{m/s}}{0.5\ \text{s}} = \langle -1, -4, 1\rangle\ \frac{\text{kg}\cdot\text{m}}{\text{s}^2}$$

1.9 *The principle of relativity

Sections marked with a "*" are optional. They provide additional information and context, but later sections of the textbook don't depend critically on them. This optional section deals with some deep issues about the "reference frame" from which you observe motion. Newton's first law of motion only applies in an "inertial reference frame," which we will discuss here in the context of the principle of relativity.

A great variety of experimental observations has led to the establishment of the following principle:

THE PRINCIPLE OF RELATIVITY

Physical laws work in the same way for observers in uniform motion as for observers at rest.

This principle is called "the principle of relativity." (Einstein's extensions of this principle are known as "special relativity" and "general relativity.") Phenomena observed in a room in uniform motion (for example, on a train moving with constant speed on a smooth straight track) obey the same physical laws in the same way as experiments done in a room that is not moving. According to this principle, Newton's first law of motion should be true both for an observer moving at constant velocity and for an observer at rest.

For example, suppose you're riding in a car moving with constant velocity, and you're looking at a map lying on the dashboard. As far as you're concerned, the map isn't moving, and no interactions are required to hold it still on the dashboard. Someone standing at the side of the road sees the car go by, sees the map moving at a high speed in a straight line, and can see that no interactions are required to hold the map still on the dashboard. Both you and the bystander agree that Newton's first law of motion is obeyed: the bystander sees the map moving with constant velocity in the absence of interactions, and you see the map not moving at all (a zero constant velocity) in the absence of interactions.

On the other hand, if the car suddenly speeds up, it moves out from under the map, which ends up in your lap. To you it looks like "the map sped up in the backwards direction" without any interactions to cause this to happen, which looks like a violation of Newton's first law of motion. The problem is that you're strapped to the car, which is an accelerated reference frame, and Newton's first law of motion applies only to nonaccelerated reference frames, called "inertial" reference frames. Similarly, if the car suddenly turns to the right, moving out from under the map, the map tends to keep going in its original direction, and to you it looks like "the map moved to the left" without any interactions. So a change of speed or a change of direction of the car (your reference frame) leads you to see the map behave in a strange way.

The bystander, who is in an inertial (non-accelerating) reference frame, doesn't see any violation of Newton's first law of motion. The bystander's reference frame is an inertial frame, and the map behaves in an understandable way, tending to keep moving with the same speed and direction when the car changes speed or direction.

The cosmic microwave background

The principle of relativity, and Newton's first law of motion, apply only to observers who have a constant speed and direction (or zero speed) relative to the "cosmic microwave background," which provides the only backdrop and frame of reference with an absolute, universal character. It used to be that the basic reference frame was loosely called "the fixed stars," but stars and galaxies have their own individual motions within the Universe and do not constitute an adequate reference frame with respect to which to measure motion.

The cosmic microwave background is low-intensity electromagnetic radiation with wavelengths in the microwave region, which pervades the Universe, radiating in all directions. Measurements show that our galaxy is moving through this microwave radiation with a large, essentially constant velocity, toward a cluster of a large number of other galaxies. The way we detect our motion relative to the microwave background is through the "Doppler shift" of the frequencies of the microwave radiation, toward higher frequencies in front of us and lower frequencies behind. This is essentially the same phenomenon as that responsible for a fire engine siren sounding at a higher frequency when it is approaching us and a lower frequency when it is moving away from us.

The discovery of the cosmic microwave background provided major support for the "Big Bang" theory of the formation of the Universe. According to the Big Bang theory, the early Universe must have been an extremely hot mixture of charged particles and high-energy, short-wavelength electromagnetic radiation (visible light, x-rays, gamma rays, etc.). Electromagnetic radiation interacts strongly with charged particles, so light could not travel very far without interacting, making the Universe essentially opaque. Also, the Universe was so hot that electrically neutral atoms could not form without the electrons immediately being stripped away again by collisions with other fast-moving particles.

As the Universe expanded, the temperature dropped. Eventually the temperature was low enough for neutral atoms to form. The interaction of electromagnetic radiation with neutral atoms is much weaker than with individual charged particles, so the radiation was now essentially free, dissociated from the matter, and the Universe became transparent. As the Universe continued to expand (the actual space between clumps of matter got bigger!), the wavelengths of the electromagnetic radiation got longer, until today this fossil radiation has wavelengths in the relatively low-energy, long-wavelength microwave portion of the electromagnetic spectrum.

Inertial frames of reference

It is an observational fact that in reference frames that are in uniform motion with respect to the cosmic microwave background, far from other objects (so that interactions are negligible), an object maintains uniform motion. Such frames are called "inertial frames" and are reference frames in which Newton's first law of motion is valid.

? Is the surface of the Earth an inertial frame?

No! The Earth is rotating on its axis, so the velocity of an object sitting on the surface of the Earth is constantly changing direction, as is a coordinate

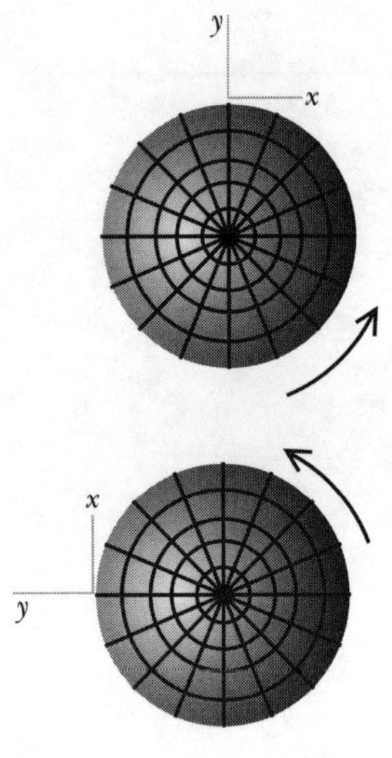

Figure 1.35 Axes tied to the Earth rotate through 90° in a quarter of a day (6 hours).

frame tied to the Earth (Figure 1.35). Moreover, the Earth is orbiting the Sun, and the Solar System itself is orbiting the center of our Milky Way galaxy, and our galaxy is moving toward other galaxies. So the motion of an object sitting on the Earth is actually quite complicated and definitely not uniform with respect to the cosmic microwave background.

However, for many purposes the surface of the Earth can be considered to be (approximately) an inertial frame. For example, it takes 6 hours for the rotation of the Earth on its axis to make a 90° change in the direction of the velocity of a "fixed" point. If a process of interest takes only a few minutes, during these few minutes a "fixed" point moves in nearly a straight line at constant speed due to the Earth's rotation, and velocity changes in the process of interest are typically much larger than the very small velocity change of the approximate inertial frame of the Earth's surface.

Similarly, although the Earth is in orbit around the Sun, it takes 365 days to go around once, so for a period of a few days or even weeks the Earth's orbital motion is nearly in a straight line at constant speed. Hence for many purposes the Earth represents an approximately inertial frame despite its motion around the Sun.

The special theory of relativity

Einstein's special theory of relativity (published in 1905) built on the basic principle of relativity but added the conjecture that the speed of a beam of light must be the same as measured by observers in different frames of reference in uniform motion with respect to each other. In Figure 1.36, observers on each spaceship measure the speed of the light c emitted by the ship at the top to be the same ($c = 3\times10^8$ m/s), despite the fact that they are moving at different velocities.

This additional condition seems peculiar and has far-reaching consequences. After all, the map on the dashboard of your car has different speeds relative to different observers, depending on the motion of the observer. Yet a wide range of experiments has confirmed Einstein's conjecture: all observers measure the same speed for the same beam of light, $c = 3\times10^8$ m/s. (The color of the light is different for the different observers, but the speed is the same.)

On the other hand, if someone on the ship at the top throws a ball or a proton or some other piece of matter, the speed of the object will be different for observers on the three ships; it is only light whose speed is independent of the observer.

Einstein's theory has interesting consequences. For example, it predicts that time will run at different rates in different frames of reference. These predictions have been confirmed by many experiments. These unusual effects are large only at very high speeds (a sizable fraction of the speed of light), which is why we don't normally observe these effects in everyday life, and why we can use nonrelativistic calculations for low-speed phenomena.

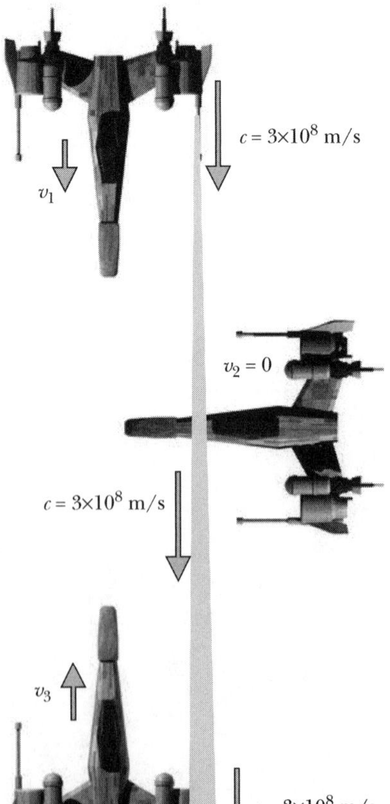

Figure 1.36 Light emitted by the top spaceship is measured to have the same speed by observers in all three ships.

1.10 Summary

Interactions (Section 1.2 and Section 1.4)

Interactions are indicated by

- change of velocity (change of direction and/or change of speed)
- change of identity
- change of shape of multiparticle system
- change of temperature of multiparticle system
- lack of change when change is expected

Newton's first law of motion (Section 1.3)

An object moves in a straight line and at constant speed except to the extent that it interacts with other objects.

Vectors (Section 1.5)

A 3D *vector* is a quantity with magnitude and a direction, which can be expressed as a triple $\langle x, y, z\rangle$. A vector is indicated by an arrow: $\vec{r}$

A *scalar* is a single number.

Legal mathematical operations involving vectors include:

- adding one vector to another vector
- subtracting one vector from another vector
- multiplying or dividing a vector by a scalar
- finding the magnitude of a vector
- taking the derivative of a vector

Operations that are *not* legal with vectors include:

- A vector cannot be added to a scalar
- A vector cannot be set equal to a scalar
- A vector cannot appear in the denominator (you can't divide by a vector)

A unit vector $\hat{r} = \vec{r}/|\vec{r}|$ has magnitude 1.
A vector can be factored using a unit vector: $\vec{F} = |\vec{F}|\hat{F}$

The symbol Δ

The symbol Δ (delta) means "change of": $\Delta t = t_f - t_i$, $\Delta\vec{r} = \vec{r}_f - \vec{r}_i$

Δ always means "final minus initial".

Velocity and change of position (Section 1.7)

DEFINITION: AVERAGE VELOCITY

$$\vec{v}_{avg} = \frac{\Delta\vec{r}}{\Delta t} = \frac{\vec{r}_f - \vec{r}_i}{t_f - t_i}$$

Velocity is a vector. $\vec{r}$ is the position of an object (a vector). t is the time. Average velocity is equal to the change in position divided by the time elapsed. SI units of velocity are meters per second (m/s).

THE POSITION UPDATE FORMULA

$$\vec{r}_f = \vec{r}_i + \vec{v}_{avg}\Delta t$$

The final position (vector) is the vector sum of the initial position plus the product of the average velocity and the elapsed time.

DEFINITION: INSTANTANEOUS VELOCITY

$$\vec{v} = \lim_{\Delta t \to 0} \frac{\Delta \vec{r}}{\Delta t} = \frac{d\vec{r}}{dt}$$

The instantaneous velocity is the limiting value of the average velocity as the time elapsed becomes very small.

Momentum (Section 1.8)

DEFINITION: MOMENTUM

$$\vec{p} = \gamma m\vec{v}$$

$$\text{where } \gamma = \frac{1}{\sqrt{1-(|\vec{v}|/c)^2}} \text{ (lower-case Greek gamma)}$$

Momentum (a vector) is the product of the relativistic factor "gamma" (a scalar), mass, and velocity.

$$\text{Combined into one equation: } \vec{p} = \frac{1}{\sqrt{1-(|\vec{v}|/c)^2}} m\vec{v}$$

APPROXIMATION: MOMENTUM AT LOW SPEEDS

$\vec{p} \approx m\vec{v}$ at speeds such that $|\vec{v}| << c$.

Useful numbers:

Radius of a typical atom: about 1×10^{-10} meter.
Radius of a proton or neutron: about 1×10^{-15} meter
Speed of light: 3×10^{8} m/s

These and other useful data and conversion factors are given on the inside back cover of the textbook.

1.11 Review questions

The purpose of review questions is to help you reflect on the most important concepts in the chapter. Try to answer the questions without flipping through the chapter looking for an answer, but think about what you know already from having done the exercises throughout the chapter. If you are stumped, look at the "Summary" on the preceding page.

If you are still stumped after looking at the chapter summary, make a note in your notebook, as a reminder that you may need to spend some extra time studying this particular concept.

Detecting interactions

RQ 1.1 Give two examples (other than those discussed in the text) of interactions that may be detected by observing:
- (a) change in velocity
- (b) change in temperature
- (c) change in shape
- (d) change in identity
- (e) lack of change when change is expected

RQ 1.2 In which of the following situations is there observational evidence for significant interaction between two objects? How can you tell?
- (a) a book rests on a table
- (b) a baseball that was hit by a batter flies toward the outfield
- (c) water freezes in an ice cube tray in the freezer
- (d) a communications satellite orbits the earth
- (e) a space probe leaves the solar system traveling at constant speed toward a distant star
- (f) a charged particle leaves a curving track in a particle detector

RQ 1.3 Which of the following observations give conclusive evidence of an interaction? (Choose all that apply.)
- (a) Change of velocity, either change of direction or change of speed.
- (b) Change of shape or configuration without change of velocity.
- (c) Change of position without change of velocity.
- (d) Change of identity without change of velocity.
- (e) Change of temperature without change of velocity.

Explain your choice.

RQ 1.4 Moving objects left the traces labelled A - F in Figure 1.37. The dots were deposited at equal time intervals (for example, one dot each second). In each case the object starts from the square. Which trajectories show evidence that the moving object was interacting with another object somewhere? If there is evidence for an interaction, what is the evidence?

RQ 1.5 A spaceship far from all other objects uses its rockets to attain a speed of 1×10^4 m/s. The crew then shuts off the power. According to Newton's first law, which of the following statements about the motion of the spaceship after the power is shut off are correct? (Choose all statements that are correct.)
- (a) The spaceship will move in a straight line.
- (b) The spaceship will travel on a curving path.
- (c) The spaceship will enter a circular orbit.
- (d) The speed of the spaceship will not change.
- (e) The spaceship will gradually slow down.
- (f) The spaceship will stop suddenly.

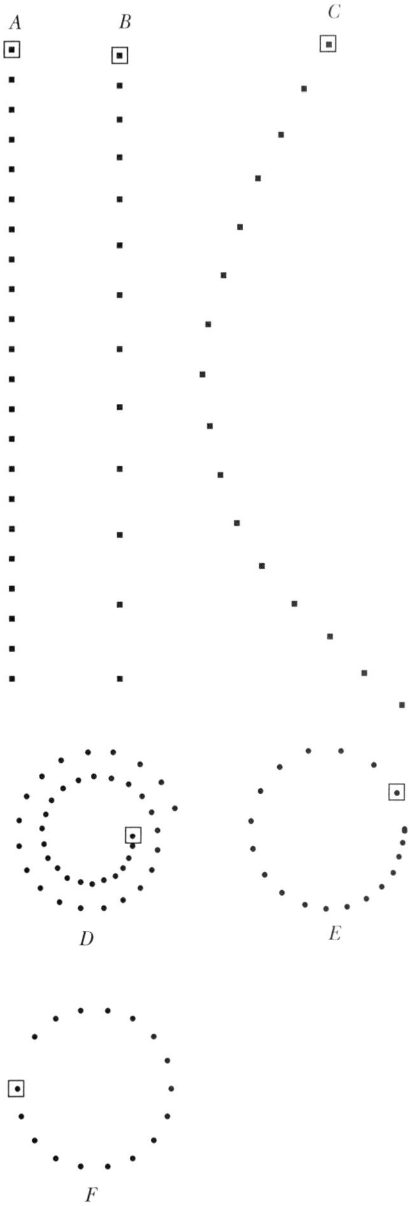

Figure 1.37 Traces left by moving objects. The dots mark the objects' positions at equal time intervals (RQ 1.4).

RQ 1.6 Why do we use a spaceship in outer space, far from other objects, to illustrate Newton's first law? Why not a car or a train? (More than one of the following statements may be correct.)

A car or train touches other objects, and interacts with them.
A car or train can't travel fast enough.
The spaceship has negligible interactions with other objects.
A car or train interacts gravitationally with the Earth.
A spaceship can never experience a gravitational force.

RQ 1.7 You slide a coin across the floor, and observe that it travels in a straight line, slowing down and eventually stopping. A sensitive thermometer shows that the coin's temperature increased. What can we conclude? (Choose all statements that are correct.)

(a) Because the coin traveled in a straight line, we conclude that it did not interact with anything.

(b) Because the coin did not change shape, we conclude that it did not interact with anything.

(c) Because the coin slowed down, we conclude that Newton's first law does not apply to objects in everyday life, such as coins.

(d) Because the coin's speed changed, we conclude that it interacted with one or more other objects.

(e) Because the coin got hot, we conclude that it interacted with one or more other objects.

RQ 1.8 Some science museums have an exhibit called a Bernoulli blower, in which a volleyball hangs suspended in a column of air blown upward by a strong fan. If you saw a ball suspended in the air but didn't know the blower was there, why would Newton's first law suggest that something must be holding the ball up?

RQ 1.9 Place a ball on a book and walk with the book in uniform motion. Note that you don't really have to do anything to the ball to keep the ball moving with constant velocity (relative to the ground) or to keep the ball at rest (relative to you). Then stop suddenly, or abruptly change your direction or speed. What does Newton's first law of motion predict for the motion of the ball (assuming the interaction between the ball and the book is small)? Does the ball behave as predicted? It may help to take the point of view of a friend who is standing still, watching you.

Velocity

RQ 1.10 How does average velocity differ from instantaneous velocity?

RQ 1.11 Start with the definition of average velocity and derive the position update formula from it. Show all steps in the derivation.

RQ 1.12 In the expression $\Delta\vec{r}/\Delta t$, what is the meaning of $\Delta\vec{r}$? What is the meaning of Δt?

Momentum

RQ 1.13 Which of the following statements about the velocity and momentum of an object are correct?

(a) The momentum of an object is always in the same direction as its velocity

(b) The momentum of an object can either be in the same direction as its velocity or in the opposite direction

(c) The momentum of an object is perpendicular to its velocity

(d) The direction of an object's momentum is not related to the direction of its velocity

(e) The direction of an object's momentum is tangent to its path

RQ 1.14 In which of these situations is it reasonable to use the approximate formula for the momentum of an object, instead of the full relativistically correct formula?

(a) A car traveling on an interstate highway
(b) A commercial jet airliner flying between New York and Seattle
(c) A neutron traveling at 2700 meters per second
(d) A proton in outer space traveling at 2×10^8 m/s
(e) An electron in a television tube traveling 3×10^6 m/s

RQ 1.15 Answer the following questions about the factor γ (gamma) in the full relativistic formula for momentum.

(a) Is γ a scalar or a vector quantity?
(b) What is the minimum possible value of γ?
(c) Does γ reach its minimum value when an object's speed is high or low?
(d) Is there a maximum possible value for γ?
(e) Does γ become large when an object's speed is high or low?
(f) Does the approximation $\gamma \approx 1$ apply when an object's speed is low or when it is high?

Change of momentum

RQ 1.16 A tennis ball of mass m traveling with velocity $\langle v_x, 0, 0\rangle$ hits a wall and rebounds with velocity $\langle -v_x, 0, 0\rangle$.

(a) What was the change in momentum of the tennis ball?
(b) What was the change in the magnitude of the momentum of the tennis ball?

Relativity

RQ 1.17 Which of the following observers might observe something that appears to violate Newton's first law of motion? Explain why.

(a) a person standing still on a street corner
(b) a person riding on a roller coaster
(c) a passenger on a starship travelling at $0.75c$ toward the nearby star Alpha Centauri
(d) an airplane pilot doing aerobatic loops
(e) a hockey player coasting across the ice

RQ 1.18 A spaceship at rest with respect to the cosmic microwave background emits a beam of red light. A different spaceship, moving at a speed of 2.5×10^8 m/s toward the first ship, detects the light. Which of the following statements are true for observers on the second ship? (More than one statement may be correct.)

(a) They observe that the light travels at 3×10^8 m/s.
(b) They see light that is not red.
(c) They observe that the light travels at 5.5×10^8 m/s
(d) They observe that the light travels at 2.5×10^8 m/s.

1.12 Problems

Vectors

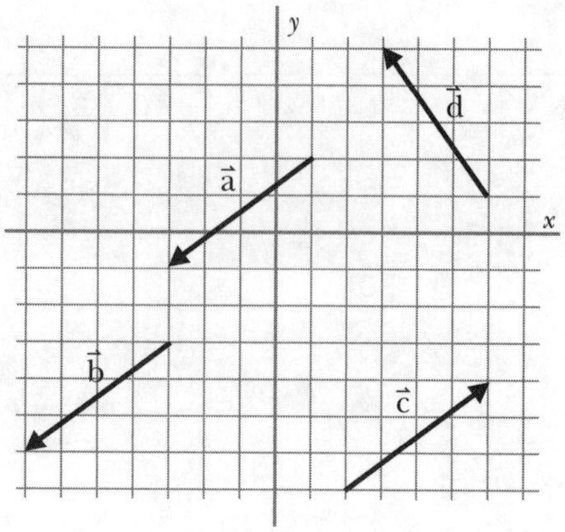

Figure 1.38 These arrows represent the vectors referred to in Problem 1.2.

Problem 1.1 On a piece of graph paper, draw arrows representing the following vectors. Make sure the tip and tail of each arrow you draw are clearly distinguishable.

(a) Placing the tail of the vector at $\langle 5, 2, 0\rangle$ draw an arrow representing the vector $\vec{p} = \langle -7, 3, 0\rangle$. Label it $\vec{p}$.

(b) Placing the tail of the vector at $\langle -5, 8, 0\rangle$ draw an arrow representing the vector $-\vec{p}$. Label it $-\vec{p}$.

Problem 1.2 The following questions refer to the vectors depicted by arrows in Figure 1.38.

(a) What are the components of the vector $\vec{a}$? (Note that since the vector lies in the xy plane, its z component is zero.)

(b) What are the components of the vector $\vec{b}$?

(c) Is this statement true or false? $\vec{a} = \vec{b}$

(d) What are the components of the vector $\vec{c}$?

(e) Is this statement true or false: $\vec{c} = -\vec{a}$?

(f) What are the components of the vector $\vec{d}$?

(g) Is this statement true or false: $\vec{d} = -\vec{c}$?

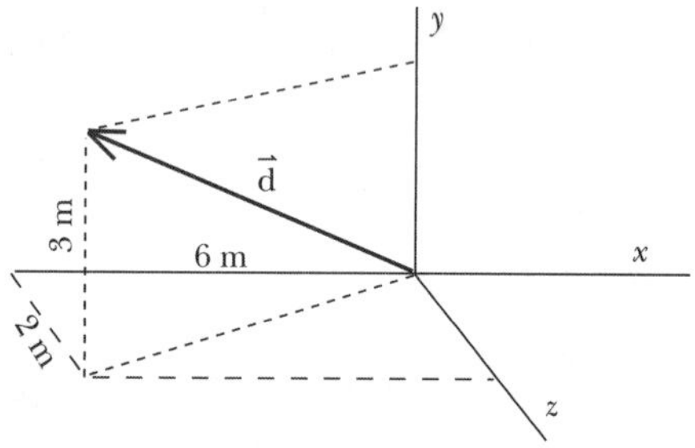

Figure 1.39 An arrow representing a position vector (Problem 1.3)

Problem 1.3

(a) What are the components of the vector $\vec{d}$, in Figure 1.39?

(b) If $\vec{e} = -\vec{d}$, what are the components of $\vec{e}$?

(c) If the tail of vector $\vec{d}$ were moved to location $\langle -5, -2, 4\rangle$ m, where would the tip of the vector be located?

(d) If the tail of vector $-\vec{d}$ were placed at location $\langle -1, -1, -1\rangle$, where would the tip of the vector be located?

Problem 1.4 Figure 1.40 shows several arrows representing vectors in the *xy* plane.

(a) Which vectors have magnitudes equal to the magnitude of $\vec{a}$?

(b) Which vectors are equal to $\vec{a}$?

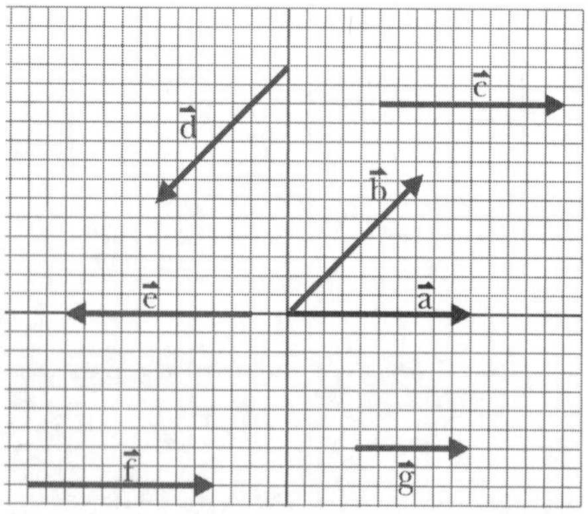

Figure 1.40 Arrows representing vectors in the *xy* plane (Problem 1.4).

Problem 1.5 Consider a vector $\vec{u} = \langle u_x, u_y, u_z\rangle$, and another vector $\vec{p} = \langle p_x, p_y, p_z\rangle$. If $\vec{u} = \vec{p}$, then which of the following statements must be true? Some, all, or none of the following may be true:

(i) $u_x = p_x$ (ii) $u_y = p_y$ (iii) $u_z = p_z$ (iv) The direction of $\vec{u}$ is the same as the direction of $\vec{p}$.

Vector operations

Problem 1.6 In the diagram in Figure 1.41 three vectors are represented by arrows in the *xy* plane. Each square in the grid represents one meter. For each vector:

(a) Write out the components of the vector.

(b) Calculate the magnitude of the vector.

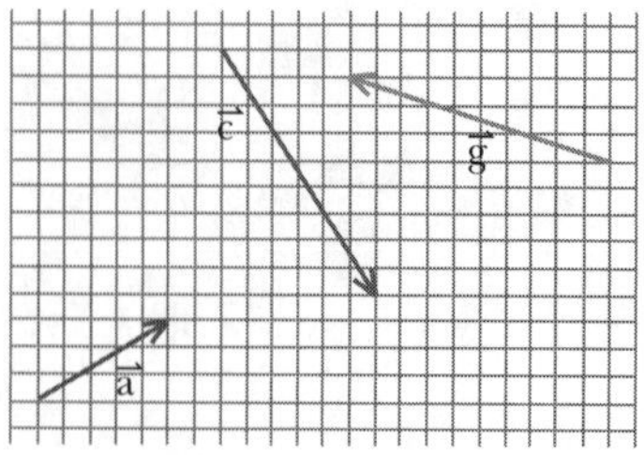

Figure 1.41 Arrows representing vectors in the *xy* plane (Problem 1.6).

Problem 1.7 Imagine that you have a baseball and a tennis ball at different locations. The center of the baseball is at $\langle 3, 5, 0\rangle$ m, and the center of the tennis ball is at $\langle -3, -1, 0\rangle$ m. On a piece of graph paper, do the following:

(a) Draw dots at the locations of the center of the baseball and the center of the tennis ball.

(b) Draw the position vector of the baseball, which is an arrow whose tail is at the origin and whose tip is at thc location of the baseball. Label this position vector $\vec{B}$. Clearly show the tip and tail of each arrow.

(c) Complete this equation: $\vec{B}$ = < ____, ____, ____ > m.

(d) Draw the position vector of the tennis ball. Label it $\vec{T}$.
(e) Complete this equation: $\vec{T}$ = < ____, ____, ____> m.
(f) Draw the relative position vector for the tennis ball relative to the baseball. The tail of this vector is at the center of the baseball, and the tip of the vector is at the center of the tennis ball. Label this relative position vector $\vec{r}$.
(g) Complete the following equation by reading the coordinates of $\vec{r}$ from your graph: $\vec{r}$ = < ____ , ____ , ____ > m
(h) Calculate the following difference: $\vec{T} - \vec{B}$ = < ____ , ____ , ____ > m.
(i) Is the following statement true? $\vec{r} = \vec{T} - \vec{B}$ ________
(j) Write two other equations relating the vectors $\vec{B}$, $\vec{T}$, and $\vec{r}$.
(k) Calculate the magnitudes of the vectors $\vec{B}$, $\vec{T}$, and $\vec{r}$.
(l) Calculate the difference of the magnitudes $|\vec{T}| - |\vec{B}|$.
(m) Does $|\vec{T}| - |\vec{B}| = |\vec{T} - \vec{B}|$?

Problem 1.8 Which of the following are vectors?
(a) 3.5 (b) 0 (c) $\langle 0.7, 0.7, 0.7\rangle$ (d) $\langle 0, 2.3, -1\rangle$ (e) -3×10^6
(f) $3 \cdot \langle 14, 0, -22\rangle$

Problem 1.9 Which of the following are vectors?
(a) $\vec{r}/2$ (b) $|\vec{r}|/2$ (c) $\langle r_x, r_y, r_z\rangle$ (d) $5 \cdot \vec{r}$

Problem 1.10 (a) What is the vector whose tail is at $\langle 9.5, 7, 0\rangle$ m and whose head is at $\langle 4, -13, 0\rangle$ m? (b) What is the magnitude of this vector?

Problem 1.11 A man is standing on the roof of a building with his head at the position $\langle 12, 30, 13\rangle$ m. He sees the top of a tree, which is at the position $\langle -25, 35, 43\rangle$ m.
(a) What is the relative position vector that points from the man's head to the top of the tree?
(b) What is the distance from the man's head to the top of the tree?

Problem 1.12 (a) On a piece of graph paper, draw the vector $\vec{f} = \langle -2, 4, 0\rangle$, putting the tail of the vector at $\langle -3, 0, 0\rangle$. Label the vector $\vec{f}$.
(b) Calculate the vector $2\vec{f}$, and draw this vector on the graph, putting its tail at $\langle -3, -3, 0\rangle$, so you can compare it to the original vector. Label the vector $2\vec{f}$.
(c) How does the magnitude of $2\vec{f}$ compare to the magnitude of $\vec{f}$?
(d) How does the direction of $2\vec{f}$ compare to the direction of $\vec{f}$?
(e) Calculate the vector $\vec{f}/2$, and draw this vector on the graph, putting its tail at $\langle -3, -6, 0\rangle$, so you can compare it to the other vectors. Label the vector $\vec{f}/2$.
(f) How does the magnitude of $\vec{f}/2$ compare to the magnitude of $\vec{f}$?
(g) How does the direction of $\vec{f}/2$ compare to the direction of $\vec{f}$?
(h) Does multiplying a vector by a scalar change the magnitude of the vector?
(i) The vector $a(\vec{f})$ has a magnitude three times as great as that of $\vec{f}$, and its direction is opposite to the direction of $\vec{f}$. What is the value of the scalar factor a?

Problem 1.13 (a) On a piece of graph paper, draw the vector $\vec{g} = \langle 4, 7, 0\rangle$ m. Put the tail of the vector at the origin.
(b) Calculate the magnitude of $\vec{g}$
(c) Calculate $\hat{g}$, the unit vector pointing in the direction of $\vec{g}$.
(d) On the graph draw $\hat{g}$. Put the tail of the vector at $\langle 1, 0, 0\rangle$ so you can compare $\hat{g}$ and $\vec{g}$.
(e) Calculate the product of the magnitude $|\vec{g}|$ times the unit vector $\hat{g}$:

$(|\vec{g}|)(\hat{g})$

Problem 1.14 A proton is located at $\langle 3\times10^{-10}, -3\times10^{-10}, 8\times10^{-10}\rangle$ m.
(a) What is $\vec{r}$, the vector from the origin to the location of the proton?
(b) What is $|\vec{r}|$?
(c) What is $\hat{r}$, the unit vector in the direction of $\vec{r}$?

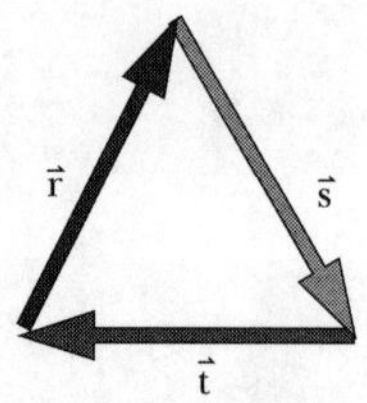

Figure 1.42 Arrows representing three vectors (Problem 1.15).

Problem 1.15 Which of the following statements about the vectors depicted by arrows in Figure 1.42 are correct?
(a) $\vec{s} = \vec{t} - \vec{r}$ (b) $\vec{r} = \vec{t} - \vec{s}$ (c) $\vec{r} + \vec{t} = \vec{s}$ (d) $\vec{s} + \vec{t} = \vec{r}$ (e) $\vec{r} + \vec{s} = \vec{t}$

Problem 1.16 Which of the following are unit vectors? (Numerical values are given only to 3 significant figures.)
(a) $\langle 0, 0, -1\rangle$ (b) $\langle 0.5, 0.5, 0\rangle$ (c) $\langle 0.333, 0.333, 0.333\rangle$ (d) $\langle 0.9, 0, 0.1\rangle$ (e) $\langle 0, 3, 0\rangle$ (f) $\langle 1, -1, 1\rangle$ (g) $\langle 0.577, 0.577, 0.577\rangle$ (h) $\langle 0.949, 0, -0.316\rangle$

Problem 1.17 Two vectors, $\vec{f}$ and $\vec{g}$, are equal: $\vec{f} = \vec{g}$. Which of the following statements are true?
$\hat{f} = \hat{g}$ (b) $g_x = f_x$ (c) $f_z = g_z$ (d) the directions of $\vec{f}$ and $\vec{g}$ may be different (e) the magnitudes of $\vec{f}$ and $\vec{g}$ may be different

Problem 1.18 A proton is located at $\vec{r}_p = \langle 2, 6, -3\rangle$ m. An electron is located at $\vec{r}_e = \langle 4, 12, -6\rangle$ m. Which of the following statements are true?
(a) $2\vec{r}_p = \vec{r}_e$ (b) $2\hat{r}_p = \hat{r}_e$ (c) $|2\vec{r}_p| = |\vec{r}_e|$

Problem 1.19 A proton is located at $\langle x_p, y_p, z_p\rangle$. An electron is located at $\langle x_e, y_e, z_e\rangle$. What is the vector pointing from the electron to the proton? What is the vector pointing from the proton to the electron?

Problem 1.20 The vector $\vec{a} = \langle 0.03, -1.4, 26.0\rangle$ and the scalar $f = -3.0$. What is $f\vec{a}$?

Problem 1.21 The vector $\vec{g} = \langle 2, -7, 3\rangle$ and the scalar $h = -2$. What is $h + \vec{g}$?
(a) $\langle 0, -9, 1\rangle$ (b) $\langle 4, -5, 5\rangle$ (c) $\langle 4, 9, 5\rangle$ (d) This is a meaningless expression.

Problem 1.22 Write each of these vectors as the product of the magnitude of the vector and the appropriate unit vector:
(a) $\langle 0, 0, 9.5\rangle$ (b) $\langle 0, -679, 0\rangle$ (c) $\langle 3.5\times10^{-3}, 0, -3.5\times10^{-3}\rangle$ (d) $\langle 4\times10^{6}, -6\times10^{6}, 3\times10^{6}\rangle$

Problem 1.23 $\vec{A} = \langle 3\times10^{3}, -4\times10^{3}, -5\times10^{3}\rangle$ and $\vec{B} = \langle -3\times10^{3}, 4\times10^{3}, 5\times10^{3}\rangle$. Calculate the following: (a) $\vec{A} + \vec{B}$ (b) $|\vec{A} + \vec{B}|$ (c) $|\vec{A}|$ (d) $|\vec{B}|$ (e) $|\vec{A}| + |\vec{B}|$

Problems on velocity and momentum

Problem 1.24 A baseball has a mass of 0.155 kg. A professional pitcher throws a baseball 90 miles per hour, which is 40 m/s. What is the magnitude of the momentum of the pitched baseball?

Problem 1.25 The position of a golf ball relative to the tee changes from <50, 20, 30> m to <53, 18, 31> m in 0.1 second. As a vector, write the velocity of the golf ball during this short time interval.

Problem 1.26 A hockey puck with a mass of 0.4 kg has a velocity of $\langle 38, 0, -27\rangle$ m/s. What is the magnitude of its momentum, $|\vec{p}|$?

Problem 1.27 A proton in an accelerator attains a speed of $0.88c$. What is the magnitude of the momentum of the proton?

Problem 1.28 The crew of a stationary spacecraft observe an asteroid whose mass is 4×10^{17} kg. Taking the location of the spacecraft as the origin, the asteroid is observed to be at location $\langle -3\times10^3, -4\times10^3, 8\times10^3\rangle$ m at a time 18.4 seconds after lunchtime. At a time 21.4 seconds after lunchtime, the asteroid is observed to be at location $\langle -1.4\times10^3, -6.2\times10^3, 9.7\times10^3\rangle$ m. Assuming the velocity of the asteroid does not change during this time interval, calculate the vector velocity $\vec{v}$ of the asteroid.

Problem 1.29 An electron with a speed of $0.95c$ is emitted by a supernova, where c is the speed of light. What is the magnitude of the momentum of this electron?

Problem 1.30 A "cosmic-ray" proton hits the upper atmosphere with a speed $0.9999c$, where c is the speed of light. What is the magnitude of the momentum of this proton?

Problem 1.31 The position of a baseball relative to home plate changes from <15, 8, −3> m to <20, 6, −1> m in 0.1 second. As a vector, write the average velocity of the baseball during this time interval.

Problem 1.32 Figure 1.43 shows the trajectory of a ball traveling through the air, affected by both gravity and air resistance. The table in Figure 1.44 gives the position of the ball at several successive times.

(a) What is the average velocity of the ball as it travels between location A and location B?

(b) If the ball continued to travel at the same average velocity during the next second, where would it be at the end of that second? (That is, where would it be at time $t = 2$ seconds?)

(c) How does your prediction from part (b) compare to the actual position of the ball at $t = 2$ seconds (location C)? If the predicted and observed locations of the ball are different, explain why.

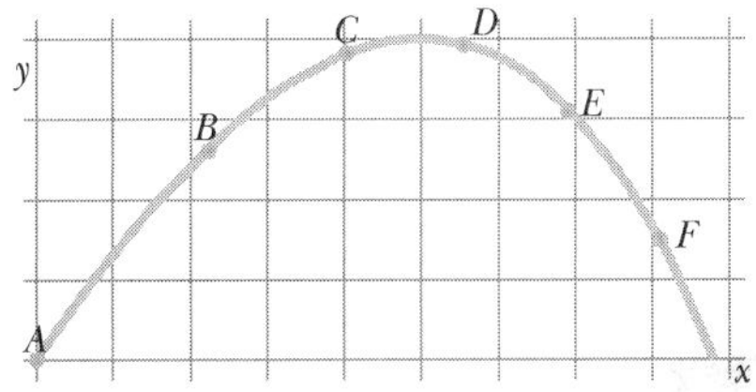

Figure 1.43 Trajectory of a ball traveling through air, affected by gravity and air resistance (Problem 1.32).

loc.	t (s)	position (m)
A	0.0	$\langle 0, 0, 0\rangle$
B	1.0	$\langle 22.3, 26.1, 0\rangle$
C	2.0	$\langle 40.1, 38.1, 0\rangle$

Figure 1.44 Table showing position of ball in Figure 1.43, at various times.

Problem 1.33 In a laboratory experiment, an electron passes location $\langle 0.02, 0.04, -0.06\rangle$ m, and 2 μs (1 microsecond = 1×10^{-6} s) later is detected at location $\langle 0.02, 1.84, -0.86\rangle$ m.

(a) What is the average velocity of the electron?

(b) If the electron continues to travel at this average velocity, where will it be in another 5 μs?

Problem 1.34 At 6 seconds after 3:00, a butterfly is observed leaving a flower whose location is $\langle 6, -3, 10\rangle$ m relative to an origin on top of a nearby tree. The butterfly flies until 10 seconds after 3:00, when it alights on a different flower whose location is $\langle 6.8, -4.2, 11.2\rangle$ m relative to the same origin.

What was the location of the butterfly at a time 8.5 seconds after 3:00? What assumption did you have to make in calculating this location?

Problem 1.35 The gray line in Figure 1.45 shows a portion of the trajectory of a ball traveling through the air. At various locations, the ball's momentum is:

$\vec{p}_B = \langle 3.03, 2.83, 0 \rangle$ kg · m/s
$\vec{p}_C = \langle 2.55, 0.97, 0 \rangle$ kg · m/s
$\vec{p}_D = \langle 2.24, -0.57, 0 \rangle$ kg · m/s
$\vec{p}_E = \langle 1.97, -1.93, 0 \rangle$ kg · m/s
$\vec{p}_F = \langle 1.68, -3.04, 0 \rangle$ kg · m/s

(a) Calculate the change in the ball's momentum between each pair of adjacent locations.

(b) On a copy of Figure 1.45, draw arrows representing each $\Delta\vec{p}$ you calculated in part (a)

(c) Between which two locations is the magnitude of the change in momentum greatest?

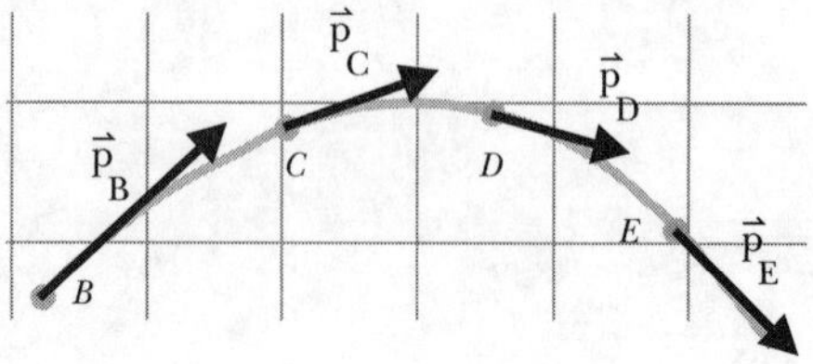

Figure 1.45 The gray line shows a portion of the path of a ball traveling through air, affected by gravity and air resistance. Arrows indicate its momentum at several locations (Problem 1.35).

Problem 1.36 A spacecraft traveling at a velocity of $\langle -20, -90, 40 \rangle$ m/s is observed to be at a location $\langle 200, 300, -500 \rangle$ m relative to an origin located on a nearby asteroid. At a later time the spacecraft is observed to be at location $\langle -380, -2310, 660 \rangle$ m.

a) How long did it take the spacecraft to travel between these locations?

b) How far did the spacecraft travel?

c) What is the speed of the spacecraft?

d) What is the unit vector in the direction of the spacecraft's velocity?

Computational problems

These problems are intended to introduce you to using a computer to model matter, interactions, and motion. You will build on these small calculations to build models of physical systems in later chapters.

Some parts of these problems can be done with almost any tool (spreadsheet, math package, etc.). Other parts are most easily done with a programming language. We recommend the free 3D programming language VPython (http://vpython.org). Your instructor will introduce you to an available computational tool and assign problems, or parts of problems, that can be addressed using the chosen tool.

Problem 1.37 Move an object across a computer screen

(a) Write a program that makes an object move from left to right across the screen at speed v. Make v a variable, so you can change it later. Let the time interval for each step of the computation be a variable dt, so that the position x increases by an amount $v*dt$ each time.

(b) Modify a copy of your program to make the object run into a wall and reverse its direction.

(c) Make a modification so that the object's speed v is no longer a constant but changes smoothly with time. Is the speed change clearly visible to an observer? Try to make one version in which the speed change is clearly noticeable, and another in which it is not noticeable.

(d) Corresponding to part (c), make a computer graph of x vs. t, where t is the time.

(e) Corresponding to part (c), make a computer graph of v vs. t, where t is the time.

Turn in your programs for parts (c), (d), and (e).

Problem 1.38 Move an object at an angle

(a) Write a program that makes an object move at an angle.

(b) Change the component of velocity of the object in the x direction but not in the y direction, or vice versa. What do you observe?

(c) Start the object moving at an angle and make it bounce off at an appropriate angle when it hits a wall.

Turn in your answer to part (b), and the final version of your computation, part (c).

Problem 1.39 Move an object, leave a trail

Write a program that makes an object move smoothly from left to right across the screen at speed v, leaving a trail of dots on the screen at equal time intervals. If the dots are too close together, leave a dot every N steps, and adjust N to give a nice display.

1.13 Answers to exercises

1.1 (page 7) a, d

1.2 (page 7) a, b, c

1.3 (page 8) Continues to move in same direction at 1×10^4 m/s.

1.4 (page 11) 3

1.5 (page 11) 1

1.6 (page 11) vector

1.7 (page 11) b, c

1.8 (page 11) 7

1.9 (page 12) scalar

1.10 (page 12) no

1.11 (page 12) 5.10 m

1.12 (page 12) no

1.13 (page 12) 2.15×10^7 **m/s**

1.14 (page 13) $\langle 6, -9, 15\rangle$ **m/s**

1.15 (page 13) $\langle 1, -1.5, 2.5\rangle$ **m/s**

1.16 (page 13) $\langle 0.04, -3.4, 60.0\rangle$

1.17 (page 13) in the opposite direction

1.18 (page 13) no

1.19 (page 13) no

1.20 (page 14) $\langle 0, 1, 0\rangle$

1.21 (page 14) $\langle -1, 0, 0\rangle$

1.22 (page 14) $\langle \frac{1}{\sqrt{3}}, \frac{1}{\sqrt{3}}, \frac{1}{\sqrt{3}}\rangle$, $\langle \frac{1}{\sqrt{3}}, \frac{1}{\sqrt{3}}, \frac{1}{\sqrt{3}}\rangle$

1.23 (page 14) $\langle 0.873, 0.436, -0.218\rangle$

1.24 (page 14) $458\langle 0.873, 0.436, -0.218\rangle \frac{\text{m}}{\text{s}^2}$

1.25 (page 16) $\langle 450, -300, -200\rangle$

1.26 (page 16) 361, 335, 577

1.27 (page 16) 696, no

1.28 (page 16) $\langle 150, 300, -200\rangle$, $\langle -150, -300, 200\rangle$

1.29 (page 16) $\langle -1, 0, -1\rangle$ **m**

1.30 (page 16) $\langle 2, 4, 0\rangle$ **m**

1.31 (page 17) $\langle -2, -4, 0\rangle$ **m**

1.32 (page 17) 2.67×10^{-3} m/s

1.33 (page 20) $\langle 8, -10, 12\rangle$ m/s, 17.55 m/s, $\langle 0.456, -0.570, 0.684\rangle$

1.34 (page 20) $\langle 25, -55, -30\rangle$ m/s, 67.45 m/s, $\langle 0.371, -0.815, -0.445\rangle$

1.35 (page 21) $\langle 2.1\times10^{5}, 1.4\times10^{5}, -2.8\times10^{5}\rangle$ m

1.36 (page 21) 0.5 s

1.37 (page 21) $\langle -2.01\times10^{5}, 5.20\times10^{4}, -1.00\times10^{3}\rangle$ m/s, 2.08×10^{5} m/s

1.38 (page 27) 650 kg · m/s

1.39 (page 27) 6.9 kg · m/s

1.40 (page 27) 2.22×10^{5} kg · m/s

1.41 (page 27) 2.415×10^{-22} kg · m/s

1.42 (page 27) $\gamma = 7.09$

1.43 (page 27) 6.71 kg · m/s

1.44 (page 29) $\vec{p}_i$ to the right ($+x$), $\vec{p}_f$ to the left ($-x$),

$\langle -5.59, 0, 0\rangle$ kg · m/s, 0.114 kg · m/s

1.45 (page 29) $\langle -1.6\times10^{28}, 0, 1.6\times10^{28}\rangle$ kg · m/s, downward to the left

1.46 (page 29) 0, 500 kg · m/s

Chapter 2

The Momentum Principle 1: Impulse and Momentum Change

Chapter 2

The Momentum Principle 1: Impulse and Momentum Change

In this chapter we introduce the Momentum Principle, the first of three fundamental principles of mechanics which together make it possible to predict and explain a very broad range of real-world phenomena (the other two are the Energy Principle and the Angular Momentum Principle). In Chapter 1 you learned how to describe positions and motions in 3D, and we discussed the notion of "interaction" where change is an indicator that an interaction has occurred. We introduced the concept of momentum as a quantity whose change is related to the amount of interaction occurring. The Momentum Principle makes a quantitative connection between amount of interaction and change of momentum.

The major topics in this chapter are:

- The Momentum Principle, relating momentum change to interaction
- Force as a quantitative measure of interaction
- The concept of a "system" to which to apply the Momentum Principle

2.1 The Momentum Principle

Newton's first law of motion, "the stronger the interaction, the bigger the change in the velocity," and therefore in the momentum, states a qualitative relationship between momentum and interaction. The Momentum Principle restates this relation in a powerful quantitative form that can be used to predict the behavior of objects. The validity of the Momentum Principle has been verified through a very wide variety of observations and experiments. It is a summary of the way interactions affect motion in the real world.

THE MOMENTUM PRINCIPLE

$$\Delta\vec{p} = \vec{F}_{net}\Delta t$$

(for a short enough time interval Δt)

In words: change of momentum (the effect) is equal to the net force acting on an object times the duration of the interaction (the cause).

As usual, the capital Greek letter delta (Δ) means "change of" (something), or "final minus initial." The "net" force $\vec{F}_{net}$ is the vector sum of all the forces acting on an object. We will study forces in detail in this chapter. Examples of forces include

- the repulsive electric force a proton exerts on another proton
- the attractive gravitational force the Earth exerts on you
- the force that a compressed spring exerts on your hand
- the force on a spacecraft of expanding gases in a rocket engine
- the force of the air on the propeller of an airplane or swamp boat

Time interval short enough

We require a "short enough time interval" for the Momentum Principle to be valid, in the sense that the net force shouldn't change very much during the time interval. If the net force hardly changes during the motion, we can use a very large time interval. If the net force changes rapidly, we need to use a series of small time intervals for accuracy. This is the same issue we met with the position update relation, $\vec{r}_f = \vec{r}_i + \vec{v}_{avg}\Delta t$, where we need to use a short enough time interval that the velocity isn't changing very much, or else we need to know the average velocity during the time interval.

Since $\Delta\vec{p} = \vec{p}_f - \vec{p}_i = \vec{F}_{net}\Delta t$ ("final minus initial"), we can rearrange the Momentum Principle like this:

UPDATE FORM OF THE MOMENTUM PRINCIPLE

$$\vec{p}_f = \vec{p}_i + \vec{F}_{net}\Delta t \text{ (for a short enough time interval } \Delta t)$$

or, written out:

$$\langle p_{fx}, p_{fy}, p_{fz}\rangle = \langle p_{ix}, p_{iy}, p_{iz}\rangle + \langle F_x, F_y, F_z\rangle(\Delta t)$$

This update version of the Momentum Principle emphasizes the fact that if you know the initial momentum, and you know the net force acting during a "short enough" time interval, you can predict the final momentum. It's an interesting fact of nature that the x component of a force doesn't affect the y or z components of momentum, as you can see from these equations.

The Momentum Principle written in terms of vectors can be interpreted as three ordinary scalar equations, for components of the motion along the x, y, and z axes:

$$p_{fx} = p_{ix} + F_{\text{net, x}}\Delta t$$

$$p_{fy} = p_{iy} + F_{\text{net, y}}\Delta t$$

$$p_{fz} = p_{iz} + F_{\text{net, z}}\Delta t$$

Note how much information is expressed compactly in the vector form of this equation, $\vec{p}_f = \vec{p}_i + \vec{F}_{net}\Delta t$. In some simple situations, for example, if we know that the y and z components of an object's momentum are not changing, we may choose to work only with the x component of the momentum update equation.

The Momentum Principle has been experimentally verified in an extremely wide range of phenomena. We will see later that it can be restated in a very general form: the change in momentum of an object plus the change in momentum of its surroundings is zero (Conservation of Momentum). In this form the principle can be applied to all objects, from the very small (atoms and nuclei) to the very large (galaxies and black holes), though understanding these systems in detail requires quantum mechanics or general relativity.

Historically, the Momentum Principle is often called "Newton's second law of motion." We will refer to it as the Momentum Principle to emphasize the key role played by momentum in physical processes.

You are already familiar with change of momentum $\Delta\vec{p}$ and with time interval Δt. The new element is the concept of "force."

2.1.1 Force

Scientists and engineers employ the concept of "force" to quantify interactions between two objects. The net force, the vector sum of all the forces acting on an object, acting for some time Δt causes changes of momentum (Figure 2.1). Like momentum or velocity, force is described by a vector, since a force has a magnitude and is exerted in a particular direction. Measuring the magnitude of the velocity of an object (in other words, measuring its speed) is a familiar task, but how do we measure the magnitude of a force?

$$\Delta\vec{\mathbf{p}} = \vec{\mathbf{F}}_{net}\Delta t$$

Figure 2.1 The bigger the net force, the greater the change of momentum.

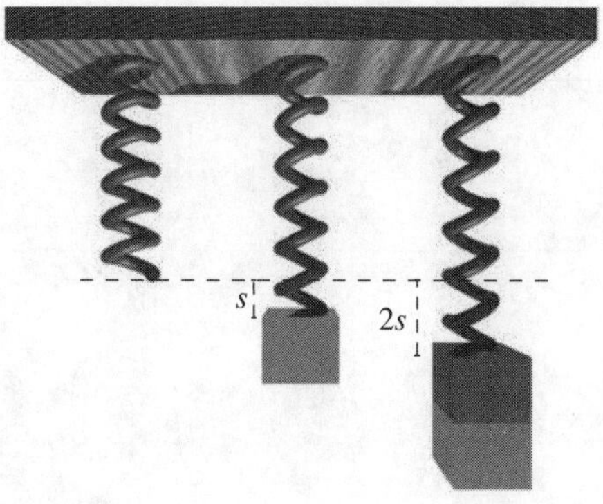

Figure 2.2 Stretching of a spring is a measure of force.

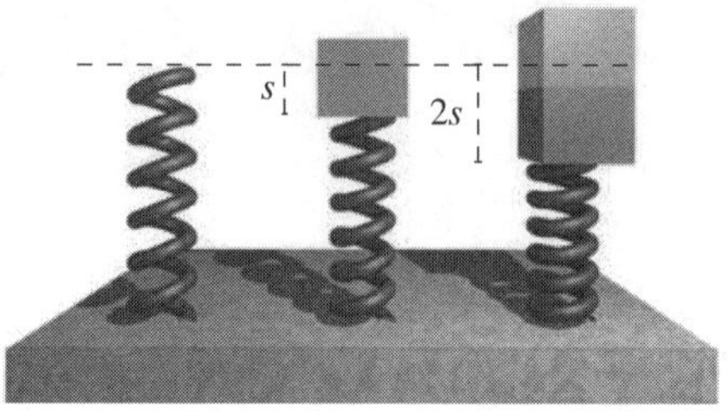

Figure 2.3 Compression of a spring is also a measure of force.

A simple way to measure force is to use the stretch or compression of a a spring. In Figure 2.2 we hang a block from a spring, and note that the spring is stretched a distance *s*. Then we hang two such blocks from the spring, and we see that the spring is stretched twice as much. By experimentation, we find that any spring made of the same material and produced to the same specifications behaves in the same way. Similarly, we can observe how much the spring compresses when the same blocks are supported by it (Figure 2.3). We find that one block compresses the spring by the same distance $|s|$, and two blocks compress it by $2|s|$. (Compression can be considered negative stretch, because the length of the spring decreases.)

We can use a spring to make a scale for measuring forces, calibrating it in terms of what force is required to produce a given stretch. The SI unit of force is the "newton," abbreviated as "N." One newton is a rather small force. A newton is approximately the downward gravitational force of the Earth on a small apple, or about a quarter of a pound. If you hold a small apple at rest in your hand, you apply an upward force of about one newton, compensating for the downward pull of the Earth.

The spring force law

A "force law" describes mathematically how a force depends on the situation. In this chapter we'll learn about various force laws, including the gravitational force law and the electric force law. For a spring, the magnitude of the force exerted by a spring on an object attached to the spring is given by the following force law:

THE SPRING FORCE LAW (MAGNITUDE)

$$|\vec{F}_{\text{spring}}| = k_s|s|$$

$|s|$ is the absolute value of the stretch; formally:

$$s = \Delta L = L - L_0$$

L_0 is the length of the relaxed spring
L is the length of the spring when stretched or compressed
k_s is the "spring stiffness"

The constant k_s is a positive number, and is a property of the particular spring: the stiffer the spring, the larger the spring stiffness, and the larger the force needed to stretch the spring. Note that *s* is positive if the spring is stretched ($L > L_0$) and negative if the spring is compressed ($L < L_0$). It is possible to write this equation as a vector equation by using unit vectors.

In the following, be sure to work through the "stop and think" activities before reading ahead. To learn the material, you need to engage actively with the questions posed here, not just read passively.

? Suppose a certain spring has been calibrated so that we know that its spring stiffness k_s is 500 N/m. You pull on the spring and observe that it is 0.01 m (1 cm) longer than it was when relaxed. What is the magnitude of the force exerted by the spring on your hand?

The force law gives $|\vec{F}_{\text{spring}}| = (500\text{ N/m})|+0.01\text{ m}| = 5\text{ N}$. Note that the total length of the spring doesn't matter; it's just the amount of stretch or compression that matters.

? Suppose that instead of pulling on the spring, you push on it, so the spring becomes shorter than its relaxed length. If the relaxed length of the spring is 10 cm, and you compress the spring to a length of 9 cm, what is the magnitude of the force exerted by the spring on your hand?

The stretch of the spring in SI units is

$$s = L - L_0 = (0.09\ \text{m} - 0.10\ \text{m}) = -0.01\ \text{m}\,.$$

The force law gives

$$\left|\vec{F}_{\text{spring}}\right| = (500\ \text{N/m})|-0.01\ \text{m}| = 5\ \text{N}\,.$$

The magnitude of the force is the same as in the previous case. Of course the direction of the force exerted by the spring on your hand is now different, but we would need to write a full vector equation to incorporate this information.

Reciprocity

In the preceding example, you pushed on a spring, compressing it, and we calculated the force exerted by the compressed spring on your hand. Of course, your hand has to exert a force on the spring in order to keep it compressed. It turns out that the force exerted by your hand on the spring is equal in magnitude (though opposite in direction) to the force exerted by the spring on your hand. This "reciprocity" of forces is a fundamental property of the electric interaction between the electrons and protons in the atoms of your hand and the electrons and protons in the atoms making up the spring. We will say more about the reciprocity of electric and gravitational forces later in this chapter.

Ex. 2.1 You push on a spring whose stiffness is 11 N/m, compressing it until it is 2.5 cm shorter than its relaxed length. What is the magnitude of the force the spring now exerts on your hand?

Ex. 2.2 A spring is 0.17 m long when it is relaxed. When a force of magnitude 250 N is applied, the spring becomes 0.24 m long. What is the stiffness of this spring?

Ex. 2.3 The spring in the previous exercise is now compressed so that its length is 0.15 m. What magnitude of force is required to do this?

2.1.2 Impulse

The amount of interaction affecting an object includes both a measure of the strength of the interaction expressed as the net force $\vec{F}_{\text{net}}$ and of the duration Δt of the interaction. Either a bigger force, or a longer application of the force, will cause more change of momentum.

The product of a force and a time interval is called "impulse" (Figure 2.4):

DEFINITION OF IMPULSE

$$\text{Impulse} \equiv \vec{F}\Delta t \ \ \text{(for small enough } \Delta t\text{)}$$

Impulse has units of N·s (newton-seconds)

With this definition of impulse we can state the Momentum Principle in words like this:

The change of momentum of an object is equal to the net impulse applied to it.

$$\Delta\vec{p} = \vec{F}_{\text{net}}\Delta t$$

Figure 2.4 Net impulse: net force times duration of the interaction. Net impulse changes momentum.

? A constant net force $\langle 3, -5, 4 \rangle$ N acts on an object for 10 s. What is the net impulse applied to the object?

$$\text{impulse} = \vec{F}_{\text{net}}\Delta t = \langle 3, -5, 4 \rangle \text{ N} \cdot (10 \text{ s}) = \langle 30, -50, 40 \rangle \text{ N} \cdot \text{s}$$

Ex. 2.4 A constant net force of $\langle -0.5, -0.2, 0.8 \rangle$ N acts on an object for 2 minutes. What is the impulse applied to the object, in SI units?

2.1.3 Predictions using the Momentum Principle

Experiments can easily be done to verify the Momentum Principle. Many introductory physics laboratories have air tracks like the one illustrated in Figure 2.5. The long triangular base has many small holes in it, and air under pressure is blown out through these holes. The air forms a cushion under the glider, allowing it to coast smoothly with very little friction. Suppose we place a block on a glider sitting on a long air track (Figure 2.5), and we attach a spring to it whose stiffness is 500 N/m, like the one discussed earlier. We will choose the x axis to point in the direction of the motion.

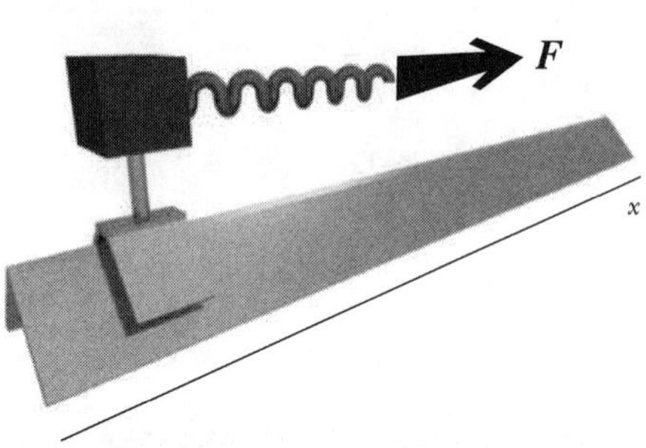

Figure 2.5 Apply a constant force to a block on a low-friction air track.

Suppose the block starts from rest, and you pull for 1 second with the spring stretched 4 cm, so that you know that you are pulling with a force of magnitude $(500 \text{ N/m})(0.04 \text{ m}) = 20 \text{ N}$ (you will have to move forward in order to keep the spring stretched).

? The block starts from rest, so $\vec{p}_i = \langle 0, 0, 0 \rangle$ kg·m/s. What would the Momentum Principle predict the new momentum of the block to be after 1 second?

Since the friction force on the glider is negligibly small, the net force on the glider is just the force with which you pull. The Momentum Principle (update form) applied to the glider is:

$$\vec{p}_f = \langle 0, 0, 0 \rangle \text{ kg·m/s} + \langle 20, 0, 0 \rangle \text{ N}(1 \text{ s})$$

Since the net force is in the x direction, we know that the y and z components of the glider's momentum will not change, and we can work with just the x component of the momentum.

$$p_{fx} = p_{ix} + F_{\text{net, x}}\Delta t = 0 + (20 \text{ N})(1 \text{ s}) = 20 \text{ kg·m/s}$$

If you do the experiment, this is what you will observe.

You keep pulling for another second, but now with the spring stretched half as much, 2 cm, so you know that you're pulling with half the original force:

$$(500 \text{ N/m})(0.02 \text{ m}) = 10 \text{ N}$$

? What would the Momentum Principle predict the new x component of the momentum of the block to be now?

The Momentum Principle would predict the following, where we take the final momentum from the first pull and consider that to be the initial momentum for the second pull:

$$p_{fx} = p_{ix} + F_{\text{net, x}}\Delta t = (20 \text{ kg·m/s}) + (10 \text{ N})(1 \text{ s}) = 30 \text{ kg·m/s}$$

This is what is observed experimentally. Note that the effects of the interactions in the two 1-second intervals add; we add the two momentum changes.

Instead of varying the net force, we could try varying the duration of the interaction (Figure 2.6). Start over with the block initially at rest and pull for 2 seconds with the spring stretched 4 cm, so the force is 20 N.

$$\Delta\vec{p} = \vec{F}_{net}\Delta t$$

Figure 2.6 Duration of interaction: the longer the net force acts, the greater the change of momentum.

? The block starts from rest ($p_{ix} = 0$). What would the Momentum Principle predict the new x component of the momentum of the block to be?

$$p_{fx} = p_{ix} + F_{\text{net},x}\Delta t = 0 + (20\text{ N})(2\text{ s}) = 40\text{ kg}\cdot\text{m/s}$$

Here the final x component of momentum was 40 kg·m/s after applying a force of 20 N for 2 s, whereas in the previous experiments we got 30 kg·m/s after applying a force of 20 N for 1 s plus a force of 10 N for 1s. In our calculations we can use big values of Δt as long as the force isn't changing much. When the force changed from 20 N for a second to 10 N for a second we had to treat the two time intervals separately.

Many different experiments have shown the validity of the Momentum Principle. If we use two springs to move the block, we find that it is indeed the vector sum of the two spring forces, the "net" force, that accounts for the change in momentum.

2.2 The superposition principle

? Drag this book across the table at constant velocity (constant speed and constant direction). You're applying a force, yet the momentum isn't changing. Explain briefly.

The force you apply to the book isn't the only force acting on the book. The table also applies a force, in the opposite direction, called a frictional force. This frictional force is due to collisions between atoms in the bottom layer of the book and atoms in the top layer of the table (see Figure 2.7 for an atomic picture of sliding friction). If the momentum isn't changing, it must be that the *net* force on the book is zero. This is why the Momentum Principle involves $\vec{F}_{\text{net}}$ rather than just the force you apply.

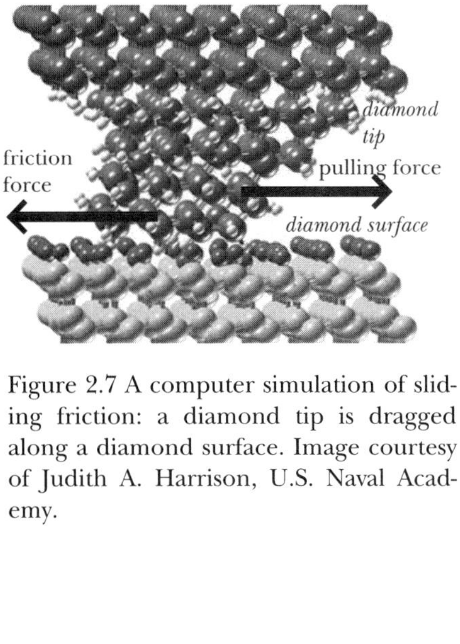

Figure 2.7 A computer simulation of sliding friction: a diamond tip is dragged along a diamond surface. Image courtesy of Judith A. Harrison, U.S. Naval Academy.

? What do you have to do to make the book go faster?

You have to apply a force that is bigger than the frictional force, so that the vector sum is nonzero.

? The Momentum Principle predicts that if the net force is zero, the momentum doesn't change but stays the same. Yet if you push the book across the table at high speed, then let go, the book doesn't keep moving but comes to a stop. Explain briefly.

After you let go, the net force on the book is just the frictional force of the table, which acts opposite to the momentum and makes the momentum decrease.

? What if you were in outer space, far from tables and other objects, and you pushed the book so that it was going fast, then let go. What would the book do?

In the absence of any forces (zero net force), the book would continue forever with the (vector) momentum you initially gave it. It would move in a straight line at constant speed: $\vec{p}_f = \vec{p}_i + \vec{F}_{\text{net}}\Delta t = \vec{p}_i$ if $\vec{F}_{\text{net}} = \vec{0}$.

Why it took so long to discover the Momentum Principle

In our everyday world friction between objects is common. As a result, for many centuries people quite naturally believed that a force was necessary to sustain motion, just as you saw when dragging the book across the table at constant speed. This made it hard to understand what made the planets keep moving. Galileo and Newton finally realized that it is the *net* force that matters, and that objects free of interactions just naturally keep moving, with no forces needed to keep them moving (Newton's first law of motion). This represented a major revolution in how humans viewed the world.

2.2.1 Net force

How do we calculate the net force acting on an object? It turns out to be surprisingly straightforward, as summarized by the superposition principle.

THE SUPERPOSITION PRINCIPLE

The net force on an object is the vector sum of the individual forces exerted on it by all other objects.

Each individual interaction is unaffected by the presence of other interacting objects.

The superposition principle is completely general. It has been found experimentally to apply to all kinds of interactions: gravitational, electromagnetic, and nuclear interactions. The implications of this principle are not always intuitively obvious. For example, this principle implies that the force exerted by the Sun on the Earth at a particular distance will always be the same, regardless of how many other planets there are that also interact with the Sun and the Earth. The presence of other objects and interactions does not block or change the interactions between each pair of objects. An interaction doesn't get "used up," and the interaction between one object and another is unaffected by the presence of a third object.

To take a silly example, hold a glass over a table and then let go (Figure 2.8). The glass falls, so evidently the table doesn't block the gravitational interaction with the Earth. In fact, the table adds a tiny additional gravitational force on the glass to that of the Earth, without changing the Earth's attraction for the glass.

Figure 2.8 The presence of the table does not change the interaction of the glass with the Earth.

Ex. 2.5 A balloon experiences a gravitational force of $\langle 0, -0.05, 0\rangle$ N and a force due to the wind of $\langle -0.03, 0, 0.02\rangle$ N. What is the net force acting on the balloon?

Ex. 2.6 A sailboat sails straight toward an island at constant speed. What objects exert forces on the boat? Is the net force due to these objects zero or nonzero?

Ex. 2.7 You watch someone carry a heavy block left to right across the room, walking at constant speed. According to the Momentum Principle, should we conclude that the net force acting on the block is upward, toward the right, or zero?

2.3 System and surroundings

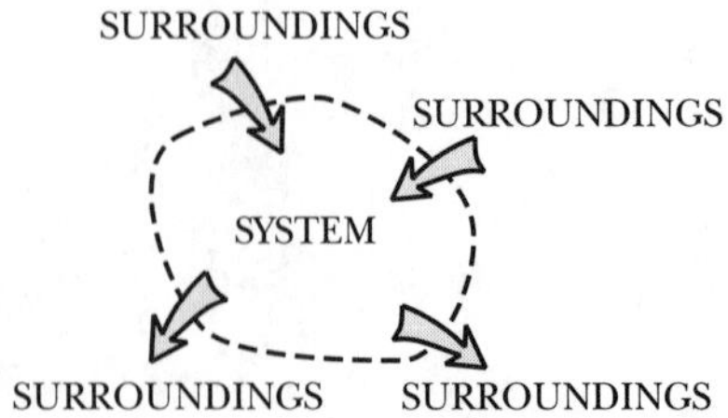

Figure 2.9 System and surroundings. Interactions in the form of impulses flow across the system boundary and change the system's momentum.

In order to use the Momentum Principle to predict motion, we must choose a "system" whose momentum change we will calculate. By the word "system" we mean some pieces of the Universe of interest to us. The rest of the Universe we call the "surroundings" (Figure 2.9).

To predict the motion of a baseball through the air, it makes sense to choose the baseball as the system whose momentum changes. The surroundings include the Earth which exerts a gravitational force on our chosen system (the baseball), and the air, which exerts a force of air resistance against the moving ball (Figure 2.10). It is true that there are changes in the momentum of the Earth and the momentum of the air molecules due to interactions with the baseball, but once we decide to choose the baseball as the system of interest we don't have to pay attention to what happens to the momentum of objects in the surroundings. The role of the surroundings is described entirely by the forces they exert on our chosen system.

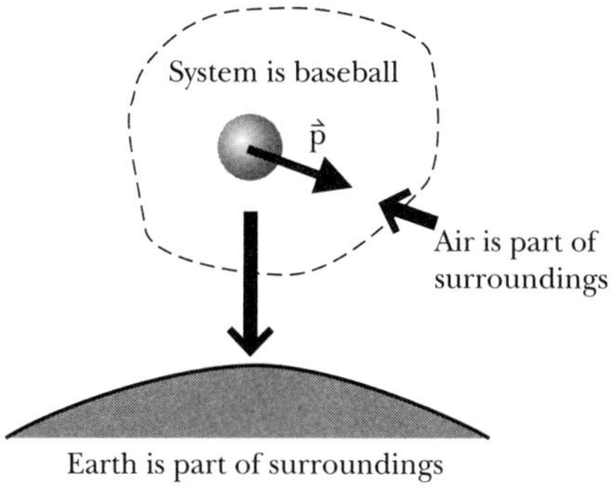

Figure 2.10 Choose a baseball as the system. The surroundings are the Earth and the air, which interact across the system boundary to change the momentum of the system.

Only external forces matter

It is an important rule that we do not include in the net force acting on the baseball any forces that the baseball exerts on itself. The atoms of the baseball do exert forces on each other, but as we'll discuss in more detail later, interatomic forces and gravitational forces come in equal and opposite pairs (Figure 2.11). Therefore these "internal" force pairs, forces between pairs of atoms internal to our chosen system, add up to zero and can safely be ignored in calculating the net force that appears in the Momentum Principle, $\vec{p}_f = \vec{p}_i + \vec{F}_{net}\Delta t$. Only "external" forces matter, forces associated with interactions between our chosen system and objects in the surroundings.

Of course objects in the surroundings experience forces exerted by objects inside the system, but when we apply the Momentum Principle just to the system, we only care about the change of momentum of the system. Also, there are equal and opposite changes of momentum in the surroundings, but they typically don't interest us.

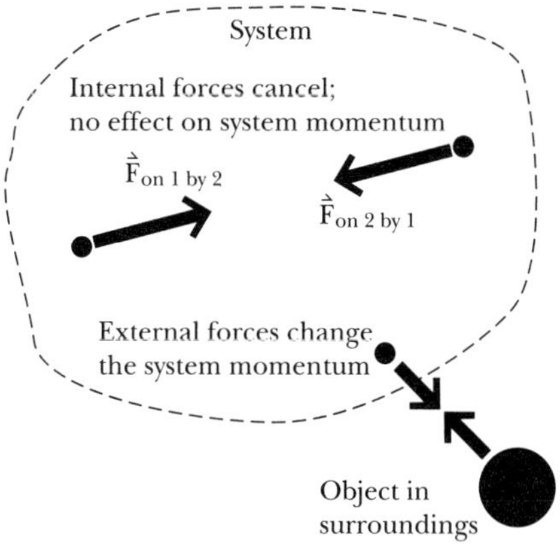

Figure 2.11 Internal force pairs cancel, so only external forces can change the momentum of a system.

Neglecting small effects

The surroundings of the system of the baseball includes the Sun, the Moon, Mars, etc. Do we have to consider all of the forces these objects exert on the baseball? In practice, no, because these forces are extremely small compared to the forces exerted on the system by the Earth and the air during the brief flight of the baseball. But if we were trying to plot an accurate course for a spacecraft going to Mars we would need to include even small forces exerted by other planets, because impulse is force times the time duration, $\vec{F}\Delta t$, and in the long time required to go to Mars even small forces can produce significant impulses, and significant changes in the momentum of the spacecraft.

Systems consisting of several objects

A system can consist of more than one object. For example, we might choose to consider a system consisting of the entire Solar System. The surroundings of this large system would include the rest of the Universe, including neighboring stars. The total momentum of the Solar System can change due to the gravitational forces exerted by stars, especially nearby stars.

Ex. 2.8 In Section 2.1.3 we applied the Momentum Principle to predict the motion of a glider on an air track. What did we choose as the system in this analysis? What objects were included in the surroundings?

2.4 Applying the Momentum Principle to a system

To apply the Momentum Principle to analyze the motion of a real-world system, several steps are required:

1. Choose a system, consisting of a portion of the Universe. The rest of the Universe is called the surroundings.

2. Make a list of the objects in the surroundings that exert significant forces on the chosen system, and make a labeled diagram showing the external forces exerted by the objects in the surroundings.

A labeled diagram (part of step 2) gives a physics view of the situation, and it defines symbols to use in writing an algebraic statement of the Momentum Principle.

3. Apply the Momentum Principle to the chosen system:

$$\vec{p}_f = \vec{p}_i + \vec{F}_{\text{net}}\Delta t$$

For each term in the Momentum Principle, substitute any values you know.

4. Apply the position update formula, if necessary:

$$\vec{r}_f = \vec{r}_i + \vec{v}_{\text{avg}}\Delta t$$

5. Solve for any remaining unknown quantities of interest.

6. Check for reasonableness (units, etc.).

2.4.1 Example: Position and momentum of a ball

Inside a spaceship in outer space there is a small steel ball of mass 0.25 kg. At a particular instant, the ball is located at position $\langle 9, 5, 0\rangle$ m and has momentum $\langle -8, 3, 0\rangle$ kg·m/s. At this instant the ball is being pulled by a string, which exerts a net force $\langle 20, 50, 0\rangle$ N on the ball. What is the ball's approximate momentum and position 3 milliseconds later $(3\times10^{-3}\text{ s})$? What approximations or simplifying assumptions did you make in your analysis?

1. Choose a system

System: the steel ball

2. List external objects that interact with the system, with diagram

the string
(a circle represents the system)

3. Apply the Momentum Principle

$$\vec{p}_f = \vec{p}_i + \vec{F}_{\text{net}}\Delta t$$

$$\vec{p}_f = (\langle -8, 3, 0\rangle\text{ kg·m/s}) + (\langle 20, 50, 0\rangle\text{ N})(3\times10^{-3}\text{ s})$$

$$\vec{p}_f = (\langle -8, 3, 0\rangle\text{ kg·m/s}) + (\langle 0.06, 0.15, 0\rangle\text{ N·s})$$

$$\vec{p}_f = \langle -7.94, 3.15, 0\rangle\text{ kg·m/s}$$

Assume that the force didn't change much during 3 ms. Or to put it another way, we assumed that a time interval of 3 ms was sufficiently short that we could update the momentum fairly well assuming a constant force during that short time interval.

4. Apply the position update formula

$$\vec{r}_f = \vec{r}_i + \vec{v}_{avg}\Delta t\text{, where } \vec{v} \approx \vec{p}/m \text{ since } v << c$$

$$\vec{r}_f = (\langle 9, 5, 0\rangle \text{ m}) + \frac{(\langle -7.94, 3.15, 0\rangle \text{ kg·m/s})}{(0.25 \text{ kg})}(3\times10^{-3}\text{ s})$$

$$\vec{r}_f = (\langle 9, 5, 0\rangle \text{ m}) + (\langle -0.0953, 0.0378, 0\rangle \text{ m})$$

$$\vec{r}_f = \langle 8.905, 5.038, 0\rangle \text{ m}$$

5. There are no remaining unknowns

6. Check

Units check (momentum: kg·m/s and position: m)

We used the momentum at the end of the interval to update the position.

Note that at very high speeds, $\vec{v} \approx \vec{p}/m$ isn't valid for updating position. See page 102.

We chose to update the position using the velocity at the end of the time interval. We could instead have used the velocity at the start of the time interval. Either value of the velocity is a good approximation to the average velocity, because the momentum (and velocity) didn't change very much.

2.4.2 Example: A fan cart (1D, constant net force)

An easy way to arrange to apply a nearly constant force is to mount an electric fan on a cart (Figure 2.12). If the fan blows backwards, the interaction with the air pushes the cart forward with a nearly constant force, making the cart's momentum continually increase. Swamp boats used in the very shallow Florida Everglades are built in a similar way, with large fans on top of the boats propelling them through the swamp.

When a fan cart or boat gets going very fast, air resistance becomes important and at high speeds is as big as the propelling force, so that the net force becomes zero, at which point the momentum doesn't increase any more, and the cart or boat travels at constant speed. For simplicity we'll consider the motion at low speed, with negligible air resistance, so we can make the approximation that the net force is due solely to the fan and is nearly constant.

(Note that the y component of the net force is zero, because the downward gravitational force on the cart is exactly balanced by the upward force exerted by the track on the cart. In a later chapter we will examine the interaction between solid objects like the cart and the track in more detail.)

Figure 2.12 A fan cart on a track.

Predict new position and new momentum

Suppose you have a fan cart whose mass is 400 grams (0.4 kg), and with the fan turned on, the net force acting on the cart, due to the air and friction with the track, is $\langle 0.2, 0, 0\rangle$ N and constant. You give the cart a shove, and you release the cart at position $\langle 0.5, 0, 0\rangle$ m with initial velocity $\langle 1.2, 0, 0\rangle$ m/s. What is the position of the cart 3 seconds later, and what is its momentum at that time?

1. Choose a system

System: the cart (including the fan)

2. List external objects that interact with the system, with diagram

the Earth, the track, the air
(represent the system by a circle)

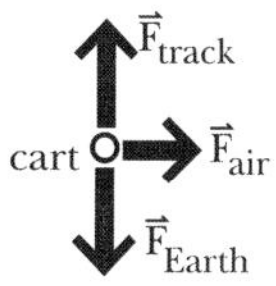

Since $v \ll c$, we can use the approximation that $\vec{p} \approx m\vec{v}$.

Since the y component of the cart's momentum does not change, we know that the y component of the net force must be zero, and $|\vec{F}_{\text{track}}| = |\vec{F}_{\text{Earth}}|$.

We can use a large time interval Δt because the force isn't changing very much in either magnitude or direction.

3. Apply the Momentum Principle

$$\vec{p}_f = \vec{p}_i + \vec{F}_{\text{net}}\Delta t = \vec{p}_i + (\vec{F}_{\text{track}} + \vec{F}_{\text{Earth}} + \vec{F}_{\text{air}})(\Delta t)$$

$$\vec{p}_f \approx \langle (0.4\text{kg})(1.2\text{m/s}), 0, 0\rangle + \langle 0.2, (|\vec{F}_{\text{track}}| - |\vec{F}_{\text{Earth}}|), 0\rangle\text{N}(3\text{ s})$$

$$\vec{p}_f \approx \langle 1.08, 0, 0\rangle\ \text{kg}\cdot\text{m/s}$$

Approximating the average velocity

In order to use the position update formula, we need the average velocity of the cart. How do we find $\vec{v}_{\text{avg}}$? We know initial and final values for $\vec{v}$:

$$\vec{v}_i = \langle 1.2, 0, 0\rangle\text{m/s}$$

$$\vec{v}_f \approx \frac{\vec{p}_f}{m} = \frac{\langle 1.08, 0, 0\rangle\ \text{kg}\cdot\text{m/s}}{0.4\ \text{kg}} = \langle 2.7, 0, 0\rangle\text{m/s}$$

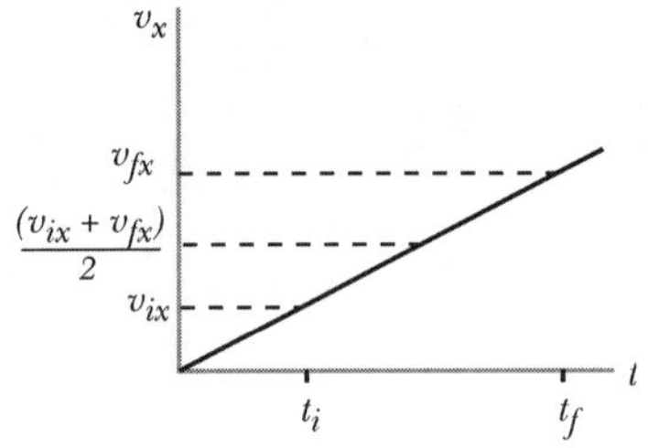

Figure 2.13 v_x is changing at a constant rate (linearly with time), so the arithmetic average is equal to $v_{\text{avg},x}$.

How can we use find $\vec{v}_{\text{avg}}$? You may have guessed that the average value of v_x is approximately the "arithmetic" (pronounced "arithMETic") average:

$$v_{\text{avg},x} \approx \frac{(v_{ix} + v_{fx})}{2} \text{ is an approximation for } v_{\text{avg},x} = \frac{\Delta x}{\Delta t}$$

The arithmetic average lies between the two extremes. For example, the arithmetic average of 6 and 8 is (6+8)/2 = 14/2 = 7, halfway between 6 and 8. The arithmetic average is often a good approximation, but it is not necessarily equal to the true average $v_{\text{avg},x} = \Delta x/\Delta t$. The arithmetic average does not give the true average unless v_x is changing at a constant rate, which is the case only if the net force is constant, as it happens to be for a fan cart (See Figure 2.13).

In contrast, if you drive 50 mi/hr for four hours, and then 20 mi/hr for an hour, you go 220 miles, and your average speed is (220 mi)/(5 hr) = 44 mi/hr, whereas the arithmetic average is (50+20)/2 = 35 mi/hr (see Figure 2.14 for another such example). In situations where the force is not constant, we have to choose short enough time intervals that the velocity is nearly constant during the brief Δt.

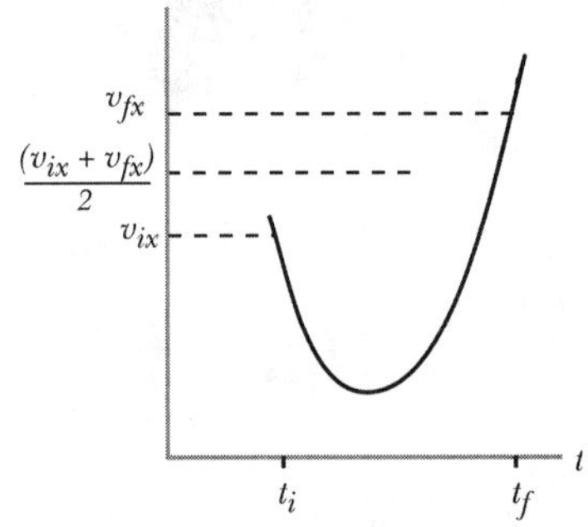

Figure 2.14 v_x is not changing linearly with time. In this case the arithmetic average is much higher than the average value of v_x.

The proof that $v_{\text{avg},x} = (v_{ix} + v_{fx})/2$ when v_x changes at a constant rate is given in optional Section 2.16 at the end of this chapter. (The proof is more complicated than one might expect.)

Now that we have a way to calculate the x component of the average velocity in this special situation (constant x component of force), we can apply the position update formula:

The net force on the cart is constant, so this calculation of average velocity gives the correct value.

4. Apply the position update formula

$$v_{\text{avg},x} = \frac{(v_{ix} + v_{fx})}{2} = \frac{(1.2 + 2.7)}{2}\text{ m/s}$$

$$\vec{v}_{\text{avg}} = \langle 1.95, 0, 0\rangle\text{ m/s}$$

$$\vec{r}_f = \vec{r}_i + \vec{v}_{\text{avg}}\Delta t = \langle 0.5, 0, 0\rangle\text{m} + \langle 1.95, 0, 0\rangle\frac{\text{m}}{\text{s}}(3\text{s})$$

$$\vec{r}_f = \langle 6.35, 0, 0\rangle\text{m}$$

5. There are no remaining unknowns

6. Check

Position: m (correct units); $x_f > x_i$, as it should be.

Ex. 2.9 A hockey puck is sliding along the ice with nearly constant momentum $\langle 10, 0, 5\rangle$ kg · m/s when it is suddenly struck by a hockey stick with a force $\langle 0, 0, 2000\rangle$ N that lasts for only 3 milliseconds (3×10^{-3} s). What is the new momentum of the puck?

Ex. 2.10 You were driving a car with velocity $\langle 25, 0, 15\rangle$ m/s. You quickly turned and braked, and your velocity became $\langle 10, 0, 18\rangle$ m/s. The mass of the car was 1000 kg. What was the (vector) change in momentum $\Delta\vec{p}$ during this maneuver? Pay attention to signs. What was the (vector) impulse applied to the car by the ground?

Ex. 2.11 In the previous exercise, if the maneuver took 3 seconds, what was the average net (vector) force $\vec{F}_{net}$ that the ground exerted on the car?

Ex. 2.12 A truck driver slams on the brakes and the momentum changes from $\langle 9\times10^4, 0, 0\rangle$ kg · m/s to $\langle 5\times10^4, 0, 0\rangle$ kg · m/s in 4 seconds due to a constant force of the road on the wheels of car. As a vector, write the force exerted by the road.

Ex. 2.13 At a certain instant, a particle is moving in the $+x$ direction with momentum +10 kg·m/s. During the next 0.1 s, a constant force $\langle -6, 3, 0\rangle$ N acts on the particle. What is the momentum of the particle at the end of this 0.1 s interval?

2.4.3 Example: A thrown ball (2D, constant net force)

A second example is the prediction of the motion of a ball thrown through the air. The Earth pulls down on the ball with a gravitational force, and the air pushes against the ball as the ball runs into air molecules. When a high-density object is thrown at low speed, this "air resistance" force is rather small compared to the gravitational force, so we may be able to neglect air resistance.

A low-density object such as a styrofoam ball experiences air resistance that is comparable to the small gravitational force on the ball, so air resistance is important unless the styrofoam ball is moving very slowly (air resistance is small at low speeds and big at high speeds, as you may have experienced if you put your hand out the window of a car). At low speeds a baseball, which has a fairly high density, moves with negligible air resistance. But at the speed that a professional pitcher can throw a baseball (about 90 mi/hr or 40 m/s), a baseball goes only about half as far in air as it would in a vacuum, because air resistance is large at this high speed (Figure 2.15).

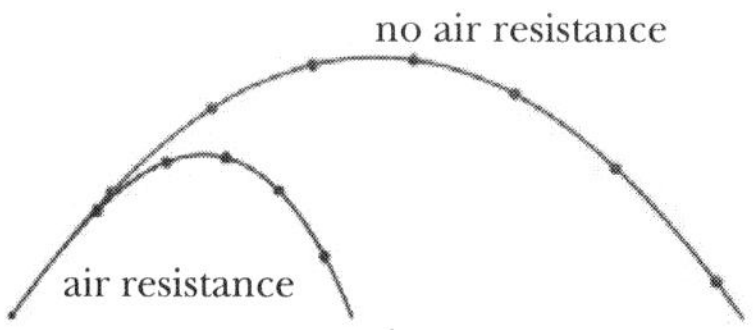

Figure 2.15 The trajectory of a baseball thrown at high speed, ignoring air resistance (top curve) and including the effect of air resistance (bottom curve). The dots indicate the ball's position at equal time intervals.

We'll analyze the flight of a ball with the assumption that we can neglect air resistance, which means we are dealing with a high-density object at low speed, or motion in a vacuum. Later we'll develop techniques that make it possible to predict the motion of a ball when air resistance is significant.

Near the Earth's surface, every kilogram of matter is attracted toward the center of the Earth by a force of approximately 9.8 N. The Earth's gravitational pull on a 2 kg block is (2 kg)(9.8 N/kg), or 19.6 N. In general, the Earth exerts a force on a mass m like this:

$$\text{Near the Earth's surface, } |\vec{F}_{grav}| \approx mg, \text{ where } g = +9.8 \text{ N/kg}$$

We'll have more to say about gravitational forces later in this chapter, but this is sufficient for analyzing the motion of a thrown ball.

Predict new velocity and new position of a thrown ball

You throw the ball so that just after it leaves your hand at location $\langle x_i, y_i, 0\rangle$ it has velocity $\langle v_{ix}, v_{iy}, 0\rangle$, with no component in the z direction. Now that it has left your hand, and we're neglecting air resistance, the net force at all times is $\langle 0, -mg, 0\rangle$, since the Earth's gravitational force acts downward, toward the center of the Earth, and we normally choose our axes so y points up. Predict the velocity and position of the ball after a time Δt.

1. Choose a system

System: the ball

2. List external objects that interact with the system, with diagram

Your hand is no longer touching the ball, so it cannot exert a force on the ball.

the Earth, the air
(represent the system by a circle)

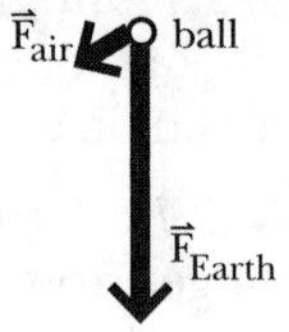

3. Apply the Momentum Principle

We make the approximation that air resistance is negligible compared to the gravitational force.

$v \ll c$ so $\vec{p} \approx m\vec{v}$

Divide both sides of the equation by m

$$\vec{p}_f = \vec{p}_i + \vec{F}_{net}\Delta t$$

$$m\langle v_{fx}, v_{fy}, 0\rangle = m\langle v_{ix}, v_{iy}, 0\rangle + \langle 0, -mg, 0\rangle\Delta t$$

$$\langle v_{fx}, v_{fy}, 0\rangle = \langle v_{ix}, v_{iy}, 0\rangle + \langle 0, -g, 0\rangle\Delta t$$

$$= \langle(v_{ix} + 0\Delta t), (v_{iy} + (-g\Delta t)), (0 + 0(\Delta t))\rangle$$

$$= \langle v_{ix}, (v_{iy} - g\Delta t), 0\rangle$$

The x and z components of velocity are not changing; this makes sense because the net force has only a y component. The y component of velocity is decreasing continuously; this makes sense because the gravitational force on the ball by the Earth affects the y component of the ball's velocity.

4. Apply the position update formula

$$\vec{v}_{avg} = \langle\left(\frac{v_{ix} + v_{fx}}{2}\right), \frac{(v_{iy} + (v_{iy} - g\Delta t))}{2}, \frac{(0+0)}{2}\rangle$$

$$\vec{v}_{avg} = \langle v_{ix}, (v_{iy} - \tfrac{1}{2}g\Delta t), 0\rangle$$

The net force is constant, so the velocity components change at a constant rate, and therefore we can use the arithmetic average to calculate the average velocity.

$$\vec{r}_f = \vec{r}_i + \vec{v}_{avg}\Delta t$$

$$\langle x_f, y_f, 0\rangle = \langle x_i, y_i, 0\rangle + \langle v_{ix}, (v_{iy} - \tfrac{1}{2}g\Delta t), 0\rangle\Delta t$$

$$\langle x_f, y_f, 0\rangle = \langle(x_i + v_{ix}\Delta t), (y_i + v_{iy}\Delta t - \tfrac{1}{2}g(\Delta t)^2), 0\rangle$$

Alternatively:

$$x_f = x_i + v_{ix}\Delta t$$

$$y_f = y_i + v_{iy}\Delta t - \tfrac{1}{2}g(\Delta t)^2$$

$$z_f = 0$$

5. There are no remaining unknowns

6. Check

The units check for the final position. Note that g is N/kg, so that $g\Delta t$ is N·s/kg, units of impulse/kg, which is kg·m/s/kg, or m/s.

? Under what circumstances can we use this result to predict the trajectory of an object?

We assumed that air resistance was negligible, that the object was traveling at a speed much less than the speed of light, and that the net force was constant. If any of these three conditions are not met, then these results do not apply, and using them will give a wrong answer.

Using this result

If a ball of mass 90 g is initially on the ground, at location $\langle 0, 0, 0 \rangle$ m , and you kick it with initial velocity $\langle 3, 7, 0 \rangle$ m/s , where will the ball be half a second later?

$$x_f = 0 + (3\text{m/s})(0.5\text{ s}) = 1.5\text{ m}$$

$$y_f = 0 + (7\text{m/s})(0.5\text{ s}) - \tfrac{1}{2}(9.8\text{N/kg})(0.5\text{ s})^2 = 2.275\text{ m}$$

? Can we use these equations to find the location of the ball 10 seconds after you kick it?

No. Our result would be that the ball was far underground, since

$$y_f = (7\text{m/s})(10\text{ s}) - \tfrac{1}{2}(9.8\text{N/kg})(10\text{ s})^2 = -420\text{ m}$$

which is not physically reasonable! (The ball would have hit the ground and stopped before 10 seconds had passed.)

? What should you remember from the preceding example?

If you understand the basic method, you can reproduce the specific results quickly and accurately. Just memorizing the results won't help you much, because you won't really understand what they mean or when they can be used.

For concreteness we analyzed the flight of a thrown ball, but this same analysis would work for any situation where an object moves in two dimensions under the influence of a constant net force. For example, just by changing the constant *y* component of the net force to be something other than $-mg$, we could analyze the two-dimensional motion of a swamp boat (if the direction of the thrust of the propeller doesn't change, and we can neglect friction with the water), or to the two-dimensional motion of an electron between two large charged plates. However, in these cases, the resulting equations would be slightly different, because the mass of the object would appear in the result (the *m* cancels only in the case of a force that is proportional to *m*, such as the gravitational force).

Graphs of the motion

In Figure 2.16 we shows graphs of position and velocity components vs. time, and the actual path (*y* vs. *x*) of the ball. The first graph, v_x vs. *t*, is simply a horizontal line, because v_x doesn't change, since there is no *x* component of force. The graph of *x* is a straight line (second graph), rising if v_x is positive. Note that the slope of the *x* vs. *t* graph is equal to v_x.

The graph of v_y is a falling straight line (third graph), because the *y* component of the force is $-mg$, which constantly makes the y component of momentum decrease. At some point the *y* component of momentum decreases to zero, at the top of the motion, after which the ball heads downward, with negative v_y. The graph of *y* vs. time *t* is an inverted parabola (fourth graph), since the equation for *y* is a quadratic function in the time.

Note that the slope of the *y* vs. *t* graph (the fourth graph) at any time is equal to v_y at that time. In particular, when the slope is zero (at the maximum y), v_y is momentarily equal to zero. Before that point the slope is positive, corresponding to $v_y > 0$, and after that point the slope is negative, corresponding to $v_y < 0$.

The actual path of the ball, the graph of *y* vs. *x*, is also an inverted parabola (the bottom graph). Since *x* increases linearly with *t*, whether we plot *y* vs. *t* or *y* vs. *x* we'll see a similar curve. The scale factor along the horizontal axis is different, of course (meters instead of seconds).

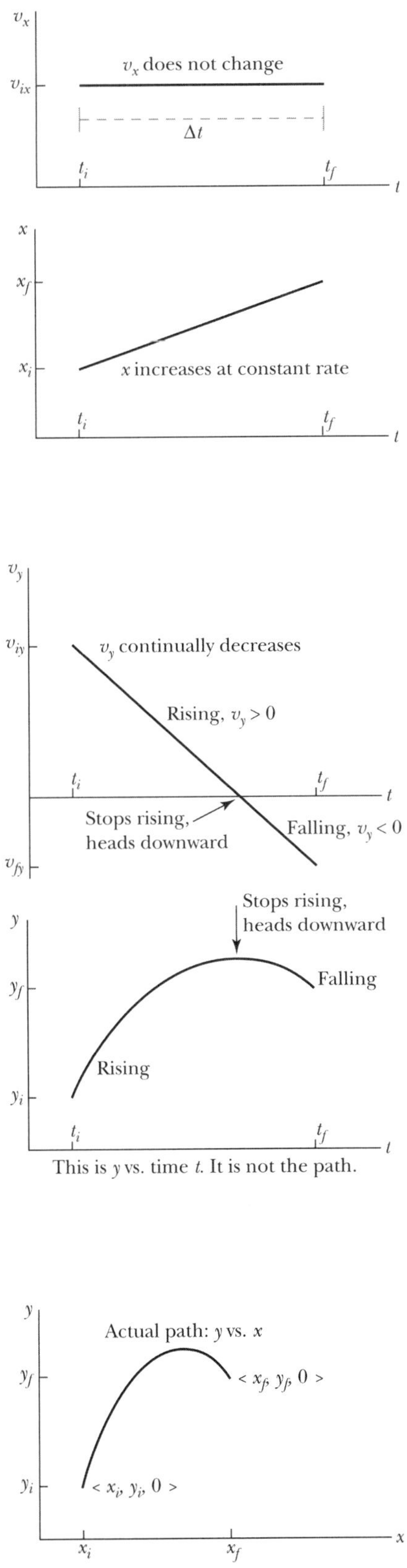

Figure 2.16 Graphs for the thrown ball.

Understanding the results

Let's explore the results a bit to understand what they tell us about the motion. The x component of the motion is very simple; it is completely unaffected by the downward-pointing gravitational force, so x simply increases at a constant rate (v_x constant). Of course, once the ball hits the ground, x no longer increases.

The y component of the motion is more interesting. The equation

$$v_{fy} = v_{iy} - g\Delta t$$

says that if you throw the ball so that v_{iy} is positive (heading up), at some time v_y will decrease to zero, and then become negative.

? At what time Δt_{top} will the ball reach its highest point?

Our result for y as a function of time:

$$y_f = y_i + v_{iy}\Delta t - \frac{1}{2}g(\Delta t)^2$$

does not help us answer this question, because we don't yet know how high the ball will go (we don't know what to use for y_f).

? Can we use the equation for v_y as a function of time to find the time Δt_{top} at which the ball reaches its highest point? Do we know the value of v_y when the ball reaches its highest point?

Just before the ball reaches its maximum height, v_y is positive. Just after it reaches its maximum height and begins to head downward, v_y is negative. At the instant the ball reaches its maximum height, $v_y = 0$. Using this information, we can solve for the elapsed time at that instant:

$$\text{If } v_{fy} = v_{iy} - g\Delta t_{\text{top}} = 0 \text{ then } \Delta t_{\text{top}} = \frac{v_{iy}}{g}$$

? How high is the ball when it turns around and heads downward?

Now that we have a value for Δt_{top}, we can use that in our general result for the final position:

$$y_f = y_i + v_{yi}\Delta t_{\text{top}} - \frac{1}{2}g(\Delta t_{\text{top}})^2 = y_i + \left[v_{iy}\left(\frac{v_{iy}}{g}\right) - \frac{1}{2}g\left(\frac{v_{iy}}{g}\right)^2\right]$$

$$y_f - y_i = \frac{v_{iy}^2}{2g} \text{ is the maximum height above your hand}$$

Check the algebra yourself to verify this result. This result makes sense, in that the bigger the initial y component of velocity, the higher the ball will go. Notice that it doesn't matter at all what x component of initial velocity you give the ball; all that matters is v_{iy}. However, it takes more effort to give the ball some v_{ix} in addition to giving it some v_{iy}, so for maximum height you want to throw the ball nearly straight up.

? Suppose you throw the ball so that it rises 2 m before falling back down. If you double v_{iy}, how far will it rise above your hand?

Since $y_f - y_i = v_{iy}^2/(2g)$, doubling v_{iy} will make the ball go up 4 times as high, to 8 m above your hand.

? Suppose you throw the ball so that it rises 2 m on Earth. If you give the ball the same v_{iy} on the Moon, where g is about one-sixth that on Earth, how far will the ball rise above your hand?

Since $y_f - y_i = v_{iy}^2/(2g)$, if g decreases by a factor of 6 but v_{iy} stays the same, the rise of the ball will increase by a factor of 6, so the ball will rise to 12 m above your hand. Because there is no air on the Moon, our analysis works well there.

? On Earth, what v_{iy} must you give the ball so that it rises 10 m above your hand?

We have $10\text{ m} = v_{iy}^2/(2(9.8\text{ N/kg}))$, and solving this we find $v_{iy} = 14\text{ m/s}$. This is a high enough speed that air resistance might be significant, and our result inaccurate.

? Suppose the ball is caught by a friend at the same height as you launched the ball. How long was the ball in the air?

In this case we have $y_f = y_i$, so $y_f = y_i + (v_{iy} - \frac{1}{2}g\Delta t)\Delta t$ gives us

$$\Delta t_{\text{up and down}} = 2\frac{v_{iy}}{g}\text{; compare with } \Delta t_{\text{top}} = \frac{v_{iy}}{g}$$

The ball takes twice as long to go up and down as to go up, so we conclude that the time to come down is the same as the time to go up. (In the presence of air resistance, the speed at every height is slower on the way down, so it takes longer to come down than to go up.) Again, notice that the time to go up and down doesn't depend at all on the x component of velocity. The x and y motions are independent of each other.

? How far from you is your friend when the ball is caught? Note that you know how long the ball was in the air, and you know the x component of velocity.

$$x_f = x_i + v_{ix}\Delta t_{\text{up and down}}\text{, so the distance is } x_f - x_i = v_{ix}\left(2\frac{v_{iy}}{g}\right)$$

This distance is called the "range" of the ball. It is the distance the ball goes in the air if it returns to the same height. It depends on v_{iy} because that determines how much time it spends in the air. The result also depends on v_{ix} because that determines how far the ball will move in the x direction while the ball is in the air.

Initial speed and angle

Sometimes you know the initial speed $|\vec{v}|$ and angle θ for the launch of a ball. Figure 2.17 shows how to calculate the sine and cosine of the angle. Use these results to solve for the velocity components:

$$v_{ix} = |\vec{v}|\cos\theta,\ v_{iy} = |\vec{v}|\sin\theta$$

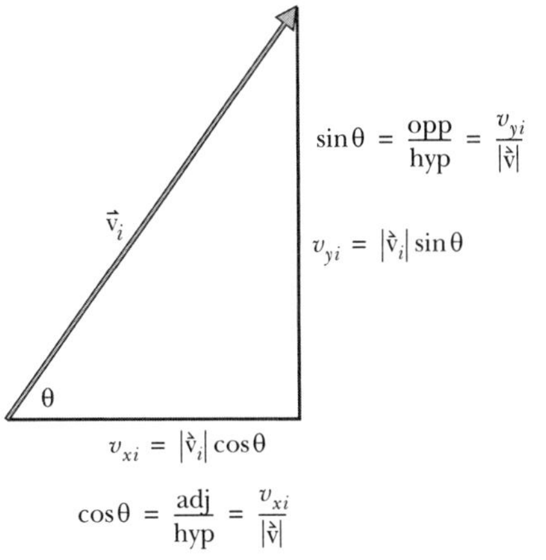

Figure 2.17 Converting speed and angle to velocity components.

Ex. 2.14 A ball is kicked from a location $\langle 9, 0, -5\rangle$ m (on the ground) with initial velocity $\langle -10, 13, -5\rangle$ m/s.
(a) What is the velocity of the ball 0.6 seconds after being kicked?
(b) What is the location of the ball 0.6 seconds after being kicked?
(c) What is the maximum height reached by the ball?
(d) At what time does the ball reach its maximum height?
(e) At what time does the ball hit the ground?
(f) What is the location of the ball when it hits the ground?

Ex. 2.15 Apply the general results obtained in the full analysis on page 60 to answer the following questions. You hold a small metal ball of mass m a height h above the floor. You let go, and the ball falls to the floor. Choose the origin of the coordinate system to be on the floor where the ball hits, with y up as usual. Just after release, what are y_i and v_{iy}? Just before hitting the floor, what is y_f? How much time Δt does it take for the ball to fall? What is v_{fy} just before hitting the floor? Express all results in terms of m, g, and h. How would your results change if the ball had twice the mass?

Ex. 2.16 A soccer ball is kicked at an angle of 60° to the horizontal with an initial speed of 10 m/s. Assume that we can neglect air resistance. For how much time is the ball in the air? How far does it go (horizontal distance along the field)? How high does it go?

2.4.4 Graphical prediction of motion

It is instructive to apply the Momentum Principle qualitatively and graphically to the thrown ball, to see visually how the Momentum Principle determines the motion. The Momentum Principle predicts that in a short time interval Δt, the momentum of the ball will change by an amount $\Delta\vec{p} = \vec{F}_{net}\Delta t$ which we can calculate, because we know the force acting on the ball. In the following diagram we show the initial momentum $\vec{p}_1$, the change in the momentum $\Delta\vec{p} = \vec{F}_{net}\Delta t$, and the new momentum $\vec{p}_2 = \vec{p}_1 + \Delta\vec{p}$. Note that this vector addition corresponds to adding the arrows $\vec{p}_1$ and $\Delta\vec{p}$ tip to tail. We approximate the average momentum in the time interval Δt by the new momentum $\vec{p}_2$, and we advance the ball in the direction of $\vec{p}_2$.

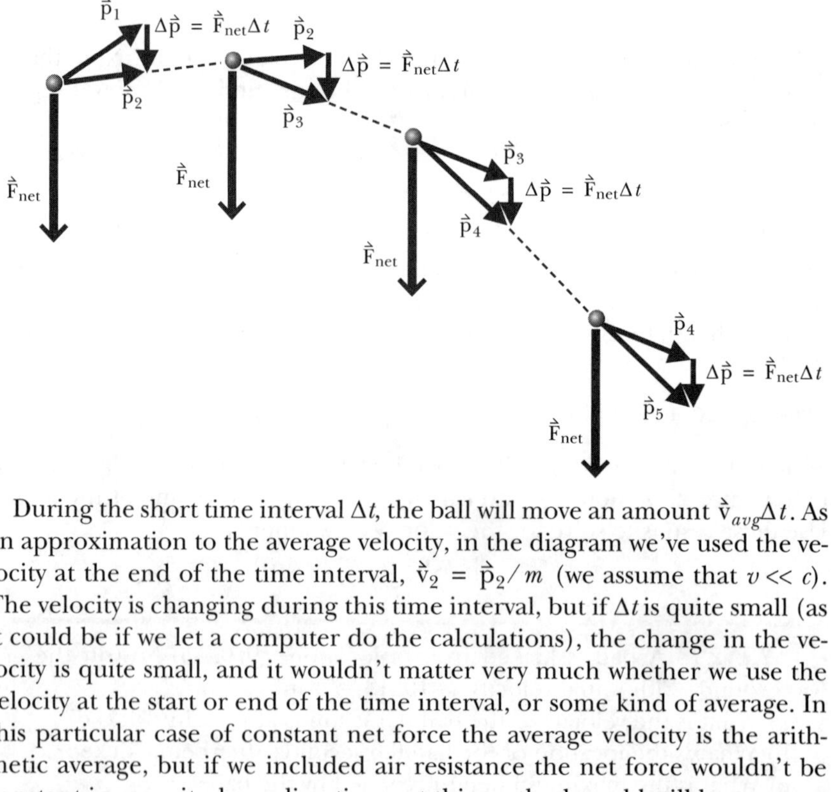

Graphical prediction of the motion of a ball, for three successive time steps.

During the short time interval Δt, the ball will move an amount $\vec{v}_{avg}\Delta t$. As an approximation to the average velocity, in the diagram we've used the velocity at the end of the time interval, $\vec{v}_2 = \vec{p}_2/m$ (we assume that $v << c$). The velocity is changing during this time interval, but if Δt is quite small (as it could be if we let a computer do the calculations), the change in the velocity is quite small, and it wouldn't matter very much whether we use the velocity at the start or end of the time interval, or some kind of average. In this particular case of constant net force the average velocity is the arithmetic average, but if we included air resistance the net force wouldn't be constant in magnitude or direction, yet this method would still be accurate as long as we use small Δt.

At the next position we repeat the procedure, graphically adding the vectors $\vec{p}_2$ and $\Delta\vec{p} = \vec{F}_{net}\Delta t$ to obtain the new momentum $\vec{p}_3$. We then advance the ball in the direction of $\vec{p}_3$. As we do this repeatedly, we trace out graphically the trajectory of the ball. You can see that the trajectory looks like what is observed in the real world, and you can even see the increasing magnitude of the ball as it falls, corresponding to increasing speed. The important point to see in the diagram is that the net impulse $\vec{F}_{net}\Delta t$ *changes* the momentum.

2.4.5 Example: Block on spring (1D, nonconstant net force)

Next we'll study how to predict motion when there is more than one force, and the net force isn't constant. The situation we'll analyze is the motion of a block bouncing on a vertical spring. The spring has a relaxed length of 20 cm (0.2 m) and the spring stiffness is 8 N/m (Figure 2.18). You place a 60 gram block (0.06 kg) on top of the spring, compressing the spring, and the block sits motionless on the spring (Figure 2.19).

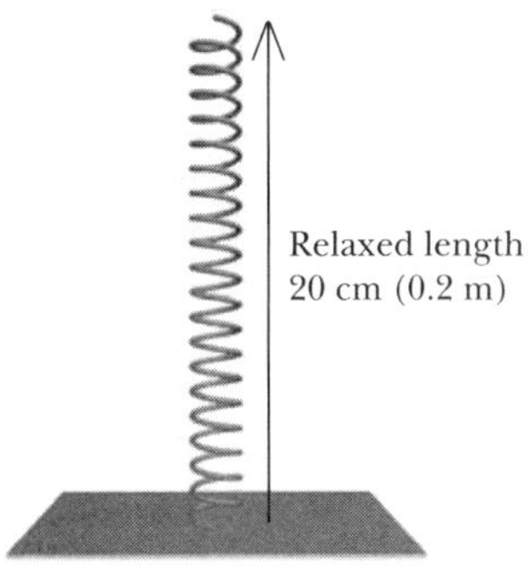

Figure 2.18 A relaxed vertical spring.

? How much is the spring compressed? Prove this using the Momentum Principle. Remember that $|\vec{F}_{spring}| = k_s|s|$, where s is the stretch of the spring (a negative number if the spring is compressed).

Since the block sits motionless over a time interval Δt, its momentum is not changing, which implies that the net force acting on the block must be zero (Figure 2.19):

$$\Delta\vec{p} = \vec{F}_{net}\Delta t$$
$$\langle 0, 0, 0\rangle = \langle 0, 0, 0\rangle(\Delta t)$$

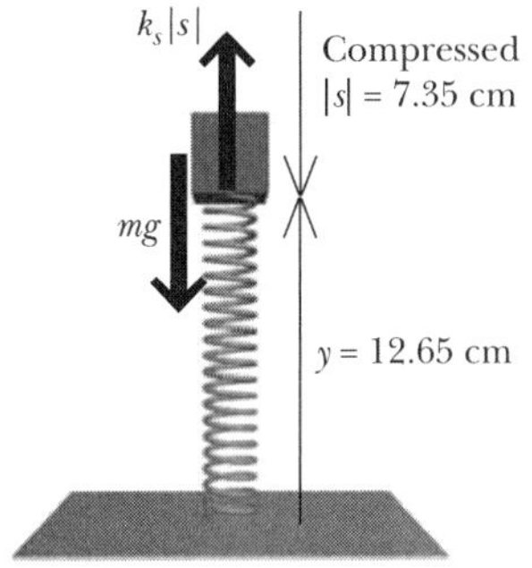

Figure 2.19 A block sits motionless on the spring. The net force must be zero.

The net force is the vector sum of all the individual forces acting on the block: the force due to the spring and the force due to the Earth.

$$\begin{aligned}\vec{F}_{net} &= \vec{F}_{spring} + \vec{F}_{Earth}\\ &= \langle 0, k_s|s|, 0\rangle + \langle 0, -mg, 0\rangle\\ &= \langle 0, k_s|s| - mg, 0\rangle\end{aligned}$$

where k_s is the spring stiffness and as defined on page 50, $s = \Delta L = L - L_0$ is the compression of the spring (the current length minus the relaxed length, which is a negative number for a compression, so we've written $|s|$). Considering just the y component, we have

$$F_{net,\,y} = 0 = (8\text{ N/m})|s| - (0.06\text{kg})(9.8\text{ N/kg})$$

$$|s| = \frac{(0.06\text{kg})(9.8\text{ N/kg})}{(8\text{ N/m})} = 0.0735\text{ m (or 7.35 cm)}$$

We'll choose the origin of our coordinate system to be at the bottom of the spring, where it is attached to the floor. As usual, we take $+y$ to be up. When the block is sitting motionless on the spring, the bottom of the block (or top of the spring) is at $y = (0.2 - 0.0735)\text{ m} = 0.1265\text{ m}$, since the relaxed length of the spring was 0.2 m.

Next you push the mass down, compressing the spring an additional amount. You release the block, and because the spring has been compressed an extra amount, the upward force of the spring is greater than the downward gravitational force exerted by the Earth, and the block will move upward with increasing momentum due to the upward net force (Figure 2.20). After you release the block, as the block passes a point 8 cm (0.08 m) above the floor, the block has an upward speed of 0.3 m/s.

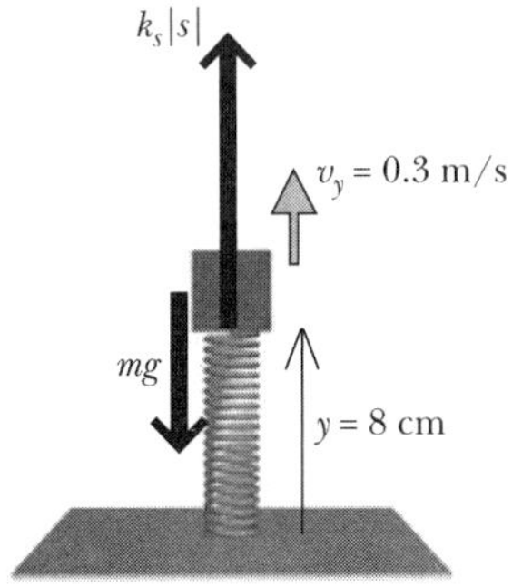

Figure 2.20 You compressed the spring an additional amount, then released the block; the block heads upward with increasing speed, because the net force is nonzero and upward.

We will predict the velocity and position of the block after 0.02 s:

1. Choose a system

System: the block

2. List external objects that interact with the system, with diagram

the Earth, the spring, the air
(see force diagram on previous page)

We make the approximation that air resistance is negligible compared to the other forces.

3. Apply the Momentum Principle

$$p_{fy} = p_{yi} + F_{net,y}\Delta t \text{ with } v << c \text{ so that } p_y \approx mv_y \text{ and } v_y \approx \frac{p_y}{m}$$

Initially the spring is compressed $(0.2 - 0.08)\text{m} = 0.12 \text{ m}$

The net force is changing as the compression changes, so the momentum update is approximate.

$$F_{net,y} = k_s|s| - mg = (8 \text{ N/kg})(0.12 \text{ m}) - (0.06 \text{ kg})(9.8 \text{ N/kg})$$

$$F_{net,y} = 0.372 \text{ N}$$

$$p_{fy} = (0.06 \text{ kg})(0.3 \text{ m/s}) + (0.372 \text{ N})(0.02 \text{ s})$$

$$p_{fy} = 0.02544 \text{ kg·m/s} \text{ ; the new } p_y$$

$$v_{fy} \approx \frac{p_{fy}}{m} = \frac{(0.02544 \text{ kg·m/s})}{(0.06 \text{ kg})} = 0.424 \text{ m/s} \text{ ; the new } v_y$$

4. Apply the position update formula

$$\vec{r}_f = \vec{r}_i + \vec{v}_{avg}\Delta t, \text{ or } y_f = y_i + v_{avg,y}\Delta t$$

Tentatively, make the approximation that the net force was nearly constant during the brief time interval 0.02 s:

$$v_{avg,y} \approx \frac{(v_{iy} + v_{fy})}{2} = \frac{(0.3 + 0.424)\text{ m/s}}{2} = 0.362 \text{ m/s}$$

$$y_f \approx ((0.08 \text{ m}) + (0.362 \text{ m/s})(0.02 \text{ s})) = 0.08724 \text{ m} \text{ ; the new } y$$

5. There are no remaining unknowns

6. Check

The units check for y_f; both v_y and y have increased as expected.

During the 0.02 s time interval the block is moving up and the length of the spring is changing, so the stretch (compression) of the spring is changing. That means that the spring force $k_s|s|$ is not constant during the 0.02 s. Therefore when we use $p_{fy} = p_{iy} + F_{net,y}\Delta t$ to update the momentum, we're making an approximation that the force doesn't change very much during that time.

We can estimate how serious this issue is by calculating the net force at the new position, $y_f = 0.08724 \text{ m}$.

? What is the new net force? (You need to determine the new compression.)

With the top of the spring now at $y_f = 0.08724 \text{ m}$, the compression of the spring is $|s| = (0.2 - 0.08724) \text{ m} = 0.11276 \text{ m}$, and the net force has become this:

$$F_{net,y} = k_s|s| - mg = (8 \text{ N/kg})(0.11276 \text{ m}) - (0.06 \text{ kg})(9.8 \text{ N/kg})$$

$$F_{net,y} = 0.314 \text{ N}$$

The net force at the start of the time interval was $F_{\text{net}, y} = 0.372\ \text{N}$. So although the net force did change during the 0.02 s, it didn't change very much, and our approximate analysis might be adequate for some purposes.

? How could our predictions be improved?

By taking shorter time steps. Instead of taking one step of 0.02 s, we could take two steps of 0.01 s, or ten steps of 0.002 s. During each of these shorter time intervals the compression would change less, so the force would be more nearly constant. Unfortunately, to achieve increased accuracy we have to do a lot more calculations.

Taking another step

To continue predicting the motion of the block into the future, we can take another step. The final values of momentum and position after the original 0.02 s step become the initial values for the next 0.02 s step:

3. Apply the Momentum Principle

$$p_{fy} = p_{iy} + F_{\text{net}, y}\Delta t \text{ with } v \ll c \text{ so that } p_y \approx mv_y \text{ and } v_y \approx \frac{p_y}{m}$$

The initial momentum and position are taken from the final momentum and position in the previous step.

Spring is compressed $(0.2 - 0.08724)\text{m} = 0.11276\ \text{m}$

$$F_{\text{net}, y} = k_s|s| - mg$$
$$= (8\ \text{N/kg})(0.11276\ \text{m}) - (0.06\ \text{kg})(9.8\ \text{N/kg})$$

$$F_{\text{net}, y} = 0.314\ \text{N}$$

$$p_{fy} = (0.02544\ \text{kg·m/s}) + (0.314\ \text{N})(0.02\ \text{s})$$

$$p_{fy} = 0.03172\ \text{kg·m/s}\text{; the new } p_y$$

$$v_{fy} \approx \frac{p_{fy}}{m} = \frac{(0.03172\ \text{kg·m/s})}{(0.06\ \text{kg})} = 0.5287\ \text{m/s}\text{; the new } v_y$$

4. Apply the position update formula

$$\vec{r}_f = \vec{r}_i + \vec{v}_{\text{avg}}\Delta t\text{, or } y_f = y_i + v_{\text{avg}, y}\Delta t$$

Make the approximation that the net force was nearly constant during the 0.02 s:

$$v_{\text{avg}, y} \approx \frac{(v_{iy} + v_{fy})}{2} = \frac{(0.424 + 0.5287)\ \text{m/s}}{2} = 0.4764\ \text{m/s}$$

$$y_f \approx ((0.08724\ \text{m}) + (0.4764\ \text{m/s})(0.02\ \text{s})) = 0.09677\ \text{m}$$

Look through this calculation and make sure you understand how the *final* results for position and momentum from the first 0.02 s time step have been used as the *initial* values for the second 0.02 s time step.

In principle we could continue, taking more and more steps to predict farther and farther into the future. Here is a summary of this iterative scheme:

- Start with initial positions and momenta of the interacting objects.
- Calculate the (vector) forces acting on each object.
- Update the momentum of each object: $\vec{p}_f = \vec{p}_i + \vec{F}_{\text{net}}\Delta t$.
- Update the positions: $\vec{r}_f = \vec{r}_i + \vec{v}_{\text{avg}}\Delta t$.
- Repeat.

Every time you repeat, the "final" momentum and position become the "initial" momentum and position for the next step. You have to use an approximate value for $\vec{v}_{avg}$, either by using the velocity at the start or end of the time interval, or by taking the arithmetic average of these two velocities as we did above.

While this scheme is very general, doing it by hand is incredibly tedious. It is possible to program a computer to do these calculations repetitively. Computers are now fast enough that it is possible to get high accuracy simply by taking very short time steps, so that during each step the net force and velocity aren't changing much. We'll talk more about computer prediction of motion later in this chapter.

Ex. 2.17 After a third time step of 0.02 seconds, what will be the position and momentum of the block?

2.4.6 Example: Fast proton (1D, constant net force, relativistic)

A proton in a particle accelerator is moving with velocity $\langle 0.96c, 0, 0\rangle$, so the speed is $0.96 \times 3\times10^8\ \text{m/s} = 2.88\times10^8\ \text{m/s}$. A constant electric force is applied to the proton to speed it up, $\vec{F}_{net} = \langle 5\times10^{-12}, 0, 0\rangle\ \text{N}$. What is the proton's speed as a fraction of the speed of light after 20 nanoseconds $(1\ \text{ns} = 1\times10^{-9}\ \text{s})$?

1. Choose a system

System: the proton

2. List external objects that interact with the system, with diagram

electric charges in the accelerator
(a circle represents the system)

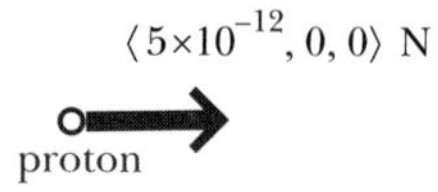

3. Apply the Momentum Principle

$$\vec{p}_f = \vec{p}_i + \vec{F}_{net}\Delta t$$

$$\langle p_{fx}, 0, 0\rangle = \langle \gamma_i m v_{ix}, 0, 0\rangle + (\langle 5\times10^{-12}, 0, 0\rangle\ \text{N})(20\times10^{-9}\ \text{s})$$

$$p_{fx} = \frac{1}{\sqrt{1-\left(\frac{0.96c}{c}\right)^2}}(1.7\times10^{-27}\ \text{kg})(0.96\times3\times10^8\ \text{m/s}) + (1\times10^{-19}\ \text{N}\cdot\text{s})$$

$$p_{fx} = (1.75\times10^{-18}\ \text{kg}\cdot\text{m/s}) + (1\times10^{-19}\ \text{N}\cdot\text{s}) = 1.85\times10^{-18}\ \text{kg}\cdot\text{m/s}$$

$$\frac{v_{fx}}{c} = \frac{\frac{p_{fx}}{mc}}{\sqrt{1+\left(\frac{p_{fx}}{mc}\right)^2}}$$

(See Section 2.20, page 102; obtaining v from p when the speed is near the speed of light.)

Evaluate $$\frac{p_{fx}}{mc} = \frac{1.85\times10^{-18}\ \text{kg}\cdot\text{m/s}}{(1.7\times10^{-27}\ \text{kg})(3\times10^8\ \text{m/s})} = 3.62\ \text{(no units)}$$

$$\frac{v_{fx}}{c} = \frac{3.62}{\sqrt{1+3.62^2}} = 0.964\ \text{(no units)}$$

The speed didn't increase very much, because the proton's initial speed, $0.96c$, was already close to the cosmic speed limit, c. Because the speed hardly changed, the distance the proton moved during the 20 ns was approximately equal to $(0.96\times3\times10^8\ \text{m/s})(20\times10^{-9}\ \text{s}) = 5.8\ \text{m}$.

2.5 Problems of greater complexity

So far all of the examples we have considered have involved finding a change in momentum (and position), given a known force acting over a known time interval. The following problems require you to find either the duration of an interaction (time interval), or the force exerted during an interaction. These large problems involve several steps in reasoning.

2.5.1 Example: Strike a hockey puck

In Figure 2.21 an 0.4 kg hockey puck is sliding along the ice with velocity $\langle 20, 0, 0\rangle$ m/s As the puck slides past location $\langle 1, 0, 2\rangle$ m on the rink, a player strikes the puck with a sudden force in the $+z$ direction, and the hockey stick breaks. Some time later, the puck's position on the rink is $\langle 13, 0, 21\rangle$ m. When we pile weights on the side of a hockey stick we find that the stick breaks under a force of about 1000 N (this is roughly 250 pounds; a force of one newton is equivalent to a force of about a quarter of a pound, approximately the weight of a small apple).

(a) For approximately how much time $\Delta t_{contact}$ was the hockey stick in contact with the puck? Evidently the contact time is quite short, since you hear a short, sharp crack. Be sure to show clearly the steps in your analysis.

(b) What approximations and/or simplifying assumptions did you make in your analysis?

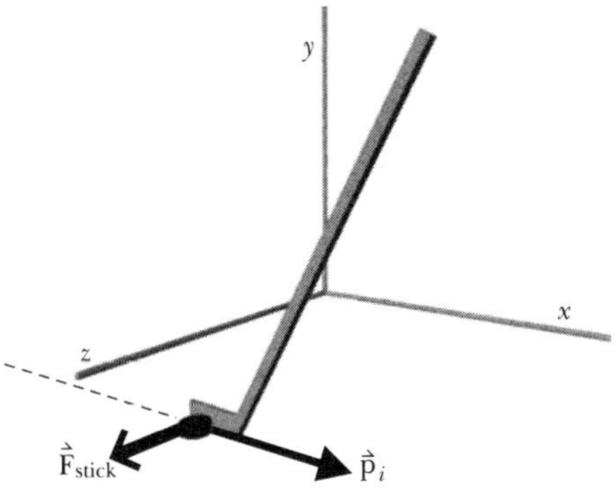

Figure 2.21 A hockey stick hits a puck as it slides by.

1. Choose a system

System: the hockey puck

2. List external objects that interact with the system, with diagram

Earth, ice, hockey stick, air

List all interacting objects, even if you think some interactions will cancel out.

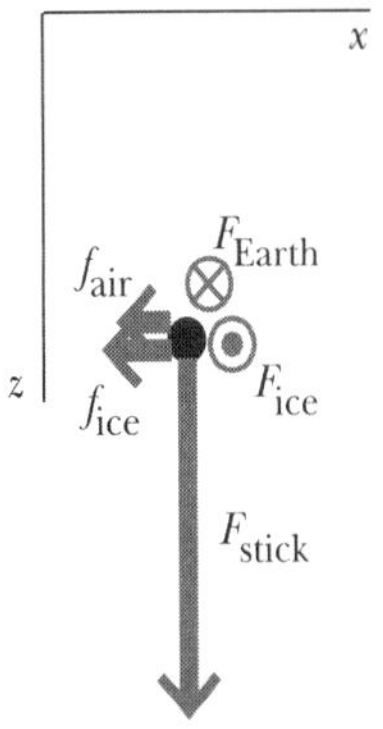

We're looking down on the ice from above. The y axis points out of the page, toward you.

The symbol ⊙ means out of the page (the tip of an arrow pointing out at you); the symbol ⊗ means into the page (tail feathers of an arrow pointing away from you)

When possible, make a 2D diagram, which is much easier than trying to draw in 3D.

3. Apply the Momentum Principle

$$\langle p_{fx}, 0, p_{fz}\rangle = \langle p_{ix}, 0, 0\rangle + \langle -f_{ice} - f_{air}, (F_{ice} - mg), F_{stick}\rangle \Delta t_{contact}$$

We use the information that the puck slides without bouncing on the ice, which means that p_y is zero at all times.

Write the three component equations separately:

x component: $p_{fx} = p_{ix} - (f_{ice} + f_{air})\Delta t_{contact}$

y component: $0 = (F_{ice} - mg)\Delta t_{contact}$ which implies $F_{ice} = mg$

z component: $p_{fz} = F_{stick}\Delta t_{contact}$

The atoms in the top layer of the ice exert upward forces on the atoms on the bottom of the puck, counteracting the downward gravitational pull of the Earth.

Find new momentum:

$f_{ice} \approx 0$ and $f_{air} \approx 0$ (negligible forces along x direction)

$p_{fx} \approx p_{ix} + (0)\Delta t_{contact} \approx p_{ix}$ so no change in p_x

$p_{fz} = F_{stick}\Delta t_{contact}$

Since the puck slides long distances with little change in velocity, friction and air resistance must be negligible.

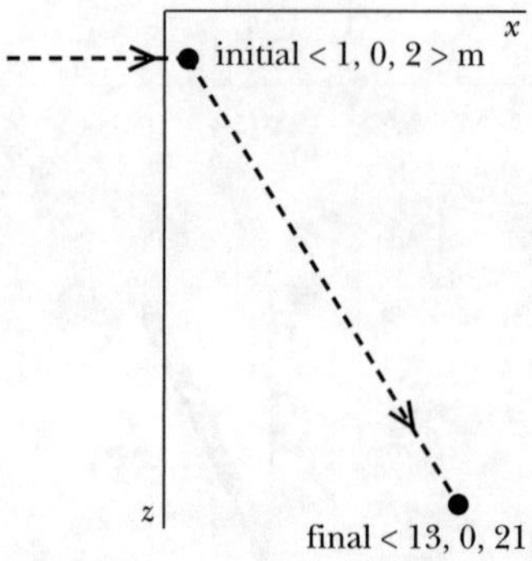

Figure 2.22 The x component of the momentum (and velocity) hardly changes, but the z component of momentum (and velocity) changes quickly from zero to some final value when the puck is hit.

? Given these results from the Momentum Principle for p_x and p_z, make a sketch of the path of the puck before and after it is hit.

The path of the puck must look something like that shown in Figure 2.22.

In the result $p_{fz} = F_{stick}\Delta t_{contact}$ there are two unknowns, p_{fz} and F_{stick}. We need another equation in order to be able to solve for the unknown contact time $\Delta t_{contact}$. We can get additional information from the position update formula $\vec{r}_f = \vec{r}_i + \vec{v}_{avg}\Delta t$.

4. Apply the position update formula

In $\vec{r}_f = \vec{r}_i + \vec{v}_{avg}\Delta t$, let Δt_{slide} be the amount of time it takes the puck to slide from where it was struck to the known final position:

$$\langle 13, 0, 21\rangle \text{ m} = (\langle 1, 0, 2\rangle \text{ m}) + \langle 20 \text{ m/s}, 0, v_z\rangle\Delta t_{slide}$$

$$x \text{ component: } (13 \text{ m}) = (1 \text{ m}) + (20 \text{ m/s})\Delta t_{slide}\text{, so}$$

$$\Delta t_{slide} = (12 \text{ m})/(20 \text{ m/s}) = 0.6 \text{ s}$$

$$y \text{ component: } 0 = 0 + 0\Delta t_{slide} \text{ so } 0 = 0 \text{ (not a surprise)}$$

$$z \text{ component: } (21 \text{ m}) = (2 \text{ m}) + v_z(0.6 \text{ s}) \text{ since } \Delta t_{slide} = 0.6 \text{ s, so}$$

$$v_z = (19 \text{ m})/(0.6 \text{ s}) = 31.7 \text{ m/s}$$

Since the puck slides long distances with little change in velocity, friction and air resistance must be negligible. We assume that after the puck is struck, the velocity is nearly constant (with a new magnitude and direction). Therefore the average velocity is the same as the velocity just after impact.

5. Solve for the unknowns

$$p_{fz} = F_{stick}\Delta t_{contact} \text{ where } p_{fz} \approx mv_{fz} \text{ (since } v << c)$$

$$(0.4 \text{ kg})(31.7 \text{ m/s}) = (1000 \text{ N})\Delta t_{contact}$$

$$\Delta t_{contact} = (0.4 \text{ kg})(31.7 \text{ m/s})/(1000 \text{ N}) = 0.013 \text{ s}$$

The contact time is very short, a small fraction of a second. As a result, you hear a short sharp crack when the stick hits the puck.

6. Check

- Units check (contact time is in seconds)
- Is the result reasonable? The contact time is very short, as expected. If for example it had come out to be 300 s (5 minutes!) we should check our calculations.

How good were our approximations?

We made the following approximations and simplifying assumptions:

- Ice exerts little force in the x or z directions; low sliding friction.
- Negligible air resistance.
- Force of stick is roughly constant during $\Delta t_{contact}$ and equal to 1000 N.
- The puck doesn't move very far during the contact time.

The neglect of sliding friction and air resistance is probably pretty good, since a hockey puck slides for long distances on ice.

We know the hockey stick exerts a maximum force of $F_{stick} = 1000$ N, because we observe that the stick breaks. We approximate the force as nearly constant during contact. Actually, this force grows quickly from zero at first contact to 1000 N, then abruptly drops to zero when the stick breaks.

The final approximation is somewhat questionable. Although 0.013 s is a short time, the puck moves (20 m/s) (0.013 s) = 0.26 m (a bit less than one foot) in the x direction during this time. Also during this time v_z increases from 0 to 31.7 m/s, with an average value of about 15.8 m/s, so the z displacement is about (15.8 m/s) (0.013 s) = 0.2 m during contact. On the other hand, these displacements aren't very large compared to the displacement from $\langle 1, 0, 2\rangle$ m to $\langle 13, 0, 21\rangle$ m, so our result isn't terribly inaccurate due to this approximation. Nevertheless, a more accurate sketch of the path of the puck should show a bend as in Figure 2.23.

It is important to see that even though our analysis of the stick contact time (0.013 s) isn't exact, it is adequate to get a reasonably good determination of this short time, something that we wouldn't know without using the Momentum Principle and the position update formula. The very short duration of the impact explains why we hear a sharp, short crack.

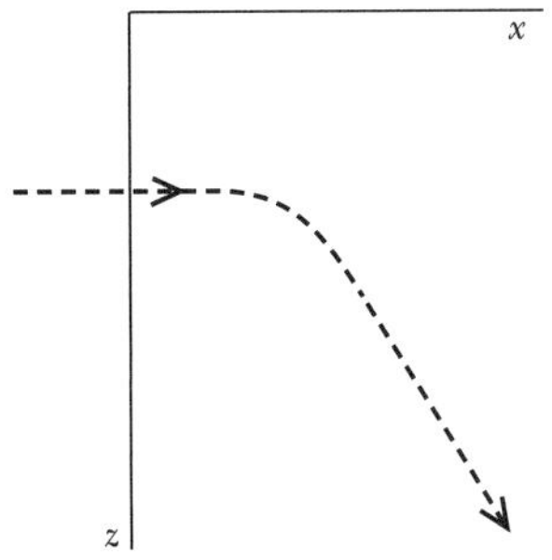

Figure 2.23 A more accurate overhead view of the path of the hockey puck, showing the bend during impact.

Choice of system

? We chose the hockey puck as the system to analyze. Why not choose the system consisting of both the hockey puck and the hockey stick?

The problem with choosing both objects as the system is that the 1000 N force is now an internal force and doesn't show up in the Momentum Principle, so we aren't able to use this information. By the reciprocity of electric forces, including the interatomic forces between stick and puck, the puck exerts a 1000 N force on the stick. (Reciprocity was mentioned earlier, page 51, and will be discussed in more detail later.) In the combined system the puck gains momentum from the stick, and the stick loses momentum to the puck. The total momentum of the system doesn't change: $\Delta\vec{p}_{\text{system}} = \vec{0}$.

Review

Let's review what we did to analyze this situation, in the form of a general scheme for attacking problems. We can summarize our work with a diagram in the shape of a diamond, which emphasizes

- starting from the Momentum Principle applied to a system,
- then expanding for the particular situation,
- then contracting down to solving for the quantity of interest,
- followed by checking for reasonableness.

Link the lines in this diamond to the hockey stick analysis:

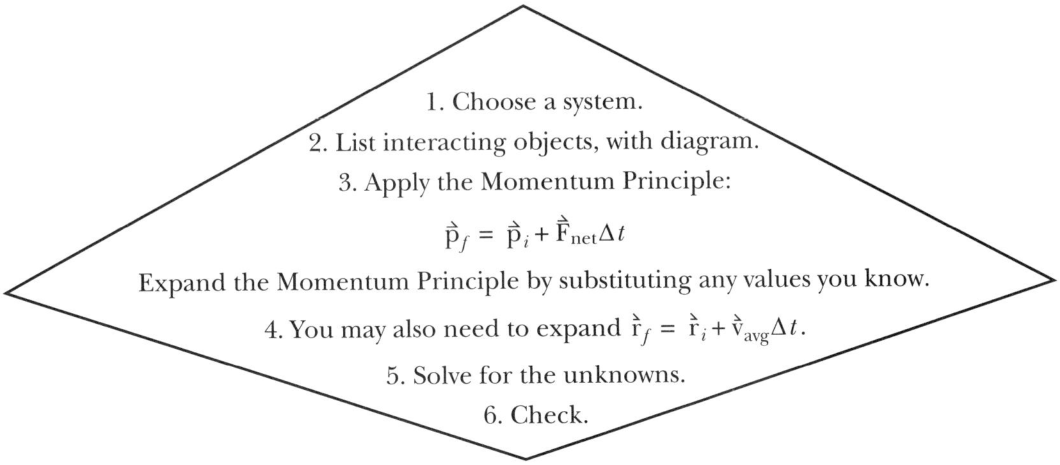

An important point implied by this diamond is that it is usually not useful to try to jump immediately to the step "Solve for the unknowns," by hunting for some formula that gives the unknown quantity directly. Very often such a special-case formula may not exist. For example, in the hockey puck problem the contact time emerged from applying the Momentum Principle and the position update formula. There wasn't some ready-made "formula for the contact time for hockey sticks and pucks" that you could use.

If you start by applying the Momentum Principle to a chosen system you can attack novel problems that you've never before encountered. The Momentum Principle is always valid, whereas special-case formulas aren't.

In previous studies you may have been taught a useful but restricted approach to solving problems, which is to start with a formula for the unknown quantity. That is, if you've trying to find v, start with a formula for v. We will help you learn a more powerful technique for solving problems, which is to start from a fundamental physics principle (in this case the Momentum Principle), expand it by substituting known values, then contract to solve for the unknown quantities. In other words, you *derive* the formula you need rather than hunt for it. This is the only technique that can give you the power to solve novel problems, ones that no one has previously encountered.

2.5.2 Example: Colliding students

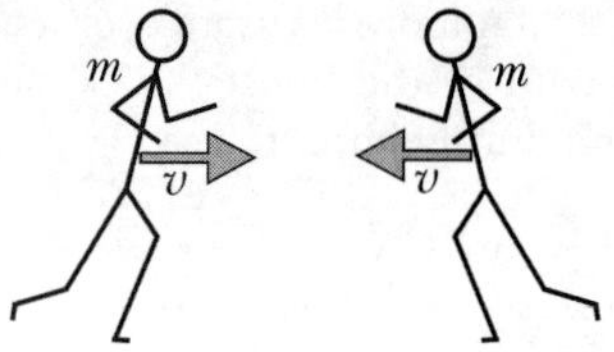

Figure 2.24 Colliding students. The speed (magnitude of velocity) of each student is assumed to be the same and is represented by v. We also make the simplifying assumption that the students have about the same mass m.

Next we'll work through in detail a messy real-world situation: Two students who are late for tests are running to classes in opposite directions as fast as they can (Figure 2.24). They turn a corner, run into each other head-on, and crumple into a heap on the ground. Using physics principles, estimate the force that one student exerts on the other during the collision. You will need to estimate some quantities; give reasons for your choices and provide checks showing that your estimates are physically reasonable.

This problem is rather ill-defined and doesn't seem much like a "textbook" problem. No numbers have been given, yet you're asked to estimate the force of the collision. This kind of problem is typical of the kinds of problems engineers and scientists encounter in their professional work. To get started, we'll make the simplifying assumption that the students have similar masses and similar speeds (Figure 2.24).

Remember the diamond scheme as a guide to how to proceed. We'll carry out the analysis symbolically and plug in estimated values of student mass and speed at the end. That way we get a general solution that can be evaluated for different values of these quantities.

1. Choose a system

System: the student on the left

2. List external objects that interact with the system, with diagram

List all interacting objects, even if you think some interactions will cancel out.

Earth, ground, other student, air

(more complex diagram shows point of application of each force)

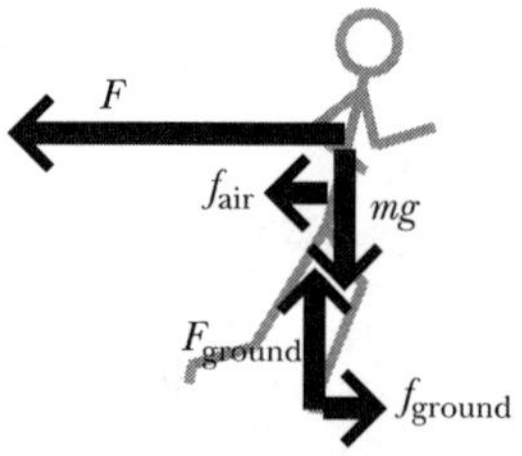

3. Apply the Momentum Principle

The student's heels are pushed to the left along the ground due to the collision, and the resisting ground pushes to the right. We'll assume that this force is much smaller than the force exerted by the other student, because the student's shoes can slip.

$$\langle 0, 0, 0\rangle = \langle p_{ix}, 0, 0\rangle + \langle -F - f_{\text{ground}} - f_{\text{air}}, (F_{\text{ground}} - mg), 0\rangle \Delta t$$

Many of the momentum components are zero. Do you see why $p_{fx} = 0$?

Write the three component equations separately:

In a vector equation, the x component on the left must equal the x component on the right; similarly for y and z.

x component: $0 = p_{ix} - (F + f_{\text{ground}} + f_{\text{air}})\Delta t$

y component: $0 = (F_{\text{ground}} - mg)\Delta t$, so $F_{\text{ground}} = mg$

z component: $0 = 0$ which is true but uninformative

Find new momentum:

Assume f_{ground} and f_{air} are negligible compared to F

$$0 = p_{ix} - F\Delta t, \text{ so } mv = F\Delta t \text{ since } v \ll c$$

We could estimate the typical mass of a student, and the likely speed of a student running at full speed, but $mv = F\Delta t$ is one equation in two unknowns, the unknown force F of the other student and the time duration Δt of the impact. We need a second equation.

Estimating the time duration

? Before trying to apply the position update formula $\vec{r}_f = \vec{r}_i + \vec{v}_{avg}\Delta t$, try to estimate Δt of the impact simply by making a guess.

It is common to guess what seems a rather short time, such as a second, or half a second. However, it is easy to show that such estimates are way off. The world record for the 100 m dash is about 10 seconds. If we assume that these students are very fast runners, they each could be running 10 m/s. An alternative way of making the estimate is to note that you can easily walk about 4 miles in an hour, which gives

$$v \approx \left(4\frac{\text{mi}}{\text{hr}}\right)\left(1.6\frac{\text{km}}{\text{mi}}\right)\left(1000\frac{\text{m}}{\text{km}}\right)\left(\frac{1\text{ hr}}{3600\text{ s}}\right) \approx 1.8\text{ m/s}$$

This moderate walking speed suggests that the students should be able to run between 5 m/s and 10 m/s. Let's estimate that they each have a speed $v \approx 5$ m/s. During the collision the speed drops quickly from v to 0, so the average speed during the collision is approximately $(v+0)/2$, or 2.5 m/s.

? If we guess that Δt is about half a second, during that half second, how far does the student move?

At an average speed of 2.5 m/s, in a half second the student would go 1.25 meters, well over a yard. This would imply that the two students pass right through each other, coming out the other side! Evidently an estimate of a half second is wildly too big and is inconsistent with the real world.

The problem is that it is very difficult to guess short time intervals accurately. On the other hand, we can estimate short distances rather well, and this gives us an indirect way to estimate short time intervals. Each student gets squeezed during the collision. Suppose each student's body gets pushed in a distance of about $\Delta x \approx 5\text{ cm} = 0.05\text{ m}$ (try pushing on your stomach and see how much deflection you can make).

4. Apply the position update formula

Apply $\vec{r}_f = \vec{r}_i + \vec{v}_{avg}\Delta t$ or equivalently $\vec{v}_{avg} = \Delta\vec{r}/\Delta t$:

$$\langle v/2, 0, 0\rangle = \langle \Delta x, 0, 0\rangle/\Delta t$$

$$x \text{ component: } v/2 = \Delta x/\Delta t, \text{ so } \Delta t = 2\Delta x/v$$

5. Solve for the unknowns

$$mv = F\Delta t = F(2\Delta x/v)$$

$$F = \frac{mv^2}{2\Delta x} = \frac{(60\text{ kg})(5\text{ m/s})^2}{2(0.05\text{ m})} = 15000\text{ N}$$

$$\Delta t = \frac{2\Delta x}{v} = \frac{2(0.05\text{ m})}{5\text{ m/s}} = 0.02\text{ s}$$

6. Check

- Units check (force is in newtons, collision time is in seconds)
- Is the result reasonable? The contact time is very short, as expected. Is the force reasonable or not? See discussion below.

Estimate that the student is brought to an abrupt stop in just 5 cm (about 2 inches).

Force has units (kg)(m/s)/s, which has the units of momentum (kg·m/s) divided by s, which is correct for a force (change of momentum divided by time).

The collision time is 25 times smaller than our original guess of 0.5 s. So the key to making a realistic estimate of a time interval is to estimate average speed and distance, and then derive a time estimate from that.

15000 N is a very large force. For example, the gravitational force on a 60 kg student (the "weight") is only about $(60\text{ kg})(9.8\text{ N/kg}) \approx 600\text{ N}$. The force of the impact is more than 25 times the weight of the student! It's like having a stack of 25 students sitting on you. If the students hit heads instead of stomachs, the squeeze might be less than 1 cm, and the force would be over 5 times as large! This is why heads can break in such a collision.

So yes, our result of 15000 N is plausible: collisions involve very large forces acting for very short times, giving impulses of ordinary magnitude.

How good were our approximations?

We made the following approximations and simplifying assumptions:

- We estimated the running speed from known 100 m dash records.
- We estimated the masses of the students; we could plug in other masses now that we have derived general results in terms of m and v.
- We assumed that the horizontal component of the force of the ground on the bottom of the student's shoe was small compared to the force exerted by the other student. Now that we find that the impact force is huge, this assumption seems quite good.
- We made the approximation that the impact force was nearly constant during the impact, so what we've really determined is an average force.
- We assumed that the students had similar masses and similar running speeds, to simplify the analysis. If this is not the case, the analysis is significantly more complicated, but we would still find that the impact force is huge.

You might object that with all these estimates and simplifying assumptions the final result of 15000 N for the impact force is not useful. It is certainly the case that we don't have a very accurate result. But nevertheless we gained valuable information, that the impact force is *very* large. Before doing this analysis based on the Momentum Principle, we had no idea of whether the force was small compared to the student's weight, comparable, or much bigger. Now we have a quantitative result that the force is on the order of 25 times the weight of a student, and we can appreciate why collisions are so dangerous.

? What if we choose the system to include both students?

In this case there is almost no external impulse during the brief time of contact. The big internal forces the students exert on each other are equal in magnitude but opposite in direction (due to the reciprocity of interatomic forces that come into play when the students make contact). Therefore these forces don't change the momentum of the combined system. Before the collision the total momentum of the combined system is zero ($mv - mv = 0$), and after the collision the total momentum is also zero (they're not moving). So there's no change in momentum, which is consistent with there being negligible external forces. Simple analysis, but with this choice of system we learn nothing about the force that one student exerts on the other. That's why we choose just one student as the system.

2.5.3 Physical models

Our model of the colliding student situation is good enough for many purposes. However, we left out some aspects of the actual motion. For example, we mostly ignored the flexible structure of the students, how much their shoes slip on the ground during the collision, etc. We have also quite sensibly neglected the gravitational force of Mars on the students, because it is so tiny compared to the force of one student on the other.

Making and using models is an activity central to physics, and the criteria for a "good" physical model will depend on how we intend to use the model.

In fact, one of the most important problems a scientist or engineer faces is deciding what interactions must be included in a model of a real physical, chemical, or biological system, and what interactions can reasonably be ignored.

Simplified (or "idealized") models

If we neglect some (hopefully small) effects, we say that we are constructing a simplified model of the situation. A useful model should omit extraneous detail but retain the main features of the real-world situation. We hope that the main results of our analysis of the simplified model will apply adequately to the complex real-world situation.

Of course if we do a poor job of modeling and make inappropriate approximations or neglect effects that are actually sizable, we may get a rather inaccurate result (though perhaps adequate for some purposes). There is therefore a certain art to making a good model. One goal of this course is to help you develop skill in formulating simplified but meaningful physical models of complex situations.

Another common way of describing a model is to say that it is an "idealization," by which we mean a simple, clean, stripped-down representation, free of messy complexities. "Ideally," a ball will roll forever on a level floor, but a real ball rolling on a real floor eventually comes to a stop. An "ideal" gas is a fictitious gas in which the molecules don't interact with each other, as opposed to a real gas whose molecules do interact, but only when they come close to each other.

The analysis of a simplified model gives an approximate result that differs somewhat from what actually happens in the real world. Alternatively, you can consider your result to be exact for a *different* real-world situation, one where the neglected influences are not actually present. For example, the calculations we made concerning the colliding students could apply almost exactly to two rubber balls moving in outer space that compress by the amount we estimated.

An important aspect of physical modeling that will engage us throughout this course is making appropriate approximations to simplify the messy, real-world situation enough to permit (approximate) analysis using Newton's laws. Actually, using Newton's laws is itself an example of modeling and making approximations, because we are neglecting the effects of quantum mechanics and of general relativity (Einstein's treatment of gravitation). Newton's laws are only an approximation to the way the world works, though frequently an extremely good one.

2.6 Fundamental forces

There are four different kinds of fundamental forces currently known to science, associated with four different kinds of interactions: gravitational, electromagnetic, nuclear (also referred to as the "strong" interaction), and the "weak" interaction.

- The *gravitational* interaction is responsible for an attraction every object exerts on every other object. For example, the Earth exerts a gravitational force on the Moon, and the Moon exerts a gravitational force on the Earth.
- The *electromagnetic* interaction includes electric forces responsible for sparks, static cling, and the behavior of electronic circuits, and magnetic forces responsible for the operation of motors driven by electric current. Protons repel each other electrically, as do electrons, whereas protons and electrons attract each other (Figure 2.25). Electric forces bind protons and electrons to each other in atoms, and are responsi-

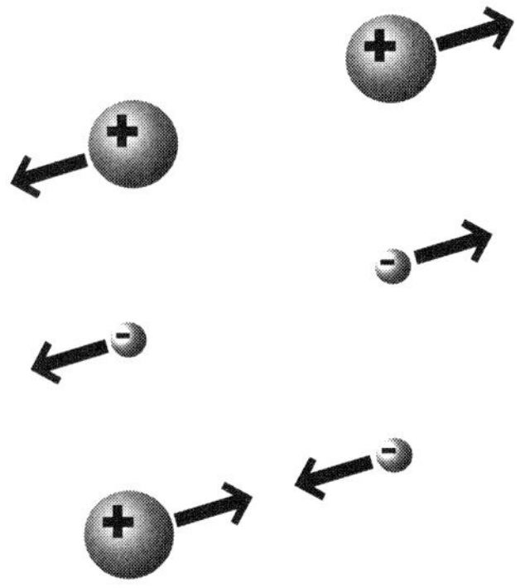

Figure 2.25 The electric force: protons repel each other; electrons repel each other; protons and electrons attract each other.

ble for the chemical bonds between atoms in molecules. The force of a stretched or compressed spring is due to electric forces between the atoms that make up the spring.

- The nuclear or *strong* interaction holds protons and neutrons together in the nucleus of an atom despite the large mutual electric repulsion of the protons (Figure 2.26). (The neutrons are not electrically charged and interact only through the strong force.)
- An example of the *weak* interaction is seen in the instability of a neutron. If a neutron is removed from a nucleus, with an average lifetime of about 15 minutes the neutron decays into a proton, an electron, and a ghostly particle called the antineutrino. This change is brought about by the weak interaction.

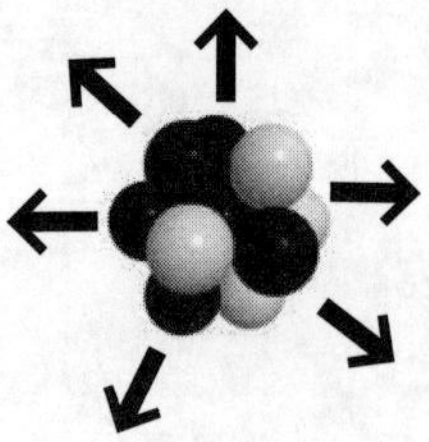

Figure 2.26 The strong force: the protons in the nucleus of an atom exert repulsive electric forces on each other, but he strong interaction holds the nucleus together despite this electric repulsion.

We will be mainly concerned with gravitational and electric interactions, but we will occasionally encounter situations where the nuclear or strong interaction plays an important role. We will have little to say about the weak interaction. Nor will we deal with magnetism, the other part of the electromagnetic interaction. The second volume of this textbook deals extensively with both electric and magnetic interactions.

2.7 The gravitational force law

To predict the motion of stars, planets, spacecraft, comets, satellites, and other massive objects traveling through space, we need to use a gravitational force law together with the Momentum Principle. As a matter of general culture, it can be interesting to understand how the motion of objects in space can be predicted by physics principles. If you are studying aerospace engineering or astrophysics this may also be a professional interest.

What if your interests are in nanotechnology or chemistry or civil engineering? Studying how to predict the motion of stars and planets is one of the most direct ways to understand in general how the Momentum Principle determines the behavior of objects in the real world. The motion of stars and planets is in important ways simpler than other mechanical phenomena, because there is no friction to worry about, so this is a good place to start your study. The basic ideas used to predict the motion of stars and planets can be applied later to a wide range of everyday and atomic phenomena.

In the 1600's Isaac Newton deduced that there must be an attractive force associated with a gravitational interaction between any pair of objects. The gravitational force acts along a line connecting the two objects (Figure 2.27), is proportional to the mass of one object and to the mass of the other object, and is inversely proportional to the square of the distance between the *centers* of the two objects (not the gap between their surfaces). Here is the formula for the gravitational force exerted on object 2 by object 1:

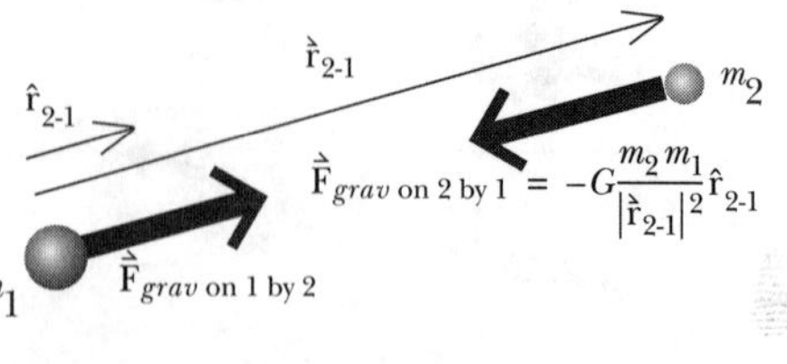

Figure 2.27 The gravitational force exerted on object 2 by object 1. (The force exerted on object 1 by object 2 has the same magnitude but opposite direction.)

THE GRAVITATIONAL FORCE LAW

$$\vec{F}_{grav\text{ on 2 by 1}} = -G\frac{m_2 m_1}{|\vec{r}_{2\text{-}1}|^2}\hat{r}_{2\text{-}1}$$

$\hat{r}_{2\text{-}1}$ points from the center of object 1 to the center of object 2

$$G \text{ is a universal constant: } G = 6.7\times10^{-11}\ \frac{\text{N}\cdot\text{m}^2}{\text{kg}^2}$$

This force law looks pretty complicated, but it could have been a lot more complicated than it is. For example, the gravitational force law does not depend on the momentum or velocity of the objects. It depends only on the masses, and on the position of one object relative to the other object.

Large spheres

An optional section at the end of this chapter (page 99) discusses the fact that uniform-density spheres interact gravitationally as though all of their mass were concentrated at the center of the sphere, so the gravitational force law applies to large uniform-density spheres as long as you use the center-to-center distance. For example, the force that the Earth exerts on you can be calculated as though all the mass of the Earth were at its center, 6.4×10^6 m away from where you are sitting.

Measurement of *G*

Another optional section at the end of this chapter (page 99) tells how Cavendish in the late 1700's measured *G* by observing the tiny gravitational forces two lead balls exerted on each other.

2.7.1 Understanding the gravitational force law

We will spend some time understanding how to calculate forces between objects using Newton's gravitational force law. Then we will use this force in the Momentum Principle to predict the future motion of objects.

The gravitational force law involves a lot of different symbols and may look pretty intimidating at first. Let's take the law apart and look at the individual pieces to try to make sense of the formula.

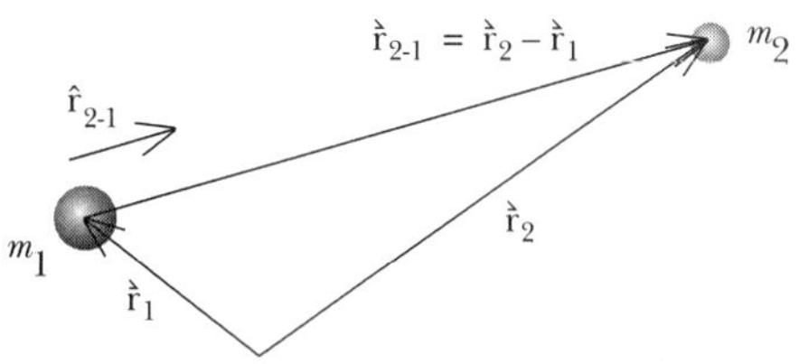

Figure 2.28 The location of object 2 relative to object 1: "final minus initial."

The relative position vector $\vec{r}_{2\text{-}1}$ in the gravitational force law points from the center of object 1 to the center of object 2, and so does the unit vector $\hat{r}_{2\text{-}1}$ ("r-hat") in Figure 2.28. Recall that in words $\vec{r}_{2\text{-}1}$ is the location of object 2 relative to object 1, "final minus initial." The magnitude of $\vec{r}_{2\text{-}1}$ is the distance between the centers of the two objects.

We say that the gravitational constant *G* is "universal" because it is the same for any pair of interacting masses, no matter how big or small they are, or where they are located. Because *G* is universal, it can be measured for any pair of objects and then used with other pairs of objects. As we describe on page 99, Cavendish was the first person to make such a measurement.

As highlighted in Figure 2.29, the gravitational force is proportional to the *product* of the two masses, $m_2 m_1$. If you double either of these masses, keeping the other one the same, the force will be twice as big. If you double both of the masses, the force will be four times as big. Since $m_2 m_1 = m_1 m_2$, the magnitude of the force exerted on object 1 by object 2 is exactly the same as the magnitude of the force exerted on object 2 by object 1 (but the direction is opposite).

Figure 2.29 The gravitational force depends on the *product* of the two masses.

The gravitational force is an "inverse square" law. As highlighted in Figure 2.30, the square of the center-to-center distance appears in the denominator. This means that the gravitational force depends very strongly on the distance between the objects. For example, if you double the distance between them, the only thing that changes is the denominator, which gets four times bigger (2 squared is 4), so the force is only 1/4 as big as before.

Figure 2.30 The gravitational force is an "inverse square" law.

? If you move the masses 10 times farther apart than they were originally, how does the gravitational force change?

The force goes down by a factor of 100. Evidently when two objects are very far apart, the gravitational forces they exert on each other will be vanishingly small: big denominator, small force.

The minus sign with the unit vector highlighted in Figure 2.31 gives the direction of the force exerted on object 2 by object 1. The vector $\vec{r}_{2\text{-}1}$ points toward object 2, as does the unit vector $\hat{r}_{2\text{-}1}$, and the force acting on object 2 points in the opposite direction.

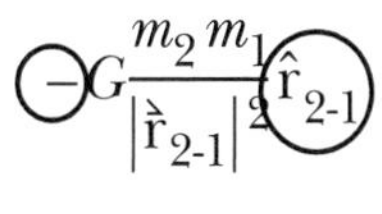

Figure 2.31 The direction of the gravitational force on object 2 is in the opposite direction to the unit vector pointing *toward* object 2.

A useful way to think about the gravitational force law is to factor it into magnitude and direction, like this:

- A vector is a magnitude times a direction: $\vec{F}_{grav} = |\vec{F}_{grav}|\hat{F}_{grav}$
  - Magnitude: $|\vec{F}_{grav}| = G\dfrac{m_2 m_1}{|\vec{r}_{2\text{-}1}|^2}$
  - Direction (unit vector): $\hat{F}_{grav} = -\hat{r}_{2\text{-}1}$

It is usually simplest to calculate the magnitude and direction separately, then combine them to get the vector force. That way you can focus on one thing at a time rather than getting confused (or intimidated!) by the full complexity of the vector force law.

Ex. 2.18 Masses M and m attract each other with a gravitational force of magnitude F. Mass m is replaced with a mass $3m$, and it is moved four times farther away. Now what is the magnitude of the force?

Ex. 2.19 A 3 kg ball and a 5 kg ball are 2 m apart, center to center. What is the magnitude of the gravitational force that the 3 kg ball exerts on the 5 kg ball? What is the magnitude of the gravitational force that the 5 kg ball exerts on the 3 kg ball?

Ex. 2.20 Measurements show that the Earth's gravitational force on a mass of 1 kg near the Earth's surface is 9.8 N. The radius of the Earth is 6400 km (6.4×10^6 m). From these data determine the mass of the Earth.

2.7.2 Calculating the gravitational force on a planet

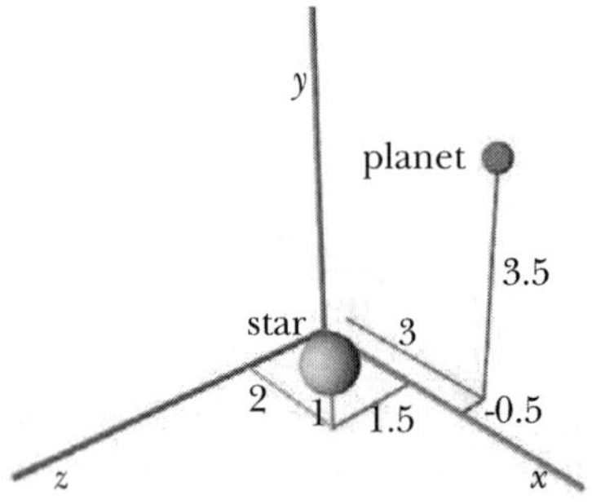

Figure 2.32 A star and a planet interact gravitationally. We will calculate the gravitational forces. The x, y, and z components are to be multiplied by 1×10^{11} m.

We'll go through a complete calculation to see in full gory detail how to evaluate the gravitational force law in the most general situation. In Figure 2.32 is a star of mass 4×10^{30} kg located at this time at position $\langle 2\times10^{11}, 1\times10^{11}, 1.5\times10^{11}\rangle$ m and a planet of mass 3×10^{24} kg located at position $\langle 3\times10^{11}, 3.5\times10^{11}, -0.5\times10^{11}\rangle$ m. These are typical values for stars and planets. Notice that the mass of the star is much greater than that of the planet.

We will calculate the gravitational force exerted on the planet by the star. In Figure 2.32 we show the x, y, and z components of the positions, to be multiplied by 1×10^{11} m. Make sure you understand how the numbers on the diagram correspond to the positions given as vectors.

Here is a summary of the steps we will take to calculate the force acting on the planet due to the star:

- Calculate $\vec{r}_{2\text{-}1}$, the position of the planet relative to the star.
- Calculate $|\vec{r}_{2\text{-}1}|$, the (scalar) distance from star to planet.
- Calculate $Gm_2m_1/|\vec{r}_{2\text{-}1}|^2$, the magnitude of the force.
- Calculate $-\hat{r}_{2\text{-}1} = -\vec{r}_{2\text{-}1}/|\vec{r}_{2\text{-}1}|$, the direction of the force.
- Multiply the magnitude times the direction to get the vector force.

The relative position vector

An important quantity in the force law is the position of the center of the planet relative to the center of the star, $\vec{r}_{2\text{-}1}$, so we start by calculating this relative position vector.

? Think about what you know about calculating relative position vectors, and try to calculate $\vec{r}_{2\text{-}1}$ before reading ahead.

As usual, we just calculate "final minus initial":

$$\vec{r}_{2\text{-}1} = \vec{r}_2 - \vec{r}_1$$

$$\vec{r}_{2\text{-}1} = \langle 3\times10^{11}, 3.5\times10^{11}, -0.5\times10^{11}\rangle \text{ m} - \langle 2\times10^{11}, 1\times10^{11}, 1.5\times10^{11}\rangle \text{ m}$$

$$\vec{r}_{2\text{-}1} = \langle 1\times10^{11}, 2.5\times10^{11}, -2\times10^{11}\rangle \text{ m}$$

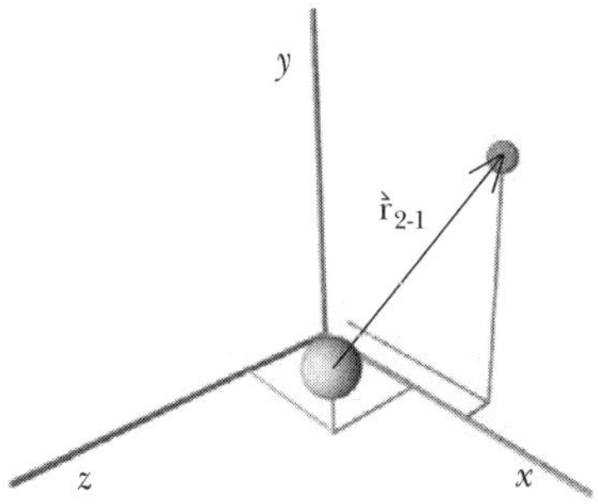

Figure 2.33 The position vector of the planet relative to the star.

In Figure 2.33, the diagram shows that the signs of the components of $\vec{r}_{2\text{-}1}$ make sense (positive x and y components, negative z component), which is an important check that we haven't made any sign errors.

The distance

In order to calculate the magnitude of the force, we'll need the distance $|\vec{r}_{2\text{-}1}|$ between the centers of the star and planet (and a bit later we'll also need $|\vec{r}_{2\text{-}1}|$ for calculating the unit vector).

? Try to calculate the magnitude of $\vec{r}_{2\text{-}1}$ before reading ahead.

As usual, we calculate the distance, which is a scalar, by using the 3D version of the Pythagorean theorem:

$$|\vec{r}_{2\text{-}1}| = \sqrt{(1\times10^{11})^2 + (2.5\times10^{11})^2 + (-2\times10^{11})^2} \text{ m}$$

$$|\vec{r}_{2\text{-}1}| = 3.35\times10^{11} \text{ m}$$

The magnitude of the force

? Before reading ahead, try to calculate the magnitude of the force, $|\vec{F}_{grav \text{ on planet by star}}|$.

The magnitude of the force on the planet by the star is this:

$$|\vec{F}_{grav \text{ on planet by star}}| = G\frac{m_2 m_1}{|\vec{r}_{2\text{-}1}|^2}$$

$$|\vec{F}_{grav \text{ on planet by star}}| = \left(6.7\times10^{-11} \frac{\text{N}\cdot\text{m}^2}{\text{kg}^2}\right)\frac{(3\times10^{24} \text{ kg})(4\times10^{30} \text{ kg})}{(3.35\times10^{11} \text{ m})^2}$$

$$|\vec{F}_{grav \text{ on planet by star}}| = 7.16\times10^{21} \text{ N}$$

This looks like a big force, but it's acting on a planet with a big mass, so it isn't obvious whether this is really a "big" force in terms of what it will do.

The direction of the force

? Can you think how to calculate the vector direction of the gravitational force, the unit vector $\hat{F}_{grav \text{ on planet by star}}$? Try it.

We've got the magnitude of the force. We also need to calculate the direction of the force, which involves the unit vector:

$$\text{Direction of force: } \hat{F}_{grav \text{ on planet by star}} = -\hat{r}_{2\text{-}1} = -\frac{\vec{r}_{2\text{-}1}}{|\vec{r}_{2\text{-}1}|}$$

$$\hat{F}_{grav \text{ on planet by star}} = -\frac{\langle 1\times10^{11}, 2.5\times10^{11}, -2\times10^{11}\rangle \text{ m}}{3.35\times10^{11} \text{ m}}$$

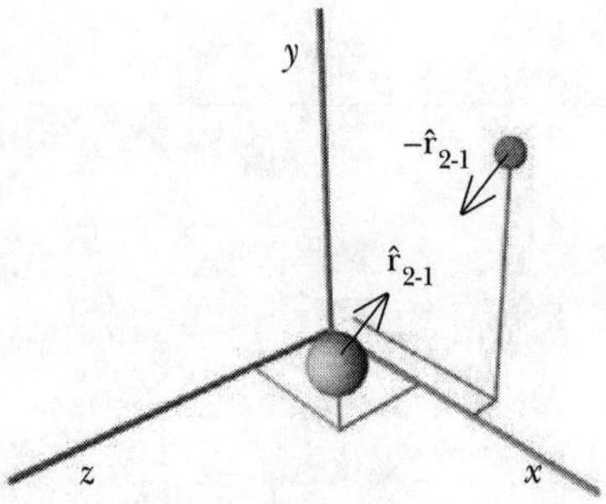

Figure 2.34 Unit vectors.

$$\hat{F}_{grav \text{ on planet by star}} = \langle -0.299, -0.746, 0.597 \rangle$$

Notice that the unit vector giving the direction has no units, because the meters in the numerator and the denominator cancel. Figure 2.34 shows the unit vectors pointing from star toward planet, and from planet toward star.

The vector gravitational force

? Try to construct the vector force. You know the magnitude, and you know the direction.

Finally we can multiply the magnitude of the force times the unit vector direction of the force to get the full vector force:

$$\vec{F}_{grav \text{ on planet by star}} = |\vec{F}_{grav \text{ on planet by star}}|\hat{F}_{grav \text{ on planet by star}}$$

$$\vec{F}_{grav \text{ on planet by star}} = (7.16\times10^{21}\ \text{N})\langle -0.299, -0.746, 0.597 \rangle$$

$$\vec{F}_{grav \text{ on planet by star}} = \langle -2.14\times10^{21}, -5.34\times10^{21}, 4.27\times10^{21} \rangle\ \text{N}$$

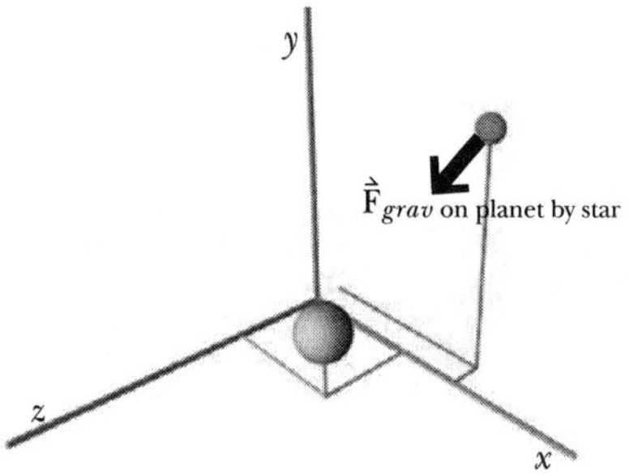

Figure 2.35 The gravitational force exerted on the planet by the star.

Checking our result

It is important to check long calculations such as this, because there are many opportunities to make a mistake along the way. There are several checks we can make.

- *Diagram:* An important check is to make a diagram like Figure 2.35 and see whether the direction of the calculated force makes sense. Pay particular attention to signs. The diagram shows that the force acting on the planet points in the $-x$, $-y$, $+z$ direction, and this agrees with our calculation, which has the same component signs.
- *Order of magnitude:* Make a rough "order of magnitude" check, dropping all the detailed numbers. In terms of "order of magnitude" the distance between star and planet is *very* roughly 1×10^{11} m, the mass of the star is *very* roughly 1×10^{30} kg, the mass of the planet is *very* roughly 1×10^{24} kg, and the gravitational constant is *very* roughly $1\times10^{-11}\ \text{N}\cdot\text{m}^2/\text{kg}^2$. Therefore we expect the magnitude of the force to be *very* roughly this:

$$\left(1\times10^{-11}\ \frac{\text{N}\cdot\text{m}^2}{\text{kg}^2}\right)\frac{(1\times10^{24}\ \text{kg})(1\times10^{30}\ \text{kg})}{(1\times10^{11}\ \text{m})^2} = 1\times10^{21}\ \text{N}$$

  In this very rough order of magnitude calculation, ignoring numerical details, we just need to add and subtract exponents, and one hardly even needs a calculator. The fact that we get within an order of magnitude of the result 7.16×10^{21} N is evidence that we haven't made any huge mistakes.
- *Units:* An important check is that the units came out correctly, in newtons.
- *Unit vector:* Check to see whether the calculated unit vector does indeed have magnitude 1:

$$|\hat{F}_{grav \text{ on planet by star}}| = \sqrt{(-0.299)^2 + (-0.746)^2 + (0.597)^2} = 1.001$$

  The magnitude isn't exactly 1 because we rounded off the intermediate calculations to three significant figures.

Our result passes all these checks. This doesn't prove we haven't made a mistake somewhere, but at least we've ruled out many possible errors.

The force on the star exerted by the planet

We've just calculated the force exerted on the planet by the star. To calculate the force exerted on the star by the planet, we could redo all these lengthy calculations in the same way, but that would be a lot of work.

> **?** Look at the form of the gravitational force law and think carefully. Do you see a way to write down immediately the gravitational force on the star by the planet, without doing any additional calculations?

Note that the magnitude is exactly the same, because it involves exactly the same quantities. The only change is the direction of the force, which is in the opposite direction. Therefore we can immediately write the new result, just by flipping the signs:

$$\vec{F}_{grav \text{ on star by planet}} = \langle 2.14\times10^{21}, 5.34\times10^{21}, -4.27\times10^{21}\rangle \text{ N}$$

Note that the signs of the force components are consistent with Figure 2.36. You might be puzzled that the planet pulls just as hard on the star as the star pulls on the planet. We'll discuss this in more detail later in this chapter.

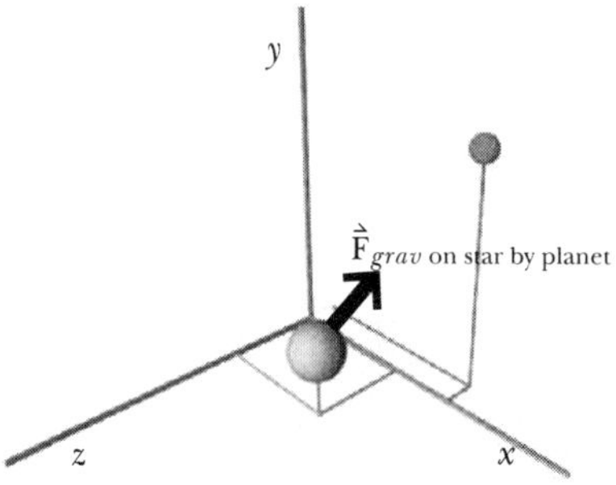

Figure 2.36 The gravitational force exerted on the star by the planet.

How can you remember all these steps?

Calculating the vector gravitational force requires many steps, but each step is pretty straight-forward. How can one remember what steps to take? Think of the equation for the gravitational force as a guide and outline for what to do.

$$\vec{F}_{grav \text{ on 2 by 1}} = -G\frac{m_2 m_1}{|\vec{r}_{2\text{-}1}|^2}\hat{r}_{2\text{-}1}$$

- When you look at the formula you see that you need $\vec{r}_{2\text{-}1}$, so you calculate it ("final minus initial").
- You also see that you need $|\vec{r}_{2\text{-}1}|$, so you calculate that.
- Now you see that you now have enough information to calculate the magnitude of the force.
- The formula tells you to calculate the negative of the unit vector, to get the direction.
- The formula says to multiply magnitude times direction. That's it!

With this scheme in mind, look over the previous few pages to review how we carried out the evaluation of the gravitational force law.

Ex. 2.21 The mass of the Earth is 6×10^{24} kg, and the mass of the Moon is 7×10^{22} kg. At a particular instant the Moon is at location $\langle 2.8\times10^{8}, 0, -2.8\times10^{8}\rangle$ m, in a coordinate system whose origin is at the center of the Earth.
(a) What is $\vec{r}_{M\text{-}E}$ the relative position vector from the Earth to the Moon?
(b) What is $|\vec{r}_{M\text{-}E}|$?
(c) What is the unit vector $\hat{r}_{M\text{-}E}$?
(d) What is the gravitational force exerted by the Earth on the Moon? Your answer should be a vector.

2.7.3 Using the gravitational force to predict motion

You now know that this star exerts this force on this planet at this instant, but what's the point in knowing this? By itself, this isn't particularly interesting. But if we use this calculated force in the Momentum Principle, we can predict the future motion of the planet.

- We know the force at this instant, and if we know the initial momentum $\vec{p}_i$ of the planet we can use the update form of the Momentum Principle, $\vec{p}_f = \vec{p}_i + \vec{F}_{net}\Delta t$, to predict the momentum of the planet a short time Δt later. We need to choose Δt short enough that the force doesn't change much during this brief time interval, due to changes in the positions of the interacting objects.
- We can also use the velocity to update the position of the planet, using the position update formula $\vec{r}_f = \vec{r}_i + \vec{v}_{avg}\Delta t$. Again, we need to choose Δt short enough that the velocity doesn't change much during that brief time interval, so that we can use the initial or final velocity as a good approximation to the average velocity.

After doing these momentum and position updates, we have successfully predicted the new momentum and position a short time into the future. Now we can repeat this process, calculating a new gravitational force based on the new position, updating the momentum, and updating the position. Step by step we can predict the motion of the planet into the future.

Although the massive star won't move very much, we could also update its momentum and position repetitively. As we've seen, it's easy to get the force on the star just by flipping signs.

Here is a summary of this iterative scheme:

- Start with initial positions and momenta of the interacting objects.
- Calculate the (vector) gravitational forces acting on each object.
- Update the momentum of each object: $\vec{p}_f = \vec{p}_i + \vec{F}_{net}\Delta t$.
- Update the positions: $\vec{r}_f = \vec{r}_i + \vec{v}_{avg}\Delta t$, where $\vec{v}_{avg} \approx \vec{p}_f/m$.
- Repeat.

In order to plan the trajectory for a spacecraft from Earth to a particular landing site on Mars, NASA has to carry out a very large number of such update calculations, taking small steps (small Δt) to achieve high accuracy. This includes calculating the net force as the vector sum of all the gravitational forces exerted on the spacecraft along the way by the Sun, Earth, Moon, Mars, and other objects in the Solar System. Doing all these force and update calculations by hand would be impossibly difficult, so people at NASA write computer programs to do these repetitive calculations. Later we will show you how to write such programs.

Updating the momentum and position of the planet

To see in detail how this iterative scheme works, we'll carry out one complete step—one update of the momentum and position. In order to start the iteration we have to know the initial positions and momenta of the star and planet. We already know their initial positions. Let's suppose we also know their initial velocities:

$$\vec{v}_{i,\text{star}} = \langle 0, 0, 0 \rangle \text{ m/s}$$

$$\vec{v}_{i,\text{planet}} = \langle 1\times10^4, 2\times10^4, 1.5\times10^4 \rangle \text{ m/s}$$

The star is initially at rest and, because it has a huge mass compared to the planet, it won't move very much. For simplicity we'll make the approximation that the star is fixed in position and never moves.

Choosing a sufficiently small time interval

We need to decide how big a value of Δt we can get away with and yet preserve adequate accuracy. The smaller the time step Δt, the more accurate the calculation, because the force and velocity won't change much during the brief time interval. However, taking smaller steps means doing more calculations to predict some time into the future.

Here is a way to choose a time interval. The initial speed is about $v \approx 3\times10^4$ m/s. Choose a Δt so that the distance the planet goes in this time interval, $v\Delta t$, is small compared to the distance d between star and planet, which is about 3×10^{11} m. As the planet changes position the gravitational force will change in magnitude and direction, as you can see in Figure 2.37. In the Momentum Principle $\vec{p}_f = \vec{p}_i + \vec{F}_{net}\Delta t$ we need for Δt to be small enough that $\vec{F}_{net}$ is nearly constant during this time interval.

Let's choose as a tentative criterion that the distance the planet moves ought to be about 0.001 times the distance from the star to the planet:

$$v\Delta t \approx 0.001\,d$$

$$\Delta t \approx \frac{0.001\,d}{v} = \frac{(0.001)(3\times10^{11}\text{ m})}{(3\times10^4\text{ m/s})} = 1\times10^4\text{ s}$$

This is about 3 hours. If instead we choose a large time interval of 1×10^7 s, say, in the first update the planet would move a long distance, about $(3\times10^4\text{ m/s})(1\times10^7\text{ s}) = 3\times10^{11}$ m, which is about as large as the distance to the star, and during that step the actual gravitational force would change a great deal in magnitude and direction. This would make our predictions very inaccurate.

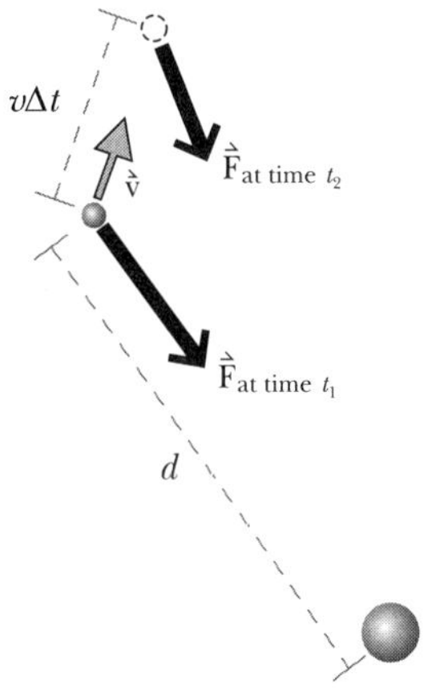

Figure 2.37 If Δt is too big, the gravitational force will change a lot in magnitude and direction during the time interval.

Whenever you choose a time interval to use for updating momentum, make sure it is sufficiently small that the net force doesn't change very much in magnitude or direction during your chosen time interval.

On the other hand, if the force happens to be constant in magnitude and direction, you can use as large a Δt as you like without making an error in the momentum update.

Update the momentum of the planet

? We know the force acting on the planet, and we've chosen a time interval to make one step in predicting the future of the planet. Try to predict the momentum the planet will have after 1×10^4 s.

The planet is moving at a high speed, about 3×10^4 m/s, but this is small compared to the speed of light, which is 3×10^8 m/s, so the initial momentum is $\vec{p}_i \approx m\vec{v}_i$.

$$\vec{p}_i \approx m\vec{v}_i = (3\times10^{24}\text{ kg})(\langle 1\times10^4, 2\times10^4, 1.5\times10^4\rangle\text{ m/s})$$

$$\vec{p}_i = \langle 3\times10^{28}, 6\times10^{28}, 4.5\times10^{28}\rangle\text{ kg}\cdot\text{m/s}$$

Use the Momentum Principle in its update form:

$$\vec{p}_f = \vec{p}_i + \vec{F}_{net}\Delta t$$

$$\vec{p}_f = \langle 3\times10^{28}, 6\times10^{28}, 4.5\times10^{28}\rangle\text{ kg}\cdot\text{m/s} + (\langle -2.14\times10^{21}, -5.34\times10^{21}, 4.27\times10^{21}\rangle\text{ N})(1\times10^4\text{ s})$$

$$\vec{p}_f = \langle 2.9979\times10^{28}, 5.9947\times10^{28}, 4.5043\times10^{28}\rangle\text{ kg}\cdot\text{m/s}$$

Notice that the momentum didn't change very much, because we deliberately chose a small time interval to ensure accuracy in the momentum update, by making sure that the force remained nearly constant during the brief time interval.

Update the position of the planet

? Think about how you would update the position of the planet.

We must use the position update formula, $\vec{r}_f = \vec{r}_i + \vec{v}_{avg}\Delta t$, but what is the average velocity $\vec{v}_{avg}$? The velocity (and momentum) changed during the 1×10^4 s time step. Should we use the initial velocity? The final velocity? The arithmetic average of these two velocities?

A key point is that the momentum changed very little, so the velocity changed very little. Therefore it hardly matters which velocity we use. To be concrete about this, and for simplicity in computer calculations, we'll normally use the final velocity (obtained from the just-calculated final momentum) to represent the average velocity. This works well as long as the time step is small. Since $\vec{p}_f \approx m\vec{v}_f$, $\vec{v}_f \approx \vec{p}_f/m$, and the position update is this:

$$\vec{r}_f \approx \vec{r}_i + \frac{\vec{p}_f}{m}\Delta t$$

$$\vec{r}_f = \left((\langle 3\times10^{11}, 3.5\times10^{11}, -0.5\times10^{11}\rangle\ \text{m}) + \frac{(\langle 2.9979\times10^{28}, 5.9947\times10^{28}, 4.5043\times10^{28}\rangle\ \text{kg}\cdot\text{m/s})}{(3\times10^{24}\ \text{kg})}(1\times10^4\ \text{s})\right)$$

$$\vec{r}_f = \langle 3.001\times10^{11}, 3.502\times10^{11}, -0.4985\times10^{11}\rangle\ \text{m}$$

What have we done?

Whew! After all that computational drudgery, what have we accomplished? We've predicted the future of the planet!

Knowing the planet's initial position and momentum, we predict that after 1×10^4 s the planet will be in a new position that we've calculated, with a new momentum that we've calculated. So what? Well, we could do this again. The second time step would involve exactly the same calculations we just did, but the "initial" position and momentum would be the "final" ones we just calculated. We could keep going, doing this over and over and over, and predict as far ahead into the future as we are willing to do the tedious calculations.

Perhaps you'd rather not do all those calculations? Right! The thing to do is to instruct a computer to do this for you. We need to learn how to write out one set of instructions for the computer to do repetitively, over and over, the calculations you have just done by hand.

2.7.4 Telling a computer what to do

This is how a computer program can be structured in order to apply the Momentum Principle repetitively to predict the future:

- Define the values of constants such as G to use in the program.
- Specify the masses, initial positions, and initial momenta of the interacting objects.
- Specify an appropriate value for Δt, small enough that the objects don't move very far during one update.
- Create a "loop" structure for repetitive calculations:
  - Calculate the (vector) forces acting on each object.
  - Update the momentum of each object: $\vec{p}_f = \vec{p}_i + \vec{F}_{net}\Delta t$.
  - Update the positions: $\vec{r}_f = \vec{r}_i + \vec{v}_{avg}\Delta t$, where $\vec{v}_{avg} \approx \vec{p}_f/m$.
  - Repeat.

The details of the actual program statements depend on what programming language or system you use, and your instructor will provide you with the details of the particular tool you will use.

A computer calculation

In the preceding sections you took one step in such a calculation, by hand. Figure 2.38 shows the orbit of the planet as predicted by a computer program written using VPython (http://vpython.org) that continually updated the momentum and position of the planet, repeatedly using the gravitational force law, the Momentum Principle, and the position update formula. The star and planet are shown much larger than they really are, for visibility in the largely empty space.

Using the gravitational force law, Newton was able to explain the motion of planets and moons in the Solar System. Later scientists have found that it is possible to explain even the motions of other stars in our own and other galaxies with the same formula, so it appears that this is truly a universal law of gravitation, applying to all pairs of masses everywhere in the universe.

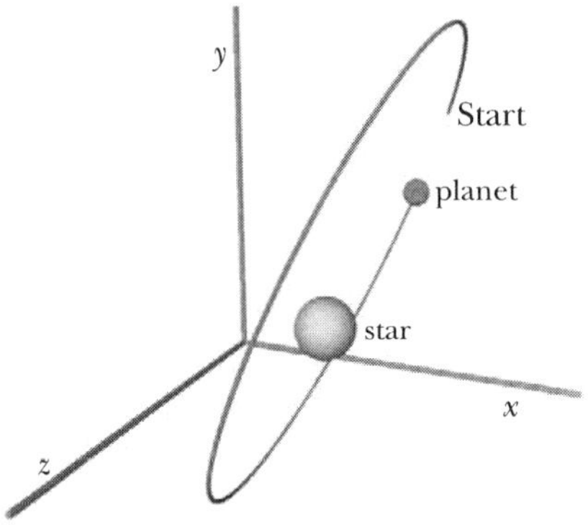

Figure 2.38 The motion of the planet around the star. The star and planet are shown much larger than they really are.

Vary the step size

After you get a program running, it is very important that you decrease the value of the step size Δt and make sure that the path doesn't change. Of course the motion will take longer, because you're doing more calculations, but the issue is that the shape of the trajectory shouldn't change. If it does change significantly, that means that Δt was originally too big, leading to the net force and velocity changing too much during the time Δt.

Keep decreasing Δt until you find that the shape of the trajectory hardly changes. Now you have an adequately small step size Δt.

Define and use constants

It is important to gather together at the start of the program the definitions of any constants such as G, and use these defined symbols everywhere else in the program (that is, use G rather than 6.7E–11 in later calculations). You should avoid entering actual numbers such as 6.7E–11 more than once in a program, since if there is a mistake in a constant you have to correct it in many places. Also, it is easier to read a program that uses G everywhere rather than the number 6.7E–11.

Why not just use calculus?

You might wonder why we don't simply use calculus to predict the motion of planets and stars. One answer is that we do use calculus. Step by step, we add up a large number of tiny increments of the momentum of a body to calculate a large change in its momentum over a long time, and this corresponds to an approximate evaluation of an integral (which is an infinite sum of infinitesimal amounts).

A more interesting answer is that the motion of most physical systems *cannot* be predicted using calculus in any way other than by this step-by-step, calculus-based approach. In a few cases calculus does give a general result without going through this procedure. For example, an object subjected to a constant force has a constant rate of change of momentum and velocity, and calculus can be used to obtain a prediction for the position as a function of time, as we saw in the example of a ball thrown through a vacuum. The elliptical orbits of two stars around each other can be predicted mathematically without an iterative approach, although the math is quite challenging. But the general motion of three stars around each other has never been successfully analyzed in this way. The basic problem is that it is usually relatively easy to take the derivative of a known function, but it is often impossible to determine in algebraic form the integral of a known function, which is what would be involved in long-term prediction.

It isn't a question of taking more math courses in order to be able to solve the "three-body" problem: there is no general mathematical solution. However, a step-by-step procedure of the kind we carried out for the planet can easily be extended to three or more bodies. Just calculate all the forces acting between pairs of bodies, update the momenta, and update the positions. This is why we study the step-by-step prediction method in detail, because it is a powerful technique of increasing importance in modern science and engineering, thanks to the availability of powerful computers to do the repetitive work for us.

Ex. 2.22 The Earth goes around the Sun in 365 days, in a nearly circular orbit. In a computer calculation of the orbit (which is actually an ellipse), approximately how big can Δt be and still get good accuracy in the prediction of the motion?

Ex. 2.23 A pendulum swings with a "period" (time for one round trip) of 2 s. In a computer calculation of the motion, approximately how big can Δt be and still get good accuracy in the prediction of the motion?

2.7.5 Approximate gravitational force near the Earth's surface

Earlier we used the expression mg to represent the magnitude of the gravitational force on an object near the Earth's surface. This is an approximation to the actual force, but it is a good one. The magnitude of the gravitational force that the Earth exerts on an object of mass m near the Earth's surface (Figure 2.39) is

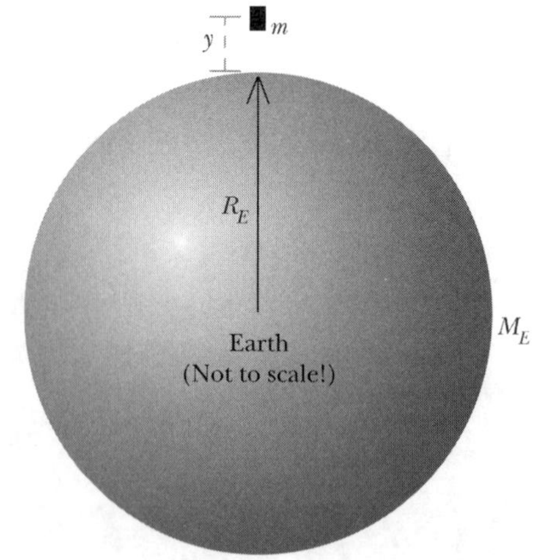

Figure 2.39 Determining the gravitational force of the Earth on an object a height y above the surface.

$$F_g = G\frac{M_E m}{(R_E + y)^2}$$

where y is the distance of the object above the surface of the Earth, R_E is the radius of the Earth, and M_E is the mass of the Earth.

A spherical object of uniform density can be treated as if all its mass were concentrated at its center. As a result, we can treat the effect of the Earth on the object as though the Earth were a tiny, very dense ball a distance $R_E + y$ away.

The gravitational force exerted by the Earth on an atom at the top of the object is slightly different from the force on an atom at the bottom of the object, because each atom is a slightly different distance from the center of the Earth. How much can this difference really matter in an analysis? Suppose the height of the object is a meter, and the bottom of the object is one meter above the surface. The radius of the Earth is about 6.4×10^6 m. Then

$$F_{top} = G\frac{M_E m}{(6400002\text{ m})^2} \text{ whereas } F_{bottom} = G\frac{M_E m}{(6400001\text{ m})^2}$$

For most purposes this difference is not significant. In fact, for all interactions of objects near the surface of the Earth, it makes sense to use the same approximate value, R_E, for the distance from the object to the center of the

Earth. This simplifies calculation of gravitational forces by allowing us to combine all the constants into a single lumped constant, g:

$$F_g \approx \left(G\frac{M_E}{R_E^2}\right)m = gm, \text{ so that } g = G\frac{M_E}{R_E^2}$$

The constant g, called "the magnitude of the gravitational field," has the value $g = +9.8$ newtons/kilogram near the Earth's surface. The "gravitational field" $\vec{g}$ at a location in space is defined to be the (vector) gravitational force that would be exerted on a 1 kg mass placed at that location. A 2 kg mass would experience a force twice as large. In general, a mass m will experience a force $m\vec{g}$ of magnitude mg.

Note that g is a *positive* number, the magnitude of the gravitational field. We can use this approximate formula for gravitational force in our analysis of any interactions occurring near the surface of the Earth.

Ex. 2.24 Show that GM_E/R_E^2 equals 9.8 N/kg. The mass of the Earth is 6×10^{24} kg.

Ex. 2.25 At what height above the surface of the Earth is there a 1% difference between the approximate magnitude of the gravitational field (9.8 N/kg) and the actual magnitude of the gravitational field at that location? That is, at what height y above the Earth's surface is $GM_E/(R_E+y)^2 = 0.99\,GM_E/R_E^2$?

2.8 The electric force law: Coulomb's law

There are electric interactions between "charged" particles such as the protons and electrons found in atoms. It is observed that two protons repel each other, as do two electrons, while a proton and an electron attract each other (Figure 2.40; where protons are said to have "positive electric charge" and electrons have "negative electric charge").

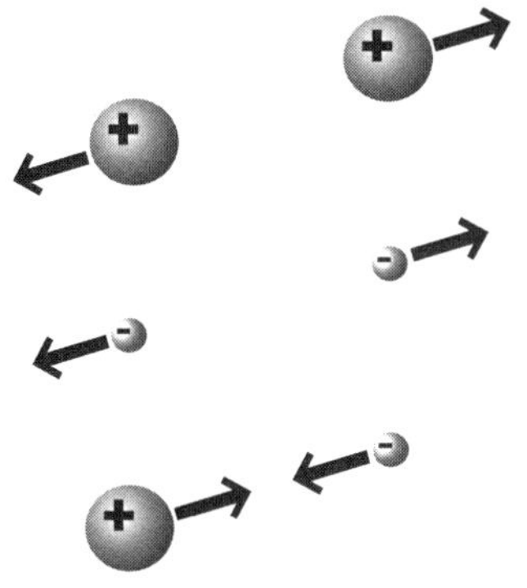

Figure 2.40 Protons repel each other; electrons repel each other; protons and electrons attract each other.

The force corresponding to this electric interaction is similar to the gravitational force law, and it is known as "Coulomb's law" to honor the French scientist who established this law in the late 1700's, during the same period when Cavendish measured the gravitational constant G. See Figure 2.41.

THE ELECTRIC FORCE LAW (COULOMB'S LAW)

$$\vec{F}_{elec\text{ on }2\text{ by }1} = \frac{1}{4\pi\varepsilon_0}\frac{q_2 q_1}{|\vec{r}_{2\text{-}1}|^2}\hat{r}_{2\text{-}1}$$

$$\text{where } \frac{1}{4\pi\varepsilon_0} \text{ is a universal constant: } \frac{1}{4\pi\varepsilon_0} = 9\times10^9\,\frac{\text{N}\cdot\text{m}^2}{\text{C}^2}$$

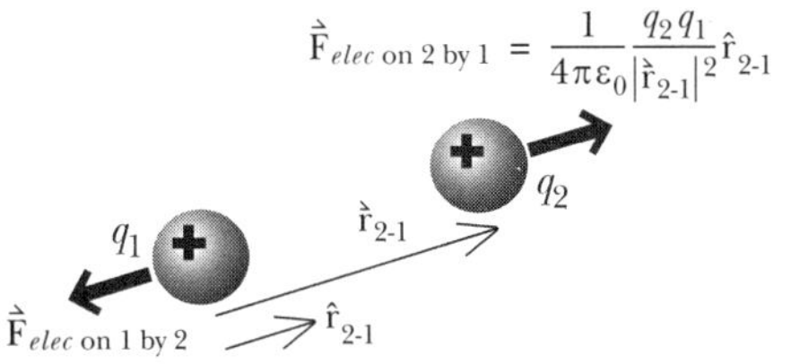

Figure 2.41 The electric force exerted on object 2 by object 1. (The force exerted on object 2 by object 1 has the same magnitude but opposite direction.)

Both the gravitational and electric forces are proportional to the inverse square of the center-to-center distance ("inverse square" laws). The universal electric constant (read as "one over four pi epsilon-zero") is very much larger than the gravitational constant, reflecting the fact that electric interactions are intrinsically much stronger than gravitational interactions. For example, consider a heavy weight that hangs from a thin metal wire. The small number of atoms interacting electrically in the wire have as big an effect on the hanging weight as a very much larger number of atoms in the huge Earth, interacting gravitationally.

The charges q_1 and q_2 must be measured in SI units called "coulombs." The proton has a charge of $+1.6\times10^{-19}$ C and the electron has a charge of -1.6×10^{-19}.

? Study the three cases in Figure 2.40. Why is no minus sign needed in the force law, unlike the case with the gravitational force law?

The difference is that two positively charged particles such as two protons repel each other, whereas masses are always positive but the gravitational force is attractive. Two negatively charged particles such as electrons also attract each other and minus times minus gives plus. Only if the two particles have opposite charges will they attract each other, and then the factor $q_2 q_1$ contributes the necessary minus sign, just as in the gravitational force law.

2.8.1 Interatomic forces

When two objects touch each other they exert forces on each other. At the microscopic level, these contact forces are due to electric interactions between the protons and electrons in one object and the protons and electrons in the other object. In a later chapter we will examine interatomic forces in more detail.

Ex. 2.26 A proton and an electron are separated by 1×10^{-10} m, the radius of a typical atom. Calculate the magnitude of the electric force that the proton exerts on the electron, and the magnitude of the electric force that the electron exerts on the proton.

2.9 Reciprocity

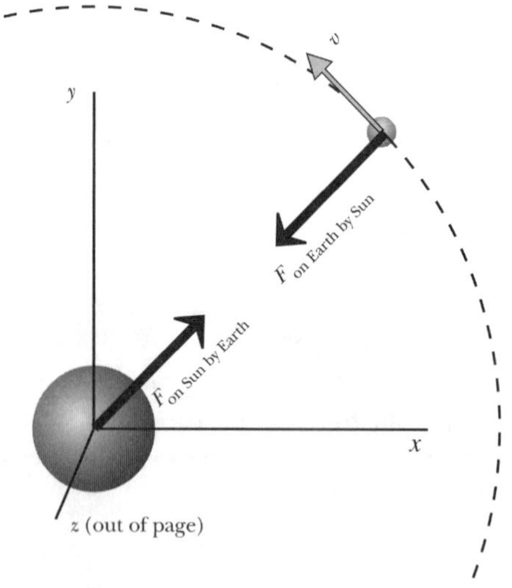

Figure 2.42 The Sun and the Earth exert equal and opposite forces on each other.

An important aspect of the gravitational and electric interactions (including the electric forces of atoms in contact with each other) is that the force that object 1 exerts on object 2 is equal and opposite to the force that object 2 exerts on object 1 (Figure 2.42). That the magnitudes must be equal is clear from the algebraic form of the laws, because $m_1 m_2 = m_2 m_1$ and $q_1 q_2 = q_2 q_1$. The directions of the forces are along the line connecting the centers, and in opposite directions.

This property is called "reciprocity" or "Newton's third law of motion":

RECIPROCITY

$$\vec{F}_{\text{on 1 by 2}} = -\vec{F}_{\text{on 2 by 1}} \text{ (gravitational and electric forces)}$$

The force that the Earth exerts on the massive Sun is just as big as the force that the Sun exerts on the Earth, so in the same time interval the momentum changes are equal in magnitude and opposite in direction ("equal and opposite" for short):

$$\Delta\vec{p}_1 = \vec{F}_{\text{on 1 by 2}}\Delta t = -\Delta\vec{p}_2$$

However, the velocity change $\Delta\vec{v} = \Delta\vec{p}/m$ of the Sun is extremely small compared to the velocity change of the Earth, because the mass of the Sun is enormous in comparison with the mass of the Earth. The mass of our Sun, which is a rather ordinary star, is 2×10^{30} kg. This is enormous compared to the mass of the Earth (6×10^{24} kg) or even to the mass of the largest planet in our Solar System, Jupiter (2×10^{27} kg). Nevertheless, very accurate measurements of small velocity changes of distant stars have been used to infer the presence of unseen planets orbiting those stars.

Magnetic forces do not have the property of reciprocity. Two electrically charged particles that are both moving can interact magnetically as well as electrically, and the magnetic forces that these two particles exert on each other need not be equal in magnitude nor opposite in direction. Reciprocity applies to gravitational and electric forces, but not in general to magnetic forces acting between individual moving charges.

Why reciprocity?

The algebraic forms of the gravitational and electric force laws indicate that reciprocity should hold. A diagram may be helpful in explaining *why* the forces behave this way.

Consider two small objects, a 3 gram object made up of three 1 gram balls, and a 2 gram object made up of two 1 gram balls (Figure 2.43). The distance between centers of the two objects is large compared to the size of either object, so the distances between pairs of 1 gram balls are about the same for all pairs.

You can see that in the 3 gram object each ball has two gravitational forces exerted on it by thc distant balls in the 2 gram object.

? How many forces act on the 3 gram object?

There is a net force of $3 \times 2 = 6$ times the force associated with one pair of balls.

? Similarly, consider the forces acting on the 2 gram object. How many forces act on the 2 gram object?

There is again a net force of $2 \times 3 = 6$ times the force associated with one pair of balls.

The effect is that the force exerted by the 3 gram object on the 2 gram object has the same magnitude as the force exerted by the 2 gram object on the 3 gram object. The same reciprocity holds, for the same reasons, for the electric forces between a lithium nucleus containing 3 protons and a helium nucleus containing 2 protons.

Figure 2.43 Reciprocity: there are 6 forces acting on the 3 gram object and 6 forces acting on the 2 gram object.

2.10 The Newtonian synthesis

Newton devised a particular explanatory scheme in which the analysis of motion is divided into two distinct parts:

1) Quantify the interaction in terms of a concept called "force." Specific examples are Newton's law of gravitation and Coulomb's law.
2) Quantify the change of motion in terms of the change in a quantity called "momentum." The change in the momentum is equal to the force times Δt.

This scheme, called the "Newtonian synthesis," has turned out to be extraordinarily successful in explaining a huge variety of diverse physical phenomena, from the fall of an apple to the orbiting of the Moon. Yet we have no way of asking whether the Universe "really" works this way. It seems unlikely that the Universe actually uses the human concepts of "force" and "momentum" in the unfolding motion of an apple or the Moon.

We refer to the "Newtonian synthesis" both to identify and honor the particular, highly successful analysis scheme introduced by Newton, but also to remind ourselves that this is not the only possible way to view and analyze the universe.

Einstein's alternative view

Newton stated his gravitational force law but could give no explanation for it. He was content with showing that it correctly predicted the motion of the planets, and this was a huge advance, the real beginning of modern science.

Einstein made another huge advance by giving a deeper explanation for gravity, as a part of his general theory of relativity. He realized that the massive Sun bends space and time (!) in such a way as to make the planets move the way they do. The equations in Einstein's general theory of relativity make it possible to calculate the curvature of space and time due to massive objects, and to predict how other objects will move in this altered space and time.

Moreover, Einstein's theory of general relativity accurately predicts some tiny effects that Newton's gravitational law does not, such as the slight bending of light as it passes near the Sun. General relativity also explains some bizarre large-scale phenomena such as black holes and the observed expansion of the space between the galaxies.

Einstein's earlier special theory of relativity established that nothing, not even information, can travel faster than light. Because Newton's gravitational force formula depends only on the distance between objects, not on the time, something's wrong with the formula, since this implies that if an object were suddenly yanked away, its force on another object would vanish instantaneously, thus giving (in principle) a way to send information from one place to another instantaneously. Einstein's theory of general relativity doesn't have this problem.

Since the equations of general relativity are very difficult to work with, and Newton's gravitational law works very well for most purposes, in this course we will use Newton's approach to gravity. But you should be aware that for the most precise calculations one must use the theory of general relativity. For example, the highly accurate atomic clocks in the satellites that make up the Global Positioning System (GPS) have to be continually corrected using Einstein's theory of general relativity. Otherwise GPS positions would be wrong by several kilometers after just one day of operation!

2.11 Systems consisting of several objects

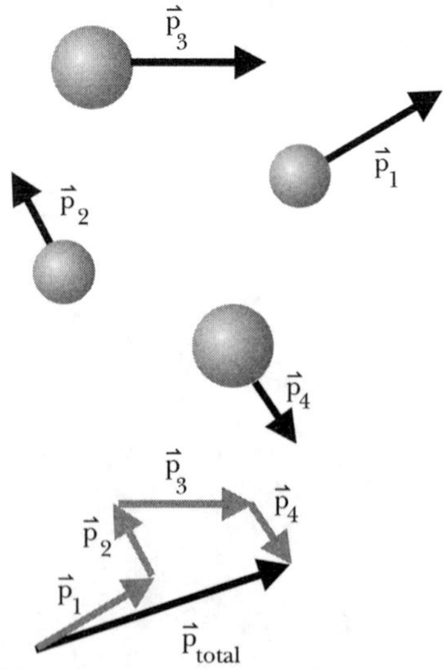

Figure 2.44 The total momentum of the system of four objects is the sum of the individual momenta of each object.

In our applications of the Momentum Principle up to this point, we have usually chosen a single object as the system of interest. In some situations, however, it can be very useful to choose a system that consists of two or more interacting objects. This is a legitimate choice of system; the Momentum Principle applied to a system of two or more objects says that the change in the *total* momentum of the system (Figure 2.44) can be determined by finding the net *external* impulse applied to the system:

MOMENTUM PRINCIPLE FOR MULTIPARTICLE SYSTEMS

$$\Delta\vec{p}_{total} = (\vec{p}_{total,f} - \vec{p}_{total,i}) = \vec{F}_{net,ext}\Delta t$$

Where:

$\vec{p}_{total} \equiv \vec{p}_1 + \vec{p}_2 + \vec{p}_3 + ...$ sum of momenta of all objects in the system

$\vec{F}_{net,ext} \equiv \vec{F}_1 + \vec{F}_2 + \vec{F}_3 + ...$ sum of all external forces on the system

This is an interesting statement of the Momentum Principle; it implies that if we know the external forces acting on a multi-object system, we can draw conclusions about the change of momentum of the system over some time interval without worrying about any of the details of the interactions of the objects with each other. This can greatly simplify the analysis of the motion of some very complex systems.

We have implicitly been using the multiparticle version of the Momentum Principle when we have treated macroscopic objects like humans, spacecraft, and planets as if they were single pointlike objects. In Section 2.15 there is a full derivation of the multiparticle version of the Momentum Principle, in which it is shown that because of reciprocity, forces between objects inside the system do not change the total momentum of the system.

2.11.1 Collisions: Negligible external forces

An event is called a "collision" if it involves an interaction that takes place in a short time and has a large effect on the momenta of the objects compared to the effects of other interactions. A collision does not necessarily involve actual physical contact between objects, which may be interacting via long distance forces like the gravitational force or the electric force. Often it can be useful to analyze collisions by choosing a system that includes all of the colliding objects.

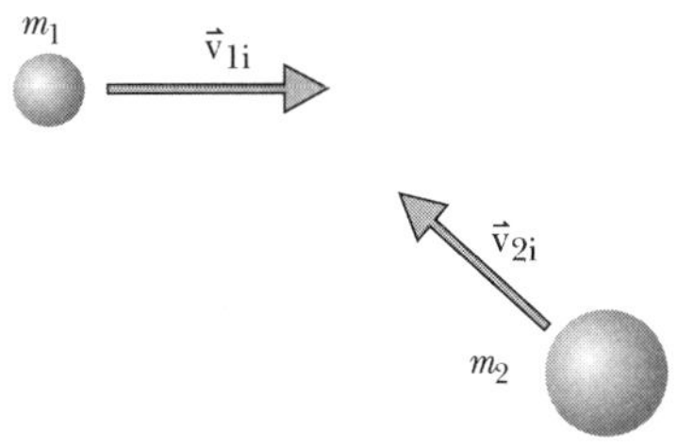

Figure 2.45 Two sticky balls traveling through the air, just before colliding.

As a simple example, consider the case of two sticky balls traveling through the air toward each other, at speeds much less than the speed of light (Figure 2.45).

? Choose both balls as the system. What is the total momentum of the system at the instant shown in Figure 2.45?

Since $v \ll c$, the total momentum of the two-ball system is

$$\vec{p}_{\text{total,i}} = m_1\vec{v}_{1i} + m_2\vec{v}_{2i}$$

? Suppose that when they collide the balls stick together. What can we conclude about their final velocity?

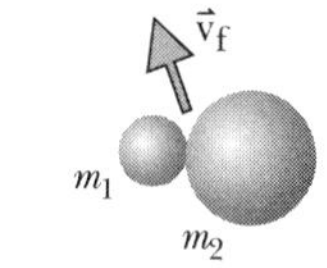

Figure 2.46 The momentum of the stuck-together balls just after the collision is equal to the sum of the initial momenta of the two balls.

If we can assume that the effects of gravity and air resistance have a negligible effect during the short time of the actual collision, we know that the momentum of the stuck-together balls (the final total momentum of the system) must be the same as the initial total momentum of the system:

$$\vec{p}_{\text{total,f}} = \vec{p}_{\text{total,i}}$$
$$(m_1 + m_2)\vec{v}_f = m_1\vec{v}_{1i} + m_2\vec{v}_{2i}$$
$$\vec{v}_f = \frac{m_1\vec{v}_{1i} + m_2\vec{v}_{2i}}{m_1 + m_2}$$

If the mass of ball 1 is 2 kg and its initial velocity is $\langle 6, 0, 0\rangle$ m/s, and the mass of ball 2 is 5 kg and its initial velocity is $\langle -5, 4, 0\rangle$ m/s, then the final velocity of the stuck-together balls will be

$$\vec{v}_f = \frac{(2\text{ kg})\langle 6, 0, 0\rangle + (5\text{ kg})\langle -5, 4, 0\rangle}{(7\text{ kg})}\text{ m/s} = \langle -1.86, 2.86, 0\rangle\text{ m/s}$$

We reached this conclusion without needing to know anything about the details of the complex forces that the balls exerted on each other during the collision.

Some problems require more than one principle

Because the balls in the previous example had the same final speed, we had enough information to solve for the final velocity of the stuck-together balls—we had one equation with only one unknown (vector) quantity. In more complex situations, we will find that we sometimes do not have enough information to solve for all of the unknown quantities by applying only the Momentum Principle.

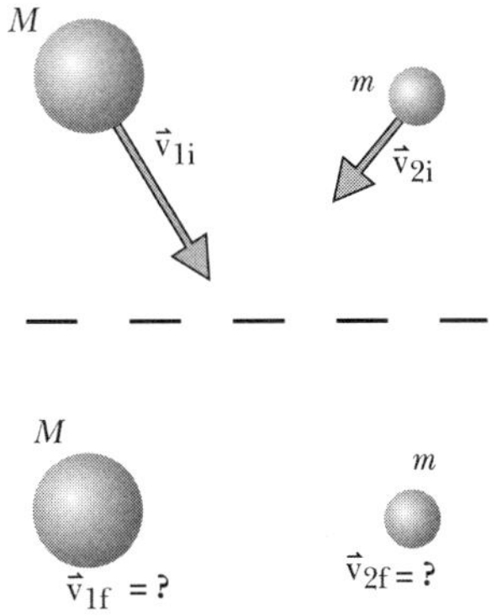

Figure 2.47 The initial momenta of the colliding objects are known, but the final momenta after the collision are unknown.

For example, consider a collision between two balls that bounce off each other, as shown in Figure 2.47. The Momentum Principle tells us that

$$\vec{p}_{\text{total,f}} = \vec{p}_{\text{total,i}}$$
$$\vec{p}_{1f} + \vec{p}_{2f} = \vec{p}_{1i} + \vec{p}_{2i}$$

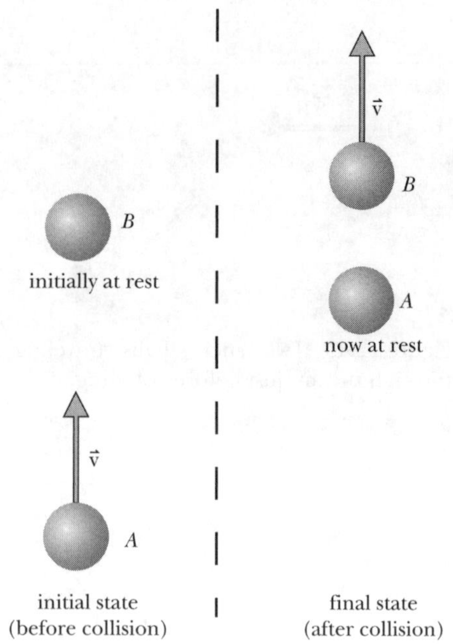

Figure 2.48 Head on collision between two balls of equal mass. Initially (left side) ball A has velocity $\vec{v}$ and B is at rest. After the collision (right side) A is at rest and B has velocity $\vec{v}$. The total momentum of the two-ball system has not changed.

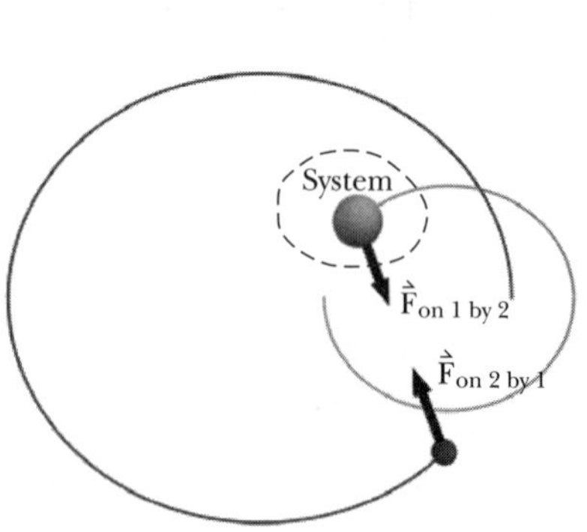

Figure 2.49 A binary star: choose just one of the stars as the system. The momentum of the system changes due to the external force.

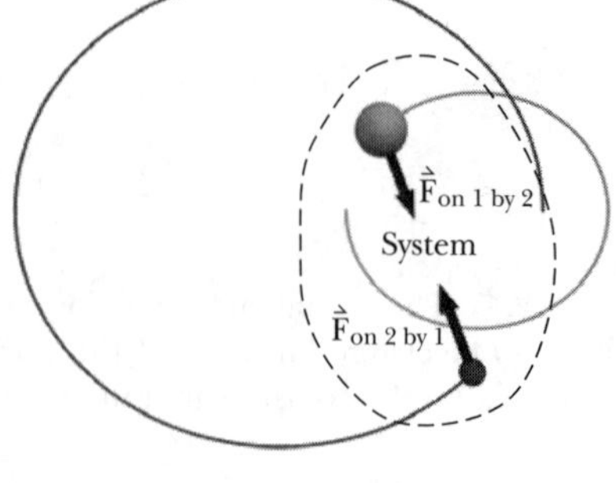

Figure 2.50 A binary star: choose both stars as the system. The momentum of the combined system doesn't change.

but we are left with two unknown (vector) quantities, $\vec{p}_{1f}$ and $\vec{p}_{2f}$, and only one equation. We will not be able to analyze problems of this kind fully until we can invoke the Energy Principle (Chapter 4) and the Angular Momentum Principle (Chapter 9), along with the Momentum Principle.

2.11.2 Momentum flow within a system

Suppose a bowling ball (ball *A*) rolls along a smooth surface at almost constant velocity and hits another bowling ball (ball *B*) of equal mass, which is initially at rest. If the collision is a head-on collision, we will see ball *A* come to a stop, and ball *B* roll along with a velocity nearly equal to the initial velocity of ball *A* (Figure 2.48).

If we choose a system including both balls, we find that the initial momentum and the final momentum of the system are the same:

$$\vec{p}_i = m\vec{v} + 0$$

$$\vec{p}_f = 0 + m\vec{v}$$

despite the fact that the momentum of each object within the system has changed. Assuming that the effects of air resistance and friction are negligible during the short time when the balls interact with each other, so the net external force applied to the system is approximately zero, this is consistent with the momentum principle:

$$\vec{p}_f - \vec{p}_i = \langle 0, 0, 0\rangle \Delta t = \langle 0, 0, 0\rangle$$

Ex. 2.27 You and a friend each hold a lump of wet clay. Each lump has a mass of 20 grams. You each toss your lump of clay into the air, where the lumps collide and stick together. Just before the impact, the velocity of one lump was $\langle 5, 2, -3\rangle$ m/s, and the velocity of the other lump was $\langle -3, 0, -2\rangle$ m/s. What was the total momentum of the lumps just before impact? What is the momentum of the stuck-together lump just after the impact? What is its velocity?

Ex. 2.28 In outer space, far from other objects, two rocks collide and stick together. Before the collision their momenta were $\langle 10, 20, -5\rangle$ kg·m/s and $\langle 8, -6, 12\rangle$ kg·m/s. What was their total momentum before the collision? What must be the momentum of the combined object after the collision?

Ex. 2.29 At a certain instant, the momentum of a proton is $\langle 3.4\times10^{-21}, 0, 0\rangle$ kg · m/s as it approaches another proton which is initially at rest. The two protons repel each other electrically, without coming close enough to touch. When they are once again far apart, one of the protons now has momentum $\langle 2.4\times10^{-21}, 1.6\times10^{-21}, 0\rangle$ kg · m/s. At this instant, what is the momentum of the other proton?

2.12 Conservation of momentum

The choice of system affects the detailed form of the Momentum Principle. Consider the case of two stars orbiting each other, a "binary star." If you choose as the system of interest just one of the stars (Figure 2.49), the other star is in the "surroundings" and exerts an external force which changes the first star's momentum.

If on the other hand you choose both stars as your system (Figure 2.50), the surroundings consist of other stars, which may be so far away as to have negligible effects on the binary star. In that case the net external force acting on the system is nearly zero.

? What happens to the total momentum of the isolated binary star as time goes by?

The Momentum Principle $\vec{p}_f = \vec{p}_i + \vec{F}_{net}\Delta t$ reduces in this case to $\vec{p}_f = \vec{p}_i$, which predicts that the total momentum of the system $\vec{p} = \vec{p}_1 + \vec{p}_2$ remains constant in magnitude and direction. This is an important special case: the total momentum of an isolated system, a system with negligible interactions with the surroundings, doesn't change but stays constant.

In a later chapter we will see that the total momentum can be expressed as $M_{total}\vec{v}_{center\ of\ mass}$, where the center of mass is a mathematical point between the two stars, closer to the more massive star. The velocity of the center of mass does not change but is constant as the binary star drifts through space (or is zero if the binary star's total momentum is zero).

One way to think about this result is to say that momentum gained by one star is lost by the other, because as we saw in the previous chapter, the gravitational forces (and impulses) are equal in magnitude but opposite in direction. The effect is that the total momentum doesn't change. Let $\vec{F}$ be the force exerted on star 1 by star 2, so $-\vec{F}$ is the force exerted on star 2 by star 1. After a short time interval Δt the new total momentum is this:

$$(\vec{p}_1 + \vec{F}\Delta t) + (\vec{p}_2 - \vec{F}\Delta t) = \vec{p}_1 + \vec{p}_2$$

Whatever momentum is lost by one star is gained by the other star. This is a simple example of an important principle, the "conservation of momentum," which says that the change of momentum in a system plus the change of momentum in the surroundings adds up to zero.

CONSERVATION OF MOMENTUM

$$\Delta\vec{p}_{system} + \Delta\vec{p}_{surroundings} = \vec{0}$$

In the case of the two stars there are no objects in the surroundings, and no external forces, so the momentum of the system doesn't change. Zero net external impulse ($\vec{F}_{net}\Delta t$), zero net momentum change. One of the stars can gain momentum (due to a force acting on it), but only if the other star loses the same amount.

Relativistic momentum conservation

Particle accelerators produce beams of particles such as electrons, protons, and pions at speeds very close to the speed of light. When these high-speed particles interact with other particles, experiments show that the total momentum is conserved, but only if the momentum of each particle is defined in the way Einstein proposed, $\vec{p} \equiv m\vec{v}/\sqrt{1-(|\vec{v}|/c)^2}$. It is found that using the low-speed approximation $\vec{p} \approx m\vec{v}$ there is no conservation of momentum when the speed v approaches the speed of light c.

Ex. 2.30 Consider the head-on collision of two identical bowling balls discussed above and shown in Figure 2.48.

(a) Choose a system consisting only of ball A. What is the momentum change of the system during the collision? What is the momentum change of the surroundings?

(b) Choose a system consisting only of ball B. What is the momentum change of the system during the collision? What is the momentum change of the surroundings?

(c) Choose a system consisting of both balls. What is the momentum change of the system during the collision? What is the momentum change of the surroundings?

2.13 Predicting the future of complex gravitating systems

We can use the Newtonian method and numerical integration to try to predict the future of a group of objects that interact with each other mainly gravitationally, such as the Solar System consisting of the Sun, planets, moons, asteroids, comets, etc. There are other, non-gravitational interactions present. Radiation pressure from sunlight makes a comet's tail sweep away from the sun. Streams of charged particles (the "solar wind") from the Sun hit the Earth and contribute to the Northern and Southern Lights (auroras). But the main interactions within the Solar System are gravitational.

The Solar System as a group is orbiting around the center of our Milky Way galaxy, and our galaxy is interacting gravitationally with other galaxies in a local cluster of galaxies (Figure 2.51), but changes in these speeds and directions take place very slowly compared to those within the solar system. To a very good approximation we can neglect these interactions while studying the inner workings of the Solar System itself.

Figure 2.51 Our Solar System orbits around the center of our galaxy, which interacts with other galaxies.

We have modeled the motion of planets and stars using the Momentum Principle and the universal law of gravitation. We can use the same approach to model the gravitational interactions of more complex systems involving three, four, or any number of gravitationally interacting objects. All that is necessary is to include the forces associated with all pairs of objects.

In principle we could use these techniques to predict the future of our entire Solar System. Of course our prediction would be only approximate, because we take finite time steps which introduce numerical errors, and we are neglecting various small interactions. We would need to investigate whether the prediction of our model is a reasonable approximation to the real motion of our real Solar System.

Our Solar System contains a huge number of objects. Most of the nine planets have moons, and Jupiter and Saturn have many moons. There are thousands of asteroids and comets, not to mention an uncounted number of tiny specks of dust. This raises a practical limitation: the fastest computers would take a very long time to carry out even one time step for all the pieces of the Solar System.

2.13.1 The three-body problem

As we have seen, in order to carry out a numerical integration we need to know the position and velocity of each object at some time t, from which we can calculate the forces on each object at that time. This enables us to calculate the position and momentum for each object at a slightly later time $t+\Delta t$. We need the net force on an object, which we get from Newton's universal law of gravitation and the superposition principle.

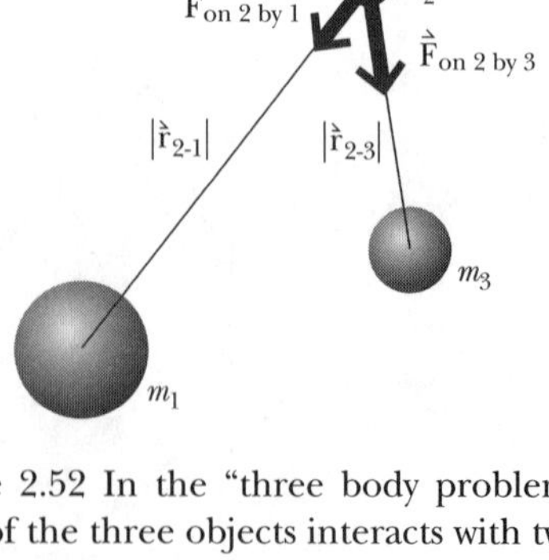

Figure 2.52 In the "three body problem" each of the three objects interacts with two other objects.

Suppose that there are three gravitating objects, as shown in Figure 2.52. The force on object m_2, which is acted upon gravitationally by objects m_1 and m_3, is a natural extension of what we did earlier for the analysis of a star and planet. We simply use the superposition principle to include the interaction with m_3 as well as the interaction with m_1:

$$\vec{F}_{\text{net on } m_2} = -G\frac{m_2 m_1}{|\vec{r}_{2\text{-}1}|^2}\hat{r}_{2\text{-}1} - G\frac{m_3 m_1}{|\vec{r}_{2\text{-}3}|^2}\hat{r}_{2\text{-}3}$$

We use this force to update the momentum of object m_2. Similar computations apply to the other masses, m_1 and m_3. These calculations are tedious, but let the computer do them! We can carry out a numerical integration for each mass: given the current locations of all the masses we can calculate the vector net force exerted on each mass at a time t, and then we can calculate the new values of momentum and position at a slightly later time $t + \Delta t$.

While there exist analytical (non-numerical) solutions for two-body gravitational motion, except for some very special cases the "three-body problem" has not been solved analytically. However, in a numerical integration, the computer doesn't mind that there are lots of quantities and a lot of repetitive calculations to be done. You yourself can solve the three-body problem numerically, by adding additional calculations to your two-body computations (see the three-body homework problem in Chapter 2).

For a single object orbiting a very massive object, or for two objects orbiting each other (with zero net momentum), the possible trajectories are only a circle, ellipse, parabola, hyperbola, or straight line. The motion of a three-body system can be vastly more complex and diverse. Figure 2.53 shows the numerical integration of a complicated three-body trajectory for one particular set of initial conditions. Adding just one more object opens up a vast range of complex behaviors. Imagine how complex the motion of a galaxy can be!

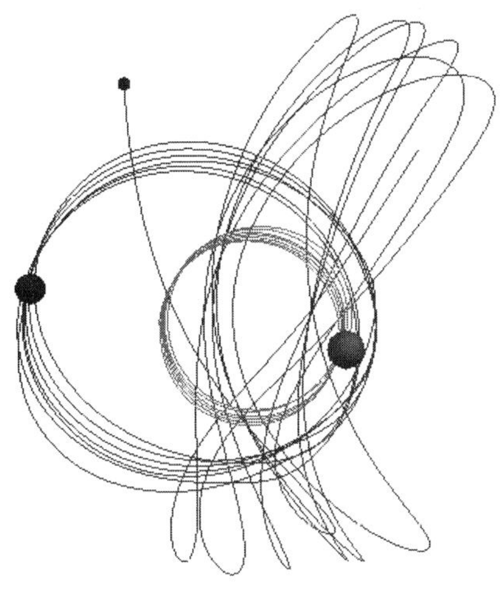

Figure 2.53 One example of three bodies interacting gravitationally.

2.13.2 Sensitivity to initial conditions

For a two-body system, slight changes in the initial conditions make only slight changes in the orbit, such as changing a circle into an ellipse, or an ellipse into a slightly different ellipse, but you never get anything other than a simple trajectory.

The situation is very different in systems with three or more interacting objects. Consider a low-mass object orbiting two massive objects which we imagine somehow to be nailed down so that they cannot move (Figure 2.54). This could also represent two positive charges that are fixed in position, with a low-mass negative charge orbiting around them.

Figure 2.54 A low mass object orbiting two massive objects.

In Figure 2.55 are two very different orbits (one shown in black, the other in gray), starting from only slightly different initial conditions. In both of these cases, the low-mass object was released from rest, but at slightly different initial locations. The trajectories are wildly different! (The mass of the object on the left is twice the mass of the object on the right in this computation.)

This sensitivity to initial conditions generally becomes more extreme as you add more objects, and the Solar System contains lots of objects, big and small. Also, we can anticipate that if small errors in specifying the initial conditions can have large effects, so too it is likely that failing to take into account the tiny force exerted by a small asteroid might make a big difference after a long integration time.

The Solar System is actually fairly predictable, because there is one giant mass (the Sun), and the other, much smaller masses are very far apart. This is unlike the example shown here, which deliberately emphasizes the sensitivity observed when there are large masses near each other.

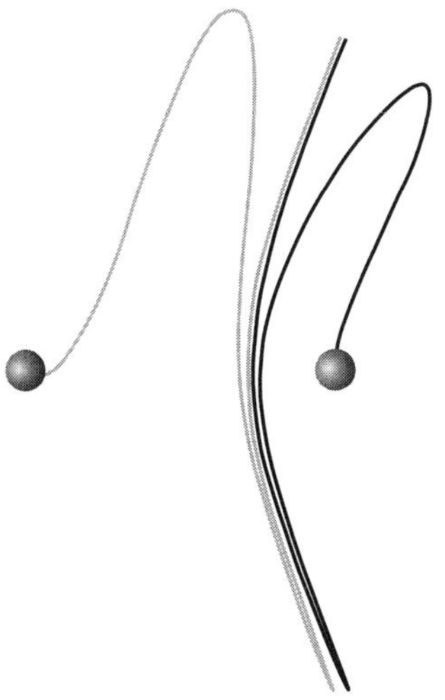

Figure 2.55 Two different orbits (one gray, the other black), starting from rest but from slightly different positions.

2.14 Determinism

If you know the net force on an object as a function of its position and momentum (or velocity), you can predict the future motion of the object by simple step-by-step calculations based on the Momentum Principle. Newton's demonstration of the power of this approach induced many seventeenth and eighteenth century philosophers and scientists to adopt the view already proposed by Descartes, Boyle, and others, that the Universe is a giant clockwork device, whose future is completely determined by the present positions and momenta of all the macroscopic and microscopic objects in it. Just turn the crank, and predict the future! This point of view is called "determinism," and taken to its extreme it raises the question of whether humans actually have any free will, or whether all of our actions are predetermined. Scientific and technological advances in the twentieth century, however, have led us to see that although the Newtonian approach can be used to predict the long-term future of a simple system or the short-term future of a complicated system, there are both practical and theoretical limitations on predictions in some systems.

2.14.1 Practical limitations

One reason we may be unable to predict the long-term future of even a simple system is a practical limitation in our ability to measure its initial conditions with sufficient accuracy. Over time, even small inaccuracies in initial conditions can lead to large cumulative errors in calculations. This is a practical limitation rather than a theoretical one, because in principle if we could measure initial conditions more accurately, we could perform more accurate calculations.

Another practical limitation is our inability to account for all interactions in our model. Every object in the Universe interacts with every other object. In constructing simplified models, we ignore interactions whose magnitude is extremely small. However, over time, even very small interactions can lead to significant effects. Even the tiny "radiation pressure" exerted by sunlight has been shown to affect noticeably the motion of an asteroid over a long time. With larger and faster computers, we can include more and more interactions in our models, but our models can never completely reflect the complexity of the real world.

A branch of current research that focuses on how the detailed interactions of a large number of atoms or molecules lead to the bulk properties of matter is called "molecular dynamics." For example, one way to test our understanding of the nature of the interactions of water molecules is to create a computational model in which the forces between molecules and the initial state of a large number of molecules—several million—are described, and then to see if running the model produces a virtual fluid that actually behaves like water. However, even with an accurate force law for the (electric) interactions between atoms or molecules, it is not feasible to compute and track the motions of the 10^{25} molecules in a glass of water. Significant efforts are underway to build faster and faster computers, and to develop more efficient computer algorithms, to permit realistic numerical integration of more complex systems.

2.14.2 Chaos

A second kind of limitation occurs in systems that display an extreme sensitivity to initial conditions. We saw something like this in the example of the three-body problem in gravitating systems—small changes in initial conditions produced large changes in behavior. In recent years scientists have discovered systems in which a change in the initial conditions, no matter how small (infinitesimal), can lead to complete loss of predictability: the difference in the two possible future motions of the system diverges exponentially with time. Such systems are called "chaotic." It is thought that over a long time period the weather may be literally unpredictable in this sense.

An interesting popular science book about this new field of research is *Chaos: Making a New Science,* by James Gleick (Penguin 1988). The issue of small changes in initial conditions is a perennial concern of time travellers in science fiction stories. Perhaps the most famous such story is Ray Bradbury's "The Sound of Thunder," in which a time traveller steps on a butterfly during the Jurassic era, and returns to an eerily changed present time.

2.14.3 Breakdown of Newton's laws

There are other situations of high interest where we cannot usefully apply Newton's laws of motion, because these classical laws do not adequately describe the behavior of physical systems. To model systems composed of very small particles such as protons and electrons, quarks, and photons, it is necessary to use the laws of quantum mechanics. To model in detail the gravitational interactions between massive objects, it is necessary to apply the principles of general relativity. Given these principles, one might assume that it would be possible to follow a procedure similar to the one we have followed in this chapter, and predict in detail the future of these systems. However, there appear to be more fundamental limitations on what we can know about the future.

2.14.4 Probability and uncertainty

Our understanding of the atomic world of quantum mechanics suggests that there are fundamental limits to our ability to predict the future, because at the atomic and subatomic levels the Universe itself is non-deterministic. We cannot know exactly what will happen at a given time, but only the probability that certain events will occur within a given time frame.

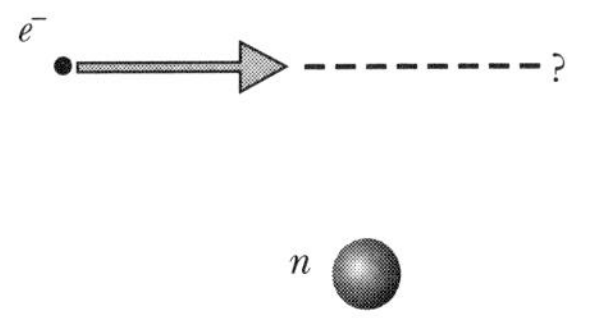

Figure 2.56 The electron may continue to travel in a straight line, but we cannot be certain that it will, because we do not know when the neutron will decay.

A simple example may make this clear. A free neutron (one not bound into a nucleus) is unstable and eventually decays into a positively charged proton, a negatively charged electron, and an electrically neutral anti-neutrino:

$$n \rightarrow p^+ + e^- + \bar{\nu}$$

The average lifetime of the free neutron is about 15 minutes. Some neutrons survive longer than this, and some last a shorter amount of time. All of our experiments and all of our theory are consistent with the notion that *it is not possible to predict when a particular neutron will decay.* There is only a (known) probability that the neutron will decay in the next microsecond, or the next, or the next.

If this is really how the Universe works, then there is an irreducible lack of predictability and determinism of the Universe itself. As far as our own predictions using the Momentum Principle are concerned, consider the following simple scenario. An electron is traveling with constant velocity through nearly empty space, but there is a free neutron in the vicinity (Figure 2.56).

We might predict that the electron will move in a straight line, and some fraction of the time we will turn out to be right (the electrically uncharged neutron exerts no electric force on the electron, there is a magnetic interaction but it is quite small, and the gravitational interaction is tiny). But if the neutron happens (probabilistically) to decay just as our electron passes nearby, suddenly our electron is subjected to large electric forces due to the decay proton and electron, and will deviate from a straight path. Moreover, the directions in which the decay products move are also only probabilistic, so we can't even predict whether the proton or the electron will come closest to our electron (Figure 2.57).

This is a simple but dramatic example of how quantum indeterminacy can lead to indeterminacy even in the context of the Momentum Principle.

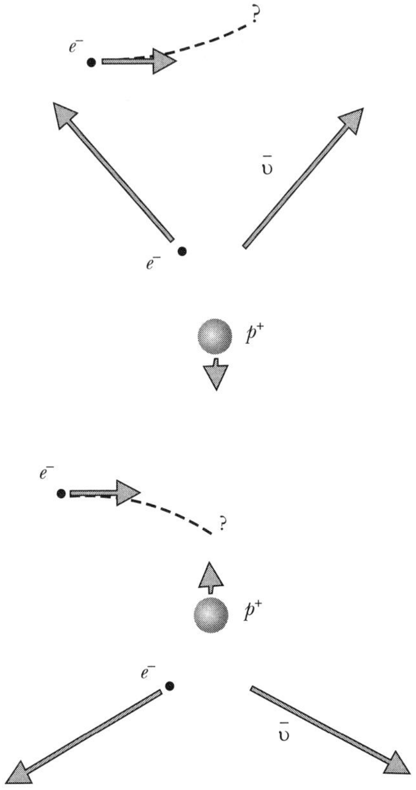

Figure 2.57 If the neutron does decay when the electron is near it, the trajectory of the electron will depend on the directions of the momenta of the decay products.

The Heisenberg uncertainty principle

The Heisenberg uncertainty principle states that there are actual theoretical limits to our knowledge of the state of physical systems. One quantitative formulation states that the position and the momentum of a particle cannot both be simultaneously measured exactly:

$$\Delta x \Delta p_x \geq h$$

This relation says that the product of the uncertainty in position Δx and the uncertainty in momentum Δp_x is equal to a constant h, called Planck's constant. Planck's constant is tiny ($h = 6.6\times10^{-34}$ kg·m^2/s), so this limitation is not noticeable for macroscopic systems, but for objects small enough to require a quantum mechanical description, the uncertainty principle puts fundamental limits on how accurately we can know the initial conditions, and therefore how well we can predict the future.

2.15 *Derivation of the multiparticle Momentum Principle

In a three-particle system (Figure 2.58) we show all of the forces acting on each particle, where the lower case $\vec{f}$'s are forces the particles exert on each other (so-called "internal" forces), and the upper case $\vec{F}$'s are forces exerted by other objects that are not shown and are not part of our chosen system (these are so-called "external" forces, such as the gravitational attraction of the Earth, or a force that you exert by pulling on one of the particles).

We will use the following shorthand notation: $\vec{f}_{1,3}$ will denote the force exerted on particle 1 by particle 3; $\vec{F}_{2,\text{ext}}$ will denote the force on particle 2 exerted by external objects, and so on. We start by applying the Momentum Principle to each of the three particles separately:

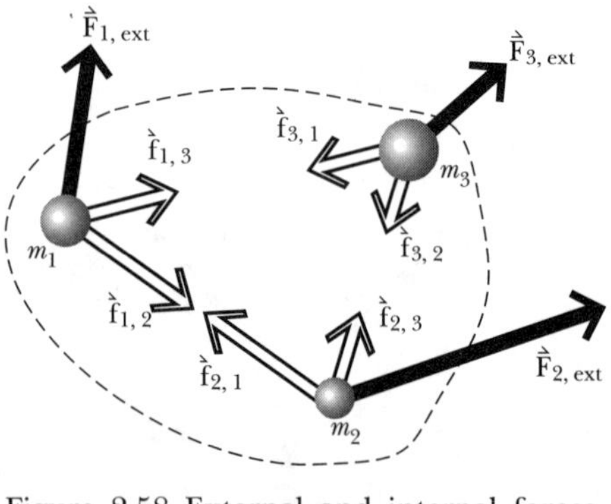

Figure 2.58 External and internal forces acting on a system of three particles.

$$\Delta\vec{p}_1 = (\vec{F}_{1,\text{ext}} + \vec{f}_{1,2} + \vec{f}_{1,3})\Delta t$$

$$\Delta\vec{p}_2 = (\vec{F}_{2,\text{ext}} + \vec{f}_{2,1} + \vec{f}_{2,3})\Delta t$$

$$\Delta\vec{p}_3 = (\vec{F}_{3,\text{ext}} + \vec{f}_{3,1} + \vec{f}_{3,2})\Delta t$$

Nothing new so far. But now we add up these three equations. That is, we create a new equation by adding up all the terms on the left sides of the three equations, and adding up all the terms on the right sides, and setting them equal to each other:

$$\Delta\vec{p}_1 + \Delta\vec{p}_2 + \Delta\vec{p}_3 = (\vec{F}_{1,\text{ext}} + \vec{f}_{1,2} + \vec{f}_{1,3} + \vec{F}_{2,\text{ext}} + \vec{f}_{2,1} + \vec{f}_{2,3} + \vec{F}_{3,\text{ext}} + \vec{f}_{3,1} + \vec{f}_{3,2})\Delta t$$

Many of these terms cancel. By the principle of reciprocity (Newton's third law of motion), which is obeyed by gravitational and electric interactions, we have the following:

$$\vec{f}_{1,2} = -\vec{f}_{2,1}$$

$$\vec{f}_{1,3} = -\vec{f}_{3,1}$$

$$\vec{f}_{2,3} = -\vec{f}_{3,2}$$

Thanks to reciprocity, all that remains after the cancellations is this:

$$\Delta\vec{p}_1 + \Delta\vec{p}_2 + \Delta\vec{p}_3 = (\vec{F}_{1,\text{ext}} + \vec{F}_{2,\text{ext}} + \vec{F}_{3,\text{ext}})\Delta t$$

The total momentum of the system is $\vec{P}_{\text{tot}} = \vec{p}_1 + \vec{p}_2 + \vec{p}_3$, so we have:

$$\Delta\vec{P}_{\text{tot}} = \vec{F}_{\text{net,ext}}\Delta t$$

The importance of this equation is that reciprocity has eliminated all of the internal forces (the forces that the particles in the system exert on each other); internal forces cannot affect the motion of the system as a whole. All that matters in determining the rate of change of (total) momentum is the net external force. The equation has exactly the same form as the Momentum Principle for a single particle.

As we mentioned earlier, in a later chapter we will see that the total momentum can be expressed as $\vec{P}_{tot} = M_{total}\vec{v}_{center\ of\ mass}$, where the center of mass is a mathematical point obtained from a weighted average of the masses in the system.

Moreover, if an object is a sphere whose density is only a function of radius, the object exerts a gravitational force on other objects as though the sphere were a point particle. We can predict the motion of a star or a planet or an asteroid as though it were a single point particle of large mass.

2.16 *Points and spheres

The gravitational force law applies to objects that are "point-like" (very small compared to the center-to-center distance between the objects). In the second volume of this textbook we will be able to show that any hollow spherical shell with uniform density acts gravitationally on external objects as though all the mass of the shell were concentrated at its center. The density of the Earth is not uniform, because the central iron core has higher density than the outer layers. But by considering the Earth as layers of hollow spherical shells, like an onion, with each shell of nearly uniform density, we get the result that the Earth can be modeled for most purposes as a point mass located at the center of the Earth (for very accurate calculations one must take into account small irregularities in the Earth's density from place to place). Similar statements can be made about other planets and stars. In Figure 2.59, the gravitational force is correctly calculated using the center-to-center distance $|\vec{r}_{2\text{-}1}|$.

This is not an obvious result. After all, in Figure 2.59 some of the atoms are closer and some farther apart than the center-to-center distance $|\vec{r}_{2\text{-}1}|$, but the net effect after adding up all the interactions of the individual atoms is as though the two objects had collapsed down to points at their centers. This is a very special property of $1/r^2$ forces, both gravitational and electric, and is not true for forces that have a different dependence on distance.

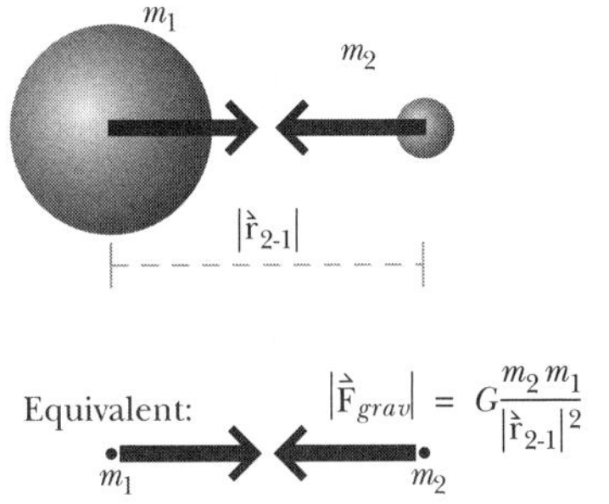

Figure 2.59 A sphere whose density depends only on distance from the center acts like a point particle with all the mass at the center.

2.17 *Measuring the universal gravitational constant G

In order to make quantitative predictions and analyses of physical phenomena involving gravitational interactions, it is necessary to know the universal gravitational constant G. In 1797-1798 Henry Cavendish performed the first experiment to determine a precise value for G (Figure 2.60). In this kind of experiment, a bar with metal spheres at each end is suspended from a thin quartz fiber which constitutes a "torsional" spring. From other measurements, it is known how large a tangential force measured in newtons is required to twist the fiber through a given angle. Large balls are brought near the suspended spheres, and one measures how much the fiber twists due to the gravitational interactions between the large balls and the small spheres.

If the masses are measured in kilograms, the distance in meters, and the force in newtons, the gravitational constant G has been measured in such experiments to be

$$G = 6.7\times10^{-11}\ \frac{\text{N}\cdot\text{m}^2}{\text{kg}^2}$$

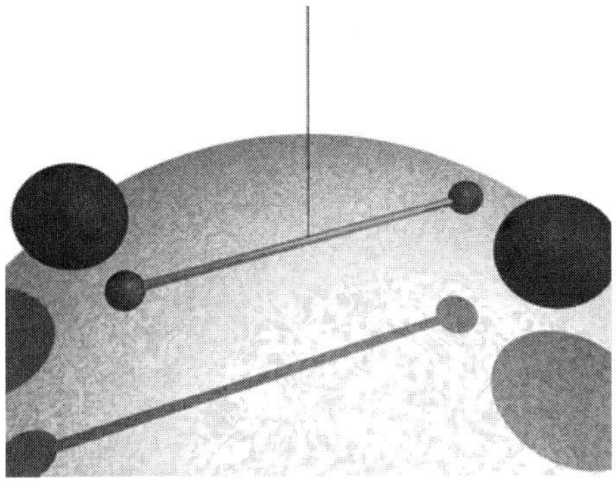

Figure 2.60 The Cavendish experiment.

This extremely small number reflects the fact that gravitational interactions are inherently very weak compared with electromagnetic interactions. The only reason that gravitational interactions are significant in our daily lives is that objects interact with the entire Earth, which has a huge mass. It takes sensitive measurements such as the Cavendish experiment to observe gravitational interactions between two ordinary-sized objects.

2.18 *The Momentum Principle is valid only in inertial frames

Newton's first law is valid only in an "inertial frame" of reference, one in uniform motion (or at rest) with respect to the pervasive "cosmic microwave background" (see optional discussion at the end of Chapter 1). Since the Momentum Principle is a quantitative version of Newton's first law, we expect the Momentum Principle to be valid in an inertial reference frame, but not in a reference frame that is not in uniform motion. Let's check that this is true.

If you view some objects from a space ship that is moving uniformly with velocity $\vec{v}_s$ with respect to the cosmic microwave background, all of the velocities of those objects have the constant $\vec{v}_s$ subtracted from them, as far as you are concerned. For example, a rock moving at the same velocity as your spacecraft would have $\vec{v}_{rock} = (\vec{v} - \vec{v}_s) = \vec{0}$ in your reference frame: it would appear to be stationary as it coasted along beside your spacecraft.

With a constant spaceship velocity, we have $\Delta\vec{v}_s = \vec{0}$, and the change of momentum of the moving object reduces to the following (for speeds small compared to c):

$$\Delta[m(\vec{v} - \vec{v}_s)] = \Delta(m\vec{v}) - \Delta(m\vec{v}_s) = \Delta(m\vec{v}) \text{ since } \Delta(m\vec{v}_s) = \vec{0}$$

$$\text{Therefore, } \Delta[m(\vec{v} - \vec{v}_s)] = \Delta(m\vec{v}) = \vec{F}_{net}\Delta t$$

If the velocity $\vec{v}_s$ of the space ship doesn't change (it represents an inertial frame of reference), the form (and validity) of the Momentum Principle is unaffected by the motion of the space ship.

However, if your space ship increases its speed, or changes direction, $\Delta\vec{v}_s \neq \vec{0}$, an object's motion relative to you changes without any force acting on it. In that case the Momentum Principle is not valid for the object, because you are not in an inertial frame. Although the Earth is not an inertial frame because it rotates, and goes around the Sun, it is close enough to being an inertial frame for many everyday purposes.

2.19 *Derivation of special average velocity result

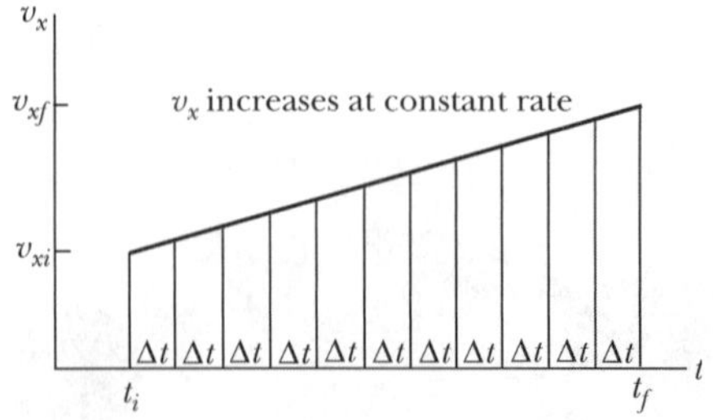

Figure 2.61 In this graph of v_x vs. t (constant force), form narrow vertical slices each of height v_x and width Δt.

Here we offer two proofs, one geometric and one algebraic (using calculus), for the following special-case result concerning average velocity:

$$v_{avg,x} = \frac{(v_{ix} + v_{fx})}{2} \text{ \textit{only} if } v_x \text{ changes at a constant rate}$$

$$\text{Similar results for } v_{avg,y} \text{ and } v_{avg,z}$$

If $F_{net,x}$ is constant, $p_{fx} = p_{ix} + F_{net,x}\Delta t$ implies that p_x changes at a constant rate. At speeds small compared to the speed of light, $v_x \approx p_x/m$, so a graph of v_x vs. time is a straight line, as in Figure 2.61. In Figure 2.61 we form narrow vertical slices, each of height v_x and narrow width Δt.

Within each narrow slice v_x changes very little, so the change in position during the brief time Δt is approximately $\Delta x = v_x\Delta t$. Therefore the change in x is approximately equal to the area of the slice of height v_x and width Δt (Figure 2.62).

If we add up the areas of all these slices, we get approximately the area under the line in Figure 2.61, and this is also equal to the total displacement $\Delta x_1 + \Delta x_2 + \Delta x_3 + ... = x_f - x_i$. If we go to the limit of an infinite number of slices, each with infinitesimal width, the sum of slices really *is* the area, and this area we have shown to be equal to the change in position. This kind of sum of an infinite number of infinitesimal pieces is called an "integral" in calculus.

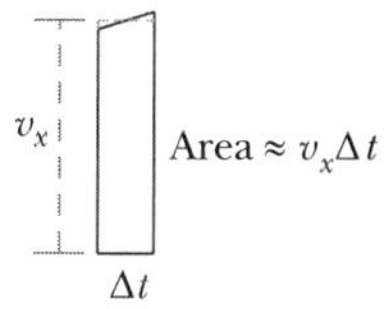

Figure 2.62 One narrow slice has an area given approximately by $v_x\Delta t$. This is equal to Δx, the displacement of the object.

The area under the line is a trapezoid, and from geometry we know that the area of a trapezoid is the average of the two bases times the altitude. Turn Figure 2.61 on its side, as in Figure 2.63, and you see that the top and bottom have lengths v_{ix} and v_{fx}, while the altitude of the trapezoid is the total time $(t_f - t_i)$. Therefore we have the following result:

$$\text{Trapezoid area} = x_f - x_i = \frac{(v_{ix} + v_{fx})}{2}(t_f - t_i)$$

Dividing by $(t_f - t_i)$, we have this:

$$\frac{x_f - x_i}{t_f - t_i} = \frac{(v_{ix} + v_{fx})}{2}$$

But by definition the x component of average velocity is the change in x divided by the total time, so we have proved that

$$v_{\text{avg},x} = \frac{(v_{ix} + v_{fx})}{2} \quad \textit{only} \text{ if } v_x \text{ changes linearly with time}$$

The proof depended critically on the straight-line ("linear") change in velocity, which occurs if $F_{\text{net},x}$ is constant (and $v << c$). Otherwise we wouldn't have a trapezoidal area. That's why the result isn't true in general; it's only true in this important but special case.

$$\text{Area} = \frac{(\text{Top} + \text{Bottom})}{2}\text{Altitude}$$

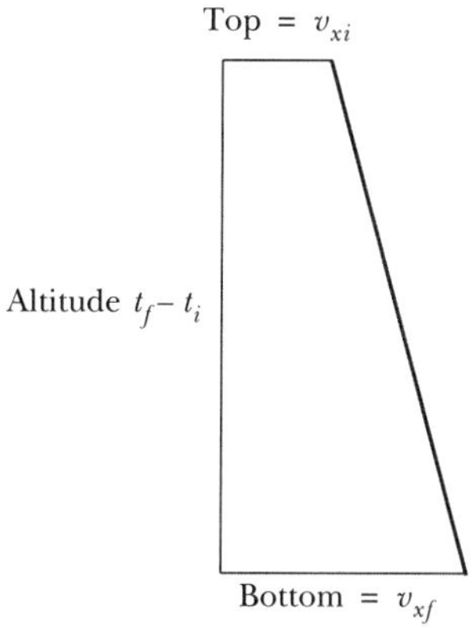

Figure 2.63 The area of the whole trapezoid is equal to the total displacement. The figure in Figure 2.61 has been turned on its side.

Algebraic proof using calculus

An algebraic proof using calculus can also be given. We will use the x component of the derivative version of the Momentum Principle (more about this in the next chapter):

$$\Delta\vec{p} = \vec{F}_{\text{net}}\Delta t \text{ implies that } \frac{\Delta\vec{p}}{\Delta t} = \vec{F}_{\text{net}}$$

$$\text{In the limit we have } \lim_{\Delta t \to 0}\frac{\Delta\vec{p}}{\Delta t} = \frac{d\vec{p}}{dt} = \vec{F}_{\text{net}} \text{ and } \frac{dp_x}{dt} = F_{\text{net},x}$$

If $F_{\text{net},x}$ is a constant, the time derivative of p_x is a constant, so we have

$$p_x = F_{\text{net},x}t + p_{xi} \text{ since } p_x = p_{ix} \text{ when } t = 0$$

You can check this by taking the derivative with respect to time t, which gives the original equation $dp_x/dt = F_{\text{net},x}$. At speeds small compared to the speed of light, $v_x \approx p_x/m$, so we can write

$$v_x = \frac{F_{\text{net},x}}{m}t + v_{ix} \text{ since } v_x = v_{ix} \text{ when } t = 0$$

But the x component of velocity is the rate at which x is changing:

$$v_x = \frac{dx}{dt} = \frac{F_{\text{net},x}}{m}t + v_{ix}$$

Now the question is, can you think of a function of x that has this time derivative? Since the time derivative of t^2 is $2t$, the following formula for x has the appropriate derivative:

$$x = \frac{1}{2}\frac{F_{\text{net},x}}{m}t^2 + v_{ix}t + x_i \text{ since } x = x_i \text{ when } t = 0$$

You can check this by taking the derivative with respect to t, which gives the equation for v_x, since $d(\frac{1}{2}t^2)/dt = t$ and $d(t)/dt = 1$.

The average velocity which we seek is the change in position divided by the total time:

$$v_{\text{avg},x} = \frac{x - x_i}{t} = \frac{1}{2}\frac{F_{\text{net},x}}{m}t + v_{ix} = \frac{1}{2}(v_{fx} - v_{ix}) + v_{xi}$$

where we have used the equation we previously derived for the velocity:

$$v_{fx} = v_x = \frac{F_{\text{net},x}}{m}t + v_{ix}$$

Simplifying the expression for $v_{\text{avg},x}$ we have the proof:

$$v_{\text{avg},x} = \frac{(v_{ix} + v_{fx})}{2} \text{ \textit{only} if } v_x \text{ changes at a constant rate}$$

2.20 *Updating position at high speed

If $v << c$, $\vec{p} \approx m\vec{v}$ and $\vec{v} \approx \vec{p}/m$. But at high speed it is more complicated to determine the velocity from the (relativistic) momentum. Here is a way to solve for $\vec{v}$ in terms of $\vec{p}$:

$$|\vec{p}| = \frac{1}{\sqrt{1 - (|\vec{v}|/c)^2}}m|\vec{v}|$$

Divide by m and square: $\dfrac{|\vec{p}|^2}{m^2} = \dfrac{|\vec{v}|^2}{1 - (|\vec{v}|/c)^2}$

Multiply by $(1 - (|\vec{v}|/c)^2)$: $\dfrac{|\vec{p}|^2}{m^2} - \left(\dfrac{|\vec{p}|^2}{m^2c^2}\right)|\vec{v}|^2 = |\vec{v}|^2$

Collect terms: $\left(1 + \dfrac{|\vec{p}|^2}{m^2c^2}\right)|\vec{v}|^2 = \dfrac{|\vec{p}|^2}{m^2}$

$$|\vec{v}| = \frac{|\vec{p}|/m}{\sqrt{1 + \left(\frac{|\vec{p}|}{mc}\right)^2}}$$

But since $\vec{p}$ and $\vec{v}$ are in the same direction, we can write this:

$$\vec{v} = \frac{\vec{p}/m}{\sqrt{1 + \left(\frac{|\vec{p}|}{mc}\right)^2}}$$

THE RELATIVISTIC POSITION UPDATE FORMULA

$$\vec{r}_f = \vec{r}_i + \frac{1}{\sqrt{1+\left(\frac{|\vec{p}|}{mc}\right)^2}}\left(\frac{\vec{p}}{m}\right)\Delta t \text{ (for small } \Delta t)$$

Note that at low speeds $|\vec{p}| \approx m|\vec{v}|$, and the denominator is $\sqrt{1+\left(\frac{|\vec{v}|}{c}\right)^2} \approx 1$, so the formula becomes the familiar $\vec{r}_f = \vec{r}_i + (\vec{p}/m)\Delta t$.

2.21 *Definitions, measurements, and units

Using the Momentum Principle requires a consistent way to measure length, time, mass, and force, and a consistent set of units. We state the definitions of the standard Système Internationale (SI) units, and we briefly discuss some subtle issues underlying this choice of units.

Units: meters, seconds, kilograms, coulombs, and newtons

Originally the meter was defined as the distance between two scratches on a platinum bar in a vault in Paris, and a second was 1/86,400th of a "mean solar day." Now however the second is defined in terms of the frequency of light emitted by a cesium atom, and the meter is defined as the distance light travels in 1/299,792,458th of a second, or about 3.3×10^{-9} seconds (3.3 nanoseconds). The speed of light is defined to be exactly 299,792,458 m/s (very close to 3×10^{8} m/s). As a result of these modern redefinitions, it is really speed (of light) and time that are the internationally agreed-upon basic units, not length and time.

By international agreement, one kilogram is the mass of a platinum block kept in that same vault in Paris. As a practical matter, other masses are compared to this standard kilogram by using a balance-beam or spring weighing scale (more about this in a moment). The newton, the unit of force, is defined as that force which acting for 1 second imparts to 1 kilogram a velocity change of 1 m/s. We could make a scale for force by calibrating the amount of stretch of a spring in terms of newtons.

The coulomb, the SI unit of electric charge, is defined in terms of electric currents. The charge of a proton is 1.6×10^{-19} coulomb.

Some subtle issues

What we have just said about SI units is sufficient for practical purposes to predict the motion of objects, but here are some questions that might bother you. Is it legitimate to measure the mass that appears in the Momentum Principle by seeing how that mass is affected by gravity on a balance-beam scale? Is it legitimate to use the Momentum Principle to define the units of force, when the concept of force is itself associated with the same law? Is this all circular reasoning, and the Momentum Principle merely a definition with little content? Here is a chain of reasoning that addresses these issues.

Measuring inertial mass

When we use balance-beam or spring weighing scales to measure mass, what we're really measuring is the "gravitational mass," that is, the mass that appears in the law of gravitation and is a measure of how much this object is affected by the gravity of the Earth. In principle, it could be that this "gravitational mass" would be different from the so-called "inertial mass"—the mass that appears in the definition of momentum. It is possible to compare the inertial masses of two objects, and we find experimentally that inertial and gravitational mass seem to be entirely equivalent.

Here is a way to compare two inertial masses directly, without involving gravity. Starting from rest, pull on the first object with a spring stretched by some amount s for an amount of time Δt, and measure the increase of speed Δv_1. Then, starting from rest, pull on the second object with the same spring stretched by the same amount s for the same amount of time Δt, and measure the increase of speed Δv_2. We *define* the ratio of the inertial masses as $m_1/m_2 = \Delta v_2/\Delta v_1$. Since one of these masses could be the standard kilogram kept in Paris, we now have a way of measuring inertial mass in kilograms. Having defined inertial mass this way, we find experimentally that the Momentum Principle is obeyed by both of these objects in all situations, not just in the one special experiment we used to compare the two masses.

Moreover, we find to extremely high precision that the inertial mass in kilograms measured by this comparison experiment is exactly the same as the gravitational mass in kilograms obtained by comparing with a standard kilogram on a balance-beam scale (or using a calibrated spring scale), and that it doesn't matter what the objects are made of (wood, copper, glass, etc.). This justifies the convenient use of ordinary weighing scales to determine inertial mass.

Is this circular reasoning?

The definitions of force and mass may sound like circular reasoning, and the Momentum Principle may sound like just a kind of definition, with no real content, but there is real power in the Momentum Principle. Forget for a moment the definition of force in newtons and mass in kilograms. The experimental fact remains that any object if subjected to a single force by a spring with constant stretch experiences a change of momentum (and velocity) proportional to the duration of the interaction. Note that it is not a change of *position* proportional to the time (that would be a constant speed), but a change of *velocity*. That's real content. Moreover, we find that the change of velocity is proportional to the amount of stretch of the spring. That too is real content.

Then we find that a different object undergoes a different rate of change of velocity with the same spring stretch, but after we've made one single comparison experiment to determine the mass relative to the standard kilogram, the Momentum Principle works in all situations. That's real content.

Finally come the details of setting standards for measuring force in newtons and mass in kilograms, and we use the Momentum Principle in helping set these standards. But logically this comes after having established the law itself.

2.22 Summary

Fundamental Physical Principles

THE MOMENTUM PRINCIPLE

$$\Delta\vec{p} = \vec{F}_{net}\Delta t \text{ (for a short enough time interval } \Delta t)$$

$$\text{Update form: } \vec{p}_f = \vec{p}_i + \vec{F}_{net}\Delta t$$

MOMENTUM PRINCIPLE FOR MULTIPARTICLE SYSTEMS

$$\Delta\vec{p}_{total} = \vec{F}_{net,ext}\Delta t$$

Where $\vec{p}_{total} \equiv \vec{p}_1 + \vec{p}_2 + \vec{p}_3 + \ldots\ldots = M_{total}\vec{v}_{center\ of\ mass}$ and

$\vec{F}_{net,ext} \equiv \vec{F}_1 + \vec{F}_2 + \vec{F}_3 + \ldots\ldots$, the sum of all external forces on the system

CONSERVATION OF MOMENTUM

$$\Delta\vec{p}_{system} + \Delta\vec{p}_{surroundings} = \vec{0}$$

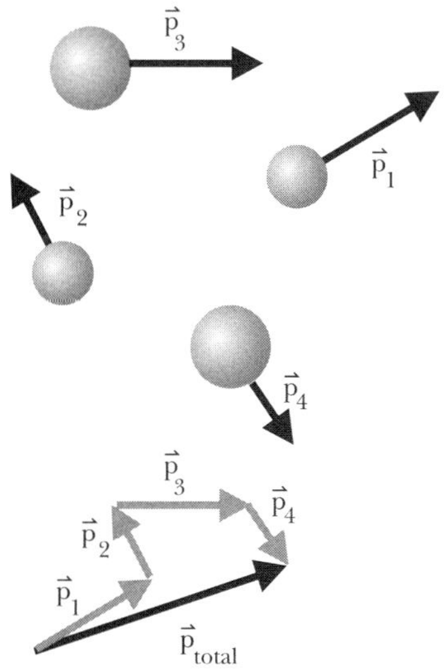

Figure 2.64 The total momentum of the system of four objects is the sum of the individual momenta of each object.

THE SUPERPOSITION PRINCIPLE

The net force on an object is the vector sum of the individual forces exerted on it by all other objects. Each individual interaction is unaffected by the presence of other interacting objects.

Major new concepts

To apply the Momentum Principle to analyze the motion of a real-world system, several steps are required:

1. Choose a system, consisting of a portion of the Universe. The rest of the Universe is called the surroundings.

2. Make a list of objects in the surroundings that exert significant forces on the chosen system, and make a labeled diagram showing the external forces exerted by the objects in the surroundings.

3. Apply the Momentum Principle to the chosen system:

$$\vec{p}_f = \vec{p}_i + \vec{F}_{net}\Delta t$$

For each term in the Momentum Principle, substitute any values you know.

4. Apply the position update formula, if necessary:

$$\vec{r}_f = \vec{r}_i + \vec{v}_{avg}\Delta t$$

5. Solve for any remaining unknown quantities of interest.

6. Check for reasonableness (units, etc.).

A labeled diagram (step 3) gives a physics view of the situation, and it defines symbols to use in writing an algebraic statement of the Momentum Principle.

System is a portion of the Universe acted on by the *surroundings*.

Force is a quantitative measure of interactions; units are newtons.

Impulse is the product of force times time $\vec{F}\Delta t$; momentum change equals net impulse (the impulse due to the net force).

Four fundamental types of interaction have been identified:

- gravitational interactions (all objects attract each other gravitationally)
- electromagnetic interactions (electric and magnetic interactions, closely related to each other); interatomic forces are electric in nature
- "strong" interactions (inside the nucleus of an atom)
- "weak" interactions (neutron decay, for example)

Physical models are tractable approximations/idealizations of the real world.

Computer prediction of motion:

- Define the values of constants such as G to use in the program.
- Specify the masses, initial positions, and initial momenta of the interacting objects.
- Specify an appropriate value for Δt, small enough that the objects don't move very far during one update.
- Create a "loop" structure for repetitive calculations:
  - Calculate the (vector) forces acting on each object.
  - Update the momentum of each object: $\vec{p}_f = \vec{p}_i + \vec{F}_{net}\Delta t$.
  - Update the positions: $\vec{r}_f = \vec{r}_i + \vec{v}_{avg}\Delta t$, where $\vec{v}_{avg} \approx \vec{p}_f / m$.
  - Repeat.

Force laws

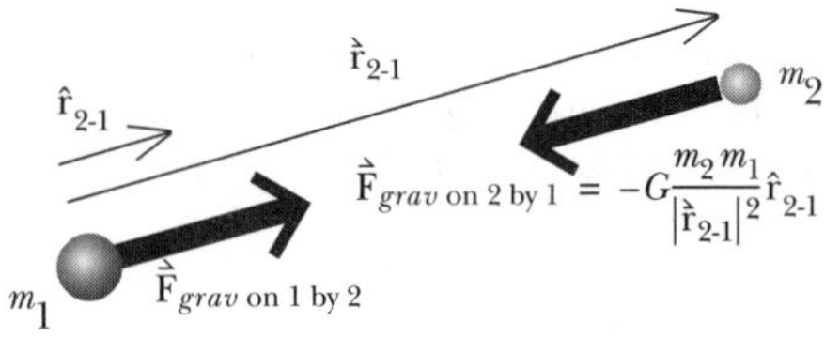

Figure 2.65 The gravitational force exerted on object 2 by object 1. (The force exerted on object 1 by object 2 has the same magnitude but opposite direction.)

THE GRAVITATIONAL FORCE LAW

$$\vec{F}_{grav\ on\ 2\ by\ 1} = -G\frac{m_2 m_1}{|\vec{r}_{2\text{-}1}|^2}\hat{r}_{2\text{-}1}$$

$\vec{r}_{2\text{-}1}$ points from the center of object 1 to the center of object 2

G is a universal constant: $G = 6.7\times10^{-11}\ \frac{N \cdot m^2}{kg^2}$

Near the Earth's surface, $|\vec{F}_{grav}| = mg$, where $g = +9.8$ N/kg

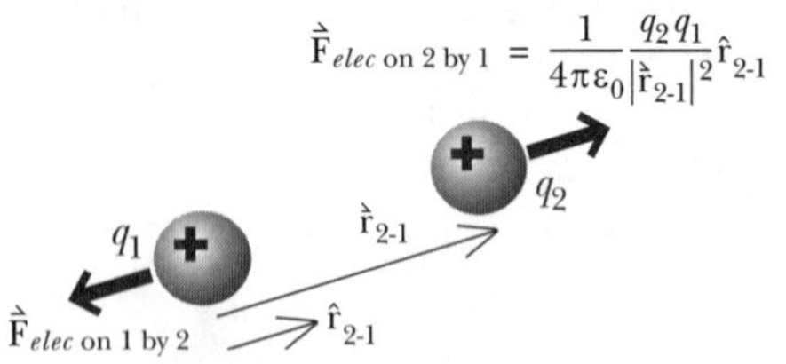

Figure 2.66 The electric force exerted on object 2 by object 1. (The force exerted on object 2 by object 1 has the same magnitude but opposite direction.)

THE ELECTRIC FORCE LAW (COULOMB'S LAW)

$$\vec{F}_{elec\ on\ 2\ by\ 1} = \frac{1}{4\pi\varepsilon_0}\frac{q_2 q_1}{|\vec{r}_{2\text{-}1}|^2}\hat{r}_{2\text{-}1}$$

where $\frac{1}{4\pi\varepsilon_0}$ is a universal constant: $\frac{1}{4\pi\varepsilon_0} = 9\times10^9\ \frac{N \cdot m^2}{C^2}$

THE SPRING FORCE LAW (MAGNITUDE)

$$|\vec{F}_{spring}| = k_s|s|$$

s is the stretch: $s = \Delta L = L - L_0$ (relaxed length - new length)

k_s is called the "spring stiffness"

RECIPROCITY

$\vec{F}_{on\ 1\ by\ 2} = -\vec{F}_{on\ 2\ by\ 1}$ (gravitational and electric forces)

This is also called "Newton's third law of motion."

Additional new concepts and results

Uniform-density spheres act as though all the mass were at the center.

Special case for average velocity:

$$v_{\text{avg},x} = \frac{(v_{ix} + v_{fx})}{2} \quad \textit{only} \text{ if } v_x \text{ changes linearly with time}$$

Similar results for $v_{\text{avg},y}$ and $v_{\text{avg},z}$

2.23 Review questions

The Momentum Principle

RQ 2.1 At a certain instant, a particle is moving in the $+x$ direction with momentum +10 kg·m/s. During the next 0.1 s, a constant force acts on the particle: $F_x = -6$ N, and $F_y = +3$ N. What is the magnitude of the momentum of the particle at the end of this 0.1 s interval?

RQ 2.2 At $t = 12.0$ seconds an object with mass 2 kg was observed to have a velocity of $\langle 10, 35, -8 \rangle$ m/s. At $t = 12.3$ seconds its velocity was $\langle 20, 30, 4 \rangle$ m/s. What was the average (vector) net force acting on the object?

RQ 2.3 A proton has mass 1.7×10^{-27} kg. What is the magnitude of the impulse required to increase its speed from $0.990c$ to $0.994c$?

The gravitational force law

RQ 2.4 Calculate the approximate gravitational force exerted by the Earth on a human standing on the Earth's surface. Compare with the approximate gravitational force of a human on another human at a distance of 3 meters. What approximations or simplifying assumptions must you make? (See data on the inside back cover.)

Reciprocity

RQ 2.5 The windshield of a speeding car hits a hovering insect. Compare the magnitude of the force that the car exerts on the bug to the force that the bug exerts on the car. Which is bigger? Compare the magnitude of the change of momentum of the bug to that of the car. Which is bigger? Compare the magnitude of the change of velocity of the bug to that of the car. Which is bigger? Explain briefly. (Note: the interatomic forces between bug and windshield are electric forces.)

The superposition principle

RQ 2.6 In order to pull a sled across a level field at constant velocity you have to exert a constant force. Doesn't this violate Newton's first and second laws of motion, which imply that no force is required to maintain a constant velocity? Explain this seeming contradiction.

2.24 Problems

Problem 2.1 In the space shuttle

A space shuttle is in a circular orbit near the Earth. An astronaut floats in the middle of the shuttle, not touching the walls. On a diagram, draw and label

(a) the momentum $\vec{p}_1$ of the astronaut at this instant;
(b) all of the forces (if any) acting on the astronaut at this instant;
(c) the momentum $\vec{p}_2$ of the astronaut a short time Δt later;
(d) the momentum change (if any) $\Delta\vec{p}$ in this time interval.
(e) Why does the astronaut seem to "float" in the shuttle?

It is ironic that we say the astronaut is "weightless" despite the fact that the only force acting on the astronaut is the astronaut's weight (that is, the gravitational force of the Earth on the astronaut).

Problem 2.2 Two books attract each other

Two copies of this textbook are standing right next to each other on a bookshelf. Make a rough estimate of the magnitude of the gravitational force that the books exert on each other. Explicitly list all quantities that you had to estimate, and all simplifications and approximations you had to make to do this calculation. Compare your result to the gravitational force on the book by the Earth.

Problem 2.3 Compare gravitational and electric forces

Use data from the inside back cover to calculate the gravitational and electric forces two protons exert on each other when they are 1×10^{-10} m apart (about one atomic radius). Which interaction between two protons is stronger, the gravitational attraction or the electric repulsion? If the two protons are at rest, will they begin to move toward each other or away from each other? Note that since both the gravitational and electric force laws depend on the inverse square distance, this comparison holds true at all distances, not just at a distance of 1×10^{-10} m.

Problem 2.4 Crash test

In a crash test, a truck with mass 2200 kg traveling at 25 m/s (about 55 miles per hour) smashes head-on into a concrete wall without rebounding. The front end crumples so much that the truck is 0.8 m shorter than before. What is the approximate magnitude of the force exerted on the truck by the wall? Explain your analysis carefully, and justify your estimates on physical grounds.

Problem 2.5 Ping-pong ball

A ping-pong ball is acted upon by the Earth, air resistance, and a strong wind. Here are the positions of the ball at several times.

Early time interval:

At $t = 12.35$ s, the position was $< 3.17, 2.54, -9.38 >$ m
At $t = 12.37$ s, the position was $< 3.25, 2.50, -9.40 >$ m

Late time interval:

At $t = 14.35$ s, the position was $< 11.25, -1.50, -11.40 >$ m
At $t = 14.37$ s, the position was $< 11.27, -1.86, -11.42 >$ m

(a) In the early time interval, from $t = 12.35$ s to $t = 12.37$ s, what was the average momentum of the ball? The mass of the ping-pong ball is 2.7 grams (2.7×10^{-3} kg). Express your result as a vector.

(b) In the late time interval, from $t = 14.35$ s to $t = 14.37$ s, what was the average momentum of the ball? Express your result as a vector.

(c) In the time interval from $t = 12.35$ s (the start of the early time interval) to $t = 14.35$ s (the start of the late time interval), what was the average net force acting on the ball? Express your result as a vector.

Problem 2.6 Kick a basketball

A 0.6 kg basketball is rolling by you at 3.5 m/s. As it goes by, you give it a kick perpendicular to its path (Figure 2.67). Your foot is in contact with the ball for 0.002 s. The ball eventually rolls at a 20° angle from its original direction. The overhead view is approximately to scale. The arrow represents the force your toe applies briefly to the basketball.

(a) In the diagram, which letter corresponds to the correct overhead view of the ball's path?

(b) Determine the magnitude of the average force you applied to the ball.

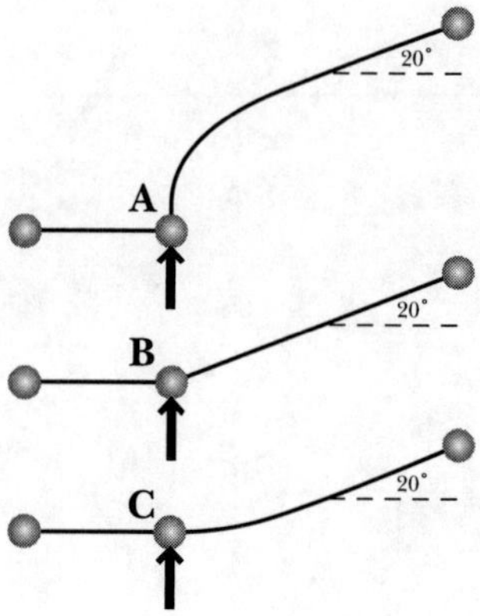

Figure 2.67 Kick a basketball (Problem 2.6).

Problem 2.7 Spacecraft and asteroid

At $t = 532.0$ s after midnight a spacecraft of mass 1400 kg is located at position $\langle 3\times10^5, 7\times10^5, -4\times10^5\rangle$ m, and an asteroid of mass 7×10^{15} kg is located at position $\langle 9\times10^5, -3\times10^5, -12\times10^5\rangle$ m. There are no other objects nearby.

(a) Calculate the (vector) force acting on the spacecraft.

(b) At $t = 532.0$ s the spacecraft's momentum was $\vec{p}_i$, and at the later time $t = 538.0$ s its momentum was $\vec{p}_f$, Calculate the (vector) change of momentum $\vec{p}_f - \vec{p}_i$.

Problem 2.8 Proton and HCl molecule

A proton interacts electrically with a neutral HCl molecule located at the origin. At a certain time t, the proton's position is $\langle 1.6\times10^{-9}, 0, 0\rangle$ m and the proton's velocity is $\langle 3200, 800, 0\rangle$ m/s. The force exerted on the proton by the HCl molecule is $\langle -1.12\times10^{-11}, 0, 0\rangle$ N. At a time $t + (2\times10^{-14}\text{ s})$, what is the approximate velocity of the proton?

Problem 2.9 A free throw in basketball

Determine two different possible ways for a player to make a free throw in basketball. In both cases give the initial speed, initial angle, and initial height of the basketball. The rim of the basket is 10 feet (3.0 m) above the floor. It is 14 feet (4.3 m) along the floor from the free-throw line to a point directly below the center of the basket.

Problem 2.10 A basketball pass

You have probably seen a basketball player throw the ball to a teammate at the other end of the court, 30 m away. Estimate a reasonable initial angle for such a throw, and then determine the corresponding initial speed. For your chosen angle, how long does it take for the basketball to go the length of the court? What is the highest point along the trajectory, relative to the thrower's hand?

Problem 2.11 The case of the falling flower pot

You are a detective investigating why someone was hit on the head by a falling flowerpot. One piece of evidence is a home video taken in a 4th-floor apartment, which happens to show the flowerpot falling past a tall window. Inspection of individual frames of the video shows that in a span of 6 frames the flowerpot falls a distance that corresponds to 0.85 of the window height seen in the video (note: standard video runs at a rate of 30 frames per second). You visit the apartment and measure the window to be 2.2 m high. What can you conclude? Under what assumptions? Give as much detail as you can.

Problem 2.12 Tennis ball hits wall

A tennis ball has a mass of 0.057 kg. A professional tennis player hits the ball hard enough to give it a speed of 50 m/s (about 120 miles per hour). The ball hits a wall and bounces back with almost the same speed (50 m/s). As indicated in Figure 2.68, high-speed photography shows that the ball is

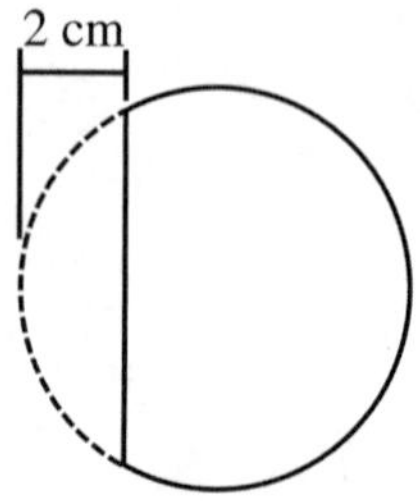

Figure 2.68 A high-speed tennis ball deforms when it hits a wall (Problem 2.12).

crushed 2 cm (0.02 m) at the instant when its speed is momentarily zero, before rebounding.

Making the very rough approximation that the large force that the wall exerts on the ball is approximately constant during contact, determine the approximate magnitude of this force. Hint: Think about the approximate amount of time it takes for the ball to come momentarily to rest. (For comparison note that the gravitational force on the ball is quite small, only about (0.057 kg)(9.8 N/kg) ≈ 0.6 N. A force of 5 N force is approximately the same as a force of one pound.)

Problem 2.13 Mars probe

A small space probe, of mass 240 kg, is launched from a spacecraft near Mars. It travels toward the surface of Mars, where it will land. At a time 20.7 seconds after it is launched, the probe is at the location $\langle 4.30\times10^3, 8.70\times10^2, 0\rangle$ m, and at this same time its momentum is $\langle 4.40\times10^4, -7.60\times10^3, 0\rangle$ kg · m/s. At this instant, the net force on the probe due to the gravitational pull of Mars plus the air resistance acting on the probe is $\langle -7\times10^3, -9.2\times10^2, 0\rangle$ N.

(a) Assuming that the net force on the probe is approximately constant over this time interval, what is the momentum of the probe 20.9 seconds after it is launched?

(b,) What is the location of the probe 20.9 seconds after launch?

Problem 2.14 Spacecraft navigation

Suppose you are navigating a spacecraft far from other objects. The mass of the spacecraft is 1.5×10^5 kg (about 150 tons). The rocket engines are shut off, and you're coasting along with a constant velocity of $\langle 0, 20, 0\rangle$ km/s. As you pass the location $\langle 12, 15, 0\rangle$ km you briefly fire side thruster rockets, so that your spacecraft experiences a net force of $\langle 6\times10^4, 0, 0\rangle$ N for 3.4 s. The ejected gases have a mass that is small compared to the mass of the spacecraft. You then continue coasting with the rocket engines turned off. Where are you an hour later? Also, what approximations or simplifying assumptions did you have to make in your analysis? Think about the choice of system: what are the surroundings that exert external forces on your system?

Problem 2.15 Electron motion in a CRT

In a cathode ray tube (CRT) used in oscilloscopes and television sets, a beam of electrons is steered to different places on a phosphor screen, which glows at locations hit by electrons (Figure 2.69). The CRT is evacuated, so there are few gas molecules present for the electrons to run into. Electric forces are used to accelerate electrons of mass m to a speed $v_0 << c$, after which they pass between positively and negatively charged metal plates which deflect the electron in the vertical direction (upward in Figure 2.69, or downward if the sign of the charges on the plates is reversed).

While an electron is between the plates, it experiences a uniform vertical force F, but when the electron is outside the plates there is negligible force on it. The gravitational force on the electron is negligibly small compared to the electric force in this situation. The length of the metal plates is d, and the phosphor screen is a distance L from the metal plates. Where does the electron hit the screen? (That is, what is y_f?)

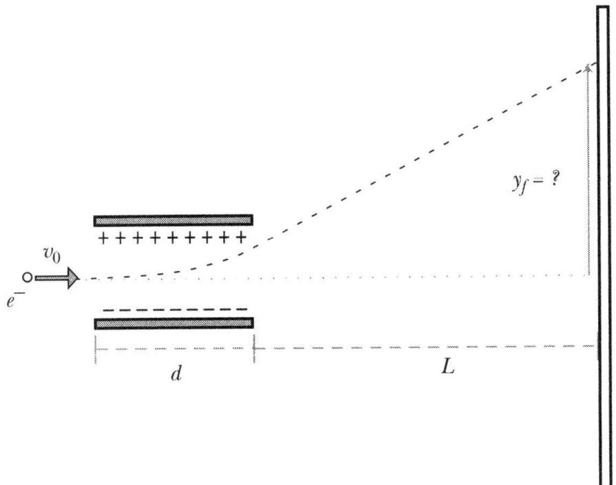

Figure 2.69 A cathode ray tube (Problem 2.15).

Problem 2.16 Star and planet
A star of mass 7×10^{30} kg is located at $\langle 5\times10^{12}, 2\times10^{12}, 0\rangle$ m. A planet of mass 3×10^{24} kg is located at $\langle 3\times10^{12}, 4\times10^{12}, 0\rangle$ m and is moving with a velocity of $\langle 0.3\times10^{4}, 1.5\times10^{4}, 0\rangle$ m/s.

(a) At a time 1×10^{6} seconds later, what is the new velocity of the planet?

(b) Where is the planet at this later time?

(c) Explain briefly why the procedures you followed in parts (a) and (b) were able to produce usable results but wouldn't work if the later time had been 1×10^{9} seconds instead of 1×10^{6} seconds after the initial time. Explain briefly how could you use a computer to get around this difficulty.

Problem 2.17 Two stars
At $t = 0$ a star of mass 4×10^{30} kg has velocity $\langle 7\times10^{4}, 6\times10^{4}, -8\times10^{4}\rangle$ m/s and is located at $\langle 2.00\times10^{12}, -5.00\times10^{12}, 4.00\times10^{12}\rangle$ m relative to the center of a cluster of stars. There is only one nearby star that exerts a significant force on the first star. The mass of the second star is 3×10^{30} kg, its velocity is $\langle 2\times10^{4}, -1\times10^{4}, 9\times10^{4}\rangle$ m/s, and this second star is located at $\langle 2.03\times10^{12}, -4.94\times10^{12}, 3.95\times10^{12}\rangle$ m relative to the center of the cluster of stars.

(a) At $t = 1\times10^{5}$ s, what is the approximate momentum of the first star?

(b) Discuss briefly some ways in which your result for (a) is approximate, not exact.

(c) At $t = 1\times10^{5}$ s, what is the approximate position of the first star?

(d) Discuss briefly some ways in which your result for (b) is approximate, not exact.

Problem 2.18 The SLAC two-mile accelerator
SLAC, the Stanford Linear Accelerator Center, located at Stanford University in Palo Alto, California, accelerates electrons through a vacuum tube two miles long (it can be seen from an overpass of the Junipero Serra freeway that goes right over the accelerator). Electrons which are initially at rest are subjected to a continuous force of 2×10^{-12} newton along the entire length of two miles (one mile is 1.6 kilometers) and reach speeds very near the speed of light.

(a) Determine how much time is required to increase the electrons' speed from $0.93c$ to $0.99c$. (That is, the quantity $|\vec{v}|/c$ increases from 0.93 to 0.99.)

(b) Approximately how far does the electron go in this time? What is approximate about your result?

Problem 2.19 Determining the mass of an asteroid
In June 1997 the NEAR spacecraft ("Near Earth Asteroid Rendezvous"; see http://near.jhuapl.edu/), on its way to photograph the asteroid Eros, passed within 1200 km of asteroid Mathilde at a speed of 10 km/s relative to the asteroid (Figure 2.70). From photos transmitted by the 805 kg spacecraft, Mathilde's size was known to be about 70 km by 50 km by 50 km. It is presumably made of rock. Rocks on Earth have a density of about 3000 kg/m^3 (3 grams/cm^3).

(a) Make a rough diagram to show qualitatively the effect on the spacecraft of this encounter with Mathilde. Explain your reasoning.

(b) Make a very rough estimate of the change in momentum of the spacecraft that would result from encountering Mathilde. Explain how you made your estimate.

(c) Using your result from part (b), make a rough estimate of how far off course the spacecraft would be, one day after the encounter.

(d) From actual observations of the location of the spacecraft one day after encountering Mathilde, scientists concluded that Mathilde is a loose ar-

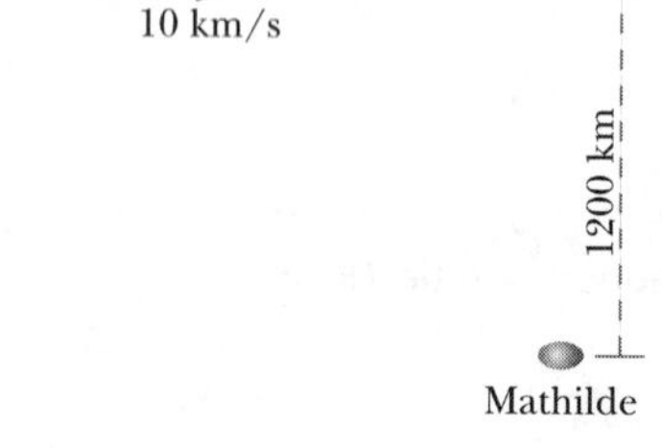

Figure 2.70 The NEAR spacecraft passes the asteroid Mathilde (Problem 2.19).

rangement of rocks, with lots of empty space inside. What was it about the observations that must have led them to this conclusion?

Experimental background: The position was tracked by very accurate measurements of the time that it takes for a radio signal to go from Earth to the spacecraft followed immediately by a radio response from the spacecraft being sent back to Earth. Radio signals, like light, travel at a speed of 3×10^8 m/s, so the time measurements had to be accurate to a few nanoseconds (1 ns = 10^{-9} s).

The following problems are intended to introduce you to using a computer to model matter, interactions, and motion. Some parts of these problems can be done with almost any tool (spreadsheet, math package, etc.). Other parts are most easily done with a programming language, for which we recommend the free 3D programming language VPython (http://vpython.org). Your instructor will introduce you to an available computational tool and assign problems, or parts of problems, that can be addressed using the chosen tool.

Problem 2.20 Planetary orbits

In this problem you will study the motion of a planet around a star. To start with a somewhat familiar situation, you will begin by modeling the motion of our Earth around our Sun. Write answers to questions either as comments in your program or on paper, as specified by your instructor.

Planning

(a) The Earth goes around the Sun in a nearly circular orbit, taking one year to go around. Using data on the inside back cover, what initial speed should you give the Earth in a computer model, so that a circular orbit should result? (If your computer model does produce a circular orbit, you have a strong indication that your program is working properly.)

(b) Estimate an appropriate value for Δt to use in your computer model. Remember that t is in seconds, and consider your answer to part (a) in making this estimate. If your Δt is too small, your calculation will require many tedious steps. Explain briefly how you decided on an appropriate step size.

Circular orbit

(c) Starting with speed calculated in (a), and with the initial velocity perpendicular to a line connecting the Sun and the Earth, calculate and display the trajectory of the Earth. Display the whole trail, so you can see whether you have a closed orbit (that is, whether the Earth returns to its starting point each time around). Include the Sun's position $\vec{r}_s = \langle x_s, y_s, z_s \rangle$ in your computations, even if you set its coordinates to zero. This will make it easier to modify the computation to let the Sun move (Problem 2.2 Binary stars).

Computational accuracy

(d) As a check on your computation, is the orbit a circle as expected? Run the calculation until the accumulated time t is the length of a year: does this take the Earth around the orbit once, as expected? What is the largest step size you can use that still gives a circular orbit and the correct length of a year? What happens if you use a much larger step size?

Noncircular trajectories

(e) Set the initial speed to 1.2 times the Earth's actual speed. Make sure the step size is small enough that the orbit is nearly unaffected if you cut the step size. What kind of orbit do you get? What step size do you need to use to get results you can trust? What happens if you use a much larger step size?

Produce at least one other qualitatively different noncircular trajectory for the Earth. What difficulties would humans have in surviving on the Earth if it had a highly noncircular orbit?

Force and momentum

(f) Choose initial conditions that give a noncircular orbit. Continuously display a vector showing the momentum of the Earth (with its tail at the Earth's position), and a different colored vector showing the force on the Earth by the Sun (with its tail at the Earth's position). You must scale the vectors appropriately to fit into the scene. A way to figure out what scale factor to use is to print the numerical values of momentum and force, and compare with the scale of the scene. For example, if the width of the scene is W meters, and the force vector has a typical magnitude F (in newtons), you might scale the force vector by a factor $0.1\,W/F$, which would make the length of the vector be one-tenth the width of the scene. Is the force in the same direction as the momentum? How does the momentum depend on the distance from the star?

Problem 2.21 Binary stars

About a half of the visible "stars" are actually systems consisting of two stars orbiting each other, called "binary stars." In your computer model of the Earth and Sun (Problem 2.20 Planetary orbits), replace the Earth with a star whose mass is half the mass of our Sun, and take into account the gravitational effects that the second star has on the Sun.

(a) Give the second star the speed of the actual Earth, and give the Sun zero initial momentum. What happens? Try a variety of other initial conditions. What kinds of orbits do you find?

(b) One of the things to try is to give the Sun the same magnitude of momentum as the second star, but in the opposite direction. If the initial momentum of the second star is in the $+y$ direction, give the Sun the same magnitude of momentum but in the $-y$ direction, so that the total momentum of the binary star system is initially zero, but the stars are not headed toward each other. What is special about the motion you observe in this case?

Problem 2.22 Other force laws

Modify your orbit computation to use a different force law, such as a force that is proportional to $1/r$ or $1/r^3$, or a constant force, or a force proportional to r^2 (this represents the force of a spring whose relaxed length is nearly zero). How do orbits with these force laws differ from the circles and ellipses that result from a $1/r^2$ law? If you want to keep the magnitude of the force roughly the same as before, you will need to adjust the force constant G.

Problem 2.23 The effect of the Moon and Venus on the Earth

In Problem 2.20 you analyzed a simple model of the Earth orbiting the Sun, in which there were no other planets or moons. Venus and the Earth have similar size and mass. At its closest approach to the Earth, Venus is about 40 million kilometers away (4×10^{10} m). The Moon's mass is about 7×10^{22} kg, and the distance from Earth to Moon is about 4×10^{8} m (400,000 km, center to center).

(a) Calculate the ratio of the gravitational forces on the Earth exerted by Venus and the Sun. Is it a good approximation to ignore the effect of Venus when modeling the motion of the Earth around the Sun?

(b) Calculate the ratio of the gravitational forces on the Earth exerted by the Moon and the Sun. Is it a good approximation to ignore the effect of the Moon when modeling the motion of the Earth around the Sun?

Problem 2.24 The three-body problem

Carry out a numerical integration of the motion of a three-body gravitational system and plot the trajectory, leaving trails behind the objects. Calculate all of the forces before using these forces to update the momenta and positions of the objects. Otherwise the calculations of gravitational forces would mix positions corresponding to different times.

Try different initial positions and initial momenta. Find at least one set of initial conditions that produces a long-lasting orbit, one set of initial conditions that results in a collision with a massive object, and one set of initial conditions that allows one of the objects to wander off without returning. Report the masses and initial conditions that you used.

Problem 2.25 The Ranger 7 mission to the Moon

The first U.S. spacecraft to photograph the Moon close up was the unmanned "Ranger 7" photographic mission in 1964. The spacecraft, shown in the illustration at the right (NASA photograph), contained television cameras that transmitted close-up pictures of the Moon back to Earth as the spacecraft approached the Moon. The spacecraft did not have retro-rockets to slow itself down, and it eventually simply crashed onto the Moon's surface, transmitting its last photos immediately before impact.

Figure 2.71 is the first image of the Moon taken by a U.S. spacecraft, Ranger 7, on July 31, 1964, about 17 minutes before impacting the lunar surface. The large crater at center right is Alphonsus (108 km diameter); above it (and to the right) is Ptolemaeus and below it is Arzachel. The Ranger 7 impact site is off the frame, to the left of the upper left corner.

You can find out more about the actual Ranger lunar missions at
http://nssdc.gsfc.nasa.gov/planetary/lunar/ranger.html

Figure 2.71 The first Ranger 7 photo of the Moon (NASA photograph).

To send a spacecraft to the Moon, we put it on top of a large rocket containing lots of rocket fuel and fire it upward. At first the huge ship moves quite slowly, but the speed increases rapidly. When the "first-stage" portion of the rocket has exhausted its fuel and is empty, it is discarded and falls back to Earth. By discarding an empty rocket stage we decrease the amount of mass that must be accelerated to even higher speeds. There may be several stages that operate for a while and then are discarded before the spacecraft has risen above most of Earth's atmosphere (about 50 km, say, above the Earth), and has acquired a high speed. At that point all the fuel available for this mission has been used up, and the spacecraft simply coasts toward the Moon through the vacuum of space.

We will model the Ranger 7 mission. Starting 50 km above the Earth's surface (5×10^4 m), a spacecraft coasts toward the Moon with an initial speed of about 10^4 m/s. Here are data we will need:

mass of spacecraft = 173 kg   mass of Earth $\approx 6\times10^{24}$ kg

mass of Moon $\approx 7\times10^{22}$ kg   radius of Moon = 1.75×10^6 m

distance from Earth to Moon $\approx 4\times10^8$ m (400,000 km, center to center)

We're going to ignore the Sun in a simplified model even though it exerts a sizable gravitational force. We're expecting the Moon mission to take only a few days, during which time the Earth (and Moon) move in a nearly straight line with respect to the Sun, because it takes 365 days to go all the way around the Sun. We take a reference frame fixed to the Earth as representing (approximately) an inertial frame of reference with respect to which we can use the Momentum Principle.

For a simple model, make the Earth and Moon be fixed in space during the mission. Factors that would certainly influence the path of the spacecraft include the motion of the Moon around the Earth, and the motion of the Earth around the Sun. In addition, the Sun and other planets exert gravitational forces on the spacecraft. As a separate project you might like to include some of these additional factors.

(a) Compute the path of the spacecraft, and display it either with a graph or with an animated image. Here, and in remaining parts of the problem, report the step size Δt that gives accurate results (that is, cutting this step size has little effect on the results).

(b) By trying various initial speeds, determine the approximate *minimum* launch speed needed to reach the Moon, to two significant figures (this is the speed that the spacecraft obtains from the multistage rocket, at the time of release above the Earth's atmosphere). What happens if the launch speed is less than this minimum value? (Be sure to check the step size issue.)

(c) Use a launch speed 10% larger than the approximate minimum value found in part (b). How long does it take to go to the Moon, in hours or days? (Be sure to check the step size issue.)

(e) What is the "impact speed" of the spacecraft (its speed just before it hits the Moon's surface)? Make sure that your spacecraft crashes on the surface of the Moon, not at the Moon's center! (Be sure to check the step size issue.)

You may have noticed that you don't actually need to know the mass m of the spacecraft in order to carry out the computation. The gravitational force is proportional to m, and the momentum is also proportional to m, so m cancels. However, nongravitational forces such as electric forces are not proportional to the mass, and there is no cancellation in that case. We kept the mass m in the analysis in order to illustrate a general technique for predicting motion, no matter what kind of force, gravitational or not.

Problem 2.26 The effect of Venus on the Moon voyage

In the Moon voyage analysis, you used a simplified model in which you neglected among other things the effect of Venus. An important aspect of physical modeling is making estimates of how large the neglected effects might be. If we take Venus into account, make a rough estimate of whether the spacecraft will miss the Moon entirely. How large a sideways deflection of the crash site will there be? Explain your reasoning and approximations.

Venus and the Earth have similar size and mass. At its closest approach to the Earth, Venus is about 40 million kilometers away (4×10^{10} m). In the real world, Venus would attract the Earth and the Moon as well as the spacecraft, but to get an idea of the size of the effects, pretend that the Earth, the Moon, and Venus are all fixed in position (Figure 2.72) and just investigate Venus's effect on the spacecraft.

Figure 2.72 The relative positions of Venus, Earth, and Moon in Problem 2.26.

The following problems deal with momentum conservation

Problem 2.27 Outer space collision

In outer space a small rock with mass 5 kg traveling with velocity $\langle 0, 1800, 0\rangle$ m/s strikes a stationary large rock head-on and bounces straight back with velocity $\langle 0, -1500, 0\rangle$ m/s. After the collision, what is the vector momentum of the large rock?

Problem 2.28 Two rocks collide

Two rocks collide in outer space. Before the collision, one rock had mass 9 kg and velocity $\langle 4100, -2600, 2800 \rangle$ m/s. The other rock had mass 6 kg and velocity $\langle -450, 1800, 3500 \rangle$ m/s. A 2 kg chunk of the first rock breaks off and sticks to the second rock. After the collision the rock whose mass is 7 kg has velocity $\langle 1300, 200, 1800 \rangle$ m/s. After the collision, what is the velocity of the other rock, whose mass is 8 kg?

Problem 2.29 Two rocks collide

Two rocks collide with each other in outer space, far from all other objects. Rock 1 with mass 5 kg has velocity $\langle 30, 45, -20 \rangle$ m/s before the collision and $\langle -10, 50, -5 \rangle$ m/s after the collision. Rock 2 with mass 8 kg has velocity $\langle -9, 5, 4 \rangle$ m/s before the collision. Calculate the final velocity of rock 2.

Problem 2.30 Two rocks stick together

In outer space two rocks collide and stick together. Here are the masses and initial velocities of the two rocks:

Rock 1: mass = 15 kg, initial velocity = $\langle 10, -30, 0 \rangle$ m/s

Rock 2: mass = 32 kg, initial velocity = $\langle 15, 12, 0 \rangle$ m/s

What is the velocity of the stuck-together rocks after the collision?

Problem 2.31 The car and the mosquito

A car of mass M moving in the x direction at high speed v strikes a hovering mosquito of mass m, and the mosquito is smashed against the windshield.

(a) What is the approximate momentum change of the mosquito? Give magnitude and direction. Explain any approximations you make.

(b) At a particular instant during the impact, when the force exerted on the mosquito by the car is F, what is the magnitude of the force exerted on the car by the mosquito?

(c) What is the approximate momentum change of the car? Give magnitude and direction. Explain any approximations you make.

(d) Qualitatively, why is the collision so much more damaging to the mosquito than to the car?

Problem 2.32 The bouncing ball

A steel ball of mass m falls from a height h onto a scale calibrated in newtons. The ball rebounds repeatedly to nearly the same height h. The scale is sluggish in its response to the intermittent hits and displays an *average* force F_{avg}, such that

$$F_{avg} T = F \Delta t$$

where $F \Delta t$ is the brief impulse that the ball imparts to the scale on every hit, and T is the time between hits.

Calculate this average force in terms of m, h, and physical constants. Compare your result with what the scale reads if the ball merely rests on the scale. Explain your analysis carefully (but briefly).

Problem 2.33 Moving the Earth

Suppose all the people of the Earth go to the North Pole and, on a signal, all jump straight up. Estimate the recoil speed of the Earth. The mass of the Earth is 6×10^{24} kg, and there are about 6 billion people (6×10^9).

Problem 2.34 Bullet embeds in block

A bullet of mass m traveling horizontally at a very high speed v embeds itself in a block of mass M that is sitting at rest on a nearly frictionless surface. What is the speed of the block after the bullet embeds itself in the block?

Problem 2.35 Meteor hits a spinning satellite

A satellite which is spinning clockwise has four low-mass solar panels sticking out as shown. A tiny meteor traveling at high speed rips through one of the solar panels and continues in the same direction but at reduced speed. Afterwards, calculate the v_x and v_y components of the center-of-mass velocity of the satellite. Initial data are provided on the diagram.

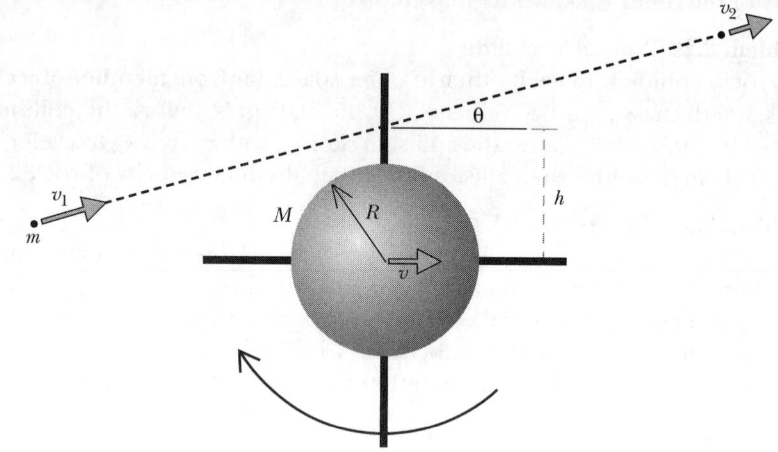

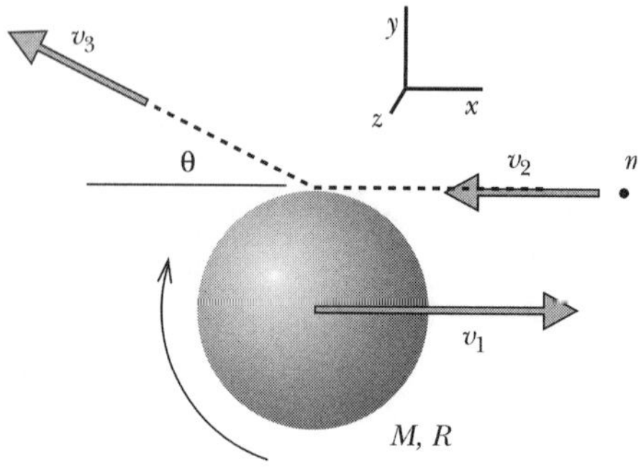

Figure 2b.73 Space junk strikes a spinning satellite (Problem 2.36).

Problem 2.36 Space junk

A tiny piece of space junk of mass m strikes a glancing blow to a spinning satellite (Figure 2b.73). After the collision the space junk is traveling in a new direction and moving more slowly. The space junk had negligible rotation both before and after the collision. The velocities of the space junk before and after the collision are shown in the diagram. The satellite has mass M and radius R,. Before the collision the satellite was moving and rotating as shown in the diagram. Just after the collision, what are the components of the center-of-mass velocity of the satellite (v_x and v_y)?

Problem 2.37 Two balls collide

A ball of mass 0.05 kg moves with a velocity $\langle 17, 0, 0 \rangle$ m/s. It strikes a ball of mass 0.1 kg which is initially at rest. After the collision, the heavier ball moves with a velocity of $\langle 3, 3, 0 \rangle$ m/s.

(a) What is the velocity of the lighter ball after impact?

(b) What is the impulse delivered to the 0.05 kg ball by the heavier ball?

(c) If the time of contact between the balls is 0.03 sec, what is the force exerted by the heavier ball on the lighter ball?

Problem 2.38 Space station

A space station has the form of a hoop of radius R, with mass M (Figure 2b.74). Initially its center of mass is not moving, but it is spinning. Then a small package of mass m is thrown by a spring-loaded gun toward a nearby spacecraft as shown; the package has a speed v after launch. Calculate the center-of-mass velocity of the space station (v_x and v_y) after the launch.

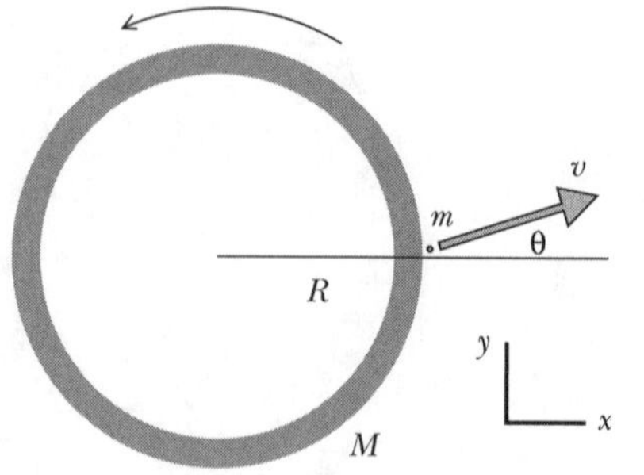

Figure 2b.74 Launch a package from a space station (Problem 2.38).

2.25 Answers to exercises

2.1 (page 51) 0.275 N

2.2 (page 51) 3571 N/m

2.3 (page 51) 71.4 N

2.4 (page 52) $\langle -60, -24, 96 \rangle$ N·s

2.5 (page 54) $\langle -0.03, -0.05, 0.02 \rangle$ N

2.6 (page 54) Earth, water, and air exert forces; net force is zero

2.7 (page 54) zero

2.8 (page 55) glider; track, spring, Earth, air

2.9 (page 59) $\vec{p}_f = \langle 10, 0, 11 \rangle$ kg·m/s

2.10 (page 59) $\Delta\vec{p} = \vec{p}_f - \vec{p}_i = \langle -15000, 0, 3000 \rangle$ kg·m/s = impulse by ground

2.11 (page 59) $\vec{F}_{net} = \langle -5000, 0, 1000 \rangle$ N

2.12 (page 59) $\vec{F}_{net} = \langle -1\times10^4, 0, 0 \rangle$ N

2.13 (page 59) $\vec{p}_f = \langle 9.4, 0.3, 0 \rangle$ kg·m/s

2.14 (page 63) (a) $\langle -10, 7.12, -5 \rangle$ m/s

(b) $\langle 3, 6.036, -8 \rangle$ m

(c) 8.6 m

(d) 1.33 s

(e) 2.66 s

(f) $\langle -17.5, 0, -18.3 \rangle$ m

2.15 (page 63) $y_i = h$, $v_{yi} = 0$, $y_f = 0$, $\Delta t = \sqrt{2h/g}$, $v_{yf} = \sqrt{2gh}$

Same results if mass is changed.

2.16 (page 64) time in air is 1.77 s; horizontal distance 8.8 m; 3.8 m high

2.17 (page 68) $y = 0.108$ m, $p_y = 0.0365$ kg·m/s

2.18 (page 78) 3/16 times as large

2.19 (page 78) 2.5×10^{-10} N, 2.5×10^{-10} N

2.20 (page 78) 6×10^{24} kg

2.21 (page 81) (a) $\langle 2.8\times10^8, 0, -2.8\times10^8 \rangle$ m

(b) 3.96×10^8 m

(c) $\langle 0.707, 0, -0.707 \rangle$ (no units; dimensionless

(d) $\langle -1.27\times10^{20}, 0, 1.27\times10^{20} \rangle$ N

2.22 (page 86) about a day or less; a small portion of the circle

2.23 (page 86) about 0.02 s; a small fraction of the round-trip time

2.25 (page 87) 3.2×10^4 m; note that Mt. Everest is about 8×10^3 m tall

2.26 (page 88) 2.3×10^{-8} N

2.27 (page 92) $\langle 0.04, 0.04, -0.1\rangle$ kg · m/s,

$\langle 0.04, 0.04, -0.1\rangle$ kg · m/s,

$\langle 1, 1, -2.5\rangle$ m/s

2.28 (page 92) $\langle 18, 14, 7\rangle$ kg · m/s, $\langle 18, 14, 7\rangle$ kg · m/s

2.29 (page 92) $\langle 1\times10^{-21}, -1.6\times10^{-21}, 0\rangle$ kg · m/s

2.30 (page 93) (a) $-m\vec{v}$, $+m\vec{v}$; (b) $+m\vec{v}$, $-m\vec{v}$; (c) $\vec{0}$, $\vec{0}$

Chapter 3

The Momentum Principle 2: Instantaneous Rate of Change

Chapter 3

The Momentum Principle 2: Instantaneous Rate of Change

In this chapter we continue to apply the Momentum Principle to a variety of systems.

The major topics in this chapter are:

- The derivative form of the Momentum Principle, with applications
- A microscopic model of solid objects and their interactions

3.1 Derivative form of the Momentum Principle

The forms of the Momentum Principle given so far, $\Delta\vec{p} = \vec{F}_{net}\Delta t$ and $\vec{p}_f = \vec{p}_i + \vec{F}_{net}\Delta t$, are particularly useful when we know the momentum at a particular time and want to predict what the momentum will be at a later time. We use these forms in repetitive computer calculations to predict future motion.

Another important form is obtained by dividing by the time interval Δt:

$$\frac{\Delta\vec{p}}{\Delta t} = \vec{F}_{net}$$

Just as we did in finding the instantaneous velocity of an object (see Section 1.7.6), we can find the instantaneous rate of change of momentum by letting the time interval Δt approach zero (an infinitesimal time interval). The ratio of the infinitesimal momentum change (written as $d\vec{p}$) to the infinitesimal time interval (written as dt) is the instantaneous rate of change of the momentum. As the time interval gets very small, $\Delta t \to 0$, the ratio of the momentum change to the time interval approaches the ratio of the infinitesimal quantities

$$\lim_{\Delta t \to 0} \frac{\Delta\vec{p}}{\Delta t} = \frac{d\vec{p}}{dt}$$

and we obtain the following form:

THE MOMENTUM PRINCIPLE (DERIVATIVE FORM)

$$\frac{d\vec{p}}{dt} = \vec{F}_{net}$$

In words, "the instantaneous time rate of change of the momentum of an object is equal to the net force acting on the object." Or in calculus terms, since the limit of a ratio of infinitesimal quantities is called a derivative, we can say that "the derivative of the momentum with respect to time is equal to the net force acting on the object."

This form of the Momentum Principle is useful when we know something about the rate of change of the momentum at a particular instant. Knowing the rate of change of momentum, we can use this form of the Momentum Principle to deduce the net force acting on the object, which is numerically equal to the rate of change of momentum. Knowing the net force, we may be able to figure out particular contributions to the net force.

Ex. 3.1 At a certain instant the z component of the momentum of an object is changing at a rate of 4 kg·m/s per second. At that instant, what is the z component of the net force on the object?

Ex. 3.2 If an object is sitting motionless, what is the rate of change of its momentum? What is the net force acting on the object?

Ex. 3.3 If an object is moving with constant momentum $\langle 10, -12, -8 \rangle$ kg·m/s, what is the rate of change of momentum $d\vec{p}/dt$? What is the net force acting on the object?

3.1.1 An approximate result: $F = ma$

In simple cases the derivative form of the Momentum Principle reduces to a form that may be familiar to you from a previous course. If the mass is constant (the usual situation), and the speed is small enough compared to the speed of light that the momentum is well approximated by $\vec{p} \approx m\vec{v}$, we have this:

$$\frac{d\vec{p}}{dt} \approx \frac{d(m\vec{v})}{dt} = m\frac{d\vec{v}}{dt} = \vec{F}_{\text{net}} \quad \text{(nonrelativistic form; constant mass)}$$

The quantity $\dfrac{d\vec{v}}{dt}$ is called *acceleration,* and is often given the symbol $\vec{a}$.

This form of the Momentum Principle (Newton's second law) says that mass times acceleration (time rate of change of velocity, $d\vec{v}/dt$) is equal to the net force, or in simplified, scalar form "$ma=F$" or "$F=ma$".

The approximate form is not valid in situations where an object's mass isn't constant. One example is a rocket with exhaust gases ejecting out the back; as a result, the rocket has decreasing mass. In such cases the momentum-based formula $d\vec{p}/dt = \vec{F}_{\text{net}}$ gives the correct results, whereas the constant-mass formula $m d\vec{v}/dt = \vec{F}_{\text{net}}$ cannot be used.

The approximate form is also not valid for objects moving at speeds close to the speed of light, because in these circumstances the approximation $\vec{p} \approx m\vec{v}$ is not valid.

When the principle is written in terms of momentum, it is valid even for objects whose mass changes or which are moving at speeds close to the speed of light, as long as we use the relativistically correct definition of momentum. So $d\vec{p}/dt = \vec{F}_{\text{net}}$ is the more general and more powerful form.

The extra symbols on $d\vec{p}/dt = \vec{F}_{\text{net}}$ compared to the simplified version $F=ma$ are important:

- The Momentum Principle is a vector principle, so the arrows over the symbols are extremely important; they remind us that there are really three separate component equations, for x, y, and z.
- It is also important to remember that we have to add up all vector forces to give the "net" force, so we write $\vec{F}_{\text{net}}$, not just F.

Ex. 3.4 The velocity of a 80 gram ball changes from $\langle 5, 0, -3 \rangle$ m/s to $\langle 5.02, 0, -3.04 \rangle$ m/s in 0.01 s, due to the gravitational attraction of the Earth and to air resistance. What is the acceleration of the ball? What is the rate of change of momentum of the ball? What is the net force acting on the ball?

3.2 Momentum not changing

We saw in Chapter 1 that if a system is in uniform motion its momentum is constant, and does not change with time. A special case of uniform motion is a situation in which object is at rest and remains at rest; this situation is called "static equilibrium."

3.2.1 Static equilibrium

? If the system never moves, what is $\frac{d\vec{p}}{dt}$?

The momentum of the system isn't changing, so the rate of change of the momentum is zero:

$$\frac{d\vec{p}}{dt} = \vec{0} = \langle 0, 0, 0 \rangle \text{ (static equilibrium)}$$

? What does this imply about the net force acting on the system?

From the derivative form of the Momentum Principle, $d\vec{p}/dt = \vec{F}_{net}$, we deduce that the net force acting on a system that never moves must be zero:

$$\vec{F}_{net} = \vec{0} = \langle 0, 0, 0 \rangle \text{ (static equilibrium)}$$

For example, consider a spring whose stiffness is $k_s = 80$ N/m. You suspend the spring vertically and hang a block of mass 0.5 kg from the bottom end. At first the block oscillates up and down but eventually air resistance brings it to a stop, and the block hangs motionless, never moving (Figure 3.1).

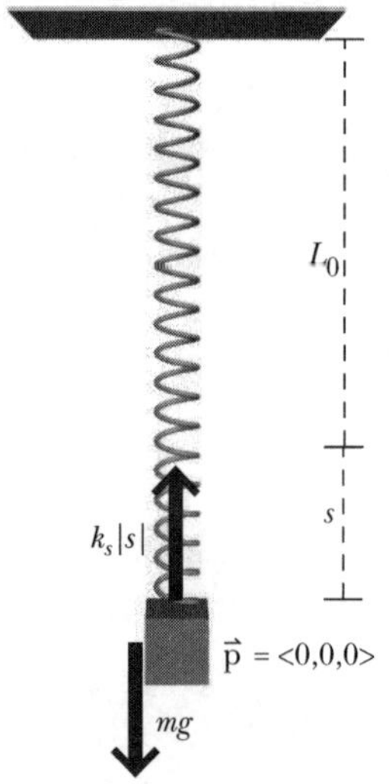

Figure 3.1 A block hangs motionless from a spring (static equilibrium).

? How much has the spring stretched? Use the Momentum Principle to prove your result rigorously.

Choose the block as the system. If the block no longer moves at all, the rate of change of its momentum $d\vec{p}/dt$ is zero. From the derivative form of the Momentum Principle we conclude that the net force acting on the block must be zero.

? What objects in the surroundings exert forces on the block?

The spring pulls up and the Earth pulls down.

? Apply the derivative form of the Momentum Principle to find the stretch.

The vector sum of these two forces must equal zero:

$$\frac{d\vec{p}}{dt} = \vec{F}_{net} = \vec{F}_{spring} + \vec{F}_{Earth}$$

$$\langle 0, 0, 0 \rangle = \langle 0, k_s s, 0 \rangle + \langle 0, -mg, 0 \rangle = \langle 0, k_s s - mg, 0 \rangle$$

$$0 = k_s s - mg \text{ which gives } s = mg/k_s$$

$$s = \frac{(0.5 \text{ kg})(9.8 \text{ N/kg})}{80 \text{ N/m}} = 0.061 \text{ m}$$

3.2.2 Uniform motion

Suppose the spring and hanging mass are on an elevator that goes up at a constant speed of 4 m/s. How does putting the apparatus into uniform motion affect our analysis?

? What is $\frac{d\vec{p}}{dt}$ for the block, which moves upward at 4 m/s?

Now the momentum of the block is not zero, but constant, and $d\vec{p}/dt$ is still zero, which means that the net force $\vec{F}_{net}$ must also be zero. The spring will be stretched 0.061 m, just as when there was no motion. This is an example of the principle of relativity, that physical laws (in this case, the Momentum Principle) work in the same in all reference frames that are in uniform motion (that is, constant velocity) with respect to the cosmic microwave background, which is the stage on which phenomena play themselves out.

3.2.3 Momentarily at rest *vs.* static equilibrium

Pull the block down a ways and release it, and it oscillates up and down at the end of the spring. During this oscillating motion, whenever the block reaches the bottom of its travel its speed is momentarily zero. At first glance, this situation may seem similar to the case of block hanging at rest which we considered in Section 3.2.1, but there are important differences.

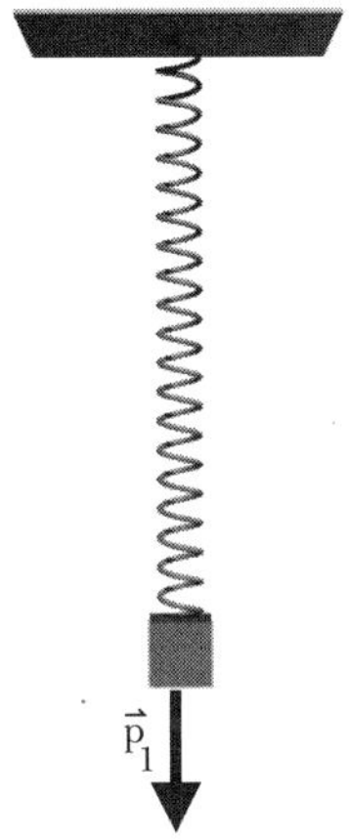

Figure 3.2 Just before the block comes to a momentary stop, its momentum is downward.

? At this instant, when the speed is momentarily zero, is the rate of change of the block's momentum zero?

No! The momentum just before hitting bottom points downward (Figure 3.2), and just after hitting bottom the momentum points upward (Figure 3.4). The momentum is changing all during this time, and the momentum is zero only momentarily (Figure 3.3).

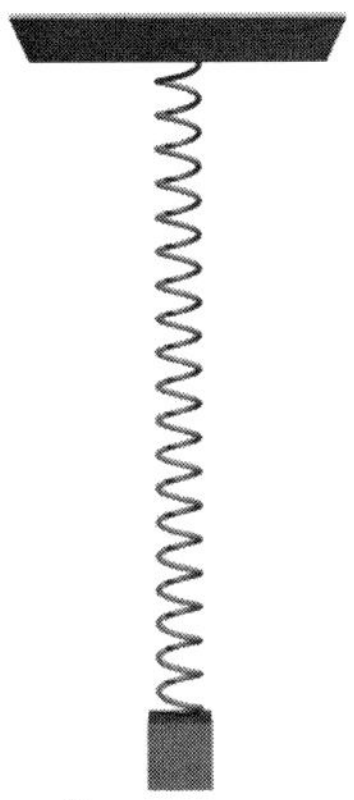

Figure 3.3 The block's momentum is zero only at the instant it reaches its lowest position.

? Apply the derivative form of the Momentum Principle. What can we deduce about the magnitude and direction of the net force on the block at the instant the block is at rest at its lowest position?

Since $d\vec{p}/dt$ is nonzero, the net force must be nonzero as well. Since the initial momentum was downward, and the final momentum is upward, the direction of $d\vec{p}/dt$ must be upward. This implies that the net force is also upward at the instant the block is at rest.

? Does an upward net force make sense physically?

If the net force at the bottom of the motion were zero, the momentum couldn't change from its current value (zero), and the block would stay forever at the bottom point of the oscillation. But that's not what happens. The block slows to a stop but then starts back upwards. In order for the y component of momentum to change from zero to a positive quantity, there must be an upward net force to cause this change.

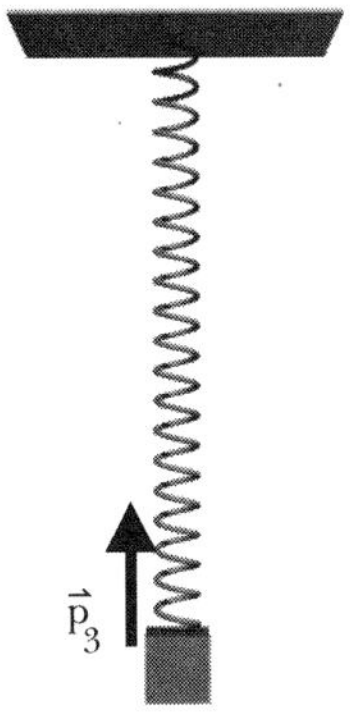

Figure 3.4 As the block starts back upward, its momentum is upward.

? At the bottom of the oscillating motion, which force is larger, the force exerted upward by the spring or the force exerted downward by the Earth? How do you know?

Since the net force must be upward in order to produce the observed change in the momentum, it must be that the spring force is larger than the gravitational force exerted by the Earth. This conclusion is also consistent with the fact that the block when headed downward slows to a stop, which means that the net force must point upward to oppose the downward motion.

? How does the stretch of the spring when the block is momentarily at rest at its lowest point compare to the stretch of the spring in the previous situation, where the block hung motionless in equilibrium?

The stretch of the spring is longer in this case (block momentarily at rest) than it is in the previous case (static equilibrium).

Let's look at this numerically. When a block of 0.5 kg hangs motionless from a spring whose stiffness is 80 N/m, we found that in static equilibrium the spring is stretched 0.061 m (6.1 cm). Now suppose the block is oscillating up and down in such a way that at the bottom of its motion the spring is stretched 0.1 m (10 cm).

? What is the net force when the block's momentum is (momentarily) zero?

The y component of the net force is $F_{\text{net},y} = k_s s - mg = 0$, which gives

$$F_{\text{net},y} = (80\text{ N/m})(0.1\text{ m}) - (0.5\text{ kg})(9.8\text{ N/kg}) = +3.1\text{ N}$$

As expected, the net force points upward (positive y component), consistent with the rate of change of momentum being upward.

Avoiding confusion

One of the hardest things to keep straight when first learning about motion and the causes of motion is that being momentarily at rest is not at all like being permanently at rest (static equilibrium). When you encounter a situation where the momentum is zero, always ask yourself whether it is staying zero permanently. If not, the momentum is changing, and the net force is nonzero.

There are many other situations in which the net force on an object is nonzero at the instant when the object's momentum is (momentarily) zero. This happens whenever an object reverses direction. Throw a ball up in the air. When the ball reaches its maximum height, p_y is momentarily zero, but the gravitational force still points downward and makes p_y decrease. If the net force dropped to zero when p_y dropped to zero, the ball would stay motionless at the top of its trajectory!

When a tennis ball hits a wall and rebounds, the momentum of the ball momentarily goes to zero, but a force due to the wall acts on the ball, changing the momentum of the ball to go in the opposite direction. If the force exerted by the wall went to zero when the speed of the ball went to zero, the ball would just remain motionless, squashed against the wall.

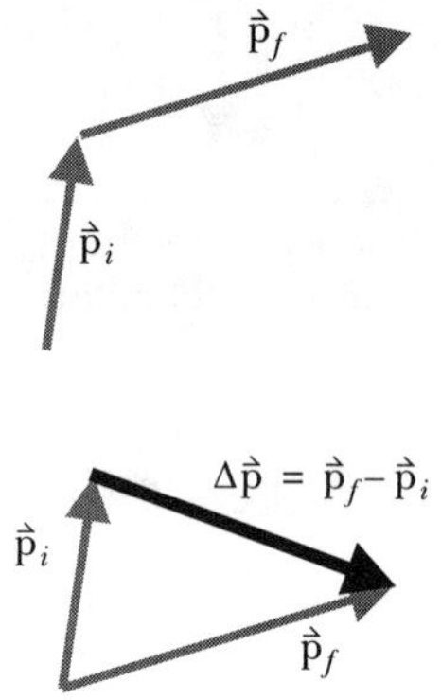

Figure 3.5 Calculation of $\Delta\vec{p}$.

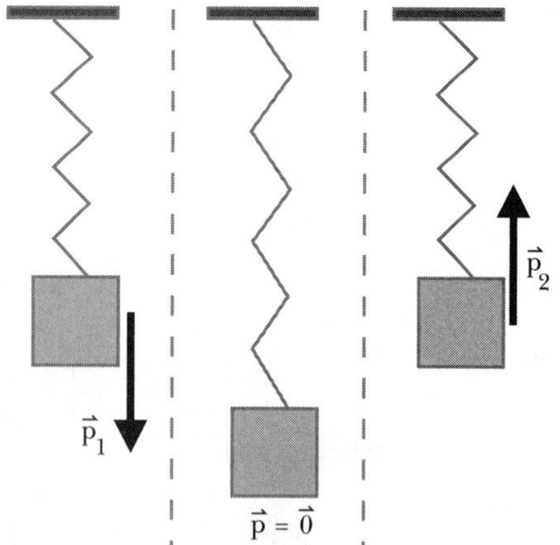

Figure 3.6 The momentum of a block oscillating on a spring, at an instant before it comes to a momentary rest at the bottom of its path (1) and an instant after it starts back upward (2).

3.2.4 Finding the rate of change of momentum

In order to apply the derivative form of the Momentum Principle, it is important to be able to decide if $d\vec{p}/dt$, the instantaneous rate of change of momentum, is zero or nonzero, and if it is nonzero, what its direction is. The procedure for finding $d\vec{p}/dt$ is based on the procedure for finding $\Delta\vec{p}$ which you practiced in Chapter 1, and requires that you draw a simple diagram, as shown in Figure 3.5.

1. Draw two arrows:

   one representing $\vec{p}_i$, the momentum of the object a short time before the instant of interest, and

   a second arrow representing $\vec{p}_f$, the momentum of the object a short time after the instant of interest

2. Graphically find $\Delta\vec{p} = \vec{p}_f - \vec{p}_i$ by placing the arrows tail to tail, and drawing the resultant arrow starting at the tail of $\vec{p}_i$ and going to the tip of $\vec{p}_f$

3. This arrow indicates the direction of $\Delta\vec{p}$, which is the same as the direction of $d\vec{p}/dt$, provided the time interval involved is sufficiently small

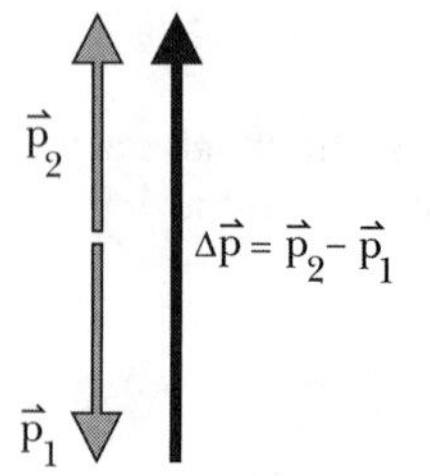

Figure 3.7 The direction of $d\vec{p}/dt$ is the direction of $\Delta\vec{p} = \vec{p}_2 - \vec{p}_1$.

The diagrams in Figure 3.6 and Figure 3.7 show the result of this procedure when applied to the oscillating block on the end of a spring, at the instant it is at rest at the bottom. The direction of $d\vec{p}/dt$ at this instant is upward.

Ex. 3.5 A block oscillating up and down on a spring comes to the top of its path, where it is momentarily at rest. Follow the procedure outlined above to answer the following questions: At the instant it is at rest, is $d\vec{p}/dt$ zero or nonzero? If nonzero, what is its direction?

Ex. 3.6 You throw a ball straight up into the air. Follow the procedure outlined above to answer the following questions: At the instant the ball reaches its highest location, is $d\vec{p}/dt$ zero or nonzero? If nonzero, what is its direction?

Ex. 3.7 A ping pong ball hits a paddle and bounces straight back. While the ball is in contact with the paddle, is $d\vec{p}/dt$ zero or nonzero? If nonzero, what is its direction?

3.2.5 Example: Hanging block (static equilibrium)

A block of mass m hangs from a string (Figure 3.8). Then you pull the block to the side by pulling horizontally with a second string. What is the magnitude F_2 of the force you must apply to the second string in order for the first string to be at an angle of θ to the vertical? What is the tension F_1 in the first string? (The "tension" in a string is the force it exerts on another object. Because a string is flexible, the force a string exerts is always in the direction of the string; a string cannot exert a force perpendicular to the string.)

1. Choose a system

System: the block

2. List external objects that interact with the system, with diagram

string 1, string 2, the Earth

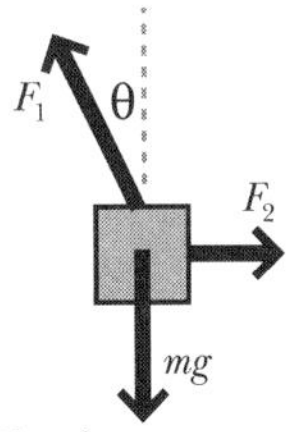

3. Apply the Momentum Principle

$$d\vec{p}/dt = \vec{F}_{net}$$
$$\langle 0, 0, 0\rangle = \langle F_2 - F_1\sin\theta, F_1\cos\theta - mg, 0\rangle$$

4. No need to apply the position update formula (no motion)

5. Solve for the unknowns (algebraic manipulation)

$$0 = F_2 - F_1\sin\theta \text{ so } F_1\sin\theta = F_2$$

$$0 = F_1\cos\theta - mg \text{ so } F_1\cos\theta = mg \text{ which gives } F_1 = \frac{mg}{\cos\theta}$$

$$F_2 = F_1\sin\theta = \frac{mg}{\cos\theta}\sin\theta = mg\tan\theta$$

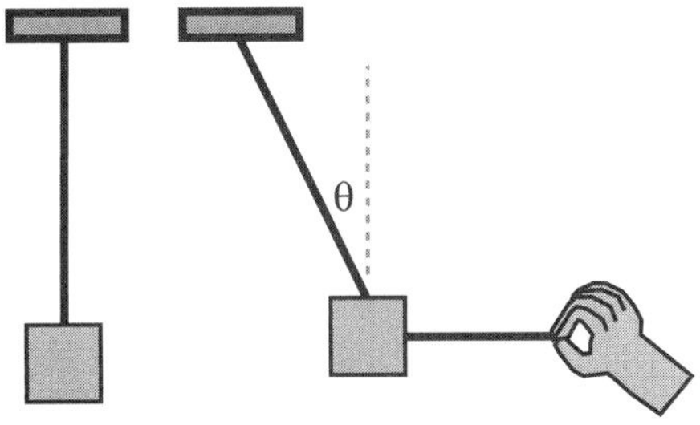

Figure 3.8 A block hangs from a string. Then you pull the block to the side with another string.

6. Check

We can make some checks on the reasonableness of the results.

If $\theta = 0°$, the first string is vertical, and we expect the tension in this string to be mg. Since $\cos(0°) = 1$, we do indeed find $F_1 = mg$. Also, $F_2 = 0$, as it should be in this case (no pulling).

On the other hand, if $\theta = 90°$, $\cos(90°) = 0$, the strings are horizontal, with $F_1 = mg/0$ and $F_2 = mg/0$, which is infinitely large and impossible. This makes sense: if both strings are horizontal, there is no y component to support the block.

3.2.6 Preview of the Angular Momentum Principle

In many static equilibrium situations, the Momentum Principle is not sufficient for carrying out a full analysis and we also need the Angular Momentum Principle, which we will discuss in a later chapter.

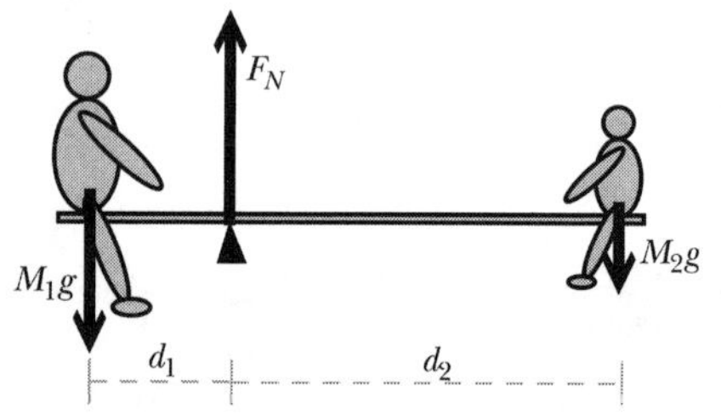

Figure 3.9 Two children sit on a seesaw in static equilibrium.

For example, consider two children sitting motionless on a seesaw (Figure 3.9). Choose as the system the two children and the (lightweight) board. For this choice of system, the objects in the surroundings that exert significant (external) forces on the system are the Earth (pulling down) and the support pivot (pushing up). The momentum of the system isn't changing, so the Momentum Principle correctly tells us that the net force must be zero, and the support pivot pushes up with a force equal to the weight of the children (we are neglecting the low-mass board).

However, the Momentum Principle alone doesn't tell us where the children have to sit in order to achieve balance and static equilibrium. The force on a child exerts a twist (the technical term is "torque") about the pivot which tends to make the board turn. The torque associated with the forces on the two children have to add up to zero (one tends to twist the board clockwise, the other counterclockwise). Torque is defined as force times lever arm, and a nonzero net *torque* causes changes in a quantity called angular momentum, just as a nonzero net *force* causes changes in the ordinary momentum we have been studying.

3.3 Rate of change of momentum along a curving path

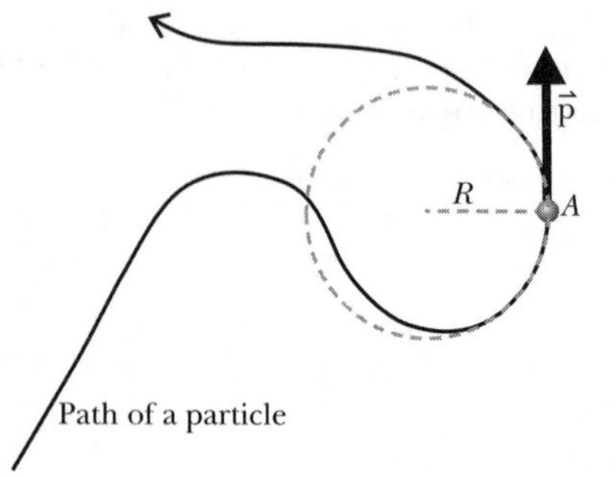

Figure 3.10 Curving path of a particle. The particle's momentum is shown at location A. At location A the particle's path has a radius of curvature R. The "kissing circle" matches the tangent and radius of curvature of the path.

When a particle moves along a curving path, such as that shown in Figure 3.10, the direction (and perhaps the magnitude) of its momentum is continuously changing. Using geometry and algebra, we can derive an algebraic expression for the rate of change of momentum of a particle at any location on a curving trajectory, by doing the following:

- Pick a short interval bracketing location A
- Graphically find $\Delta\vec{p}$ over this interval
- Inscribe a "kissing" circle inside the curve
- Use similar triangles to relate $\Delta\vec{p}$ to the radius of the circle

We will find that the rate of change of momentum at a location depends on:

- the momentum of the particle when it is at that location
- the particle's speed
- the radius of a circle "kissing" the inside of the curve at that location

We will show that when the particle is (momentarily) at a particular location (such as location A in Figure 3.10),

$$\left|\frac{\Delta\vec{p}}{\Delta t}\right| = \frac{|\vec{v}|}{R}|\vec{p}| \text{ in the special case where the speed is not changing,}$$

where $|\vec{v}|$ is the speed of the particle at the moment that it passes location A, and R is the radius of the kissing circle. Since R is in the denominator, the larger the radius of curvature, the smaller the rate of change of the momentum. This is reasonable, because when the radius of curvature gets extremely large, the path of the particle is essentially a straight line, and the direction of the momentum isn't changing. On the other hand, a tight turn (small radius of curvature R) means that the direction of the momentum is changing very rapidly, and $\Delta\vec{p}/\Delta t$ is large.

Simplest case: magnitude of momentum constant

To begin, let's assume that the magnitude of the momentum isn't changing, only the direction is changing. Consider motion along the curving path shown in Figure 3.10. We want to find the rate of change of momentum of a particle traveling along this path, at the instant when the particle is at location A.

In Figure 3.10 a dashed circle has been drawn inside the path, tangent to the path at location A. This dashed circle of radius R is called the "osculatory" or "kissing" circle, because it just "kisses" the curving path at location A, fitting into the trajectory as smoothly as possible, with the circle and the trajectory sharing the same tangent and same radius of curvature R at location A. We pick a short interval centered on location A, and draw arrows representing the momentum of the particle at the beginning and end of that interval, as shown in Figure 3.11.

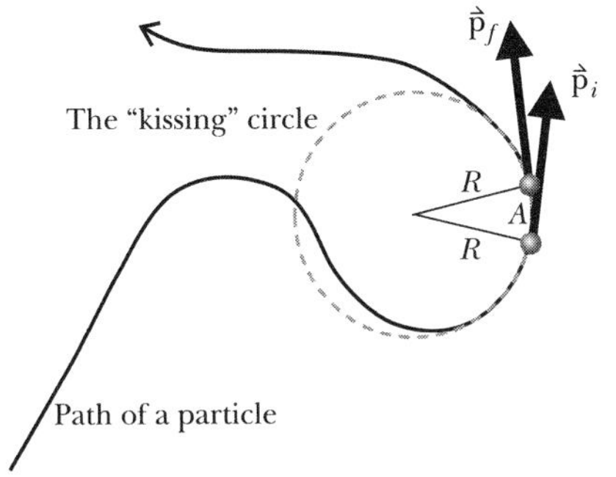

Figure 3.11 The particle's momentum shown at two locations before and after location A.

In Figure 3.12 we find $\Delta\vec{p}$ by putting the initial and final momentum vectors tail to tail and drawing the vector $\Delta\vec{p} = \vec{p}_f - \vec{p}_i$ (final minus initial). The important thing to notice is that the *change* in momentum is not in the direction of the momentum but is perpendicular to it. $\Delta\vec{p}$ points to the left, toward the center of the kissing circle. In order to make such a change in the momentum, you would have to push toward the left.

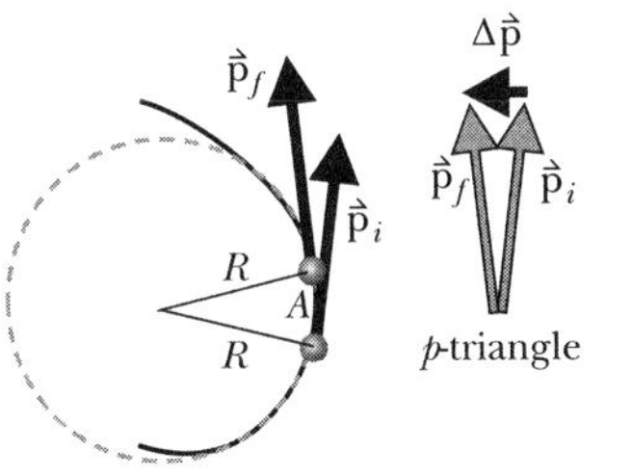

Figure 3.12 We find $\Delta\vec{p} = \vec{p}_f - \vec{p}_i$ graphically, by putting the vectors tail to tail.

We can use geometry to calculate the rate at which the momentum changes due to the changing direction. In Figure 3.14 the *p*-triangle and the *R*-triangle are similar to each other, with the same acute angles, because the radius vectors in the *R*-triangle and the momentum vectors in the *p*-triangle are at right angles to each other. To say it another way, you can rotate one of the triangles through 90 degrees and the small acute angles will clearly be the same. We're assuming that the magnitude of the momentum isn't changing, only the direction, so $|\vec{p}_f| = |\vec{p}_i|$.

For short times (small angles), the length of the short side of the *R*-triangle is approximately equal to the arc length, which is the distance the particle goes in the short time, which is equal to $|\vec{v}|\Delta t$. Since the two triangles are similar, the ratio of the side opposite the acute angle to the side adjacent to the acute angle must be the same for both triangles:

$$\frac{|\Delta\vec{p}|}{|\vec{p}|} = \frac{|\vec{v}|\Delta t}{R} \text{ (constant speed)}$$

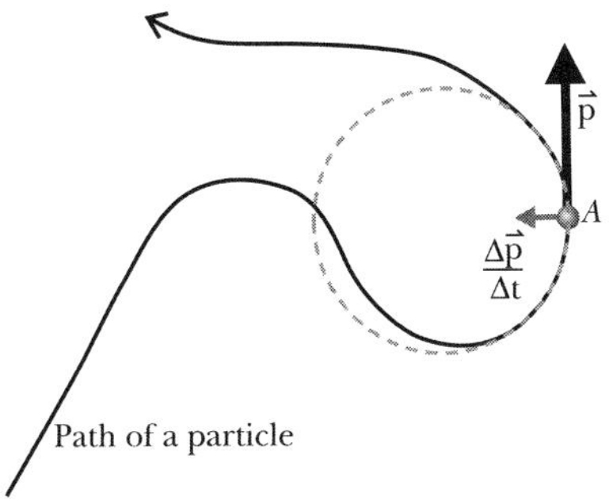

Figure 3.13 For a particle moving at constant speed along a curving path, $\Delta\vec{p}/\Delta t$ is perpendicular to the particle's momentum, and points to the inside of the curve.

We have used the fact that we're assuming that the magnitude of the momentum isn't changing, only the direction, so that we can write $|\vec{p}|$ for the magnitude of either the initial or final momentum near location A. We are interested in the rate of change of the momentum, so rearrange the formula like this:

$$\frac{|\Delta\vec{p}|}{\Delta t} = \frac{|\vec{v}|}{R}|\vec{p}| \text{ (constant speed)}$$

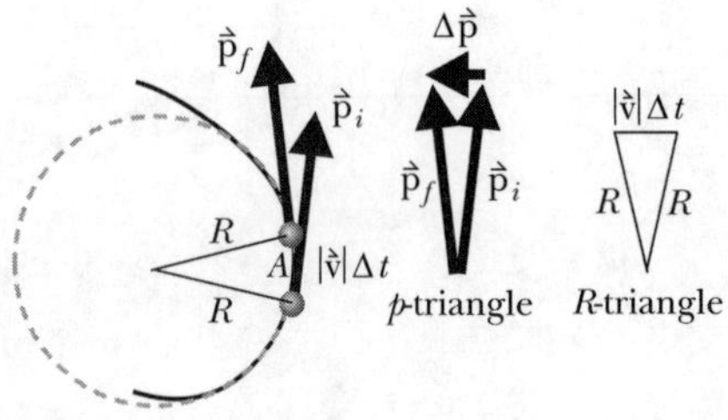

Figure 3.14 The two triangles are similar, with the same acute angle, because the sides of one triangle were originally perpendicular to the side of the other.

Since the time is increasing, Δt is always positive and therefore the same as $|\Delta t|$, so we can express the magnitude of the rate of change of the momentum like this:

$$\left|\frac{\Delta\vec{p}}{\Delta t}\right| = \frac{|\vec{v}|}{R}|\vec{p}| \text{ (constant speed)}$$

The direction of $\Delta\vec{p}/\Delta t$ is the same as the direction of $\Delta\vec{p}$: perpendicular to the particle's momentum, toward the center of the kissing circle (Figure 3.13).

Checking units

A simple check on the results of a calculation or derivation is a units check.

? Do the units on both sides of the preceding equation match?

$$\text{Left hand side: } \frac{\left(\frac{\text{kg}\cdot\text{m}}{\text{s}}\right)}{\text{s}} = \frac{\text{kg}\cdot\text{m}}{\text{s}^2}$$

$$\text{Right hand side: } \frac{\left(\frac{\text{m}}{\text{s}}\right)}{\text{m}}\left(\frac{\text{kg}\cdot\text{m}}{\text{s}}\right) = \frac{\text{kg}\cdot\text{m}}{\text{s}^2}$$

The units match. (If they had not matched, the equation could not be correct, and we would have to look for errors in our derivation.)

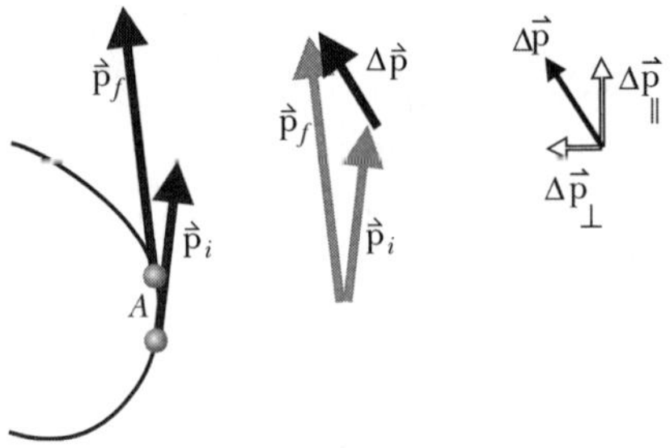

Figure 3.15 If both the magnitude and the direction of the momentum change, then $\Delta\vec{p}$ has a component parallel to the motion as well as a component perpendicular to the motion.

General case: magnitude of momentum changing

Despite the fact that the preceding derivation hinged on the assumption that the magnitude of momentum was constant, we can extend this result to the general case in which both direction and magnitude of momentum are changing. In Figure 3.15 we see what happens if both the speed and direction of motion are changing. As usual, we move the initial and final momentum vectors to be tail to tail, and the difference vector (final minus initial) is the change in the momentum. The initial and final momentum vectors have different lengths (magnitudes). As a result, the momentum change $\Delta\vec{p}$ has a component parallel to the motion as well as a component perpendicular to the motion.

In this general case, $\Delta\vec{p}$ can be separated into two components: one component perpendicular to the momentum, $\Delta\vec{p}_{\perp}$, along the radius of the kissing circle (this component indicates change of direction), and another component parallel to the momentum, $\Delta\vec{p}_{||}$, along the tangent to the curving trajectory (this component indicates change of speed). If we consider only the perpendicular component, then the result obtained earlier still applies.

Instantaneous rate of change of momentum

In the limit as Δt becomes infinitesimally small we have a derivative, and we obtain this important result:

PERPENDICULAR COMPONENT OF $d\vec{p}/dt$
FOR MOTION ALONG A CURVING PATH

$$\text{Perpendicular component of } \frac{d\vec{p}}{dt} \text{ is given by } \left|\left(\frac{d\vec{p}}{dt}\right)_{\perp}\right| = \frac{|\vec{v}|}{R}|\vec{p}|$$

The direction is toward the center of the kissing circle of radius R.

? Is this result valid for high speeds, when $|\vec{v}| \approx c$?

We did not use the approximate formula for momentum, so our result should be valid even if a particle is traveling at a high speed. Given the full definition of momentum, we can expand the result like this:

$$\left|\left(\frac{d\vec{p}}{dt}\right)_{\perp}\right| = \left(\frac{|\vec{v}|}{R}\right)\frac{1}{\sqrt{1-(|\vec{v}|/c)^2}}(m|\vec{v}|) = \frac{1}{\sqrt{1-(|\vec{v}|/c)^2}}\left(\frac{m|\vec{v}|^2}{R}\right)$$

If the particle's speed is small compared to the speed of light, we have the following approximate result for the perpendicular component of $d\vec{p}/dt$:

APPROXIMATE RESULT

Perpendicular component: $\left|\left(\frac{d\vec{p}}{dt}\right)_{\perp}\right| \approx \frac{m|\vec{v}|^2}{R}$ if $v << c$

It makes sense to remember the more general result, $|(d\vec{p}/dt)_{\perp}| = (|\vec{v}|/R)|\vec{p}|$, since it is not really any more complicated to use, and it is valid at all speeds, not just low speeds.

Ex. 3.8 Which of the dashed circles in Figure 3.16 best represents the "kissing circle" tangent to the path of the particle, with the same radius of curvature as that of the path at the location marked by "x"?

Ex. 3.9 Assume that the particle whose path is shown in Figure 3.16 is traveling at constant speed. At the location marked "x", what is the direction of $d\vec{p}/dt$ for the particle?

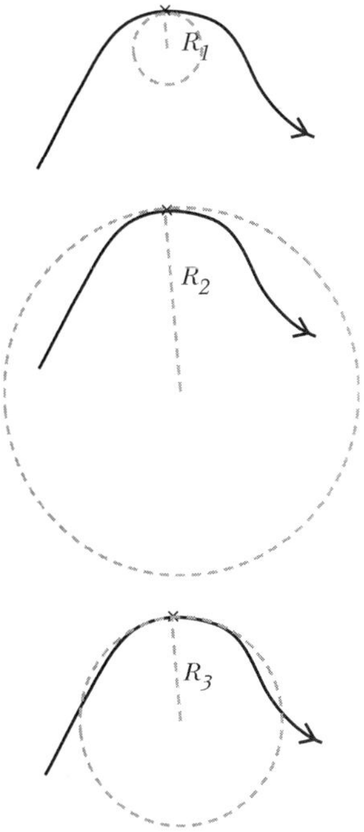

Figure 3.16 Which circle best represents the kissing circle? (Exercise 3.8)

3.3.1 Example: The Moon around the Earth (curving motion)

The Moon, which has a mass of 7×10^{22} kg, orbits the Earth once every 28 days (a lunar month), following a path which is nearly circular (Figure 3.17). The distance from the Earth to the Moon is 4×10^{8} m. What is the magnitude of the rate of change of the Moon's momentum? Calculate the gravitational force exerted on the Moon by the Earth, whose mass is 6×10^{24} kg. Compare these two results.

$$|\vec{v}| = \frac{2\pi(4\times10^8\ \text{m})}{(28\ \text{days})(24\ \text{hr/day})(60\ \text{min/hr})(60\ \text{s/min})} = 1\times10^3\ \text{m/s}$$

$$|\vec{p}| \approx m|\vec{v}| = (7\times10^{22}\ \text{kg})(1\times10^3\ \text{m/s}) = 7.3\times10^{25}\ \text{kg}\cdot\text{m/s}$$

$$\left|\left(\frac{d\vec{p}}{dt}\right)_{\perp}\right| = \frac{|\vec{v}|}{R}|\vec{p}| = \frac{(1\times10^3\ \text{m/s})}{(4\times10^8\ \text{m})}(7.3\times10^{25}\ \text{kg}\cdot\text{m/s}) = 1.8\times10^{20}\ \frac{\text{kg}\cdot\text{m}}{\text{s}^2}$$

The Moon's speed is not changing, so $d\vec{p}/dt$ is perpendicular to $\vec{p}$, as shown in Figure 3.17.

The magnitude of the gravitational force of the Earth on the Moon is

$$|\vec{F}_{\text{net}}| = G\frac{Mm}{R^2} = \left(6.7\times10^{-11}\ \frac{\text{N}\cdot\text{m}^2}{\text{kg}^2}\right)\frac{(6\times10^{24}\ \text{kg})(7\times10^{22}\ \text{kg})}{(4\times10^8\ \text{m})^2}$$

$$|\vec{F}_{\text{net}}| = 1.8\times10^{20}\ \text{N}$$

The net force is equal in magnitude to the observed magnitude of the rate of change of the Moon's momentum. The directions are also the same: toward the center of the kissing circle.

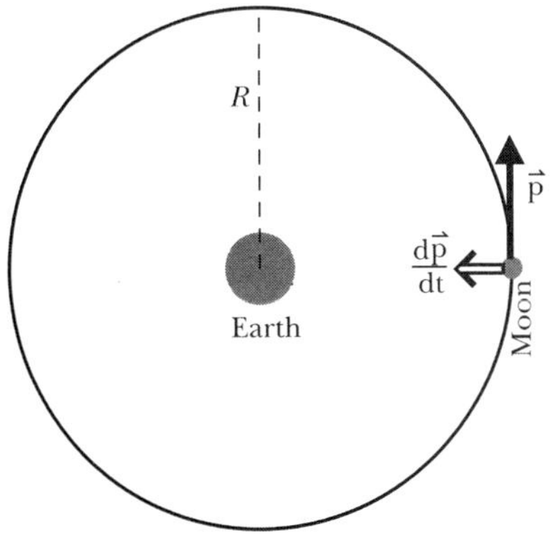

Figure 3.17 The Moon's orbit around the Earth is nearly circular. Not to scale: the sizes of the Earth and Moon are exaggerated.

This comparison was first made by Newton, who realized that the fall of an apple and the orbit of the Moon were due to the same fundamental cause, the gravitational attraction of the Earth for the apple and for the Moon. Until Newton had this insight no one had made this connection.

Ex. 3.10 A child of mass 40 kg sits on a wooden horse on a carousel. The wooden horse is 5 meters from the center of the carousel, which completes one revolution every 90 seconds. What is the rate of change of the momentum of the child, both magnitude and direction? What is the net force acting on the child?

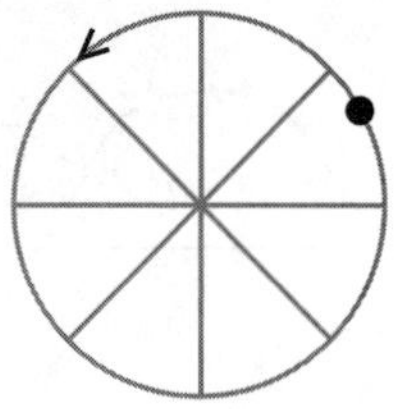

Figure 3.18 A person rides on a Ferris wheel at constant speed (Exercise 3.11). When the person is at the location shown, what is the direction of the net force acting on the person?

Ex. 3.11 A person rides on a Ferris wheel at constant speed (Figure 3.18). When the person is at the location shown, what is the direction of the person's rate of change of momentum, $d\vec{p}/dt$? What must be the direction of the net force acting on the person? (The net force is due to the Earth's gravitational pull plus the force of the seat on the person.)

Ex. 3.12 The orbit of the Earth around the Sun is approximately circular, and takes one year to complete. The Earth's mass is 6×10^{24} kg, and the distance from the Earth to the Sun is 1.5×10^{11} m. What is the magnitude of the rate of change of the Earth's momentum? What is the direction of the rate of change of the Earth's momentum? What is the magnitude of the gravitational force the Sun (mass 2×10^{30} kg) exerts on the Earth? What is the direction of this force?

3.3.2 Example: Tarzan swings from a vine (curving motion)

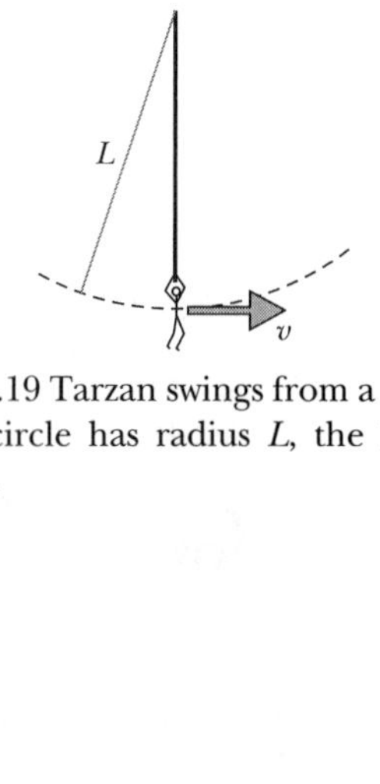

Figure 3.19 Tarzan swings from a vine. The kissing circle has radius *L*, the length of the vine.

Tarzan swings from a vine, and the radius of his circular motion is *L*, the length of the vine (Figure 3.19). We will calculate how strong the vine must be so that at the bottom of the swing (with the vine momentarily vertical) the vine doesn't break. Tarzan's mass is *M* and his speed at the bottom of the swing is *v*. Because Tarzan's motion is circular, we know that the component of the rate of change of his momentum that is perpendicular to his motion has magnitude $(|\vec{v}|/L)|\vec{p}|$ and is directed toward the center of the kissing circle (the top of the vine).

1. Choose a system

System: Tarzan

2. List external objects that interact with the system, with diagram

the vine (tension F_V), the Earth, and the air (represent the system by a circle)

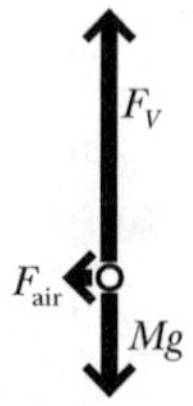

3. Apply the Momentum Principle

$$d\vec{p}/dt = \vec{F}_{net}$$

We'll neglect air resistance.

$$\langle 0, \frac{|\vec{v}|}{L}|\vec{p}|, 0\rangle = \langle 0, F_V - Mg, 0\rangle$$

4. No need to apply the position update formula

5. Solve for the unknowns

$$\frac{|\vec{v}|}{L}|\vec{p}| = F_V - Mg \text{ so } F_V = \left(\frac{v}{L}\right)Mv + Mg$$

As usual, we should check that these results make sense:

6. Check

Mg has units of N, and $(v/L)Mv$ has units of (1/s)(kg)(m/s) or kg·m/s^2, which are the units of the rate of change of momentum, which are equivalent to N. So the units check.

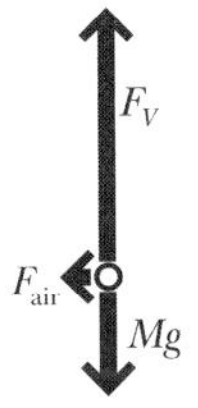

Is the result reasonable? The result shows that the vine has to exert an upward force that is greater than Tarzan's weight Mg. That makes sense. If Tarzan were hanging motionless (static equilibrium), the vine would only have to exert an upward force of Mg (net force would be zero). But to make Tarzan's momentum turn from horizontal to upward requires that the net force be upward.

As a numerical example, Tarzan's mass might be 100 kg (about 220 pounds), and the vine might be 10 m long. Suppose his speed at the bottom of the motion is 15 m/s. His weight Mg is 980 N, but the vine has to provide a much larger force:

$$F_V = \frac{(100\ \text{kg})(15\ \text{m/s})^2}{10\ \text{m}} + 980\ \text{N}$$

$$F_V = (2250\ \text{N}) + (980\ \text{N}) = 3230\ \text{N}$$

If the maximum tension this vine can support without breaking is less than 3230 N, Tarzan is in trouble. Note that the vine has to support a force about three times as big as Tarzan's weight.

3.3.3 The Momentum Principle relates different things

The equation $d\vec{p}/dt = \vec{F}_{net}$ seems to imply that $d\vec{p}/dt$ and $\vec{F}_{net}$ are the same thing. But in fact,

$$\frac{d\vec{p}}{dt} \text{ and } \vec{F}_{net} \text{ are } \textit{very different things!}$$

In the analysis of Tarzan and the vine, note carefully that we knew the rate of change of the momentum, $d\vec{p}/dt$, because we knew what kind of curving path Tarzan was following. This result, $d\vec{p}/dt = \langle 0, (v/L)Mv, 0\rangle$, does not involve the strength of the vine nor g, the magnitude of the gravitational field; it just depends on the way Tarzan's motion is curving.

$$\frac{d\vec{p}}{dt} \text{ tells about the motion, no matter what the forces are}$$

On the other hand, we know that the net force is $\langle 0, F_V - Mg, 0\rangle$ (the vector sum of all the external forces). This expression for the net force doesn't contain anything about how Tarzan is moving.

$$\vec{F}_{net} \text{ tells about the forces, no matter what the motion is}$$

It is the derivative form of the Momentum Principle which links these two kinds of information together, telling us that the (vector) rate of change of the momentum is numerically equal to the (vector) net force. We set the two vectors equal to each other and solve for the quantity of interest.

The result $|(d\vec{p}/dt)_\perp| = (|\vec{v}|/R)|\vec{p}|$ is essentially a geometrical result, which we proved by using a geometrical argument. It is important to understand that it doesn't involve anything at all concerning the particular force that is causing the change of direction. For example, the gravitational force law constant G does not appear in this result, nor is there anything about electric charges or electric forces in the result.

Yet the Momentum Principle says that $d\vec{p}/dt$ is equal to $\vec{F}_{net}$. These two quantities are very different in kind. They even have different units: $(\text{kg} \cdot \text{m/s})/\text{s}$ and newtons (although the Momentum Principle indicates that these two physical units must be equivalent in some sense).

The Momentum Principle, $d\vec{p}/dt = \vec{F}_{net}$, is powerful *precisely because it relates two very different things to each other.* If you know something about the net force, you know how rapidly the momentum must be changing. Or if you know how rapidly the momentum is changing, you know what the net force must be. The Momentum Principle relates an *effect* (rate of change of momentum) to its *cause* (net force due to interactions with other objects).

This is very different from a mere definition such as the definition of momentum, $\vec{p} \equiv m\vec{v}/\sqrt{1-(|\vec{v}|/c)^2}$. We introduce the symbol $\vec{p}$ and let it stand for the quantity $m\vec{v}/\sqrt{1-(|\vec{v}|/c)^2}$ as a convenience, but there is no *physics* in this, because there is no difference between what we mean by $\vec{p}$ and what we mean by $m\vec{v}/\sqrt{1-(|\vec{v}|/c)^2}$.

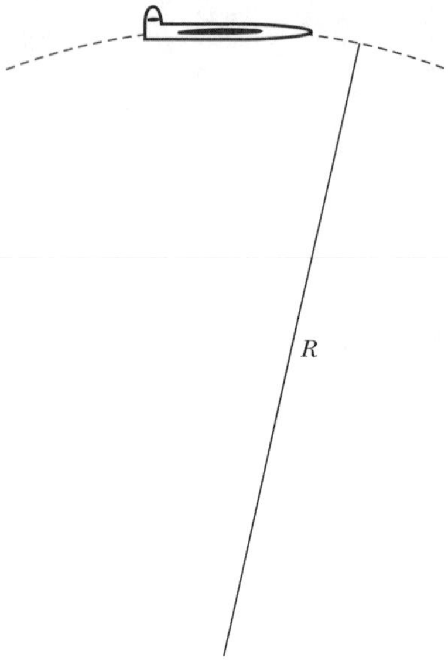

Figure 3.20 An airplane goes over the top of its path with speed v. The kissing circle has radius R at the top of the path.

3.3.4 Example: Sitting in an airplane (curving motion)

An airplane makes a curving motion at constant speed v as shown in Figure 3.20, where the radius of the kissing circle at the top of the path is R. Choose a passenger as the system of interest and determine the force that the seat exerts on the passenger's bottom.

1. Choose a system

   System: passenger in airplane

2. List external objects that interact with the system, with diagram

   the seat and the Earth
   (represent the system by a circle)

3. Apply the Momentum Principle

$$d\vec{p}/dt = \vec{F}_{net}$$

$$\langle 0, -\left(\frac{|\vec{v}|}{R}\right)|\vec{p}|, 0\rangle = \langle 0, F_S - Mg, 0\rangle$$

The rate of change of the momentum $d\vec{p}/dt$ as usual points toward the center of the kissing circle, so the y component of $d\vec{p}/dt$ is negative.

4. No need to apply the position update formula

5. Solve for the unknowns

$$-\left(\frac{|\vec{v}|}{R}\right)|\vec{p}| = F_S - Mg \text{ so } F_S = Mg - \left(\frac{v}{R}\right)Mv$$

6. Check

   The units check; see previous example on these units.

Let's explore the physical significance of this result. There is an interesting and important minus sign in our result. The upward force of the seat F_S on the passenger's bottom is *less* than the passenger's weight Mg.

? That means that the net force points downward. Does that make sense?

Yes, because the net force should point toward the center of the kissing circle, which is below the airplane.

? But since $F_S = Mg - Mv^2/R$, if the airplane's speed v is big enough, the force exerted by the seat could be zero. Does that make any sense?

Yes, that's the case where the airplane changes direction so quickly that the seat is essentially jerked out from under the passenger, and there is no longer contact between the seat and the passenger. No contact, no force of the seat on the passenger.

In this case, what does the situation look like to the passenger? From the passenger's point of view it is natural to think that the passenger has been thrown upward, away from the seat. But actually the seat has been yanked out from under the passenger, in which case the passenger is falling toward the Earth (being no longer supported by the seat), but changing velocity less rapidly than the airplane and so the ceiling comes closer to the passenger.

In the extreme case, if the airplane's maneuver is very fast, it may seem as though the passenger is "thrown up against the cabin ceiling." In actual fact, however, it is the cabin ceiling that is yanked down and hits the passenger!

You can see the value of wearing a seat belt, to prevent losing contact with the seat. When the airplane yanks the seat downward, the seat belt yanks the passenger downward. This may be uncomfortable, but it is a lot better than having the cabin ceiling hit you hard in the head.

? How would this analysis change if the airplane goes through the bottom of a curve rather than the top (Figure 3.21)?

Now the center of the kissing circle is above the airplane, so $d\vec{p}/dt$ points upward. That means that the net force must also point upward.

? Which is larger, the force of the seat on the passenger, or the gravitational force of the Earth on the passenger?

The force of the seat is greater than Mg, since the net force is upward. The passenger sinks deeper into the seat. Or to put it another way, the airplane yanks the seat up harder against the passenger's bottom, squeezing the seat against the passenger.

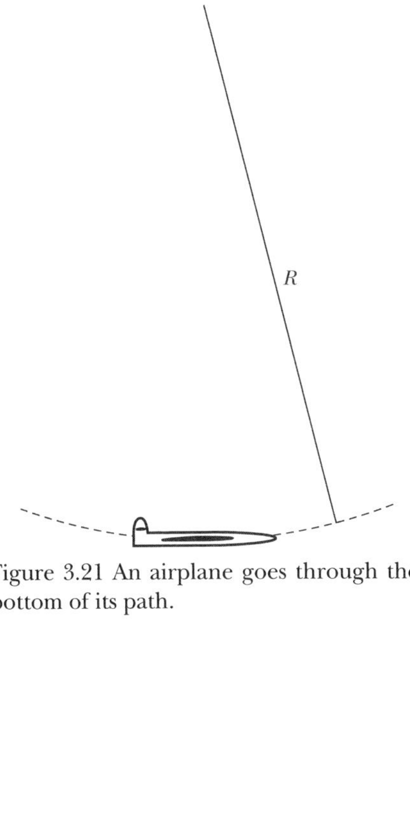

Figure 3.21 An airplane goes through the bottom of its path.

The feeling of weight, and of weightlessness

As you sit reading this text, you feel a sensation that you associate with weight. If someone suddenly yanks the chair out from under you, you feel "weightless," which feels funny and odd. It also feels scary, because you know from experience that whenever you lose the support of objects under you (chair, airplane seat, floor, mountain path), something bad is about to happen.

In reality, what we perceive as "weight" is actually not the gravitational force at all but rather the forces of atoms in the chair (or airplane seat or floor or mountain path) on atoms in your skin. You have nerve endings that sense the compression of interatomic bonds in your skin, and you interpret this as evidence for gravity acting on you. But if you were placed in a spaceship that was accelerating, going faster and faster, its floor would squeeze against you in a way that would fool your nerve endings and brain into thinking you were subjected to gravity, even if there are no stars or planets nearby.

If you lose contact with the seat in an airplane that is going rapidly over the top of a curving path, your nerve endings no longer feel any contact forces, and you feel "weightless." Yet this is a moment when the only force acting on you is in fact the Earth's gravitational force. Weightlessness near the Earth paradoxically is associated with being subject only to your weight *Mg*.

Nor is it just nerve endings in your skin that give you the illusion of "weight." As you sit here reading, your internal organs press upward on other organs above them, making the net force zero. You feel these internal contact forces. If you are suddenly "weightless" a main reason for feeling funny is that the forces one organ exerts on another inside your body are suddenly gone.

To train astronauts, NASA has a cargo plane that is deliberately flown in a curving motion over the top, so that people in the padded cargo bay lose contact with the floor and seem to float freely (in actual fact, they are accelerating toward the Earth due to the *Mg* forces acting on them, but the plane is also deliberately accelerating toward the Earth rather than flying level, so the people don't touch the walls and appear to be floating). All the nerve endings are crying out for the usual comforting signals and not getting them. This airplane is sometimes called the "vomit comet" in honor of the effects it can have on trainees.

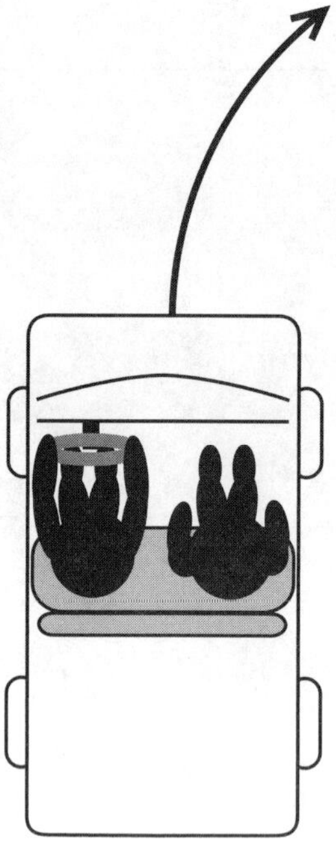

Figure 3.22 You are a passenger in a car that makes a sudden right turn.

3.3.5 Example: A turning car (curving motion)

Suppose you're riding as the passenger in a convertible with left hand drive (American or continental European), so you are sitting on the right (Figure 3.22). Assume you have foolishly failed to fasten your seat belt. The driver makes a sudden right turn. Before studying physics you would probably have said that you are "thrown to the left," and in fact you do end up closer to the driver.

> **?** There's something wrong with this "thrown to the left" idea. What object in the surroundings exerted a force to the left to make you move to the left?

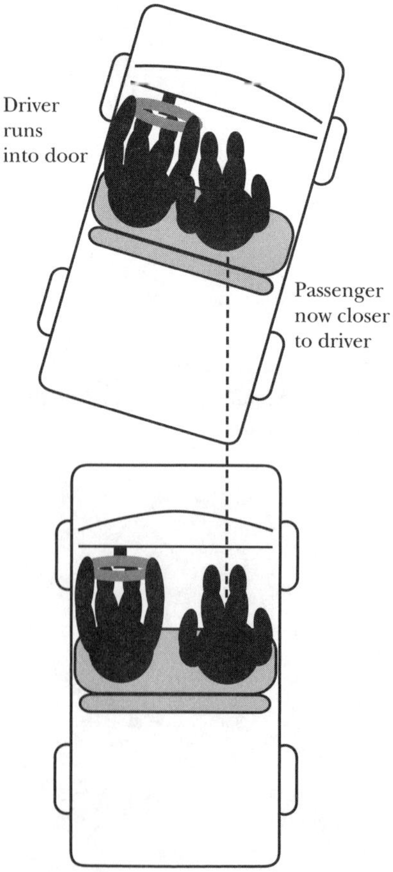

Figure 3.23 The passenger moves straight ahead and is now closer to the driver, who was thrown to the right. The driver was thrown to right by the door pushing the driver to the right.

There is no such object! To understand what's happening, watch the turning car from a fixed vantage point above the convertible, and observe carefully what happens to the passenger. In the absence of forces to change the passenger's direction of motion, the passenger keeps moving ahead, in a straight line at the same speed as before (Figure 3.23). It is the car that is yanked to the right, out from under the passenger. The driver moves closer to the passenger; it isn't that the passenger moves toward the driver.

The driver may also feel "thrown to the left" against the door. But it is the door that runs into the driver, forcing the driver to the *right*, not to the left.

Sometimes people invent fictitious forces such as the so-called "centrifugal" force to account for the passenger and driver being "thrown to the left." But this just adds a second confusion to the first. The passenger and driver are thrown to the right in a right turn, not to the left, and there are real forces to the right that make them go to the right (for example, the force of the door on the driver's left shoulder).

Moreover, forces are associated with interactions between objects in a chosen system and objects in the surroundings. The "centrifugal" force is just made up and is not associated with any real object.

3.4 A special case: Circular motion at constant speed

Motion in a circle at constant speed is a common kind of motion. This is approximately the situation for the planets in our Solar System.

Although the orbits are actually ellipses, except for Pluto these elliptical orbits are very close to circular. In the case of the Earth, the ellipse is so close to a circle that on a drawing the size of this page you couldn't tell the difference. The orbit of our Moon around the Earth is also nearly circular at constant speed. Also, each day you are carried around in a circle at constant speed around the Earth's axis by the rotation of the Earth.

A Ferris wheel or a merry-go-round moves its passengers in a circle at constant speed. Some physical properties of the hydrogen atom can be understood by a simple model of an electron going around in a circle around the proton at constant speed (though this model is too simple to explain many properties of hydrogen atoms, and quantum mechanics is needed for a full analysis).

For a planet of mass m moving around a star of mass M in a circular orbit at constant speed, we can write

$$\left|\frac{d\vec{p}}{dt}\right| = |\vec{F}_{net}|$$

$$\frac{|\vec{v}|}{R}|\vec{p}| = |\vec{F}_{grav}| = G\frac{Mm}{R^2}$$

Planets move at very high speeds compared to baseballs, but at speeds that are very small compared to the speed of light. So approximately, $\vec{p} \approx m\vec{v}$, and we can write

$$\frac{|\vec{v}|}{R}m|\vec{v}| \approx G\frac{Mm}{R^2}$$

Canceling the common factors m and R, we have the following result:

$$|\vec{v}|^2 \approx G\frac{M}{R} \quad \text{(special case; see following paragraph)}$$

This is such a special-case formula that it is not a good idea to memorize it. If you do, you are likely to misuse it, using it in situations where it does not apply. It applies only to

- circular motion at constant speed
- with a gravitational force
- between a massive nearly motionless star and
- a comparatively low-mass planet.

You should be able to derive this result from the Momentum Principle and the gravitational force law, as done above. Only then can you say that you really understand the result. Note how short the derivation is!

The gravitational force is unusual because the force is proportional to the mass m of the object, so when the gravitational force appears in the Momentum Principle, the mass m cancels. This is not the case when the force is an electric force, which depends on electric charge, not on mass.

In Einstein's general theory of relativity, which describes gravity in a very different way than Newton did, the star's mass alters the space around the star, and all small masses move in that space in the same way, no matter what their (small) mass. We see an echo of this Einstein view in the cancellation of m in the result above.

3.4.1 The initial conditions required for a circular orbit

What can we do with our result, $|\vec{v}|^2 \approx GM/R$ for a circular orbit at constant speed? We know the mass M of the Sun, and we know the gravitational constant G, so for a circular orbit whose radius of curvature is R we can calculate the speed that a planet must have in such a circular orbit. Conversely, for a particular speed, we can calculate the radius that will give a circular orbit.

? What if the planet starts out in a direction perpendicular to the line between Sun and planet, but the planet has less than the calculated speed for a circular orbit?

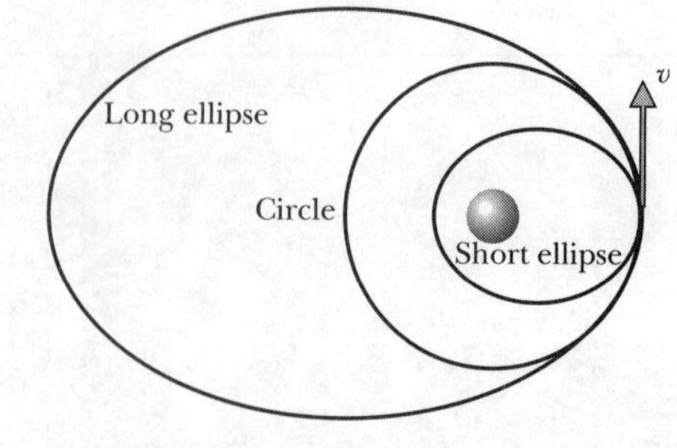

Figure 3.24 The initial speed v has to be just right to initiate a circular orbit. If the speed is more or less than this value, you get an ellipse (or for very high speeds, a parabola or hyperbola).

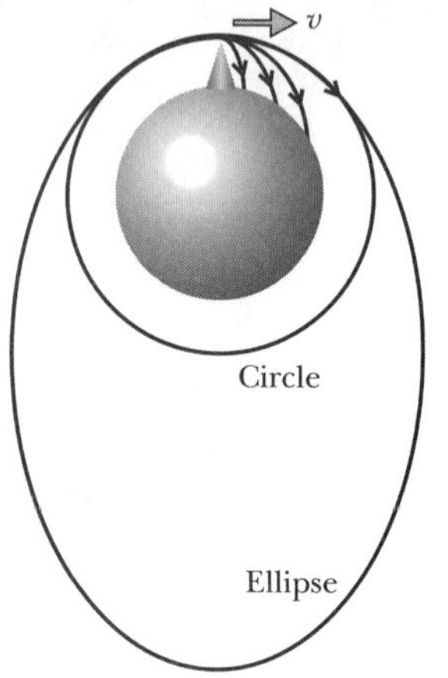

Figure 3.25 Newton throws a rock horizontally from a high mountain. Only one particular speed makes the rock move in a circular orbit.

In that case the planet will not go in a circle; its orbit will be a short ellipse (Figure 3.24). If the planet has more than this calculated speed, the orbit will be a long ellipse (or a parabola or a hyperbola), not a circle. The circular orbit is very special and requires just the right combination of radius of curvature and speed.

For a given radius, only one value of the speed gives a circular orbit. Likewise, if you know the speed of an object in a circular orbit, you can figure out what the radius of its orbit has to be.

Newton himself published a diagram that shows very clearly how the initial speed determines the type of orbit. He imagined standing on a very high mountain on the Earth, so high as to be above the atmosphere, and throwing a rock horizontally at various speeds. In Figure 3.25 you see that if Newton throws the rock with a low speed the rock simply falls toward the Earth and hits the ground near the base of the mountain. If he throws a bit faster, the rock goes further as it falls, hitting the ground a bit farther from the base of the mountain. With a somewhat higher initial speed the rock goes a significant distance around the Earth, falling all the time, before hitting the ground. But if Newton throws the rock just right, the rock falls, and falls, and falls, and falls in a circle, and comes around and hits him in the back of the head!

This is the magic, special speed for circular motion, and there is only one speed that will produce a circle. It is a speed which you can calculate (as Newton did) from what you now know, and it is the speed of those satellites that are in circular orbits just above our atmosphere. The speed is thousands of meters per second, so this is a "thought experiment" as far as an unaided human thrower is concerned.

If Newton throws the rock even faster than the magic speed for circular motion, the rock moves in a long elliptical orbit. And for even higher speeds (still small compared to the speed of light), the orbit is a parabola or hyperbola, and the rock never comes back.

Example: The Earth's elliptical orbit can be well approximated for many purposes by a circular orbit of radius 1.5×10^{11} m. What would we predict for the orbital speed of the Earth? Start from the Momentum Principle as applied to circular motion. Solar system data are given in the inside back cover of the textbook.

Solution: $\left|\frac{d\vec{p}}{dt}\right| = |\vec{F}_{net}|$ (Momentum Principle)

$$\left(\frac{|\vec{v}|}{R}\right)|\vec{p}| = G\frac{Mm}{R^2} \quad \text{(circular motion; gravitational force)}$$

$$\left(\frac{|\vec{v}|}{R}\right)m|\vec{v}| \approx G\frac{Mm}{R^2} \quad (v << c)$$

$$\frac{|\vec{v}|^2}{R} = G\frac{M}{R^2} \quad \text{(mass } m \text{ of the Earth cancels)}$$

$$|\vec{v}|^2 = G\frac{M}{R} \quad \text{(solve for speed squared)}$$

$$|\vec{v}| = \sqrt{G\frac{M}{R}} = \sqrt{(6.7\times10^{-11}\ \text{N}\cdot\text{m}^2/\text{kg}^2)\frac{2\times10^{30}\ \text{kg}}{1.5\times10^{11}\ \text{m}}} = 3\times10^4\ \text{m/s}$$

Ex. 3.13 What is the speed of a satellite in a circular orbit near the Earth? You can use the radius of the Earth as the approximate radius of the orbit, because the atmosphere is only about 50 km thick, whereas the radius of the Earth is 6.4×10^6 m. The mass of the Earth is 6×10^{24} kg. Follow the analysis in the example above, but think about the meaning of M and R for this situation.

3.4.2 Nongravitational situations

Our analysis is not limited to gravitational forces. Any time an object moves at constant speed in a circular orbit when subjected to a net force of constant magnitude, directed radially inward, we have from the Momentum Principle

$$\left|\frac{d\vec{p}}{dt}\right| = \left(\frac{|\vec{v}|}{R}\right)|\vec{p}| = |\vec{F}_{net}| \quad \text{(circular motion at constant speed)}$$

For example, if you swing a rock at the end of a string in a horizontal circular orbit at constant speed, the net force is the horizontal component of the tension in the string.

For this reason, rather than memorizing the very special result $|\vec{v}|^2/R \approx GM/R^2$ for the case of circular motion due to a gravitational force, it is better to memorize the more general result for the time rate of change of the perpendicular component of momentum, $|(d\vec{p}/dt)_\perp| = (|\vec{v}|/R)|\vec{p}|$, and then equate this perpendicular component of the rate of change of momentum to whatever the perpendicular component of the net force happens to be.

Ex. 3.14 In outer space, a ball of mass 5 kg at the end of a spring moves in a circle of radius 2 m at a constant speed of 3 m/s. What is the magnitude of the force exerted by the spring on the rock? In words, in what direction does the force act?

3.4.3 Period

Often we know or would like to calculate the time it takes a planet to go around its star once, called the "period" T. For example, the period of the Earth's orbit around the Sun is one calendar year.

? If a particle goes all the way around a circle of radius R in a time T at constant speed, what is the speed?

The distance around once is the circumference $2\pi R$, so the speed is this distance divided by the time it takes:

$$|\vec{v}| = \frac{2\pi R}{T}$$

This makes it possible to predict a relationship between the period T of a planet and its distance R from the Sun. Just put this value for the speed into the result we obtained above.

This relationship was first discovered empirically by Johannes Kepler, who lived before Newton. Kepler produced his formula by fitting mathematical curves to accurate observations of planetary orbits made by Tycho Brahe (without telescopes, which hadn't been invented yet). The later explanation of Kepler's equation in terms of the Momentum Principle and the gravitational force law was a triumph of the Newtonian approach.

Ex. 3.15 On the previous page we predicted that the speed of the Earth around the Sun should be 3×10^4 m/s. Do you get the same result for the speed from the simple geometrical formula $|\vec{v}| = (2\pi R)/T$? (Hint: You need the period in seconds. How many seconds in a year?) Compare with our prediction from the Momentum Principle.

3.4.4 Example: Circular pendulum (curving motion)

As another, rather complex example of an analysis using the derivative form of the Momentum Principle, we will analyze a ball hanging at the end of a string, but with the ball in motion. There are many possible motions of this system. The ball could swing back and forth in a plane; this is the motion of a "simple pendulum." The ball could go around in a path something like an ellipse, in which case the ball's height would vary, just as with simple pendulum motion. The ball could move so violently that the string goes slack for part of the time, with abrupt changes of momentum every time the string suddenly goes taut. If the string can stretch noticeably, there can be a sizable oscillation superimposed on the swinging.

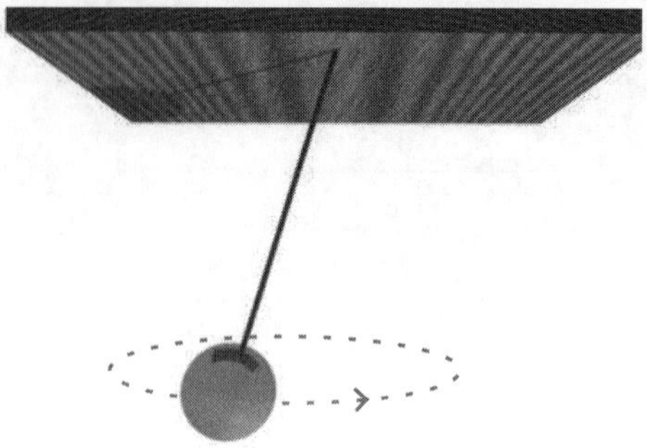

Figure 3.26 A circular pendulum.

All of these motions are difficult to analyze. In principle, we could use a computer to predict the general motion of the ball hanging from the string if we had a formula for the tension force in the string, as a function of its length. Essentially, we would need the effective spring stiffness for this stiff "spring." However, the string may be so stiff that even a tiny stretch implies a huge tension (a nearly "inextensible" string). The observable length of the string is nearly constant, but the tension force that the string applies to the ball can vary a great deal. This makes it very difficult to do a computer numerical integration, because a tiny error in position makes a huge change in the force.

There is however one particular motion of the ball that is simple enough to model and to analyze without a lot of mathematics—circular motion with a fixed stretch of the string. This is a "circular pendulum" (Figure 3.26). As you discovered when studying planetary orbits, circular motion doesn't just happen. To get the ball moving in a circle you have to start it moving with just the right initial conditions. We tackle a problem that we can solve, by imposing the requirement that we only consider the case of circular motion.

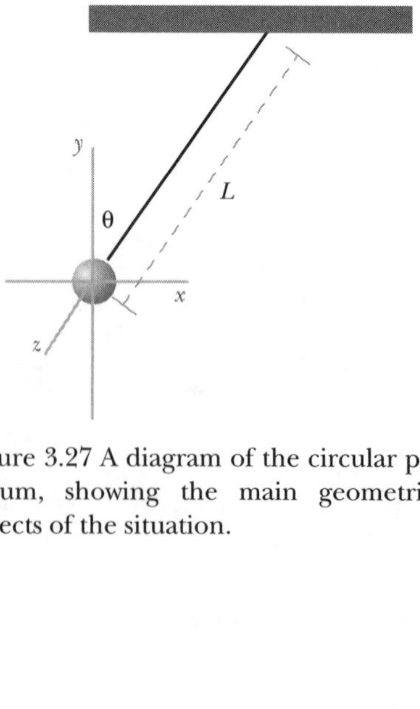

Figure 3.27 A diagram of the circular pendulum, showing the main geometrical aspects of the situation.

A large circular pendulum was used by Newton to measure *g*, the magnitude of the gravitational field, with considerable accuracy. You could observe and measure the motion of a circular pendulum yourself, and your analysis of the motion would let you deduce a value for *g*, which you could compare with the accepted value.

A physics diagram (Figure 3.27) shows a snapshot of a side view of the circular pendulum, at an instant when the string (of length L to the center of the ball) lies in the *xy* plane (at an angle θ to the vertical), and the ball is at the leftmost point in its circular path of radius $r = L\sin\theta$. Let's assume that the ball is moving counter-clockwise as seen from above. Here is the analysis:

1. Choose a system

   System: the ball

2. List external objects that interact with the system, with diagram

   the string, the Earth, and the air (but neglect air resistance) (see Figure 3.28)

3. Apply the Momentum Principle

$$d\vec{p}/dt = \vec{F}_{net}$$

$$\left\langle \frac{|\vec{v}|}{r}|\vec{p}|, 0, 0 \right\rangle = \langle F_T \sin\theta, F_T\cos\theta - mg, 0\rangle$$

4. No need to apply the position update formula

5. Solve for the unknowns

$$\frac{|\vec{v}|}{r}|\vec{p}| = F_T\sin\theta \text{ so } F_T = \frac{mv^2}{r\sin\theta} \text{ where } r = L\sin\theta$$

$$0 = F_T\cos\theta - mg \text{ so } F_T = \frac{mg}{\cos\theta}$$

   From the two expressions for the tension in the string,

$$F_T = \frac{mv^2}{L(\sin\theta)^2} = \frac{mg}{\cos\theta} \text{ so } g = \frac{v^2\cos\theta}{L(\sin\theta)^2}$$

   Measuring the speed v, length L and angle θ determines g.

   If the time to go around once is T, a distance $2\pi r = 2\pi L\sin\theta$, the speed is $v = (2\pi L\sin\theta)/T$, so you can measure L, θ, and T to get g.

6. Check

   The units can be shown to be correct. Also, note that if $\theta = 0$ the string hangs vertically and there is no motion, so the tension in the string ought to be mg, which is consistent with $F_T = \frac{mg}{\cos 0} = mg$.

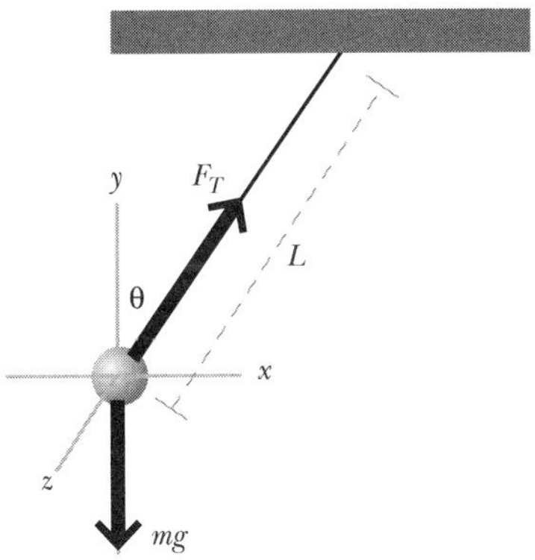

Figure 3.28 We show known and unknown forces to create a useful force diagram.

No "centrifugal force"

We point out that in this analysis there is no role for a radially outward "centrifugal force" (Figure 3.29). There is no interaction associated with such a fictitious force. Moreover, if the net force were zero, Newton's first law says that the ball would have to move in a straight line!

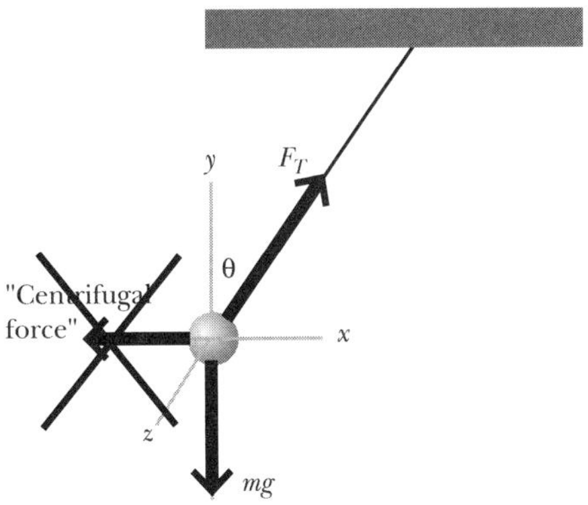

Figure 3.29 There is no "centrifugal force." There is no interaction associated with such a fictitious force. Moreover, if the net force were zero, the ball would have to move in a straight line!

3.4.5 Comment: Forward reasoning

A common technique for solving problems is to start by considering what you want to know (or what you have been asked to determine), and to work backward from that to the solution. We have approached the analysis of physical systems in the opposite way. We have begun by organizing our knowledge about the system according to a fundamental principle—the Momentum Principle—and systematically recording our knowledge in equations and diagrams. After we have recorded all the information we have, whatever is missing is what we need to figure out. At this point it is usually quite clear how to figure out this missing information.

This way of working problems, sometimes called "forward reasoning," is typical of the way expert physicists solve problems of this kind—starting each new problem from the fundamental principles. This approach works! For the student, however, approaching problems in this way may require a bit of faith—it is not necessarily obvious how the desired answer will emerge from the analysis. If you practice approaching problems in this way, you will develop confidence in the approach and will gain important experience in explaining complex phenomena starting from fundamental principles.

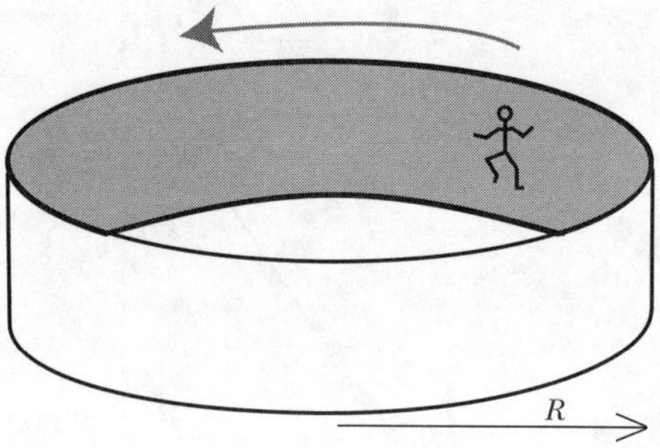

Figure 3.30 An amusement park ride.

3.4.6 Example: An amusement park ride (curving motion)

There is an amusement park ride that some people love and others hate in which a bunch of people stand against the wall of a cylindrical room of radius R, and the room starts to rotate at higher and higher speed (Figure 3.30). When a certain critical speed is reached, the floor drops away, leaving the people stuck against the whirling wall.

Explain why the people stick to the wall without falling down. Include a carefully labeled force diagram of a person (Figure 3.31), and discuss how the person's momentum changes, and why.

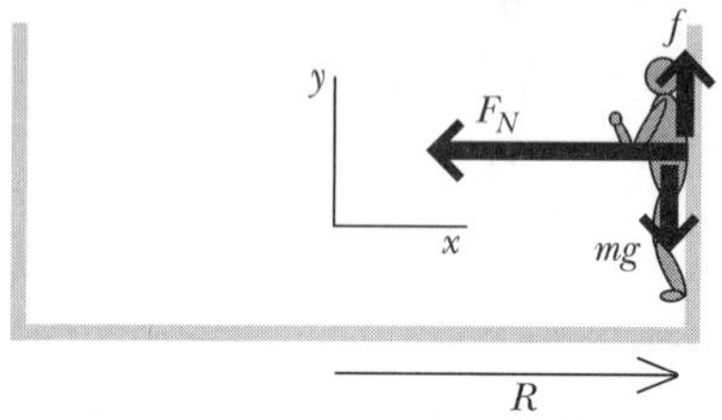

Figure 3.31 Physics diagram of the person. At this instant the person is moving in the $-z$ direction.

1. Choose a system

   System: person

2. List external objects that interact with the system, with diagram

   the wall, the Earth, and the air (neglect air resistance)
   (represent the system by a circle)

3. Apply the Momentum Principle

$$d\vec{p}/dt = \vec{F}_{net}$$

$$\langle -\left(\frac{|\vec{v}|}{R}\right)|\vec{p}|, 0, 0\rangle = \langle -F_N, f - mg, 0\rangle$$

4. No need to apply the position update formula

5. Solve for the unknowns

$$-\left(\frac{|\vec{v}|}{R}\right)|\vec{p}| = -F_N \text{ so } F_N = \left(\frac{v}{R}\right)mv$$

$$0 = f - mg \text{ so } f = mg$$

6. Check

   The units check. Also, the faster the ride, the bigger the inward force of the wall, which sounds right.

The wall exerts an unknown force which must have a y component $+f$ (because the person isn't falling) and an x component $-F_N$ normal to the wall (because the person's momentum is changing direction). Note that there is no fictitious outward-going "centrifugal" force! There is a momentum change inward; if the net force were zero, the person would move in a straight line.

The vertical component f of the wall force is a frictional force. If the wall has friction that is too low, the person won't be supported. In Chapter 5 we will see that $f \le \mu F_N$, where the "coefficient of friction" μ has a value for many materials between about 0.1 and 1.0. Basically, the more strongly two objects are squeezed together, the harder it is to slide one along the other.

The speed of the ride has to be large enough that $f \le \mu F_N$, so we have

$$mg \le \mu\left(\frac{mv^2}{R}\right)$$

$$\mu \ge \frac{gR}{mv^2}$$

This predicts that the lower the speed v, the greater the coefficient of friction must be to hold the person up on the wall, which sounds right. Conversely, if the ride spins very fast, the wall doesn't have to have a very large coefficient of friction to hold the people against the wall.

3.5 A model of a solid: Balls connected by springs

The situations we have analyzed in the preceding sections of this chapter have involved the interaction of solid objects, such as a person and a wall, an airplane seat and a passenger, or a vine and Tarzan. As a result of our analysis, we have found that the magnitude of the force exerted by a wall on a person, or by a string on a weight, can vary, depending on the motion of the person or the weight.

? You may have wondered how an inanimate object like a wall or a seat "knows" how big a force to exert on an object touching it.

To answer this question, we need to understand what is happening inside the wall or the string, at the microscopic level. Up until now we have emphasized the great power of the Momentum Principle, which allows us to predict and understand motion. However, the Momentum Principle alone (or even combined with other powerful fundamental principles, the Energy Principle, and the Angular Momentum Principle), is not sufficient to let us understand some important aspects of the real world. We also need some simple models of real matter to help us explain many kinds of phenomena.

The great 20th-century American physicist Richard Feynman said, "If, in some cataclysm, all of scientific knowledge were to be destroyed, and only one sentence passed on to the next generations of creatures, what statement would contain the most information in the fewest words? I believe it is the *atomic hypothesis* (or the atomic *fact*, or whatever you wish to call it) that *all things are made of atoms—little particles that move around in perpetual motion, attracting each other when they are a little distance apart, but repelling upon being squeezed into one another.* In that one sentence, you will see, there is an *enormous* amount of information about the world, if just a little imagination and thinking are applied." (*The Feynman Lectures on Physics* by R. P. Feynman, R. B. Leighton, & M. Sands, 1965; Palo Alto: Addison-Wesley.)

In the quote given above, Feynman has summarized the basic properties of atoms and of interatomic forces. The main properties of atoms that will be important to us in this course are these:

- All matter consists of atoms, whose typical radius is about 10^{-10} meter.
- Atoms attract each other when they are close to each other but not too close.
- Atoms repel each other when they get too close to each other.
- Atoms move around in perpetual motion.

These properties were established during a century of intense study of atoms by physicists and chemists, using a wide variety of experimental techniques. You can see for yourself some of the evidence for these properties. For example, consider a solid block of aluminum, which we have good reason to believe is made up of a large number of tiny aluminum atoms.

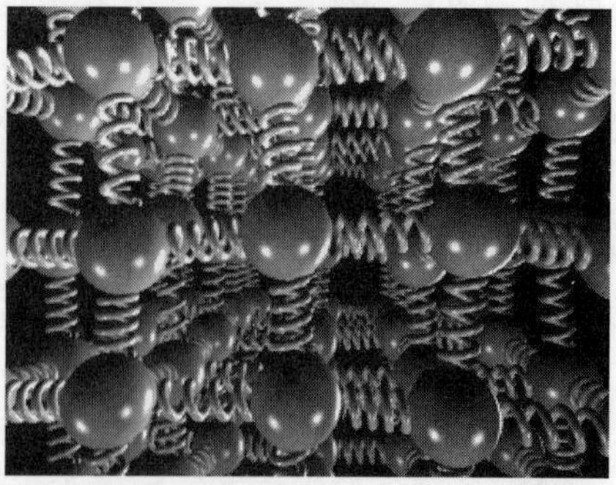

Figure 3.32 A simple model of a solid: tiny balls in constant motion, connected by springs.

? What properties of this block of aluminum, observable without special equipment, support the claims that "atoms attract each other when they are not too close to each other" and "atoms repel each other when they get too close to each other"?

Since the block of metal does not spontaneously fall apart or evaporate, the atoms in the block must be attracting each other. It is very difficult to compress the block, so the atoms must resist an attempt to push them closer together. Evidently the atoms in a block of aluminum are normally at just the right distance from each other, not too close and not too far away. This just-right distance is called the "equilibrium" distance between atoms.

Given these properties of solids, a very useful model for a block of aluminum is an array of tiny balls in constant motion, connected by springs (Figure 3.32). If two balls (models of atoms) are pulled apart and then released, the spring connecting the balls will pull them back toward their original equilibrium positions. But if the balls are moved closer together, the spring will push them back toward the equilibrium positions. In general, the balls will continually oscillate short distances away from and back toward their equilibrium positions.

In a solid at room temperature, the atoms are in motion, continually oscillating around their equilibrium positions. If we heat the solid, these atomic oscillations become more vigorous. One of the most important outcomes of research on the atomic nature of matter was the realization that for many materials, the temperature of an object, as measured by an ordinary thermometer, is just an indicator of the average energy of the atoms; the higher the temperature, the more vigorous the atomic motion. During this course we will have much more to say about the relationship between temperature and energy. Figure 3.32 is a "snapshot" of the atoms in a model solid at a particular instant.

One can think of a ball in Figure 3.32 as representing the atomic nucleus plus the inner electrons. Although almost all of the mass of an atom is concentrated in the nucleus, the nucleus is only about $1/100000$ as large as the electron cloud surrounding it, so nuclei would not be visible on the scale of this picture. If the balls represented the entire electron cloud of the atoms, including the outer electrons, the balls would of course touch each other. A spring in the picture can be thought of as representing the interatomic forces associated with the outer electrons, which are involved in the bonds between atoms. Compare the ball-and-spring model with the actual scanning tunneling microscope (STM) images we first showed in Chapter 1, where the electron clouds of the atoms on the surface of solid silicon touch each other (Figure 3.33).

Figure 3.33 STM images of atoms on two different surfaces of crystalline silicon.

Solids in which the atoms are arranged in regular "lattices" like those shown here are called crystals, including metals, quartz, diamond, and table salt (NaCl), but not most organic solids such as wood.

Limitations of the ball-and-spring model

The simple ball-and-spring model of a block of aluminum is not perfect. Some properties of a block of aluminum, such as its thermal conductivity, electrical conductivity, and color can be fully explained only with quantum mechanics. This branch of physics is called "condensed matter physics" (as opposed to the study of individual atoms or subatomic particles). Nevertheless, many properties can be explained fully or at least approximately in terms of the simple ball-and-spring model. Even professional condensed matter physicists often use this simplified model in their day-to-day work, both to get an approximate idea of what will happen in a particular situation, and to check approximately the results of full quantum-mechanical calculations.

A solid vs. a single atom

A block of aluminum has very different properties than a single atom of aluminum. A block of aluminum looks metallic, melts at high temperature, can be bent or dented, is a good conductor of electric current, etc. A single atom of aluminum has none of these properties! When atoms bind to each other to form a solid, macroscopic object, that object has special properties arising from the interactions of the individual atoms. Aggregates of molecules also have different properties from those of individual molecules; a single H_2O molecule is not "wet."

Diameter of an atom in a solid

The diameter of an atom in a solid is one of the important properties of matter which plays a role in interactions. We can calculate atomic diameters for crystals by making a "macro-micro" connection; that is, by relating a macroscopic property of a large block of crystalline material to the same property at the microscopic level. If we calculate the density of the material in both a macro and micro way, we can determine the diameter of an atom in a solid.

The density of a macroscopic object is its mass divided by its volume. For example, one cubic centimeter of water has a mass of one gram, so its macro density is 1 gram/cm^3. For comparison, the macro density of aluminum is 2.7 grams/cm^3 and the macro density of lead is 11.4 grams/cm^3. That is, one cubic centimeter of aluminum has a mass of 2.7 grams, and one cubic centimeter of lead has a mass of 11.4 grams. Density can also be expressed in SI units. For example, here is the density of water in SI units:

$$\left(1\frac{\text{gram}}{\text{cm}^3}\right)\left(\frac{(100\text{ cm})^3}{1\text{ m}^3}\right)\left(\frac{1\text{ kg}}{1000\text{ grams}}\right) = 1000\text{ kg/m}^3$$

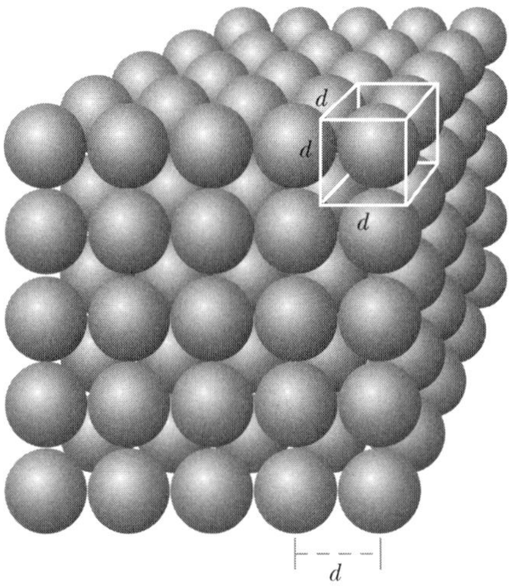

Figure 3.34 A simple cubic arrangement of atoms. The volume of space associated with each atom is a tiny cube *d* by *d* by *d*.

Consider a simple "cubic" arrangement of atoms as shown in Figure 3.34 (this is called a cubic lattice). The volume of space taken up by each atom is a tiny cube *d* on a side, where *d* is the diameter of an atom in the solid and is also the distance from the center of one atom to the center of a neighboring atom. The micro density is the mass of one atom divided by the volume of the tiny cube associated with each atom. This micro density is the same as the macro density, mass per volume:

$$\text{macro density} = \text{micro density} = \frac{\text{mass of 1 atom}}{\text{volume of 1 atom}} = \frac{M_{1\text{ atom}}}{d^3} \quad \text{(cubic lattice)}$$

Therefore if you know the macro density (for example, 2.7 grams/cm^3 for aluminum), and you can calculate the mass of one atom, you can determine the diameter of an atom in the material. The mass of one atom can be determined using the mass of one mole and the knowledge that one mole contains 6×10^{23} atoms (Avogadro's number):

$$M_{1\text{ atom}} = \frac{\text{mass of one mole}}{6\times10^{23}\text{ atoms/mole}}$$

It might be tempting to put these concepts together into a formula of the form "$d = ...$", but try to avoid the temptation. It is much more memorable and much safer to think through these physical relationships each time you make the calculation, to avoid serious mistakes. The key point to understand is that the macro density is equal to the micro density, and the micro density is equal to the mass of one atom divided by the volume d^3 of the tiny cube-formed space associated with each atom. You probably remember from chemistry how to calculate the mass of one atom, as the mass of one mole divided by Avogadro's number.

In the following exercises, you can do the calculations in terms of grams and centimeters, then convert the final result from centimeters to meters. Or you can use SI units from the beginning, in which case you need to be careful to convert macro densities from grams/cm^3 to kg/m^3 and molar masses from grams to kilograms.

Few solid elements actually have cubic lattices, because a cubic lattice is rather unstable. An example of more common and more stable arrangement is "body-centered cubic" in which there are 8 atoms at the corners of a cube with another atom at the center of the cube. However, it is adequate for our purposes to assume that crystalline solids have cubic lattices.

Example: One mole of copper (6×10^{23} atoms) has a mass of 64 grams (see the periodic table on the inside front cover). The density of copper is 8.94 grams/cm^3. What is the approximate diameter of a copper atom in solid copper?

Solution: $$\text{(macro) } 8.94\frac{\text{grams}}{\text{cm}^3} = \frac{\left(\dfrac{64 \text{ grams/mole}}{6\times10^{23} \text{ atoms/mole}}\right)}{d^3} \text{ (micro)}$$

By taking a cube root, you can solve for $d = 2.3\times10^{-8}$ cm, which in SI units is $d = 2.3\times10^{-10}$ m. Alternatively, you could convert density and mass to SI units and obtain d in meters directly.

Ex. 3.16 The density of aluminum is 2.7 grams/cm^3. What is the approximate diameter of an aluminum atom in solid aluminum?

Ex. 3.17 The density of lead is 11.4 grams/cm^3. What is the approximate diameter of a lead atom in solid lead? You may be surprised by the comparisons among these very different atoms, copper and aluminum and lead.

Figure 3.35 A thin wire is stretched by a heavy ball.

3.6 Static properties of a solid: The interatomic "spring"

We begin our study of a solid by looking at its "static" properties, when the solid is not moving. We will use our simple ball-and-spring model of a solid to analyze how much the interatomic "springs" inside a thin wire are stretched when we hang a heavy ball from the end of the wire (Figure 3.35). The upper end of the wire is fastened to a support whose base is firmly embedded in the ground.

We need a quantitative description of springs in general and a quantitative description of the gravitational forces acting on the ball and the wire. Then we will be able to determine the strength of the interatomic "spring."

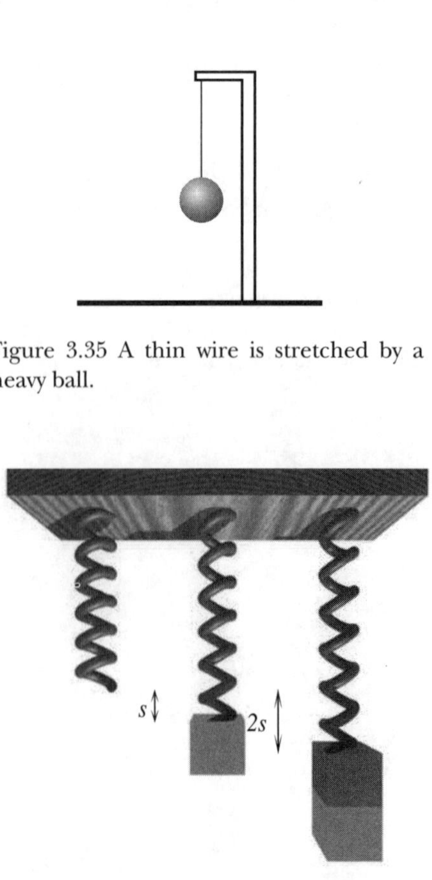

Figure 3.36 Double the force acting on a spring, double the stretch.

3.6.1 Force exerted by a spring

We need a quantitative formulation of the force exerted by a spring on an object touching it. As we discussed in Chapter 2, the magnitude of the force required to stretch or compress a spring is observed to be proportional to the magnitude of the stretch or compression s. For example, if we double the force acting on the spring, we get double the stretch or compression s (Figure 3.36 and Figure 3.37). We write this:

$$|\vec{F}_{\text{spring}}| = k_s|s|$$

$|s|$ is the absolute value of the stretch; formally:

$$s = \Delta L = L - L_0$$

L_0 is the length of the relaxed spring
L is the length of the spring when stretched or compressed

k_s is the "spring stiffness"

This is an approximate relationship, because it is true only over a limited range. If you stretch a spring so much that it turns into a straight wire, it suddenly becomes extremely difficult to stretch it any further. If you compress a spring so much that its coils come into contact, it suddenly becomes extremely difficult to compress the spring any further.

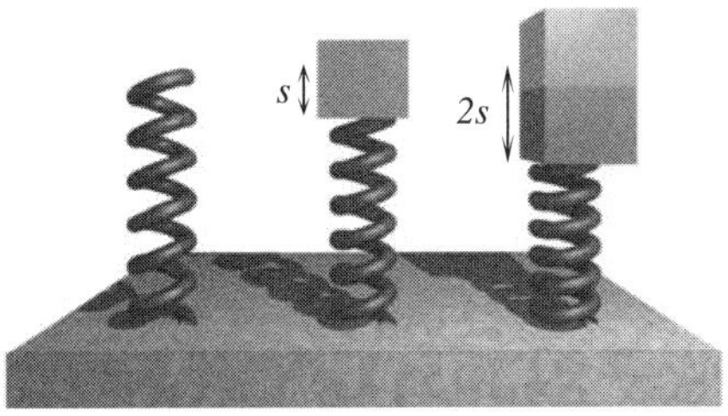

Figure 3.37 Double the force acting on a spring, double the compression.

Ex. 3.18 A spring has a length of 8 cm when relaxed, and a length of 10 cm when stretched by a force of 60 N. What is the stiffness k_s of this spring?

Ex. 3.19 This same spring is then compressed by an unknown force so that the length of the spring becomes 7 cm. What is the magnitude of this force?

Ex. 3.20 A different spring of length L has stiffness k_s. It is cut into two springs, each of length $L/2$. What is the stiffness of one of these shorter springs? (Note that when you stretch the full spring, each half of the spring stretches half the total amount of stretch.)

3.6.2 A microscopic view: Interactions within the wire

For a moment consider a wire that supports only its own weight; nothing is hanging from it. For simplicity, Figure 3.38 shows a segment near the bottom of a wire that is only one atom wide. The atom at the very bottom of the wire experiences a gravitational force mg due to its interaction with the Earth. Since its momentum is not changing, the net force on the atom must be zero ($d\vec{p}/dt = \vec{F}_{net}$), so the "spring" between the atoms must exert an upward force $k_s s_1$. The second atom feels a gravitational force mg, and also a downward force mg due to the first spring (since the force exerted by a spring is proportional to its stretch, the magnitude of the force it exerts on both atoms connected to it must be the same).

Since the second atom's momentum is not changing, the next spring must exert an upward force of $k_s(2s_1) = 2mg$, so it must be stretched twice as much as the bottom "spring," the next spring must be stretched three times as much, and so on. Clearly the interatomic bonds at the top of the wire are stretched more than the bonds at the bottom of the wire, so the tension in the wire is not actually uniform. You can see a similar effect if you hold up a long spring such as a Slinky. The coils at the top of the spring are stretched considerably more than the coils at the bottom.

In Figure 3.38 the amount of stretch of the interatomic springs is greatly exaggerated. In a thin wire supporting only its own weight, even the interatomic bond at the top is stretched very little, because the mass of all the atoms in the wire itself is small.

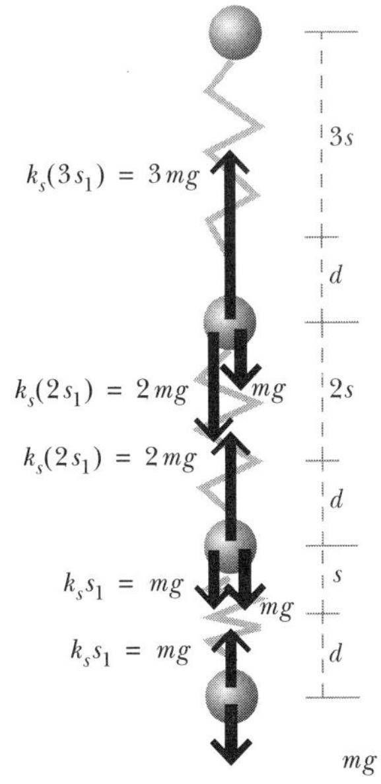

Figure 3.38 The bottom of a hanging wire, with nothing hanging from the wire. d is the equilibrium interatomic distance (for a wire lying on a table). The stretch of the interatomic bonds is greatly exaggerated in this figure.

Wire supporting a ball

When a large ball is hung from the wire, it will have a very large effect on the stretch of the interatomic springs (Figure 3.39). Now the interatomic bond at the bottom of the wire must support the large ball. The mass of the

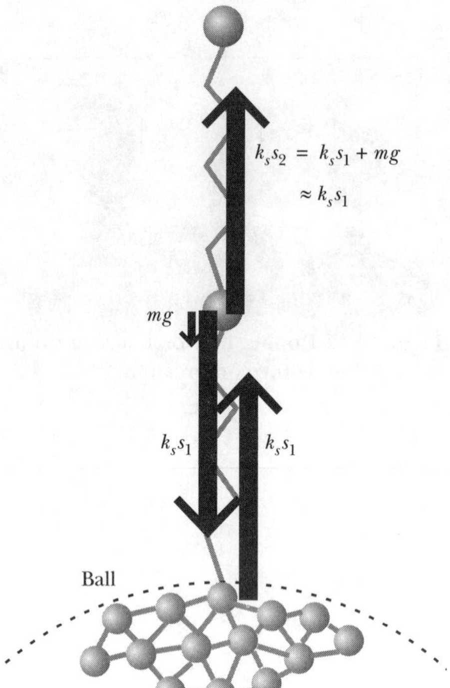

Figure 3.39 Forces on atoms in the wire with a heavy ball hanging from the wire.

ball is much greater than the mass of one of the atoms, so the interatomic "spring" stretch will be much greater than it was without the ball. As before, the "springs" at the top of the wire will be stretched slightly more than the "springs" at the bottom of the wire, due to the accumulated weight of the atoms in the wire. However, if the mass of the wire is very small compared to the mass of the ball, this small increase in stretch will be very small compared to the stretch caused by the ball, and all the stretches will be similar.

Note that for a wire whose mass is very small, it is a good approximation to neglect the effect of the mass of the wire and attribute the interatomic stretch entirely to the ball. However if the wire's mass is comparable to that of the ball, this is a poor approximation.

You may have seen the phrase "a massless wire" or a "massless spring" in textbook physics problems. Of course there is no such thing as a wire or a spring with no mass. This phrase refers to a frequently used model or "idealization" of a thin wire, a string, or a rope. An "ideal" string has no mass at all and does not stretch at all when under tension. A "massless, inextensible string" is a useful model of a real string or wire in situations where the mass of the string is very small and the effective spring stiffness of the interatomic bonds in the string is very large (that is, the "springs" are very stiff), so the string lengthens very little.

3.6.3 A macroscopic view: Tension

Instead of taking an atomic viewpoint, suppose we consider the wire as a macroscopic object with mass m_w, and we mostly ignore its internal (atomic) structure. The Earth exerts a downward gravitational force on the wire, $m_w g$, which is the sum of the gravitational forces on the individual atoms in the wire (Figure 3.40). Although the wire interacts gravitationally with other objects such as the support, the ball, and the Moon, we will neglect these forces because they are very small.

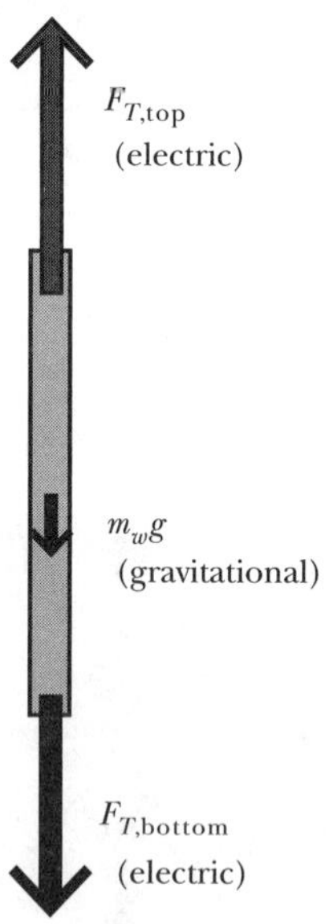

Figure 3.40 Tension force in the wire.

The hanging ball exerts a downward force on the bottom of the wire, of magnitude $F_{T,\text{bottom}}$. This is an interatomic electric force. The "T" is for "tension," which is what we call the force internal to the wire that is associated with the stretch of the interatomic "springs" in our simple ball-and-spring model of a solid.

The support exerts a force $F_{T,\text{top}}$ on the top end of the wire. This is an interatomic electric force.

The wire is at rest, so according to the Momentum Principle (Newton's second law of motion) the net force should be zero. In the y direction, we have

$$\frac{dp_y}{dt} = 0 = F_{T,\text{top}} - F_{T,\text{bottom}} - m_w g.$$

We can solve this equation to get $F_{T,\text{top}} = F_{T,\text{bottom}} + m_w g$. In other words, the "tension" in the top of the wire is equal to the total downward force acting on the wire, which is the (electric) force exerted by the ball on the bottom of the wire plus the gravitational force of the Earth on the wire. If the mass of the thin wire is very small compared to that of the hanging mass, we can use the "massless, inextensible wire" idealization, and say that the tension $F_{T,\text{top}}$ in the topmost portion of the wire is approximately equal to the tension $F_{T,\text{bottom}}$ in the bottom of the wire. We then do not have to worry about the different values for the tension along the length of the wire, and we can use the approximate value $F_T \approx F_{T,\text{top}} \approx F_{T,\text{bottom}}$ for the tension everywhere in the wire. Notice that "tension" is defined as the force F_T the string exerts on the ball, not $2F_T$.

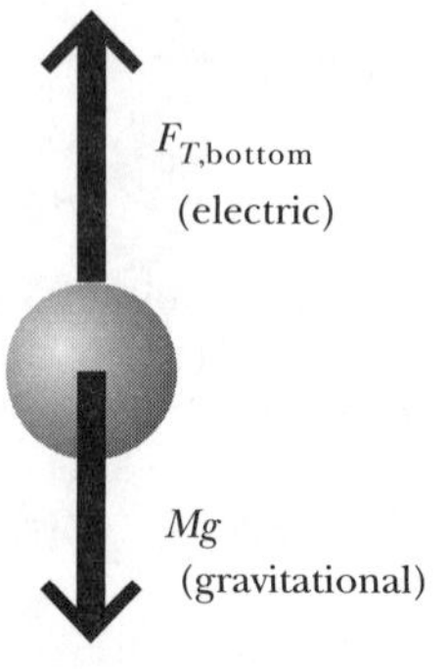

Figure 3.41 Forces on the hanging ball.

Now consider the equilibrium of the ball, of mass M (Figure 3.41). According to the principle of reciprocity (Newton's third law), the wire must exert an upward force with the same magnitude $F_{T,\,bottom}$ on the ball that the ball exerts on the wire. This force must equal the gravitational force Mg exerted by the Earth on the ball. If the ball has much more mass than the wire ($M >> m_w$), we have $F_T \approx F_{T,\,top} \approx F_{T,\,bottom} \approx Mg$; the tension F_T in the wire is approximately equal to the weight Mg of the ball, everywhere throughout the wire. (By "weight" we mean the gravitational force on an object.)

3.6.4 How much does the wire stretch?

In the real, non-ideal world the thin wire will stretch somewhat due to the tension F_T. We are now in a position to discuss quantitatively how much the wire will stretch, and then deduce the effective stiffness of the interatomic "springs" in our ball-and-spring model.

If we knew the spring stiffness k_s for the wire we could calculate the stretch s from the relation $F_T = k_s s$. Usually a different relationship is used to describe the stretching of straight elements such as a wire. If the wire has a length L, we call the amount of stretch of the wire ΔL (a small increment in the length, what we have called s in a spring), and the fractional stretch $\Delta L/L$ is called the "strain." As long as the strain isn't too big, the strain is proportional to the tension per unit area, F_T/A, where A is the cross-sectional area of the wire (Figure 3.42). The tension per unit area is called the "stress."

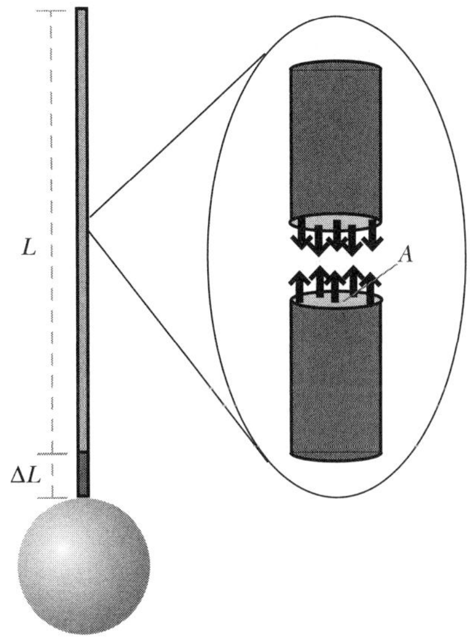

Figure 3.42 The heavy ball stretches the wire an amount ΔL. We are interested in the stretch of the interatomic "springs." A is the cross-sectional area—the area of a slice through the wire, perpendicular to its length.

$$\text{strain} = \frac{\Delta L}{L};\ \text{stress} = \frac{F_T}{A}$$

At the atomic level, the stress F_T/A can be related to the force that neighboring atoms exert on each other, and the strain $\Delta L/L$ can be related to the stretch of the interatomic bond. The ratio of stress to strain is a property of the material and differs for steel, aluminum, etc. A quantitative form of the relation between stress and strain in a wire is written as follows:

$$\frac{F_T}{A} = Y\frac{\Delta L}{L}$$

The parameter Y is called "Young's modulus" and is a property of the material. The stiffer the material, the larger is Young's modulus. Note the similarity to the spring formula, $F_T = k_s s$ $(= k_s\Delta L)$. At this point it is appropriate to measure Young's modulus experimentally (Problem 3.1).

Limit of applicability of Young's modulus

If you apply too large a stress, the wire "yields" (stretches a great deal) or breaks, and the proportionality of stress and strain is no longer even approximately true. Figure 3.43 shows a graph of the strain $\Delta L/L$ that results as a function of the applied stress F_T/A for a particular aluminum alloy.

You see that for moderate stress the resulting strain is proportional to the applied stress: double the stress, double the strain. But once you reach the yield stress, a further slight increase in the stress leads to a very large increase in the length of the metal. This can be quite dramatic: as you add more weights to the end of a wire the wire gets slightly longer, then suddenly it starts to lengthen a lot, very rapidly, and then the wire breaks. (In materials reference manuals stress is usually plotted against strain, so that the slope of the line is equal to Young's modulus. We plot strain vs. stress to emphasize that stress is the *cause* and strain the *effect*.)

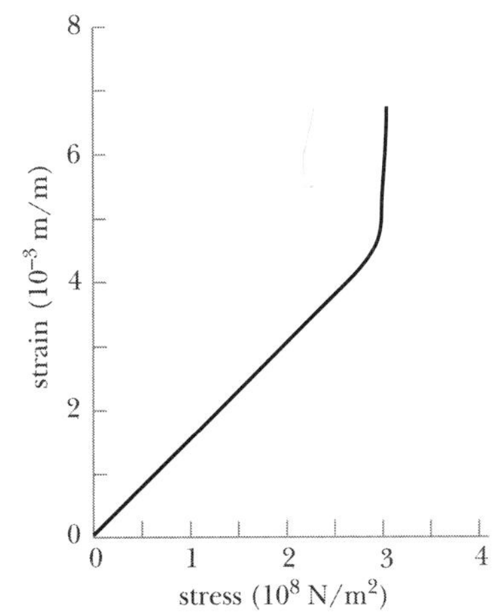

Figure 3.43 Strain vs. stress for a particular aluminum alloy.

Ex. 3.21 In Figure 3.43, what is the approximate value of Young's modulus for this aluminum alloy?

Ex. 3.22 Suppose we hang a heavy ball with a mass of 10 kg (about 22 pounds) from a steel wire 3 m long that is 3 mm in diameter. Steel is very stiff, and Young's modulus for steel is unusually large, $2\times10^{11}\ \mathrm{N/m^2}$. Calculate the stretch ΔL of the steel wire. This calculation shows why in many cases it is a very good approximation to pretend that the wire doesn't stretch at all ("ideal non-extensible wire").

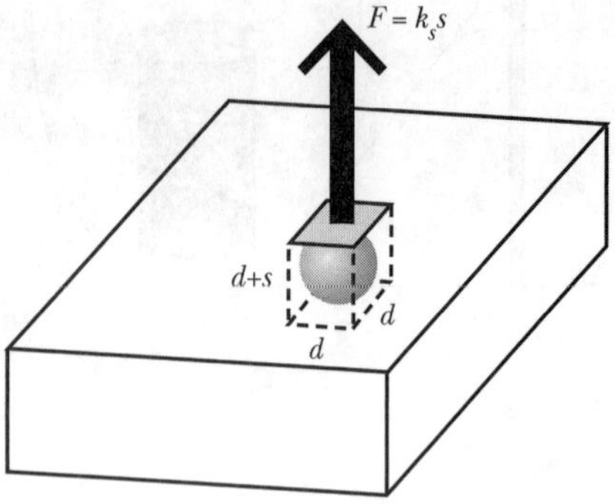

Figure 3.44 One atom occupies a portion d^2 of the total cross-sectional area, and the force acting on that portion is $k_s s$.

3.6.5 Effective interatomic spring stiffness

From macroscopic measurements of Young's modulus we can calculate an approximate spring stiffness k_s for the interatomic force by again making a "macro-micro" connection.

Consider a slice through a wire that is being stretched, with a force uniformly distributed across the cross-sectional area. Let d represent the diameter of one atom, so that each atom occupies a portion d^2 of the total cross-sectional area (Figure 3.44).

? What is the magnitude of the stress on one atom in terms of the interatomic force $k_s s$?

The atomic stress (force per unit area) is $k_s s/d^2$.

? In terms of the stretch s of the interatomic "spring," what is the strain for one atom (change of height of the atom, divided by its normal height)?

The strain for one atom is s/d, because its height was originally d, and the change in its height is s. Therefore we can express Young's modulus in terms of atomic quantities:

$$\text{macro } Y = \text{micro } Y = \frac{(k_s s/d^2)}{(s/d)} = \frac{k_s}{d}$$

If you know Young's modulus for some metal, you can calculate the effective stiffness of the interatomic bond, modeled as a spring: $k_s = Yd$, where d is the atomic diameter in the metal. This is another example of an important theme in this course, that of relating macroscopic properties to microscopic (atomic-level) properties.

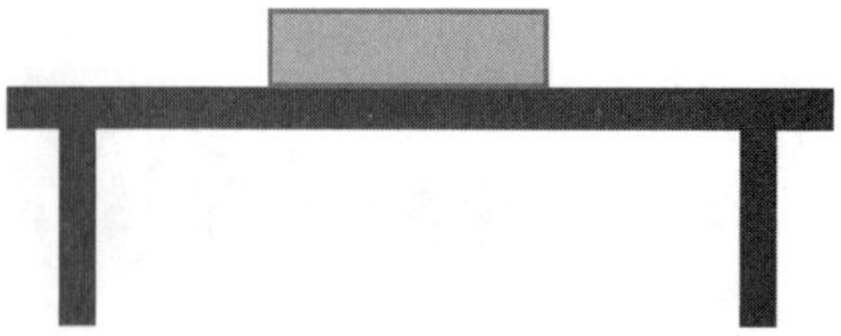

Figure 3.45 A brick lies on a table.

3.7 Brick lying on a table: Compression

A closely related situation is that of a lead brick lying at rest on a table (Figure 3.45), except that the brick is compressed while the wire was stretched.

Take the lead brick as the system of interest (Figure 3.46). The Earth pulls down on the brick of mass M, and the magnitude of the gravitational force is Mg. The table pushes up on the brick with a force whose magnitude we'll call F_N, a "normal" force—that is, a force that is normal (perpendicular) to the table. What kind of a force is this, exactly?

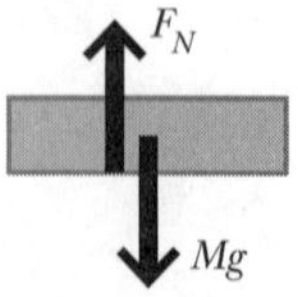

Figure 3.46 Forces acting on the brick.

If the lead brick were extremely heavy, we might not be surprised to observe the table sagging under its weight. Even if the table does not sag visibly, the molecules in the top layer of the table are pushed down by molecules in the bottom layer of the brick. Both objects deform somewhat (their interatomic "springs" are compressed).

Atoms in the table and the brick are squeezed closer than their normal equilibrium positions, like compressing a spring. Figure 3.47 shows a tiny contact region. The bottom layers of atoms in the brick are compressed the most. Higher layers support less of the total weight of the brick. Young's modulus relates stress and strain for compression as well as tension.

It would be appropriate to call the upward force of the table F_C for "compression force," but it is common practice to label this kind of force as a "normal" force. The word is used in the mathematics sense, meaning "perpendicular to." The table exerts a "normal" force—that is, perpendicular to its surface.

Note: Neither "normal" force F_N nor "tension" force F_T
is a different kind of fundamental force!
These are merely names for interatomic electric interactions.

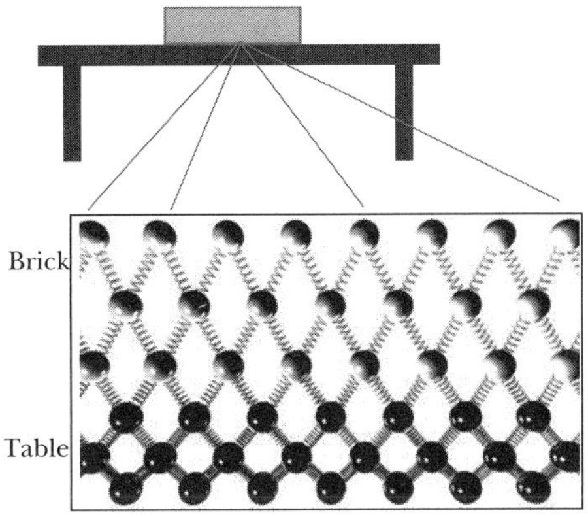

Figure 3.47 Interatomic "springs" in a tiny contact region are compressed by the weight of atoms above.

Since the brick is at rest, the vertical component of the net force must be zero:

$$\frac{dp_y}{dt} = 0 = +F_N - Mg$$

We conclude that the normal force F_N exerted by the table is equal to the weight Mg of the brick, an unsurprising result. This neglects the effects of the air on the brick, which we will discuss later in this chapter.

3.8 Dynamic properties of a solid: A spring-mass system

We have modeled a solid as a collection of atoms connected to their neighbors by spring-like interatomic forces, and we have studied some "static" aspects of these forces. Next we will study the "dynamics" of masses and springs.

We have commented that the atoms in a solid are in constant motion around their equilibrium positions, and that there is more such motion at higher temperatures. In the following sections we'll study the motion of a mass on a spring. This will help us visualize the behavior of atoms in a solid, and this model will be important later when we attempt to understand the thermal properties of solids. Many of the concepts associated with the motion of a mass on a spring have remarkably broad application to other kinds of phenomena, including phenomena as disparate as the behavior of certain kinds of electric circuits, and the vibrational motion of a diatomic molecule such as oxygen, O_2.

At this point it is useful to observe a spring-mass system. See Problem 3.2, in which you study experimentally a vertical spring-mass system like the one shown in Figure 3.48.

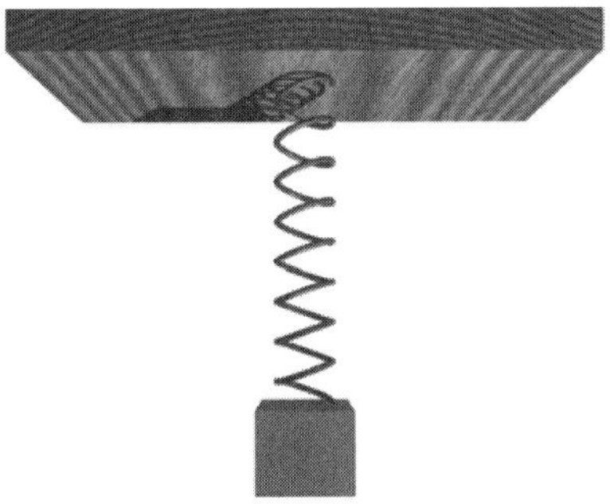

Figure 3.48 A vertical spring-mass system.

3.8.1 Applying the Momentum Principle to a spring-mass system

Consider a block of mass m connected to a spring whose stiffness is k_s (that is, the force exerted by the spring is $k_s s$ when the spring is stretched or compressed an amount s). The block slides with almost no friction on an air table, where it is supported on a cushion of air. The other end of the spring is attached to the wall (Figure 3.49). The horizontal motion of this horizontal spring-mass system is very similar to but easier to analyze than the motion of the vertical spring-mass system that is the subject of the experimental study in Problem 3.2. It is easier to do a theoretical study of the horizontal system because the Earth's gravitational force plays no important role. It is easier to do experiments with a the vertical system because there is very little friction.

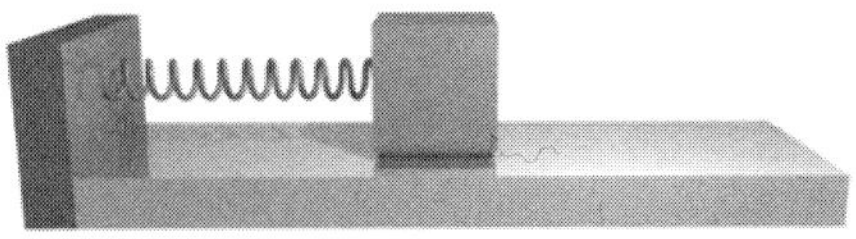

Figure 3.49 A block connected by a spring to a wall slides back and forth with little friction.

In order to write the equations corresponding to the Momentum Principle we need to select an origin with respect to which we will measure the position of the block. We can simplify our calculations by choosing to measure the position of the block with respect to an origin chosen to be at the equilibrium position, the location of the block when the spring is relaxed, so that the stretch s of the spring is equal to the position x of the block.

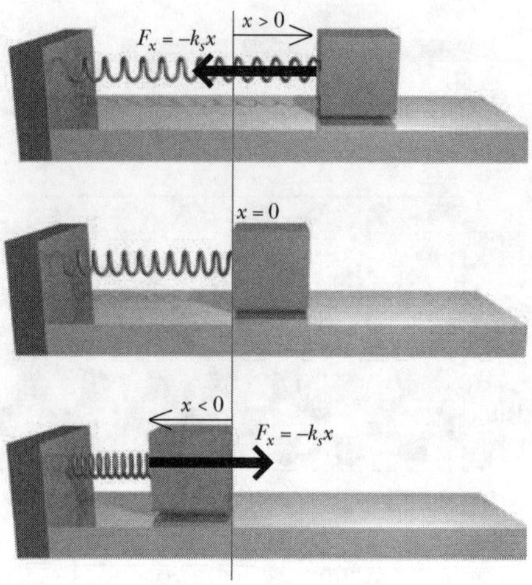

Figure 3.50 The x component of the force that the spring exerts on the block is equal to $-k_s x$.

In the middle of Figure 3.50, the spring is relaxed, at the equilibrium position, with zero stretch ($x = L - L_{\text{relaxed}} = 0$). Therefore the force that the spring exerts on the block is zero. At the top of Figure 3.50, the spring has been stretched, and the position of the block is $x > 0$. The length of the spring is more than its equilibrium length by an amount x, and the quantity $k_s x$ is positive. The x component of the force that the spring exerts on the block, F_x, is given by $F_x = -k_s x$, which is a negative quantity and correctly predicts an x component of force on the block to the left, as we would expect.

At the bottom of Figure 3.50, the spring is compressed, and the position of the block is $x < 0$. The length of the spring is less than its equilibrium length by an amount x, and the quantity $k_s x$ is negative. The x component of the force that the spring exerts on the ball, F_x, is given by $F_x = -k_s x$, which is a positive quantity and correctly predicts an x component of force on the block to the right, as we would expect.

Therefore the equation $F_x = -k_s x$ describes the x component of the force exerted on the block by the spring no matter whether the spring is stretched or compressed. The direction of the force is opposite to the direction of the displacement from the equilibrium position.

? What would be the formula for the spring force in terms of the position of the block if the origin were located at the wall, and the relaxed length of the spring is L?

In that case we would write $F_x = -k_s(x - L)$, because what matters in this formula is the stretch (change in length of the spring), which is $(x - L)$. A check on this is the fact that when the block is at location $x = L$, the stretch $(x - L)$ is zero.

1. Choose a system

System: block

2. List external objects that interact with the system, with diagram

the spring, the air table, the Earth, and the surrounding air (see force diagrams in Figure 3.50; the table pushes up with F_N)

3. Apply the Momentum Principle

We'll neglect sliding friction and air resistance in the diagram and analysis.

$$d\vec{p}/dt = \vec{F}_{\text{net}}$$

$$\langle \frac{dp_x}{dt}, 0, 0 \rangle = \langle -k_s x, F_N - mg, 0 \rangle$$

$$\frac{dp_x}{dt} = -k_s x \text{ and } F_N = mg$$

3.8.2 Predicting the motion of a spring-mass system

In order to predict the horizontal motion, we can apply the same method to the block that we applied to computing the motion of the planet in Chapter 2. At each step in the numerical integration we can evaluate the known force law $-k_s x$ in terms of the current position x. (Later we will also discuss a way to obtain an analytical solution.)

Suppose that at some instant in time you know the position of the block and its momentum exactly. We call these data the "initial conditions" x_0 and p_0 at time $t = 0$. Using the Momentum Principle in a computer program as you did in Chapter 2, you can predict what the position x and momentum p of the block will be at some future time t. At this point it is appropriate to do Problem 3.3, which deals with the details of a numerical integration of a block connected to a spring.

Your calculations should show that the block oscillates back and forth along the x axis, producing a graph of the position of the block *vs.* time that should look something like Figure 3.51.

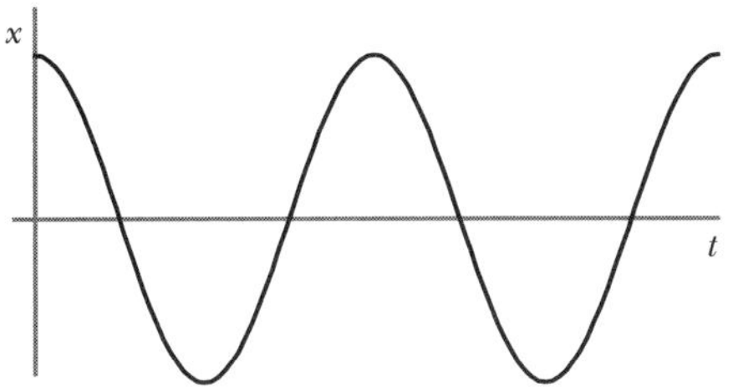

Figure 3.51 A graph of the position of the block as a function of time.

Oscillating back and forth is familiar behavior for a spring. The graph looks suspiciously like a cosine function, which is somewhat intriguing, since no trigonometry is used in the numerical integration in Problem 3.3. Later we will show that the function is indeed a cosine, but for now the most important point is not what "standard" math function this graph may represent, but rather that it is possible to predict the future one step at a time. This illustrates the great power of Newtonian analysis. If you know the net force on an object as a function of its position and velocity (in this case, $F_x = -k_s x$), you can predict the motion of the object far into the future by simple step-by-step calculations based on the Momentum Principle.

3.8.3 Harmonic oscillators

? In your experiments and in your program how does the period change when you start out with an initial stretch twice as big as before?

Doubling the initial stretch apparently just doubles the size of the oscillations without changing the period of the oscillations.

A system that oscillates sinusoidally is called a "harmonic oscillator," and a spring-mass system is an example of a harmonic oscillator. Other kinds of oscillators are sometimes called "anharmonic" oscillators. A simple example is a bouncing ball, whose motion is repetitive but not sinusoidal; its height as a function of time involves t^2, not a sine or cosine of the time.

A very special property of harmonic oscillators is that the period does not depend on the amplitude. If you double the amplitude, the round-trip time doesn't change despite having to go twice as far. The bigger force due to the larger stretch makes the mass go just enough faster to compensate for the longer distance. This makes the harmonic oscillator extremely important in measuring time, because even if the amplitude changes somewhat (for example, decreases slowly due to friction), the period stays the same. This is not true for other kinds of oscillators, as you can see in part (e) of Problem 3.3.

Here is another example of an anharmonic oscillator:

Ex. 3.23 A bouncing ball is an example of an anharmonic oscillator, and its "ticking" makes a poor clock. On each bounce, it bounces not quite as high as before. As a result, what happens to the period as this "clock" winds down?

3.8.4 A vertical spring-mass system

To reduce friction, it was easier to experiment with a mass oscillating vertically up and down on a spring (Figure 3.52) rather than sliding on an air table. We will show that we expect the same period whether the spring is vertical or horizontal.

We choose a coordinate system with the positive y axis pointing downward (Figure 3.52). We'll measure y downward from the location where the end of the relaxed spring would be (the relaxed length of the spring is L).

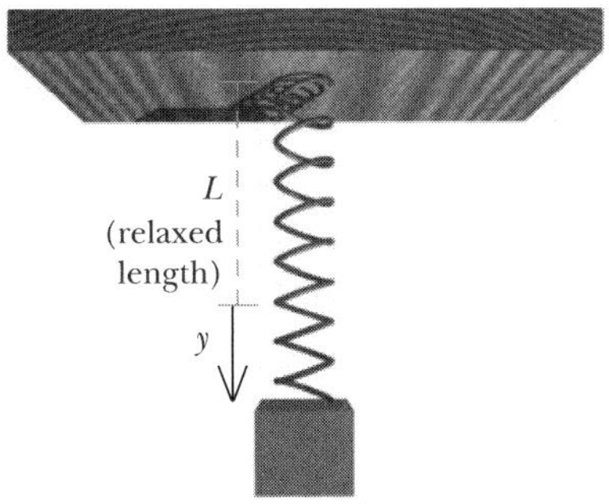

Figure 3.52 A vertical spring-mass system minimizes sliding friction.

Neglecting air resistance and gravitational interactions with objects other than the Earth, the physics diagram of the forces acting on the hanging block is that shown in Figure 3.53. We then have the following:

$$\frac{dp_y}{dt} = -k_s y + mg$$

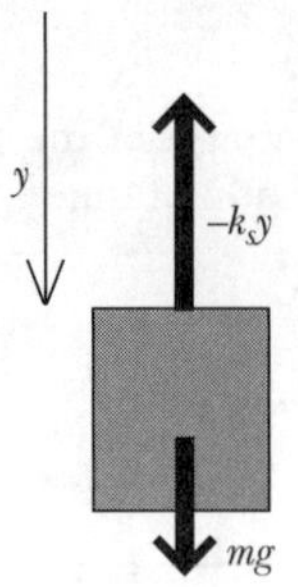

Figure 3.53 Physics diagram showing the forces acting on the hanging block.

We choose to analyze only simple vertical motion, with no bouncing side to side, so we know this:

$$p_x = 0 \text{ at all times, so } \frac{dp_x}{dt} = 0$$

We can reduce the p_y equation to a familiar form if we measure a new coordinate, w, from the equilibrium position, when the block is hanging motionless, rather than from the unstretched spring position (Figure 3.54).When the mass hangs motionless from the spring, the spring is stretched an amount $k_s s_0 = mg$, so $s_0 = mg/k_s$. We'll measure w downward from the end of the spring, so

$$y = (s_0 + w) = \frac{mg}{k_s} + w \text{, and therefore}$$

$$\frac{dp_y}{dt} = -k_s\left(\frac{mg}{k_s} + w\right) + mg = -k_s w$$

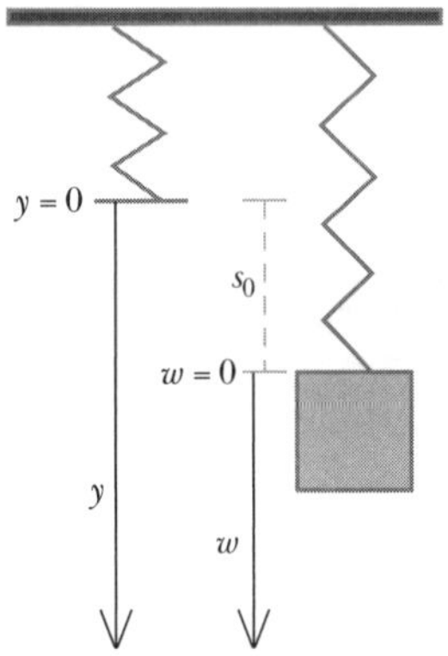

Figure 3.54 Measure w from the place where the block can hang motionless.

This says that if we measure from the end of the spring at its equilibrium location, mg/k_s below the bottom of the unstretched spring, the equation for the motion of the mass is the same as in a horizontal spring-mass system, whose motion we already know. The moving mass will oscillate up and down around this equilibrium position, as you could observe in your own experiments.

3.8.5 Discussion of modeling

At this point it is useful to have a discussion about the validity and accuracy of our calculational attempt to predict the future motion of an actual spring-mass system. Think about the following issues and be prepared to discuss them:

1) Does your actual spring-mass system behave *exactly* as predicted? How does the observed motion differ from the prediction?

2) A wide range of observations are consistent with the Momentum Principle (at least in the domain where we need not invoke quantum mechanics or general relativity), and we expect the Momentum Principle to be valid for predicting the motion of a spring-mass system. Therefore it is prudent to look at failings of our *model* of the spring-mass system. What assumptions or approximations did we make that aren't entirely appropriate for your own mass oscillating on your own spring?

3) What could be done to make your own spring-mass system correspond more closely to the simple model? What could be done to make your own spring-mass system behave even more differently from the simple model?

4) On the other hand, how could we make the *model* more realistic? That is, could we remove some of the simplifying approximations, and then apply our step-by-step analysis to this more realistic model?

5) Is it conceivable that you could model your spring-mass system *completely*, taking into account *all* interactions, no matter how small?

6) Can we know the initial conditions *exactly*? If we don't know them exactly, can we predict the future exactly?

3.8.6 An analytical prediction for a spring-mass system

Even if one uses a high-accuracy numerical integration scheme, and takes very short time steps (with correspondingly lots of calculations), there is some unavoidable error associated with finite time steps in a numerical integration. Presumably Nature uses *infinitesimal* time steps! In some cases we can use calculus to predict the future using infinitesimal time steps, and eliminate entirely the errors introduced by a numerical integration. Even so, our prediction of the future is still constrained by the inevitable failings of our physics model to encompass all of the messiness of the real world.

The spring-mass model is one of the (few!) systems for which we can use calculus to obtain an "analytical" prediction for our model, with the prediction in the form of a symbolic expression rather than a set of numbers. As mentioned already, the graph of our numerical results on page 153 looks suspiciously like a cosine function. In a moment we will show that a cosine function is indeed compatible with the Momentum Principle, and that an analytical solution to the motion is given by this:

AN ANALYTICAL SOLUTION FOR SPRING-MASS

$$x = A\cos(\omega t)\text{, with } \omega = \sqrt{\frac{k_s}{m}}$$

A (the "amplitude") is a constant that depends on the initial conditions.

Angular frequency, period, and frequency

The constant ω in $x = A\cos(\omega t)$ is called the "angular frequency" and is measured in radians per second (ω is lower-case Greek omega). When the argument ωt of the cosine increases by 2π radians, the motion repeats (one complete cycle), so if we call the "period" T (the time required for one complete cycle, Figure 3.55), we have $\omega T = 2\pi$.

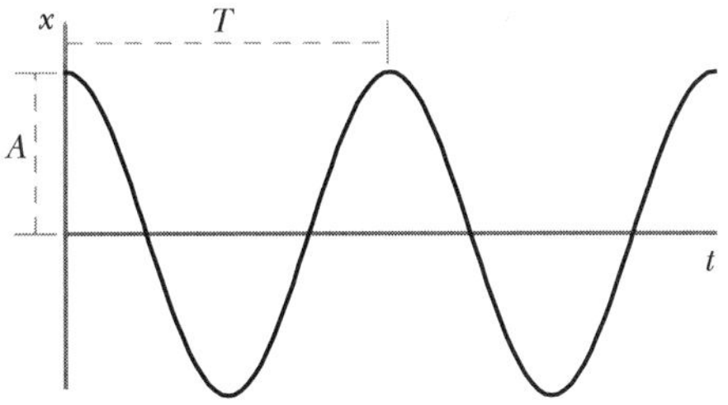

Figure 3.55 Amplitude A and period T.

One other quantity often used to describe oscillating systems is the "frequency" f, which is the number of complete cycles per second: $f =$ (1 cycle) per T seconds. Cycles per second are also called "hertz."

$$\text{Angular frequency } \omega = \frac{2\pi}{T} = \sqrt{\frac{k_s}{m}}\text{, radians per second}$$

$$\text{Period } T = \frac{2\pi}{\omega}\text{, seconds}$$

$$\text{Frequency } f = \frac{1}{T} = \frac{\omega}{2\pi}\text{, cycles per second or "hertz"}$$

In Chapter 2 you calculated the period of a planet orbiting a star. Both the oscillation of a mass on a spring and the orbital motion of a planet are called "periodic" motion. If you project the motion of a body in a circular orbit onto a line (for example, by shining a light edge-on to the orbital plane and observing the resulting shadow), this component of the circular motion will look like the motion of a harmonic oscillator.

Ex. 3.24 The constant A is called the "amplitude." What was the amplitude in your numerical integration? What are the minimum and maximum values of x during a cycle?

Ex. 3.25 Calculate the period T from the analytical solution and compare with the result obtained from your numerical integration.

3.8.7 Checking the analytical solution

In this section we verify that the cosine function given above does indeed satisfy the Momentum Principle for the spring-mass system. If you have not yet learned in calculus the derivatives of sines and cosines, or have not yet learned what is usually called the "chain rule," consult the appendix on basic calculus at the end of this volume before studying this section.

If we think we have an expression for $x(t)$, we can check it by differentiating it twice to see if it is consistent with our equation for the Momentum Principle, because nonrelativistically the time derivative of momentum that appears in the Momentum Principle is actually the mass times the second derivative of position with respect to time:

$$\frac{dp_x}{dt} \approx \frac{d(mv_x)}{dt} = m\frac{d}{dt}\left[\frac{dx}{dt}\right]$$

Based on our observations of a spring-mass system, we make the guess that

$$x = A\cos(\omega t),$$

where A and ω are initially unknown constants, and we check whether this function works when we plug it into the derivative form of the Momentum Principle, $dp_x/dt = F_{\text{net},x}$. Since we need the time derivative of the momentum p_x, we start by differentiating x to get v_x:

? $v_x = \dfrac{dx}{dt} = \dfrac{d[A\cos(\omega t)]}{dt} = ?$

Using the chain rule of differentiation, we have the following:

$$\frac{d[A\cos(\omega t)]}{dt} = -A\sin(\omega t)\frac{d(\omega t)}{dt} = -A\sin(\omega t)\omega$$

So $v_x = -A\sin(\omega t)\omega$. Next we take the derivative of v_x:

? $\dfrac{dv_x}{dt} = \dfrac{d[-A\sin(\omega t)\omega]}{dt} = ?$

Apply the chain rule a second time:

$$\frac{d[-A\sin(\omega t)\omega]}{dt} = -A\cos(\omega t)\omega\frac{d(\omega t)}{dt} = -A\cos(\omega t)\omega^2$$

Therefore $\dfrac{dv_x}{dt} = -A\cos(\omega t)\omega^2$. We know from the Momentum Principle that

$$\frac{dp_x}{dt} = m\frac{dv_x}{dt} = -k_s x$$

Therefore for *any* time t that you care to choose, the following equation must be correct for our guessed solution, $x = A\cos(\omega t)$, to be valid:

$$m[-A\cos(\omega t)\omega^2] = -k_s A\cos(\omega t)$$

We can indeed satisfy this equation for any and all time t if (and only if) we choose ω^2 to be equal to k_s/m. In that case we have proven that

$$x = A\cos(\omega t)$$

satisfies the differential equation (the Momentum Principle) *for all values of the time t.* We have an "analytical" solution of the prediction problem. If someone gives you a time t, you can predict what the stretch x will be at that time, without having to carry out a numerical integration step by step from the present time to the specified time.

3.8.8 Real springs have mass

Real springs have mass, but we have ignored this fact of the real world. It is not easy to include the effect of the mass of the spring, either in a numerical integration or in an analytical treatment. One way to model a real spring with mass is with a chain of point masses connected by massless springs.

? Under what circumstances is it likely to be a good approximation to neglect the mass of the spring?

In many cases the spring has much less mass than the objects that are connected to it, in which case the effect of the spring's mass is likely to be small. A contrasting example is an oscillating Slinky, in which the only mass is the mass of the spring.

3.8.9 Existence of analytical solutions

It is fortunate that we can obtain an analytical solution—a prediction for all time of the motion of our model spring-mass system not subject to the additional inaccuracies introduced by a finite time step. Unfortunately, for many systems the differential equation corresponding to the Momentum Principle has a mathematical form that cannot be solved analytically and must be solved numerically. The power of numerical integration applied to the Newtonian scheme is that it can be used even in those situations where no analytical solution seems possible.

? As a simple example, show that $x = A\cos(\omega t)$ *cannot* solve the following "nonlinear" differential equation for any value of ω:

$$\frac{dp_x}{dt} = m\frac{d}{dt}\left[\frac{dx}{dt}\right] = -k_s x^3$$

There may or may not exist an analytical solution of this differential equation in terms of well-known mathematical functions, but *any* differential equation can be solved numerically, just as we did for the spring-mass equation. Given x and p at some instant in time, we can calculate the force $(-k_s x^3)$ and determine the new values of x and p at a time Δt later, then repeat as often as necessary to reach the time of interest. In fact, you may have explored this possibility in part (e) of Problem 3.3.

It is important to keep clearly in mind that either an analytical or a numerical prediction of the future applies only to our *model* of the actual real-world situation. If our model is a good approximation to the real world, our prediction will be a good approximation to what will actually happen, but even an analytical prediction cannot be exact because of the practical impossibility of knowing exactly the initial conditions and the net force due to all the other objects in the Universe.

One might think that analytical solutions are better than numerical ones, but note that one of the ways to calculate a table of cosine values is to do a numerical integration of the equation that applies to spring-mass systems! In general, a fruitful way to describe a mathematical function is by specifying the differential equation for which it is a solution, because then a numerical integration of that equation yields numerical values of the function.

3.8.10 General solution for the spring-mass system

It is easy to show that the function $x = A\cos(\omega t + \phi)$ is a general solution for a spring-mass system with arbitrary k_s, m, and initial conditions, where $\omega = \sqrt{k_s/m}$, and where the amplitude A and the "phase shift" ϕ (measured in radians and denoted by the Greek letter phi) are constants determined by the initial conditions, as is explained below. The "phase shift" essentially

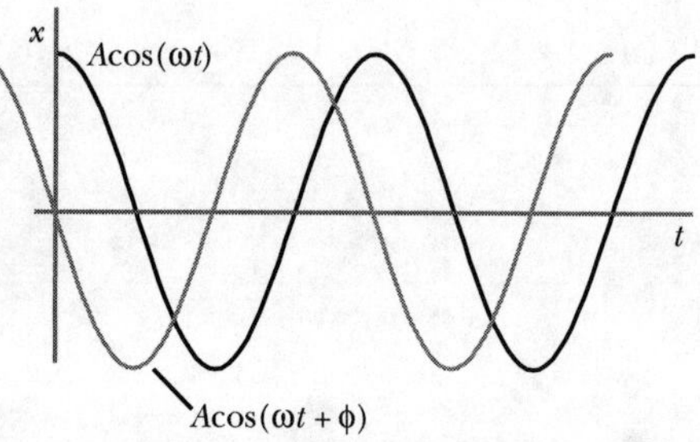

Figure 3.56 Phase shift ϕ.

shifts the starting point of the oscillation. For example, if $\phi = \pi/2$, the cosine function actually represents a negative sine function (Figure 3.56):

$$x = A\cos\left(\omega t + \frac{\pi}{2}\right) = -A\sin(\omega t)$$

Determining the amplitude and phase shift

The amplitude A and phase shift ϕ can be determined from the initial values at time t_0 of position x_0 and velocity v_0:

$$x_0 = A\cos\left(\sqrt{\frac{k_s}{m}}t_0 + \phi\right)$$

$$v_0 = -\sqrt{\frac{k_s}{m}}A\sin\left(\sqrt{\frac{k_s}{m}}t_0 + \phi\right)$$

Using the important trigonometric identity $\sin^2\theta + \cos^2\theta = 1$ (which is the Pythagorean theorem applied to a triangle whose hypotenuse has length 1), we can obtain an equation that can be solved for the amplitude A:

$$x_0^2 + \frac{m}{k_s}v_0^2 = A^2\text{, so we have } A = \sqrt{x_0^2 + \frac{m}{k_s}v_0^2}$$

For example, if the initial velocity is zero, the amplitude is simply equal to the initial stretch.

The value for A can be plugged back into either the x_0 equation or the v_0 equation to obtain the phase shift ϕ. Alternatively, by dividing the x_0 equation by the v_0 equation we obtain an equation that can be solved for the phase shift ϕ:

$$\frac{v_0}{x_0} = -\sqrt{\frac{k_s}{m}}\tan\left(\sqrt{\frac{k_s}{m}}t_0 + \phi\right)$$

$$\sqrt{\frac{k_s}{m}}t_0 + \phi = \arctan\left(-\sqrt{\frac{m}{k_s}}\frac{v_0}{x_0}\right)$$

$$\phi = \arctan\left(-\sqrt{\frac{m}{k_s}}\frac{v_0}{x_0}\right) - \sqrt{\frac{k_s}{m}}t_0$$

Ex. 3.26 Suppose the system is not oscillating and you strike it with a hammer, applying a large force F for a *very* short time Δt. What is the initial speed v_0? Explain briefly. What will be the amplitude (the maximum displacement away from the equilibrium position)?

Ex. 3.27 In the case you just analyzed (striking with a hammer at time $t_0 = 0$), what is the phase shift ϕ? (Does this make sense?)

Comment: Two constants for a second-order differential equation

Note that whether we specify the starting conditions in terms of x_0 and v_0 or in terms of the amplitude A and the phase shift ϕ, two constants are required. You can see why two constants are needed by looking at the start of the numerical integration. You needed the initial position x_0 in order to be able to calculate the initial force (which lets you step the velocity), and you also needed the initial momentum p_0 or velocity v_0 in order to be able to step the position.

We have been working with what is called a "second-order differential equation." That is, the highest derivative in the equation is a second derivative—the acceleration $dv/dt = d(dx/dt)/dt$. A second-order differential equation normally requires two constants to specify the starting conditions completely.

3.8.11 *Linear systems

We saw that doubling the amplitude of a spring-mass system simply multiplies the solution by a factor of two without changing the time dependence of the motion. This is an extremely important property of systems described by "linear" differential equations (equations in which the variables appear only to the first power, not squared, for example). More formally, let x be replaced by the quantity Rx, where R is a constant:

$$m\frac{d}{dt}\left(\frac{d(Rx)}{dt}\right) = -k_s(Rx)$$

The R's cancel, leaving us with the original equation. Therefore starting with a different initial stretch will simply scale up the vertical axis of the graph of x vs. t. Also, since the motion is periodic, a change in the initial velocity corresponds to starting the motion at a different time. For example, instead of starting with an initial stretch and zero velocity, you could start with zero stretch and some initial velocity, and this would correspond to a point on the original graph where the function crosses the time axis.

In contrast, consider a differential equation in which there is an x^3, and try scaling by R:

$$m\frac{d}{dt}\left(\frac{d(Rx)}{dt}\right) = -b(Rx)^3 = -bR^3x^3$$

In this case the R's do *not* cancel. The solutions to this "nonlinear" equation do not scale up in the simple way that solutions to a linear equation do.

3.9 Example: A vertical steel rod

Steel has a rather high Young's modulus, $2\times10^{11}\ \text{N/m}^2$.

(a) A thin steel rod is 4 m long with diameter 2 mm (2×10^{-3} m). You hang a 50 kg mass from the bottom of the vertical rod, and the rod does not sag catastrophically. How much does the rod stretch? Explain your work clearly.

(b) You pull the 50 kg mass down very slightly and release it. In the resulting vertical oscillations of the large mass, how long does it take for the mass to movc from its lowest position to its highest position? Explain your work, including a brief explanation of the model of the rod-mass system you used.

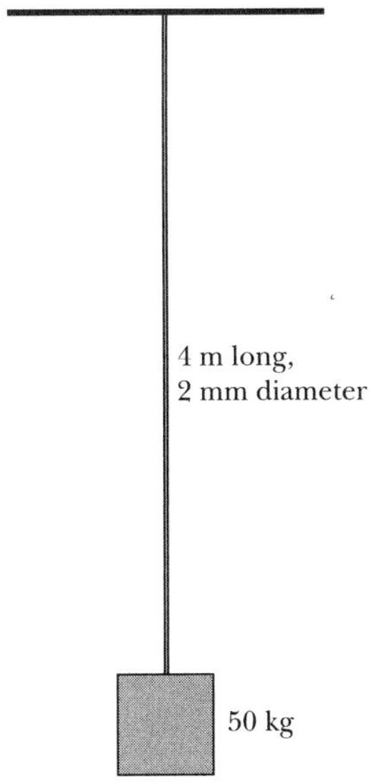

Figure 3.57 A thin steel rod supports a 50 kg mass.

Solution

Start from a fundamental principle: the Momentum Principle. This is a case of known motion (no motion) and unknown forces.

(a) The rod pulls up on the 50 kg mass with a force F_T, the tension in the rod. Since the momentum of the 50 kg mass isn't changing, the net force on it must be zero, so

$$F_T - mg = 0\text{, and the tension } F_T = mg$$

Since the rod doesn't sag catastrophically, we can assume that the stress-strain relation in terms of Young's modulus is valid:

$$\frac{F_T}{A} = Y\frac{\Delta L}{L}$$

$$\Delta L = \frac{F_T L}{AY} = \frac{(50\ \text{kg})(9.8\ \text{N/kg})(4\ \text{m})}{\pi(10^{-3}\ \text{m})^2(2\times10^{11}\ \text{N/m}^2)} = 3.1\times10^{-3}\ \text{m} = 3.1\ \text{mm}$$

(b) We can consider the rod to be a very stiff spring, and model the system as a mass on a spring, since the stretch of the rod, $s = \Delta L$, is proportional to the force $F_T = mg$ that the 50 kg mass exerts on the rod. (By the principle of reciprocity, the interatomic force that the mass exerts on the bottom of the rod is equal and opposite to the interatomic force the rod exerts on the top of the mass.)

$$F_T = \left(\frac{AY}{L}\right)\Delta L = k_s s$$

Therefore the effective spring stiffness of the rod is this:

$$k_s = \frac{AY}{L}$$

We have a known solution for the motion of a mass acted on by a force proportional to displacement, $y = A\cos(\omega t)$. Since $\omega = \sqrt{k_s/m} = 2\pi/T$, we find that the period is

$$T = 2\pi\sqrt{\frac{m}{k_s}} = 2\pi\sqrt{\frac{mL}{AY}} = 2\pi\sqrt{\frac{(50\ \text{kg})(4\ \text{m})}{\pi(10^{-3}\ \text{m})^2(2\times10^{11}\ \text{N/m}^2)}} = 0.112\ \text{s}$$

Moving from the lowest position to the highest position takes half a period, which is

$$\frac{T}{2} = 0.056\ \text{s}$$

3.10 The speed of sound in a solid

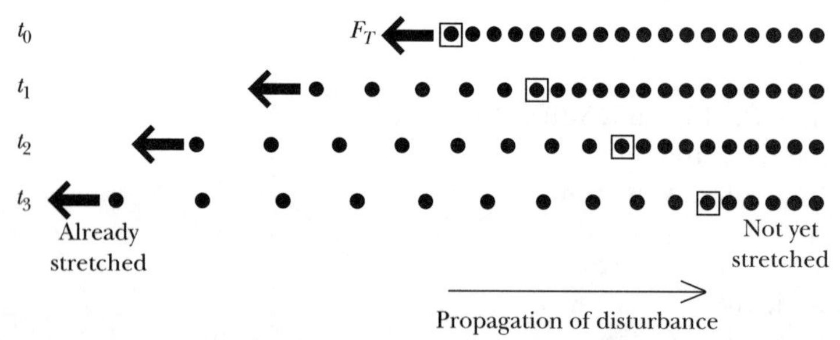

Figure 3.58 You start pulling to the left on a metal rod with a force F_T. A disturbance propagates to the right (boundary atoms are marked by boxes).

Our ball-and-spring model of a solid should allow us to predict real properties of matter—otherwise the model is of no value. One of the properties we might try to predict is the speed of propagation of a disturbance in a solid.

When you first start pulling to the left on the end of a metal rod, you very slightly lengthen the interatomic bonds between neighboring atoms in the nearby section of the rod. As a result of their leftward displacement, these atoms stretch the bonds of their neighbors to the right, and very quickly this new interatomic bond-stretching propagates to the right, all the way to the other end of the rod, and the whole rod is then in tension.

On Figure 3.58 we have marked atoms on the boundary between the already stretched region of the rod and the newly stretched region. It is this boundary that moves to the right, along the length of the rod. Individual atoms move extremely short distances, though for clarity in the diagram we have enormously exaggerated the interatomic stretches and the distance any individual atom moves.

This process is very fast, but it is not instantaneous. The boundary (or the the place where there is a change in the average length of the interatomic bond) propagates at a rate that we will try to determine, using our ball-and-spring model and our understanding of spring-mass motion.

The rate of propagation of the disturbance is called the "speed of sound" in the material, and we should explain what the propagation has to do with sound. When you clap your hands, a pulse of sound propagates through the air to your ear (Figure 3.59). This pulse starts as a momentary compression of the air (an increase in the air density) made by your hands. This compression is passed on to neighboring regions of the air, and the air near the hands relapses to its original density.

Eventually your eardrum is hit by the pulse and moves in response to the sudden increase in density (and pressure). The motion of the eardrum is detected by your inner ear and passed on to the brain, which interprets the signal as meaning that you clapped your hands. Note that no individual air molecule moves from near your hands to your ear. Rather it is the disturbance that moves from one place to another.

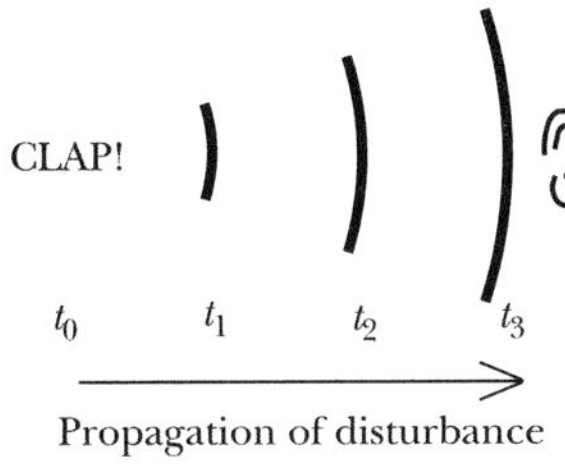

Figure 3.59 Clap your hands—sound travels to your ear as a pulse of air compression.

Similarly, a burst of sound hitting a solid surface compresses the atoms at the surface of the solid very slightly, and this compression propagates into the solid at a rate called the "speed of sound" in the solid. The sudden change in tension caused by your pulling on one end of the rod also propagated at this speed.

Dimensional analysis

? How would you expect the speed of sound in a solid to depend on the interatomic spring stiffness k_s? On the mass m of one of the atoms?

Our ball-and-spring model of a solid correctly suggests that the larger the interatomic spring stiffness k_s, the faster the propagation, due to quicker response of a neighboring atom. Also, the larger the mass m of the atoms, the slower the propagation (slower response of a neighboring atom). It is reasonable to guess that the speed of sound is proportional to the angular frequency $\omega = \sqrt{k_s/m}$ of atomic oscillations.

The angular frequency ω has units of 1/s. If we multiply by some atomic distance to get units of m/s, we might have a rough prediction for the speed of sound. A plausible guess is to say that ω is related to the rate at which one atom hits its neighbor, propagating a change through a distance of one interatomic spacing. Therefore multiplying by the interatomic spacing d (the distance between nuclei) might give us the speed of sound:

$$v = \omega d \text{ ?}$$

This informal reasoning is an example of "dimensional analysis," in which we identify meaningful quantities in the situation and combine them to guess a formula for some other quantity of interest. While this can be a fruitful procedure, we shouldn't be surprised if we find that the speed of sound is actually given by $v = 2\omega d$, or $v = \omega d/\pi$, for example. Moreover, note that atomic oscillations in the x direction involve two atomic "springs" on each side of an atom, not one, so we can't even be sure that $\omega = \sqrt{k_s/m}$ is the right way to estimate ω. Nevertheless, we can guess that $(\sqrt{k_s/m})d$ is proportional to the actual speed of sound, with the correct dependence on k_s and m. As it happens, it can be shown that this *is* the correct formula. We omit the difficult derivation.

Speed of sound in different materials

In aluminum, which is very stiff (large k_s and small atomic mass m), the speed of sound is very high, about 5000 m/s. If a wire is half a meter long, it takes about

$$(0.5\text{ m})/(5000\text{ m/s}) = 10^{-4}\text{ s (a tenth of a millisecond)}$$

from the time you pull on one end for the other end to become tense. On the other hand, lead is quite soft (small k_s and large atomic mass m), and the speed of sound in lead is only about 1200 m/s, much lower than in aluminum. (The speed of sound in solids is much higher than the speed of sound in air, which is about 340 m/s. Because air molecules are usually far from one another, the speed of sound in air is determined by the average speed of the air molecules, not by interatomic forces.)

If possible, it is interesting to measure the speed of sound in a metal bar directly, by hitting one end of a long bar with a short pulse and observing on an oscilloscope when sensors at two locations along the bar detect the sound.

Ex. 3.28 Using your results from Exercise 3.16, predict the speed of sound in aluminum and compare with the measured value (about 5000 m/s). Also, using your results from Exercise 3.17, predict the speed of sound in lead and compare with the measured value (about 1200 m/s).

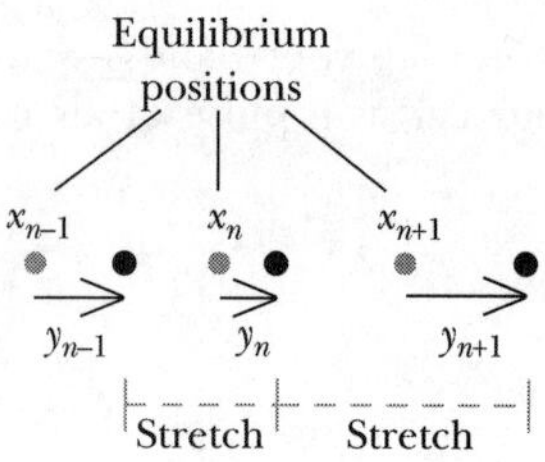

Figure 3.60 Figuring out the interatomic stretches to the left and the right of the *nth* atom, in terms of the displacements y_n away from the equilibrium positions.

3.10.1 Computing and visualizing propagation

Problem 3.25 involves modeling the propagation of a disturbance in a metal. From the computer modeling you have already done, you have most of the tools required to study the speed of sound. We discuss two important aspects of the modeling that may be less familiar.

For every atom in a long chain of atoms, you need to take into account interatomic spring forces exerted by atoms both to the left and to the right of the atom. Consider the *nth* atom in the chain, located at equilibrium position x_n (Figure 3.60). When there is no disturbance, the distance between the centers of the atoms is *d*, and the interatomic spring forces are zero. If atoms *n*–1, *n*, and *n*+1 are displaced from their equilibrium positions by amounts y_{n-1}, y_n, and y_{n+1}, the interatomic forces are computed from the stretches:

$$k_s(y_{n+1} - y_n) - k_s(y_n - y_{n-1})$$

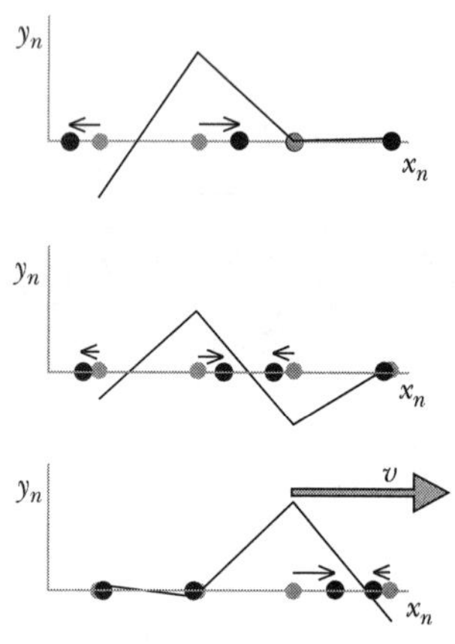

Figure 3.61 Visualizing the propagation by plotting the displacements y_n away from the equilibrium positions x_n.

Knowing the current displacements of all the atoms (y_1, y_2, etc.), you can compute the forces on each atom and determine how they will move in the next time step.

Visualizing the propagation of the disturbance can be done in various ways. A particularly useful visualization is to plot the horizontal displacements (y_n) vertically, at the locations of the horizontal equilibrium locations (x_n). An example is shown in Figure 3.61, where you see that a displacement to the left is plotted below the axis, and a displacement to the right is plotted above the axis. For further details, see Problem 3.25 (page 176) Computer model of the propagation of sounds in a metal.

3.11 Buoyancy

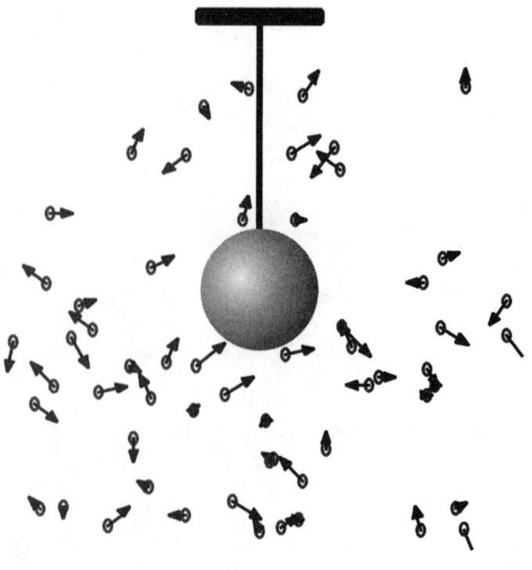

Figure 3.62 Air molecules in motion near a hanging ball.

Until now we have neglected the force that the air exerts on objects. Consider the surrounding air that is in contact with a ball hanging from a wire, as shown in Figure 3.62. You might think that the air has no effect on the ball, especially since the ball is not moving, so we don't need to worry about air resistance. To be certain, let's consider this issue at the microscopic level.

Air is a gas, composed of about 80% nitrogen and 20% oxygen, with very small amounts of other gases. The molecules in the air are continuously in motion, moving in random directions. Some of the molecules in the air near the ball will occasionally bump into the ball, exerting a small force on it. If there is no wind, the motion of the gas molecules is random, so in a short time period there should be just as many molecules colliding with the right hand side of the ball as with the left hand side, and there should on average be no sideways force on the ball.

However, the situation is not quite the same for collisions with the top and bottom of the ball. The directions of molecular velocities are still random, but there are more air molecules per cubic meter below the ball than there are above it. The density of the Earth's atmosphere decreases as one moves upward away from the Earth's surface, until finally at a sufficient distance (approximately 50 km) there is essentially no air left. You may have observed this variation yourself if you have ever gone from sea level to a location several thousand meters higher; you may feel light-headed if you exercise at high altitude because the density of oxygen is significantly lower.

This variation in air density occurs over kilometers; can it possibly be important over a distance of a few centimeters? Interestingly enough, the surprising answer is yes! There is actually a "buoyant force" upward on the ball, because the number of air molecules hitting the bottom of the ball per second is slightly greater than the number of molecules hitting the top of the ball per second.

A macroscopic view of buoyancy

How can we determine the magnitude of the buoyant force on the ball due to the air? Let's consider a macroscopic viewpoint. The following discussion, based on the Momentum Principle, applies equally well to the buoyant force exerted by any fluid (air, water, other liquids, other gases) on a floating object. The result we will obtain is called the Archimedes principle, for the Greek thinker who first understood it.

Imagine a box filled with air (or water). Consider a ball-shaped region that a ball will eventually occupy, at a moment when the ball is not yet there, so the ball-shaped region is filled with the fluid. Let's choose this ball-shaped mass of fluid as our system, and consider what forces must be acting on it (Figure 3.63). This may seem odd, but it is a perfectly legitimate choice of system.

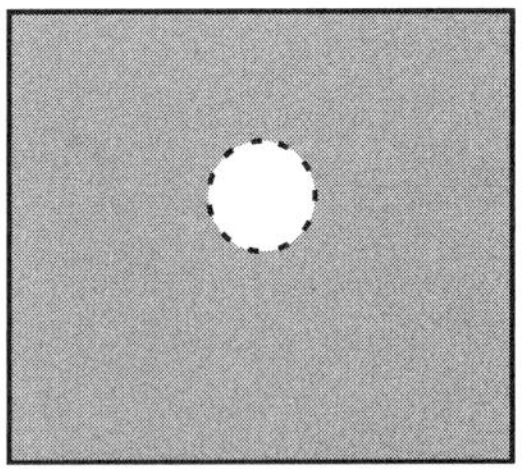

Figure 3.63 A system consisting of a ball-shaped spherical region of fluid, surrounded by more fluid.

? Is the momentum of the indicated sphere of fluid changing? What must the net force on this sphere be?

The momentum of the indicated sphere of fluid is not changing. The Earth exerts a downward gravitational force on it, but the sphere does not sink. Because its momentum isn't changing, the net force on the sphere must be zero, so there must be an upward force that balances the gravitational force.

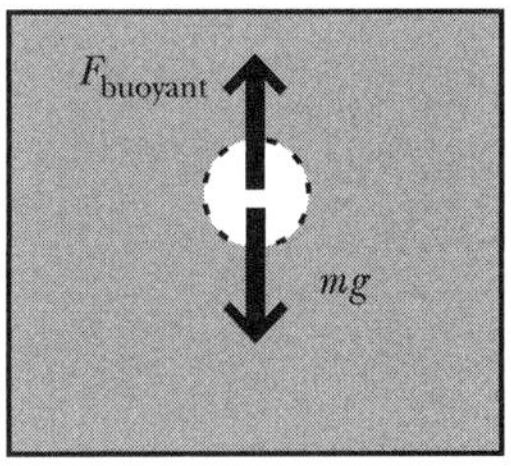

Figure 3.64 Forces on the sphere of fluid. Here m is the mass of the fluid in the sphere.

? What exerts this upward force on the sphere?

The only object in contact with this sphere is the rest of the fluid, so it must be that the rest of the fluid exerts an upward force F_b ("buoyant force") whose magnitude is mg, where m is the mass of the sphere of fluid (Figure 3.64).

Now remove the sphere of fluid and replace it with the ball. It must be that initially the rest of the fluid still exerts the same upward force F_b of magnitude mg (Figure 3.65). This force may not be large enough to counteract the downward gravitational force on the ball; in the case of a ball hanging in air it clearly is not, since if the ball were not supported by a wire it would begin accelerating downward.

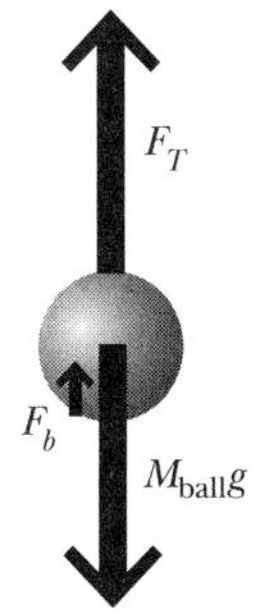

Figure 3.65 The tension force is actually less than the weight of the ball, because the buoyant force F_b due to the air is also in the upward direction.

The upward force is called a "buoyant" force but is simply an interatomic contact force due to fluid molecules striking atoms in the surface of the ball. The key point is that despite the complexity of the interatomic interactions, the net effect is simply an upward force whose magnitude is the mass of an equivalent volume of fluid, times g. Often it is reasonable to neglect buoyant forces in air (though usually not in water). To see why, let's compare the buoyant force to the gravitational force F_g on a ball hanging in air, where we write V for the volume of the ball:

$$\frac{F_b}{F_g} = \frac{m_{\text{air}}g}{M_{\text{ball}}g} = \frac{m_{\text{air}}}{M_{\text{ball}}} = \frac{m_{\text{air}}/V}{M_{\text{ball}}/V} = \frac{\rho_{\text{air}}}{\rho_{\text{ball}}}$$

The ratio of the forces is equal to the ratio of ρ_{air} (the density of air) to ρ_{ball} (the density of the ball; ρ is Greek lower-case letter rho). The density of air at STP ("standard temperature and pressure," 0°C and 1 atmosphere) can be calculated if we remember from chemistry that one mole of a gas at STP occupies 22.4 liters. The molecular mass of N_2 is 28 and that of O_2 is 32, which gives an average molecular mass of about 29 for air (which is about 80% nitrogen and 20% oxygen):

$$\rho_{\text{air}} \approx \frac{29 \text{ grams/mole}}{22.4\times10^3 \text{ cm}^3 \text{ / mole}} \approx 1.3\times10^{-3} \text{ grams/cm}^3$$

What is the density of the ball? We can estimate it by noting that the density of water is 1 gram/cm^3, and that most solids are more dense than water. Using 1 g/cm^3 as the density of the ball, we find that $F_b/F_g \approx 1.3\times10^{-3}/1$.

The buoyant force $m_{\text{air}}g$ of air on a solid object is therefore very small compared to the gravitational force mg. In many cases we can safely choose to neglect it. Since the upward force exerted by the air on the ball is small compared to the gravitational force or the tension force exerted by the wire, we can choose to neglect it in an analysis of the forces on a hanging ball. However, it is not always appropriate to neglect the buoyant force; in accurate analytical chemistry it is necessary to correct for buoyancy when weighing a sample of liquid or low-density solid on an analytical balance.

A final note: the buoyant force being equal to $m_{\text{air}}g$ is really a time average, because the buoyant force is due to random collisions with air molecules and therefore has an intermittent character. However, the collision rate is so high that the force seems nearly continuous and constant. If, however, the ball is of microscopic size, it may not be possible to ignore the intermittent nature of the collisions. In a microscope it is possible to observe small particles being jostled about by the random collisions with the water molecules. This effect is called "Brownian motion."

Ex. 3.29 Calculate the buoyant force in air on a kilogram of iron (whose density is about 8 grams per cubic centimeter). Compare with the weight mg of this much iron.

3.12 Pressure and suction

There is an interesting case where we cannot neglect the effect of the air. Suppose that an object lies on a table and the surfaces of the table and object are so smooth that upon squeezing them together all air is expelled. Such a situation occurs when a suction cup is pressed down onto a smooth surface. Now there are no longer any air molecules striking the bottom surface of the suction cup, while air molecules continue to strike the top surface. The force of the air on the top surface is quite large, about 10^5 newtons on each square meter at sea level! This large force per unit area is called the "pressure" P of the air. The pressure can be thought of as due to the weight of a one-square-meter column of the entire atmosphere, extending upward many kilometers (although with diminishing density), as illustrated in Figure 3.66.

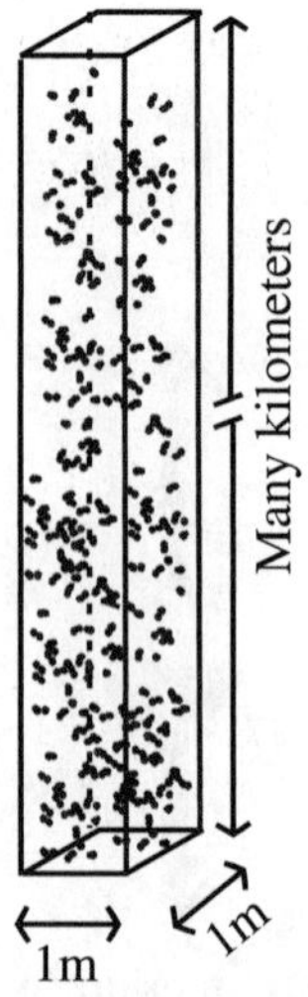

Figure 3.66 A one-square-meter column of air extending upward through the entire atmosphere.

Pressure $P = F/A$ (force per unit area)

If the atmosphere had a constant density ρ like that near the surface (1.3 grams per cubic centimeter), we could calculate the height h of the atmosphere in the following way. The volume of a column of air of height h and cross-sectional area A is Ah, and its mass is $M_{\text{air}} = \rho(Ah)$. The pressure, which is about 10^5 newtons per square meter, is the weight $M_{\text{air}}g$ divided by the cross-sectional area: $P = M_{\text{air}}g/A = \rho gh$.

The height of a constant-density atmosphere, $h = P/(\rho g)$, would be about 8000 meters (about 5 miles), about the height of Mount Everest. In reality air density is not constant but decreases as you go higher, so there is still some low-density air at much higher altitudes than this.

A physics diagram for a suction cup looks like Figure 3.67, where PA represents the force of the air on the area A of the top surface of the suction cup. The normal force exerted upward by the table and the downward force of the air are huge compared to the weight of the suction cup mg.

The normal force $F_N = PA + mg \approx PA$, since the force of the air is so much larger than the weight mg of the suction cup. For example, a rubber suction cup with a diameter of 4 cm has an area of $\pi(0.02\text{ m})^2 = 1.3\times10^{-3}\text{ m}^2$, so the force of the air on the suction cup is

$$PA = (10^5\text{ N/m}^2)(1.3\times10^{-3}\text{ m}^2) = 130\text{ newtons}$$

whereas its own weight mg might be about 0.01 kg (10 grams) times 9.8 N/kg, or only about 0.1 newton. Note that a force of 130 newtons is the weight of an object whose mass is about 13 kg, or about 29 pounds, which is why it isn't easy to pull even a small suction cup off a flat surface.

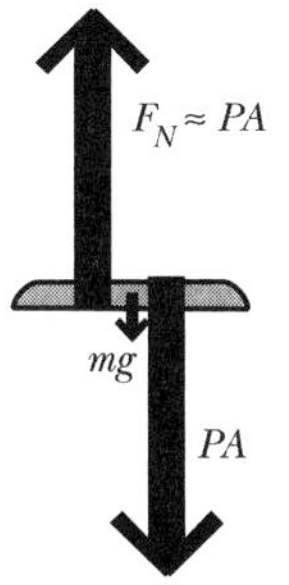

Figure 3.67 The weight of the suction cup is very small compared to the downward force by the air or the normal force by the table.

Every solid object in air, such as a hanging ball, is subjected to large compression forces by the pressure of the surrounding air (Figure 3.68). The upward buoyant force is the tiny difference between these huge forces. A hollow object must be very strong to support this compression, if air is removed from the inside. There is no problem however if air can flow into and out of the container, because then it is just the solid walls that are compressed, since the air exerts comparable pressure on the inner and outer surfaces.

All of this is even more relevant under water, because the density of water is about 1000 times greater than the density of air, and the compression forces are about 1000 times as strong. A hollow submersible vehicle that dives very deep must have an extremely strong hull. When a scuba diver descends to greater depth where the water pressure is larger, the breathing apparatus automatically increases the pressure of the air supplied to the diver to equal the increased pressure of the outside water at the new depth, in order to prevent the lungs from being crushed.

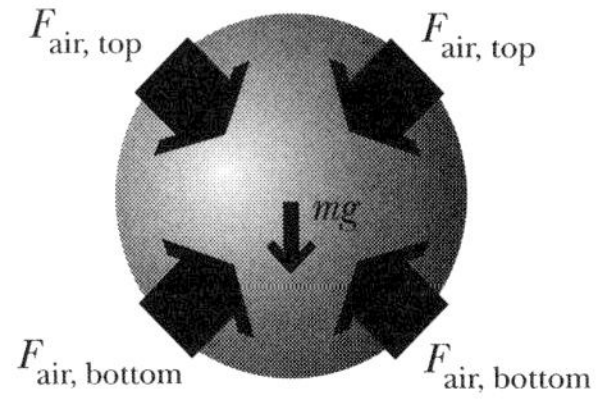

Figure 3.68 A ball hanging in air is subject to large compression forces due to the surrounding air.

Ex. 3.30 It is hard to imagine that there can be enough air between a book and a table so that there is a net upward (buoyant) force on the book despite the large downward force on the top of the book. About how many air molecules are there between a textbook and a table, if there is an average distance of about 0.01 mm between the uneven surfaces of the book and table?

3.13 Summary

THE MOMENTUM PRINCIPLE (DERIVATIVE FORM)

$$\frac{d\vec{p}}{dt} = \vec{F}_{net}$$

In words, "the instantaneous time rate of change of the momentum of an object is equal to the net force acting on the object." Or in calculus terms, since the limit of a ratio of infinitesimal quantities is called a derivative, we can say that "the derivative of the momentum with respect to time is equal to the net force acting on the object."

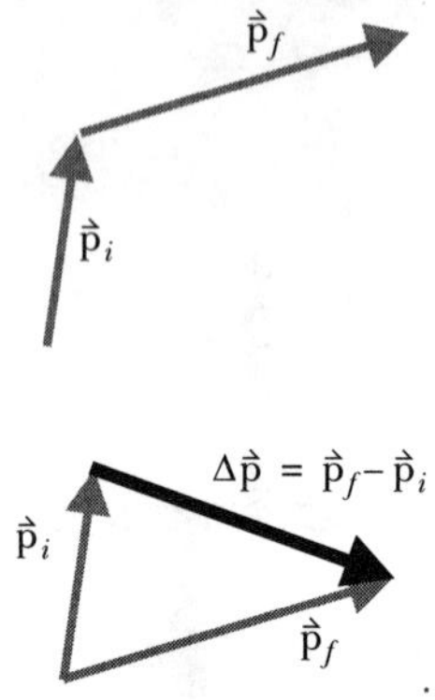

Figure 3.69 Calculation of $\Delta\vec{p}$.

Finding $d\vec{p}/dt$ (Section 3.2.4)

1. Draw two arrows (Figure 3.69),

   one representing $\vec{p}_i$, the momentum of the object a short time before the instant of interest, and

   a second arrow representing $\vec{p}_f$, the momentum of the object a short time after the instant of interest

2. Graphically find $\Delta\vec{p} = \vec{p}_f - \vec{p}_i$ by placing the arrows tail to tail, and drawing the resultant arrow starting at the tail of $\vec{p}_i$ and going to the tip of $\vec{p}_f$

3. This arrow indicates the direction of $\Delta\vec{p}$, which is the same as the direction of $d\vec{p}/dt$, provided the time interval involved is sufficiently small

Rate of change of momentum along a curving path (Section 3.3.3)

PERPENDICULAR COMPONENT OF $d\vec{p}/dt$ FOR MOTION ALONG A CURVING PATH

Perpendicular component of $\frac{d\vec{p}}{dt}$ is given by $\left|\left(\frac{d\vec{p}}{dt}\right)_\perp\right| = \frac{|\vec{v}|}{R}|\vec{p}|$

The direction is toward the center of the kissing circle of radius R.

New concepts

Macroscopic spring: $|\vec{F}| = k_s|s|$, where $s = L - L_{relaxed}$
Forces describing interatomic interactions (contact forces):
Tension force in a wire or a string due to stretching of interatomic bonds in the object
"Normal" force due to compression of object

$\frac{F_T}{A} = Y\frac{\Delta L}{L}$; atomic version $Y = \frac{(k_s s/d^2)}{(s/d)} = \frac{k_s}{d}$

Concepts of amplitude A, angular frequency ω, frequency f, and period T for oscillator.

Buoyancy force due to variation in air density.

Results

Spring-mass motion

Numerical solution of the Momentum Principle $\frac{dp_x}{dt} = -k_s x$ for a horizontal mass on a spring produces an oscillatory motion. The idealized spring-mass system has an analytical prediction:

$$x = A\cos\left(\sqrt{\frac{k_s}{m}}t\right)$$

Angular frequency $\omega = \sqrt{\frac{k_s}{m}}$, Period $T = \frac{2\pi}{\omega}$, Frequency $f = \frac{1}{T} = \frac{\omega}{2\pi}$

Speed of sound in a solid

$v = d\sqrt{k_s/m}$, where d is distance between nuclei, m is atomic mass

Buoyancy and pressure

Upward buoyancy force = $m_{\text{fluid}}g$, where m_{fluid} is mass of equivalent volume of fluid.

Pressure is force per unit area; air pressure at sea level is about 10^5 N/m^2.

3.14 Review questions

Circular motion

RQ 3.1 You are driving an American car, sitting on the left side of the front seat. You make a sharp right turn. You feel yourself "thrown to the left" and your left side hits the left door. Is there a force that pushes you to the left? What object exerts that force? What really happens? Draw a diagram to illustrate and clarify your analysis.

Circular motion

RQ 3.2 A 30 kg child rides on a playground merry-go-round, 2 m from the center. The merry-go-round makes one complete revolution every 4 seconds. How large is the net force on the child? In what direction does the net force act?

RQ 3.3 The radius of a merry-go-round is 7 meters, and it takes 12 seconds to make a complete revolution.

(a) What is the speed of an atom on the outer rim?

(b) Which statement below (i, ii, or iii) best describes the direction of the momentum of this atom?

(c) Which statement below (i, ii, or iii) best describes the direction of the rate of change of the momentum of this atom?

(i) Inward, toward the center

(ii) Outward, away from the center

(iii) Tangential

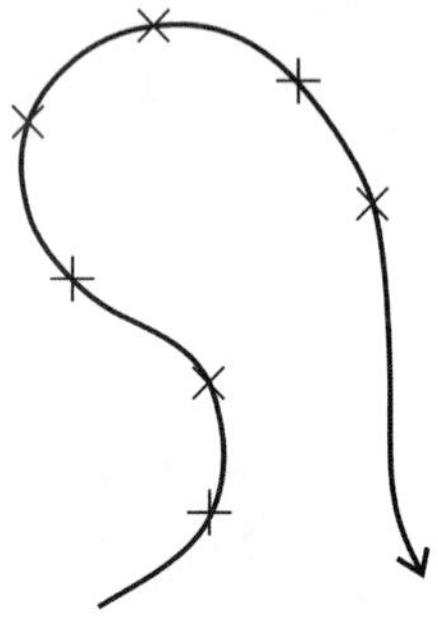

Figure 3.70 Path of a particle traveling at constant speed (RQ 3.4).

RQ 3.4 Figure 3.70 shows the path of a particle. Assuming the particle's speed does not change as it travels along this path, at each location labeled "x" draw an arrow showing the direction of $d\vec{p}/dt$ for the particle as it passes that location.

Determinism

RQ 3.5 List as many limitations as you can on our ability to predict the future.

The nature of solids

RQ 3.6 Suppose you attempt to pick up a very heavy object. Before you tried to pick it up, the object was sitting still—its momentum was not changing. You pull very hard, but do not succeed in moving the object. Isn't this a violation of the Momentum Principle? How can you be exerting a large force on the object without causing a change in its momentum? What does change when you apply this force?

Spring-mass system without friction

For the following review questions, assume that friction is negligible.

RQ 3.7 Write the Momentum Principle for a horizontal spring-mass system with spring stiffness k_s and mass m, and write an analytical solution.

RQ 3.8 In terms of k_s and m, what is the angular frequency? The frequency? The period? The amplitude?

RQ 3.9 If you double the amplitude, by what factor does the period change? If you double the mass, by what factor does the period change? If you double the spring stiffness, by what factor does the period change?

RQ 3.10 How should you start the system going at $t = 0$ in order for the motion to be $A\cos(\omega t)$? How should you start the system going at $t = 0$ in order for the motion to be $A\sin(\omega t)$?

Other oscillators

RQ 3.11 Describe two examples of oscillating systems that are *not* harmonic oscillators.

The atmosphere

RQ 3.12 Calculate the height that the atmosphere would have if the air density were constant.

Pressure

RQ 3.13 Air pressure at the surface of a fresh water lake near sea level is about 10^5 N/m^2. At approximately what depth below the surface does a diver experience a pressure of 2×10^5 N/m^2? How would this be different in sea water, which has higher density than fresh water?

3.15 Problems

Problem 3.1 Experiment—study of strain vs. stress

Measure and graph the strain vs. stress curve for a long metal wire by hanging weights from the end of the wire and carefully measuring the stretch as a function of the hanging weight. If possible, go beyond the "yield" stress and observe the rapid, large increase in the strain. BE CAREFUL! Once you exceed the yield stress, the wire may break and dump the weights on whatever is underneath.

Problem 3.2 Experiment—a vertical spring-mass system

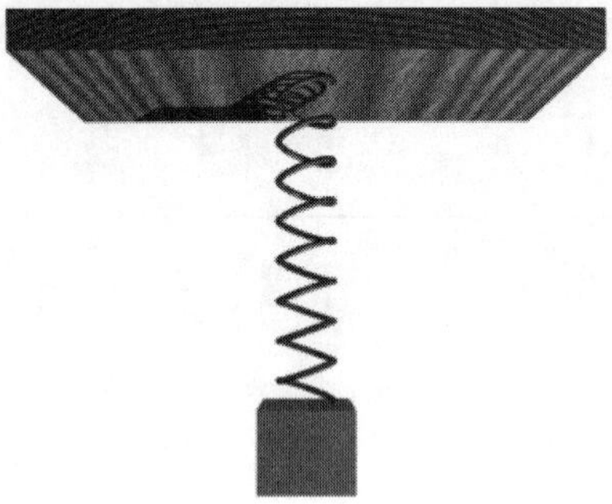
Figure 3.71 A vertical spring-mass system.

You will be given a spring and one or more masses with which to study the motion of a mass hanging from a spring (Figure 3.71).

(a) Measure the value of the spring stiffness k_s of the spring in N/m. The magnitude of a spring force is $k_s s$, where s is the stretch of the spring (change from the unstretched length). Explain briefly how you measured k_s. Include in your report the unstretched length of the spring.

(b) Measure the period (the round-trip time) of a mass hanging from the vertical spring. Report the mass that you use, and the amplitude of the oscillation. Amplitude is the maximum displacement, plus or minus, from the equilibrium position (the position where the mass can hang motionless).

(c) With twice the amplitude that you reported in part (b), measure the period again. Since the mass has to move twice as far, one might expect the period to lengthen. Does it?

How to measure the period accurately

There are unavoidable fluctuations in starting and stopping the timing, but you can minimize the error this contributes by timing many complete cycles so that the starting and stopping fluctuations are a small fraction of the total time measured.

It is good practice to count out loud *starting from zero, not one.* To count five cycles, say out loud "Zero, one, two, three, four, five." If on the other hand you say "One, two, three, four, five," you have actually counted only four cycles, not five.

Since the motion is periodic, you can start (say "zero") at any point in the motion. It is best to start and stop when the mass is moving fast past some marker near the equilibrium point, because it is difficult to estimate the exact time when the mass reaches the very top or the very bottom, because it is moving slowly at those turnaround points. Be sure to measure full round-trip cycles, not the half-cycles between returns to the equilibrium point (but going in the opposite direction).

Problem 3.3 Numerical integration of a spring-mass system without friction

Carry out a numerical integration of the motion of a horizontal spring-mass system. Use the spring stiffness k_s and mass m that you measured experimentally for a vertical spring-mass system. Predict the motion of this spring-mass system sufficiently far into the future to observe several oscillations.

(a) Initial conditions: Make the initial position of the block be such that the stretch of the spring is equal to the amplitude of the oscillations in your experiment, and release the block with zero initial momentum. Display the motion in two ways:

(1) Plot a graph of the position of the block as a function of time. Label the scales on both axes, so your numerical results are clear.

(2) If possible using your computer tool, make an animation of the motion of the block. (You don't need to draw a real spring; it is sufficient to draw a line connecting the block to the wall.)

(Continued on next page.)

Very briefly, what are the most important aspects of the predicted motion?

(b) Vary the time step and report the maximum step size that gives reasonable results, in the sense that a smaller step size has little effect.

(c) Read the "period" (the round-trip time) of the oscillation off the graph. How does this compare with the period you measured in your experiment?

(d) Repeat the calculation with double the initial stretch. What happens to the period? Is this what you observed in your experiment?

(e) Temporarily change the force law to be $-k_s x^3$, and observe the period with the amplitude of part (a) and with double this amplitude. What happens to the period when you change the amplitude?

(f) How long would the spring-mass system theoretically continue to oscillate? Would a real block connected to a spring behave that way? What *would* happen?

Problem 3.4 Experiment—determine *g* with a circular pendulum

Use a circular pendulum to determine g. You can increase the accuracy of the time it takes to go around once by timing N revolutions and then dividing by N. This minimizes errors contributed by inaccuracies in starting and stopping the clock. It is wise to start counting from zero (0, 1, 2, 3, 4, 5) rather than starting from 1 (1, 2, 3, 4, 5 represents only 4 revolutions, not 5). It also improves accuracy if you start and stop timing at a well-defined event, such as when the mass crosses in front of an easily visible mark.

This was the method used by Newton to get an accurate value of g. Newton was not only a brilliant theorist but also an excellent experimentalist. For a circular pendulum he built a large triangular wooden frame mounted on a vertical shaft, and he pushed this around and around while making sure that the string of the circular pendulum stayed parallel to the slanting side of the triangle.

Problem 3.5 Relating radius to period for circular orbits

The planets in our Solar System have orbits around the Sun that are nearly circular, and $v << c$. Calculate the period T (a "year"—the time required to go around the Sun once) for a planet whose orbit radius is r. This is the relationship discovered by Kepler and explained by Newton. (It can be shown by advanced techniques that this result also applies to elliptical orbits if you replace r by the "semi major axis," which is half the longer, "major" axis of the ellipse.) Use this analytical solution for circular motion to predict the Earth's orbital speed, using the data for Sun and Earth on the inside back cover of the textbook.

Problem 3.6 Circular motion in a magnetic field

When a particle with electric charge q moves with speed v in a plane perpendicular to a magnetic field B, there is a magnetic force at right angles to the motion with magnitude qvB, and the particle moves in a circle of radius r (Figure 3.72). This formula for the magnetic force is correct even if the speed is comparable to the speed of light. Show that

$$p = \frac{mv}{\sqrt{1-(|\vec{v}|/c)^2}} = qBr$$

even if v is comparable to c. Remember that $\vec{F} = m\vec{a}$ is *not* valid for high speeds.

This result is used to measure relativistic momentum: if the charge q is known, we can determine the momentum of a particle by observing the radius of a circular trajectory in a known magnetic field.

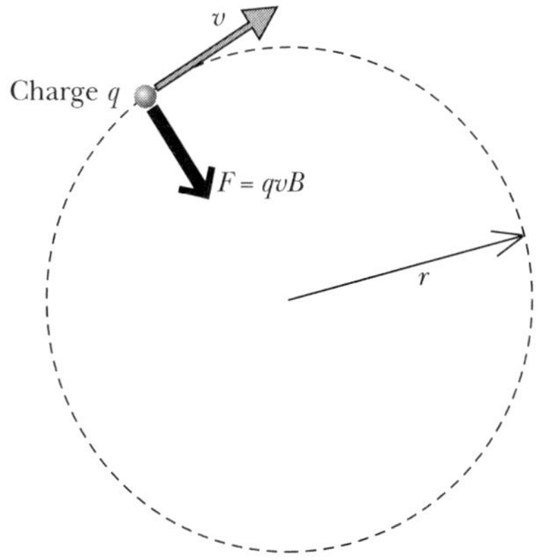

Figure 3.72 A charge moving in a circle due to a magnetic force (Problem 3.6).

Problem 3.7 Mass on a spring with circular motion
A small block of mass m is attached to a spring with stiffness k_s and relaxed length L. The other end of the spring is fastened to a fixed point on a low-friction table. The block slides on the table in a circular path of radius $R > L$. How long does it take for the block to go around once?

Problem 3.8 Dark matter in the Universe
In the 1970's the astronomer Vera Rubin made observations of distant galaxies that she interpreted as indicating that perhaps 90% of the mass in a galaxy is invisible to us ("dark matter"). She measured the speed with which stars orbit the center of a galaxy, as a function of the distance of the stars from the center. The orbital speed was determined by measuring the "Doppler shift" of the light from the stars, an effect which makes light shift toward the red end of the spectrum ("red shift") if the star has a velocity component away from us, and makes light shift toward the blue end of the spectrum if the star has a velocity component toward us. She found that for stars farther out from the center of the galaxy, the orbital speed of the star hardly changes with distance from the center of the galaxy, as is indicated in the diagram below. The visible components of the galaxy (stars, and illuminated clouds of dust) are most dense at the center of the galaxy and thin out rapidly as you move away from the center, so most of the visible mass is near the center.

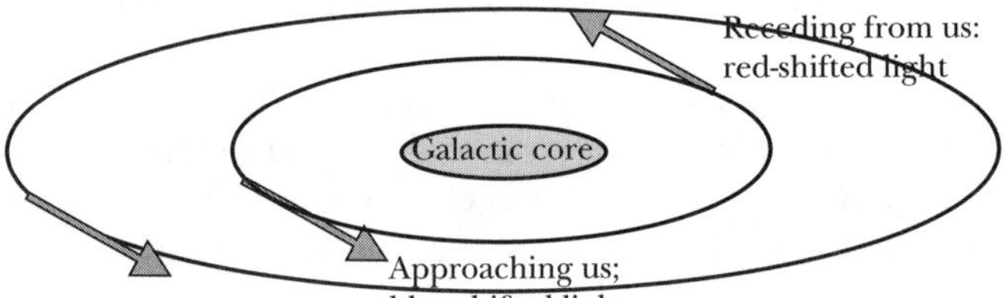

(a) Predict the speed v of a star going around the center of a galaxy in a circular orbit, as a function of the star's distance r from the center of the galaxy, assuming that almost all of the galaxy's mass M is concentrated at the center.

(b) Construct a logical argument as to why Rubin concluded that much of the mass of a galaxy is not visible to us. Reason from principles discussed in this chapter, and your analysis of part (a). Explain your reasoning. You need to address the following issues:

- Rubin's observations are not consistent with your prediction in (a).
- Most of the *visible* matter is in the center of the galaxy.
- Your prediction in (a) assumed that most of the mass is at the center.

This issue has not yet been resolved, and is still a current topic of astrophysics research. Here is a reference to the original work: "Dark Matter in Spiral Galaxies" by Vera C. Rubin, *Scientific American*, June 1983 (96-108). You can find several graphs of the rotation curves for spiral galaxies on page 101 of this article.

Problem 3.9 Orbital periods
(a) Many communication satellites are placed in a circular orbit around the Earth at a radius where the period (the time to go around the Earth once) is 24 hours. If the satellite is above some point on the equator, it stays above that point as the Earth rotates, so that as viewed from the rotating Earth the satellite appears to be motionless. That is why you see dish antennas pointing at a "fixed" point in space. Calculate the radius of the orbit of such a "synchronous" satellite. Explain your calculation in detail.

(b) Electromagnetic radiation including light and radio waves travels at a speed of 3×10^{8} m/s. If a phone call is routed through a synchronous satellite to someone not very far from you on the ground, what is the minimum delay between saying something and getting a response? Explain. Include in your explanation a diagram of the situation.

(c) Some human-made satellites are placed in "near-Earth" orbit, just high enough to be above almost all of the atmosphere. Calculate how long it takes for such a satellite to go around the Earth once, and explain any approximations you make.

(d) Calculate the orbital speed for a near-Earth orbit, which must be provided by the launch rocket. (Recently large numbers of near-Earth communications satellites have been launched. Their advantages include making the signal delay unnoticeable, but with the disadvantage of having to track the satellites actively and having to use many satellites to ensure that at least one is always visible over a particular region.)

(e) When the first two astronauts landed on the Moon, a third astronaut remained in an orbiter in circular orbit near the Moon's surface. During half of every complete orbit, the orbiter was behind the Moon and out of radio contact with the Earth. On each orbit, how long was the time when radio contact was lost?

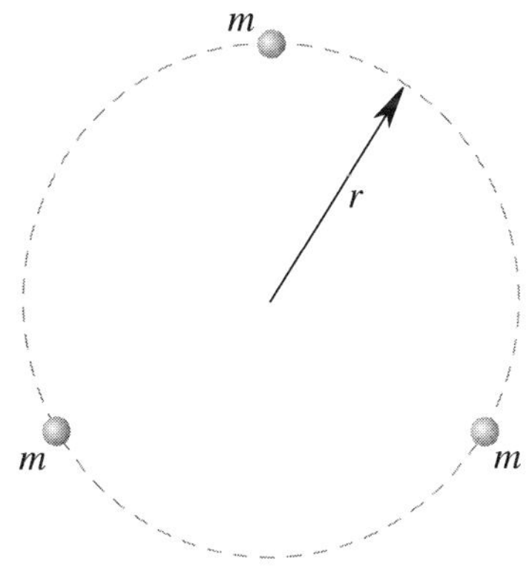

Figure 3.73 Three stars in a circular orbit (Problem 3.10).

Problem 3.10 Analytical 3-body orbit

There is no general analytical solution for the motion of a 3-body gravitational system. However, there do exist analytical solutions for very special initial conditions. Figure 3.73 shows three stars, each of mass m, which move in the plane of the page along a circle of radius r. Calculate how long this system takes to make one complete revolution. (In many cases 3-body orbits are not stable: any slight perturbation leads to a break-up of the orbit.)

Problem 3.11 The center of our galaxy

Remarkable data indicate the presence of a massive black hole at the center of our Milky Way galaxy. One of the two W. M. Keck 10 meter diameter telescopes in Hawaii was used by Andrea Ghez and her colleagues to observe infrared light coming directly through the dust surrounding the central region of our galaxy (visible light is multiply scattered by the dust, blocking a direct view). Stars were observed for several consecutive years to determine their orbits near a motionless center that is completely invisible ("black") in the infrared but whose precise location is known due to its strong output of radio waves, which are observed by radio telescopes. The data were used to show that the object at the center must have a mass that is huge compared to the mass of our own Sun, whose mass is a "mere" 2×10^{30} kg.

(a) The speeds of the stars as a function of distance from the center were consistent with Newtonian predictions for orbits around a massive central object. How should the speed of each star depend on its distance from the center, and the mass of the central object? Express your result symbolically.

(b) The star nearest the center of our galaxy was observed to be about 1×10^{14} m from the center, with a period of about 15 years. What is the speed of this star in m/s? Also express the speed as a fraction of the speed of light. (The actual orbit is an ellipse as reproduced in Figure 3.74, but for purposes of this analysis you may approximate it as being circular.)

(c) This is an extraordinarily high speed for a star. Is it reasonable to approximate the star's momentum as mv?

(d) Calculate the mass of the massive black hole about which this star is orbiting.

(e) How many of our Suns does this represent?

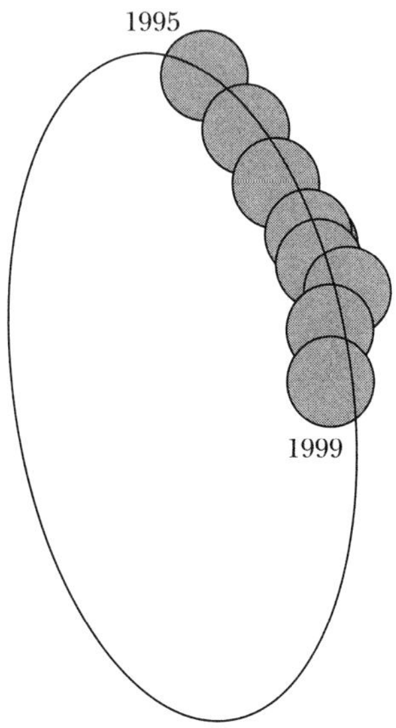

Figure 3.74 Observed positions of a star from 1995 to 1999 near the center of our Milky Way Galaxy, and an orbit fit to the data (Problem 3.11).

It is thought that all galaxies may have such a black hole at their centers, as a result of long periods of mass accumulation. When many bodies orbit each other, sometimes an object happens in an interaction to acquire enough speed to escape from the group, and the remaining objects are left closer together. Some simulations show that over time, as much as half the mass may be ejected, with agglomeration of the remaining mass. This could be part of the mechanism for forming massive black holes.

For more information, see the home page of Andrea Ghez, http://www.astro.ucla.edu/faculty/ghez.htm. The papers available there refer to "arcseconds," which is an angular measure of how far apart objects appear on the sky, and to "parsecs," which is a distance equal to 3.3 light-years (a light-year is the distance light goes in one year).

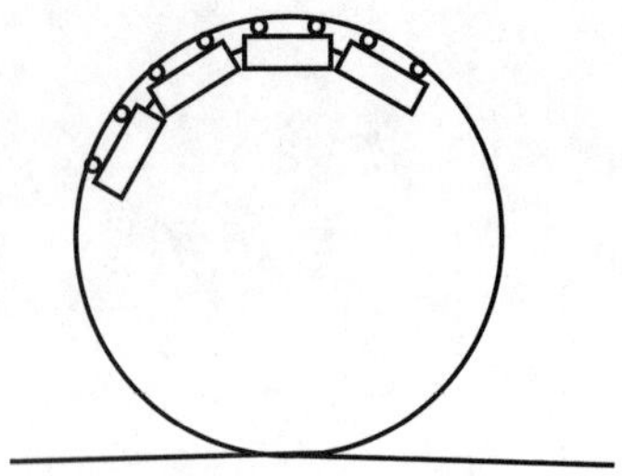

Figure 3.75 Make a loop in a roller coaster (Problem 3.12).

Problem 3.12 Looping roller coaster
What is the minimum speed v that a roller coaster car must have in order to make it around an inside loop and just barely lose contact with the track at the top of the loop (Figure 3.75)? The center of the car moves along a circular arc of radius R. Include a carefully labeled force diagram. State briefly what approximations you make. Design a plausible roller coaster loop, including numerical values for v and R.

Problem 3.13 Your weight on a bathroom scale
If you stand on a bathroom scale at the North Pole, the scale shows your "weight" as an amount Mg (actually, it shows the force F_N that the scale exerts on your feet). At the North Pole you are 6357 km from the center of the Earth. If instead you stand on the scale at the equator, the scale reads a different value due to two effects: 1) The Earth bulges out at the equator (due to its rotation), and you are 6378 km from the center of the Earth. 2) You are moving in a circular path due to the rotation of the Earth (one rotation every 24 hours). Taking into account both of these effects, what does the scale read at the equator?

Problem 3.14 Space station
An engineer whose mass is 70 kg holds onto the outer rim of a rotating space station whose radius is 14 m and which takes 30 s to make one complete rotation. What is the magnitude of the force the engineer has to exert in order to hold on? What is the magnitude of the net force acting on the engineer?

Problem 3.15 NEAR spacecraft orbits Eros
After the NEAR spacecraft passed Mathilde, on several occasions rocket propellant was expelled to adjust the spacecraft's momentum in order to follow a path that would approach the asteroid Eros, the final destination for the mission. After getting close to Eros, further small adjustments made the momentum just right to give a circular orbit of radius 45 km (45×10^3 m) around the asteroid. So much propellant had been used that the final mass of the spacecraft while in circular orbit around Eros was only 500 kg. The spacecraft took 1.04 days to make one complete circular orbit around Eros. Calculate what the mass of Eros must be.

Problem 3.16 At the bottom of a Ferris wheel
A Ferris wheel is a vertical, circular amusement ride. Riders sit on seats that swivel to remain horizontal as the wheel turns. The wheel has a radius R and rotates at a constant rate, going around once in a time T. At the bottom of the ride, what are the magnitude and direction of the force exerted by the seat on a rider of mass m? Include a diagram of the forces on the rider.

Problem 3.17 Weightlessness
By "weight" we usually mean the gravitational force exerted on an object by the Earth. However, when you sit in a chair your own perception of your own "weight" is based on the contact force the chair exerts upward on your rear end rather than on the gravitational force. The smaller this contact force is, the less "weight" you perceive, and if the contact force is zero, you feel peculiar and "weightless" (an odd word to describe a situation when the only force acting on you is the gravitational force exerted by the Earth!). Al-

so, in this condition your internal organs no longer press on each other, which presumably contributes to the odd sensation in your stomach.

(a) How fast must a roller coaster car go over the top of a circular arc for you to feel "weightless"? The center of the car moves along a circular arc of radius R (Figure 3.76). Include a carefully labeled force diagram.

(b) How fast must a roller coaster car go through a circular dip for you to feel three times as "heavy" as usual, due to the upward force of the seat on your bottom being three times as large as usual? The center of the car moves along a circular arc of radius R (Figure 3.77). Include a carefully labeled force diagram.

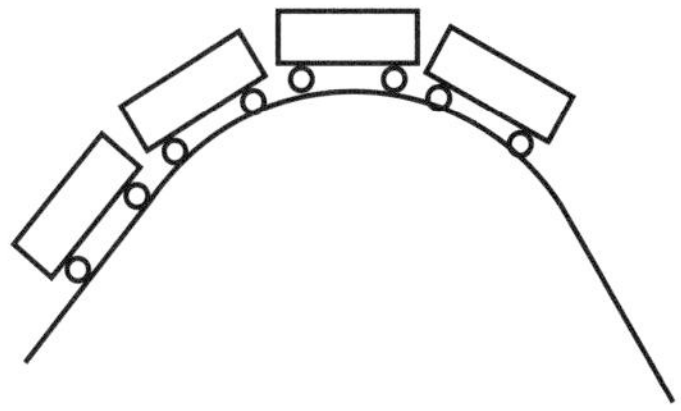

Figure 3.76 Go over the top in a roller coaster (Problem 3.17a).

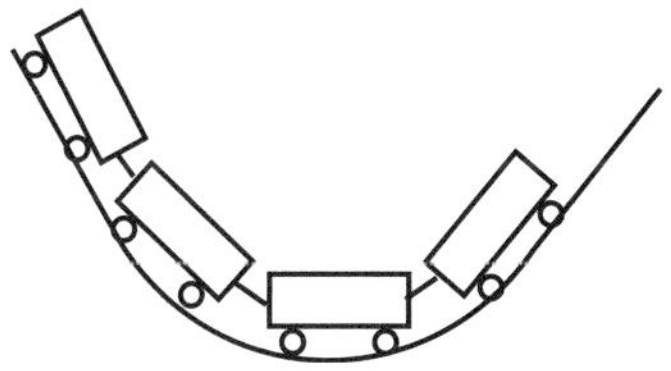

Figure 3.77 Go through a dip in a roller coaster (Problem 3.17b).

Problem 3.18 Swing a ball on a spring

A ball of unknown mass m is attached to a spring. In outer space, far from other objects, you hold the other end of the spring and swing the ball around in a circle of radius 1.5 m at constant speed.

(a) You time the motion and observe that going around 10 times takes 6.88 seconds. What is the speed of the ball?

(b) Is the momentum of the ball changing or not? How can you tell?

(c) If the momentum is changing, what interaction is causing it to change? If the momentum is not changing, why isn't it?

(d) The relaxed length of the spring is 1.2 m, and its stiffness is 1000 N/m. While you are swinging the ball, since the radius of the circle is 1.5 m, the length of the spring is also 1.5 m. What is the magnitude of the force that the spring exerts on the ball?

(e) What is the mass m of the ball?

Problem 3.19 Car going over a hill

A sports car (and its occupants) of mass M, is moving over the rounded top of a hill of radius R. At the instant when the car is at the very top of the hill, the car has a speed v. You can safely neglect air resistance.

(a) Taking the sports car as the system of interest, what object(s) exert non-negligible forces on this system?

(b) At the instant when the car is at the very top of the hill, draw a diagram showing the system as a dot, with force vectors whose tails are at the location of the dot. Label the force vectors (that is, give them algebraic names). Try to make the lengths of the force vectors be proportional to the magnitudes of the forces.

(c) Starting from the Momentum Principle, calculate the force exerted by the road on the car.

(d) Under what conditions will the force exerted by the road on the car be zero? Explain.

Problem 3.20 Ferris wheel

A person of mass 70 kg rides on a Ferris wheel whose radius is 4 m. The person's speed is constant at 0.3 m/s. The person's location is shown by a dot in the diagram in Figure 3.78.

(a) What is the magnitude of the rate of change of the momentum of the person at the instant shown?.

(b) What is the direction of the rate of change of momentum of the person at the instant shown?

(c) What is the magnitude of the *net* force acting on the person at the instant shown? Draw the *net* force vector on the diagram at this instant, with the tail of the vector on the person.

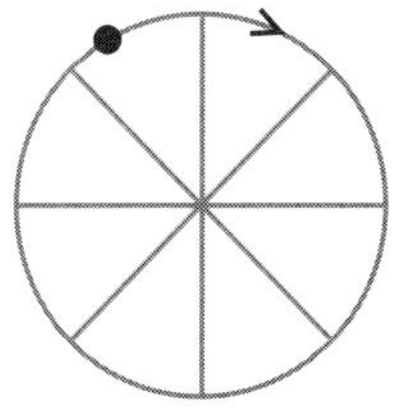

Figure 3.78 A person (represented by a dot) riding on a Ferris wheel (Problem 3.20).

Problem 3.21 A spring-like device

A certain spring-like device with uneven windings has the property that to stretch it an amount s from its relaxed length requires a force that is given by $F = bs^3$. You suspend this device vertically, and its unstretched length is 20 cm.

(a) You hang a mass of 15 grams from the device, and you observe that the length is now 24 cm. What is b, including units? Start your analysis from the Momentum Principle.

(b) You hold the 15 gram mass and throw it downward, releasing it when the length of the spring-like device is 27 cm and the speed of the mass is 4 m/s. One millisecond later (10^{-3} s), what is the approximate change in the stretch of the device, and what is the approximate change in the speed of the mass?

Problem 3.22 Pushing a rod through space

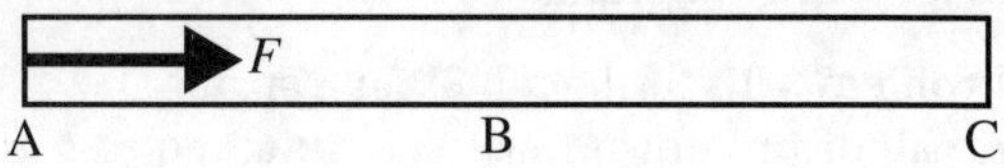

(a) In outer space, a rod is pushed to the right by a constant force F. Describe the pattern of interatomic distances along the rod. Include a specific comparison of the situation at locations A, B, and C. Explain briefly in terms of fundamental principles. Hint: Consider the motion of an individual atom inside the rod, and various locations along the rod.

(b) After the rod in part (a) reaches a speed v, the object that had been exerting the force on the rod is removed. Describe the subsequent motion of the rod, and the pattern of interatomic distances inside the rod. Include a specific comparison of the situation at locations A, B, and C. Explain briefly.

Problem 3.23 Pushing on a block

Two blocks of mass m_1 and m_3, connected by a rod of mass m_2, are sitting on a low-friction surface, and you push to the left on the right block (mass m_1) with a constant force (Figure 3.79).

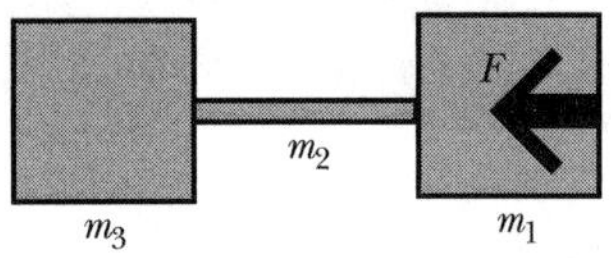

Figure 3.79 Push on a block (Problem 3.22).

(a) What is the acceleration dv_x/dt of the blocks?

(b) What is the compression force in the rod (mass m_2) near its right end? Near its left end?

(c) How would these results change if you *pull* to the left on the *left* block (mass m_3) with the same force, instead of pushing the right block?

Problem 3.24 Interatomic spring stiffness

Young's modulus for aluminum is 6.2×10^{10} N/m^2. The density of aluminum is 2.7 grams/cm^3, and the mass of one mole is 27 grams. If we model the interactions of neighboring aluminum atoms as though they were connected by springs, determine the approximate spring stiffness of such a spring. Repeat this analysis for lead: Young's modulus for lead is 1.6×10^{10} N/m^2, the density of lead is 11.4 grams/cm^3, and the mass of one mole is 207 grams. Make a note of these results, which we will use for various purposes later on. Note that aluminum is a rather stiff material, whereas lead is quite soft.

Problem 3.25 Computer model of the propagation of sounds in a metal

Create a computer model of a long metal bar as many atomic masses connected by interatomic springs, all in a straight line. Give the first atom a sudden push and compute the motions of all the atoms, one time step after another. Calculate all of the forces before using these forces to update the momenta and positions of the objects. Otherwise the calculations of the forces would mix positions corresponding to different times.

After each time step, display the displacements of all the atoms away from their equilibrium positions, and determine the speed of sound (the distance between the first and last atoms, divided by the amount of time it takes for the pulse to reach the last atom). Compare with the prediction that $v = (\sqrt{k_s/m})d$.

Problem 3.26 The interatomic spring stiffness for copper
A hanging copper wire with diameter 2 mm (2×10^{-3} m) is initially 3 m long. When a 5 kg mass is hung from it, the wire stretches an amount 0.425 mm, and when a 10 kg mass is hung from it, the wire stretches an amount 0.85 mm. A mole of copper has a mass of 63 grams, and its density is 9 grams/cm^3. Find the approximate value of the effective spring stiffness of the interatomic force, and explain your analysis.

Problem 3.27 Observing spring-mass systems
Using masses and springs similar to those you used, it was found that a 20 gram mass hanging from a spring had an oscillation period of 1.2 seconds.

(a) When two 20 gram masses are hung from this spring, what would you predict for the period in seconds? Explain briefly.

(b) When one 20 gram mass is supported by two of these vertical, parallel springs (Figure 3.80), what would you predict for the period in seconds? Explain briefly.

(c) Suppose you cut one spring into two equal lengths, and you hang one 20 gram mass from this half spring. What would you predict for the period in seconds? Explain briefly.

(d) Suppose you take a single (full-length) spring and a single 20 gram mass to the Moon and watch the system oscillate vertically there. Will the period you observe on the Moon be longer, shorter, or the same as the period you measured on Earth? (The gravitational field strength on the Moon is about one-sixth that on the Earth.) Explain briefly.

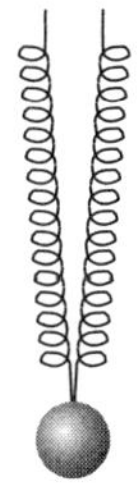

Figure 3.80 Hang the mass from two springs (Problem 3.27).

Problem 3.28 Vibration of diatomic molecules
In Exercise 3.16 and Exercise 3.17 we found the effective spring stiffness corresponding to the interatomic force for aluminum and lead. Let's assume for the moment that, *very* roughly, other atoms have similar values.

(a) What is the (very) approximate frequency f for the vibration of H_2, a hydrogen molecule?

(b) What is the (very) approximate frequency f for the vibration of O_2, an oxygen molecule?

(c) What is the approximate vibration frequency f of D_2, a molecule both of whose atoms are deuterium atoms (that is, each nucleus has one proton and one neutron)?

(d) Explain why the *ratio* of the deuterium frequency to the hydrogen frequency is quite accurate, even though you have estimated each of these quantities very approximately, and the effective spring stiffness is normally expected to be significantly different for different atoms. (Hint: what interaction is modeled by the effective "spring"?)

Problem 3.29 Speed of sound in uranium
Uranium-238 (U^{238}) has three more neutrons than uranium-235 (U^{235}). Compared to the speed of sound in a bar of U^{235}, is the speed of sound in a bar of U^{238} higher, lower, or the same? Explain your choice, including justification for assumptions you make.

Problem 3.30 Speed of sound in nickel
One mole of nickel (6×10^{23} atoms) has a mass of 59 grams, and its density is 8.9 grams per cubic centimeter. You have a bar of nickel 2.5 m long, with a square cross section, 2 mm on a side. You hang the rod vertically and attach a 40 kg mass to the bottom, and you observe that the bar becomes 1.2 mm longer.

Next you remove the 40 kg mass, place the rod horizontally, and strike one end with a hammer.

How much time T will elapse before a microphone at the other end of the bar will detect a disturbance? Be sure to show clearly the steps in your analysis.

Problem 3.31 Falling through the Earth

In the approximation that the Earth is a sphere of uniform density, it can be shown that the gravitational force it exerts on a mass m inside the Earth at a distance r from the center is $mg(r/R)$, where R is the radius of the Earth. (Note that at the surface, the force is indeed mg). Suppose that there were a hole drilled along a diameter straight through the Earth, and the air were pumped out of the hole. If an object is released from one end of the hole, how long will it take to reach the other side of the Earth? Include a numerical result.

Problem 3.32 Floating objects

(a) A block of wood 20 cm long by 10 cm wide by 6 cm high has a density of 0.7 grams/cm^3 and floats in water (density 1.0 grams/cm^3). How far below the surface of the water is the bottom of the block? Explain your reasoning.

(b) An advertising blimp consists of a gas bag, the supporting structure for the gas bag, and the gondola hung from the bottom, with its cabin, engines, and propellers. The gas bag is about 30 meters long with a diameter of about 10 meters, and it is filled with helium (density about 4 grams per 22.4 liters under these conditions). Estimate the total mass of the blimp, including the helium.

Problem 3.33 The simple pendulum

A "simple" pendulum (Figure 3.81; next page) consists of a small mass of mass m swinging at the end of a low-mass string of length L (in contrast to a pendulum whose mass is distributed, such as a rod swinging from one end).

(a) Show that the momentum of the small mass obeys the following equation, where s is the arc length $L\theta$:

$$\frac{dp}{dt} = -mg\sin\theta = -mg\sin\left(\frac{s}{L}\right)$$

Figure 3.81 A simple pendulum (Problem 3.33).

(b) What form does this equation take for small-amplitude swings? (Note that if θ is measured in radians, $\sin\theta \approx \theta$ for small angles. For example, even for an angle as large as $\theta = 30°$, $\theta = \pi/6 = 0.524$ radians, which is close to $\sin(30°) = 0.5$, and the approximation gets better for smaller angles.)

(c) Compare the form of the approximate equation you obtained in part (b) with the form of the Momentum Principle for a spring-mass system. By comparing these equations for different systems, determine the period of a simple pendulum for small-amplitude swings. (Note that any time you can approximate the Momentum Principle for some system by an equation that looks like the equation for a mass on a spring, you know that the system will move approximately sinusoidally.)

(d) Make a simple pendulum and predict its period, then measure the period. Do this for a long pendulum and for a short pendulum. Report your experimental data and results, with comparisons to your theoretical predictions.

(e) For large-amplitude swings, the small-angle approximation is not valid, but you can use numerical integration to calculate the motion. For a mass on a string, the largest possible amplitude is 90°, but if the string is replaced by a lightweight rod, the amplitude can be as large as 180° (standing upside-down, with the mass at the top). It turns out that such a pendulum can be modeled by the same equation as the equation derived in part (a), but the concepts of "torque" and "angular momentum" are required to prove it. Assume that the equation of part (a) is valid, and use numerical integration to plot the position and velocity as a function of time for a pendulum whose amplitude is nearly 180°. (Continued on next page.)

Note how different these plots are from sines or cosines, because the system is not well approximated by a "spring-mass" equation. Also note that for this anharmonic oscillator the period increases with increasing amplitude, unlike the situation with a harmonic oscillator.

3.16 Answers to exercises

3.1 (page 123) 4 N

3.2 (page 123) $\vec{0}$, $\vec{0}$

3.3 (page 123) $\vec{0}$, $\vec{0}$

3.4 (page 123) $\langle 2, 0, -4\rangle$ m/s^2, $\langle 0.16, 0, -0.32\rangle$ (kg · m/s)/s, $\langle 0.16, 0, -0.32\rangle$ N

3.5 (page 127) nonzero, downward

3.6 (page 127) nonzero, downward

3.7 (page 127) nonzero, in direction of final momentum

3.8 (page 131) The third (bottom) one (R_3)

3.9 (page 131) downward on the page

3.10 (page 132) 0.975 (kg · m/s)/s toward center

0.975 N toward center

3.11 (page 132) Both $d\vec{p}/dt$ and $\vec{F}_{net}$ point toward the center

3.12 (page 132) 3.57×10^{22} (kg · m/s)/s toward the Sun

3.57×10^{22} N toward the Sun

3.13 (page 138) 8000 m/s (8 km/s)

3.14 (page 139) 22.5 N, toward center of circle

3.15 (page 139) 3×10^{4} m/s

(page 146) 2.6×10^{-10} **m**

3.17 (page 146) 3.1×10^{-10} **m**

3.18 (page 147) 3000 N/m

3.19 (page 147) 30 N

3.20 (page 147) $2k_s$

3.21 (page 149) $Y \approx 6\times10^{10}$ N/m2

3.22 (page 150) $\Delta L \approx 0.2$ millimeter

3.23 (page 153) The period gets shorter and shorter.

3.24 (page 155) $A = 0.05$ m. Minimum and maximum values of x are –0.05 m and + 0.05 m.

3.25 (page 155) $T = 2\pi\sqrt{m/k}$

3.26 (page 158) $v_0 = \frac{F}{m}\Delta t$; $A = v_0\sqrt{\frac{m}{k}}$

3.27 (page 158) 90° ($\pi/2$ radians), corresponding to a sine rather than a cosine (start at x = 0)

3.29 (page 164) $F_{buoyant} = 1.6\times10^{-3}$ N

$$\frac{F_{buoyant}}{F_{weight\ of\ iron}} = 1.6\times10^{-4}$$

3.30 (page 165) For a book the size of this book, there are approximately 10^{19} air molecules between the book and the table.

Chapter 4

The Energy Principle

Chapter 4

The Energy Principle

If your numerical integration of a spring-mass system showed the maximum displacement (the amplitude) growing with time, why would you know that there is something wrong with your calculation? If your computer model of the Earth going around the Sun showed the Earth spiraling outward with greater and greater speed, why would you suspect something is wrong with your calculations? How would you predict how far a car could travel by burning one gallon of gasoline?

These are questions that take us up to and beyond the boundaries of what we can analyze using only the Momentum Principle. The concept of "energy" greatly expands the scope of our models of physical systems. Conservation of energy is a powerful principle that allows us to decide what is possible and what is impossible in the physical world. Although we will introduce the concept of energy conservation in the context of Newtonian mechanics, we will see later that energy conservation is a very general principle, which applies even in situations where Newton's laws do not apply.

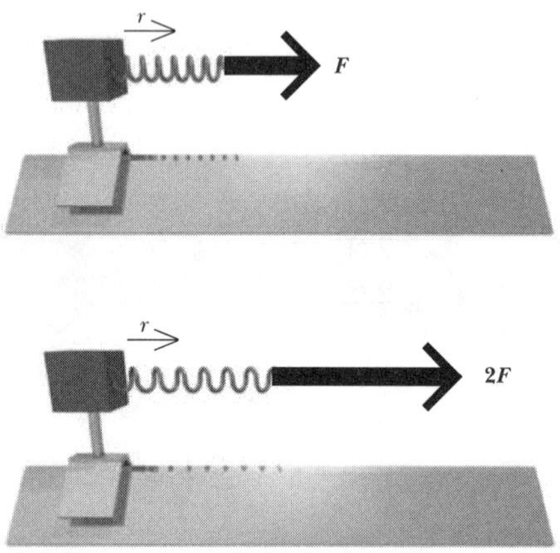

Figure 4.1 Pulling on a spring connected to an object.

4.1 Work: Mechanical energy transfer

We already have a good feel for some aspects of energy from our everyday experience. It is clear that energy is associated with motion—the faster an object is moving, the more energy it has. It is also clear that energy must be expended to get the object moving rapidly—as the old saying goes, there ain't no such thing as a free lunch. Suppose you apply a force on an object and increase its speed. The object gains "kinetic energy" (energy associated with motion). You have lost "chemical" energy, because you have expended some of the energy stored in your body. The net effect is that energy has been transferred from you to the object.

To be quantitative, suppose you apply a constant force F in a constant direction to an object of mass m that was initially at rest, pulling it through a displacement r (Figure 4.1). The object moves faster and faster, acquiring some kinetic energy. If you were to pull twice as hard ($2F$) through the same displacement r, the object would presumably acquire twice as much kinetic energy (and you would expend twice as much chemical energy).

If you lift a heavy weight to a height h above the ground, you will have expended a certain amount of chemical energy (Figure 4.2). If you lift the weight twice as high (a height $2h$ above the ground), you will presumably have had to expend twice as much energy.

Because the amount of energy expended appears to be proportional to the magnitude of the force exerted, and to the distance through which an object is moved, let's tentatively define the amount of energy transfer between you and the object as the product of force times displacement, Fr. This type of mechanical energy transfer is called "work." Here we use the word "work" as a technical term in the context of physics. This technical meaning is only loosely connected with everyday uses of the word.

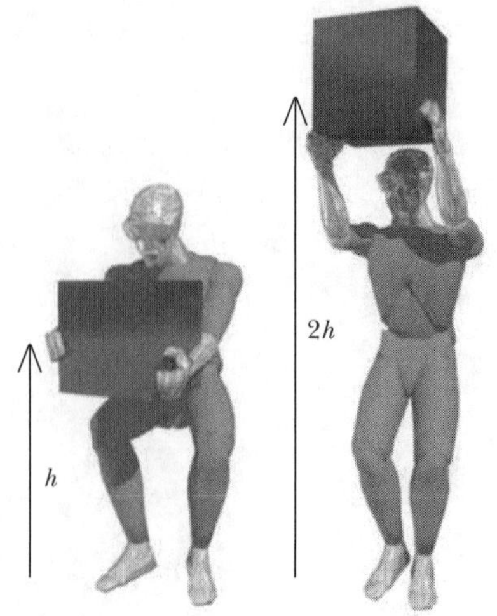

Figure 4.2 You expend more energy when lifting the weight higher.

Contrast between distance and time

Doubling the displacement is not the same as doubling the time. If you pull the block on the air track through twice the displacement, you do twice as much work, and the block acquires twice as much kinetic energy. But because the block is moving faster in the second half of the doubled displacement, the time interval is less than doubled (and the impulse is less than doubled), so the momentum change is not doubled.

Note carefully: work and kinetic energy change are proportional to displacement, while impulse and momentum change are proportional to time interval. In symbols: work $F\Delta r$ is not the same as impulse $F\Delta t$.

Force at an angle

Let's consider the motion of a block pulled through a small displacement Δr by a spring stretched a constant amount s, held at an angle θ to the horizontal, as shown in Figure 4.3. We resolve the spring force into vertical and horizontal components, as shown in Figure 4.4. We pull at an angle such that the block does not move in the vertical direction.

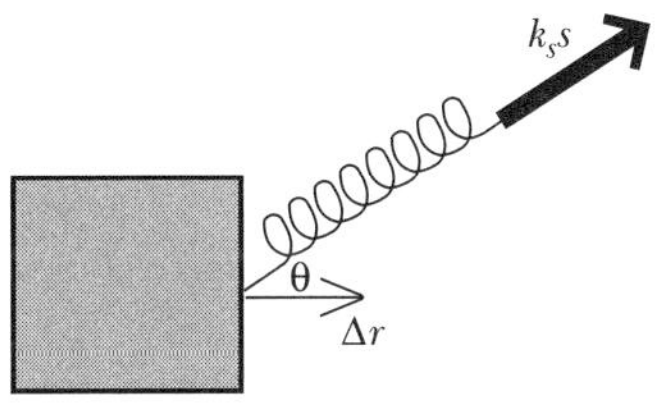

Figure 4.3 Exerting a force at an angle to the motion.

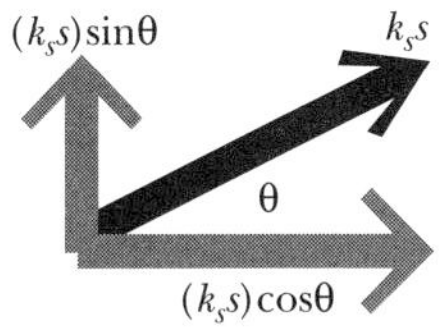

Figure 4.4 *x* and *y* components of the force.

? In calculating the amount of work done to pull the block, should we use the magnitude of the force ($k_s s$), or just the horizontal component of the force ($k_s s\cos\theta$)?

Because there is no vertical displacement, the vertical component of the force does no work:

$$F_y \Delta r_y = F_y \cdot 0 = 0$$

Only the component $k_s s\cos\theta$ in the direction of the motion does work. If the block moves to the right through a displacement r, the amount of work done by the spring force is:

$$F_x \Delta r_x = (k_s s\cos\theta)\Delta r$$

Because only the component of a force parallel to the direction of motion does any work, the work done is given by the product of the component of the force parallel to the displacement times the small displacement:

$$\text{work} \equiv F_{||}\Delta r$$

The unit of energy, 1 newton · meter, is called a joule, abbreviated as J. Mechanical energy transfer ("work") is also measured in joules.

Work can be positive or negative. If work is positive, the object gains energy and the surroundings lose energy. If work is negative, the object loses energy and the surroundings gain energy.

Example: You throw a 250 gram ball straight up, and it rises 8 meters into the air. Neglecting air resistance, how much work is done by the gravitational force?

Solution: $F_{||} = mg\cos\theta = -mg$, because $\theta = 180°$.

$\text{work} \equiv F_{||}\Delta r = [-(0.25\ \text{kg})(9.8\ \text{N/kg})](8\ \text{m}) = -19.6\ \text{joule}$

Negative work is done on the ball, and the ball slows down, losing some energy.

Ex. 4.1 You pull your little sister across a flat snowy field on a sled. Your sister plus the sled have a mass of 20 kg. The rope is at an angle of 35 degrees to the ground. You pull a distance of 50 m with a force of 30 N. How much work do you do?

Ex. 4.2 About how much mechanical work is done by the Sun on the Earth during six months? (A separate kind of energy transfer is solar radiation, which warms up the Earth.)

Ex. 4.3 Contrast work and impulse. If the magnitude of the Earth's momentum is *P*, what is the magnitude of the impulse that the Sun applies to the Earth during six months?

4.1.1 The dot product

As we saw above, the component $F_{||}$ of the force vector $\vec{F}$ in the direction of the displacement $\vec{r}$ is equal to $F\cos\theta$, where θ is the angle between the two vectors $\vec{F}$ and $\vec{r}$. The work done by this force is therefore $(F\cos\theta)\Delta r$, or $F\Delta r\cos\theta$.

A more compact notation involves the "dot product." This mathematical operation is defined for any two vectors in terms of their magnitudes and the cosine of the angle between them, as shown in Figure 4.5.

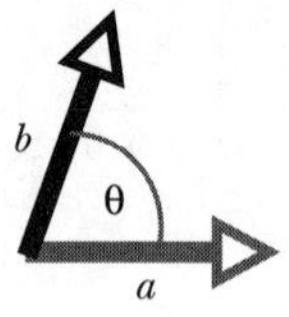

Figure 4.5 θ is the angle between two vectors whose magnitudes are *a* and *b*.

DOT PRODUCT: FORM USING ANGLE θ

$$\vec{a} \bullet \vec{b} \equiv ab\cos\theta \text{ (a scalar quantity)}$$

Note that the result of the dot product is a *scalar* quantity, a single number, and not a vector involving components.

Using this definition, we can write the following compact expression for the work W done by a force $\vec{F}$ acting through a small displacement $\Delta\vec{r}$:

$$W = \vec{F} \bullet \Delta\vec{r}$$

? Using the definition of the dot product, what is the work done by a force of magnitude F acting through a small displacement of magnitude Δr if the angle is 0°? 90°? 180°?

If the force is in the same direction as the displacement (0°), the work is $F\Delta r\cos(0°) = F\Delta r$; if the force is perpendicular to the displacement (90°), the work is $F\Delta r\cos(90°) = 0$; and if the force is opposite to the displacement (180°), the work is $F\Delta r\cos(180°) = -F\Delta r$.

The dot product can also be expressed as the sum of products involving the individual components. This form is often more useful in computations:

DOT PRODUCT: FORM USING COMPONENTS

$$\vec{F} \bullet \Delta\vec{r} = F\Delta r\cos\theta = F_x\Delta r_x + F_y\Delta r_y + F_z\Delta r_z \text{ (a scalar quantity)}$$

To confirm this for yourself, write each vector using unit vectors (vectors of unit length along the *x*, *y*, and *z* axes):

$$\vec{F} = F_x\hat{\imath} + F_y\hat{\jmath} + F_z\hat{k} \text{ and } \Delta\vec{r} = \Delta r_x\hat{\imath} + \Delta r_y\hat{\jmath} + \Delta r_z\hat{k}$$

When you multiply out the product $\vec{F} \bullet \Delta\vec{r}$, note that $\hat{\imath} \bullet \hat{\imath} = 1$ (because $\cos 0° = 1$), and $\hat{\imath} \bullet \hat{\jmath} = 0$ (because $\cos 90° = 0$).

? Writing each vector in terms of its components, again calculate the work done by a force of magnitude F acting through a displacement of magnitude Δr if the angle is 0°, 90°, or 180°. (Choose axes so that $\Delta\vec{r}$ points in the x direction.)

If the force is in the x direction, the work is

$$F_x\Delta r_x + F_y\Delta r_y + F_z\Delta r_z = F\Delta r + (0)(0) + (0)(0) = F\Delta r$$

If the force is in the y direction, the work is

$$F_x\Delta r_x + F_y\Delta r_y + F_z\Delta r_z = (0)\Delta r + F(0) + (0)(0) = 0$$

And if the force is in the $-x$ direction (angle is 180°), the work in that case is

$$F_x\Delta r_x + F_y\Delta r_y + F_z\Delta r_z = (-F)\Delta r + (0)(0) + (0)(0) = -F\Delta r$$

Example: You pull a block 1.5 m across a table in the $-x$ direction, while exerting a force of $(-0.3\text{ N})\hat{\imath} + (0.25\text{ N})\hat{\jmath}$. How much work do you do on the block?

Solution: $W = \vec{F} \bullet \Delta\vec{r} = [(-0.3\text{ N})\hat{\imath} + (0.25\text{ N})\hat{\jmath}] \bullet [(-1.5\text{ m})\hat{\imath}]$

$W = (-0.3\text{ N})(-1.5\text{ m}) = 0.45\text{ J}$

Ex. 4.4 A 2 kg ball rolls off a 30 m high cliff, and lands 25 m from the base of the cliff. Express the displacement and the gravitational force in terms of unit vectors and calculate the work done by the gravitational force. Note that the gravitational force is $-mg\hat{\jmath}$, where g is a *positive* number (+9.8 N/kg).

4.1.2 Non-constant force

If the force changes in magnitude or direction during the process, we can't simply multiply a constant force times the net displacement. The nonconstant force acts on the object along a path, and we split the path into small increments $\Delta\vec{r}$ (Figure 4.6). This is similar to the way in which we took small time steps Δt when computing the trajectories of objects.

If our increments along the path are sufficiently small, the force is approximately constant in magnitude and direction within one increment. We can write the total work done as a sum along the path:

$$W = \vec{F}_1 \bullet \Delta\vec{r}_1 + \vec{F}_2 \bullet \Delta\vec{r}_2 + \vec{F}_3 \bullet \Delta\vec{r}_3 + \ldots \text{ (small steps)}$$

More compactly, we can use the symbol Σ (Greek capital sigma) to mean "sum":

WORK

$$W = \sum \vec{F} \bullet \Delta\vec{r} \text{ (joules; a sum of scalar quantities)}$$

It is worth noting that this technical definition of "work" by itself is merely a definition, with no significance for physics unless we can show that it is related to something else. However we will soon show that, as we suspected from the beginning, work is equal to the amount of energy transferred from the surroundings to an object, through the actions of forces.

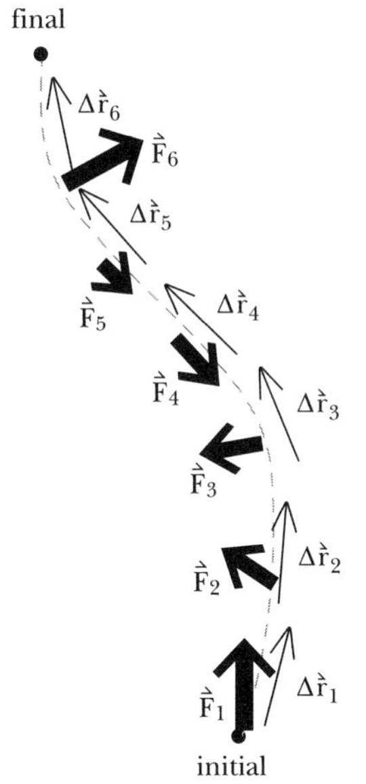

Figure 4.6 A varying force is exerted on an object moving along a curved path.

Ex. 4.5 You push a heavy crate out of a carpeted room and down a hallway with a waxed linoleum floor. While pushing the crate 2.3 m out of the room you exert a force of 30 N; while pushing it 8 m down the hallway you exert a force of 15 N. How much work do you do in all?

Ex. 4.6 You pull on a spring whose spring constant is 20 N/m, and stretch it from its equilibrium length of 0.2 m to a length of 0.3 m. Estimate the work done by dividing the stretching process into two stages and using the average force you exert to calculate work done during each stage.

4.1.3 Comment: Work as an integral

If we go to the limit where the increments are infinitesimal ($\Delta\vec{r} \to d\vec{r}$), the sum used for calculating work turns into a sum of an infinite number of infinitesimal contributions. Such a summation is called a "definite integral." It is written as a distorted "S" for "sum," with an indication of the initial position "*i*" and the final position "*f*":

WORK

$$W = \int_i^f \vec{F} \bullet d\vec{r} \text{ (joules)}$$

If you have already studied integration in a calculus course, you already know some techniques for evaluating definite integrals such as this; if not, you will learn them soon in calculus. In many cases it is not possible to find an analytical form for evaluating an integral, in which case one approximates the infinite sum by a finite sum, with finite increments $\Delta\vec{r}$. This is what you did in the computer analyses to calculate the sums of $\Delta\vec{r}$ and $\Delta\vec{p}$ to find $\vec{r}_f$ and $\vec{p}_f$ from $\vec{r}_i$ and $\vec{p}_i$, and the procedure is called "numerical integration." For example, you calculated the net change in the momentum like this:

$$\vec{p}_{\text{final}} - \vec{p}_{\text{initial}} = \vec{F}_1\Delta t_1 + \vec{F}_2\Delta t_2 + \vec{F}_3\Delta t_3 + ...$$

And you calculated the net change in the position like this:

$$\vec{r}_{\text{final}} - \vec{r}_{\text{initial}} = (\vec{p}_1/m)\Delta t_1 + (\vec{p}_2/m)\Delta t_2 + (\vec{p}_3/m)\Delta t_3 + ...$$

Much of our work in this course will involve finite sums such as these, rather than integrals, because many integrals cannot be evaluated analytically.

4.2 Energy of a particle

Remember that we began our study of work in the hopes of understanding something about energy: when you do work on an object, its energy changes. The question we will now address is, "How much will the speed of a single particle change if we do a specific amount of work on it?"

What is a "particle"?

It is important to be clear about what we mean by a "particle." A particle is an object that does not undergo any internal changes during the process of interest. Some examples may be helpful. A proton may often be treated as a particle, as long as there is no change in the configuration of the quarks that make up the proton. A single water molecule may often be treated as a particle, but not if it rotates and the rotation rate changes, nor if the molecule vibrates and the mode of vibration changes. A baseball or even the whole Earth may often be treated as a particle, in situations where there is no significant change in shape or rotation or vibration or temperature of the object.

Sometimes the notion of particle is emphasized by speaking of a "point particle," which is a particle so small that its size is negligible compared to other distances in the situation. On the scale of our galaxy, the Sun can often be treated as though it were a "point particle."

The best example of a point particle is the electron. The most precise measurements are consistent with the radius of the electron being zero! Of course even more precise measurements might discover a nonzero radius, but we have good reasons to suspect that the radius really is zero. And yet the electron also behaves in some ways like a spinning ball, despite its zero size.

Looking for the right formula for particle energy

We expect the energy of a particle to depend on its speed v and its mass m. For a definition of energy E to be consistent with our definition of work, the following should be true for a particle:

$$\Delta E = W$$

What we mean by the notation ΔE is a *change* in particle energy, which is the final energy minus the initial energy, $E_{\text{final}} - E_{\text{initial}}$. If the particle energy increases from 5 joules to 7 joules, the *change* is +2 joules (final minus initial). If the particle energy decreases from 9 joules to 8 joules, the *change* is –1 joule (final minus initial).

Consider a single point-like particle of mass m. For convenience in the reasoning we are about to do, choose coordinate axes so that the x axis is in the direction of the motion at this instant, so the displacement in the next short time interval is simply Δx (Figure 4.7). There is a net force $\vec{F}$ (magnitude F) acting at an angle θ to the displacement. The physical results we will obtain do not depend on our choice of coordinate axes; we are free to choose axes that simplify our calculations.

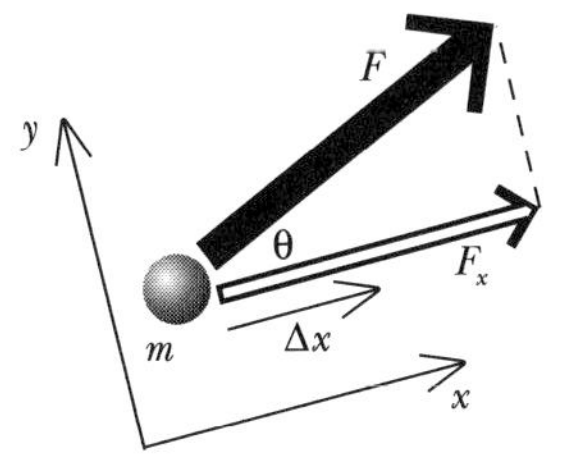

Figure 4.7 A particle of mass m. We choose the x axis to be in the direction of the motion. The net force acting on the particle has a component F_x in the direction of motion.

The component of the net force that is parallel to the displacement, $F_x = F\cos\theta$, affects the speed and hence the kinetic energy of the particle. The perpendicular component F_y merely changes the direction of the momentum.

In the very short displacement Δx, the energy E of the particle changes by a small amount:

$$\Delta E = \vec{F} \bullet \Delta\vec{r} = F_x \Delta x$$

We can rewrite the product $F_x \Delta x$ using the Momentum Principle:

$$\Delta E = F_x \Delta x = \left(\frac{\Delta p_x}{\Delta t}\right)\Delta x$$

Dividing by Δx, we have this:

$$\frac{\Delta E}{\Delta x} = \frac{\Delta p_x}{\Delta t}$$

In the limit as $\Delta t \to 0$, this relates derivatives in the direction of the motion:

$$\frac{dE}{dx} = \frac{dp_x}{dt}$$

In words, this says that the change in energy of the particle per unit distance (in the direction of the motion) is equal to the change in the particle's momentum in that direction per unit time. What we are looking for is a formula for the energy E that satisfies this relationship between the two derivatives. In a section at the end of this chapter we show that the following formula has the right properties:

ENERGY OF A PARTICLE MOVING AT SPEED *v*

$$E = \frac{mc^2}{\sqrt{1 - v^2/c^2}} \quad \text{(single point particle with rest mass } m\text{)}$$

As usual, c is the speed of light (3×10^8 m/s). Like the formula for momentum, this was predicted by Einstein in his special theory of relativity, and it has been confirmed in a very wide range of experiments with high-speed particles.

Note the similarities with and differences from the formula for momentum, $\vec{p} = m\vec{v}/\sqrt{1 - v^2/c^2}$. Both have the same square root in the denominator, and both are proportional to the rest mass m. The energy formula has a scalar value c^2 where the momentum formula has the vector value $\vec{v}$.

Interpreting the energy formula—low speed

Let's see what the energy formula reduces to when the speed v is small compared to the speed of light.

? If a particle is at rest, what is its energy?

Setting $v = 0$ in the energy formula we find the famous formula $E = mc^2$. This has the striking interpretation that a particle has energy even when it is sitting still. We will return to this important point in a moment.

? If the particle is moving with speed v, is its energy greater than mc^2?

If $v > 0$, $\sqrt{1 - v^2/c^2}$ is less than 1, so $1/\sqrt{1 - v^2/c^2}$ is greater than 1, and the particle energy $E = mc^2/\sqrt{1 - v^2/c^2}$ increases with increasing speed. As a result of its motion, the particle has additional, "kinetic" energy K:

KINETIC ENERGY *K* OF A PARTICLE

$$E = \frac{mc^2}{\sqrt{1 - v^2/c^2}} = mc^2 + K$$

$$K = E - mc^2 = \frac{mc^2}{\sqrt{1 - v^2/c^2}} - mc^2$$

If a particle is at rest, the particle energy is just its "rest energy" mc^2. If the particle moves, it has not only its rest energy mc^2 but also an additional amount of energy which we call the kinetic (motional) energy K.

A speed limit

Doing more and more work on a particle increases its energy more and more, and there is no theoretical limit on how much energy a particle can acquire. As the speed approaches the speed of light, the denominator $\sqrt{1 - v^2/c^2}$ becomes very small, and the energy becomes very large.

However, even though you can increase the energy, it becomes very difficult to make large increases in the speed v, because a sizable speed increase would mean a huge energy increase. There is a kind of speed limit in effect: it would require an infinite amount of work to raise the speed v all the way to the speed of light c.

The nonrelativistic approximation

Next consider the case of a particle moving with a speed v that is small compared to the speed of light ($v << c$). An approximate formula for the energy at low speeds can be obtained through use of the mathematics of the "binomial expansion," as is shown in a section at the end of this chapter:

$$\frac{mc^2}{\sqrt{1 - v^2/c^2}} = mc^2\left[1 + \frac{1}{2}\frac{v^2}{c^2} - \frac{3}{8}\left(\frac{v^2}{c^2}\right)^2 - \frac{5}{16}\left(\frac{v^2}{c^2}\right)^3 + ...\right]$$

$$\frac{mc^2}{\sqrt{1 - v^2/c^2}} \approx mc^2 + \frac{1}{2}mv^2 \text{ if } v/c << 1$$

$E = mc^2 + K$, so at low speeds the kinetic energy is approximately this:

NONRELATIVISTIC KINETIC ENERGY *K* OF A PARTICLE

$$K \approx \frac{1}{2}mv^2 = \frac{p^2}{2m} \quad (\text{if } v << c)$$

Example: A proton has a mass of 1.7×10^{-27} kg. What is its rest energy in joules? If its total energy is 3 times its rest energy, what is its kinetic energy K in joules and its speed v in m/s?

Solution: $mc^2 = (1.7\times10^{-27}\text{ kg})(3\times10^8\text{ m/s})^2 = 1.5\times10^{-10}\text{ J}$

$$E = mc^2 + K = 3mc^2\text{, so } K = 2mc^2 = 3\times10^{-10}\text{ J}$$

Because $E = mc^2/\sqrt{1 - v^2/c^2}$, $1 - v^2/c^2 = (mc^2/E)^2 = 1/9$

$v/c = \sqrt{8/9}$, and $v = (0.943)(3\times10^8\text{ m/s}) = 2.82\times10^8\text{ m/s}$

Note that we cannot use $K \approx \frac{1}{2}mv^2$, because v is large.

Ex. 4.7 It is not very difficult to accelerate an electron to a speed that is 99% of the speed of light, because it has such a very small mass. What is the ratio of the kinetic energy K to the rest energy mc^2 in this case? In the definition of what we mean by kinetic energy K, $E = mc^2 + K$, you must use the full relativistic formula for E, because v/c is not small compared to 1.

Ex. 4.8 A pitcher can throw a baseball at about 100 miles/hour (about 44 m/s). What is the ratio of the kinetic energy to the rest energy mc^2? (Can you use $K \approx \frac{1}{2}mv^2$?)

4.2.1 Relativistic energy and momentum

Because both relativistic energy and momentum have the factor

$$\sqrt{1 - v^2/c^2}$$

in the denominator, it seems likely that they can be related in some way. As you can show in Ex. 4.9 on page 190, the relationship turns out to be:

RELATIVISTIC ENERGY AND MOMENTUM OF A PARTICLE

$$E^2 - (pc)^2 = (mc^2)^2 \quad \text{(rest mass } m\text{)}$$

This relation is true in all reference frames. Both energy and momentum will change if a phenomenon is viewed in a different reference frame, but this relation will still be true. The quantity $E^2 - (pc)^2$ is therefore called an "invariant" quantity—a quantity that does not vary when the reference frame is changed.

This relation turns out to be $E^2 - (pc)^2 = 0$ for those particles such as photons (and probably neutrinos) which always travel at the speed of light. Even though they can never be at rest, it is as though they had zero rest mass. The relation for so-called "massless" particles reduces to $E^2 = (pc)^2$, or $E = pc$. This relation is also approximately true for any particle that is traveling at very high speed, because in that case mc^2 is small compared to E.

Example: A proton has a mass of 1.7×10^{-27} kg. If its total energy is 3 times its rest energy, what is its momentum?

Solution: $mc^2 = (1.7\times10^{-27}\text{ kg})(3\times10^8\text{ m/s})^2 = 1.5\times10^{-10}\text{ J}$

$$E = mc^2 + K = 3mc^2 \text{ and } (pc)^2 = E^2 - (mc^2)^2$$

$$pc = \sqrt{E^2 - (mc^2)^2} = \sqrt{(3mc^2)^2 - (mc^2)^2} = \sqrt{8(mc^2)^2} = \sqrt{8}(mc^2)$$

$$p = \sqrt{8}(mc) = \sqrt{8}(1.7\times10^{-27}\text{ kg})(3\times10^8\text{ m/s}) = 1.4\times10^{-18}\text{ kg}\cdot\text{m/s}$$

Ex. 4.9 Show the validity of the relation $E^2 - (pc)^2 = (mc^2)^2$ in the case where $m \neq 0$, by making the substitutions:

$$E = \frac{mc^2}{\sqrt{1 - v^2/c^2}} \text{ and } p = \frac{mv}{\sqrt{1 - v^2/c^2}}$$

4.3 The Energy Principle for a particle

So far we have discussed the work done by a force acting through a small displacement, and the energy of a particle. We now return to the connection between work and energy. Here again is the small energy change ΔE that results from a small amount of work:

$$\Delta E = F_{\|}\Delta r = \vec{F} \bullet \Delta\vec{r}$$

If we add up lots of small amounts of work done on a particle as it moves from some initial position to a final position (Figure 4.8), we have this:

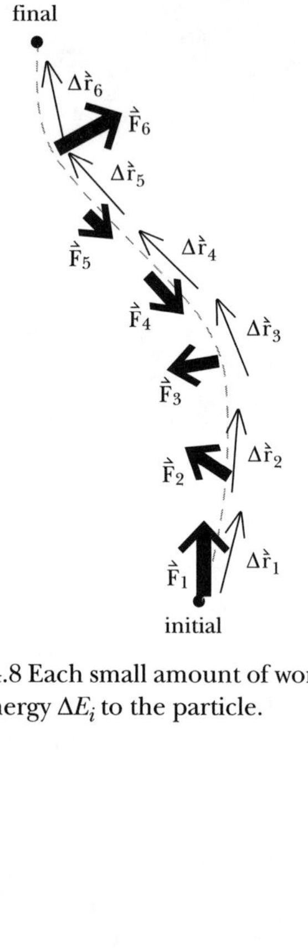

Figure 4.8 Each small amount of work adds some energy ΔE_i to the particle.

$$\Delta E_1 + \Delta E_2 + \Delta E_3 + \ldots = \vec{F}_1 \bullet \Delta\vec{r}_1 + \vec{F}_2 \bullet \Delta\vec{r}_2 + \vec{F}_3 \bullet \Delta\vec{r}_3 + \ldots$$

But the left side of this equation, the sum of all the energy changes, is simply the difference between the initial and final values of the particle energy. Therefore we have this result:

$$\Delta E = \vec{F}_1 \bullet \Delta\vec{r}_1 + \vec{F}_2 \bullet \Delta\vec{r}_2 + \vec{F}_3 \bullet \Delta\vec{r}_3 + \ldots$$

Using the summation symbol, we have the following important result, that the total work W done on a particle is equal to the change in its energy:

THE ENERGY PRINCIPLE FOR A PARTICLE

$$\Delta E = \Delta\left(\frac{mc^2}{\sqrt{1 - v^2/c^2}}\right) = \sum \vec{F}_{net} \bullet \Delta\vec{r} \text{ (for a particle)}$$

If there is no change in the identity of the particle, the mass doesn't change, in which case $\Delta E = \Delta(mc^2 + K)$ is the same as ΔK, the change in the kinetic energy, so we can write $\Delta K = W$.

The Energy Principle for a particle is true for any net force, and does not depend on the net force being constant. Even if the force varies (requiring us to add up the various contributions in a sum), the change in the kinetic energy of a particle is equal to the work done on the particle.

The work done by the net force on the particle can be calculated as a finite sum of small quantities, or as an infinite sum of infinitesimal quantities:

$$W = \sum \vec{F}_{net} \bullet \Delta\vec{r} \rightarrow \int_i^f \vec{F}_{net} \bullet d\vec{r} \text{ (for a particle)}$$

The nonrelativistic limit

At low speeds we have the following approximate form:

NONRELATIVISTIC ENERGY PRINCIPLE FOR A PARTICLE

$$\Delta K = \Delta\left(\frac{p^2}{2m}\right) = \frac{p_f^2}{2m} - \frac{p_i^2}{2m} = \sum \vec{F}_{net} \bullet \Delta\vec{r} \text{ } (v << c\text{, no change in identity})$$

We have developed the relationship between work and particle energy for the general, relativistic case, in order to establish the right foundation. Much of what we will do next will deal with the nonrelativistic approximation, but we will return to relativistic energy later in this chapter.

Example: If you pull on a 3 kg block and change its speed from 4 m/s to 5 m/s, what is the minimum change of chemical (food) energy in you? Give magnitude, sign, and units. (Your temperature rises when you do work like this, involving an increase in thermal energy, so there is more actual chemical cost than the minimum amount you calculate.)

Solution: The speed is tiny compared to c, so the change in kinetic energy of the block is

$$\Delta K = \frac{1}{2}m(v_f^2 - v_i^2) = \frac{1}{2}(3\text{ kg})[(5\text{ m/s})^2 - (4\text{ m/s})^2]$$

$$\Delta K = 13.5\text{ J}$$

Because $\Delta K = W$, you did an amount of work equal to +13.5 J, which means your store of chemical (food) energy has decreased at least this much; the change is at least -13.5 J.

Ex. 4.10 Show that the units of $\frac{p^2}{2m}$ and mc^2 are indeed joules. (Note that 1 N is 1 $\text{kg}\cdot\text{m/s}^2$.)

Ex. 4.11 An automobile traveling on a highway has an average kinetic energy of 1.1×10^5 J. Its mass is 1.5×10^3 kg; what is its average speed? Convert your answer to miles per hour to see if it makes sense.

Ex. 4.12 If you could use all of the mc^2 rest energy of some fuel to provide the car in the previous exercise with its kinetic energy of 1.1×10^5 J, how much mass of fuel would you need?

4.4 Particle interactions: Potential energy

For isolated, individual point particles, there is only one kind of energy, the relativistic particle energy:

$$E = \frac{mc^2}{\sqrt{1 - v^2/c^2}} \quad \text{(single point particle, rest mass } m\text{)}$$

It is often convenient to split this into two parts, $E = mc^2 + K$, though this is only a convenience. We call mc^2 the rest energy, and we call K the kinetic energy (the additional energy the particle has if it is moving).

In systems of interacting particles, such as galaxies of stars interacting gravitationally, or atoms in which the protons and electrons interact electrically, there is energy associated with the interactions. This interaction energy is in addition to the energies of the individual particles. As you might expect, changes in interaction energy are associated with changes in shape of the multiparticle system.

4.4.1 Choosing a system

The word "system" has a special scientific meaning. When we talk about a "system of three particles," as in Figure 4.9, we are implicitly dividing the Universe into two parts: the objects within the system, and everything in the rest of the Universe. Having created this division, it makes sense to talk about interactions that are internal to the system (that is, interactions between the particles within the system) and interactions with objects external to the system.

When considering a problem, we are free to define any collection of objects we wish as the "system" of interest. Once we have done this, however, we must stick strictly to this definition in our analysis. If we decide to include an object in a system, it must be considered this way throughout our analysis.

4.4.2 Interaction energy in a multiparticle system

First we will consider a generic multiparticle system to see how to calculate interaction energy in general. Then we'll be able to calculate the interaction energy for several specific systems of interest to us.

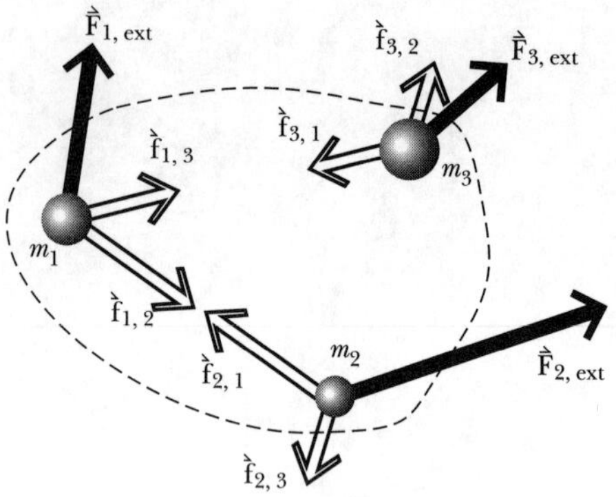

Figure 4.9 A system consisting of three particles. They exert "internal" forces on each other, and additional "external" forces are exerted by objects outside the system.

Consider a system of three particles that exert forces on each other, and which are also acted upon by objects outside this system, as shown in Figure 4.9. For example, these could be three charged particles that exert electric forces on each other but also have springs attached to them that exert additional forces. The forces that are exerted by other particles in the system are called "internal" forces. The forces exerted by objects outside the system are called "external" forces.

The work done by these internal and external forces on each particle changes the particle energy E of that particle. We can write the work-energy relation for each particle:

$$\Delta E_1 = (\vec{f}_{1,\,2} + \vec{f}_{1,\,3} + \vec{F}_{1,\,\text{ext}}) \bullet \Delta\vec{r}_1 = W_{1,\,\text{int. forces}} + W_{1,\,\text{ext. forces}}$$

$$\Delta E_2 = (\vec{f}_{2,\,1} + \vec{f}_{2,\,3} + \vec{F}_{2,\,\text{ext}}) \bullet \Delta\vec{r}_2 = W_{2,\,\text{int. forces}} + W_{2,\,\text{ext. forces}}$$

$$\Delta E_3 = (\vec{f}_{3,\,1} + \vec{f}_{3,\,2} + \vec{F}_{3,\,\text{ext}}) \bullet \Delta\vec{r}_3 = W_{3,\,\text{int. forces}} + W_{3,\,\text{ext. forces}}$$

If we add up the changes in the particle energy for each of the three particles (the left sides of the equations) and the work done on each particle (the right sides of the equations) we can write this:

$$\Delta(E_1 + E_2 + E_3) = W_{\text{int. forces}} + W_{\text{ext. forces}}$$

where $W_{\text{int. forces}}$ is the sum of the work done by internal forces on each of the particles, and $W_{\text{ext. forces}}$ is the sum of the work done by external forces on each of the particles. We emphasize that this represents the sum of the changes in the individual particle energies. Note that the work done by each individual external force involves the displacement of that force (equivalently, the displacement of the particle on which that particular force acts).

We now make a simple algebraic rearrangement that has major conceptual consequences. We move the work done on the particles by the internal forces, $W_{\text{int. forces}}$, to the other side of the equation:

$$\Delta(E_1 + E_2 + E_3) + (-W_{\text{int. forces}}) = W_{\text{ext. forces}}$$

We relabel the term $(-W_{\text{int. forces}})$ as a change of what is called "potential" energy U, energy associated with the interactions of the particles with each other:

$$\Delta(E_1 + E_2 + E_3) + \Delta U = W_{\text{ext. forces}}$$

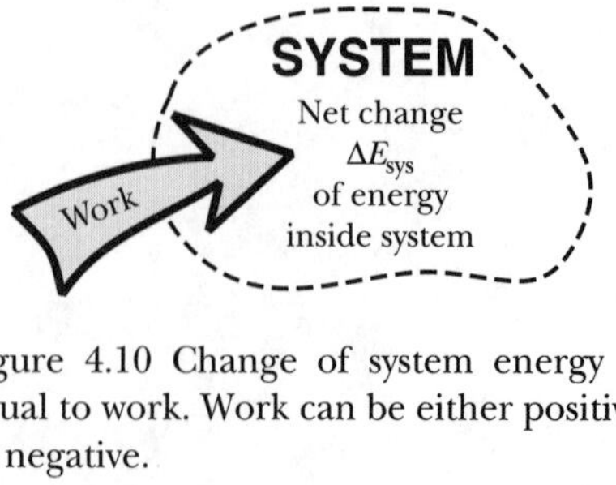

Figure 4.10 Change of system energy is equal to work. Work can be either positive or negative.

We define the total energy of the three-particle system, E_{system}, to be the sum of all the particle energies in the system, plus the potential energy (interaction energy):

$$E_{\text{system}} \equiv (E_1 + E_2 + E_3) + U$$

With this definition, we can write an extremely important equation for how the energy of a multiparticle system changes as a result of work done by external forces (Figure 4.10):

THE ENERGY PRINCIPLE FOR A MULTIPARTICLE SYSTEM

$$\Delta E_{\text{system}} = W_{\text{ext. forces}}$$

$$E_{\text{system}} = (E_1 + E_2 + \ldots) + U$$

E_1, E_2, etc. are the particles' energies in the system

U is the potential energy of the interacting particles in the system

The importance of this way of describing energy changes is that to the left of the equal sign are quantities that are internal to the system, while to the right of the equal sign are the external effects on the system. The interpretation is, "if you do work on a system, you change its energy." The crucial difference between the single-particle work-energy relation and this broader Energy Principle for multiparticle systems is the role of the potential energy U associated with interactions inside the system. Note that external work can affect both particle energy and potential energy inside the system.

You get to choose what system you want to analyze, but once you have specified what portion of the Universe you're considering to be the system of interest, you must be consistent about what effects act on the system (external work done by external forces), versus what energy changes occur inside your system (possible changes in particle energies or potential energy).

It is customary to drop the subscript and write $W_{\text{ext. forces}}$ simply as W, with the understanding that the work W is mechanical energy transfer from the surroundings to the system, with a resulting change in the energy of the system:

$$\Delta E_{\text{system}} = W$$

Remember that W can be positive or negative, resulting in either an energy gain or an energy loss in the system of interest.

Ex. 4.13 Consider an isolated system, by which we mean a system free of external forces. If the sum of the particle energies in the system increases by 50 joules, what must have been the change ΔU in the system's potential energy?

4.4.3 Conservation of energy

A *conserved* quantity is defined to be a quantity that is constant for a system plus its surroundings, where by "surroundings" we mean everything but the system of interest. An example of a conserved quantity is momentum: If an object gains momentum due to forces acting on it, the surroundings lose the same amount of momentum. Thanks to the reciprocity principle, electric or gravitational forces on the object and the surroundings are equal in magnitude and opposite in direction, so the impulses are equal and opposite.

In the 1800's it was discovered experimentally that energy is a conserved quantity. If a system gains energy, the surroundings lose the same amount of energy. If a system loses energy, the surroundings gain the same amount of energy. When new forms of energy such as nuclear energy were discovered in the 1900's, it was found experimentally that energy conservation was still valid. Here is a statement of the conservation of energy:

CONSERVATION OF ENERGY

$$\Delta E_{\text{system}} + \Delta E_{\text{surroundings}} = 0$$

Many systems of particles of practical interest are nearly isolated from their surroundings and can be considered to be nearly free of external influences. For isolated systems we can say that the energy of the system stays constant with time, and the energy is called "a constant of the motion."

In an isolated system the total energy doesn't change, but energy can be swapped back and forth among particle energies and potential energy. For example, in an isolated binary star system the kinetic energies of the stars can increase if there is a corresponding decrease in potential energy. In collisions in an isolated container of helium, some helium atoms gain particle energy and others lose, but the total energy of the gas doesn't change.

A system can exchange energy with its surroundings, and one form of energy exchange is mechanical energy transfer, called work *W*. In that case we have this:

$$\Delta E_{\text{system}} = W$$

$$\Delta E_{\text{surroundings}} = -W$$

$$\Delta E_{\text{system}} + \Delta E_{\text{surroundings}} = 0$$

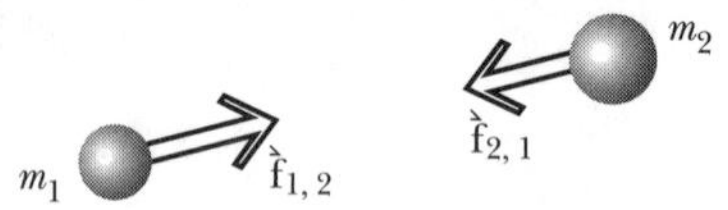

Figure 4.11 A system consisting of two stars interacting gravitationally.

4.4.4 The potential energy of pairs of particles

The reciprocity of gravitational forces and electric forces makes it useful to consider pair-wise contributions to potential energy. Consider the sum of the work done by internal gravitational forces in a short time interval on the individual stars in a system consisting of two stars (Figure 4.11):

$$\Delta U = -W_{\text{int. forces}} = -(\vec{f}_{1,2} \bullet \Delta\vec{r}_1 + \vec{f}_{2,1} \bullet \Delta\vec{r}_2)$$

Because $\vec{f}_{1,2} = -\vec{f}_{2,1}$, we have

$$\Delta U = -(-\vec{f}_{2,1} \bullet \Delta\vec{r}_1 + \vec{f}_{2,1} \bullet \Delta\vec{r}_2) = -\vec{f}_{2,1} \bullet (\Delta\vec{r}_2 - \Delta\vec{r}_1)$$

Because $\Delta\vec{r}_2 - \Delta\vec{r}_1 = \Delta(\vec{r}_2 - \vec{r}_1)$, we can rewrite the position changes:

$$\Delta U = -\vec{f}_{2,1} \bullet \Delta(\vec{r}_2 - \vec{r}_1)$$

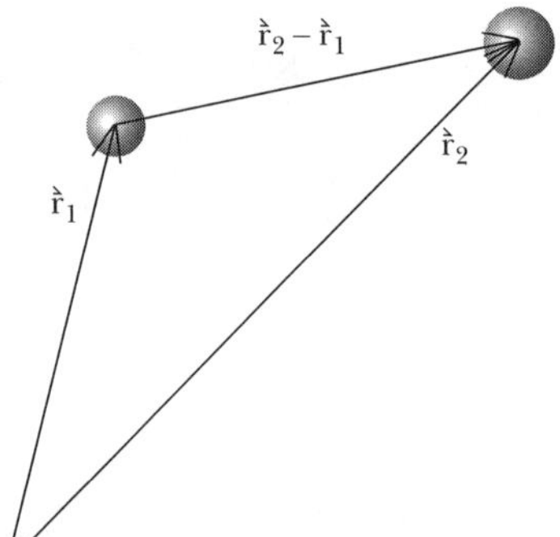

Figure 4.12 The difference in the position vectors of the two stars is the position of star 2 relative to star 1.

But $(\vec{r}_2 - \vec{r}_1)$ is the position of star 2 relative to star 1 (Figure 4.12). Therefore we can calculate the amount of change of potential energy ΔU simply in terms of the relative displacements between pairs of particles. We will express the energy of a multiparticle system in terms of the particle energies plus the pair-wise potential energies. We write U_{12} for the potential energy of the pair consisting of particles 1 and 2, U_{23} for pair 2 and 3, etc.

Ex. 4.14 Write the energy of a system containing four particles, including the relativistic particle energies E_1 etc. plus the potential energy pairs U_{12} etc. How many potential energy pairs are there?

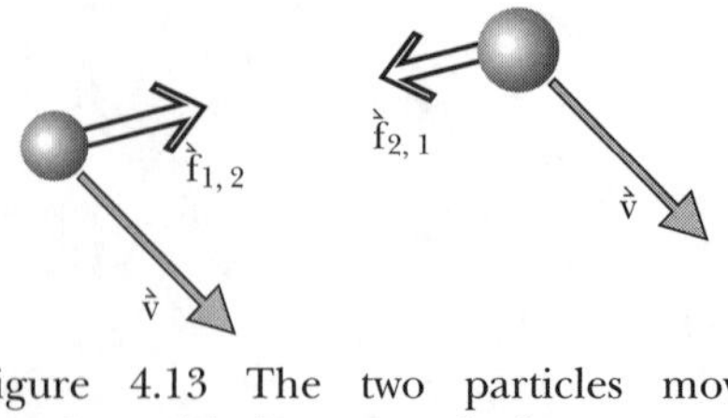

Figure 4.13 The two particles move together, with the same velocity.

4.4.5 Potential energy change involves a change of shape or size

We found that in a short time interval, $\Delta U_{12} = -\vec{f}_{2,1} \bullet \Delta\vec{r}_{21}$, where $\Delta\vec{r}_{21}$ is the change in the position of particle 2 relative to particle 1.

? Suppose that both particles move together, with the same velocity (Figure 4.13). In that case, what is $\Delta\vec{r}_{21}$? What is ΔU_{12}?

In this case the relative positions don't change, $\Delta\vec{r}_{21}$ is zero, and there is no change in the pair-wise potential energy.

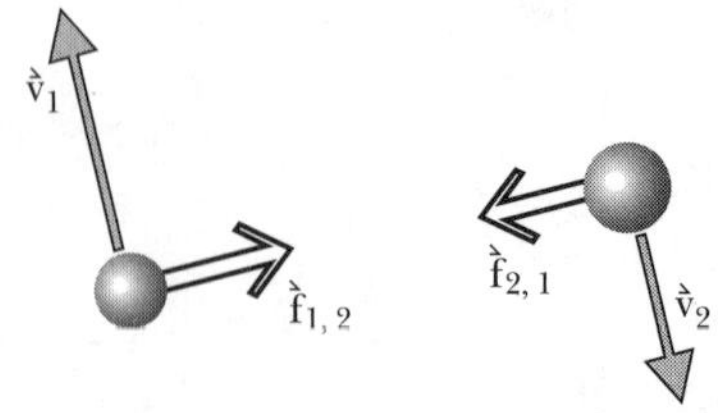

Figure 4.14 The two particles rotate around each other.

? Suppose instead that the two particles rotate around each other (Figure 4.14). What can you say about ΔU_{12}?

This is a bit trickier, because if there is rotation, $\Delta\vec{r}_{21}$ isn't zero, because $\vec{r}_{21}$ is now a rotating vector. But in rotation, $\Delta\vec{r}_{21}$ is perpendicular to $\vec{r}_{21}$, and to the force $\vec{F}_{2,1}$, so the dot product $\vec{F}_{2,1} \bullet \Delta\vec{r}_{21}$ is zero. So again there is no change in the pair-wise potential energy.

? What can you conclude about potential energy for a rigid system?

Evidently U is constant for a rigid system, one whose shape doesn't change. For the potential energy to change, there must be a change of shape or size. Conversely, a change of shape or size is an indicator that U may have changed.

4.4.6 The sign of potential energy

The total energy of a multiparticle system is $E_{\text{system}} = (E_1 + E_2 + \dots) + U$, where $E_1 = m_1c^2/\sqrt{1 - v_1^2/c^2}$ is the particle energy of particle 1, etc. When the particles are very far apart, we must define U to be zero, so that the total energy of these far-apart particles is simply the sum of the particle energies:

U = 0 WHEN THE PARTICLES ARE VERY FAR APART

This is necessary for consistency with particle energy.

If the particles start to come closer together, we find that U starts to become negative if the interactions are attractive, while U starts to become positive if the interactions are repulsive. To see this, first consider an isolated system consisting of two stars that are very far apart, initially at rest, so the initial energy of the system is just the sum of the rest energies of the two stars. At first slowly, then more and more rapidly, their kinetic energies increase due to their mutual gravitational attractions (Figure 4.15).

Figure 4.15 Initially two stars are very far apart, at rest, but they attract each other.

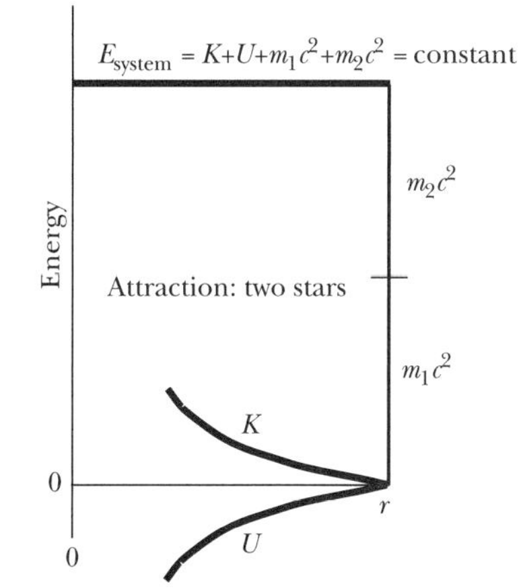

Figure 4.16 A graph of the energies of two stars as a function of separation r. U must be zero at large r. As r decreases, K increases and U decreases. The total energy of the two-star system is constant.

? U was initially zero, so what must be the sign of U when the stars have acquired significant kinetic energies?

If the system is isolated, there are no external forces and hence no external work, so E_{system} does not change. If the relativistic particle energies increase (so the total kinetic energy increases), the potential energy must decrease. Because U was defined to be zero initially, U must become negative. A graph of the energies is shown in Figure 4.16. (In this graph the kinetic and potential energies are greatly exaggerated for clarity. We will soon see that K and U are normally extremely small compared to the rest energies.)

Does this negative sign make sense in terms of the definition of potential energy change, $\Delta U = -W_{\text{int. forces}}$? The attractive gravitational forces that the stars exert on each other certainly do a positive amount of work on each star, so the potential energy change must indeed be negative.

Next consider an isolated system consisting of two protons that are very far apart, and suppose that initially they have high speeds, heading inward (Figure 4.17). Their mutual electric repulsions make the protons slow down.

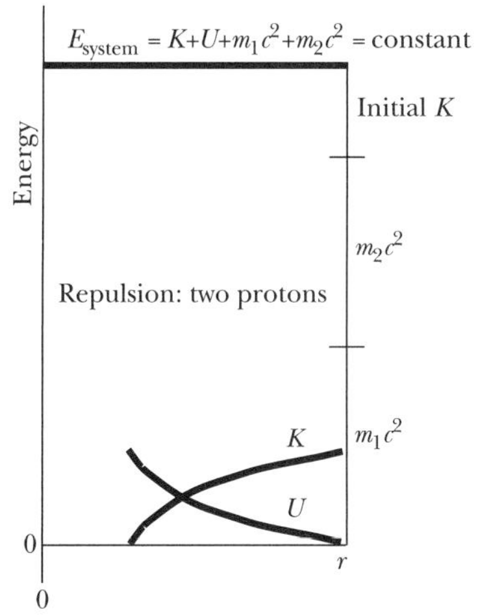

Figure 4.18 A graph of the energies of two protons as a function of separation. U must be zero at large r. As r decreases, K decreases and U increases. The total energy of the two-proton system is constant.

? U was initially zero, so what must be the sign of U when the protons have slowed down somewhat?

Again, there is no external work done on this isolated multiparticle system, so E_{system} does not change. The relativistic particle energies decrease, so the potential energy must increase; U becomes positive (Figure 4.18). Also, the repulsive electric forces do negative work on each other, so $\Delta U = -W_{\text{int. forces}}$ is positive. (Again, in Figure 4.18 we have greatly exaggerated the kinetic and potential energies in comparison with the rest energies.)

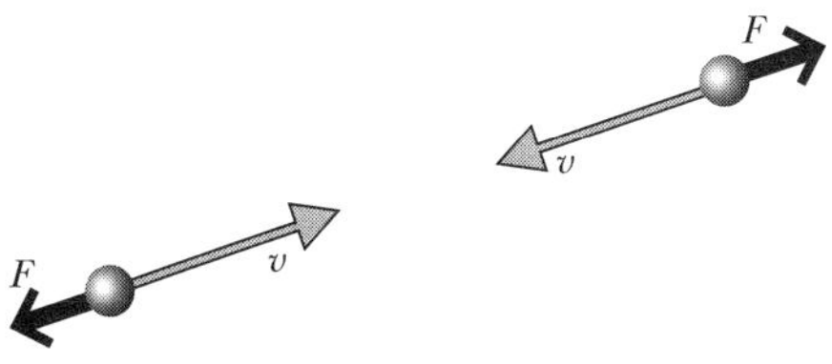

Figure 4.17 Initially two protons are very far apart, heading inward. The mutual repulsions slow the protons down.

More precisely, we will soon show that it is the *change* of U, positive or negative, that indicates whether the interaction is attractive or repulsive. But we wanted to give you a qualitative feel for the issues right away, and in particular to emphasize that there is nothing wrong with a negative value of U; this is expected for an attractive force.

Note that the change in potential energy for the two stars or the two protons was associated with a change in shape of the two-particle system. If there is no change in shape or size, there is no change in potential energy.

4.4.7 A specific example: Gravitational potential energy

We have investigated some general properties of potential energy. Next we will show how to find a specific formula for potential energy, starting from a force law. We will apply this method to find a formula for gravitational potential energy.

Remember that we found that in a short time interval the change in potential energy depends on the relative displacement $\Delta\vec{r}_{21}$ of object 2 relative to object 1:

$$\Delta U_{12} = -\vec{f}_{2,1} \bullet \Delta\vec{r}_{21}$$

It is the relative displacement that matters, so we can pretend that object 1 remains stationary, and just calculate the work done on object 2. The force is in a line with $\vec{r}_{21}$, so we can evaluate the dot product and write the following, where $F_{g,r}$ is the component of the force on object 2 in the direction of $\vec{r}_{21}$, and r is the distance from object 1 to object 2:

$$\Delta U_g = -F_{g,r}\Delta r$$

Therefore if we already knew the formula for potential energy, we could calculate the associated force like this: $F_{g,r} \approx -\Delta U_g/\Delta r$. If we let Δr become very small, we have the following:

FORCE IS NEGATIVE GRADIENT OF POTENTIAL ENERGY

$$F_{g,r} = -\frac{dU_g}{dr}$$

Figure 4.19 The gradient is the rate of change of y with respect to x.

A rate of change of a quantity with respect to position is called a "gradient." For example, if a hill is said to have "a 7% grade," this means that it rises 7 m vertically for every 100 meters horizontally—the tangent of the angle of the hill (the slope) is 0.07 (Figure 4.19).

All that remains to be done to find a formula for gravitational energy is to think of an expression whose (negative) derivative is the component of the gravitational force acting on m_2 in the direction of $\vec{r}_{21}$. From Figure 4.20 we see that this component of the gravitational force is negative: $F_{g,r} = -G(m_1 m_2/r^2)$. The minus sign reflects the fact that the attractive gravitational force on object 2 points back toward object 1.

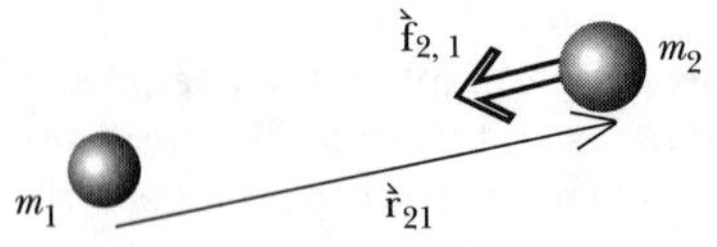

Figure 4.20 The component of the gravitational force, $F_{g,r}$

? Can you think of a function of r whose (negative) derivative with respect to r has this value of $F_{g,r}$? (If necessary, review the appendix on basic calculus at the end of this book.)

The derivative of r^n with respect to r is nr^{n-1}. We have a r^{-2} factor in the gravitational force, so $n = -1$, and the "antiderivative" of r^{-2} (the function whose derivative gives r^{-2}) must be $-r^{-1}$.

? Therefore, what is the formula for gravitational energy?

Being careful about the signs, we find that $U_g = -Gm_1 m_2/r$

? Check that the negative gradient of this gravitational energy is indeed the force, by differentiating with respect to r.

What about the minus sign on U_g? Is it correct? Yes, because the potential energy decreases as the particles get closer together. When stars fall toward each other, their kinetic energies increase and the pair-wise potential energies must decrease to more and more negative values. In summary,

GRAVITATIONAL POTENTIAL ENERGY U_g

$$U_g = -G\frac{m_1 m_2}{r}, \text{ where } r \text{ is the separation of } m_1 \text{ and } m_2$$

A possible constant?

It is true that the gradient of $U_g = (-Gm_1 m_2/r + \text{constant})$ would also give us the correct gravitational force, because the derivative of a constant is zero. However, remember that when the objects are very far apart the total energy of the system must be equal just to the sum of the particle energies, and the potential energy must be zero.

? When the particle separation r is very large, what is the value of $-Gm_1 m_2/r$?

When r is very large, $-Gm_1 m_2/r$ is nearly zero. Therefore, the value of the constant in $U_g = (-Gm_1 m_2/r + \text{constant})$ must be zero, in order for U_g to be zero at large separations. We have shown that the pair-wise gravitational potential energy must be simply $U_g = -Gm_1 m_2/r$, with no added constant.

The minus sign

It may still seem odd that the expression for gravitational potential energy has a minus sign in it. Figure 4.21 shows that although U_g is negative, it does increase with increasing separation r, which means that you have to do work to move two gravitationally attracting objects farther apart.

Here are two little exercises that will help you see why the minus sign is appropriate:

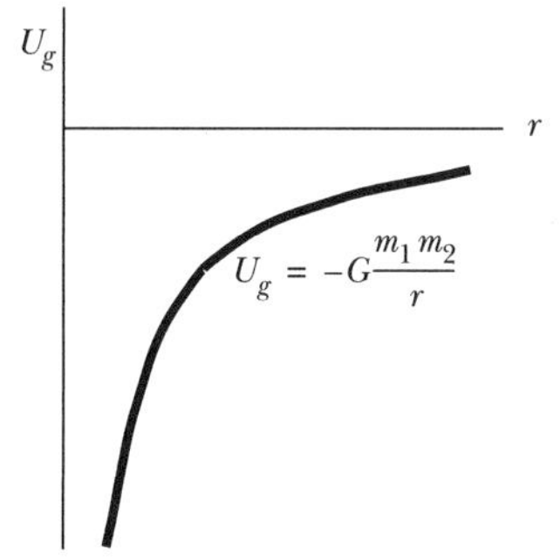

Figure 4.21 Gravitational potential energy is negative. It increases with increasing separation r.

Ex. 4.15 If a block falls toward the Earth, the kinetic energy increases and the gravitational potential energy must decrease. In this process, what is the sign of $\Delta(-GMm/r)$ (the change in the potential energy), in which the final separation r_f is less than the initial separation r_i?

Ex. 4.16 If a block rises away from the Earth, the kinetic energy decreases and the gravitational potential energy must increase. In this process, what is the sign of $\Delta(-GMm/r)$ (the change in the potential energy), in which the final separation r_f is greater than the initial separation r_i?

4.4.8 More about force as gradient of potential energy

Because the (negative) gradient of U is equal to the force, the force is the (negative) slope of a graph of potential energy. This means that you can mentally read the force right off a potential energy graph.

? Where the curve on a potential energy curve is steep, is the force large or small? How can you determine the direction of the force?

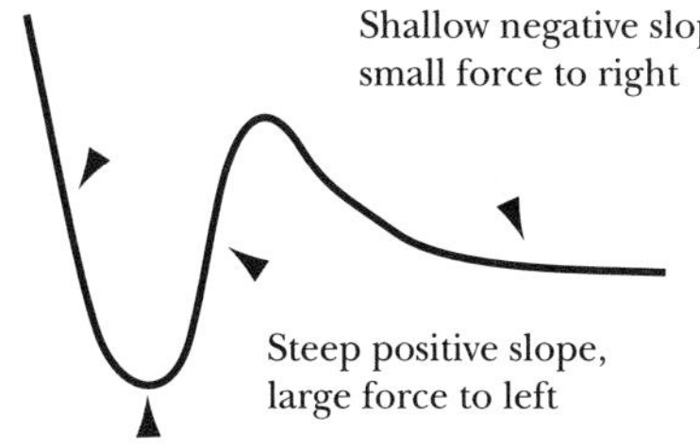

Figure 4.22 Force is (negative) gradient of potential energy, so force is the (negative) slope of the potential energy graph.

Where the curve is steep, the force is large (large gradient), and where the curve is shallow, the force is small (Figure 4.22). Where the slope of the curve is horizontal, the force is zero, and this represents a possible equilibrium position for the system. The component of the force is given by the negative gradient, so a negative slope corresponds to a force to the right, while a positive slope corresponds to a force to the left.

Actually, in determining the x component of the force, we differentiate with respect to x while holding y and z constant. This kind of derivative is called a "partial derivative" and has a special symbol to emphasize that everything else is held constant:

$$F_x = -\frac{\partial U}{\partial x}$$

To get all three components of the force we take partial derivatives of the potential energy with respect to x, y, and z:

FORCE IS NEGATIVE GRADIENT OF POTENTIAL ENERGY

$$F_x = -\frac{\partial U}{\partial x},\ F_y = -\frac{\partial U}{\partial y},\ F_z = -\frac{\partial U}{\partial z}$$

$$\vec{F} = \left\langle -\frac{\partial U}{\partial x}, -\frac{\partial U}{\partial y}, -\frac{\partial U}{\partial z} \right\rangle$$

(Another standard mathematical notation for this relation is $\vec{F} = -\vec{\nabla} U$.)

Ex. 4.17 With y pointing upward from the center of the Earth, find the y component of the gravitational force from $-\partial U/\partial y$, where $U = -GMm/y$. Find the x component of the force from $-\partial U/\partial x$.

4.4.9 Energy diagrams

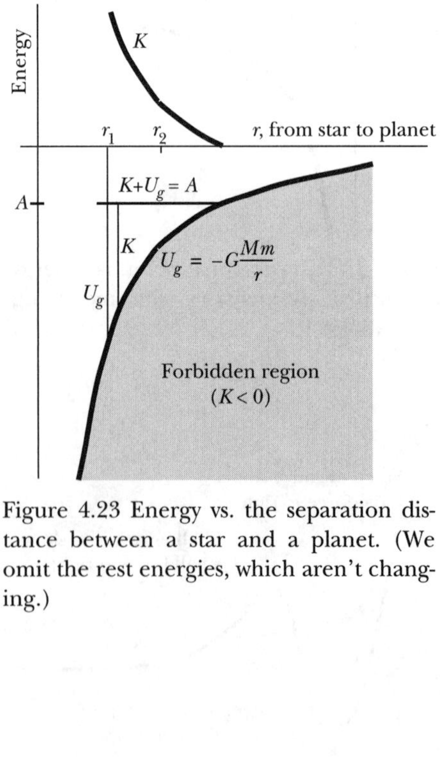

Figure 4.23 Energy vs. the separation distance between a star and a planet. (We omit the rest energies, which aren't changing.)

A graphical representation of energy can be exceptionally helpful in reasoning about a process. Consider an isolated system of a star of mass M and a planet of mass m. As a function of star-planet separation r, we plot a graph of the potential energy $-GMm/r$ (Figure 4.23). We omit the rest energies, which are not changing during the motion.

We add to the graph a horizontal line whose height above or below the axis represents the total kinetic plus potential energy of the two-particle system. The line for $K+U_g$ equal to the value A is horizontal because no external work is done on this isolated system, and there is no change of the rest masses (no change of identity), so $K+U_g$ is constant during the motion.

? Is A positive or negative for the system shown in (Figure 4.23)?

The horizontal line representing $K+U_g = A$ is below the axis representing $U_g = 0$, so A is negative. Go down to the curve by an amount U_g, then up by an amount K, to arrive at the horizontal line that represents the fact that $K+U_g$ is a constant (A). By mentally drawing a vertical line from the potential energy curve up to the horizontal total energy line, you can visualize how the total kinetic energy of the star and planet changes during the motion.

The gray area in (Figure 4.23) represents a forbidden region, where the kinetic energy K would be negative, no matter what the value of $K+U_g$ (more about this in the next section).

Ex. 4.18 The star-planet separation increases from r_1 to r_2 (Figure 4.23). Does the kinetic energy of the star-planet system increase, decrease, or remain constant?

4.4.10 Limits on possible motion

The energy graph shows at a glance the limits on the motion for a given energy. In Figure 4.24, a star and planet with $K+U_g < 0$ cannot get farther away from each other than r_3, a bound that is set by the potential-energy curve. At a larger separation r_4, the kinetic energy would have to be negative, which is "classically forbidden" (that is, impossible according to the laws of classical, prequantum mechanics). Physically, because kinetic energy is the energy a particle has as a result of being in motion (in addition to its rest energy), kinetic energy cannot be negative.

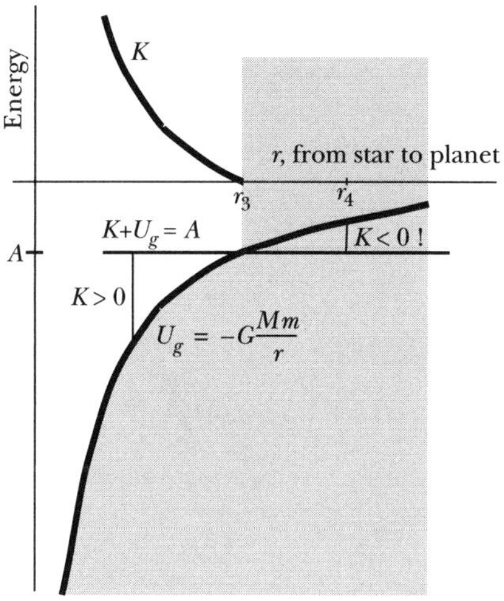

Figure 4.24 The star and planet cannot get farther away from each other than r_3; at r_4 the kinetic energy would be negative. (We omit the rest energies, which aren't changing.)

If the system has the total $K+U_g$ shown ($A < 0$), the separation of the star and planet can never be larger than r_3. An energy graph shows at a glance the range of possible positions achievable during the motion, for a given total $K+U$. We should mention, however, that in the world of atoms where a full analysis requires quantum mechanics, in some cases a system can "tunnel through" a region that is classically forbidden!

An energy graph can show at a glance whether a system is "bound" or "unbound." An example of a bound system is a planet in circular or elliptical orbit around a star: the planet cannot escape. Another example is an electron bound to an atom.

An example of an unbound system is the Earth plus a spacecraft whose initial speed is great enough that the spacecraft will get away from the Earth. For another example, if an amount of energy greater than or equal to the ionization energy is supplied to an atom, an electron can become unbound and escape from the atom. An unbound system has total $K+U_g \geq 0$ (which is easy to see on an energy graph). Only when $K+U_g \geq 0$ can the system become separated by large distances, because K is never negative, and U_g goes to zero at large separations.

Ex. 4.19 In this energy graph for some system (Figure 4.25), consider the various energy states indicated. Which of these values for the energy $K+U$ (A, B, or C) represent a bound state? Which represent an unbound state (the particle can escape)? Which represents a trapped state with enough energy to be unbound but with a barrier that (classically) prevents escape?

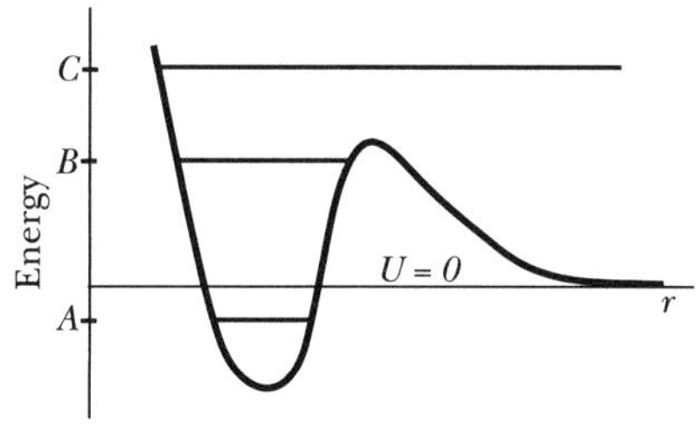

Figure 4.25 An energy graph for some system. Which states are bound? (Exercise 4.19)

4.5 Applications of gravitational potential energy

In the next sections we apply the Energy Principle to the analysis of a variety of situations involving gravitational energy. We will frequently use energy diagrams as a powerful tool.

4.5.1 Application: Energy graphs and types of orbits

In Figure 4.26 are plotted graphs of potential energy, and potential plus kinetic energy, for three different orbits of a planet and star. The following exercise asks you to interpret the meaning of these graphs.

Ex. 4.20 In Figure 4.26, what kind of motion is represented by the situation with $K+U_g = A$? B? C? Think about the range of r in each situation. For example, C represents a circular orbit (constant r).

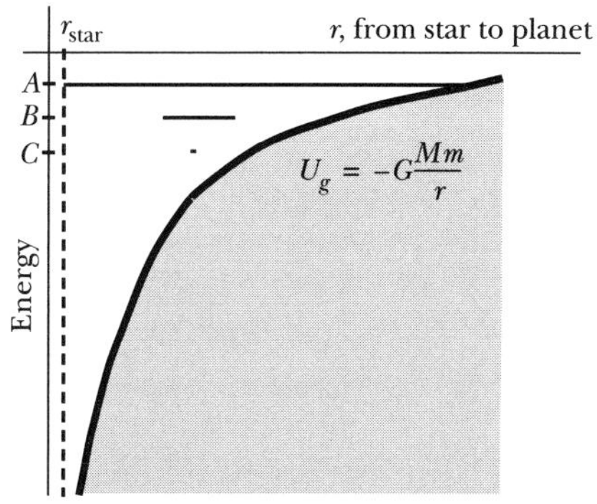

Figure 4.26 What kind of motion is represented by each of these situations? (Exercise 4.20).

4.5.2 Application: Escape speed

An analysis based on the Energy Principle will allow us to answer the following question: With what minimum speed v must a spacecraft leave the surface of an airless planet (no air resistance) of mass M and radius R in order that it can coast away without ever coming back (Figure 4.27)? Assume there are no other objects near by.

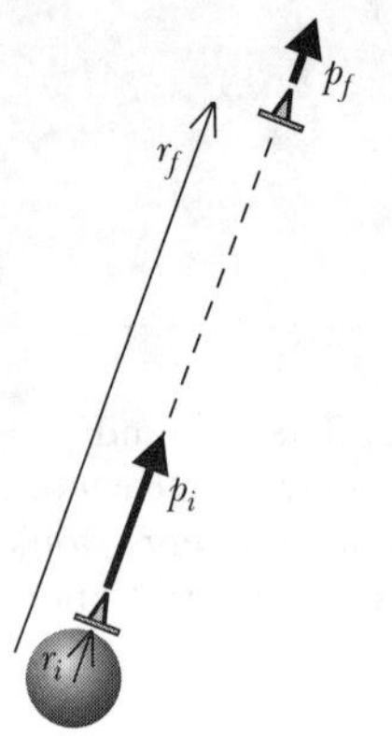

Figure 4.27 A spacecraft is launched with momentum $\vec{p}_i$ from near an airless planet's surface; it has momentum $\vec{p}_f$ when it has traveled far away.

If you did the Moon voyage problem (Problem 2.7 The Ranger 7 mission to the Moon) you found that there was a minimum initial speed required to reach the Moon. With an initial speed smaller than this, the spacecraft fell back to Earth without reaching the Moon. Your determination of the minimum initial speed was done by trial and error. However, we can now use energy relationships to calculate such minimum speeds directly.

As is often the case, an energy diagram can be very helpful. Consider Figure 4.28 (page 200). Motions with $K+U_g$ = A, B, C, or D all start from the surface of an airless planet, with a separation R between the center of the planet and the spacecraft. (Remember that the gravitational force outside a uniform sphere is exactly the same as it would be if all the mass collapsed to the center of the sphere.)

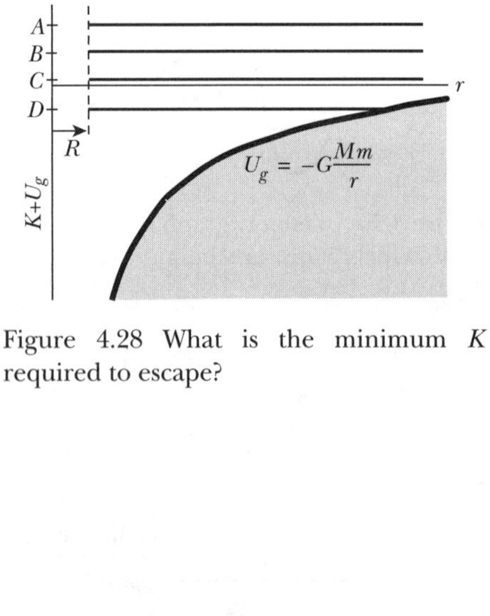

Figure 4.28 What is the minimum K required to escape?

? Which of these motions starts with the largest kinetic energy K? Which starts with the least K?

Each of the horizontal lines represents motion with $K+U_g$ constant. Evidently A is the highest value of $K+U_g$, and therefore involves the largest initial kinetic energy. D is the lowest value of $K+U_g$, and starts with the least kinetic energy.

? For which of these motions does the spacecraft "escape" and never come back?

For motions with $K+U_g = A$, B, or C, no matter how far from the planet the spacecraft gets, there is still some kinetic energy, so the spacecraft will never return. This is another example of an unbound state being associated with a positive value of $K+U_g$.

In contrast, with motion $K+U_g = D$ the spacecraft will reach a maximum separation from the planet, at which point its kinetic energy has fallen to zero. The spacecraft will fall back to the planet.

? How would you describe the motion that requires the least possible initial kinetic energy to achieve escape?

Evidently the least costly escape is to have $K = 0$ when the spacecraft has reached a distance very far from the planet. At a large separation, $U_g = 0$, so $K+U_g = 0$. This corresponds on the diagram to a horizontal line lying on the axis. Because $K+U_g$ doesn't change (no external work on the planet-spacecraft system), it must also be true that $K+U_g = 0$ at the start of the motion:

MINIMAL CONDITION FOR ESCAPE

$$K+U_g = 0$$

The minimal initial kinetic plus potential energy of the system composed of planet plus spacecraft is this, assuming that the kinetic energy of the planet is negligible:

$$K_i + U_i = \frac{p^2}{2m} + \left[-G\frac{Mm}{r}\right]_{r=R} = 0 \text{ (for } v \ll c)$$

This lets us calculate the minimum speed, which is called the "escape speed" v_{esc}, from a planet with mass M and radius R.

$$\frac{p_{esc}^2}{2m} = \frac{1}{2}mv_{esc}^2 = G\frac{Mm}{R}, \text{ or } v_{esc} = \sqrt{\frac{2GM}{R}}$$

Though we speak of "escape speed," it is wrong to say that the spacecraft has "escaped from gravity." As long as the spacecraft is a finite distance from the planet, it feels a finite gravitational attraction to the planet.

If you give the spacecraft more than the minimum kinetic energy, it not only escapes but has nonzero kinetic energy when it is far away (for example, see motions A and B in Figure 4.28).

Don't memorize the formula for escape speed!

You should not memorize the formula for the escape speed. Instead, you should study the analysis leading to this result, especially the graphical analysis, and be able to rederive the formula whenever you need it. Memorization of such easily derivable formulas turns them into dead knowledge, with no context, and carries the very real danger that the formulas will be misapplied outside their range of validity. For example, the formula above does not apply if the spacecraft has some nonzero speed when it is far from the planet.

Acquiring the ability to obtain such results starting from fundamental principles makes the result come alive, gives you practice in reasoning from fundamental principles, and helps you avoid using the formula in situations where it isn't valid.

Initial direction

It is interesting to note that in this analysis, the initial direction of motion doesn't matter. An energy analysis is insensitive to direction. If the initial velocity points directly away from the planet, the path is a straight line. If the initial velocity points in some other direction, the path can be shown to be a parabola if the initial speed is exactly equal to escape speed, and a hyperbola for higher speeds.

If the speed is below escape speed, there is no escape. The various possible orbits (Figure 4.29) can be classified on the basis of their total kinetic plus potential energy, with initial speed v_i:

$\frac{1}{2}mv_i^2 + \left(-G\frac{Mm}{r}\right) > 0$: escape with $v_\infty > 0$; straight line or hyperbola

$\frac{1}{2}mv_i^2 + \left(-G\frac{Mm}{r}\right) = 0$: escape with $v_\infty = 0$; straight line or parabola

$\frac{1}{2}mv_i^2 + \left(-G\frac{Mm}{r}\right) < 0$: no escape; straight line, circular, or elliptical

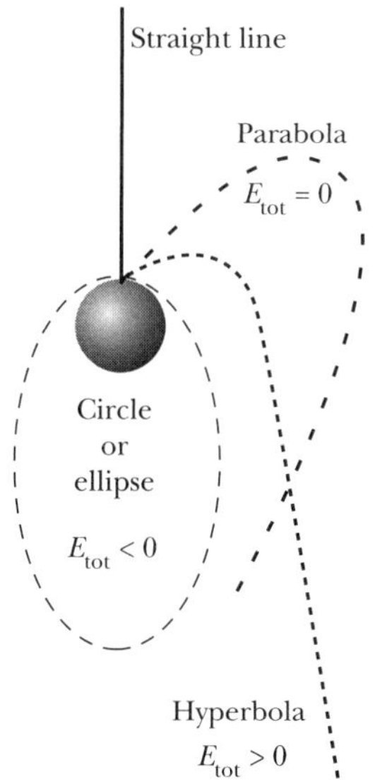

Figure 4.29 Possible orbital trajectories.

Bound vs. unbound states

This is a specific example of an important general principle. If the kinetic plus potential energy is negative, the system is "bound": the objects cannot become widely separated, because that would require that the kinetic energy become negative, which is impossible (classically). If the kinetic plus potential energy is positive, the system is "unbound" or "free": the objects can become widely separated, with net kinetic energy (and zero potential energy).

BOUND AND UNBOUND STATES

If $K + U < 0$, the system is in a bound state.

If $K + U \geq 0$, the system is unbound (free).

Example: What is escape speed from Earth?

Solution: Consider the system of Earth + spacecraft.
Minimum K for escape means $K+U_g = 0$:

$$\frac{1}{2}mv_{esc}^2 + (-GMm/r) = 0\text{, so } v_{esc} = \sqrt{2GM/r}$$

$$v_{esc} = \sqrt{\frac{2(6.7\times10^{-11}\ \text{N}\cdot\text{m}^2/\text{kg}^2)(6\times10^{24}\ \text{kg})}{(6.4\times10^{6}\ \text{m})}} = 1.12\times10^{4}\ \text{m/s}$$

Ex. 4.21 This escape speed is slightly larger than the launch speed you found was necessary to reach the Moon in Problem 2.7 The Ranger 7 mission to the Moon. Explain why.

Ex. 4.22 Turn the argument around. If an object falls to Earth starting from rest a great distance away, what is the speed with which it will hit the upper atmosphere? (Actually, a comet or asteroid coming from a long distance away might well have an even larger speed, due to its interaction with the Sun.) Small objects vaporize as they plunge through the atmosphere, but a very large object can penetrate and hit the ground at very high speed. Such a massive impact is thought to have killed off the dinosaurs (see *T. Rex and the Crater of Doom*, Walter Alvarez, Princeton University Press, 1997).

Ex. 4.23 The radius of the Moon is 1750 km, and its mass is 7×10^{22} kg. What is escape speed from the Moon? Why was a small rocket adequate to lift the lunar astronauts back up from the surface of the Moon?

4.5.3 Application: The Moon voyage

We have been applying energy considerations to two-particle systems. In our simple model for the Moon voyage (Problem 2.25 The Ranger 7 mission to the Moon), there were three "particles": the spacecraft with mass m, the Earth with mass M_{Earth}, and the Moon with mass M_{Moon}.

? How many interaction pairs are there in this three-particle system?

In a three-particle system there are three interaction pairs: U_{12}, U_{13}, and U_{23}. We can write the gravitational potential energy like this:

$$U_g = \left[-G\frac{M_{Earth}m}{r_{to\ Earth}}\right] + \left[-G\frac{M_{Moon}m}{r_{to\ Moon}}\right] + \left[-G\frac{M_{Earth}M_{Moon}}{r_{Earth\ to\ Moon}}\right]$$

In our simple model, the Earth and Moon were fixed in space, so the Earth-Moon term of this expression doesn't change. Also, there is no change of identity of the three particles, so their rest masses do not change. Therefore the energy equation for the system can be written like this, where p is the momentum of the spacecraft, the only moving particle in the model system (and $v << c$):

$$\Delta\left(\frac{p^2}{2m}\right) + \Delta\left[-G\frac{M_{Earth}m}{r_{to\ Earth}}\right] + \Delta\left[-G\frac{M_{Moon}m}{r_{to\ Moon}}\right] = 0$$

The change in the energy of the system is zero because no external work is done on the system (we're ignoring the effects of other objects, including the Sun). We can rearrange the terms and write the energy equation in terms of initial and final energies, which must be equal:

$$\frac{p_f^2}{2m}+\left[-G\frac{M_{\text{Earth}}m}{r_{\text{to Earth},f}}\right]+\left[-G\frac{M_{\text{Moon}}m}{r_{\text{to Moon},f}}\right]=\frac{p_i^2}{2m}+\left[-G\frac{M_{\text{Earth}}m}{r_{\text{to Earth},i}}\right]+\left[-G\frac{M_{\text{Moon}}m}{r_{\text{to Moon},i}}\right]$$

That is, the final (kinetic plus potential) energy of the system is equal to the initial (kinetic plus potential) energy of the system, because no external work is done on the system (and the rest masses don't change; no change of identity).

4.5.4 Gravitational potential energy near the Earth's surface

An important special case of gravitational potential energy arises in projectile motion near the surface of the Earth. Suppose you throw a baseball upward (Figure 4.30). When it has reached a height h above you, what is the change in gravitational energy? Let M be the mass of the Earth and m the mass of the ball:

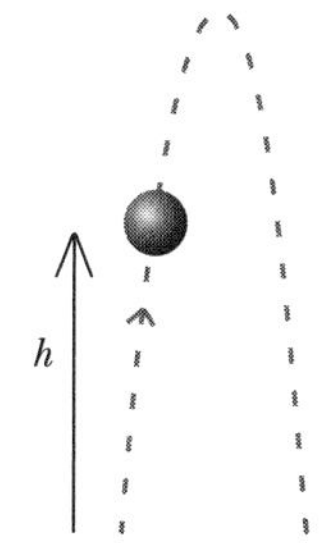

Figure 4.30 A baseball thrown upward has reached a height h.

$$\Delta U_g=\Delta\left[-G\frac{Mm}{r}\right]=\left[-G\frac{Mm}{R+h}\right]-\left[-G\frac{Mm}{R}\right]=GMm\left[\frac{(R+h)-R}{R(R+h)}\right]$$

Because h is very small compared to R, the product $R(R+h)$ is nearly the same as R^2, so

$$\Delta U_g\approx GMm\left[\frac{h}{R^2}\right]=\left(G\frac{Mm}{R^2}\right)h$$

Note that while we can neglect h in the denominator, we cannot neglect it in the numerator, where it is the largest (indeed the only) term. Remembering that $GMm/R^2\approx mg$, and writing $h=\Delta y$, we can rewrite the change in the gravitational potential energy like this:

CHANGE IN GRAVITATIONAL POTENTIAL ENERGY NEAR THE SURFACE OF THE EARTH

$$\Delta U_g\approx\Delta(mgy)$$

Notice that there is no minus sign in this (approximate) expression for the change in gravitational potential energy. As you go up (farther from the center of the Earth), the change $\Delta(mgy)$ is positive as expected; if you go down, it is negative as expected. In the full formula there is a $(-1/r)$ because moving to larger distance r means an increase (gravitational potential energy *less negative*), as can be seen in Figure 4.31.

In the formula $\Delta U_g=\Delta(mgy)$, Δy must be small compared to the radius of the Earth (6400 kilometers), because the gravitational force gets smaller as you go farther away from the Earth, inversely proportional to the square of the distance from the center of the Earth. But if y doesn't change much, neither does gravity. If you go up 0.1 km (100 meters, the length of a football field), the gravitational force decreases by a tiny factor:

$$\left(\frac{6400\text{ km}}{6400.1\text{ km}}\right)^2\approx 0.99997$$

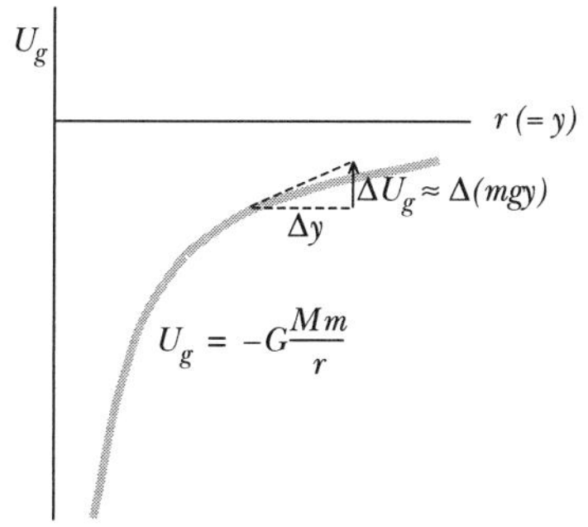

Figure 4.31 A small increase Δy in height y corresponds to a small increase in the (negative) gravitational energy.

Example: At a certain instant a ball is falling with a speed of 6 meters/second, and its position is $y = 35$ meters above the Earth's surface. How fast is the ball falling when it has fallen to a position $y = 20$ meters above the Earth's surface, assuming that we can neglect air resistance?

Solution: Choose (ball + Earth) as the system. Neglect external forces on the system, such as air resistance. There are no changes in rest mass. Then

$$\Delta K + \Delta U = 0$$

$$(K_f - K_i) + (U_f - U_i) = 0\text{, so } K_f = K_i - (U_f - U_i)$$

$$\frac{1}{2}mv_f^2 = \frac{1}{2}mv_i^2 - (mgy_f - mgy_i)$$

$$v_f^2 = v_i^2 - 2g(y_f - y_i)$$

$$v_f = \sqrt{v_i^2 - 2g(y_f - y_i)}$$

$$v_f = \sqrt{(6\text{ m/s})^2 - 2(9.8\text{ N/kg})[(20\text{ m}) - (35\text{ m})]} = 18.2\text{ m/s}$$

We did not have to say whether the ball was initially heading upward or downward. In either case, if we neglect air resistance, the speed will be 18.2 m/s when the ball reaches a height of 20 m above the ground. (If it was initially headed upward, it will again have a speed of 6 m/s when it returns to a height of 35 m.)

Ex. 4.24 Suppose in this situation we measure y from 25 meters above the surface, so that $y_{\text{initial}} = +10$ meters, and $y_{\text{final}} = -5$ meters. Why do we get the same value for the final speed?

Ex. 4.25 Use energy conservation to find the approximate final speed of a basketball dropped from a height of 2 meters (roughly the height of a professional basketball player). Why don't you need to know the mass of the basketball?

Ex. 4.26 Suppose a pitcher can throw a ball straight up at 100 miles per hour (about 45 m/s). Use energy conservation to calculate how high the baseball goes. (See the Answers section for a hint, if necessary.) Explain your work. Actually, a pitcher probably can't attain this high a speed when throwing straight up, so your result will be an overestimate of what a human can do; air resistance also reduces the achievable height.

4.5.5 Including horizontal motion

Moving vertically up or down involves a change in gravitational energy approximately equal to $\Delta(mgy)$. What about sideways, horizontal movements? There is negligible change in the gravitational energy of the Universe when an object is moved a short distance horizontally (that is, tangent to the Earth's surface). Consider moving a block along a surface with it rolling very easily on low-friction wheels, or a hockey puck sliding easily along the ice. You hardly have to use any energy at all to move the block or puck horizontally, as long as there is little friction.

What if a block is moved horizontally, but at a constant height of 3 meters above the surface? You could support the block with a tall cart, and again there would be little effort involved in moving the block horizontally. Or you could pile up dirt 3 meters deep and place a slick surface on top, and then you could easily move the block horizontally. We conclude that horizontal position doesn't affect the gravitational energy. (Again, we're considering movements over distances small compared with the radius of the Earth, so we can ignore the Earth's curvature.)

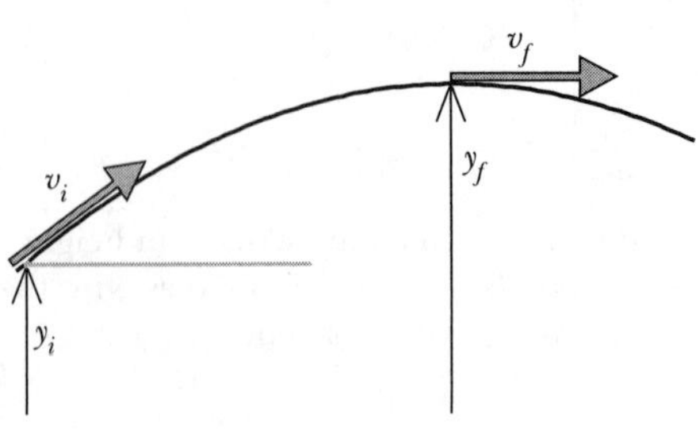

Figure 4.32 Path of a ball thrown at an angle to the horizontal.

Suppose you throw a ball at an angle to the horizontal, and just after it leaves your hand at a height y_i it has a speed v_i (Figure 4.32). At the top of its trajectory, at the instant that it is momentarily traveling horizontally, it has a speed v_f.

Ex. 4.27 What is the height y_f at the top of the trajectory, in terms of the other known quantities, assuming that we can neglect air resistance? Start by picking a system. Choose an initial and a final state, and apply the Energy Principle. If necessary, see the Answers section for a hint.

4.6 Electric potential energy

In this section we will study the potential energy associated with the electric interactions of charged particles, and the potential energy associated with the electric interactions of neutral atoms, which contain charged particles.

4.6.1 Electric potential energy involving charged particles

The electric force law for two charged particles is similar to the gravitational force law:

THE ELECTRIC FORCE

$$F_{\text{electric}} = \frac{1}{4\pi\varepsilon_0}\frac{|q_1 q_2|}{r^2}, \text{ where } \frac{1}{4\pi\varepsilon_0} = 9\times10^9\frac{\text{N}\cdot\text{m}^2}{\text{coulomb}^2}$$

The constant $1/(4\pi\varepsilon_0)$ is read as "one over four pi epsilon-zero". The quantities q_1 and q_2 represent the amount of charge on the two particles (measured in coulombs), and r is the distance between the two particles. Because this is an inverse square force, like gravitation, the electric potential energy has a formula very similar to that for gravitational potential energy:

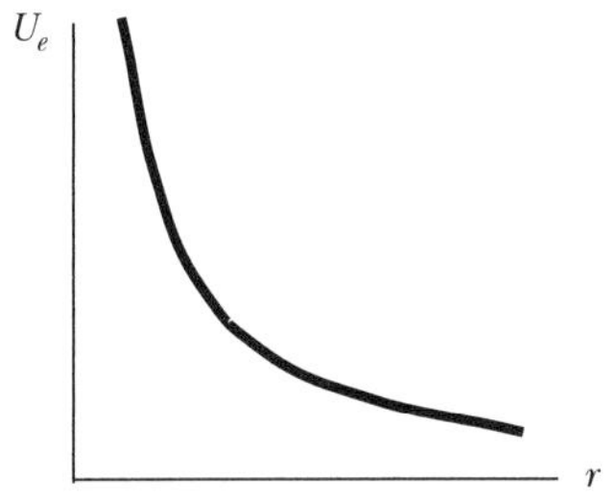

Figure 4.33 Electric potential energy is positive when like charged objects repel each other.

ELECTRIC POTENTIAL ENERGY U_e

$$U_e = \frac{1}{4\pi\varepsilon_0}\frac{q_1 q_2}{r}$$

There is no minus sign in the formula for electric potential energy, because when the electric interaction is attractive, like gravitation, q_1 and q_2 have different signs, and the minus sign shows up automatically. However, if the two charges have the same sign, the interaction is repulsive, and there is a plus sign in the formula for the electric energy (Figure 4.33). If the two charges have opposite signs, the interaction is attractive, and there is a minus sign in the formula for electric energy (Figure 4.34).

A certain amount of energy is required to remove an outer electron from a neutral atom, leaving behind a positive "ion" that is missing one electron and whose charge is $+e$, where $e = 1.6\times10^{-19}$ coulomb. (The electron's charge is of course $-e$.) The minimum energy to remove one outer electron is called the "ionization" energy.

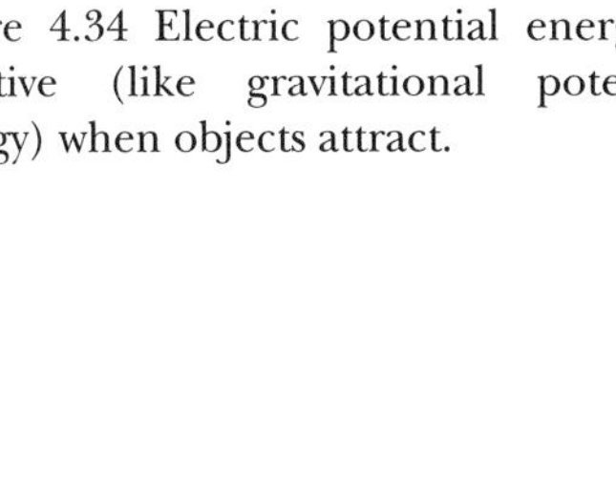

Figure 4.34 Electric potential energy is negative (like gravitational potential energy) when objects attract.

Example: A typical atom has a radius of about 10^{-10} m . Calculate a typical ionization energy in joules. Then convert your result to "electron volts," a unit of energy equal to 1.6×10^{-19} joules. This small energy unit is frequently used in discussions of atomic phenomena.

Solution: As the system we choose two particles: the outer electron with charge $-e$, and the remaining part of the atom, the ion with charge $+e$. Assume that no external forces act on this isolated system. In the initial state, the electron is bound to the atom, at a distance r_{atom} from the center; in the final state the two particles are very far from each other, at rest.

Make the rough approximation that initially the electron has negligible kinetic energy, so $K_f = K_i = 0$. Then, applying the Energy Principle:

$$\Delta E_{sys} = W$$

$$\Delta U_e + \Delta K = \Delta U_e + 0 = 0$$

$$\Delta U_e = \frac{1}{4\pi\varepsilon_0}\frac{q_1 q_2}{r_\infty} - \frac{1}{4\pi\varepsilon_0}\frac{q_1 q_2}{r_{atom}} = -\frac{1}{4\pi\varepsilon_0}\frac{(+e)(-e)}{r_{atom}}$$

$$\Delta U_e \approx \left(9\times10^9 \frac{\mathrm{N\cdot m^2}}{\mathrm{C^2}}\right)\frac{(1.6\times10^{-19}\ \mathrm{C})^2}{(10^{-10}\ \mathrm{m})} = 2.3\times10^{-18}\ \mathrm{J}$$

$$\Delta U_e = (2.3\times10^{-18}\ \mathrm{J})\left(\frac{1\ \mathrm{eV}}{1.6\times10^{-19}\ \mathrm{J}}\right) = 14\ \mathrm{eV}$$

This is a rough estimate, but typical ionization energies are indeed a few electron volts.

Ex. 4.28 Two protons are hurled straight at each other, each with a kinetic energy of 0.1 MeV, where 1 mega electron volt is equal to $10^6 \times (1.6\times10^{-19})$ joules. What is the separation of the protons from each other when they momentarily come to a stop?

4.6.2 Electric potential energy involving neutral atoms

In the preceding section we obtained the formula for electric potential energy for pairs of charged particles. In Chapter 3 we studied the spring-like forces between electrically *neutral* atoms. These interatomic forces are the superposition of electric attractions and repulsions among the positively charged protons and negatively charged electrons of which the atoms are made.

When two electrically neutral atoms come very near each other (Figure 4.35), they attract each other. If you try to push the two atoms closer together, you eventually encounter a rapidly increasing repulsion. Both of these properties are due to the superposition of the electric forces of the various protons and electrons in the atoms.

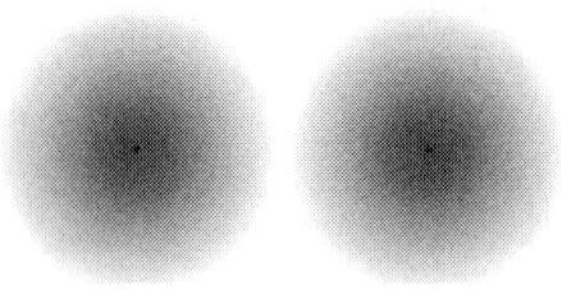

Figure 4.35 Two atoms attract each other at intermediate distances, but repel each other at very short distances.

In Chapter 3 we found that a good approximation to the net force over a limited range of separations is $F_s = -k_s s$; that is, the component of the force in the direction of the interatomic bond is opposite to the stretch s. When s is positive (longer bond), the force is attractive and pulls the atoms toward each other (negative component). When the atoms are forced closer together (s negative), the force is repulsive (positive component).

Remember that the (negative) gradient of potential energy is equal to the associated force. So we're looking for a function whose derivative is $+k_s s$.

? Can you think of a function of s whose derivative with respect to s is $+k_s s$? (You might review the Appendix on basic calculus at the end of this volume.)

The derivative with respect to s of $\frac{1}{2}k_s s^2$ is $+k_s s$, and the negative gradient of $\frac{1}{2}k_s s^2$ is the force $-k_s s$:

CHANGE IN INTERATOMIC POTENTIAL ENERGY

$$\Delta U_s = \Delta[\tfrac{1}{2}k_s s^2]$$

Note that the stretch s appears squared in the formula for spring energy. This is due to the fact that either a lengthening or a compression of the interatomic bond involves increased potential energy.

As usual, change of the interatomic potential energy involves a change of shape of the two-atom system.

What about large separations?

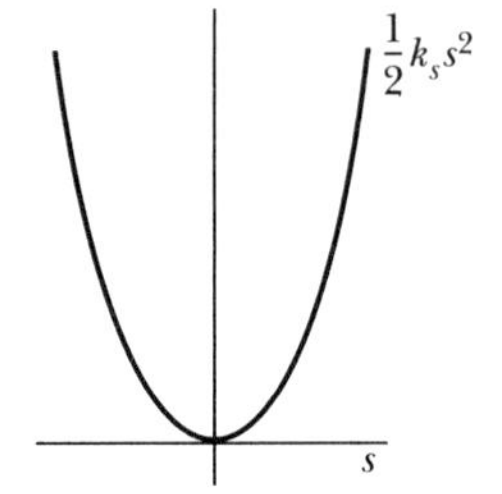

Figure 4.36 The predicted form of the potential energy becomes *infinite!*

Our analysis is fine as far as it goes, but there's a serious problem. The potential energy U_s of the two atoms should go to zero when they are far apart, so that the total energy is just the sum of their particle energies, but $\frac{1}{2}k_s s^2$ becomes *infinite!* (See Figure 4.36.) We need to go deeper into the details of the interatomic force.

Figure 4.37 shows a function called the Morse potential energy U_M that approximately represents the potential energy of two neutral atoms, as a function of the distance r between their nuclei:

$$U_M = E_0[1 - e^{-\alpha(r - r_{\text{eq}})}]^2 - E_0$$

The Morse approximation is designed to model the basic behavior of the interatomic force, and is in approximate agreement with measurements of the interatomic force. The parameter E_0 gives the depth of the potential energy where it is a minimum, and the parameter r_{eq} (on the order of 10^{-10} m) is the distance at which the potential energy curve is a minimum. The parameter α is adjusted to make the curve have the right width. When the interatomic separation r is very large, $U_M = 0$, as it should.

The depth of the potential well, E_0, is typically a few electron volts, where 1 eV is equal to 1.6×10^{-19} joule, a tiny energy unit often used in describing the world of atoms. For example, the potential energy well is about 5 eV deep for oxygen, O_2. (As you saw in the example on page 205, a typical ionization energy is also in the range of electron volts). However, for the noble gases such as helium and argon the well is so shallow that these atoms don't form stable diatomic molecules at room temperature, because collisions easily provide enough energy to break the molecules apart.

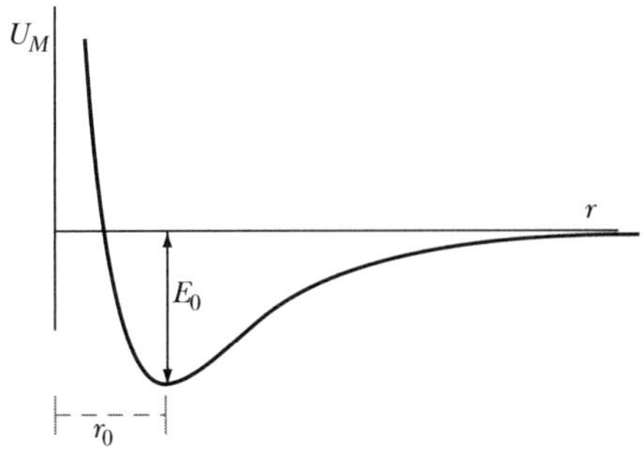

Figure 4.37 The Morse potential energy function U_M, representing the potential energy of two neutral atoms.

? What is the physical significance of the minimum of the curve?

At the minimum of the potential energy curve, the slope is zero, so the force between the two atoms is zero. This is the position of stable equilibrium.

? What is the direction of the force on an atom at a distance smaller than the equilibrium distance?

The left side of the curve has a large negative slope, so the sign of the force will be positive. This corresponds to a force to the right (in the $+r$ direction), away from the other atom. This makes sense; when the atoms are too close together they push each other apart.

Likewise, if the atom moves to the right (farther away) there should be an attractive restoring force to the left. The slope of this region of the curve is positive, so the sign of the force is negative, meaning its direction is to the left, as expected.

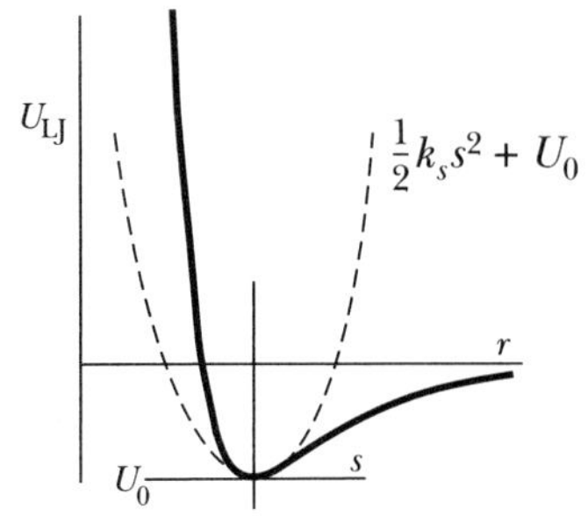

Figure 4.38 The bottom of the interatomic potential energy curve may be approximated by a harmonic oscillator energy curve.

As indicated by the dashed curve in Figure 4.38, the bottom of the interatomic potential energy curve can be approximated like this:

THE "SPRING" APPROXIMATION FOR INTERATOMIC *U*

$\frac{1}{2}k_s s^2 + U_0$ (U_0 is a negative number; s measured from equilibrium)

For the Morse potential energy formula, $U_0 = -E_0$

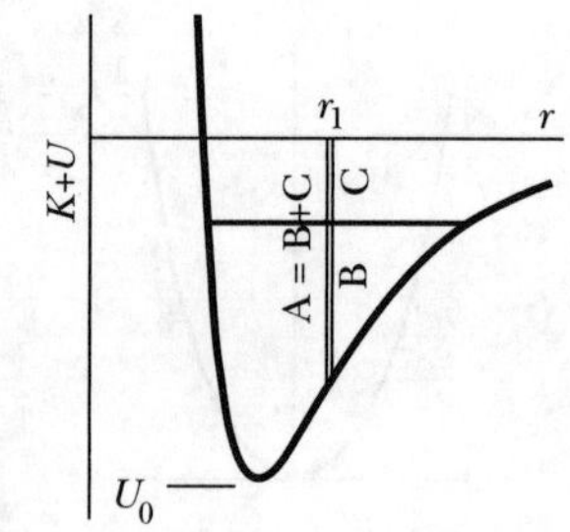

Figure 4.39 A horizontal line represents constant kinetic plus potential energy.

Because we can make this approximate fit, we know that for small oscillations around the equilibrium point the two atoms will act as though they are connected by a spring with a $-k_s s$ force. This is the rationale for modeling a solid as a collection of balls connected by springs.

Bound and unbound states with interatomic potential energy

In Figure 4.39, a horizontal line represents motion in which the sum of the kinetic energies and the potential energy of the two-atom system does not change. A horizontal line corresponds to a possible state of the system when it is isolated from external forces.

Ex. 4.29 At separation r_1 in Figure 4.39, what is the physical significance of the quantity A? Of the quantity B? Of the quantity C?

Ex. 4.30 In Figure 4.40, which of the states are bound states? Which are unbound states?

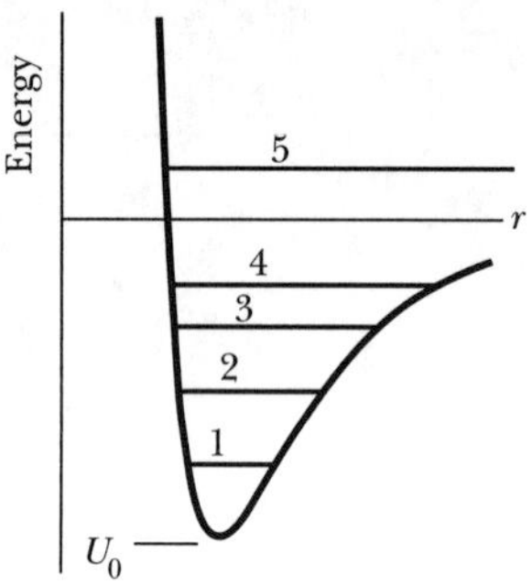

Figure 4.40 Which are bound states? Which are unbound states?

4.7 Rest energy in a multiparticle system

If the individual particles in a multiparticle system don't change their identity, both the initial and the final energies of the system include mc^2 rest-energy terms for each particle, and these cancel out in the energy equation:

$$m_1c^2 + m_2c^2 + K_i + U_{s,\,i} + U_{g,\,i} = m_1c^2 + m_2c^2 + K_f + U_{s,\,f} + U_{g,\,f}$$

$$K_i + U_{s,\,i} + U_{g,\,i} = K_f + U_{s,\,f} + U_{g,\,f}$$

This is an important special case. Next we consider the interesting effects observed when the rest masses of the particles can change due to change of identity.

Neutron decay is a good example to illustrate the issues. A free neutron (one not bound into a nucleus) is unstable, with an average lifetime of about 15 minutes. As shown in Figure 4.41, it decays into a proton, electron, and a (nearly) massless antineutrino, which travels at (nearly) the speed of light. All three of these particles have kinetic energy. That is, they have energy above and beyond their rest energy; the energy of the antineutrino is (nearly) all kinetic, because it has (almost) no rest energy. Here is a situation with a change of particle identity, and a change in particle rest mass.

$$n \rightarrow p^+ + e^- + \bar{\nu}$$

$$m_nc^2 = (m_pc^2 + K_p) + (m_ec^2 + K_e) + E_\nu$$

Figure 4.41 A stationary neutron decays into a proton, an electron, and a massless antineutrino (Exercise 4.31).

The rest energy of a proton is 938.3 MeV (1 MeV is a million electron volts, where 1 electron volt is 1.6×10^{-19} joule). One mole of atomic hydrogen has a mass of 10^{-3} kg, so we have approximately

$$mc^2 \approx \frac{\left(\dfrac{10^{-3}\text{ kg}}{6\times10^{23}\text{ protons}}\right)(3\times10^8\text{ m/s})^2}{1.6\times10^{-19}\text{ J/eV}}\text{; accurate value is }938.3\times10^6\text{eV}$$

We may use the unit MeV/c^2 as a unit of mass, because dividing energy by c^2 gives mass. The rest mass of a proton is 938.3 MeV/c^2. The rest mass of a neutron is 939.6 MeV/c^2. The rest mass of the electron is 0.511 MeV/c^2. The antineutrino is nearly massless.

Ex. 4.31 In the decay of the free neutron (Figure 4.41), how much kinetic energy (in electron volts) do the decay products (proton, electron, and antineutrino) share? Why can't a proton decay into a neutron?

4.7.1 The mass of a multiparticle system

A surprising prediction of Einstein's special theory of relativity (1905) was that mass and energy are actually the same thing. This prediction has been confirmed in many different kinds of experiments. The most dramatic example of mass-energy equivalence is the annihilation reaction that occurs between matter and anti-matter, as in the case of the annihilation of an electron and a positron (a positive electron) into two photons (high-energy quanta of light denoted by lower-case Greek gamma, γ):

$$e^- + e^+ \rightarrow \gamma + \gamma$$

Before the reaction, an electron and a positron can be almost at rest far from each other, a situation we would naturally describe as one where there is mass but seemingly no energy. After the reaction there is light (in the form of two photons), a situation which we would naturally describe as pure energy but seemingly no mass. But from the point of view of special relativity, the mass of the electron-positron system *is* energy, and the energy of the two-photon system *is* mass.

The equivalence of mass and energy is of course visible in Einstein's formula for the energy of a particle:

$$E = \frac{mc^2}{\sqrt{1 - v^2/c^2}} \equiv mc^2 + K \quad (\textit{definition} \text{ of kinetic energy } K)$$

In this equation m is called the rest mass of the particle. and is a constant. The particle has "rest energy" mc^2 even if it is not moving. If it is moving, it has additional, "kinetic" energy K and additional mass. Remember that a valid way of thinking about a fast-moving particle is that it has more mass, because mass is equivalent to energy, so the mass is E/c^2.

What is the mass of a multiparticle system consisting of interacting particles? Consider the simple case where the multiparticle system is overall "at rest." The individual particles are moving around, but the system as a whole is not moving (technically, as we will study in Chapter 7, the total momentum of the system is zero, and its center of mass is at rest). An example would be a box containing air molecules that are moving around rapidly in random directions, but the box isn't going anywhere. Another example is a diatomic molecule such as O_2 which is vibrating but has no overall translational motion.

It would seem natural to suppose that the total mass of such a system is just the sum of the rest masses of the particles: $M = m_1 + m_2 + \ldots$ However, this is not quite right. Rather, as predicted by the theory of relativity, the energy of the multiparticle system is Mc^2, and therefore its mass is this:

$$M = \frac{E_{\text{system}}}{c^2} = \frac{m_1c^2 + K_1 + m_2c^2 + K_2 + \ldots + U}{c^2}$$

$$M = (m_1 + m_2 + \ldots) + \left(\frac{K_1 + K_2 + \ldots + U}{c^2}\right)$$

The mass M of the system may be greater or less than the sum of the individual particle rest masses, because the final term in this equation may be positive or negative.

In most everyday situations the kinetic energies of the individual particles are very small compared to their rest energies (that is, the speeds of the particles are small compared to the speed of light), and the interaction energy U (the sum of all the pair-wise interactions) is also small compared to the rest energies. In that case, the mass of the multiparticle system is very nearly equal to the sum of the individual rest masses. However, this is really only an approximation, and it is often not a good approximation in nuclear and particle physics.

4.7.2 The total energy of a multiparticle system

Consider an O_2 molecule at rest. Instead of plotting only the kinetic and potential energy for this two-atom system, let's include the rest energy of the atoms, too, as we attempt to show in Figure 4.42. We can't draw the full vertical scale, because the rest energies are enormous compared to the kinetic and potential energies. In order to show the rest energy to the same scale as K and U, we would need a page stretching from the Earth to the Moon!

The total energy of the system (indicated by the thick horizontal line near the top of the graph) is the sum of the rest energies of the two atoms, the potential energy U (which is negative), and K, the kinetic energy of the two atoms. For the bound state drawn on the graph you can see that the rest energy of the two-atom O_2 system is very slightly less than the sum of the individual atomic rest energies, because $K + U < 0$. When the diatomic molecule O_2 forms, about 5 eV of energy are given off (for example, by emitting photons, which carry energy). Correspondingly, it takes an input of about 5 eV of energy to break O_2 apart, if the molecule's energy is near the bottom of the well. This is called the "binding energy."

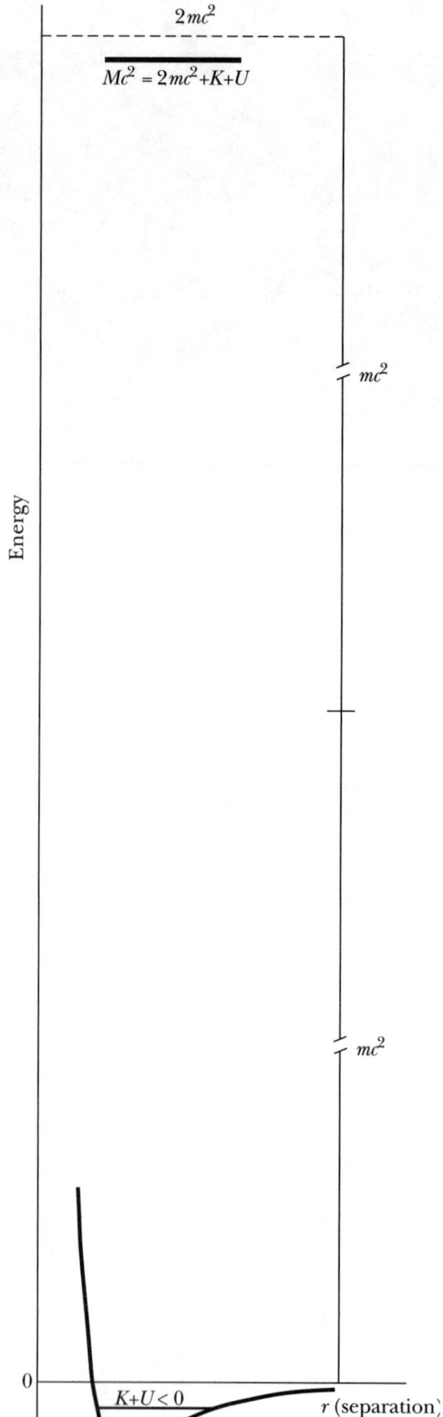

Figure 4.42 A plot of rest energy, kinetic energy, potential energy, and total energy (Mc^2) for O_2. Note that the total energy is less than the sum of the rest energies of the two individual atoms.

Two ways of thinking about the energy of the system

Given these aspects of particle energy, there are two rather different ways to think about the energy of a multiparticle system:

- The energy of a multiparticle system consists of the individual particle energies of the particles that make up the system, plus their pair-wise interaction energies.
- A multiparticle system itself has energy, rather like a particle, and if the system is at rest its energy E is simply Mc^2, where M is the mass of the system.

The mass of the system

The mass M may be greater or smaller than the sum of the rest masses of the individual particles, depending on whether the total K+U of the particles is positive or negative. A vertical line on the graph indicates that in this case the quantity Mc^2 (where M is the mass of the whole molecule) is less than $2mc^2$. From the graph we can see that:

$$Mc^2 = 2mc^2 + K + U < 2mc^2$$

Mass of a hot object

This way of thinking about mass leads us to conclude that a hot block has a greater mass than a cold block. As we'll discuss in more detail in the next chapter, a hot metal block has more thermal energy than a cold metal block, with more kinetic energy of the atoms and more potential energy in the interactions with neighboring atoms (represented as springs in our simple model of a solid). Although the effect is much too small to measure directly, the equivalence of mass and energy embodied in the special theory of relativity, $M = E/c^2$, leads to the conclusion that a hot block is actually more massive than a cold block! The same impulse $\vec{F}\Delta t$ applied to the two blocks should change the momentum of the cold block very slightly more than the momentum of the hot block.

Ex. 4.32 Calculate the approximate ratio of the binding energy of O_2 (about 5 eV) to the rest energy of O_2. Most oxygen nuclei contain 8 protons and 8 neutrons, and the rest energy of a proton or neutron is about 940 MeV. Do you think you could use a laboratory scale to detect the difference in mass between a mole of molecular oxygen (O_2) and two moles of atomic oxygen?

4.7.3 Application: Nuclear binding

In chemical reactions it is nearly impossible to measure the change of rest mass directly, but in nuclear interactions the changes are easily measurable. Consider the nuclear binding energy of an iron nucleus, which contains 26 protons and 30 neutrons, a total of 56 nucleons. When the protons and neutrons bind to each other through the "strong" force, the resulting iron nucleus has a mass that is about 1% smaller than the total mass of its individual constituents when separated. There is a sizable negative potential energy U associated with the strong force, which makes the multiparticle system's rest energy measurably smaller than the sum of the individual rest energies.

The average binding energy per nucleon is largest in iron. Lighter nuclei (hydrogen through manganese) are less tightly bound and can in principle fuse together to form the more tightly bound iron, with the release of energy. In fact, fusion reactions in stars build up from hydrogen to heavier and heavier nuclei, stopping at iron (heavier elements are made in supernova explosions).

Heavier nuclei (cobalt through uranium and beyond) are also less tightly bound than iron and can in principle fission to form the more tightly bound iron. However, only a few nuclei such as uranium and plutonium do actually spontaneously fission.

Figure 4.43 shows an energy diagram (not to scale) for one of the protons in an iron nucleus, subjected to the forces of all the other protons and neutrons. There is a deep potential well corresponding to the strong or nuclear force, which is essentially the same for protons and neutrons. This potential well has a very steep side corresponding to the very strong nuclear force (force is negative gradient of potential energy). The well is so steep it can be described approximately as a "square well." This deep well extends only out to the radius R of the nucleus, because the nuclear force has a very short range. A proton (but not a neutron) in addition experiences the repulsive electric force of the other protons (the electric U curve outside the nucleus). The total rest mass of the nucleus is about 1% less than the sum of the rest masses of the individual nucleons (protons and neutrons).

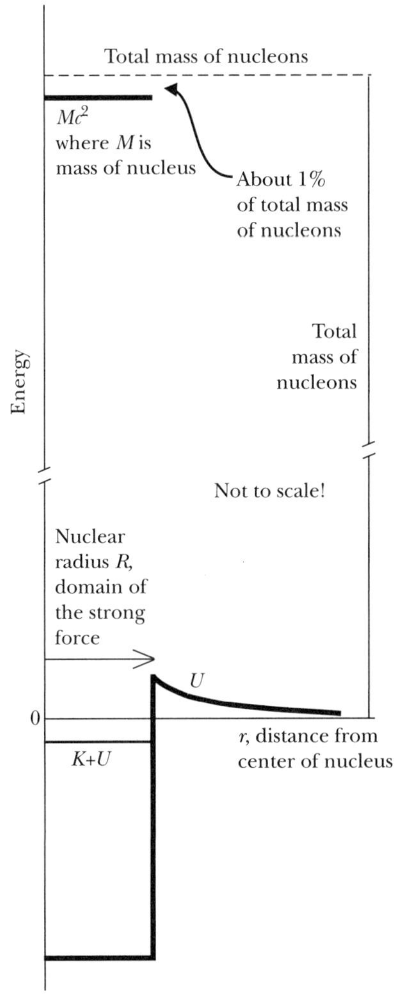

Figure 4.43 An energy diagram for a proton in an iron nucleus, not to scale. The mass of the nucleus is about 1% less than the sum of the masses of the individual nucleons (protons and neutrons).

Example: The rest energy of an iron nucleus (26 protons and 30 neutrons) is 52107 MeV. Show that the average binding energy per nucleon is about 1% of the rest energy of a nucleon, which is easily measurable. The rest energy of a proton is 938.3 MeV; of a neutron, 939.6 MeV.

Solution: As shown in Figure 4.42, the binding energy is essentially the difference between the rest energy of the multiparticle system (the nucleus) and the sum of the rest energies of its constituent particles.

$$\text{binding energy} = ((26)(938)\text{ MeV} + (30)(939)\text{ MeV}) - 52107\text{ MeV}$$
$$= 481\text{ MeV}$$

$$\text{energy per nucleon} = \frac{481\text{ MeV}}{26+30} = 8.6\text{ MeV}$$

This is about 1% of the rest energy of one nucleon.

Ex. 4.33 The deuteron, the nucleus of the deuterium atom ("heavy" hydrogen), consists of a proton and a neutron. It is observed experimentally that a high-energy photon ("gamma ray") with a minimum energy of 2.2 MeV can break up the deuteron into a free proton and a free neutron; this process is called "photo-dissociation." About what fraction of the deuteron rest energy corresponds to its binding energy? The result shows that the deuteron is very lightly bound compared to the iron nucleus.

4.7.4 Application: Nuclear fission

The nuclear fission of uranium also leads to an easily measured change in rest masses. The average binding energy per nucleon in uranium is about 7.8 MeV, so the uranium nucleus is less tightly bound than an iron nucleus, whose binding is 8.5 MeV per nucleon. This suggests the possibility of obtaining energy from the fission of a uranium nucleus, corresponding to roughly 0.1% (0.7 MeV per nucleon) of its rest energy (about 1000 MeV per nucleon).

In other words, the "mass deficiency" of a medium-mass nucleus is about 1%, with a variation of about a tenth of that. Even though there is only a small difference in average binding energy between iron and uranium, the energy involved is vast compared to the energy of a chemical reaction, which only involves the outer electrons of an atom, not the nucleus.

A uranium-235 nucleus contains 92 protons and 143 neutrons. The 92 protons exert strong electric repulsions on each other, but the large number of neutrons spreads the protons throughout a larger volume and reduces the electric repulsions. The neutrons and the protons also attract neighboring protons with strong nuclear forces. As discussed in Chapter 1, uranium-235 has an unusual property that few other nuclei have. If you hit a U-235 nucleus with a slowly moving neutron, the nucleus may deform into a dumbbell-like form and then split apart because the electric repulsion of the protons can now overcome the short-range attraction of the nucleons for each other (Figure 4.44). (In addition to the two main fission fragments there are typically one or more free neutrons in the final state.)

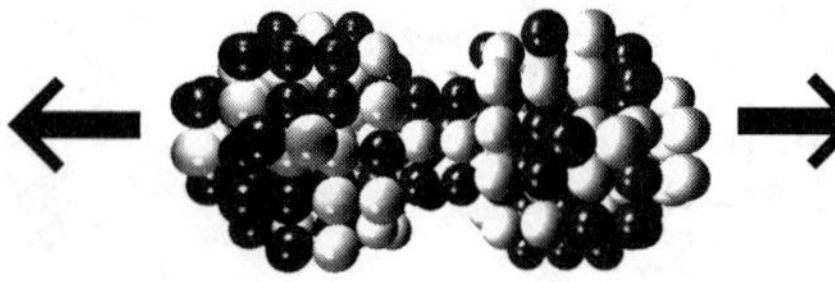

Figure 4.44 A U-235 nucleus in the process of fissioning.

The fission fragments quickly acquire very high speed, and they leave the original atom's 92 electrons behind. These highly charged nuclei storm through the block of uranium metal, tearing up the material and violently heating the metal. In a nuclear power reactor the reaction is controlled in such a way that the metal merely gets hot enough to boil water. With many uranium-235 nuclei fissioning, the metal can boil water in a nuclear power plant to produce steam and power a turbine to drive an electric generator.

The rest masses of the two fission-fragment nuclei total less than the rest mass of the uranium-235 nucleus. The rest mass difference Δm is only about one-tenth of one percent of the original rest mass, but this rest mass change is easily measurable, and c^2 times this rest mass change is indeed found to be equal to the observed kinetic energy of the fission fragments. This 0.1% change is about ten million times as big an effect as the rest mass changes that occur in chemical reactions, which only involve rearrangements of the outer electrons in the atoms, and do not affect the nucleus.

4.8 Reflection: Why energy?

It is appropriate at this point to reflect on why we introduced the topic of energy. Energy methods often make it possible to analyze some aspect of a phenomenon with much less effort than is required by direct application of the Momentum Principle. A striking example is the ability to calculate escape speed, or the impact speed of the Ranger spacecraft on the Moon, without carrying out a lengthy numerical integration.

Limits on the possible

Another powerful capability of energy analyses is that they can easily predict whether some process can occur or not. For example, if there is insufficient energy, the rocket can't escape from the planet, or the atom cannot be ionized, or the biochemical reaction cannot proceed.

Less detail

However, energy analyses give less detailed information than is given by a step-by-step numerical integration of the Momentum Principle. In particular, energy analyses tell us nothing about how fast a process will proceed; time does not appear in the statement of conservation of energy.

Here is an example that illustrates the strengths and weaknesses of energy analyses. Consider snowboarding down the three hills shown in Figure 4.45, starting from rest. If we can neglect friction, the final speed at the bottom should be the same in all three cases, because the final kinetic energy should be equal to *mgh*, the amount by which the gravitational energy decreases.

? But how will the amount of *time* compare for the ride down the hill?

On the first hill you get going very fast very quickly, and the trip is short. On the second hill you move very slowly for a long time before dropping to the bottom, and the total trip takes a long time. On the third hill you never get to the bottom, despite the fact that the change in potential energy would be negative, because you would have to pass through a region that is energetically forbidden. So energy methods are powerful, but at the cost of losing some information, including the amount of time required for a process.

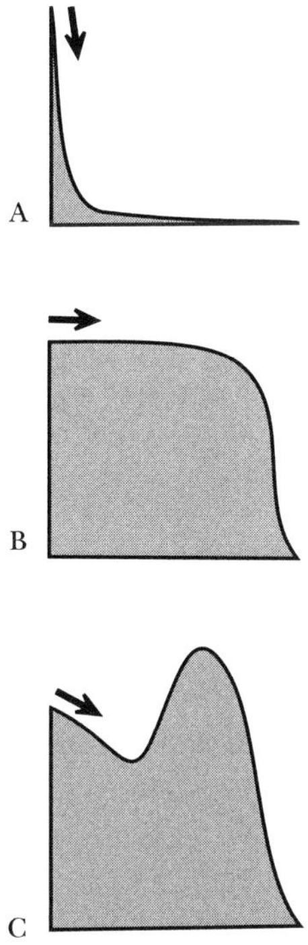

Figure 4.45 The final speed of the snowboarder will be the same at the bottom of each hill.

4.9 Solving energy problems

Problems involving energy can appear quite complex and intimidating, since they may involve many objects, many interactions, and even changes in identity. Nonetheless, the following simple way of organizing your solution will always work!

First: Explicitly choose a system. Write it down so you will not get confused later.

Second: Identify an initial state and a final state of this system. Some problems may involve more than two states. Since you can only apply the Energy Principle to two states, this means you will have to solve such a problem in more than one step.

Third: Write down the Energy Principle as it applies to your chosen system. Be careful to note whether or not there are external energy inputs to your system.

You can write the Energy Principle in two different ways:

$$\Delta E_{sys} = W, \text{ or}$$

$$E_{final} = E_{initial} + W$$

These are of course equivalent, but sometimes one or the other is more easily applied to a particular problem.

4.9.1 Example problem: Collision of alpha and carbon nucleus

An alpha particle (the nucleus of a helium atom, containing two protons and two neutrons) moving with momentum p_1 is shot toward a carbon-12 nucleus (containing six protons and six neutrons) that is moving with the same momentum p_1 toward the alpha particle. The speeds are nonrelativistic.

(a) What is the minimum momentum p_1 necessary to allow the alpha particle and carbon-12 nucleus to come in contact, so that a nuclear reaction can occur? (With enough energy it is possible to produce oxygen-16.) The radius of a nucleus is approximately $(1.3\times10^{-15}\text{ m})A^{1/3}$, where A is the total number of protons and neutrons. Briefly explain your analysis.

(b) What is the kinetic energy of the alpha particle in part (a), in MeV (1 MeV = 10^6 eV)?

(c) There is a head-on collision of the alpha particle and the carbon-12 nucleus, and an excited state of an oxygen-16 nucleus is formed at rest. A very short time later the oxygen nucleus drops to its ground state, emitting a photon whose energy is 10.352 MeV (1 MeV = 10^6 eV). The oxygen nucleus recoils with negligible kinetic energy. What is the minimum (nonrelativistic) momentum p_2 that the alpha particle must have in order for this sequence of phenomena to occur? Masses of the ground states of selected nuclei, where 1 u = 1.7×10^{-27} kg:

He-4 (2p+2n) = 4.00040868 u
C-12 (6p+6n) = 11.99670852 u
O-16 (8p+8n) = 15.99052636 u

(d) For the process in part (a), as a function of the separation r between the alpha particle and the carbon-12 nucleus, plot K_1+K_2 (the sum of the kinetic energies of the two nuclei), plot the potential energy, and plot the sum of the kinetic energies and the potential energy. Label each of the three plots clearly.

Solution

Figure 4.46 Initial state (part a).

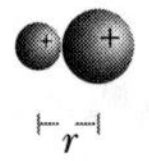

Figure 4.47 Final state (part a).

(a) *System:* alpha particle and carbon nucleus. No external forces.

Initial state: Very far apart, particles moving toward each other (Figure 4.46).

Final state: Particles at rest, touching each other (Figure 4.47).

Energy principle: $\Delta K + \Delta U_e + \Delta(mc^2) + \Delta(Mc^2) = 0$

$$U_i = 0 \text{ (particles very far apart) and } K_f = 0 \text{ (particles at rest)}$$

$$\text{no change of identity, so } \Delta(mc^2) + \Delta(Mc^2) = 0$$

$$\text{so } \left[0 - \left(\frac{p_1^2}{2m} + \frac{p_1^2}{2M}\right)\right] + \left(\frac{1}{4\pi\varepsilon_0}\frac{(2e)(6e)}{r} - 0\right) = 0$$

where m is the mass of the alpha particle and M is the mass of the carbon nucleus.

$$p_1 = \sqrt{\frac{\frac{1}{4\pi\varepsilon_0}\frac{12e^2}{r}}{\frac{1}{2}\left(\frac{1}{m} + \frac{1}{M}\right)}}$$

$$\text{final separation } r = (1.3\times10^{-15}\text{ m})(4^{1/3} + 12^{1/3}) = 5.04\times10^{-15}\text{ m, so}$$

$$p_1 = \sqrt{\frac{\left(9\times10^{9}\,\frac{\text{N}\cdot\text{m}^2}{\text{C}^2}\right)\frac{12(1.6\times10^{-19}\text{ C})^2}{(5.04\times10^{-15}\text{ m})}}{\frac{1}{2}\left(\frac{1}{4}+\frac{1}{12}\right)\left(\frac{1}{1.66\times10^{-27}\text{ kg}}\right)}} = 7.39\times10^{-20}\text{ kg}\cdot\text{m/s}$$

$$\text{(b) } K_\alpha = \frac{p_1^2}{2m} = \frac{(7.39\times10^{-20}\text{ kg}\cdot\text{m/s})^2}{2(4\times1.66\times10^{-27}\text{ kg})}\left(\frac{10^{-6}\text{ MeV}}{1.6\times10^{-19}\text{ J}}\right) = 2.57\text{ MeV}$$

Figure 4.48 Initial state (part c).

(c) *System:* alpha particle and carbon nucleus. No external forces.

Initial state: Very far apart, particles moving toward each other (Figure 4.48).

Final state: Oxygen nucleus and photon (Figure 4.49).

Energy principle: $E_{\text{initial}} = E_{\text{final}}$

Figure 4.49 Final state (part c).

since $K_{\text{oxygen}} \approx 0$,

$$\left(mc^2 + \frac{p_2^2}{2m}\right) + \left(Mc^2 + \frac{p_2^2}{2M}\right) = M_{\text{oxygen}}c^2 + E_\gamma$$

$$p_2 = \sqrt{\frac{(M_{\text{oxygen}} - m - M)c^2 + E_\gamma}{\frac{1}{2}\left(\frac{1}{m}+\frac{1}{M}\right)}}$$

To calculate accurately the quantity $(M_{\text{oxygen}} - m - M)$ we must use very accurate values of the nuclear masses, because the differences in the rest masses are very slight:

$$M_{\text{oxygen}} - m - M = (15.99052636 - 4.00040868 - 11.99670852)u$$

$$M_{\text{oxygen}} - m - M = -0.00659084\text{ u}$$

$$(-0.00659084\text{ u})\left(1.66\times10^{-27}\frac{\text{kg}}{\text{u}}\right)\left(3\times10^{8}\frac{\text{m}}{\text{s}}\right)^2 = -9.85\times10^{-13}\text{ J}$$

$$E_\gamma = (10.352\text{ MeV})\left(\frac{1.6\times10^{-19}\text{ J}}{10^{-6}\text{ MeV}}\right) = 16.56\times10^{-13}\text{ J}$$

We don't have to be so careful in calculating the sum $\left(\frac{1}{m}+\frac{1}{M}\right)$:

$$p_2 = \sqrt{\frac{(16.56\times10^{-13} - 9.85\times10^{-13})\text{ J}}{\frac{1}{2}\left(\frac{1}{4}+\frac{1}{12}\right)\left(\frac{1}{1.66\times10^{-27}\text{ kg}}\right)}} = 8.18\times10^{-20}\text{ kg}\cdot\text{m/s}$$

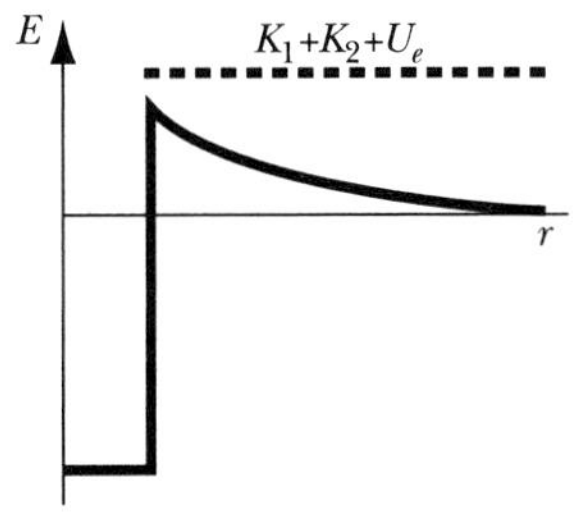

Figure 4.50 The initial kinetic energies are a bit greater than what is needed to overcome the electric potential energy barrier.

We find that the incoming momentum p_2 required to create the excited oxygen nucleus is greater than the momentum p_1 required just to overcome the electric potential energy barrier and bring the alpha particle and carbon nucleus in contact (Figure 4.50 shows the energies involved). It is necessary to bring nuclei into contact for a reaction to proceed, since the nuclear force has a very short range. However, in this reaction just making contact is insufficient. A bit more energy is needed in order to produce the larger-mass oxygen nucleus.

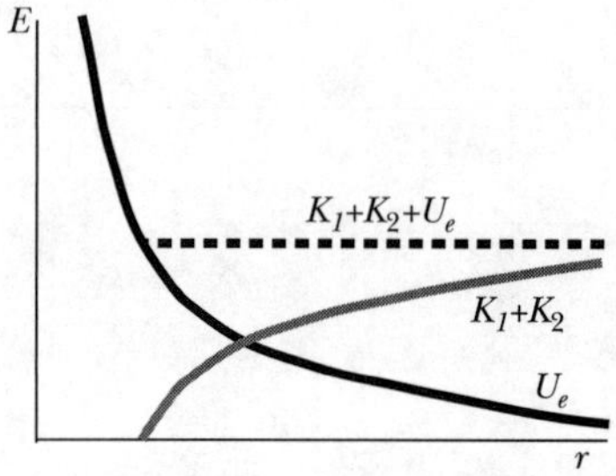

Figure 4.51 Energies as a function of separation between alpha particle and carbon nucleus.

(d) See Figure 4.51. The sum of the kinetic and potential energies doesn't change, since no work is done on the two-particle system. The smallest separation is associated with the kinetic energy going to zero.

4.10 *A puzzle

The principle of conservation of energy states that when a system's energy increases, the energy of the surroundings decreases by the same amount. For example, take a block as the system of interest. If you push the block away from you and raise its kinetic energy, your store of chemical energy decreases by the same amount. The flow of energy from you (the "surroundings") to the block takes the form of mechanical energy transfer we call "work." We can demonstrate that there is a puzzle concerning the energy in the surroundings of a star whose kinetic energy increases due to an interaction with another star.

Consider two stars of equal mass initially at rest, far apart (Figure 4.52). The two stars do work on each other:

$$\Delta E_1 = W_{\text{on }1} \quad \text{star 1 as system}$$

$$\Delta E_2 = W_{\text{on }2} \quad \text{star 2 as system}$$

Figure 4.52 Initially the stars were very far apart, at rest, but they attract each other.

If we add these two energy equations for the two stars, we have this:

$$\Delta E_1 + \Delta E_2 = W_{\text{on }1} + W_{\text{on }2} \quad \text{sum of two energy equations}$$

Moving $(W_{\text{on }1} + W_{\text{on }2})$ to the left side of the equation and renaming the negative of this quantity ΔU, we have this familiar result:

$$\Delta E_1 + \Delta E_2 + \Delta U = 0 \quad \text{energy equation for two stars}$$

Our interpretation of this equation is that the gravitational potential energy U of the two-star system (calculated from the work on the individual particles) decreases as the particle energies increase. This point of view is valid, yields physically correct results, and is the view that we will take in this introductory course.

However, it is not difficult to show that there is a puzzle concerning energy flow between one of the stars and its surroundings. Consider star 1 as the system. Since it is a single object, there is no potential energy U—no pairwise interaction energy. The energy of star 1 increases due to work done on it: $\Delta E_1 = W_{\text{on }1}$. There are very strong reasons to believe in energy conservation, so we expect that when star 1 gains an amount of energy ΔE_1 the surroundings of star 1 should lose that much energy.

? Part of the surroundings is star 2. Does star 2 lose energy?

No! Star 2 doesn't lose energy, it *gains* energy, and of the same amount $\Delta E_2 = \Delta E_1$, due to the symmetry of the situation, with the two stars having equal mass.

What else is there in the surroundings that could experience a change in energy? The only other matter is star 2, but its energy change is inconsistent with the principle of energy conservation. Evidently our model of the world is incomplete. To be able to analyze this situation fully, we need to introduce the abstract idea of a "field." Fields, which are the subject of Volume II of this textbook, are associated with objects having mass (gravitational fields) or charge (electric and magnetic fields), and they extend throughout all space. It is energy stored in the gravitational field that accounts for the energy in the surroundings of star 1. Fields can also have momentum, as we will see in Volume II.

Despite the fact that we haven't yet defined what a "field" is, we can determine indirectly how much the field energy must change. Since the energy in the surroundings of star 1 changes by an amount $-\Delta E_1$, and the energy of star 2 increases by an amount $+\Delta E_1$, it must be that the energy stored in the gravitational field changes by an amount $-2\Delta E_1$. (It is an indication of the power of the principle of conservation of energy that we can calculate the energy change of an entity that we know nothing about!) This energy change is precisely the amount of change in the potential energy of the two-star system: $\Delta U = -(\Delta E_1 + \Delta E_2) = -2\Delta E_1$.

Because the potential energy U correctly accounts for the energy changes, we don't have to consider field energy in our calculations. Our analyses in terms of U that focus on particles and their interactions, not their surroundings, are completely correct. However, the difficulty in accounting for the energy in the surroundings of a star suggests that eventually we will need to include fields in our models of the world.

4.11 *Integrals and antiderivatives

In determining the formula for gravitational potential energy, we started from these complementary formulas:

$$\Delta U_g = -F_r \Delta r \text{ and } F_r = -\frac{dU_g}{dr}$$

We calculated the change in potential energy from the negative of the internal work:

$$U_{s,f} - U_{s,i} = \Delta\left(-G\frac{m_1 m_2}{r}\right) = -\Sigma F_r \Delta r$$

But if we had let Δr approach zero, the finite sum would have turned into an integral:

$$\Delta\left(-G\frac{m_1 m_2}{r}\right) = -\int_i^f F_r dr = \int_i^f G\frac{m_1 m_2}{r^2} dr$$

You can see that a definite integral (the infinitesimal version of a finite sum) can be evaluated in terms of the antiderivative. The antiderivative of $Gm_1 m_2 / r^2$ is $-Gm_1 m_2 / r$, and the change in the value of the antiderivative turns out to be equal to the definite integral (the infinite sum of infinitesimal elements).

The reason why this works this way can be traced back to the fact that the force is the derivative of the energy, so the energy must be the antiderivative of the force, yet the energy is also the sum of lots of $F_r\Delta r$'s. Hence the sum is given by the change in the antiderivative.

When the quantity to be integrated (here, the force) is a simple function, it may be easy to find the antiderivative. However, in many practical cases there is no known simple function that is the antiderivative, in which case an integral must be evaluated by numerical integration, as a finite sum.

4.12 *Approximation for particle energy

An approximation for the kinetic energy of a particle at low speeds can be obtained through use of the mathematics of the "binomial expansion":

$$(1+\varepsilon)^n = 1 + n\varepsilon + \frac{n(n-1)}{2}\varepsilon^2 + \frac{n(n-1)(n-2)}{3 \cdot 2}\varepsilon^3 + \ldots..$$

This formula comes from working through what happens when you multiply the quantity $(1+\varepsilon)$ times itself n times. The first term is 1 multiplied by itself n times (1^n); the next term is made up of all those products, n of them, in which ε appears only once [$n\varepsilon$]; etc.

Now for the important point that makes the binomial expansion useful in obtaining approximate results. If ε is a number that is very small compared to 1, each term in the expansion is much smaller than the preceding term:

$$n\varepsilon \text{ is much smaller than } 1, \quad \frac{n(n-1)}{2}\varepsilon^2 \text{ is much smaller than } n\varepsilon\text{, etc.}$$

Therefore we can get a good approximation if we keep just the first few terms and ignore the rest.

Although we have developed the binomial expansion for the case where n is an integer, it can be shown that the formula is valid even for non-integer values of n. In particular,

$$\frac{1}{\sqrt{1-\varepsilon}} = (1-\varepsilon)^{-1/2} = 1 + (-\tfrac{1}{2})(-\varepsilon) + \frac{(-\frac{1}{2})(-\frac{3}{2})}{2}\varepsilon^2 + \frac{(-\frac{1}{2})(-\frac{3}{2})(-\frac{5}{2})}{6}\varepsilon^3 + ...$$

If v/c is small compared to 1, then v^2/c^2 is even smaller compared to 1. For example, 1/10 is small compared to one, but $(1/10)^2 = 1/100$ is very small indeed. Therefore we can write this for low speeds:

$$\frac{mc^2}{\sqrt{1-v^2/c^2}} = mc^2\left[1 + \frac{1}{2}\frac{v^2}{c^2} - \frac{3}{8}\left(\frac{v^2}{c^2}\right)^2 - \frac{5}{16}\left(\frac{v^2}{c^2}\right)^3 + ...\right] \approx mc^2 + \frac{1}{2}mv^2$$

This last element can be rewritten in terms of momentum $p \approx mv$:

$$\frac{1}{2}mv^2 = \frac{1}{2}\frac{(mv)^2}{m} = \frac{p^2}{2m}$$

4.13 *Determining the formula for particle energy

If we choose our axes so that x is in the direction of the motion at this instant (Figure 4.53), we have

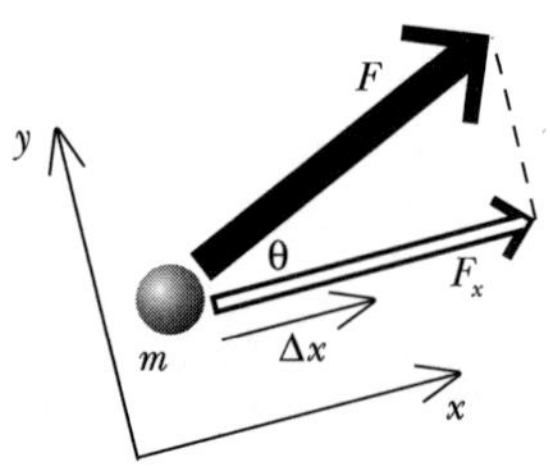

Figure 4.53 A particle of mass m. We choose the x axis to be in the direction of the motion. The net force acting on the particle has a component F_x in the direction of motion.

$$\Delta E = F_x\Delta x = \left(\frac{\Delta p_x}{\Delta t}\right)\Delta x$$

for the change in energy of a point particle acted on by a net force. The energy E should satisfy the following relationship between two derivatives:

$$\frac{dE}{dx} = \frac{dp_x}{dt}$$

We will show that if we choose the following function for E, the relationship between the two derivatives will be satisfied:

$$E = \frac{mc^2}{\sqrt{1-v^2/c^2}}$$

The proof proceeds by calculating dE/dx, calculating dp_x/dt, and showing that the two derivatives are equal to each other. First, here is dE/dx:

$$\frac{dE}{dx} = \frac{d}{dx}\left(\frac{mc^2}{\sqrt{1-v^2/c^2}}\right) = \frac{mc^2}{(1-v^2/c^2)^{3/2}}\left(-\frac{1}{2}\right)\left(-\frac{2v}{c^2}\right)\frac{dv}{dx}$$

$$\frac{dE}{dx} = \frac{mv}{(1-v^2/c^2)^{3/2}}\frac{dv}{dx}$$

Next we calculate dp_x/dt, keeping in mind that $v_x = v$, because we chose our axes so that x is in the direction of the motion at this instant:

$$\frac{dp_x}{dt} = \frac{d}{dt}\left(\frac{mv}{\sqrt{1-v^2/c^2}}\right) = \frac{m}{\sqrt{1-v^2/c^2}}\frac{dv}{dt} + \frac{mv}{(1-v^2/c^2)^{3/2}}\left(-\frac{1}{2}\right)\left(-\frac{2v}{c^2}\right)\frac{dv}{dt}$$

$$\frac{dp_x}{dt} = \frac{m}{(1 - v^2/c^2)^{3/2}}\left[\left(1 - \frac{v^2}{c^2}\right) + \frac{v^2}{c^2}\right]\frac{dv}{dt} = \frac{m}{(1 - v^2/c^2)^{3/2}}\frac{dv}{dt}$$

This is almost in the same form as the derivative dE/dx. Now note this:

$$\frac{dv}{dt} = \left(\frac{dv}{dt}\right)\left(\frac{dx}{dx}\right) = \left(\frac{dv}{dx}\right)\left(\frac{dx}{dt}\right) = \left(\frac{dv}{dx}\right)v\text{, because } v = \frac{dx}{dt}$$

Substituting this result for dv/dt into the derivative dp_x/dt, we find that this derivative is indeed equal to the derivative dE/dx if E is defined as

$$E = \frac{mc^2}{\sqrt{1 - v^2/c^2}}$$

Hence the relativistic formula for energy does follow from the relativistic formula for momentum.

It might seem that we could define the particle energy with an additive constant, which would drop out when we took the derivative dE/dx. However, when one considers a variety of particle reactions, it is found that to account for energy in a consistent way requires that the additive constant be zero.

This approach to solving a differential equation (proposing a solution and showing that it satisfies the equation) is a legitimate technique of solving differential equations, and in fact is used quite frequently.

4.14 Summary

Fundamental Principles

The Energy Principle:

$$\Delta E_{\text{system}} = W \text{ where } E_{\text{system}} = (E_1 + E_2 + \ldots) + U$$

New Concepts

Work W_F done by a single force $\vec{F}$:

$$W_F = \vec{F} \bullet \Delta\vec{r} = F\Delta r\cos\theta = F_x\Delta r_x + F_y\Delta r_y + F_z\Delta r_z \text{ (constant force)}$$

$$W_F = \sum \vec{F} \bullet \Delta\vec{r} \text{ (variable force)}$$

Energy of a particle with rest mass m is $E = \dfrac{mc^2}{\sqrt{1 - v^2/c^2}} = mc^2 + K$

Kinetic energy $K = \dfrac{mc^2}{\sqrt{1 - v^2/c^2}} - mc^2$; $K \approx \frac{1}{2}mv^2 = \dfrac{p^2}{2m}$ if $v << c$

Energy and momentum: $E^2 - (pc)^2 = (mc^2)^2$ (rest mass m)

For so-called "massless" photon or neutrino, $E = pc$

Potential energy: $\Delta U \equiv -W_{\text{internal}}$

Force is negative gradient of potential energy: $F_x = -\dfrac{dU}{dx}$

Results

Gravitational potential energy $U_g = -G\dfrac{Mm}{r}$

$\Delta U_g \approx \Delta(mgy)$ near Earth's surface

Electric potential energy $U_e = \dfrac{1}{4\pi\varepsilon_0}\dfrac{q_1 q_2}{r}$,

where $\dfrac{1}{4\pi\varepsilon_0} = 9\times10^9 \text{ N}\cdot\text{m}^2/\text{coulomb}^2$

Approx. interatomic potential energy: $U_M = E_0[1 - e^{-\alpha(r - r_{eq})}]^2 - E_0$

Near bottom of well: $U_s \approx \frac{1}{2}k_s s^2 + U_0$, where U_0 is negative.

Nuclear binding: heavy nuclei can fission, light nuclei can fuse, with energy released in both cases.

Electron volt: $1 \text{ eV} = 1.6\times10^{-19}$ J

Problem Solving Techniques

Use energy calculations to place limits on what is possible. For example:
- How high a ball can go given its initial velocity
- Escape speed for a spacecraft
- Whether a particle reaction can proceed

4.15 Review questions

Work

RQ 4.1 You push a crate 3 m across the floor with a 40 N force whose direction is 30° below the horizontal. How much work do you do?

RQ 4.2 Give brief explanations for your answers to each of the following questions:

(a) You hold a 1 kg book in your hand for 1 minute. How much work do you do on the book?

(b) In a circular pendulum, how much work is done by the string on the mass in one revolution?

(c) For a mass oscillating horizontally on a spring, how much work is done by the spring on the mass in one complete cycle? In a half cycle?

Kinetic energy

RQ 4.3 Approximately what is the kinetic energy of a baseball pitcher's fast ball? Of a basketball passed from one player to another?

RQ 4.4 What is the speed of an electron whose total energy is equal to the total energy of a proton that is at rest? What is the kinetic energy of this electron?

Work versus impulse, kinetic energy versus momentum

RQ 4.5 You pull a block of mass m across a frictionless table with a constant force. You also pull with an equal constant force a block of larger mass M. The blocks are initially at rest. If you pull the blocks through the same distance, which block has the greater kinetic energy, and which block has the greater momentum? If instead you pull the blocks for the same amount of time, which block has the greater kinetic energy, and which block has the greater momentum?

Energy conservation

RQ 4.6 You throw a ball straight up, and it reaches a height of 20 meters above your hand before falling back down. What was the speed of the ball just after it left your hand?

Gravitational energy

RQ 4.7 The escape speed from an asteroid whose radius is about 10 km is only 10 m/s. If you throw a rock away from the asteroid at a speed of 20 m/s, what will be its final speed?

Graphs of potential energy

RQ 4.8 The gravitational force inside the Earth is approximately $(GMm/R^3)r = gr/R$, where r is the distance from the center and R is the radius of the Earth; see Problem 4.11 (page 224) for additional discussion. Draw the potential energy as a function of r, from $r = 0$ to very large r ($>> R$). Note that the potential energy curve must be continuous, with no breaks.

Mass and energy

RQ 4.9 In positron-emission tomography (PET) used in medical research and diagnosis, compounds containing unstable nuclei that emit positrons are introduced into the brain, destined for a site of interest in the brain. When a positron is emitted, it goes only a short distance before coming nearly to rest. It forms a bound state with an electron, called "positronium," which is rather similar to a hydrogen atom. The binding energy of positronium is very small compared to the rest energy of an electron. After a short time the positron and electron annihilate. In the annihilation, the positron and the electron disappear, and all of their rest energy goes into two photons (particles of light) which have zero mass; all their energy is kinetic energy. These high energy photons, called "gamma rays," are emitted at nearly 180° to each other. What energy of gamma ray (in MeV, million electron volts) should each of the detectors be made sensitive to? (The mass of an electron or positron is 9×10^{-31} kg.)

RQ 4.10 One often hears the statement, "Nuclear energy production is fundamentally different from chemical energy production (such as burning of coal), because the nuclear case involves a change of mass." Critique this statement. Discuss the similarities and differences of the two kinds of energy production.

Additional exercises

Ex. 4.34 A jar of honey with a mass of 0.5 kg is knocked off the kitchen counter and falls one meter to the floor. What force acts on the jar during its fall? How much work is done by this force?

Ex. 4.35 An electron traveling through a wire in an electric circuit experiences a constant force of 5×10^{-19} N, always in the direction of its motion. How much work is done on the electron by this force as it travels through 0.5 m of the wire?

Ex. 4.36 A 0.5 kg teddy bear is knocked off a windowsill and falls 2 m to the ground. What is its kinetic energy at the instant it hits the ground? What is its speed? What assumptions or approximations did you make in this calculation?

Ex. 4.37 A 1.0 kg flowerpot is knocked off a windowsill and falls 2 m to the ground. What is its kinetic energy at the instant it hits the ground? What is its speed? How do the speed and kinetic energy compare to that of the teddy bear in the previous exercise?

Ex. 4.38 A commercial airliner flew 20,000 km nonstop from Seattle to Kuala Lumpur at an average speed of 900 km/hr. If its average mass was 140,000 kg what was its average kinetic energy? (We take an average because the mass of an airplane changes as it exhausts burnt fuel.)

4.16 Problems

Problem 4.1 Work done in Moon shot

If you have already done Problem 2.7 The Ranger 7 mission to the Moon, you can use the program you wrote for that problem, and go directly to part (b). Otherwise:

(a) Write a program to model the journey of a spacecraft coasting from the Earth to the Moon. Start the spacecraft at a height of 50 km above the Earth's surface with a speed of 1.3×10^4 m/s. This is approximately the speed it would have after all the rockets have fired, and is high enough to be above most of the Earth's atmosphere. Include the spacecraft's interactions with both the Earth and the Moon. Use a *dt* of 5 seconds. Make sure you stop the program when the spacecraft reaches the Moon's surface (not its center!). The data you need may be found on page 68, or on the inside back cover of this book. If you are interested in more of the details of this trip, see Problem 2.7 The Ranger 7 mission to the Moon, on page 68.

(b) Add a calculation of the work done by the gravitational forces of the Earth and the Moon to your analysis of sending a spacecraft to the Moon. You need to approximate the work by adding up the amount of work done by gravitational forces along each step of the path:

$$W = \sum \vec{F} \bullet \Delta\vec{r} = \vec{F}_1 \bullet \Delta\vec{r}_1 + \vec{F}_2 \bullet \Delta\vec{r}_2 + \vec{F}_3 \bullet \Delta\vec{r}_3 + \ldots$$

Compare the numerical value of the work with the change in the kinetic energy (final kinetic energy just before crashing on the Moon, minus initial kinetic energy when released above the Earth's atmosphere). Try shorter time steps to make sure that the errors introduced by a finite time step are not significant. Report your results.

Problem 4.2 Energy in the Moon shot

Energy conservation is a powerful check on the accuracy of a numerical integration. Plot a graph of kinetic energy, of gravitational potential energy, and of the sum of the kinetic energy and the gravitational potential energy, *vs.* position during the Moon voyage, Problem 4.1. Does the kinetic plus potential energy remain constant? What if you vary the step size (which varies the accuracy of the numerical integration)? Vary the launch speed, and explain the effect that this has on your graphs.

Problem 4.3 Calculating the impact speed

Use energy conservation to calculate analytically (that is, without doing a numerical integration) the final speed of the spacecraft just before it hits the Moon. Include the gravitational effect of the Moon. Use a launch speed of 1.3×10^4 m/s. Modify your program for Problem 4.2 to print out the speed of the spacecraft when it hits the surface of the Moon, and compare this value to your analytical result. What questions that could be addressed in the numerical integration are you *not* able to answer by doing this energy calculation?

Problem 4.4 Boosting a satellite orbit

Calculate the speed of a satellite in an orbit near the Earth (just above the atmosphere). If the mass of the satellite is 200 kg, what is the minimum energy required to move the satellite very far away from the Earth?

Problem 4.5 Comet orbit

A comet is in an elliptical orbit around the Sun. Its closest approach to the Sun is a distance of 4×10^{10} m (inside the orbit of Mercury), at which point its speed is 8.17×10^4 m/s. Its farthest distance from the Sun is far beyond the orbit of Pluto. What is its speed when it is 6×10^{12} m from the Sun? (This is the approximate distance of Pluto from the Sun.)

Problem 4.6 An electron in an accelerator

An electron is traveling at a speed of $0.95c$ in an electron accelerator. An electric force of 1.6×10^{-13} N is applied in the direction of motion while the electron travels a distance of 2 m. What is the new speed of the electron?

Problem 4.7 Four protons

Four protons, each with mass M and charge $+e$, are initially held at the corners of a square that is d on a side. They are then released from rest. What is the speed of each proton when the protons are very far apart?

Problem 4.8 Throwing from an asteroid

You stand on a spherical asteroid of uniform density whose mass is 2×10^{16} kg and whose radius is 10 km (10^4 m). These are typical values for small asteroids, although some asteroids have been found to have much lower average density and are thought to be loose agglomerations of shattered rocks; see Problem 2.11 (page 70). The asteroid is rotating a bit faster than once per day, so that objects on the surface have a speed of 1 m/s, including you and a rock you are holding.

(a) How fast (relative to you) do you have to throw the rock so that it never comes back to the asteroid and ends up traveling at a speed of 3 m/s when it is very far away? Explain briefly.

(b) Sketch graphs of the kinetic energy of the rock, the gravitational potential energy of the rock plus asteroid, and their sum, as a function of separation (distance from center of asteroid to rock). Label the graphs clearly.

Problem 4.9 Collision with Jupiter

This problem is closely related to the spectacular impact of the comet Shoemaker-Levy with Jupiter in July 1994 (http://www.jpl.nasa.gov/sl9/sl9.html). A rock far outside our Solar System is initially moving very slowly relative to the Sun, in the plane of Jupiter's orbit around the Sun. The rock falls toward the Sun, but on its way to the Sun it collides with Jupiter.

Calculate the rock's speed just before colliding with Jupiter. Explain your calculation and any approximations that you make.

$M_{\text{Sun}} = 2\times10^{30}$ kg $\qquad M_{\text{Jupiter}} = 2\times10^{27}$ kg

Distance, Sun to Jupiter $= 8\times10^{11}$ m $\qquad$ Radius of Jupiter $= 1.4\times10^{8}$ m

Problem 4.10 Particle-antiparticle annihilation

Figure 4.54 Plot various energies (Problem 4.10).

(a) A particle with mass M and charge $+e$ and its antiparticle (same mass M, charge $-e$) are initially at rest, far from each other. They attract each other and move toward each other. On axes like those in Figure 4.54, graph and label the various energies involved in this process, as a function of the distance r between the two particles. Be sure to include the rest energy of the particles.

(b) When the particle and antiparticle collide, they annihilate and produce a different particle with rest mass m (much smaller than M) and charge $+e$ and its antiparticle (same rest mass m, charge $-e$). When these two particles have moved far away from each other, how fast are they going? Is this speed large or small compared to c?

(c) Now take the specific case of a proton and antiproton colliding to form a positive and negative pion. Each pion has a rest mass of 2.5×10^{-28} kg. When the pions have moved far away, how fast are they going?

(d) How far apart must the two pions be (in meters) for their electric potential energy to be negligible compared to their kinetic energy? Be explicit and quantitative about your criterion and your result.

Problem 4.11 Gravitational energy inside the Earth

In the rough approximation that the density of the Earth is uniform throughout its interior, the gravitational field strength (force per unit mass) inside the Earth at a distance r from the center is gr/R, where R is the radius of the Earth. (In actual fact, the outer layers of rock have lower density than the inner core of molten iron.) Using the uniform-density approximation, calculate the amount of energy required to move a mass m from the center of the Earth to the surface. Compare with the amount of energy required to move the mass from the surface of the Earth to a great distance away.

Problem 4.12 Two neutral atoms

Figure 4.55 shows a potential energy curve for the interaction of two neutral atoms. The two-atom system is in a vibrational state indicated by the heavy solid horizontal line.

(a) At $r = r_1$, what are the approximate values of the kinetic energy K, the potential energy U, and the quantity $K+U$?

(b) What minimum energy must be supplied to cause these two atoms to separate?

(c) In some cases, when r is large, the interatomic potential energy can be expressed approximately as $U = -a/r^6$. For large r, what is the algebraic form of the magnitude of the force the two atoms exert on each other in this case?

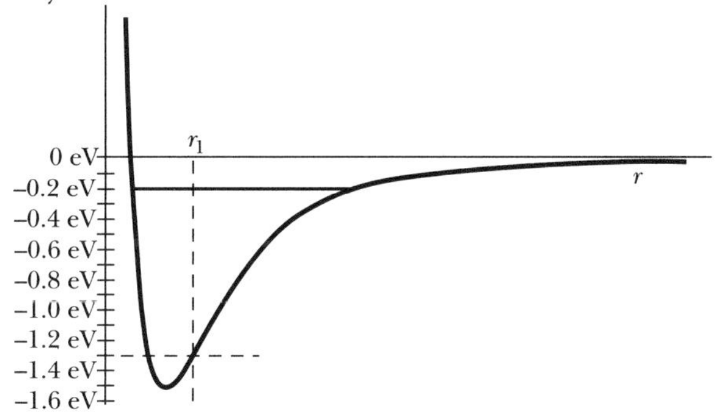

Figure 4.55 Potential energy diagram for Problem 4.12.

Problem 4.13 Alpha decay

Many heavy nuclei undergo spontaneous "alpha decay," in which the original nucleus emits an alpha particle (a helium nucleus containing two protons and two neutrons), leaving behind a "daughter" nucleus that has two fewer protons and two fewer neutrons than the original nucleus. Consider a Radium-220 nucleus that is at rest before it decays to Radon-216 by alpha decay.

The mass of the Radium-220 nucleus is 219.96274 u (unified atomic mass units) where 1 u = 1.6603×10^{-27} kg (approximately the mass of one nucleon). The mass of a Radon-216 nucleus is 215.95308 u, and the mass of an alpha particle is 4.00151 u. Radium has 88 protons, Radon 86, and an alpha particle 2.

(a) Make a diagram of the final state of the Radon-216 nucleus and the alpha particle when they are far apart, showing the momenta of each particle to the same relative scale. Explain why you drew the lengths of the momentum vectors the way you did.

(b) Calculate the final kinetic energy of the alpha particle. For the moment, assume that its speed is small compared to the speed of light.

(c) Calculate the final kinetic energy of the Radon-216 nucleus.

(d) Show that the nonrelativistic approximation was reasonable.

Problem 4.14 The SLAC two-mile accelerator

SLAC, the Stanford Linear Accelerator Center, located at Stanford University in Palo Alto, California, accelerates electrons through a vacuum tube two miles long (it can be seen from an overpass of the Junipero Serra freeway that goes right over the accelerator). Electrons which are initially at rest are subjected to a continuous force of 2×10^{-12} newton along the entire length of two miles (one mile is 1.6 kilometers) and reach speeds very near the speed of light. The analysis in Problem 2.12 (page 70) required numerical integration, but with the new techniques of this chapter you can analyze the motion analytically.

(a) Calculate the final energy, momentum, and speed of the electron.

(b) Calculate the approximate time required to go the two-mile distance.

Problem 4.15 Nuclear fission

Uranium-235 fissions when it absorbs a slow-moving neutron. The two fission fragments can be almost any two nuclei whose charges Q_1 and Q_2 add up to $92e$ (where e is the charge on a proton, $e = 1.6\times10^{-19}$ coulomb), and whose nucleons add up to 236 protons and neutrons (U-236; U-235 plus a neutron). One of the possible fission modes involves nearly equal fragments, palladium nuclei with $Q_1 = Q_2 = 46e$. The rest masses of the two palladium nuclei add up to less than the rest mass of the original nucleus. (In addition to the two main fission fragments there are typically one or more free neutrons in the final state; in your analysis make the simplifying assumption that there are no free neutrons, just two palladium nuclei.)

The rest mass of the U-236 nucleus is 235.996 u (unified atomic mass units), and the rest mass of each Pd-118 nuclei is 117.894 u, where 1 u = 1.7×10^{-27} kg (approximately the mass of one nucleon).

(a) Calculate the final speed v, when the palladium nuclei have moved far apart (due to their mutual electric repulsion). Is this speed small enough that $p^2/(2m)$ is an adequate approximation for the kinetic energy of one of the palladium nuclei? (It is all right to go ahead and make the nonrelativistic assumption first, but you then must check that the calculated v is indeed small compared to c.)

(b) Using energy considerations, calculate the distance between centers of the palladium nuclei just after fission, when they are starting from rest.

(c) A proton or neutron has a radius of roughly 10^{-15} m, and a nucleus is a tightly packed collection of nucleons. Experiments show that the radius of a nucleus containing N nucleons is approximately $(1.3\times10^{-15}\text{ m}) \times N^{1/3}$. What is the approximate radius of a palladium nucleus? Draw a sketch of the two palladium nuclei in part (b), and label the distances you calculated in parts (b) and (c).

Problem 4.16 Nuclear fusion

One of the thermonuclear or fusion reactions that takes place inside a star such as our Sun is the production of helium-3 (^{3}He, with two protons and one neutron) and a gamma ray (high-energy photon) in a collision between a proton (^{1}H) and a deuteron (^{2}H, the nucleus of "heavy" hydrogen, consisting of a proton and a neutron):

$$^1\text{H} + {}^2\text{H} \rightarrow {}^3\text{He} + \gamma$$

The rest mass of the proton is 1.0073 u (unified atomic mass unit, 1.7×10^{-27} kg), the rest mass of the deuteron is 2.0136 u, the rest mass of the helium-3 nucleus is 3.0155 u, and the gamma ray is massless.

(a) The strong interaction has a very short range and is essentially a contact interaction. For this fusion reaction to take place, the proton and deuteron have to come close enough together to touch. The approximate radius of a proton or neutron is about 10^{-15} m. What is the approximate initial total kinetic energy of the proton and deuteron required for the fusion reaction to proceed, in joules and electron volts (1 eV = 1.6×10^{-19} joule)?

(b) Given the initial conditions found in part (a), what is the kinetic energy of the ^{3}He plus the energy of the gamma ray, in joules and in electron-volts?

(c) The net energy released is the kinetic energy of the ^{3}He plus the energy of the gamma ray found in part (b), minus the energy input that you calculated in part (a). What is the net energy release, in joules and in electron volts? Note that you do get back the energy investment made in part (a).

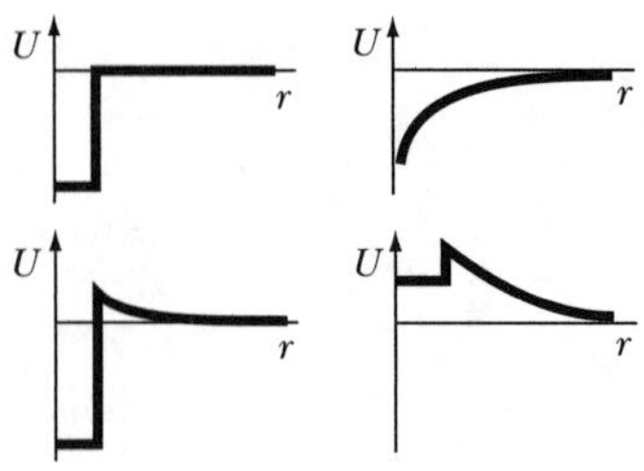

Figure 4.56 Potential energy curves for Problem 4.16 part (c).

(d) Which of the potential energy curves in Figure 4.56 is a reasonable representation of the interaction in this fusion reaction? Why?

As we will study later, the average kinetic energy of a gas molecule is $\frac{3}{2}kT$, where k is the "Boltzmann constant," 1.4×10^{-23} joule/Kelvin degree (J/K), and T is the absolute or Kelvin temperature, measured from absolute zero (so that the freezing point of water is 273 K). The approximate temperature required for the fusion reaction to proceed is very high. This high temperature, required because of the electric repulsion barrier to the reaction, is the main reason why it has been so difficult to make progress toward thermonuclear power generation. Sufficiently high temperatures are found in the interior of the Sun, where fusion reactions take place.

Problem 4.17 Determining the precise mass of the deuteron
A proton (1.6726×10^{-27} kg) and a neutron (1.6749×10^{-27} kg) at rest combine to form a deuteron, the nucleus of deuterium or "heavy hydrogen." In this process, a gamma ray (high-energy photon) is emitted, and its energy is measured to be 2.2 MeV (2.2×10^{6} eV).

(a) Keeping all five significant figures, what is the mass of the deuteron? Assume that you can neglect the small kinetic energy of the recoiling deuteron.

(b) Momentum must be conserved, so the deuteron must recoil with momentum equal and opposite to the momentum of the gamma ray. Calculate approximately the kinetic energy of the recoiling deuteron and show that it is indeed small compared to the energy of the gamma ray.

Problem 4.18 Potential energy and a pendulum
A pendulum consists of a very light but stiff rod of length L hanging from a nearly frictionless axle, with a mass m at the end of the rod (Figure 4.57).

(a) Calculate the gravitational potential energy as a function of the angle θ, measured from the vertical.

(b) Sketch the potential energy as a function of the angle θ, for angles from $-210°$ to $+210°$.

(c) Let $s = L\theta$ = the arc length away from the bottom of the arc. Calculate the tangential component of the force on the mass by taking the (negative) gradient of the energy with respect to s. Does your result make sense?

(d) Suppose you hit the stationary hanging mass so that it has an initial speed v_i. What is the minimum initial speed in order that the pendulum go over the top (θ = 180°)? On your sketch of the potential energy (part b), draw and label energy levels for the case where the initial speed is less than, equal to, or greater than this critical initial speed.

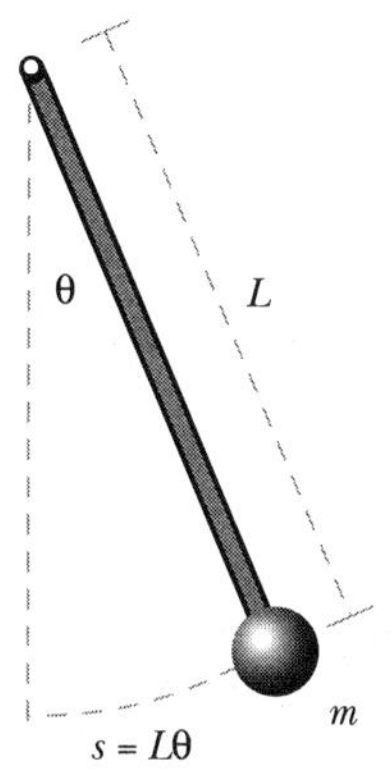

Figure 4.57 Pendulum (Problem 4.18).

4.17 Answers to exercises

4.1 (page 183)	1230 J
4.2 (page 183)	zero
4.3 (page 184)	$2P$
4.4 (page 185)	$(25\text{ m})\hat{\imath} - (30\text{ m})\hat{\jmath}$, $(-19.6\text{ N})\hat{\jmath}$, 588 J
4.5 (page 185)	189 J
4.6 (page 185)	0.1 J
4.7 (page 189)	about 6
4.8 (page 189)	about 10^{-14}
4.9 (page 190)	Make the indicated substitution; after that it's just algebra.
4.10 (page 191)	$\text{kg}\cdot\text{m}^2/\text{s}^2 = (\text{kg}\cdot\text{m}/\text{s}^2)\cdot\text{m} = \text{N}\cdot\text{m} = \text{J}$
4.11 (page 191)	12 m/s, 27 m.p.h.
4.12 (page 191)	1.2×10^{-12} kg (an extremely small amount of mass!)
4.13 (page 193)	$\Delta U = -50$ joules
4.14 (page 194)	$E_1 + E_2 + E_3 + E_4 + U_{12} + U_{13} + U_{14} + U_{23} + U_{24} + U_{34}$
4.15 (page 197)	– (more negative)
4.16 (page 197)	+ (less negative)
4.17 (page 198)	$-GMm/y^2$; 0
4.18 (page 198)	K has decreased
4.19 (page 199)	A is bound; C is unbound; B is trapped. In a quantum system in state B there is some probability of "tunneling" through the barrier and getting out of the trap!
4.20 (page 199)	A: go out in a straight line and stop, then fall back B: elliptical (motion with varying separation) C: nearly circular (nearly constant separation)
4.21 (page 202)	v to reach Moon < v_{esc}, because didn't go to infinity
4.22 (page 202)	1.1×10^4 m/s
4.23 (page 202)	2.4×10^3 m/s, much smaller than v_{esc} from Earth
4.24 (page 204)	same ΔU_g
4.25 (page 204)	6.3 m/s; the mass cancels from both sides of the energy equation
4.26 (page 204)	Hint: At top of trajectory the ball's vertical speed is momentarily zero. Answer on next page.
4.27 (page 205)	Hint: $E_f = E_i$. Answer on next page.
4.28 (page 206)	7.2×10^{-15} m Note: the radius of the proton is about 1×10^{-15} m.
4.29 (page 208)	A is potential energy, B is kinetic energy, C is $K+U$
4.30 (page 208)	1, 2, 3, 4 are bound states; 5 is an unbound state
4.31 (page 208)	0.789 MeV; proton mass is less than neutron mass
4.32 (page 210)	About 1.5×10^{-10}; accuracy of laboratory scale is far from what would be needed to detect the difference
4.33 (page 212)	About 0.1%

Additional exercises

4.34 (page 222)	4.9 N downward, 4.9 J Positive work makes the jar speed up.
4.35 (page 222)	2.5×10^{-19} J
4.36 (page 222)	9.8 J, 6.3 m/s, neglect air friction

4.37 (page 222) 19.6 J, 6.3 m/s; same speed, twice the kinetic energy
4.38 (page 222) 4.4×10^9 J

Additional answers
4.26 (page 204) 101 m
4.27 (page 205) $y_f = y_i + (v_i^2 - v_f^2)/2g$

Chapter 5

Energy in Macroscopic Systems

Chapter 5

Energy in Macroscopic Systems

In this chapter we consider energy in macroscopic systems, and we make connections between microscopic and macroscopic views of energy. We also show how the choice of system affects the form of the energy equation. The last part of the chapter deals with energy dissipation in macroscopic systems when there is friction.

5.1 Potential energy of macroscopic springs

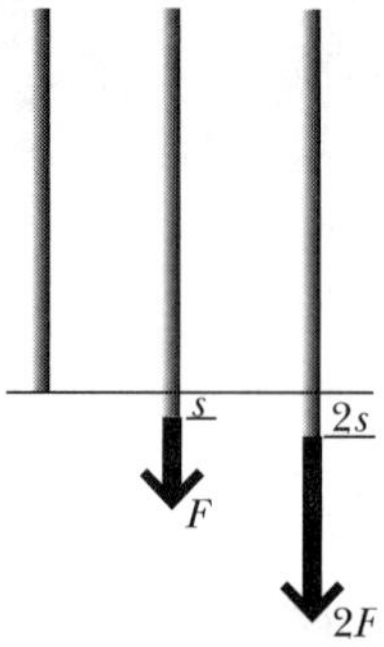

Figure 5.1 A bar of metal can be treated as a spring. For a limited range, double the force gives double the stretch.

A bar of metal can be treated as a spring, since stretching or compressing the bar stretches or compresses the interatomic bonds (Figure 5.1). Double the force produces double the change in length, as long as we're in the range where the spring approximation to the interatomic force is adequate. This is usually described for a bar in terms of Young's modulus, which we studied in Chapter 3, where the tension force F_T (per unit area) is proportional to the stretch ΔL (per unit length):

$$\frac{F_T}{A} = Y\frac{\Delta L}{L}$$

Writing this as $F_T = (YA/L)\Delta L$, we see that if we think of the bar as a (very stiff) spring, its spring stiffness is YA/L.

If you stretch the bar too much, two things can make the bar stop behaving in a spring-like manner. You might exceed the interatomic stretch for which $k_s s$ is a good approximation to the magnitude of the interatomic force. Also, the regular array of atoms may be disrupted by dislocations of the crystal structure, leading to large-scale slippage of crystal planes. Suddenly the bar "yields" and grows very much longer with little applied force.

A straight object is not usually used as a longitudinal spring but may be used in a bending mode. For example, when a diving board bends under a load, atomic bonds in the upper part of the board are stretched, and atomic bonds in the lower part are compressed (Figure 5.2). The net effect is a spring-like behavior of the diving board, with the amount of bend proportional to the applied force.

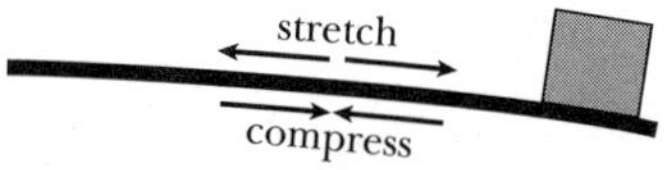

Figure 5.2 The bending of a diving board stretches and compresses interatomic bonds.

Helical springs

We will be most interested in helical (spiral) springs. A modest force applied to a helical spring gives a substantial change in the length of the spring, even though the total length of the coiled wire hardly changes. A large stretch of the spring with a small force implies a much smaller spring stiffness than the wire would have when straightened out.

If you stretch a helical spring far enough, it straightens out into a long straight wire, and it suddenly becomes very difficult to lengthen the wire any further. If you continue to stretch the straight wire, eventually it yields and then breaks. If you compress a helical spring so far that its coils run into each other, it suddenly gets extremely hard to compress the spring any further. Therefore, the linear relation between stretch and force, $F_s = -k_s s$, is valid only over a limited region of stretch, although this linear range is much wider than for the microscopic "spring."

In many situations the entire phenomenon takes place within the valid range of the linear approximation. Or one can speak of an "infinite" well as shown in Figure 5.3, which is an idealization of the real situation. Within the linear range where $F_s = -k_s s$ is the negative gradient of $\frac{1}{2}k_s s^2$, we have this:

SPRING POTENTIAL ENERGY

$$U_s = \frac{1}{2}k_s s^2 + U_0, \ s \text{ measured from equilibrium point}$$

U_0 is a negative number representing how far below $U_s = 0$ the bottom of the potential "well" is located. An amount of work $W_{external} = |U_0|$ is needed to stretch the wire out straight, and make it deform.

In Figure 5.3 (which is an idealization of the real situation) there are very steep sides of the potential well, corresponding to the large forces required to compress a spring whose coils are in contact, and the large forces required to stretch the spring after it has straightened out into a straight wire. Remember that force is the negative gradient of the potential energy, so a steep slope corresponds to a large force.

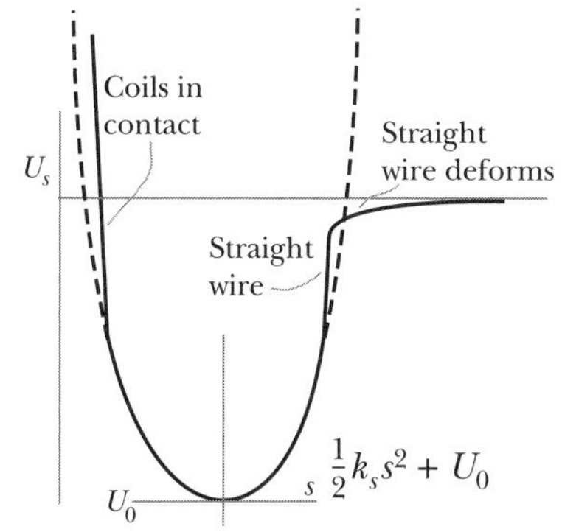

Figure 5.3 Spring potential energy, within the linear range where $F_s = -k_s s$. The dotted line represents an ideal spring, while the solid line represents a real macroscopic spring.

5.1.1 Application: Energy in an oscillating spring-mass system

Consider a block connected by a helical spring to a wall, whose oscillations we treated in Chapter 3. If you compress the spring and then let go, the block acquires kinetic energy (Figure 5.4), with a loss of spring potential energy. If you stretch the spring and then let go, the block also acquires kinetic energy, again with a loss of spring potential energy (Figure 5.5).

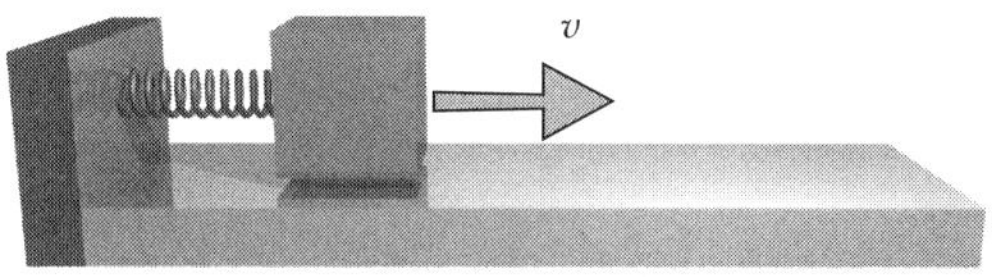

Figure 5.4 As the compressed spring expands the kinetic energy of the block increases.

We'll apply the energy principle to this spring-mass system. For the system consisting of the block of mass m and the spring of stiffness k_s, the sum of the changes in the kinetic energy K of the block and the spring potential energy U_s, must be zero, because there is no external work done on this system, if we neglect friction (the rest energy mc^2 of the block doesn't change). The wall exerts a force on the left end of the spring, but this force acts through no distance and does no external work; it only affects the momentum of the system, which changes continuously. The vertical forces act at right angles to the displacement and so do no work.

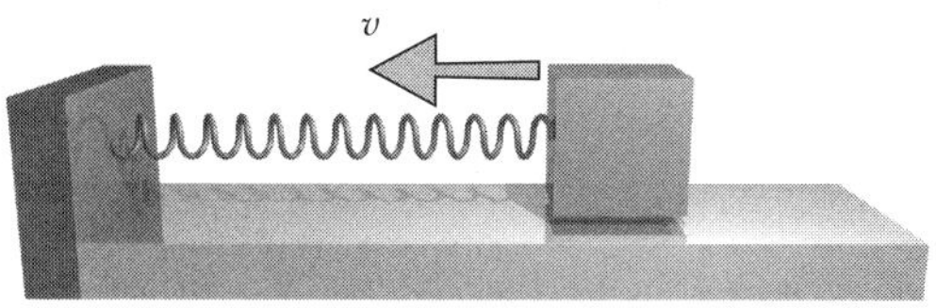

Figure 5.5 As the stretched spring contracts the kinetic energy of the block increases.

$$\Delta(K + U_s) = 0 \ \text{(neglecting friction)}$$

$$\Delta\left(\frac{p^2}{2m} + \frac{1}{2}k_s s^2 + U_0\right) = 0$$

Consider a mass and spring oscillating on a horizontal air track with very little friction. We start the oscillation by stretching the spring a distance s and releasing it from rest.

? At the moment that the spring is released, what is the kinetic energy of the spring-mass system? What is the total energy of the system?

Since the block is released from rest, its kinetic energy is zero. The energy of the system is equal to the energy stored in the spring, $\frac{1}{2}k_s s^2 + U_0$.

? At what point in the oscillation is the kinetic energy of the system highest? At that moment, what is the potential energy of the system?

When the contracting spring reaches its equilibrium length, the energy stored in the spring is a minimum (U_0). The maximum possible amount of spring potential energy has now been converted to kinetic energy, so this is the instant when the kinetic energy of the system is highest.

? At what point in the oscillation will the kinetic energy of the system be lowest?

At either endpoint of the oscillation—spring fully compressed or spring fully extended—the instantaneous speed of the mass is zero. The kinetic energy of the system is lowest here, and all the energy has been momentarily converted back into spring potential energy.

We can write the energy equation for the closed system of mass plus spring to emphasize that over a time interval the total energy (kinetic plus spring) doesn't change:

$$\Delta\left(\frac{p^2}{2m} + \frac{1}{2}k_s s^2 + U_0\right) = 0$$

Or we can rearrange the terms and write the energy equation in a somewhat different form:

$$\frac{p_f^2}{2m} + \frac{1}{2}k_s s_f^2 + U_0 = \frac{p_i^2}{2m} + \frac{1}{2}k_s s_i^2 + U_0$$

That is, the final kinetic plus potential energy of the system is equal to the initial kinetic plus potential energy of the system, no matter what we decide to choose as initial and final states. Note that the potential energy U_0 at the bottom of the "well" cancels out in either view of energy conservation.

5.1.2 Shifting the zero of potential energy

For consistency with special relativity, it is essential that potential energy be zero when the particles of a system are widely separated. However, when analyzing a macroscopic spring-mass system we can conveniently ignore the negative U_0 term and pretend that the potential energy is zero when the stretch is zero: $U_s = \frac{1}{2}k_s s^2$. The reason why this works is that in a process we only care about *changes* in energy, and the change in the constant U_0 term is zero.

Example: A mass of 0.2 kg is attached to a horizontal spring whose stiffness is 12 N/m. Friction is negligible. At $t = 0$ the spring has a stretch of 3 cm and the mass has a speed of 0.5 m/s. What is the amplitude (maximum stretch) of the oscillation? What is the maximum speed?

Solution: System: mass and spring

Initial state: $s = 3$ cm, $v = 0.5$ m/s

Final state: $v = 0$

Energy principle: $E_f = E_i$ (neglecting friction)

The initial energy, ignoring changes in rest mass and in U_0, is

$$\frac{1}{2}mv_i^2 + \frac{1}{2}k_s s^2 = (\tfrac{1}{2})(0.2\text{ kg})(0.5\text{ m/s})^2 + (\tfrac{1}{2})(12\text{ N/m})(0.03\text{ m})^2$$

$$E_i = 0.03\text{J}$$

With negligible friction, the energy will be 0.03 J at all times. The stretch is maximum when the speed and kinetic energy are zero:

$$\frac{1}{2}k_s s_{max}^2 = 0.03\text{J}$$

$$s_{max} = \sqrt{2(0.03\text{J})/(12\text{ N/m})} = 0.07\text{ m}$$

Now consider different final state: $s = 0$

The speed is maximum when the stretch and potential energy are zero:

$$\frac{1}{2}mv_{max}^2 = 0.03\text{J}$$

$$v_{max} = \sqrt{2(0.03\text{J})/(0.2\text{ kg})} = 0.55\text{ m/s}$$

Ex. 5.1 About how many joules of energy can you store in a spring like the one you used in Chapter 3, starting from a relaxed spring? Assume $k_s \approx 0.6$ N/m and a stretch of about 20 cm.

Ex. 5.2 A horizontal spring with stiffness 0.5 N/m has a relaxed length of 15 cm. A mass of 20 grams is attached and you stretch the spring to a total length of 25 cm. The mass is then released from rest. What is the speed of the mass at the moment when the spring returns to its relaxed length of 15 cm?

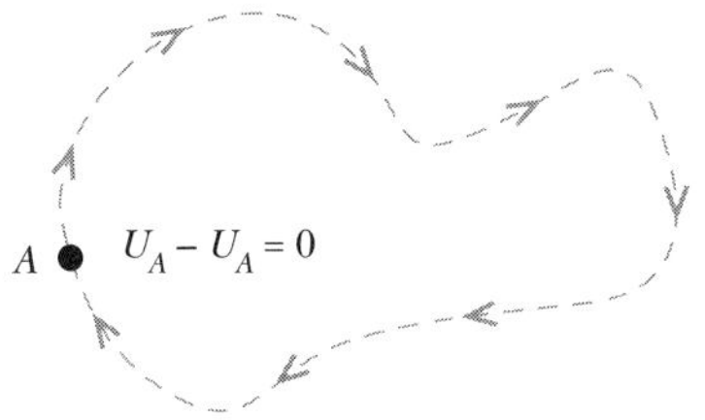

Figure 5.6 A particle moves from *A* along a path and returns to *A*.

5.2 Path independence of potential energy

In situations describable by a potential energy, the associated interaction depends solely on position, not on speed or direction. For example, spring energy depends on the stretch *s*, and gravitational and electric energy depend on the separation *r*. It is easy to prove that the following must be true:

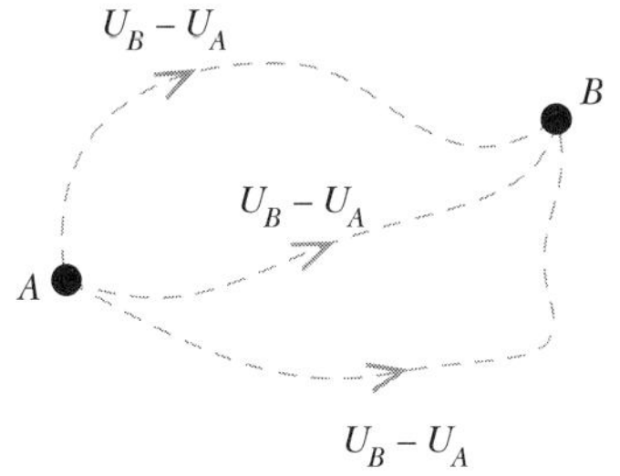

Figure 5.7 Potential energy difference is independent of the path between the initial and final locations.

PATH INDEPENDENCE OF POTENTIAL ENERGY

In a round trip, the potential energy doesn't change.

Change in potential energy doesn't depend on the path taken.

Here is the proof: In Figure 5.6, if one of the particles in a multiparticle system starts at point *A*, moves around and comes back to the starting point, the change in potential energy *U* of the multiparticle system must be zero, since the initial and final potential energy depends on pair-wise distances *r* or *s*, and these return to their original values in a round trip. Similarly, in Figure 5.7 change in potential energy *U* of the system when one of its particles moves along some path from point *A* to point *B* depends solely on the initial and final values of *r* or *s*, so it doesn't matter what path the particle follows to get from *A* to *B*.

A simple example is projectile motion when air resistance can be neglected (Figure 5.8). You throw a ball up in the air with speed *v*, and when it returns to your hand it has regained the same speed *v* that you gave it (the gravitational energy change is zero, and the change in the kinetic energy is zero). Another example is the motion of a mass on a spring in the absence of friction: every time the mass passes a particular position, it has the same speed (same *s* implies same *v*).

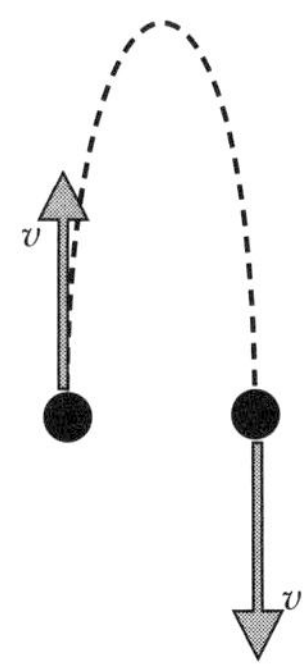

Figure 5.8 A ball goes up and down with negligible air resistance.

In both of these examples, although the speed doesn't change, the velocity may change. The ball was headed up but is now headed down; a mass on a spring may be going to the left and then going to the right with the same speed. The key point is this:

Potential energy depends only on position,
not on the direction of motion.

There is a connection between the fact that change of potential energy is independent of path, and that a round trip gives zero. Suppose that along some path from *A* to *B* to *C* in Figure 5.9 the change in potential energy is +5 J.

? What will the change in potential energy be along the same path in the opposite direction, from *C* to *B* to *A*?

Going backwards from *C* to *B* to *A* along this same path the change would be –5 J. Since the change is independent of the path, you could take *any* path to return from *C*, such as *C* to *D* to *A*, with a change of –5 J. Therefore any round trip from *A* to *B* to *C* to *D* and back to *A* will yield a change of zero.

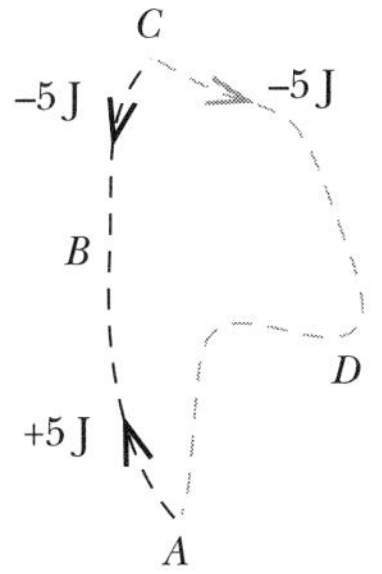

Figure 5.9 Forward and reversed paths, and a round trip.

Ex. 5.3 During one complete oscillation of a mass on a spring (one period), what is the change in potential energy of the mass + spring system, in the absence of friction?

5.3 Example: A rebounding block

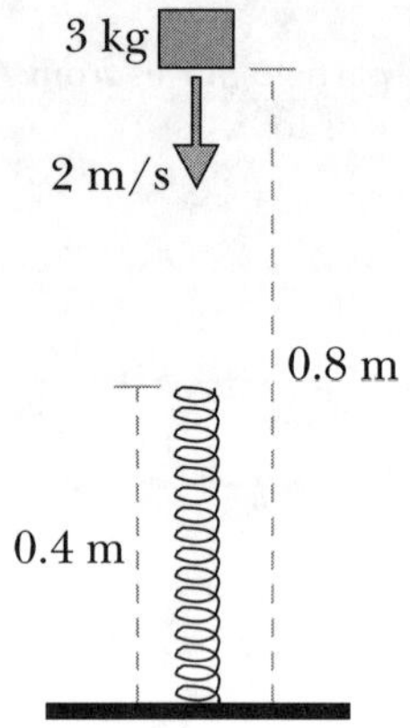

Figure 5.10 A block rebounds from a vertical spring.

A metal block of mass 3 kg is moving downward with speed 2 m/s when the bottom of the block is 0.8 m above the floor (Figure 5.10). When the bottom of the block is 0.4 m above the floor, it strikes the top of a relaxed vertical spring 0.4 m in length. The stiffness of the spring is 2000 N/m.

(a) The block continues downward. When the bottom of the block is 0.3 m above the floor, what is its speed?

(b) The block eventually heads back upwards, loses contact with the spring, and continues upward. What is the maximum height reached by the bottom of the block above the floor?

(c) What approximations did you make?

Solution

(a) Take the Earth, block, and spring as the system to which we'll apply the energy principle. There are no energy transfers to this system from the surroundings, so the energy of the system is a constant. We will calculate E_i and E_f, and set them equal to each other. In the initial state the energy is gravitational potential energy plus kinetic energy; in the final state the energy is gravitational plus kinetic plus spring potential energy (we omit the rest energy, which does not change).

$$E_i = E_f$$

$$mgy_i + \frac{1}{2}mv_i^2 = mgy_f + \frac{1}{2}mv_f^2 + \frac{1}{2}k_s s_f^2$$

Solve for the final speed:

$$v_f^2 = v_i^2 + 2g(y_i - y_f) - \frac{k_s}{m}s_f^2$$

$$v_f = \sqrt{(2\text{ m/s})^2 + 2(9.8\text{ N/kg})(0.5\text{ m}) - \frac{(2000\text{ N/m})}{(3\text{ kg})}(0.1\text{ m})^2}$$

$$v_f = 2.7\frac{\text{m}}{\text{s}}$$

(b) Here we need to consider a different final state. Take as the initial state the original initial state, and as final state the instant when the block is at the top of its trajectory, where $v = 0$ and spring stretch $s = 0$:

$$mgy_i + \frac{1}{2}mv_i^2 = mgy_f$$

Divide by mg:

$$y_f = y_i + \frac{v_i^2}{2g} = (0.8\text{ m}) + \frac{(2\text{ m/s})^2}{2(9.8\text{ N/kg})} = 1.0\text{ m}$$

(c) We assumed negligible air resistance and dissipation in the spring. We used the approximate expression $U_g = mgy$ near Earth's surface. We neglected the extremely tiny kinetic energy change of the Earth. The gravitational and spring forces on the Earth and on the block are equal and opposite, so the momentum changes are equal and opposite. But since $K = p^2/(2m)$, for the same momentum p the kinetic energy of the Earth is extremely small compared to the kinetic energy of the block.

Another way to look at this is to say that while the forces on the Earth and the block have the same magnitudes, the Earth hardly moves, so the force does extremely little work on the Earth and gives it little kinetic energy.

5.4 Thermal energy

In our idealized models of spring-mass systems, the sum of kinetic plus spring potential energy is constant. In real spring-mass systems, however, the oscillations die away ("damp down") with time, which means that the kinetic plus spring potential energy must be decreasing. The decrease in the amplitude of the oscillations is mainly due to air resistance or to friction with a solid, and some energy is transferred to the surroundings, to the air or to a solid along which the mass slides. On the microscopic level, if we could look inside the solid, we would see increased atomic motion—the atoms in the solid have gained energy. A macroscopic indication of this increased atomic motion is an increase in the temperature of the surface.

We can measure "macroscopic" spring potential energy and kinetic energy with simple measurements that don't require a microscope. But when the temperature of a solid object increases, the increased energy inside the solid is not visible to the naked eye, and we face the problem of how to evaluate the increase in the "thermal" energy.

Microscopic kinetic and spring potential energy

The increased "microscopic" energy in a solid is in two forms. First, the atoms on the average are moving around faster, with increased kinetic energy $p^2/(2m)$. Second, there is increased spring potential energy $\frac{1}{2}k_s s^2 + U_0$ in the interatomic bonds ("springs" in our model of solids), where k_s is the stiffness of an interatomic spring (Figure 5.11).

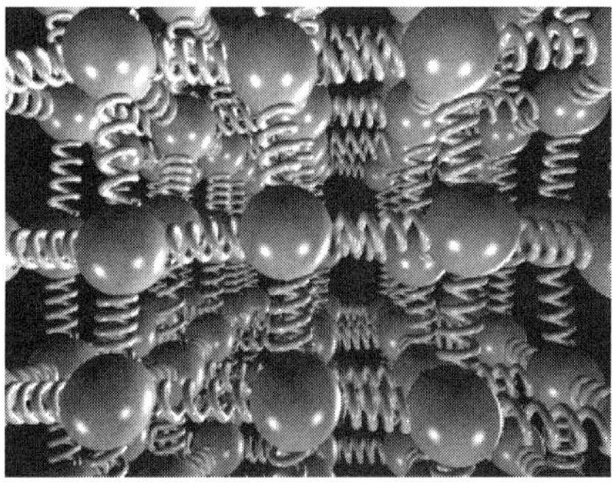

Figure 5.11 In our model solid we must account for the kinetic energy of every ball and the spring energy of every spring.

In principle the microscopic, thermal energy in a solid could be evaluated by simultaneously measuring at some instant the momenta of *all* of the atoms, and the stretches or compressions of *all* of the springs (that is, changes in atomic positions away from their equilibrium positions, together with knowledge of the spring stiffness). Since there are about 10^{23} atoms (and even more "springs") in an ordinary-sized object, it is in practice impossible to evaluate the energy this way.

Perhaps we could make microscopic measurements of just one atom (and its attached "springs"), assume that this is the average energy of each atom, and then multiply by the number of atoms there are in the object to get the total energy. That doesn't work either, because at some instant we may find an atom that is momentarily sitting motionless at its equilibrium position (zero energy), and an instant later it is displaced away from its equilibrium position and moving rapidly. Energy keeps getting passed back and forth among the many atoms and springs, and measuring the energy of just one atom and its attached springs at one instant doesn't give us the correct average energy we would need to determine the total energy of the object.

Temperature

In the 1800's it was realized that for many systems the temperature is essentially a measure of the average energy of the atoms in the system:

Temperature and the average energy of atoms are related

Given the impossibility of evaluating the energy of each individual atom, we can simply use a thermometer to measure thermal energy. A familiar example is the mercury thermometer (silver colored) or the alcohol thermometer (tinted red for visibility). In these thermometers, a very thin column of liquid expands when heated. By tradition, marks on the thermometer are placed to represent "degrees of temperature," but we could just as well

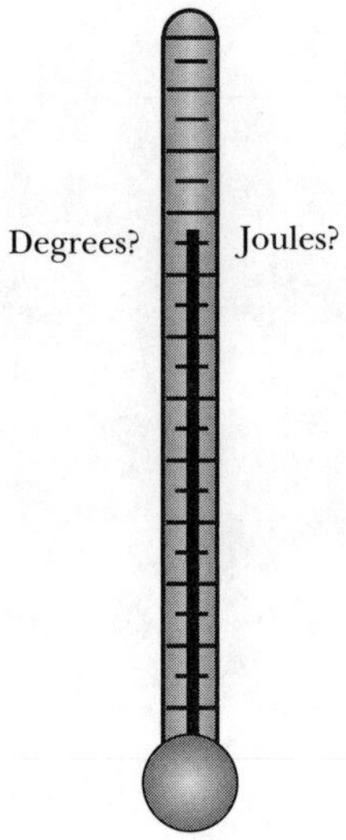

Figure 5.12 A thermometer might be calibrated in joules instead of in degrees.

place marks to indicate the average energy of the molecules in the thermometer, in joules (Figure 5.12). One Kelvin (or one Celsius degree of temperature) is equivalent to an average molecular energy of about 10^{-23} joules.

This is not the whole story. Later in this course we will make more precise the relationship between temperature and energy, and we will see that temperature is more directly related to a quantity called "entropy" than it is to energy. Nevertheless, for most ordinary systems at room temperature it is true that temperature is a good measure of the average energy of the atoms.

The usefulness of a thermometer lies in the important observation that when two objects are in good contact with each other, the average kinetic energies of the molecules in the two objects slowly come to be equal. Where the two objects touch each other, molecules collide with each other, and the faster molecules on average lose kinetic energy to the slower molecules in a collision. Once the average kinetic energies of molecules in both objects have come to be equal, additional collisions on average make no further change. To put it in terms of temperature (which is related to the average molecular kinetic energy), the hotter object gets cooler, and the cooler object gets hotter. Everyday examples abound: a cold drink warms up in a hot room, and warm water placed in a freezer cools down and freezes.

To measure the microscopic energy of an object, we place a thermometer in contact with the object. Then we wait for the temperatures (average molecular kinetic energies) of the object and thermometer to equilibrate, and we read the thermometer. It is of practical importance that the thermometer have a relatively small mass compared to the object of interest, so that attaching the thermometer doesn't add or subtract much energy to or from the object. Another advantage of a small thermometer is that it reaches thermal equilibrium quicker than a large thermometer does.

How was the calibration between one Kelvin (or one Celsius degree) and energy established? In one of a series of classic experiment performed by Joule in the 1840's, a paddle wheel in water was turned by a falling weight. The work done by the Earth's gravitational force on the falling weight (mgh), was accompanied by an increase in the thermal energy of the water (presumably also mgh, assuming that energy is conserved), with a measurable rise in the water temperature measured in Kelvins (after the paddle wheel had stopped turning). It was found for water that the energy required to raise the temperature of one gram of water by one Kelvin (1 K) is 4.2 joules. This is often summarized by saying that the "heat capacity" of water is 4.2 joules per gram per Kelvin (Figure 5.13). Other materials have been measured to have different heat capacities. We will study an atomic theory of heat capacity later in the course.

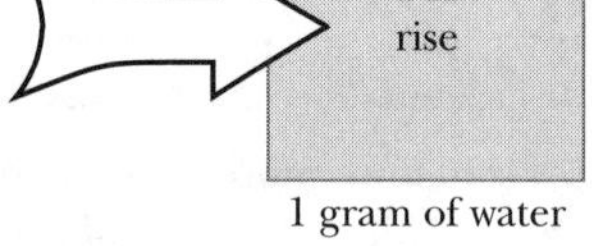

Figure 5.13 Energy input of 4.2 joules into a gram of water raises the temperature by 1 K. We say that the heat capacity of water is 4.2 J/gram/K.

Volume and temperature

An ordinary thermometer depends on the effect that higher temperature makes the material expand. Why does this happen?

For large oscillations of atoms whose interaction is described by a potential energy such as that shown in Figure 5.14, the average stretch increases. Because the actual potential energy curve is not really symmetric around the equilibrium point, with increasing energy the average interatomic bond length gets slightly longer. This is why an object typically expands at higher temperature. Higher temperature implies larger amplitude of oscillations and higher energy, and there is a slight shift in the center of the oscillations at higher energy.

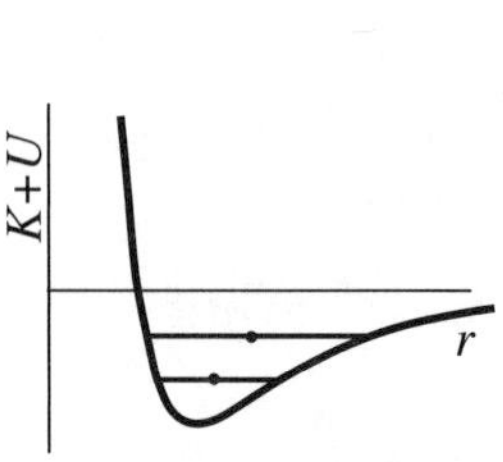

Figure 5.14 For larger oscillations the average stretch increases.

In the interior of the solid the net potential curve for an atom is symmetric, not asymmetric, because the potential curve is associated with forces exerted by atoms to the left and to the right. But since the bond lengthening can start at the surface and propagate into the interior, a solid ultimately expands at higher temperatures.

Other kinds of thermometers

There are many other kinds of thermometers. All materials change in some way when they get hotter or colder, and some of these effects are the basis for useful temperature indicators. A new kind of thermometer that is increasingly used is the liquid crystal thermometer. In a liquid crystal there is some molecular order over long distances in the liquid, unlike a true liquid in which there is hardly any long-range order. In one form of liquid crystal thermometer, helical molecules remain roughly parallel to each other, and the distance between coils in the helix changes with temperature. Because the distance between coils is comparable to the wavelength of visible light, this distance affects the way light is reflected from the liquid crystal. The effect is that the material changes color with changing temperature.

Calculations using heat capacity

Joule found that it takes 4.2 J of energy input to a gram of water to raise its temperature 1 K. If such experiments are done on other materials, the temperature rise is found to be different. For example, the heat capacity of ethanol is found to be 2.4 J/gram/K, and the heat capacity of copper is only 0.4 J/gram/K. One of the many unusual and useful properties of water is that it has a very large heat capacity on a per-gram basis, which means that it is difficult to change its temperature. Here is the meaning of heat capacity on a per-gram basis, where $\Delta E_{\text{thermal}}$ is the rise in the energy of the system, in the form of increased atomic kinetic and potential energy:

HEAT CAPACITY *C* ON A PER-GRAM BASIS

$$\Delta E_{\text{thermal}} = mC\Delta T, \text{ or } C = \frac{\Delta E_{\text{thermal}}}{m\Delta T} \quad (m \text{ in grams})$$

In principle, it should be possible to calculate the heat capacity of a material if something is known about its atomic structure, since a temperature rise is a measure of increased energy at the atomic level. Comparisons of calculations with experimental heat capacity data are a good test of our understanding of atomic models. In later chapters you will learn how to calculate the heat capacity of a gas, and of a solid, based on the ball-and-spring model of a solid.

If the heat capacity of a material is known, the amount of energy transfer into the material can be determined by observing the temperature rise.

Example: You stir 12 kg of water vigorously, doing 36000 joules of work. If the water is well insulated (so that all of your energy input goes into increasing the energy of the water), what temperature rise would you expect?

Solution: $$\left(\frac{36000\text{ J}}{12000\text{ grams}}\right)\left(\frac{1}{4.2\text{ J/gram/K}}\right) = 0.7\text{ K}$$

Including units helps avoid making calculational mistakes.

Ex. 5.4 Niagara Falls is about 50 meters high. What is the temperature rise in Kelvins of the water from just before to just after it hits the rocks at the bottom of the falls, assuming negligible air resistance during the fall and that the water doesn't rebound but just splats onto the rock?

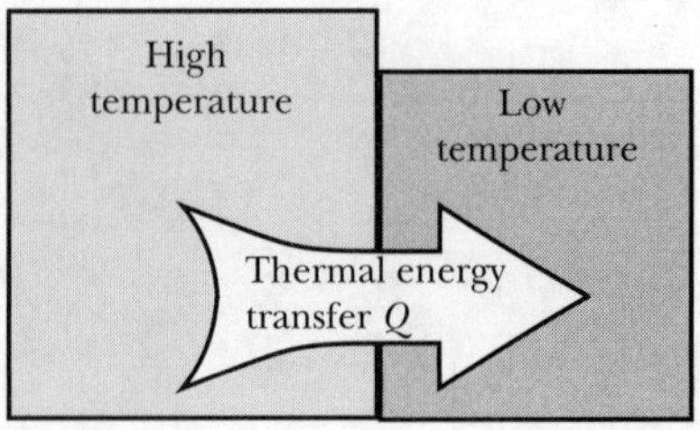

Figure 5.15 A hot block in contact with a cold block.

5.4.1 Thermal transfer of energy

Work is mechanical energy transfer into or out of a system involving forces acting through macroscopic displacements. When a hot object is placed in contact with a cold object, energy is transferred from the hot object to the cold one, but there are no macroscopic forces or displacements, so we don't refer to this kind of energy transfer as work.

Instead, we speak of "thermal transfer of energy" Q, a term which originated at a time before the atomic model of matter was prevalent. At the microscopic level thermal transfer of energy actually does involve work; when a hot block is placed in contact with a cold block (Figure 5.15), at the interface the atoms in the two blocks collide with each other, and do work on each other. The atoms in the hot block have greater average kinetic energy than the atoms in the cold block, so in an individual collision it is likely that a fast-moving atom in the hot block loses energy to a slow-moving atom in the cold block.

It can happen that the atom in the hot block happens to be moving slowly and gains energy when it is hit by an atom in the cold object that happens to be moving fast, but this is less likely. On average there is energy flow, "thermal transfer of energy" (microscopic work), from the hot block to the cold block. These energy changes will diffuse throughout the blocks. If left to themselves, the two blocks will eventually come to the same temperature, intermediate between their two initial temperatures.

THERMAL TRANSFER OF ENERGY

> Thermal transfer of energy is the name we give to a process in which energy moves between a system and the surroundings with which it is in contact, due to a temperature difference.

If we choose just the initially cold block as our system of interest, this system gains energy (at the expense of the surroundings, the initially hot block). At the atomic level, this energy increase is the result of work done on the atoms at the surface of the cold block. Usually we are unable to observe these atomic-level interactions directly. We can infer the amount of thermal transfer of energy into the system by observing the temperature rise of that system, if we know the heat capacity of the material.

It is common practice to denote thermal transfer of energy by the letter Q and work (macroscopic, mechanical energy transfer) by the letter W. The energy equation for an "open" (not isolated) system is then

$$\Delta E_{\text{system}} = Q + W$$

That is, thermal transfer of energy and mechanical energy transfers into or out of the system produce a change in the energy *of* the system. The energy of the system includes particle energy and potential energy, at the microscopic or macroscopic level.

Note that Q and/or W can be negative. If there is thermal transfer of energy out of a system, we give Q a negative sign, because the change in the energy of the system is a decrease. Typically such transfer would be due to the system having a higher temperature than the surroundings. Similarly, if the system does work on its surroundings rather than the other way round, the sign of W is negative. (Some older textbooks on thermodynamics reverse the sign of W, counting energy outputs from the system as positive.)

We believe that for the Universe as a whole Q and W are zero, and the energy of the Universe does not change. But even a small system may be effectively closed, with its energy unchanging, if it is thermally well insulated (to prevent thermal transfer of energy into or out of the system) and if there are no other energy transfers from the surroundings.

Terminology issues

The technical meaning of the word "heat" in science is not supposed to be the same as its everyday meaning, as is also the case with the word "work." In science the word "heat" is supposed to be reserved to refer to thermal transfer of energy (microscopic work) across a system boundary. Rather than saying "there is heat in the object," one says there is energy in the object. The amount of energy in the object might increase due to "heat" (energy transfer into the system due to a temperature difference). However, even professional scientists sometimes slip in their usage of the word "heat"!

To avoid confusion, we will avoid the use of the word "heat" as a noun. Instead, we will speak of "thermal transfer of energy" between system and surroundings due to a temperature difference, and change of "thermal energy" inside the system (in the form of increased atomic kinetic energy and potential energy).

Ex. 5.5 Suppose you warm up 500 grams of water (half a liter, or about a pint) on a stove while doing 5×10^4 J of work on the water with an electric beater. The temperature of the water is observed to rise from 20° C to 80° C. What was the change in the thermal energy of the water? Taking the water as the system, how much thermal transfer of energy Q was there across the system boundary? What was the energy change of the surroundings?

Other input: Electromagnetic radiation

Work and heat are very common kinds of energy transfers into or out of an open system. Also, electromagnetic radiation, including visible light and infrared, can carry energy into or out of an open system. Sunlight absorbed by the Earth raises the temperature (and the thermal energy) of the daylight side of the Earth. We will typically account for this kind of energy transfer in terms of photons (packets of energy) crossing the system boundary, which is a kind of change of identity for the system.

5.5 Example: Heat capacity

A 300 gram block of aluminum at temperature 500 K is placed on a 650 gram block of iron at temperature 350 K in an insulated enclosure. At these temperatures the heat capacity of aluminum is approximately 1.0 J/K/gram, and the heat capacity of iron is approximately 0.42 J/K/gram. Within a few minutes the two metal blocks reach the same common temperature T. Calculate T.

Solution

System: the two blocks
Initial state: different temperatures, not in contact
Final state: blocks have come to thermal equilibrium, same T.
Energy principle: $\Delta E_{Al} + \Delta E_{Fe} = 0$
(The total energy of the two blocks does not change, because there is no energy transfer from or to the surroundings.)

$$(300\text{ gram})(1.0\text{ J/K/gram})(T-500) + (650\text{ gram})(0.42\text{ J/K/gram})(T-350) = 0$$

Solving for the final temperature, we find $T = 429\text{ K}$.

In words, what happens in this process is that the aluminum temperature falls from 500 K to 429 K, and the iron temperature rises from 350 K to 429 K. The thermal energy decrease in the aluminum is equal to the thermal energy increase in the iron.

5.6 Reflection: Forms of energy

There are two fundamental kinds of energy in a multiparticle system:

- particle energy (expressible as rest energy plus kinetic energy)
- potential energy

All other forms of energy are examples of these basic forms. For example, chemicals in a system can react with each other and raise the temperature (and thermal energy) of the system, so we speak of "chemical energy," of which food is an example, with the molecule ATP providing energy storage in the body. The "chemical energy" is the kinetic energy and electric potential energy of molecules. Change of shape (configuration) is associated with change of potential energy, and there may also be changes of the kinetic energy of the molecules.

Nuclear energy is similar. Nucleons can be more or less tightly bound to particular nuclei, associated with different amounts of nuclear potential energy (change of shape) and different amounts of kinetic energy of the nucleons. Also, a change of identity in a nuclear reaction can be associated with a (huge) kinetic energy change associated with a change of rest mass.

5.7 Power: Energy per unit time

In technical usage, the word "power" is defined to mean "energy per unit time." From the point of view of energy usage, it makes no difference whether you take a minute or a month to climb a stairs, yet from a practical point of view you certainly do notice the *rate* of energy usage, called power. The units of power are joules per second, or watts (honoring James Watt, the developer of the first efficient steam engine).

POWER

Energy per unit time (J/s or watts)

There is a useful formula for the instantaneous power associated with the work done by a force:

INSTANTANEOUS POWER

$$\text{power} = \frac{dW}{dt} = \frac{\vec{F} \bullet d\vec{r}}{dt} = \vec{F} \bullet \frac{d\vec{r}}{dt} = \vec{F} \bullet \vec{v}$$

Example: How much energy is required to run a 100-watt light bulb for an hour?

Solution: $(100\text{J/s})(1\text{ hr})\left(60\frac{\text{min}}{\text{hr}}\right)\left(60\frac{\text{s}}{\text{min}}\right) = 3.6\times10^5\text{ J}$

Ex. 5.6 In the Niagara Falls hydroelectric generating plant, the energy of falling water is converted into electricity. The height of the falls is about 50 meters. Assuming that the energy conversion is highly efficient, approximately how much energy is obtained from one kilogram of falling water? Therefore, approximately how many kilograms of water must go through the generators every second to produce a megawatt of power (10^6 watts)?

Ex. 5.7 A vehicle with a mass of 1000 kg has an engine whose maximum power output is 50 kilowatts (about 67 horsepower; one horsepower is 746 watts). At a speed of 20 m/s (about 45 miles/hour), what is the maximum acceleration possible?

5.8 Open and closed systems

We have already had some experience with the difference between a closed system and an open system. Because of the importance of this distinction, in this section we will go more deeply into the issues. Consider this diagram showing transfers of money into and out of a bank during a particular time period, and the corresponding change in accounts inside the bank (Figure 5.16).

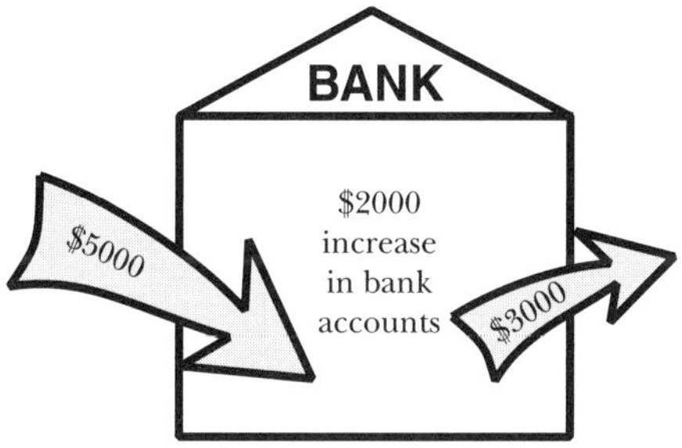

Figure 5.16 A bank is an open system.

? We have shown a situation where more money is transferred into the bank than is transferred out. During these transactions, does the total amount of money inside the bank remain unchanged?

We say that the bank is an "open system"—a portion of the world open to transfers in and out, and therefore subject to changes in its internal amount of money. During times when the bank does not permit transfers, the bank is temporarily a "closed system" and its total internal amount of money is unchanged, although there may be changes in the form of money inside the bank such as transfers between checking accounts and savings accounts.

Now consider Figure 5.17 which shows energy transfers during a particular time period in the winter, into and out of a house that is heated by gas, and the corresponding change in the energy inside the house. This is a situation where more energy comes into the house than goes out, and as a result the temperature inside the house rises.

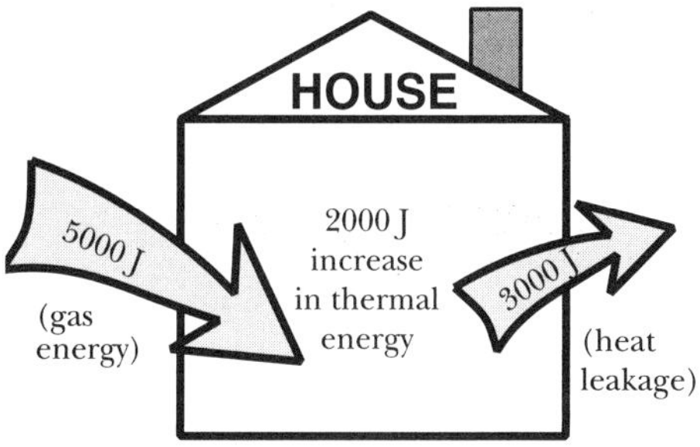

Figure 5.17 A house is an open system.

? During these energy transactions, the total amount of energy in the whole Universe is unchanged, but is the amount of energy inside the house unchanged?

The amount of energy inside the house increased by 2000 joules. We say that the house is an "open system"—a portion of the Universe open to energy transfers, and therefore subject to changes in its internal amount of energy. If the house could be perfectly insulated against heat leakage (and solar radiation), and the gas (and electricity) turned off, the house would be a "closed system" with respect to energy and its total internal amount of energy would be unchanged, although there may be changes in the form of energy inside the house such as the family dog converting chemical energy into kinetic energy when chasing its tail.

These considerations lead to the following scheme for keeping track of energy. Choose some portion of the Universe and mentally surround it by a dashed line marking the boundary (Figure 5.18). Then the energy transfer into the system across the system boundary minus the energy transfer out of the system across the system boundary is equal to the change of the energy inside the system.

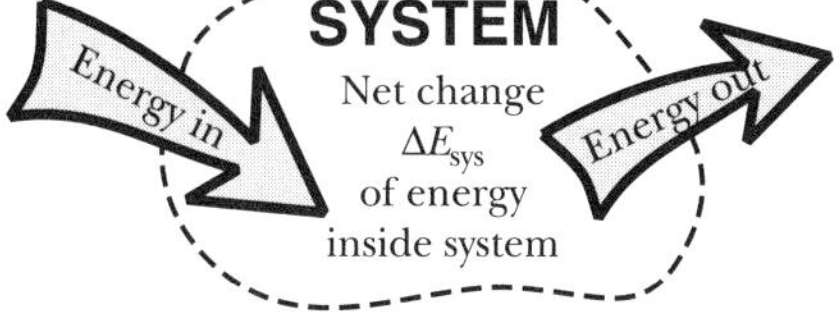

Figure 5.18 The change of energy of the system is equal to energy transferred into the system across the system boundary minus energy transferred out of the system across the system boundary.

The change in energy *inside* the system, ΔE_{sys}, can be either a positive or a negative amount of energy. For example, take an electric car as the system of interest:

Charge the battery: ΔE_{sys} is positive (more energy stored in the battery).
Do work on the car by pushing it: ΔE_{sys} is positive (more kinetic energy).
Run the headlights (radiate electromagnetic energy): ΔE_{sys} is negative (less energy stored in the battery).

The car is an open system whose total amount of energy changes due to energy transfers into or out of the system: the energy of an *open* system is *not* constant.

In contrast, the total energy of the Universe remains constant, because there are compensating energy changes in the surroundings of the car: the electric company lost a store of chemical energy by running its generators to charge the car, you used up some chemical energy to push the car, and the light from the headlights is absorbed by the surroundings and leads to a rise in temperature of the surroundings. The measurements that we are able to make confirm the premise that the Universe is a closed system whose total amount of energy never changes, although the form of the energy may change.

For any closed system, *inflow* = *outflow* = 0; $\Delta E_{sys} = 0$.

The energy of a closed system does not change.

The Universe as a whole is the most important example of a closed system, but it is often the case that a portion of the Universe can be considered to be a closed system, at least approximately. For example, put hot water and ice cubes into a very well-insulated container. During the short time that the ice melts and the water gets somewhat cooler, we can neglect the small amount of energy leakage through the walls of the insulated container. There is an increase in the energy of the ice cubes (which changed from solid to liquid), and a decrease in the energy of the hot water, but negligible net change in the energy inside the container, which is approximately a closed system whose total energy is (approximately) unchanged.

5.8.1 The choice of system affects energy accounting

What you choose to include in your chosen system affects the form of the energy equation for that system. For example, when we analyze a system consisting of an airless Earth plus a falling ball, the energy of the system does not change, because there are no energy transfers into or out of our chosen system (Figure 5.19). The ball falls a distance h, starting from rest:

$$\Delta E_{sys} = \Delta K + \Delta U_g = 0 \text{ (system is Earth + ball)}$$

$$\Delta K + (-mgh) = 0$$

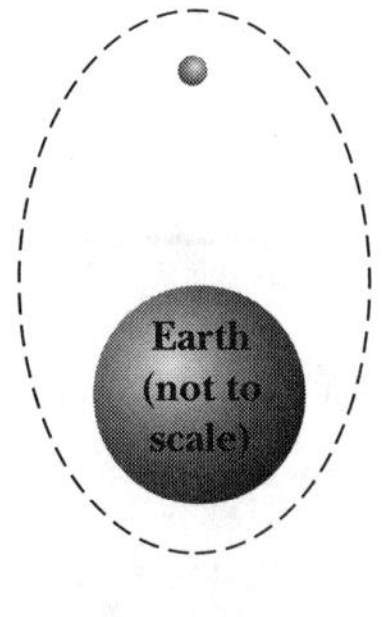

Figure 5.19 System: Earth + ball.

On the other hand, when we analyze a system consisting of just the falling ball (Figure 5.20), the Earth (which is now outside our chosen system) does work on the ball, and the energy of the ball increases:

$$\Delta E_{sys} = \Delta K = +mgh \text{ (system is ball only)}$$

Energy terms move back and forth across the equal sign in the energy equation, depending on our choice of the system of interest. It is extremely important to be clear about the choice of system, because this affects the form of the energy equation. In fact, if one is vague about the choice of system, it is possible to make a serious mistake.

Figure 5.20 System: ball only.

Warning: avoid double counting!

? A student said, "The Earth exerts a force mg through a distance h and does an amount of work mgh, and there is also a decrease in gravitational potential energy $-mgh$, so the kinetic energy increases by an amount $2mgh$." What is wrong with the student's statement?

This is an easy mistake to make. The problem here is double counting due to a lack of clarity about the choice of system. If the Earth is part of the system, the force mg is internal to the system and does no external work on the system; rather, there is work done by internal forces on both the Earth and the ball, and the negative of this work by internal forces is the change in potential energy. On the other hand, if the ball alone is the system, there is no shape change of the ball and no change of gravitational potential energy, but there is an external force *mg* which does an amount of work *mgh*.

In the next section we will illustrate the issue of choice of system by analyzing the same problem three times, using three different choices of system. We should find that the form of the energy principle will be different in each case, but that we should obtain the same physical results each time.

The woman, the barbell, and the Earth

To help avoid double-counting mistakes, it is useful practice to write the energy equation for various choices of system in a slightly more complicated situation, that of a woman applying a constant force F to lift a barbell of mass m from rest through a distance h, at which point the barbell is not only higher above the Earth but has also acquired some speed v. We're considering a time when the barbell is still headed upward, before the woman has brought the barbell to a stop above her.

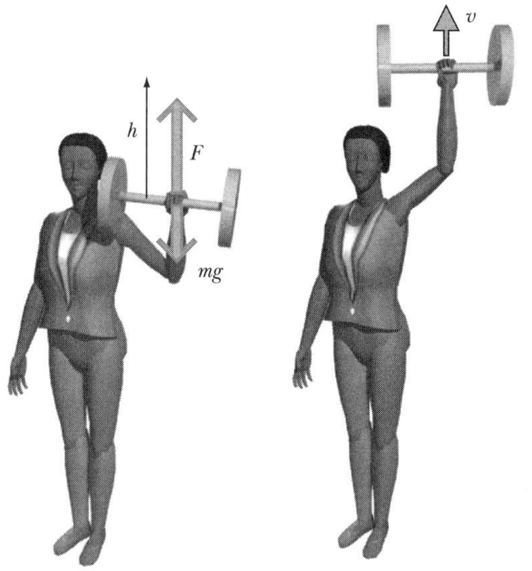

For convenience in the following discussion, let E_w represent the following energy terms:

E_w = chemical energy of woman +
kinetic energy of woman (moving arm) +
gravitational energy of woman and Earth +
thermal energy of woman

Woman + barbell + Earth:

If we choose the system to consist of the woman, the barbell, and the Earth (Figure 5.21), there are no energy transfers into or out of our system, so energy does not change, and the energy equation looks like this as the woman lifts the barbell a distance h, starting from rest (we neglect the tiny kinetic energy of the recoiling Earth):

$$\Delta E_{\text{sys}} = \Delta E_w + \Delta U_{\text{g, barbell and Earth}} + \Delta K_{\text{barbell}} = 0$$

Figure 5.21 System: woman+barbell+Earth.

? Write the energy equation for the system of woman + barbell + Earth, in terms of the known quantities m, g, h, the constant force F, p, and E_w.

Here is the energy equation for the combined system:

$$\Delta E_w + (+mgh) + \left(\frac{p^2}{2m}\right) = 0$$

The energy term ΔE_w represents changes in energy associated with the woman, mainly a decrease in her stored chemical energy, because she must perform chemical reactions with food (or with chemicals stored in body tissues after eating food) in order to be able to move her arm. This term also includes an increase in the kinetic energy of her arm, an increase in gravitational potential energy of woman and Earth associated with her arm being higher, and an increase in her thermal energy (because her temperature rises a bit when she exerts herself).

? The energy term ΔE_w includes one decrease and some increases. What is the net sign of ΔE_w, positive or negative?

The net effect must be a decrease in her energy, because the barbell and the Earth are farther apart, and the barbell goes faster, both of which represent energy increases. The woman has to burn more energy than the rest of the system gets, since she has to supply energy both to the rest of the system and to supply her own energy increases (increased kinetic energy and gravitational potential energy associated with her arm, and increased thermal energy).

Barbell only:

If we instead choose the system to consist just of the barbell (Figure 5.22), the woman does (positive) work $+Fh$ on the system, and the Earth does (negative) work $-mgh$ on the system (because mg is down and the displacement h is up)

Figure 5.22 System: barbell only.

? Write the energy equation for the system consisting.solely of the barbell.

Here is what we get for this choice of the barbell only as the system:

$$\Delta E_{\text{sys}} = \Delta K_{\text{barbell}} = W_{\text{done by woman}} + W_{\text{done by Earth}}$$

$$\left(\frac{p^2}{2m}\right) = (+Fh) + (-mgh)$$

Compare this equation to the equation we obtained when considering the system of woman+barbell+Earth. We see that energy terms have moved to the other side of the equation as a result of our different choice of system.

? Which is bigger, F or mg? Why?

Since the barbell's upward momentum increases, F must be larger than mg, since the net force in the upward direction is $F-mg$.

? Compare the equations for system 1 (woman + barbell + Earth) and system 2 (barbell only). In terms of F, calculate ΔE_w.

By comparing the two equations, we see that the change in the energy associated with the woman must be $-Fh$. The sign seems right: we had concluded earlier that her energy change must be negative.

Warning! It is tempting to include a term for the change in the gravitational energy of the barbell and to write the energy equation for the barbell alone like this:

$$\left(\frac{p^2}{2m}\right) + (mgh) = (+Fh) + (-mgh) \quad \textit{(Wrong!)}$$

? Why is this wrong (in addition to giving the wrong result)?

The error here is in assigning some gravitational energy to the barbell alone. Gravitational potential energy is associated with a change in the shape of a multi object system; in particular, to changes in the separations of pairs of interacting objects. When we choose the barbell alone as the system, the system does not change shape. Instead of a change of gravitational potential energy we have an external force exerted by the Earth on our barbell system. Instead of a $+mgh$ on the left of the equation we have a $-mgh$ on the right. We shouldn't double-count and include both terms!

Barbell + Earth:

If we choose the system to consist of the barbell plus the Earth (Figure 5.23), this system changes shape as the woman pushes the two pieces of the system apart. It is almost as though she is stretching an invisible spring connecting the two objects.

Figure 5.23 System: barbell+Earth.

? Write the energy equation for this choice of system (barbell + Earth).

Here is the energy equation for the barbell + Earth:

$$\Delta E_{sys} = \Delta U_{g,\text{ barbell and Earth}} + \Delta K_{barbell} = W_{\text{done by woman}}$$

$$(+mgh) + \left(\frac{p^2}{2m}\right) = (+Fh)$$

? Does the woman do any work *on the Earth*? Why or why not?

The woman's feet push down on the Earth but do negligible work, because there is essentially no displacement there (the Earth hardly moves). Work done by a force is calculated by taking into account the displacement of the point where the force is applied. In this case, the point of application hardly moves, so the force does hardly any work.

Reflection

Looking back over the equations obtained for different choices of system, you can see how energy terms that represent transfers across the system boundary for one choice of system become changes of energy *inside* the system for a different choice of system. Comparing equations for different choices of system can be useful in determining an unknown quantity (such as the energy change in the woman, $-Fh$, in the example you just worked through). Also, analyzing a process for more than one choice of system is a good check on your calculations and your understanding.

Ex. 5.8 Consider a harmonic oscillator (mass on a spring without friction). Taking the mass alone to be the system, how much work is done on the system as the spring contracts from its maximum stretch A to its relaxed length? What is the change in kinetic energy of the system during this motion? For what choice of system does energy remain constant during this motion?

5.9 Energy dissipation

The total energy of the Universe does not change, but useful energy is often "dissipated" into forms less useful to us. Push a chair across the floor, and some of the work that you do goes into heating the floor and the chair, rather than into kinetic energy of the chair. Throw a ball up into the air, and some of the initial energy is dissipated into increased microscopic energy of the air. Sliding friction, air resistance, viscous friction—all of these phenomena are examples of energy dissipation discussed in the rest of the chapter.

5.10 Air resistance

We begin our study of energy dissipation by considering air resistance. Air resistance isn't a major factor when you drop a metal ball a short distance. A video sequence of a falling metal ball shows that the ball moves faster and faster as it falls, an effect that is hard to observe by eye alone. The time between adjacent frames in Figure 5.24 is 1/15 second, and the increasing distances between heights of the ball in adjacent frames show that the speed is getting faster and faster. (Also note the increasing blur due to faster motion of the ball.) The gravitational attraction of the Earth acting on the ball makes the momentum of the ball continually increase. A curve is drawn along the tops of the ball images. The major vis-

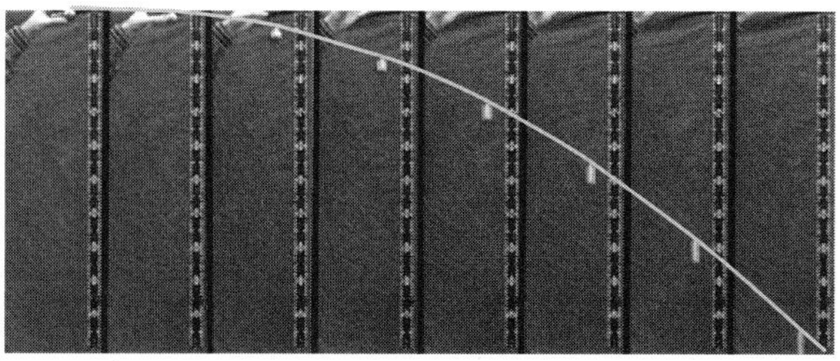

Figure 5.24 A falling metal ball goes faster and faster.

ible marks on the vertical meter sticks are 10 cm apart. In 7/15 s the ball falls about 1 meter.

Terminal speed

In contrast, it is observed that although a sky diver initially speeds up due to the gravitational force acting downward, the sky diver's speed does *not* keep getting bigger and bigger. Rather the sky diver eventually reaches a "terminal speed" and falls thereafter at constant speed despite the gravitational force. This terminal speed for falling humans is so high (about 60 m/s, or about 135 miles per hour) that hitting the ground without a parachute is normally fatal, although there have been cases of people hitting deep snow at terminal speed and surviving.

Similarly, if you drop a bowl-shaped paper coffee filter, a video sequence of the falling filter shows that the filter's speed does increase at first, but instead of continuing to gain speed it quickly reaches a constant terminal speed despite the gravitational force acting on it. In the video sequence below the time between adjacent frames is again 1/15 second. In 16/15 s the filter falls about 1 meter. A curve is drawn through the centers of the coffee filter images. The nearly straight line in the later part of the motion indicates motion at constant speed.

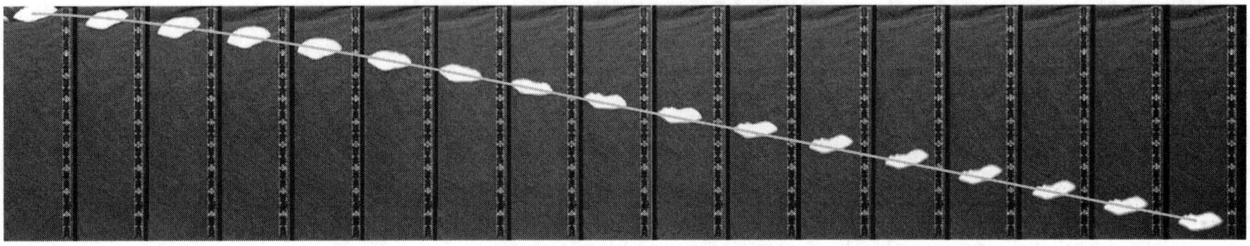

Figure 5.25 A falling coffee filter speeds up briefly, then falls at constant speed.

The restraining effect of the air is called "air resistance" or "drag." You yourself have probably observed that it is harder to move something quickly through a fluid such as water or air than to move it slowly. This suggests that air resistance acting on a falling object might depend on the speed of the object.

> **?** Consider a falling coffee filter at two different times: 1) while the filter's speed is still increasing, and 2) after terminal speed has been reached. How do you know the air resistance is larger at the second instant than at the first instant?

While the filter's speed is increasing there must be a nonzero net force, which means that the air resistance is smaller than the gravitational force (Figure 5.26). But later the net force is zero, because the speed is no longer increasing, which means that the force of the air has grown to be as large as the gravitational force (which is nearly constant at mg and in fact increases very slightly as the coffee filter gets closer to the Earth). Since the change in the density of the air is small over the distance the filter falls, and the change in the gravitational force is very small, it is reasonable to conclude that the change in the air resistance is due to a speed dependence of that force.

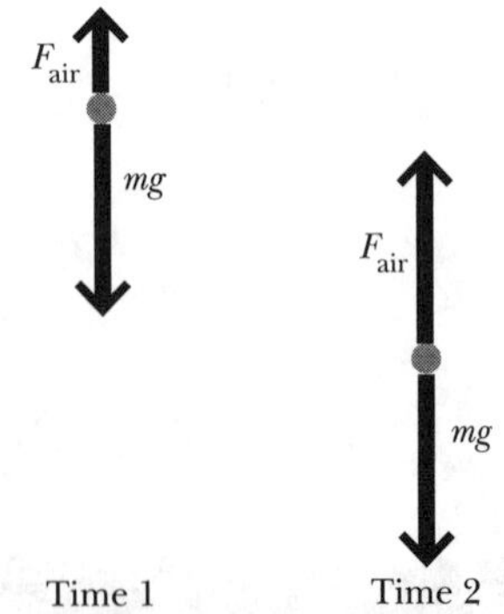

Figure 5.26 Forces acting on a falling coffee filter at Time 1 when the filter's speed is increasing and at Time 2 when terminal speed has been reached.

Dependence on cross-sectional area

We observe that the effect of air resistance increases as the cross sectional area of an object increases—a sky diver falls much more slowly with an open parachute than with a closed parachute. The low-density paper coffee filter has a large cross-sectional area, and a small gravitational force acts on it, so air resistance plays a major role. In contrast, a high-density metal ball has a small cross-sectional area, but a large gravitational force acts on it, and air resistance may be negligible in comparison with gravity.

Applications

Drop a ball. At the instant when the ball has nearly reached the ground, the gravitational potential energy of the Universe has decreased, with a corresponding increase in the kinetic energy of a portion of the Universe (the ball). In the presence of air, kinetic plus gravitational potential energy is only approximately constant. As the ball falls, it has to push aside the air underneath it, and the air molecules acquire additional kinetic energy K_{air}.

Ex. 5.9 Therefore, in air will the *actual* final speed of the ball be greater or less than you would predict from the constancy of kinetic plus gravitational energy? Why?

Ex. 5.10 A coffee filter of mass 1.4 grams dropped from a height of 2 m reaches the ground with a speed of 0.8 m/s. How much kinetic energy K_{air} did the air molecules gain from the falling coffee filter?

5.10.1 Mechanism of air resistance

We would like to understand the details of the interaction of the air with a falling object. In particular, we'd like to see if we can find a reason that the air resistance would depend on the speed of the object.

Here is a possible microscopic model to explain air resistance. When you hold a coffee filter stationary, air molecules hit it nearly equally from above and below, as seen in Figure 5.27. (Actually there are slightly more collisions per second on the lower surface than on the upper surface, leading to a small upward buoyant force which is very small compared to air resistance.) But when the filter is moving downward due to the gravitational attraction of the Earth, the bottom side of the filter is running into the air molecules, while the top side is moving away from the air molecules.

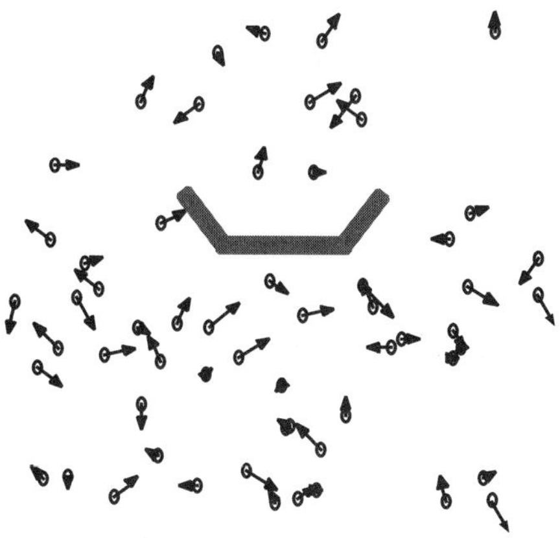

Figure 5.27 Air molecules collide with a falling coffee filter.

On average, the bottom side will have a significantly increased number of collisions per second with air molecules, and with greater impact. On the top side there will be a reduced number of collisions per second, and with less impact. There is a net upward push on the filter. This air resistance force increases with increasing downward speed of the filter, because higher speed increases the rate and impact of collisions on the bottom, and decreases the rate and impact of collisions on the top.

Eventually, when the downward speed of the coffee filter has become big enough, the net upward push by the air molecules becomes as large as the downward gravitational pull of the Earth. From then on the filter falls at a constant speed (terminal speed).

While this model captures the qualitative aspects of the situation, a simple quantitative calculation along these lines (using techniques discussed in a later chapter on gases) predicts a terminal speed that is *a thousand times smaller* than is observed! This is a spectacular failure of a simple model. The problem is that the falling coffee filter establishes macroscopic "wind"-like motions in the air, and the random motions of the air molecules at any particular location are relative to the bulk wind motion at that location. The techniques of "fluid dynamics" are capable of analyzing such phenomena, but these techniques are mathematically quite difficult and well beyond the scope of this introductory course. (Our later study of gases will address simpler situations, where there is no large-scale wind motion.)

5.10.2 Dependence of air resistance on speed

We established that the force of air resistance is dependent on the speed of a falling object.

> **?** Did we establish that the force F_{air} of the air is proportional to the speed v of the falling object?

All we know is that the air resistance increases with increasing speed, but it need not be simply proportional to v. For all we know the force could be proportional to $\sqrt{v}$ or v^2 or v^3. There is a simple experiment you can do to determine the form of the speed dependence, at least for low speeds. This is the topic of Problem 5.2 Experiment—falling coffee filters .

Other aspects of air resistance

> **?** If you could double the cross-sectional area without changing the mass or the shape of the filter (by making the filter out of thinner paper), what change would you expect in the magnitude of the air resistance force at a particular speed? If you double the density of the air through which the filter falls, what change would you expect?

Measurements of a variety of ordinary-sized objects moving through air or other fluids show that the friction force is in fact proportional to the cross-sectional area A (hitting more molecules per second.) Air resistance is also proportional to the density ρ of the air (again, hitting more molecules per second), so there is less air resistance at higher altitudes. For ordinary-sized blunt objects such as a coffee filter moving at ordinary speeds, the force is found to be approximately proportional to the square of the speed, which is probably what you found in your own experiment. Combining these effects, the air resistance is approximately proportional to $\rho A v^2$.

> **?** There is another important effect. From your own experience, would you expect larger air resistance (at a particular speed) for a pointed object or a blunt object?

As you might expect, the air resistance force depends not only on density, cross-sectional area, and speed, but also on the shape of the object. There is a larger air resistance force on a blunt object such as a coffee filter than on a ball or a streamlined object such as an arrow. This effect, and the dependence on v^2, are not easily explainable in quantitative detail in terms of a molecular model without performing a large molecular dynamics calculation.

The shape effect is usually taken into account by measuring the air resistance force (for example, by measuring the terminal speed) and determining an experimental "drag coefficient" C such that the following formula expresses the experimental results for the friction force:

APPROXIMATE AIR RESISTANCE FORCE (EMPIRICAL)

$$F_{air} \approx \frac{1}{2} C \rho A v^2 \text{, where } 0.3 \leq C \leq 1.0 \text{ (blunt objects, ordinary speeds)}$$

The direction is opposite to the velocity: $\vec{F}_{air} \approx -\frac{1}{2} C \rho A v^2 \hat{v}$

Blunter objects have a higher value of C.

It is important to keep in mind that this is *not* a fundamental force law like the law of gravitation. Air resistance is the average result of a huge number of momentary contact electric interactions of air molecules hitting atoms on the surfaces of the falling object, and the average net air resistance is very difficult to calculate from fundamental principles at the molecular level.

Motion through Earth's atmosphere

How important is the effect of air resistance in the everyday world? In Chapter 2 we analyzed the motion of a ball thrown through the air. However, we neglected air resistance, which can have a sizable effect. In Problem 5.3 you are asked to include the air resistance force given above. You find that a baseball thrown at high speed by a professional pitcher goes only about half as far in air as it would go in a vacuum.

A spinning ball

Even when we include air resistance in our calculations, we still have ignored a force that can have a major effect on the trajectory. If a ball has spin, there is an effect of fluid flow around the ball that raises the air pressure on the side where the rotational motion is in the same direction as the ball's velocity, and lowers the air pressure on the other side, where the rotational motion is in the opposite direction to the velocity. If the force points to the side, the ball curves; this is an important effect in baseball pitching. If the force points upward (due to "backspin," as in Figure 5.28), it lifts the ball and extends the range. If the force points downward (due to "topspin") it decreases the range. The calculations involved in modeling this effect are quite complex, and involve "fluid dynamics" (the study of fluid flow).

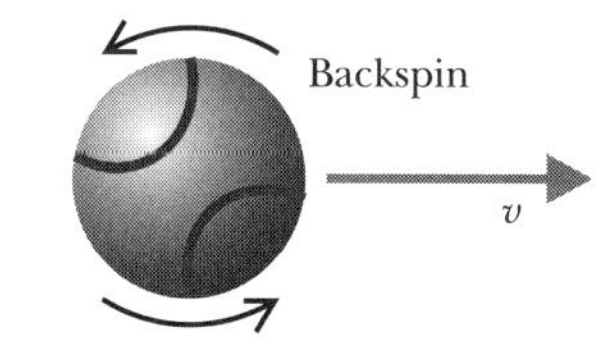

Figure 5.28 A spinning ball experiences additional forces.

Viscous friction

Very small particles such as dust or droplets of fog falling slowly through air, or small objects moving through a thick liquid such as honey, experience a friction force that is proportional to v rather than v^2. This is called "viscous" friction. The approximate dependence on v or v^2 for air resistance is roughly valid only for specific situations. The formulas for friction do not have the wide applicability of such major physical laws as the law of gravitation. However, even approximate formulas for friction make it possible to make much better predictions than we could make if we ignore friction completely.

5.11 Motion across a surface: Sliding friction

Sliding friction between two solid objects is a familiar interaction that involves energy dissipation. We'll consider it in some detail. Connect a spring to a block and stretch the spring an amount s to the right without moving the block (Figure 5.29). We need to identify interactions and draw forces on the block on a physics diagram.

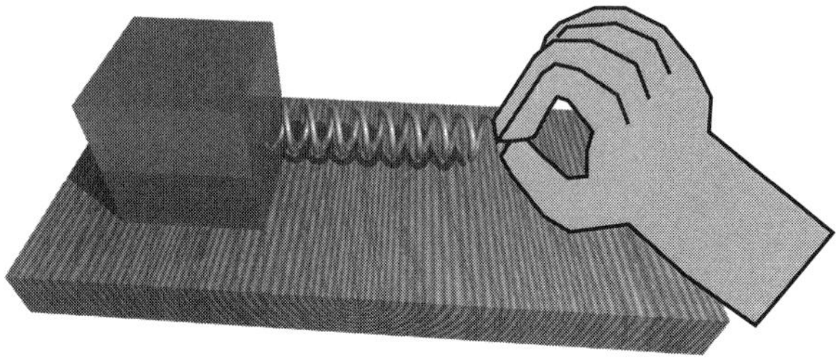

Figure 5.29 Stretch the spring without moving the block.

Non-contact forces

? What objects exert non-contact forces on the block?

The Earth exerts a force on the block straight down, in the $-y$ direction (Figure 5.30). Your hand doesn't touch the block, so if it exerts a force on the block it must exert a non-contact force. How could it do this? The hand can exert a very small gravitational force on the block, but this force is so small that we can neglect it. The hand and block are not electrically charged. There is no mechanism for a significant interaction between the hand and the block, so we can't list the hand as an object that interacts with the block.

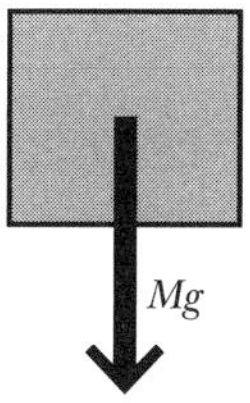

Figure 5.30 Physics diagram (incomplete) showing all non-contact forces.

Contact forces

? What objects exert contact forces on the block?

The spring touches the block, and exerts a force to the right (in the $+x$ direction) of magnitude $k_s s$. The table touches the block. What is the direc-

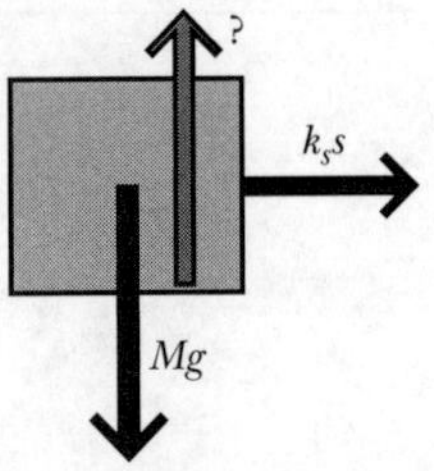

Figure 5.31 Physics diagram (incomplete) showing some contact forces as well as non-contact forces.

tion of the force exerted on the block by the table? Clearly the table exerts a force on the block upward, in the $+y$ direction (Figure 5.31). In Chapter 3 we examined the interactions between a brick and a table, and found that the table exerted a force upward on the brick as the atoms in the table were compressed by the brick. A similar interaction will happen between the block and the table, and the upward force by the table on the block can balance the downward gravitational force exerted on the block by the Earth.

Write the momentum principle and motion information

? Is the momentum of the block changing? Is your answer consistent with the system diagram in Figure 5.31?

We are left with a problem. Although the force exerted on the block by the table is upward, we can understand why the y component of the block's momentum is not changing, because there is a downward gravitational force. However, there appears to be a component of the net force on the block to the right, yet the block is at rest—the block's momentum is not changing. This is not consistent with the momentum principle. We must have failed to identify a force.

What could be exerting a force to the left on the block? The only object that can possibly be exerting a force to the left to balance the spring force is the table. How does it do that? Atoms of the block "run into" atoms in the table, and the resulting interactions stretch and compress interatomic bonds in the objects. Figure 5.32 provides an atomic-level picture of such an interaction.

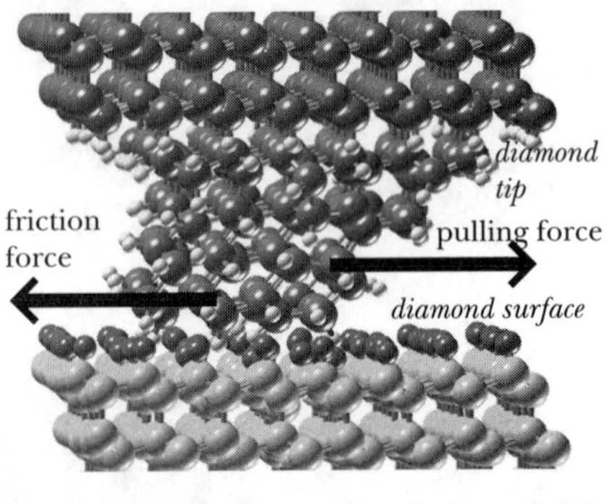

Figure 5.32 A computer simulation of sliding friction between a diamond tip and diamond surface. Image courtesy of Judith A. Harrison, U.S. Naval Academy.

This image is a single frame from a molecular dynamics computer simulation of sliding friction between two objects, performed by Judith A. Harrison and coworkers at the United States Naval Academy in Annapolis, Maryland. The computations are similar to the numerical integrations you have done, but involving a large number of objects (atoms). The forces between the atoms are modeled by spring-like forces.

In this image, a diamond tip is dragged to the right across a diamond surface. The large spheres represent carbon atoms, and the small ones are hydrogen atoms on the carbon surfaces. Note that the projecting tip has "caught" on the lower surface, causing interatomic bonds in the two objects to deform. The stretching and compression of interatomic bonds produces the horizontal component of the force exerted on the moving object by the lower object. For more information about Harrison's work, see

http://www.chemistry.usna.edu/jah/ResearchStuff/Friction_page.html

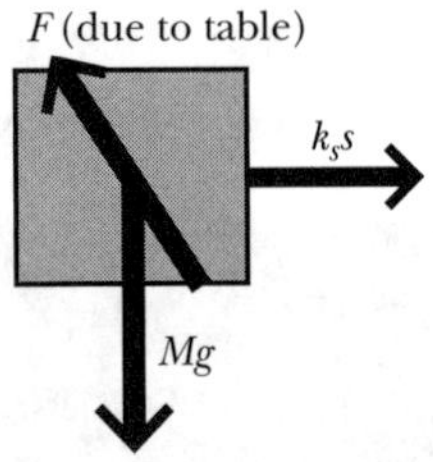

Figure 5.33 A complete physics diagram for the block.

Figure 5.33 shows a complete physics diagram for the block. The force exerted by the table acts at an angle. It is common practice to replace the force vector exerted by the table with two equivalent vectors, a compression force F_N normal to the surface and a friction force f parallel to the surface, as shown in Figure 5.34. This turns out to be very useful for many reasons, but it is important to remember that an atom in the table exerts one interatomic force on a neighboring atom in the block, not two. It is for our own convenience that we split this force into two equivalent forces.

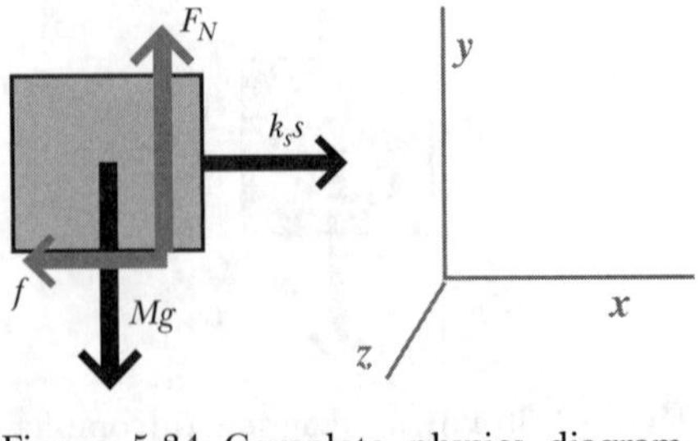

Figure 5.34 Complete physics diagram for the block; the force on the block by the table has been resolved into components normal to and parallel to the surface.

We can construct the component equations for the momentum principle by reading the forces off of our physics diagram. We choose axes with x to the right and y up (z is outward). We know that the block remains at rest even though we are pulling on it, so we can write motion information in addition to the momentum principle:

$$\frac{d\vec{p}}{dt} = \langle (k_s s - f), (F_N - Mg), 0 \rangle$$

$$p_x = 0 \text{ at all times, so } dp_x/dt = 0$$

$$p_y = 0 \text{ at all times, so } dp_y/dt = 0$$

Solve equations

? What is the magnitude of the compression force F_N? What is the magnitude of the friction force f?

From $(F_N - Mg) = 0$ we deduce that the compression force is just Mg; the table supports the block. From $(k_s s - f) = 0$ we deduce that the friction force must be equal to the spring force. If you reduce the stretch of the spring, the friction force f gets smaller, too, because the sideways interatomic forces are smaller. As long as the block remains at rest, the friction force has a range of values equal to the range of spring forces applied to the block. However, this is not the case when the block is in motion.

Note that this is another example of a situation where we don't have a force law for one of the forces, the force exerted by the table (which we split into two components). We are able to deduce that $F_N = Mg$ and $f = k_s s$ thanks to the motion information that supplements our incomplete force information.

5.11.1 Block in motion with sliding friction

Now slowly stretch the spring more and more, and eventually the block moves to the right, which means that during the instant when it began moving, its momentum changed. By the momentum principle, that implies that the spring force $k_s s$ to the right became greater than the friction force f to the left. Evidently there must be a maximum possible friction force f_{max} for this object on this table (it is different for different surfaces).

If we apply a larger spring force than this maximum friction force, we break temporary interatomic bonds between the block and the table, and we can speed up the block. Measurements on sliding objects show that the friction force remains approximately equal to f_{max}, independent of the speed of sliding. This implies that the friction force is related to the strength of the interatomic bonds, which are continually made and continually broken as the block slides along.

In contrast, remember that air resistance depends on the speed of the moving object. The mechanism for air resistance is very different from that of sliding friction and leads to a dependence on the speed.

Coefficient of friction

It is approximately true for many objects on many surfaces that the maximum friction force f_{max} is proportional to the amount of compression F_N, with a proportionality constant μ that is a property of the two surfaces that are in contact (Figure 5.35):

$$f_{max} \approx \mu F_N$$

Typical values for the "coefficient of friction" μ for wooden blocks sliding on wooden tables are in the range of 0.25 to 0.5 (μ is Greek lowercase mu).

These approximate properties of sliding friction do not have the status of fundamental physical laws such as the law of gravitation. For example, the details of motion with friction depend critically on how clean the surface is. Very flat, very clean metal or glass plates may bind to each other so strongly that it is almost impossible to slide them.

When two very clean and flat metal surfaces are brought in contact, the interface region looks much like the interior of the metals, and the two metal objects may behave pretty much like a single block. Also, when surfaces are very flat, it is possible that air will be squeezed out from between the objects, in which case air pressure on the outer surfaces raises the compression force F_N to a very large value, and the maximum friction force μF_N is then

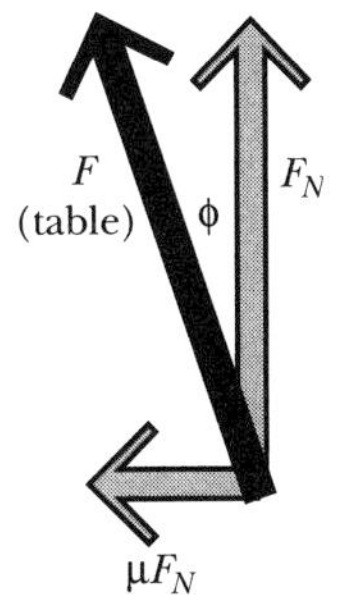

Figure 5.35 Sliding friction approximately independent of the speed.

$$\tan\phi \approx \frac{\mu F_N}{F_N} = \mu$$

very large. This is an example of a situation where F_N is much greater than the weight Mg of the block. What matters in the formula $f_{max} = \mu F_N$ is how much the two surfaces are squeezed together, not the weight of the block.

Dragging with constant speed

You can experience some of the vagaries of sliding friction by trying to drag a block at a constant speed. You will need to keep the stretch of the spring constant, so $k_s s = f_{max}$, and there should be no acceleration—no change in the constant velocity. You may find that it is very difficult to maintain a constant stretch of the spring (and a constant velocity of the block), in part because the block skips and loses some contact with the table. It can be easier to maintain constant velocity if the object is quite massive and therefore less subject to sudden large changes in velocity.

You may observe that you have to pull harder to break the block loose from the table and start it moving than you do to keep it moving at a constant speed. For some surfaces the maximum friction force for a block at rest is larger than the maximum friction force for a moving block. We will look at sliding friction in more detail later in the course, but a full understanding of this complex phenomenon is the subject of ongoing contemporary research.

An intriguing recent experimental result is that with extremely clean surfaces ("atomically" clean), it matters whether the interatomic spacings of the two objects are commensurate or not; that is, whether the grid of atoms of one surface fit into the grid of atoms of the other surface. If they do fit, like nested egg cartons, it takes a sizable force to pull the block up out of the atomic grid, and a smaller force to keep it going. But if there isn't a good fit, the slightest force will move the block.

Dragging with continually increasing speed

Figure 5.36 Pulling with a force greater than the maximum friction force.

To drag with continually increasing speed, it is necessary for the pulling force to be greater than the friction force (Figure 5.36). Try dragging a block across a table by stretching the spring enough that $k_s s$ is greater than f_{max}, and try to maintain a constant stretch s in order to obtain a constant nonzero dv_x/dt. You may find that it is quite difficult to maintain a constant acceleration for very long. A practical difficulty is that as the speed increases, you need to move your hand faster and faster to maintain a constant large stretch in the spring, and you also may run out of room on the table. In Figure 5.36 we have assumed, based on experiments, that during sliding the frictional force remains approximately equal to f_{max}.

Ex. 5.11 If the mass of the block is M, what is the acceleration $d\vec{v}/dt$ of the block? (Apply the momentum principle along both the x and y axes.)

5.11.2 Energy dissipation

If you pull the block with a constant force F at constant speed through a distance d, you do an amount of work Fd, but the block's speed doesn't increase. The making and breaking of interatomic bonds at the sliding interface increases the oscillation of atoms around their equilibrium positions. This increased energy is then shared with many other atoms in the block and table, with the result that the average energy of atoms in both objects increases. This is the same thing as saying that the temperature rises in the block and in the table. You can observe such a rise in temperature by rubbing your hands together.

The energy is dissipated throughout the block and table, and cannot easily be recovered. There is a kind of irreversibility in a process where friction plays a role, because the dissipated or dispersed energy can't be brought back to the interface region and used to move the block.

5.11.3 A paradox involving friction

If we are careful in our energy bookkeeping for a system involving motion with friction, we encounter a paradox that cannot easily be resolved. Consider the situation shown in Figure 5.37.

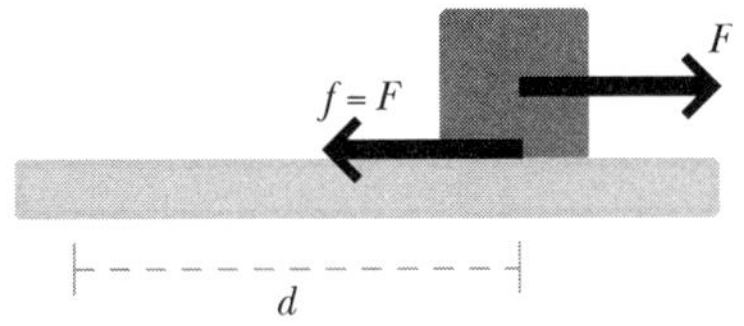

Figure 5.37 Forces in the x direction on a block pulled with constant speed across a table. (Forces in the y direction are not shown).

You exert a constant force F, and pull a block a distance d across a table. Because there is friction between the block and the table, the block travels at constant speed. Since the momentum of the block does not change, we conclude correctly that the net force on the block is zero, and that therefore the magnitude of the friction force must also be F.

In terms of energy inputs, you have done an amount of work Fd on the block. The friction force, acting in the opposite direction, has apparently done an amount of work $-Fd$. The net work done on the block therefore would seem to be zero. This is consistent with the fact that the kinetic energy of the block did not change.

It is not, however, consistent with the fact that both the block and the table get hotter as they rub together! If the net work done on the block was zero, where did that increased thermal energy come from?

This is not a "trick question," but a genuine puzzle. We will return to this issue in Chapter 7, when we will deal with it in enough detail to resolve the paradox.

5.12 Including friction in modeling a spring-mass system

One of the most obvious things about the motion of a real macroscopic spring-mass system is that the amplitude gets smaller and smaller with time. Our model of a spring-mass system that we developed in Chapter 3 is too simple: we should include the effects of friction. See Problem 5.4 Numerical integration of spring-mass system with friction .

In the case of "viscous" friction, where the friction force is $\vec{f} = -c\vec{v}$, it is possible with elaborate math (or by guessing the solution and plugging it into the momentum principle for the system) to find an analytical solution, in which a cosine function is multiplied by an exponential factor:

$$Ae^{-\frac{c}{2m}t}\cos(\omega t)$$

The amplitude dies away exponentially with time. The angular frequency is

$$\omega = \sqrt{\frac{k_s}{m} - \left(\frac{c}{2m}\right)^2}$$

which is approximately $\sqrt{k_s/m}$ for low friction.

There is no such simple analytical solution if the friction force is independent of the speed (sliding friction) or proportional to v^2 (air resistance for large blunt objects).

5.13 Friction and irreversibility

When there is friction, processes are not reversible, in contrast with situations describable in terms of potential energy. If energy dissipation is negligible, a bouncing ball continually returns to nearly its original position and momentum, and the process would run just as well backwards as forwards. A movie of the bouncing ball could be played backwards and it would look the same as when the movie is played forwards. But in the real world, energy is dissipated in the collision with the floor, and due to air resistance, so that the bouncing ball does *not* return to its initial position and momentum, and the process is not reversible. A movie of the process played backward would look quite odd, because the ball would bounce higher and higher.

Energy dissipation

In section 5.2 we saw that potential energy is independent of path. In the absence of air resistance, a ball thrown up into the air will return to your hand with exactly the same kinetic energy it had when it left your hand. The situation is entirely different in the presence of friction. You throw the ball up in the air, and as it goes up it loses some energy to the air (air resistance), and on the way down it loses some more energy to the air. When it returns to your hand the net change in the kinetic energy is negative despite the fact that the net change in the gravitational energy is zero (same position). The round trip difference in the kinetic energy is *not* zero.

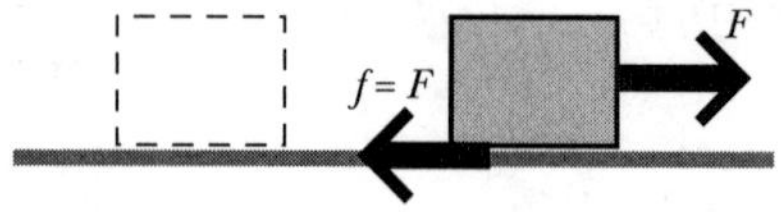

Figure 5.38 Pull to the right; the block and table get hot.

Similarly, in the presence of friction, when a mass on a spring returns to the same s, it has a smaller v. Again, the change in the kinetic energy of the mass during the round trip is not zero, even though the spring potential energy has not changed.

A particularly compelling case involving friction is that of pulling a block at constant speed to the right across a table with friction and then to the left back to its starting point. On the way to the right the block and table get hot, and on the way back to the left the block and table get even hotter (Figure 5.38 and Figure 5.39). In order to have no change in the round trip, the block and table would have had to cool down as you pulled the block to the left; this would be a very surprising effect indeed.

Note that the friction force exerted by the table points to the left on the way out, and then to the right on the way back. The direction of the friction force depends on the direction of the motion, not merely on position. An interaction that we can associate with potential energy would not behave this way. The force would have the same direction on the way out and on the way back ($F_x = -dU/dx$); one energy change would be positive and the other negative, making a net change of zero. Friction is a good example of an interaction for which we cannot define a potential energy, because the friction force depends on the direction of motion.

Figure 5.39 Pull back to the left; the block and table get even hotter.

Forces which depend only on position, and for which it is possible to define an associated potential energy (with the force equal to the negative gradient of the potential energy), are called "conservative forces." Friction is not a conservative force.

A key issue is that we define potential energy in terms of the interactions of point particles, or particle-like objects whose internal state doesn't change. But a block being dragged across a table is a complicated multiparticle system. Individual atoms interact through electric forces associated with the interatomic potential energy, and at the atomic level these interactions are reversible. But energy spreads throughout the block and floor, and macroscopically this looks to us like energy dissipation.

Irreversibility has far-reaching consequences for the statistical or probabilistic behavior of macroscopic objects consisting of large numbers of atoms. We will consider issues of reversibility and irreversibility in more detail later in the course.

Ex. 5.12 You move a block slowly to the right a distance of 0.2 m on a table. You apply a constant force of 20 N. How much work do you do? You then move slowly back to the starting point; how much work do you do in this second move? What change has there been in the thermal energy of the block and table?

Ex. 5.13 A block is attached to a spring whose stiffness is 40 N/m. You move the block slowly from $s = 0$ to $s = 0.2$ m, with negligible friction present. How much work do you do? You then move slowly back to the starting point; how much work do you do in this second move? What change has there been in the thermal energy of the block?

5.14 *Resonance

Imagine pushing somebody in a swing (Figure 5.40). If you time your pushes to match the natural frequency of the swing, you can build up an increasingly large amplitude of the swing. But larger amplitude means higher speed and a higher rate of energy dissipation.

? You keep pushing with constant (small) amplitude, and the swing eventually reaches a "steady state" with constant (large) amplitude. What determines the size of this steady-state amplitude?

The steady state is established when the energy dissipated through air resistance on each cycle has grown to be equal to the energy input you make on each cycle. The large steady-state oscillation needs only little pushes at the right times from you to make up for energy dissipation, and the amplitude of the oscillation can be much larger than the amplitude of the pushes.

On the other hand, if you deliberately push at the wrong times, not matching the natural frequency of the swing, the amplitude of the swing won't continually build up. To get a large response from an oscillating system, inputs should be made at the natural, free-oscillation frequency of that system. This important feature of driven oscillating systems is called "resonance."

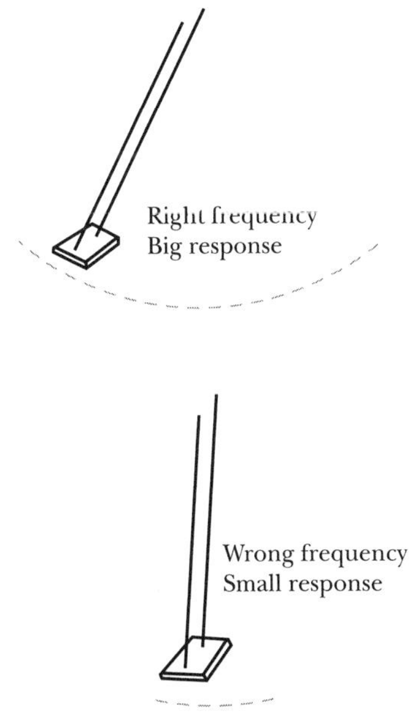

Figure 5.40 When you push a swing, you get a big response if you match the natural frequency of the swing.

5.14.1 *Analytical treatment of resonance

Problem 5.22 is a numerical integration of a sinusoidally driven spring-mass oscillator (Figure 5.41). This problem lets you study the buildup of the steady state as well as the steady state itself. For the steady-state portion of the motion, it is possible to obtain an analytical solution, if the energy dissipation is due to viscous friction. Here is the momentum principle for the driven oscillator in Figure 5.41, with viscous friction force $-cv$:

$$\frac{dp}{dt} = -k_s[x - D\sin(\omega_D t)] - cv$$

where x is the position of the mass; D is the amplitude of the sinusoidal motion and ω_D is the variable angular frequency of the driven end of the spring. For more details on how this equation was obtained, see the discussion in Problem 5.22.

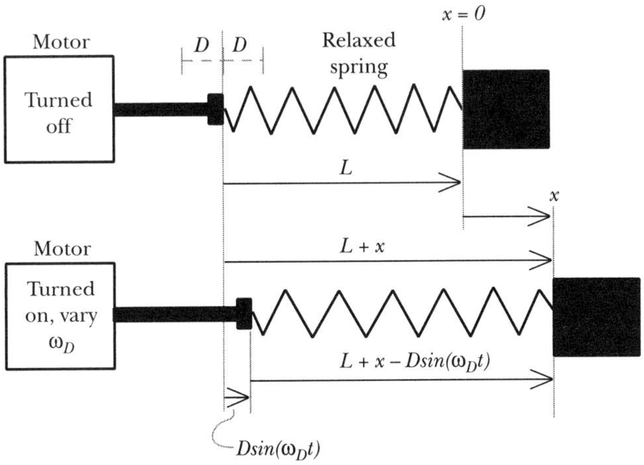

Figure 5.41 Sinusoidal driving of a spring-mass system; ω_D can be varied.

In the steady state x must be a sinusoidal function, with amplitude A and phase shift ϕ depending strongly on the driving angular frequency ω_D:

$$x = A\sin(\omega_D t + \phi)$$

By taking derivatives of this expression to obtain $v = dx/dt$ and dv/dt as a function of time t, and plugging these derivatives into the momentum principle, it can be shown that this expression for x as a function of time is a solution in the steady state if the amplitude A and phase shift ϕ have values determined by the following expressions:

$$A = \frac{\omega_F^2}{\sqrt{(\omega_F^2 - \omega_D^2)^2 + \left(\frac{c}{m}\omega_D\right)^2}} D$$

$$\cos\phi = \frac{(\omega_F^2 - \omega_D^2)}{\sqrt{(\omega_F^2 - \omega_D^2)^2 + \left(\frac{c}{m}\omega_D\right)^2}}$$

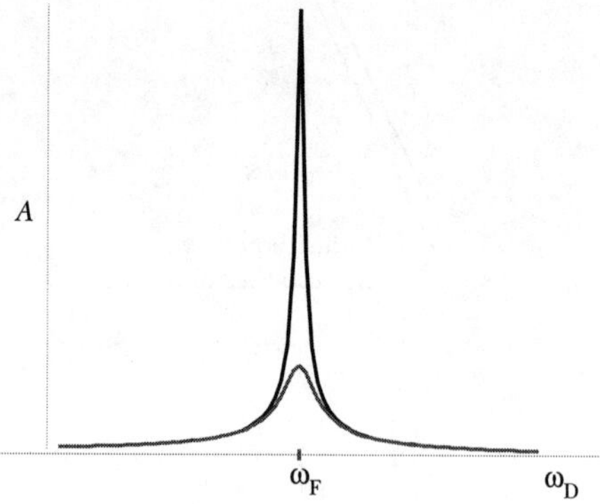

Figure 5.42 Amplitude as a function of driving frequency. The damping in the lower curve is five times greater than in the upper curve.

The detailed proof is rather complicated, so we omit it. Instead, in the exercises below we give you the opportunity to see that these expressions for amplitude and phase shift are in agreement with your experimental observations and the computer modeling of Problem 5.22. Figure 5.42 is a graph of the amplitude A as a function of the driving angular frequency ω_D, for two different values of the viscous friction parameter c.

Ex. 5.14 Using the formula for the amplitude A, show that if the viscous friction is small, the amplitude is large when ω_D is approximately equal to ω_F.

Ex. 5.15 Show that with small viscous friction, the amplitude A drops to $1/\sqrt{2}$ of the peak amplitude when the driving angular frequency differs from resonance by this amount:

$$|\omega_F - \omega_D| \approx \frac{c}{2m}\omega_F.$$

(Hint: note that near resonance $\omega_D \approx \omega_F$, so $\omega_F + \omega_D \approx 2\omega_F$.)

Ex. 5.16 Given the results of the previous exercise, how does the width of the resonance peak depend on the amount of friction? What would the resonance curve look like if there were very little friction?

Ex. 5.17 Using the formula involving the phase shift ϕ, show that the phase shift is approximately 0° for very low driving frequency ω_D, approximately 180° for very high driving frequency ω_D, and 90° at resonance, consistent with your experiment.

5.14.2 *Resonance in other systems

We have discussed resonance in the context of a particular system—a mass on a spring driven by a motor. We study resonance because it is an important phenomenon in a great variety of situations.

When you choose a radio or television station you adjust the parameters of an electrically oscillating circuit so that the circuit has a narrow resonance at the chosen station's main frequency, and other stations with significantly different frequencies have little effect on the circuit.

Magnetic resonance imaging (MRI) is based on the phenomenon of nuclear magnetic resonance (NMR) in which nuclei acting like toy tops precess in a strong magnetic field. The precessing nuclei are significantly affected by radio waves only at the precession frequency. Because the precession frequency depends slightly on what kind of atoms are nearby, resonance occurs for different radio frequencies, which makes it possible to identify different kinds of tissue and produce remarkably detailed images.

5.15 Summary

Fundamental physical principles

No new fundamental principles were introduced in this chapter.

New concepts

Spring (elastic) energy $U_s = \frac{1}{2}k_s s^2 + U_0$

Thermal energy =average energy of atoms in an object, proportional to temperature of the object

Power = $\frac{\text{energy}}{\text{time}}$ (J/s = watts); also calculable as $\vec{F} \bullet \vec{v}$

Thermal transfer of energy across the system boundary, Q

Heat capacity = energy per gram required to raise temperature 1K

$$\Delta E_{\text{thermal}} = mC\Delta T;\ C = \frac{\Delta E_{\text{thermal}}}{m\Delta T}\ \ (C \text{ in J/K per gram, } m \text{ in grams})$$

Open system: energy flows into and/or out of system.

Closed system: no energy flows into or out of system.

The choice of system (open or closed) affects energy accounting.

Dissipation of energy

Terminal speed

Air resistance: collisions between air molecules and moving object

Sliding friction force due to forming and breaking of bonds between two solid objects

Path independence of potential energy
  Impossibility of describing friction in terms of potential energy

Resonance: sensitivity of driven oscillator to driving frequency

Results

Air resistance

Air resistance is approximately proportional to speed squared, and in a direction opposite to the velocity:

$$\vec{F}_{\text{air}} \approx -\frac{1}{2}C\rho A v^2 \hat{v}$$

Viscous friction

For small particles ("viscous" friction), fluid friction is approximately proportional to v.

Sliding friction

Sliding friction force $f \le \mu F_N$; the coefficient of friction μ depends on the surfaces. F_N is the normal force compressing the two surfaces together. Sliding friction force does not depend on speed (approximate).

**Driven spring-mass system*

Analytical solution for driven spring-mass system: $x = A\sin(\omega_D t + \phi)$, where $\omega_F = \sqrt{k_s/m}$; viscous friction cv, driven at ω_D:

$$A = \frac{\omega_F^2}{\sqrt{(\omega_F^2 - \omega_D^2)^2 + \left(\frac{c}{m}\omega_D\right)^2}}D,\quad \cos\phi = \frac{(\omega_F^2 - \omega_D^2)}{\sqrt{(\omega_F^2 - \omega_D^2)^2 + \left(\frac{c}{m}\omega_D\right)^2}}$$

5.16 Review questions

Definitions

RQ 5.1 An oil company included in its advertising the following phrase: "Energy—not just a force, it's power!" In technical usage, what are the differences among the terms energy, force, and power?

Energy conservation

RQ 5.2 50 grams of water whose temperature is 25° Celsius are added to a thin glass containing 800 grams of water at 20° Celsius (about room temperature). What is the final temperature of the water? What simplifying assumptions did you have to make in order to determine your approximate result?

Spring potential energy

RQ 5.3 A spring has a relaxed length of 6 cm and a stiffness of 100 N/m. How much work must you do to change its length from 5 cm to 9 cm?

Power

RQ 5.4 Electricity is billed in kilowatt-hours. Is this energy or power? How much is one kilowatt-hour in standard physics units? (The typical cost of one kilowatt-hour is 5 to 10 cents.)

Open and closed systems

RQ 5.5 Which of the following are open systems with respect to energy, and which are closed? Explain why.

a car; a person; an insulated picnic chest; the Universe; the Earth

RQ 5.6 A horse whose mass is M gallops at constant speed v up a long hill whose vertical height is h, taking an amount of time t to reach the top. The horse's hooves do not slip on the rocky ground, so the work done by the force of the ground on the hooves is zero (no displacement of the force). When the horse started running, its temperature rose quickly to a point at which from then on, heat transferred from the horse to the air keeps the horse's temperature constant.

(a) First consider the horse as the system of interest: What objects exert forces on the horse? How much work is done on the horse by each of these forces? What non-work energy transfers are there? What is the change in the energy of the horse? In what forms?

(b) Next consider the Universe as the system of interest: What objects exert forces on this system? How much work is done on the system by each of these forces? What non-work energy transfers are there? What is the change in the energy of the system? In what forms?

Heat capacity

RQ 5.7 Substance A has a large heat capacity (per gram), while substance B has a smaller heat capacity. If the same amount of energy is put into a 100 gram block of each substance, and if both blocks were initially at the same temperature, which one will now have the higher temperature?

Energy dissipation

RQ 5.8 When a falling object reaches terminal speed, its kinetic energy reaches a constant value. However, the gravitational energy of the system consisting of object plus Earth continues to decrease. Does this violate the principle of conservation of energy? Explain why or why not.

Air resistance

RQ 5.9 Describe a situation in which it would be appropriate to neglect the effects of air resistance.

RQ 5.10 Describe a situation in which neglecting the effects of air resistance would lead to significantly wrong predictions.

RQ 5.11 List all the approximations that are made in order to find a simple analytical solution for the motion of a projectile near the Earth's surface.

RQ 5.12 You are standing at the top of a 50 m cliff. You throw a rock as hard as you can, in the horizontal direction. Approximately where does it hit on the flat plain below? Is your answer too large or too small?

RQ 5.13 Write the computational statements required at the heart of a numerical integration of projectile motion that includes air resistance.

RQ 5.14 If you let a mass at the end of a string start swinging, at first the maximum swing decreases rather quickly, but once the swing has become small it takes a long time for further significant decrease to occur. Try it! Explain this simple observation.

Sliding friction

RQ 5.15 You drag a block with constant speed v across a table with friction. Explain in detail what you have to do in order to change to a constant speed of $2v$ on the same surface. (That is, the puzzle is to explain how it is possible to drag a block with sliding friction at different constant speeds.)

Damped oscillators

RQ 5.16 Sketch a graph of position vs. time for a spring-mass oscillator subject to friction (not over-damped).

Potential energy

RQ 5.17 A particle moves inside a circular glass tube under the influence of a tangential force of constant magnitude F (Figure 5.43). Explain why we cannot associate a potential energy with this force (which is therefore a "nonconservative" force). How is this situation different from the case of a block on the end of a string, which is swung in a circle?

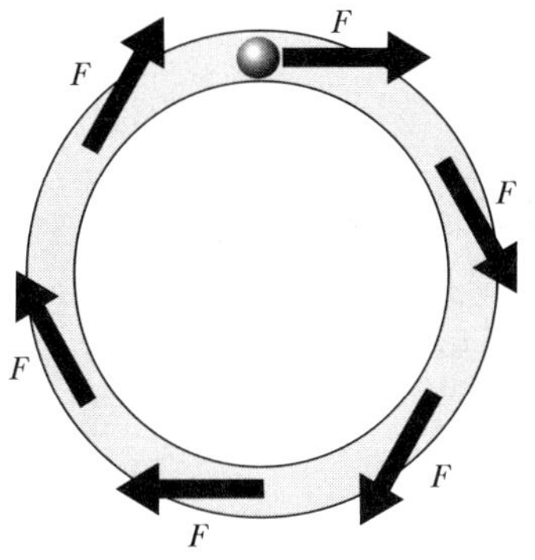

Figure 5.43 A particle moves in a circular glass tube (RQ 5.17).

RQ 5.18 In Figure 5.43, calculate the change in electric energy along two different paths in moving charge q away from charge Q from A to B along a radial path, then to C along a circle centered on Q, then to D along a radial path. Also calculate the change in energy in going directly from A to D along a circle centered on Q. Specifically, what are $U_B - U_A$, $U_C - U_B$, $U_D - U_C$, and there sum? What is $U_D - U_A$? Also, calculate the round-trip difference in the electric energy when moving charge q along the path from A to B to C to D to A.

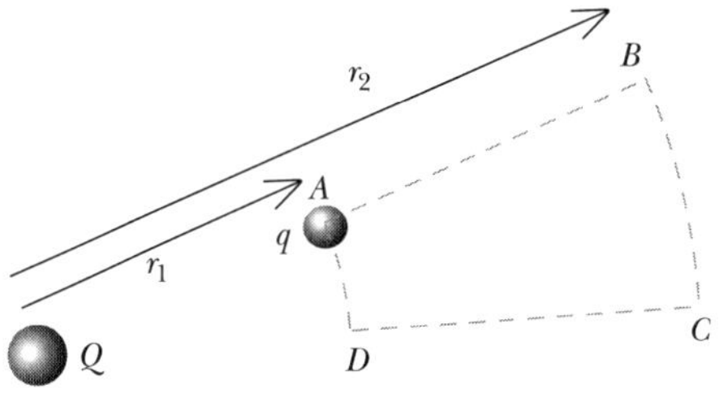

Figure 5.44 Change in electric energy along two different paths (RQ 5.18).

5.17 Problems

Problem 5.1 Energy in a spring-mass system

Energy calculations provide a powerful check on the accuracy of a numerical integration. Modify your numerical integration of the motion of a mass on a spring that you did in Chapter 3, Problem 3.3. In addition to plotting graphs of position versus time, also plot graphs of kinetic energy, of spring energy, and of the sum of the kinetic energy and the spring energy, versus time during the motion. It is up to you to choose some value for the potential energy U_0 at the bottom of the well. Label the energy graphs with numerical values and units. Make sure that the energy values are reasonable for your system.

(a) Does the sum of kinetic and potential energy remain constant as a function of time? What if you vary the step size (which varies the accuracy of the numerical integration)?

(b) Increase the energy by increasing the initial speed. How does the amplitude A change? How does the angular frequency ω change?

Problem 5.2 Experiment—falling coffee filters

A paper coffee filter in the shape of a truncated cone falls in a stable way and quickly reaches terminal speed because it has a large area (big air resistance) but low mass (small gravitational force). Use a stopwatch to time the drop of coffee filters from as high a starting position as you can conveniently manage. If a stairwell is available, drop the coffee filters there to time a longer fall. This experiment is easier to do with a partner.

(a) By stacking coffee filters you can change the mass of a falling object without changing the shape. With this scheme you can explore how the terminal speed depends on the mass. Time the fall for different numbers of stacked filters, taking some care that the shape of the bottom filter is always the same (the filters tend to flatten out when removed from a stack). Start the measurement after the filters have fallen somewhat, to allow them to reach terminal speed. Average the results of several repeated measurements of time and height. Can you think of a simple experiment you can do to verify that the coffee filters do in fact reach terminal speed before you start timing?

(b) We want to know quantitatively how air resistance depends on speed. Plot your data for the air resistance force *vs.* the terminal speed. (The air resistance force is equal to the gravitational force when terminal speed has been reached, so it is proportional to the number of filters in a stack). How does the air resistance force depend on v?

There is an important constraint on the graph of speed dependence: Should the curve of the air resistance force vs. terminal speed go through the origin (force and terminal speed both very small), or not?

Analysis hint: If you suspect that the force is proportional to v^3, you might plot force vs. v^3 and see whether a straight line fits the data.

Problem 5.3 Projectile motion with air resistance

Use a computer to carry out the following step-by-step quantitative calculations for a baseball. Remember that the direction of the air resistance force on an object is opposite to the direction of the object's velocity.

A baseball has a mass of 155 grams and a diameter of 7 cm. The drag coefficient C for a baseball is about 0.35, and the density of air at sea level is about 1.3×10^{-3} grams/cm^3, or 1.3 kg/m^3.

(a) Including the effect of air resistance, compute and plot the trajectory of a baseball thrown or hit at an initial speed of 100 miles per hour, at an angle of 45° to the horizontal. How far does the ball go? Is this a reasonable distance?

(Continued on next page.)

(A baseball field is about 400 feet from home plate to the fence in center field. An outfielder cannot throw a baseball in the air from the fence to home plate.)

(b) Plot a graph of $(K + U_g)$ *vs.* time.

(c) Temporarily neglect air resistance (set $C = 0$ in your computations) and determine the range of the baseball. Check your computations by using the analytical solution for motion of a projectile without air resistance. Compare your result with the range found in part (a), where the effect of air resistance was taken into account. Is the effect of air resistance significant? (The effect is surprisingly large—about a factor of 2!)

(d) Neglecting air resistance, plot a graph of $(K + U_g)$ *vs.* time.

(e) In Denver, a mile above sea level, the air is about 83% as dense as the air at sea level. Including the effect of air resistance, use your computer model to predict the trajectory and range of a baseball thrown in Denver. How does the range compare with the range at sea level?

Problem 5.4 Numerical integration of spring-mass system with friction
Modify your numerical integration of the spring-mass system to include the effects of friction. Such a system is called a "damped oscillator."

Simulate the effect of all three kinds of friction: sliding friction (independent of v), viscous friction (proportional to v), and air resistance for large blunt objects (proportional to v^2). Choose friction to be small enough that you get at least two full cycles of oscillation, but large enough that you see an effect. If the friction is so large that the graph never crosses the axis, the system is said to be "over-damped." For each kind of friction, display the following:

(1) Display an animation of the motion of the system
(2) Plot a graph of position *vs.* time
(3) Plot graphs of K, U, and $(K+U)$ *vs.* time

On the energy graphs, when is the energy loss rate large and small, and why?

Problem 5.5 Experiment—observing resonance
You can observe the main effects of resonance with very simple experiments. Hold a spring vertically with a mass suspended at the other end, and observe the frequency of "free" oscillations with your hand kept still. Then stop the oscillations, and move your hand *extremely* slowly up and down in a kind of slow sinusoidal motion. You will see that the mass moves up and down with the same very low frequency.

(a) How does the amplitude (plus or minus displacement from the center location) of the mass compare with the amplitude of your hand? (Notice that the phase shift of the oscillation is 0°; the mass moves up when your hand moves up.)

(b) Next move your hand up and down at a significantly higher frequency than the free-oscillation frequency. How does the amplitude of the mass compare to the amplitude of your hand? (Notice that the phase shift of the oscillation is 180°; the mass moves down when your hand moves up.)

(c) Finally, move your hand up and down at the free-oscillation frequency. How does the amplitude of the mass compare with the amplitude of your hand? (It is hard to observe, but the phase shift of the oscillation is 90°; the mass is at the midpoint of its travel when your hand is at its maximum height.)

(d) Change the system in some way so as to increase the air resistance significantly. For example, attach a piece of paper to increase drag. At the free-oscillation frequency, how does this affect the size of the response?

A strong dependence of the amplitude and phase shift of the system to the driving frequency is called resonance.

Problem 5.6 Energy in a spring-mass system

A horizontal spring-mass system has low friction, spring stiffness 200 N/m, and mass 0.4 kg. The system is released with an initial compression of the spring of 10 cm and an initial speed of the mass of 3 m/s.

(a) What is the maximum stretch during the motion?

(b) What is the maximum speed during the motion?

(c) Now suppose that there is energy dissipation of 0.01 J per cycle of the spring-mass system. What is the average power input in watts required to maintain a steady oscillation?

Problem 5.7 Vertical spring-mass system

Write an equation for the total energy of a system consisting of a mass suspended vertically from a spring, and include the Earth in the system. Place the origin for gravitational energy at the equilibrium position of the mass, and show that the changes in energy of a vertical spring-mass system are the same as the changes in energy of a horizontal spring-mass system.

Problem 5.8 Hot water

400 grams of boiling water (temperature 100° C, heat capacity 4.2 J/gram/K) are poured into an aluminum pan whose mass is 600 grams and initial temperature 20° C (the heat capacity of aluminum is 0.9 J/gram/K). After a short time, what is the temperature of the water? Explain your calculation. What simplifying assumptions did you have to make?

Problem 5.9 Place a mass on a vertical spring

A spring with stiffness k_s and relaxed length L stands vertically on a table. You hold a mass M just barely touching the top of the spring.

(a) You *very slowly* let the mass down onto the spring a certain distance, and when you let go, the mass doesn't move. How much did the spring compress? How much work did *you* do?

(b) Now you again hold the mass just barely touching the top of the spring, and then let go. What is the maximum compression of the spring? State what approximations and simplifying assumptions you made.

(c) Next you push the mass down on the spring so that the spring is compressed an amount s, then let go, and the mass starts moving upward and goes quite high. When the mass is a height of $2L$ above the table, what is its speed?

Problem 5.10 Bungee jumping

Design a "bungee jump" apparatus. A bungee jumper falls from a high platform with two elastic cords tied to the ankles. The jumper falls freely for a while, with the cords slack. Then the jumper falls an additional distance with the cords increasingly tense. Assume that you have cords that are 10 m long, and that the cords stretch in the jump an additional 20 m for an average jumper.

(a) Make a series of simple diagrams, like a comic strip, showing the platform, the jumper, and the two cords at various times in the fall and the rebound. On each diagram, draw and label vectors representing the forces acting on the jumper, and the jumper's velocity. Make the relative lengths of the vectors reflect their relative magnitudes.

(b) At what instant is there the greatest tension in the cords? How do you know?

(c) What is the jumper's speed at this instant?

(d) Is the jumper's momentum changing at this instant or not? (That is, is dp/dt nonzero or zero?) Explain briefly.

(e) Focus on this instant, and use the principles of this chapter to determine the spring stiffness k_s for each cord. Explain your analysis. Give numerical values for your design.

(Continued on next page.)

(f) What is the maximum tension that each cord must support without breaking? Give numerical values for your design.

(g) What is the maximum acceleration (in "g's") that the jumper experiences? What is the direction of this maximum acceleration?

(h) State clearly what approximations and estimates you have made in your design.

Problem 5.11 Heating a house

During three hours one winter afternoon, when the outside temperature was 0° C (32° F), a house heated by electricity was kept at 20° C (68° F) with the expenditure of 45 kwh (kilowatt·hours) of electric energy. What was the average energy leakage in joules per second through the walls of the house to the environment (the outside air and ground)?

The rate at which heat is transferred between two systems is often proportional to their temperature difference. Assuming this to hold in this case, if the house temperature had been kept at 25° C (77° F), how many kwh of electricity would have been consumed?

Problem 5.12 Experiment—your power capability

(a) If you follow a diet of 2000 food calories per day (2000 kilocalories), what is your average power consumption in watts? (A food or "large" calorie is a unit of energy equal to 4.2×10^3 J; a regular or "small" calorie is equal to 4.2 J.) Compare with the power consumption in a table lamp.

(b) You can produce much higher power for short periods. Make appropriate measurements as you run up some stairs, and report your measurements. Use these measurements to estimate your power output (this is in addition to your basal metabolism—the power needed when resting). Compare with a horsepower (which is about 750 watts) or a toaster (which is about 1000 watts).

(c) How many days of a diet of 2000 large calories are equivalent to the gravitational energy difference for you between sea level and the top of Mount Everest, 8848 m above sea level? (However, the body is not anywhere near 100% efficient in converting chemical energy into change in altitude. Also note that this is in addition to your basal metabolism.)

Problem 5.13 Human energy consumption

(a) A typical candy bar provides 280 calories (one "food" or "large" calorie is equal to 4.2×10^3 joules). How many candy bars would you have to eat to replace the chemical energy you expend doing 100 sit-ups? Explain your work, including any approximations or assumptions you make. (In a sit-up, you go from lying on your back to sitting up.)

(b) Humans have about 60 milliliters (60 cm^3) of blood per kilogram of body mass, and blood makes a complete circuit in about 20 seconds, to keep tissues supplied with oxygen. Make a crude estimate of the *additional* power output of your heart (in watts) when you are standing compared with when you are lying down. Note that we're not asking you to estimate the power output of your heart when you are lying down, just the change in power when you are standing up. You will have to estimate the values of some of the relevant parameters. Because we're only looking for a crude estimate, try to construct a model that is as simple as possible. Describe what approximations and estimates you made.

Problem 5.14 Graph of energy with dissipation

Figure 5.45 is a portion of a graph of energy terms vs. time for a mass on a spring, subject to air resistance. Identify and label the three curves as to what kind of energy each represents. Explain briefly how you determined which curve represented which kind of energy.

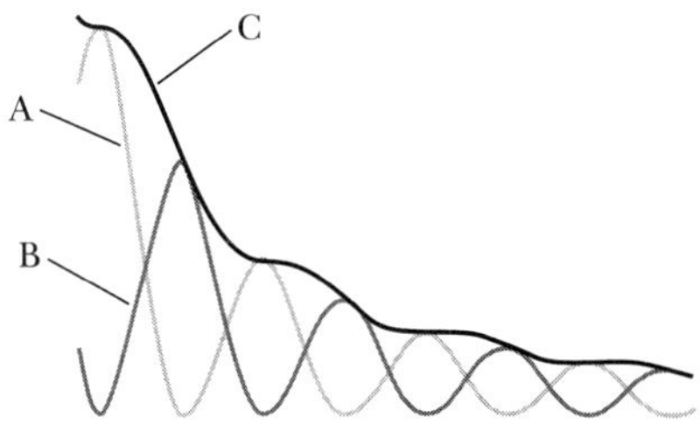

Figure 5.45 Which curve represents which kind of energy (Problem 5.14)?

Problem 5.15 Drop coffee filters from a tall building

You drop a single coffee filter from a very tall building, and it takes 52 seconds to reach the ground. Next you drop a stack of 3 of these coffee filters. About how long will they take to hit the ground? Explain briefly, including any approximations or simplifying assumptions you had to make.

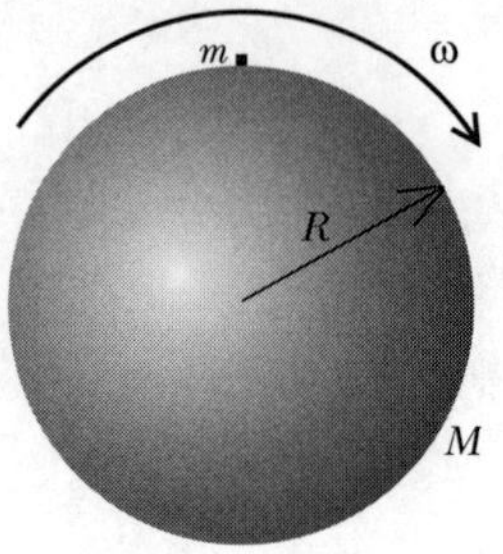

Figure 5.46 Launch a package from an asteroid (Problem 5.16).

Problem 5.16 Launch a package from an asteroid

A package of mass m sits at the equator of an airless asteroid of mass M, radius R, and spin angular speed ω (Figure 5.46). We want to launch the package in such a way that it will never come back, and when it is very far from the asteroid it will be traveling with speed v. We have a powerful spring whose stiffness is k_s. How much must we compress the spring?

Problem 5.17 A mass hangs from a spring

A mass of 0.3 kg hangs motionless from a vertical spring whose length is 0.8 m and whose unstretched length is 0.65 m. Next the mass is pulled down to where the spring has a length of 0.9 m and given an initial speed upwards of 1.2 m/s. What is the maximum length of the spring during the motion that follows? What approximations or simplifying assumptions did you make?

Problem 5.18 A sky diver

Terminal speed for a falling sky diver has been measured to be about 60 m/s. Using your computer model, determine how far (in meters) and how long (in seconds) a sky diver falls before reaching terminal speed. (You will need to think about the meaning of "terminal speed" and how to test for it.)

Plot graphs of the sky diver's speed *vs.* time and position *vs.* time.

Problem 5.19 Experiment—coefficient of friction

Design and carry out an experiment to determine the coefficient of sliding friction for a block on a flat plate.

Problem 5.20 Experiment—speed dependence of sliding friction

For this problem you will need measurements of the position vs. time of a block sliding on a table, starting with some initial velocity, slowing down, and coming to rest. If you do not have an appropriate laboratory setup for making these measurements, your instructor will provide you with such data. Analyze these data to see how well they support the assertion that the force of sliding friction is essentially independent of the speed of sliding. (Recall that a reason was given for this assertion—the friction force is determined by the strength of temporary interatomic bonds that must be broken.)

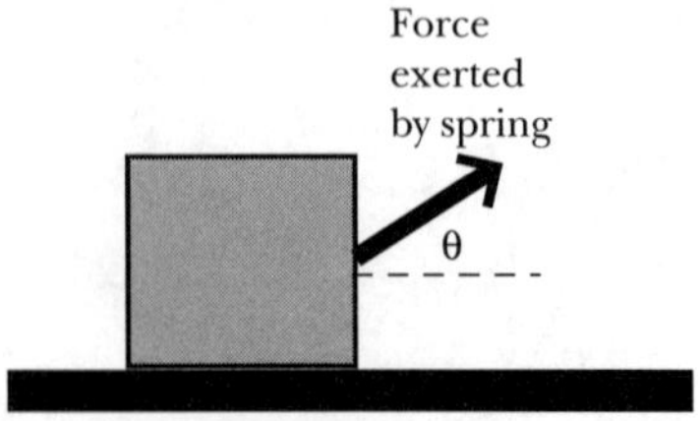

Figure 5.47 A block is dragged by a force at an angle to the horizontal (Problem 5.21).

Problem 5.21 Block dragged at an angle

A block of mass M is dragged along a table by a force applied at an angle θ to the horizontal by a spring of stiffness k_s, stretched a constant amount s (Figure 5.47). The coefficient of friction between block and table is μ. At $t = 0$, the speed of the block was v_0. What is its speed v at time t?

Problem 5.22 Computer modeling of resonance

You can study resonance in a driven oscillator in detail by modifying your computation for the spring-mass system with friction. Let one end of the spring be moved back and forth sinusoidally by a motor, with a motion given by $D\sin(\omega_D t)$ (Figure 5.48). Here D is the amplitude of the motor motion and ω_D is the angular frequency of the motor, which can be varied. (The free-oscillation angular frequency $\omega_F = \sqrt{k_s/m}$ of the spring-mass system has a fixed value, determined by the spring stiffness k_s and the mass m.)

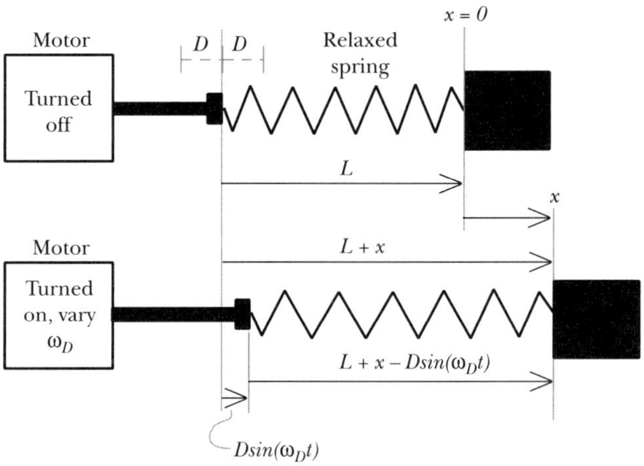

Figure 5.48 A driven spring-mass system (Problem 5.22).

We need an expression for the stretch s of the spring in order to be able to calculate the force $-k_s s$ of the spring on the mass. We have to do a bit of geometry to figure out the length of the spring when the mass is displaced a distance x from the equilibrium position, and the motor has moved the other end of the spring by an amount $D\sin(\omega_D t)$.

On the diagram, the spring gets longer when the mass moves to the right ($+x$), but the spring gets shorter when the motor moves to the right ($-D\sin(\omega_D t)$). The new length of the spring is $L + x - D\sin(\omega_D t)$, and the net stretch of the spring is this quantity minus the unstretched length L, yielding $s = x - D\sin(\omega_D t)$. A check that this is the correct expression for the net stretch of the spring is that if the motor moves to the right the same distance as the mass moves to the right, the spring will have zero stretch. Replace x in your computer computation with the quantity $[x - D\sin(\omega_D t)]$.

(a) Use viscous damping (friction proportional to v). The damping should be small. That is, without the motor driving the system ($D = 0$), the mass should oscillate for many cycles. Set ω_D to $0.9\omega_F$ (that is, 0.9 times the free-oscillation angular frequency, $\omega_F = \sqrt{k_s/m}$), and let x_0 be 0. Graph position x vs. time for enough cycles to show that there is a transient buildup to a "steady state." In the steady state the energy dissipation per cycle has grown to exactly equal the energy input per cycle.

Vary ω_D in the range from 0 to $2.0\omega_F$, with closely spaced values in the neighborhood of $1.0\omega_F$. Have the computer plot the position of the motor end as well as the position of the mass as a function of time. For each of these driving frequencies, record for later use the steady-state amplitude and the phase shift (the difference in angle between the sinusoids for the motor and for the mass). If a harmonic oscillator is lightly damped it has a large response only for driving frequencies near its own free-oscillation frequency.

(b) Is the steady-state angular frequency equal to ω_F or ω_D for these various values of ω_D? (Note that during the transient buildup to the steady state the frequency is not well-defined, because the motion of the mass isn't a simple sinusoid.)

(c) Sketch graphs of the steady-state amplitude and phase shift vs. ω_D. Note that when $\omega_D = \omega_F$ the amplitude of the mass can be much larger than D, just as you observed with your hand-driven spring-mass system. Also note the interesting variation of the phase shift as you go from low to high driving frequencies—does this agree with the phase shifts you observed with your hand-driven spring-mass system?

(d) Repeat part (c) with viscous friction twice as large. What happens to the resonance curve (the graph of steady-state amplitude vs. angular frequency ω_D)?

5.18 Answers to exercises

5.1 (page 235) If $k_s = 0.6$ N/m and $s = 0.2$ m, $U_s = 0.012$ J

5.2 (page 235) 0.5 m/s

5.3 (page 236) No change; same stretch

5.4 (page 239) 0.12 K (that is, 0.12 Kelvin)

5.5 (page 241) 1.26×10^5 J; 7.6×10^4 J; -1.26×10^5 J

5.6 (page 242) About 500 J; about 2000 kg per second

5.7 (page 242) 2.5 m/s^2

5.8 (page 247) $\frac{1}{2}kA^2$; $\frac{1}{2}kA^2$; spring + mass

5.9 (page 249) less; some of the initial energy is now in the air

5.10 (page 249) $[2.7\times10^{-2} - 4.5\times10^{-4}]\,\text{J} \approx 2.7\times10^{-2}$ J

5.11 (page 254) $\dfrac{dv_x}{dt} = \dfrac{(k_s s - f_{max})}{M}$ and $\dfrac{dv_y}{dt} = 0$.

5.12 (page 256) +4 J; +4J; 8 J

5.13 (page 257) +0.8 J; –0.8 J; 0 J

5.14 (page 258) Denominator gets very small when $\omega_D = \omega_F$.

5.15 (page 258) When $\omega_D = \omega_F$, denominator = $\dfrac{c}{m}\omega_D$.

Denominator = $\sqrt{2}\dfrac{c}{m}\omega_D$ when $\left|(\omega_F^2 - \omega_D^2)\right| = \dfrac{c}{m}\omega_D$.

But since near resonance $\omega_D \approx \omega_F$, we have

$(\omega_F^2 - \omega_D^2) = (\omega_F + \omega_D)(\omega_F - \omega_D) \approx 2\omega_D(\omega_F - \omega_D)$

Hence when $\left|\omega_F - \omega_D\right| \approx \dfrac{c}{2m}\omega_D \approx \dfrac{c}{2m}\omega_F$, amplitude down by a factor of $1/\sqrt{2}$.

5.16 (page 258) Larger friction, wider resonance peak. If friction very small, resonance peak is a very narrow spike.

5.17 (page 258) Low frequency, $\cos\phi \approx \dfrac{(\omega_F^2)}{\sqrt{(\omega_F^2)^2 + \left(\dfrac{c}{m}\omega_D\right)^2}} \approx 1$, so $\phi \approx 0°$.

High frequency, $\cos\phi \approx \dfrac{(-\omega_D^2)}{\sqrt{(-\omega_D^2)^2 + \left(\dfrac{c}{m}\omega_D\right)^2}} \approx -1$, so $\phi \approx 180°$.

Resonance frequency, $\cos\phi = 0$, so $\phi \approx 90°$.

Chapter 6

Energy Quantization

Chapter 6

Energy Quantization

In the early 20th century, as scientists learned more about the detailed behavior of atoms they were excited to discover that small systems such as atoms can have only certain energies, called energy "levels," and cannot have energies in between these discrete levels. The energy changes associated with interactions therefore come in finite packets, or "quanta."

6.1 Energy quantization

Discrete energy levels were not noticed for a long time because the energy differences between these levels are quite small. A typical separation between energy levels in an atom is about 10^{-19} joules. The energies of macroscopic objects are on the order of a joule; an energy increase of 10^{-19} joules is completely undetectable even by the most sensitive measurements of motion or temperature. However, it is possible to detect these tiny differences by observing the interaction of light and matter, as we will see.

The discovery of discrete energy levels is analogous to the discovery of the atomic nature of matter. We now know that matter is not continuous but lumpy, being made up of atoms. For most of human history this lumpiness was not noticed because the atoms are so tiny. Similarly, the noncontinuous, lumpy nature of energy was not noticed because the quantized energies are so very close together that energy seemed continuous.

6.1.1 Electronic energy levels

An example of discrete energy levels is found in the "electronic" energy associated with the motion of electrons around the nucleus of an atom. You probably know from your study of chemistry that electronic energy levels are discrete—an atom can have only certain energies. The electronic energy levels in atomic hydrogen are shown in Figure 6.1 (atomic hydrogen is a single H atom, composed of one proton and one electron, not the usual diatomic molecule H_2).

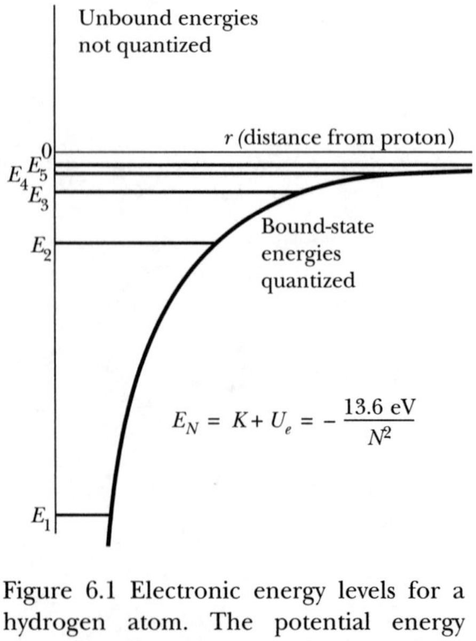

Figure 6.1 Electronic energy levels for a hydrogen atom. The potential energy curve describes the potential energy of the electron-proton system as a function of distance.

The diagram shows the familiar electric potential energy of the proton-electron system. The horizontal lines represent the possible bound energy states for the hydrogen atom; bound states with other values of $(K+U)$ are not observed. Unbound states however $(K + U \geq 0)$ are not quantized.

The electric potential energy of an electron and a proton bound together in a hydrogen atom is negative for all separations:

$$U_e = \frac{1}{4\pi\varepsilon_0}\frac{q_1 q_2}{r} = \frac{1}{4\pi\varepsilon_0}\frac{(+e)(-e)}{r} = -\frac{1}{4\pi\varepsilon_0}\frac{e^2}{r}$$

A quantum mechanical calculation of the energy levels of a system involves solving a differential equation called the Schrödinger equation, which includes a potential energy term. Solving this equation for the hydrogen atom predicts that the discrete energy levels for bound states of hydrogen (ignoring the rest energies of the proton and the electron) will be these:

ENERGY LEVELS FOR THE HYDROGEN ATOM

$$E_N = K + U_e = -\frac{13.6\text{ eV}}{N^2},\ N = 1, 2, 3, \text{etc.}$$

An electron-volt (eV) is an energy unit described below.

This arrangement of hydrogen energy levels has been verified by observations of the interaction of light and hydrogen, as we will see.

These energy levels denote different bound states of the electron+proton system. Recall that a bound state has a negative value of $K+U$, since a potential energy of zero is associated with a very large separation between the electron and the proton. We have seen that bound gravitational orbits are also associated with negative values of $K+U$. As you can see from the energy diagram, in higher energy states the average location of the electron is farther from the proton. Note that there is a lowest energy state; a hydrogen atom cannot have $K+U<-13.6$ eV!

Before the development of the fully quantum mechanical Schrödinger equation, Niels Bohr made a simple model for the hydrogen atom which also yielded the energy formula given above. We will study the Bohr model later in Chapter 9.

6.1.2 Electron volts (eV)

In our analysis of quantum systems we will frequently be considering very small changes in energy, on the order of 10^{-19} joules. A common unit of energy used in descriptions of quantum phenomena is the "electron volt," abbreviated "eV." In terms of joules:

$$1 \text{ eV} = 1.6\times10^{-19} \text{ J}$$

Here are some typical energies measured in electron-volts:

Typical chemical reaction: about 1 eV.

Typical nuclear reaction: about 1 MeV (mega-electron-volt, or 10^6 eV).

Rest mass of the electron: about 0.5 MeV.

Rest mass of the proton: about 1 GeV (giga-electron-volt, or 10^9 eV).

Example: How much energy in electron-volts is required to raise the energy of a hydrogen atom from its lowest level ($N = 1$) to the next lowest level ($N = 2$)?

Solution: Calculate the required increase in energy:

$$\Delta E = E_2 - E_1 = \left(-\frac{13.6 \text{ eV}}{2^2}\right) - \left(-\frac{13.6 \text{ eV}}{1^2}\right) = 10.2 \text{ eV}$$

Note that you cannot add less than 10.2 eV to a hydrogen atom in its lowest energy level; any smaller amount fails to be absorbed.

Ex. 6.1 How much energy in electron-volts is required to ionize a hydrogen atom (that is, remove the electron from the proton), if initially the atom is in its lowest energy level?

6.2 The quantum model of interaction of light and matter

What is the evidence for energy quantization? A key piece of evidence is that atomic systems emit and absorb light only with quantized packets of energy, implying a quantized loss or gain of energy in the system. The study of the interaction of light and matter is called spectroscopy.

6.2.1 A particle model of light

Early in the 20th century it was discovered that light can be considered to be lumpy, being made up of packets of energy called photons. As with atoms and energy, the noncontinuous, lumpy nature of light was not noticed for a long time because the amount of energy carried by one photon in an ordinary beam of light is extremely small compared to the total energy of the beam. It has now become a routine matter to detect individual photons in a very weak beam of light, and to measure the energy of an individual photon.

You may already be familiar with the classical model of light, in which light is described as a travelling wave. The classical wave model of light makes it possible to account for important aspects of the behavior of light, including interference and diffraction phenomena. To account for other properties of light, including the observation that packets of light corresponding to certain fixed energies can be absorbed or emitted by quantum systems, we need a quantum model of light that describes light as particles (photons). In this chapter we will emphasize the quantum model of light, and we will be concerned with the energy of photons emitted or absorbed as quantum systems go from one energy state to another.

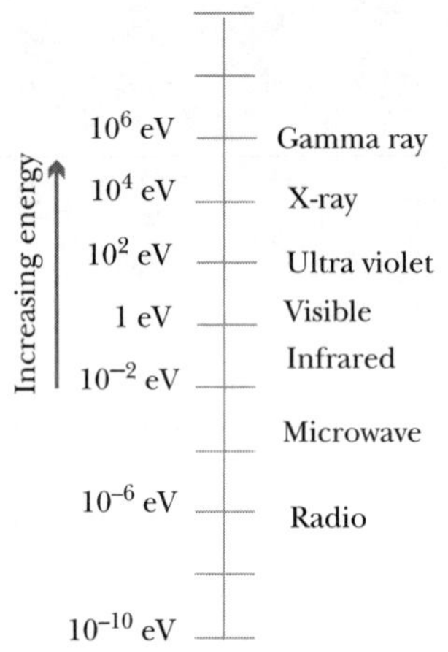

Figure 6.2 The electromagnetic spectrum.

The electromagnetic spectrum

Visible light is an example of electromagnetic radiation, a phenomenon associated with the production and detection of electric and magnetic fields of a particular kind. The instruments needed to detect and measure electromagnetic radiation depend on the photon energies involved. The spread of observable energies is called the electromagnetic spectrum (Figure 6.2).

That part of the electromagnetic spectrum for which your eye is a good detector is called "visible" light (Figure 6.3). Its energy ranges from around 1.8 eV per photon (red light) to 3.1 eV per photon (violet light). Photon energies greater than 3 eV include ultraviolet, x-rays, and gamma rays. Electromagnetic radiation with photon energies less than 1.8 eV includes infrared, radio waves, and microwaves. Sometimes the term "light" is used to refer just to visible electromagnetic radiation, that part of the spectrum detectable by human eyes.

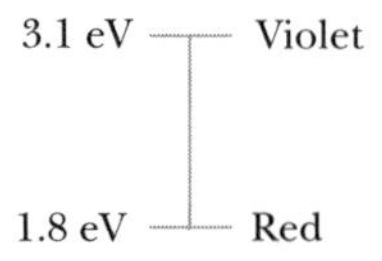

Figure 6.3 The visible portion of the electromagnetic spectrum.

Ex. 6.2 The photon energy for green light lies between that for red and violet light. What is the approximate energy of the photons in green light?

Ex. 6.3 The intensity of sunlight is about 1400 watts (J/s) per square meter. That is, when sunlight hits perpendicular to a square meter of area, about 1400 watts of energy can be absorbed. Using the average photon energy found in the previous exercise, about how many photons per second strike an area of one square meter? (This is why the lumpiness of light was not noticed for so long.)

6.2.2 A spectrum

A beam of light can be composed of photons of many different energies. When the photons in the beam are separated according to their energies, the resulting distribution is called the spectrum of that beam of light. One way to separate a beam of visible light into its component energies is to pass a narrow beam of light through a prism. You have probably seen the continuous rainbow spectrum seen when ordinary white light goes through a prism, as in Figure 6.4. Your eye reacts differently to photons of different energies, which is why you see a colorful spectrum.

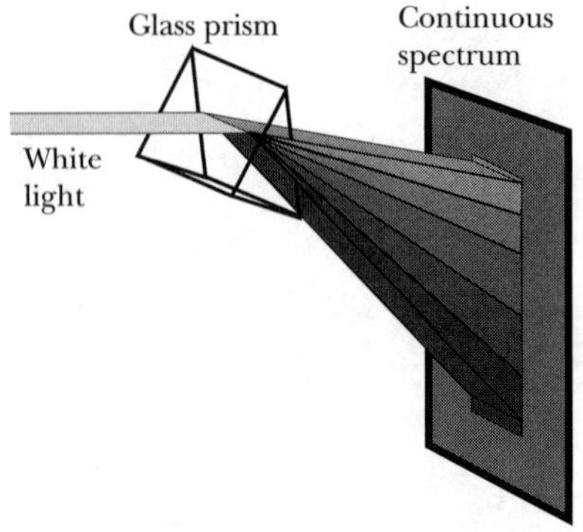

Figure 6.4 A prism separates a narrow beam of white light into a continuous rainbow of colors.

6.2.3 Emission spectra

In contrast to white light, when a narrow beam of light emitted by electrically excited atomic hydrogen is passed through a prism it does not look like a continuous spectrum but rather appears as a few widely spaced bright lines of particular colors (Figure 6.5). Such discontinuous spectra are strong evidence for energy quantization in the systems that are emitting the light.

When a hydrogen atom drops from a higher energy level E_4 to a lower energy level E_1 it emits a photon with the associated energy $E_4 - E_1$ (Figure 6.6). A collection of hydrogen atoms can emit photons whose energies correspond to the differences in energy between any two electronic energy levels of the hydrogen atom. The light emitted by excited atomic hydrogen is called the "hydrogen emission spectrum," and the bright lines of light of particular energies in the spectrum are called "spectral lines." The observed energies of light emitted by excited atomic hydrogen are consistent with the energy level differences predicted by quantum mechanics. Other atoms have different quantized energy levels, and a line spectrum can be used to identify the emitting atoms by the pattern of emitted photon energies.

If an isolated hydrogen atom is initially in a high energy level, over time it will spontaneously emit photons and drop to lower and lower states, eventually ending up in the lowest energy level, which is called the "ground state." Once an atom has dropped to the ground state, it cannot lose any more energy, nor emit photons, because there is no lower energy level than the ground state. This is quite different from the classical (prequantum) picture, in which a system can have an arbitrarily low energy.

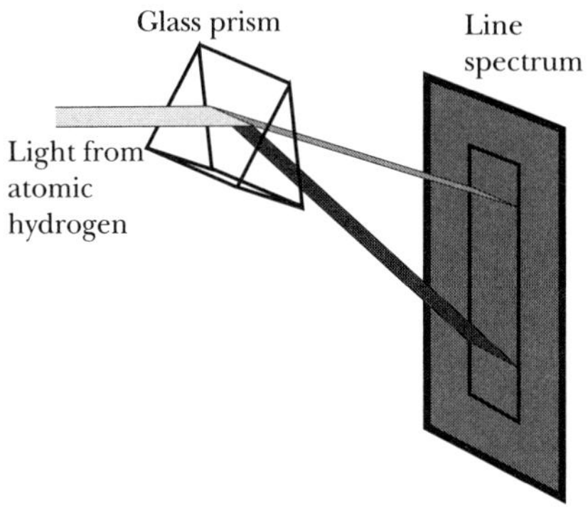

Figure 6.5 The light emitted by excited atomic hydrogen contains only a few separate colors. Light in a spectrum of a particular color and energy is called a "spectral line."

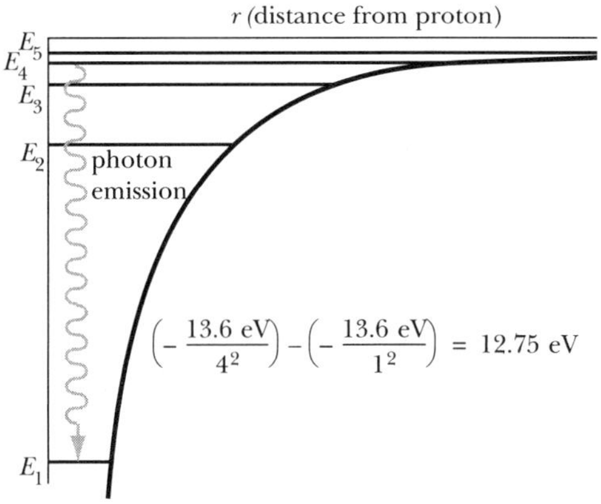

Figure 6.6 An atom can drop to a lower energy level by emitting a photon whose energy corresponds to the difference between the energies of the levels. By convention a squiggly line with an arrowhead indicates a photon.

Example: Consider a collection of hydrogen atoms. Initially each individual atom is in one of the lowest three energy states ($N = 1, 2, 3$). What will be the energies of photons emitted by this collection of hydrogen atoms?

Solution: There are three possibilities: transitions from $N = 3$ to $N = 2$, from $N = 2$ to $N = 1$, or from $N = 3$ to $N = 1$. The photon energies for the three transitions are these:

$$N = 3 \text{ to } N = 2: \quad \left(-\frac{13.6\text{ eV}}{3^2}\right) - \left(-\frac{13.6\text{ eV}}{2^2}\right) = 1.9\text{ eV}$$

$$N = 2 \text{ to } N = 1: \quad \left(-\frac{13.6\text{ eV}}{2^2}\right) - \left(-\frac{13.6\text{ eV}}{1^2}\right) = 10.2\text{ eV}$$

$$N = 3 \text{ to } N = 1: \quad \left(-\frac{13.6\text{ eV}}{3^2}\right) - \left(-\frac{13.6\text{ eV}}{1^2}\right) = 12.1\text{ eV}$$

Ex. 6.4 How many different photon energies would emerge from a collection of hydrogen atoms which occupy the lowest four energy states ($N = 1, 2, 3, 4$)? (You need not calculate the energies of these transitions.)

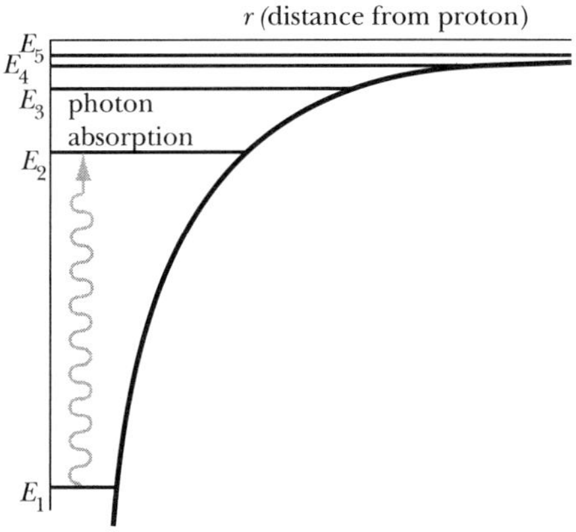

Figure 6.7 An atom can gain energy by absorbing a photon whose energy corresponds to the difference between two energy levels.

6.2.4 Absorption spectra

An atom can drop from a higher energy level to a lower energy level and emit a photon. An atom can also absorb a photon of an appropriate energy and jump from a lower energy level to a higher energy level (Figure 6.7). If an atom is in its ground state, only photon absorption is possible; photon emission cannot occur. Energy absorption can occur if the photon energy corresponds exactly to the difference between two energy levels of the atom. If we shine white light (containing a mixture of different photon energies) through a collection of atoms (such as a container of hydrogen gas), only photons whose energy is the difference of atomic energy levels will be strongly absorbed, while other photons can pass right through the material. In this case, the spectral lines are dark lines (missing energies) on a light background, instead of the bright lines on a dark background observed in an emission spectrum. This is called an "absorption spectrum."

Photon absorption is followed almost immediately by emission from the excited state to a lower state, so at any instant there are few atoms in excited states which could absorb photons. The result is that dark-line absorption spectra only involve absorption from the ground state, not from higher-level states.

The atoms that have jumped to higher energy levels through photon absorption will eventually drop to lower energy levels with the emission of photons. However, these photons are emitted in all directions, not just in the direction of the original beam of light, so the light that has passed through the material will have little intensity at photon energies corresponding to the strongly absorbed energies (Figure 6.8).

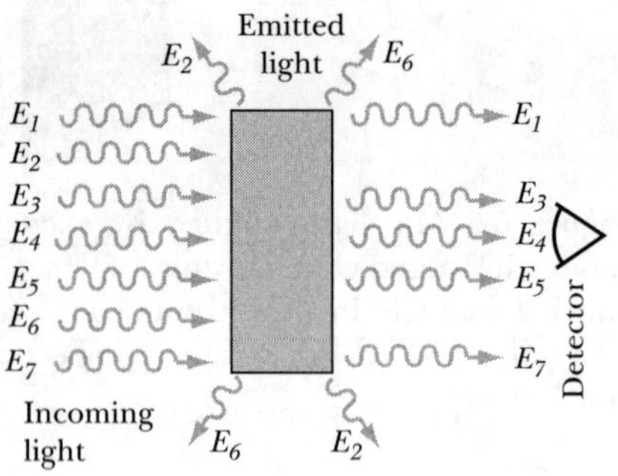

Figure 6.8 Measuring an absorption spectrum. A sample of a material is placed in a transparent container. Incoming light is absorbed, leading to emission. The outgoing light is depleted of the absorption energies.

Example: At room temperature in a gas of atomic hydrogen almost all of the atoms are in the ground state. Why might you expect the gas to be completely transparent to visible light?

Solution: As we saw in the example on page 271, the energy required to go from the ground state of atomic hydrogen to the first excited state is 10.2 eV, but visible light consists of photons whose energies lie between 1.8 eV and 3.1 eV. So visible light cannot be absorbed by ground-state hydrogen atoms. (In a collision between a photon of visible light and an atom, the atom can acquire some kinetic energy, but no internal energy change is possible.)

Ex. 6.5 Suppose you had a collection of hypothetical quantum objects whose individual energy levels were –4.0 eV, –2.3 eV, and –1.6 eV. If nearly all of the individual objects were in the ground state, what would be the energies of dark spectral lines in an absorption spectrum if visible white light (1.8 to 3.1 eV) passes through the material?

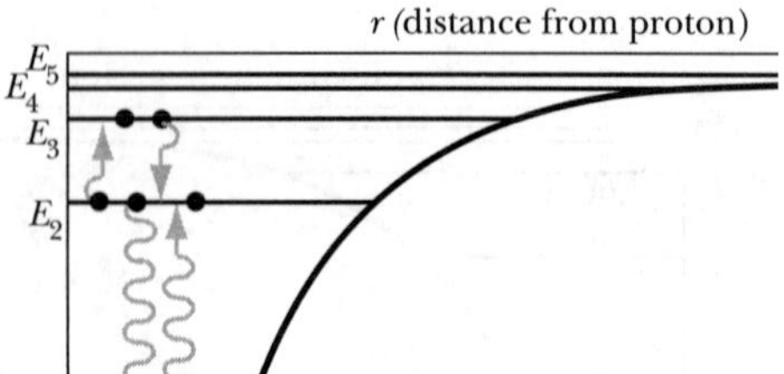

Figure 6.9 A dot represents the energy state of one hydrogen atom in a group of atoms. Energy is exchanged with the surroundings through photon emission and absorption.

6.2.5 The effects of temperature

If an atom absorbs radiation coming from elsewhere, or absorbs energy from being jostled by a neighboring atom, it may jump to a higher energy level, and eventually emit photons and drop again to a lower level (Figure 6.9). In thermal equilibrium with its surroundings, an atom's energy continually goes up and down, but with an average energy determined by the temperature of the collection of atoms.

In Figure 6.9 each dot represents one hydrogen atom in a container of atomic hydrogen. If three dots are on a line, this means three atoms are in this particular state. Each hydrogen atom has the same allowed energy levels, and at any particular instant some of the atoms will be in the ground state, some in the first excited state, etc.

For a particular temperature, at any instant there is a certain probability that one particular atom will be in state 1, 2, 3, etc. High temperature is associated with a significant probability that the various atoms are in various states, from the ground state up to a high energy level. Very low temperature is associated with small probability that atoms are in any state other than the ground state (Figure 6.10). If the temperature is near absolute zero, the atoms will nearly always be found in the ground state. As we will see in Chapter 10 on "statistical mechanics," the population of a level whose energy is E above the ground state is proportional to the "Boltzmann factor" $\exp(-E/kT)$, where k is the Boltzmann constant (1.4×10^{-23} J/K) and T is the absolute temperature in kelvins.

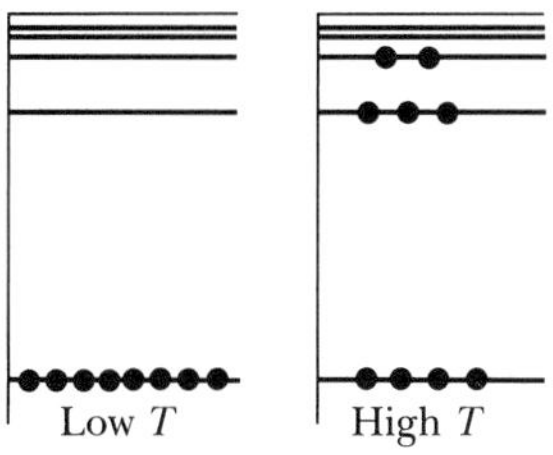

Figure 6.10 At a low temperature (system on the left) all atoms are in low energy states. At a high temperature (system on the right) some atoms are in higher energy states.

? Can you observe an emission spectrum if the material is very cold? Can you observe an absorption spectrum if the material is very cold?

Very cold matter will normally be in the ground state and will not emit, but an absorption spectrum will show gaps (dark lines) corresponding to photons being absorbed and lifting an atom above its ground state. (In order to observe an absorption spectrum you would of course need to use a beam of photons whose energies are sufficiently high to raise the system energy at least to the first excited state above the ground state.)

Room temperature is "very cold" for transitions involving visible light

In a sample at room temperature (about 300 K), almost all atoms are in their atomic ground state. For example, if the first excited state is 1 eV above the ground state, the Boltzmann factor is

$$\exp(-E/kT) = \exp\left(-\frac{1.6\times10^{-19}\text{ J}}{(1.4\times10^{-23}\text{ J/K})(300\text{ K})}\right) = \exp(-38) = 3\times10^{-17}$$

This extremely tiny number represents the fact that at room temperature hardly any atoms will be excited thermally to 1 eV above the ground state, and all the atoms in an ordinary room-temperature absorption measurement can be considered to be in the ground state.

On the other hand, the surface temperature of the Sun is about 5000 K, and with $E = 1$ eV the Boltzmann factor is $\exp(-E/kT) = \exp(-2.3) = 0.1$, so a large fraction of atoms can be in an excited state 1 eV above the ground state, due to thermal excitation.

6.2.6 Electron excitation

In order for an atom to emit a photon, it first has to absorb enough energy to raise the atom to an excited state, above the ground state. A common way to excite an atoms in a gas is to run high energy electrons through the gas. Because electrons are (negatively) charged electrically, they interact strongly with atoms, and an electron that passes near an atom may excite the atom from the ground state to an excited state, with a corresponding loss of kinetic energy of the electron.

Unlike photon absorption, which almost always occurs only if the photon energy exactly matches the difference between energy levels, an electron can give up some kinetic energy and have some kinetic energy left over.

? For example, suppose an electron with kinetic energy of 11 eV collides with a hydrogen atom which is in the ground state. What can happen?

The first excited state of hydrogen is 10.2 eV above the ground state. This 11 eV electron has enough kinetic energy to excite the hydrogen atom to the first excited state, giving up 10.2 eV of energy. The electron moves away with 0.8 eV of kinetic energy remaining. The hydrogen atom will quickly drop back down to the ground state, emitting a 10.2 eV photon. If a beam containing a large number of energetic electrons passes through a gas of atomic hydrogen, there will be continual electron excitation and continual photon emission. But as soon as you turn off the electron beam, the hydrogen will stop emitting photons.

? What happens if you send a beam of electrons with kinetic energy of 7 eV through the gas of atomic hydrogen?

7 eV isn't enough energy to go from the ground state to the first excited state in atomic hydrogen, so the electrons will pass right through without losing any kinetic energy.

? What happens if you send a beam of electrons with kinetic energy of 25 eV through the gas of atomic hydrogen?

Because the ground state of hydrogen has $K + U = -13.6\text{ eV}$, only 13.6 eV is needed to raise a hydrogen atom to any of its excited states, or even to break the atom apart into a proton and an electron. Therefore a beam containing lots of 25 eV electrons passing through the atomic hydrogen gas will keep many excited states populated, and you'll observe many different photon energies being emitted as atoms drop from excited states to lower excited states or to the ground state.

6.3 Summary of three important processes

- Electron excitation: A beam containing many electrons with high kinetic energy can maintain a gas of atoms in many excited states.
- Photon emission: Atoms in excited states emit photons as the atoms drop from higher energy states to lower energy states.
- Photon absorption: Atoms in cold material, in the absence of an electron beam, can absorb photons and go from the ground state to an excited state, but only if the photon's energy exactly matches the energy difference between the excited state and the ground state. Absorption leads to a "dark-line" spectrum.

6.4 Example: Observed photon emission

You observe photon emissions from a collection of quantum objects, each of which is known to have just four quantized energy levels. The collection is kept at a high temperature, and you detect emitted photons using a detector sensitive to photons in the energy range from 2.5 eV to 30 eV. With this detector, you observe photons emitted with energies of 3 eV, 6 eV, 8 eV, and 9 eV, but no other energies.

(a) It is known that the ground state energy of each of these objects is -10 eV. Propose two possible arrangement of energy levels that are consistent with the experimental observations. Explain in detail, using diagrams.

(b) You obtain a second detector that is sensitive to photon energies in the energy range from 0.1 eV to 2.5 eV. What additional photon energies do you observe to be emitted? Explain briefly.

(c) You cool the collection of objects to a very low temperature and send a beam of photons with a wide range of energies through the material. Using both detectors to determine the absorption spectrum, what are the photon energies of the dark lines? How can this information be used to choose between your two proposed energy level schemes? Explain briefly.

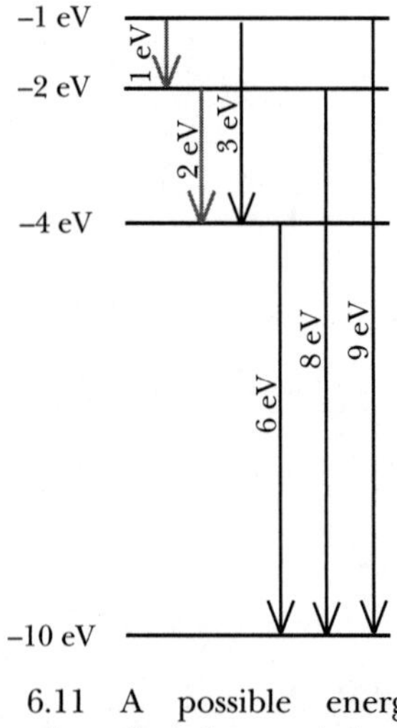

Figure 6.11 A possible energy-level scheme that fits the emission data. Lighter arrows indicate photons not detectable by this detector.

Solution

Fundamental principle: the energy principle.
System: some unknown object in an excited energy state

Initial state: object in excited state
Final state: object in lower energy state, plus photon
Energy principle: $E_{obj,i} = E_{obj,f} + E_{photon}$
So the energy of any emitted photon must be equal to the difference in energy between an excited state and a lower energy state.

(a) Figure 6.11 shows a possible energy-level scheme that fits the emission data. We found this scheme essentially by trial and error. We knew the energy differences between levels, so we tried various arrangements. The arrows are labeled with the energies of photons emitted in transitions between the two states indicated. The 1 eV and 2 eV emissions would not be detected by a detector that is sensitive to the photon energy range 2.5 eV to 30 eV.

Figure 6.12 shows another possible scheme that fits the emission data. This is somewhat less likely, because it is more common to find the energy level spacing decreasing at higher energies, as shown in Figure 6.11.

(b) With this detector we observe the 1 eV and 2 eV photon emissions.

(c) At very low temperature almost all of the objects are in the ground state. If the energy levels are as shown in Figure 6.11, the dark lines will be at 6 eV, 8 eV, and 9 eV, corresponding to absorption in the ground state. If the energy levels are as shown in Figure 6.12, the dark lines will be at 1 eV, 3 eV, and 9 eV. Therefore the low temperature absorption spectrum allows us to distinguish between the two possible energy level schemes.

Comment: Since this was an energy analysis, we didn't need to know what kind of object this was, or whether the energy levels were electronic, vibrational, or rotational energy levels.

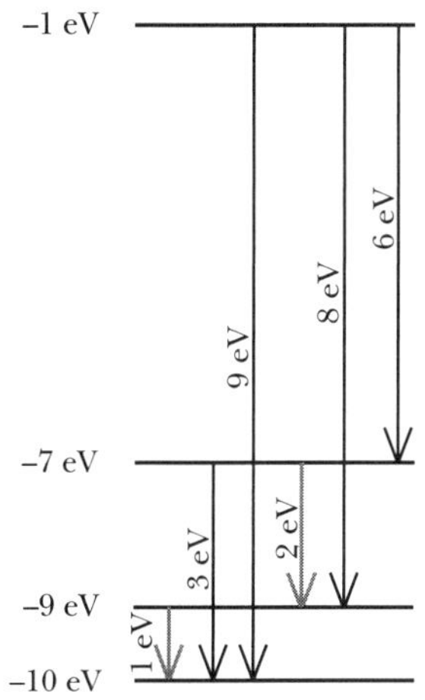

Figure 6.12 Another possible energy-level scheme that fits the emission data.

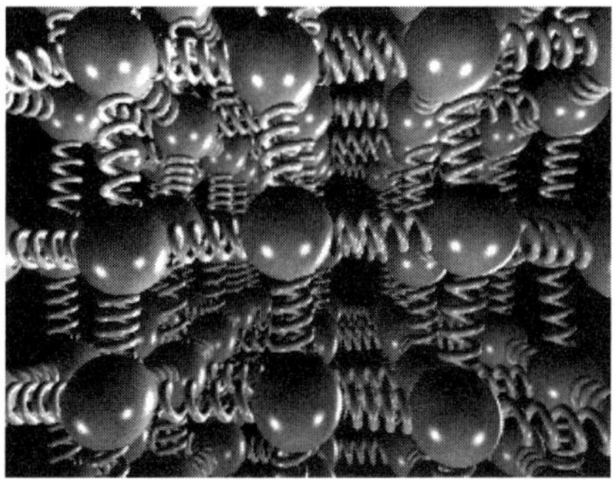
Figure 6.13 Model of a solid as a network of balls and springs.

6.5 Quantized vibrational energy levels

For concreteness we have described energy quantization for the electronic states of atomic hydrogen. Many other kinds of microscopic systems have quantized energy. We will next discuss vibrational energy quantization in atomic oscillators.

6.5.1 Vibrational energy in a classical oscillator

We have modeled a solid as a network of classical harmonic oscillators, balls and springs (Figure 6.13). Although this classical model explains many aspects of the behavior of solids, the predictions it makes regarding the thermal properties of solids do not correspond to experimental measurements at low temperatures. The classical model also fails to explain some aspects of the interaction of light with solid matter. To explain these phenomena we need to make our model of a solid more sophisticated, by incorporating the quantum nature of these atomic oscillators into our model.

The vibrational energy of a classical (non-quantum) spring-mass system can have *any* value (the gray region in Figure 6.14). Equivalently, the amplitude A (maximum stretch) can have *any* value. We can express the energy as $E = \frac{1}{2}k_sA^2$, because at the maximum stretch A the speed is zero, so the kinetic energy is zero, and the energy K+U is just the potential energy at that instant. Classically, the amplitude A can have any value whatsoever, and the energy is continuous, not quantized.

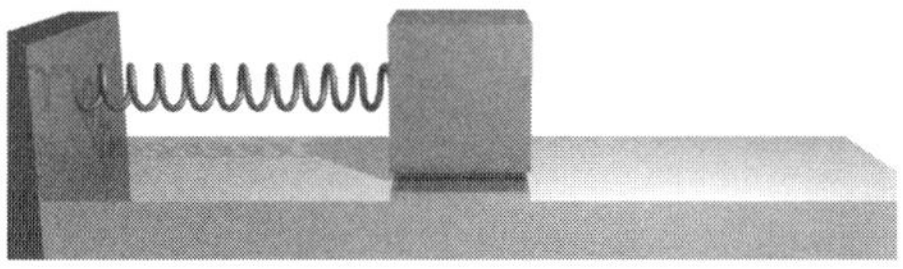

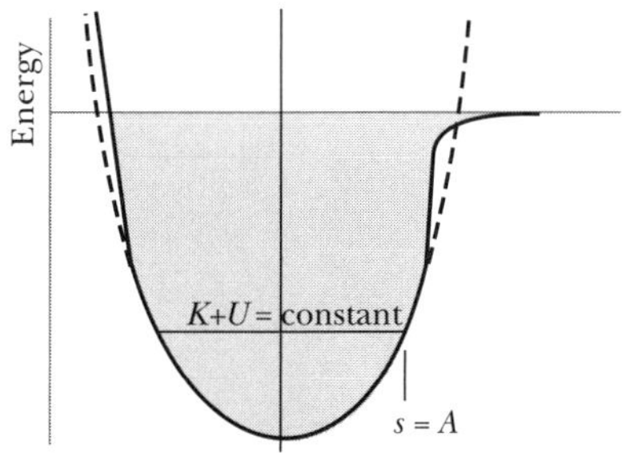

Figure 6.14 A classical harmonic oscillator (spring-mass system) can vibrate with any amplitude, and hence can have any energy. K+U of the system could have any value in the gray region.

6.5.2 Vibrational energy in a quantum mechanical oscillator

Historically, the first hint that energy might be quantized came from an attempt by Max Planck in 1900 to explain some puzzling features of the spectrum of light emitted through a small hole in a hot furnace (so-called "blackbody radiation"). Planck found that he could correctly predict the properties of blackbody radiation if he assumed that transfers of energy in the form of electromagnetic radiation were quantized—that is, an energy transfer could take place only in fixed amounts, rather than having a continuous range of possible values.

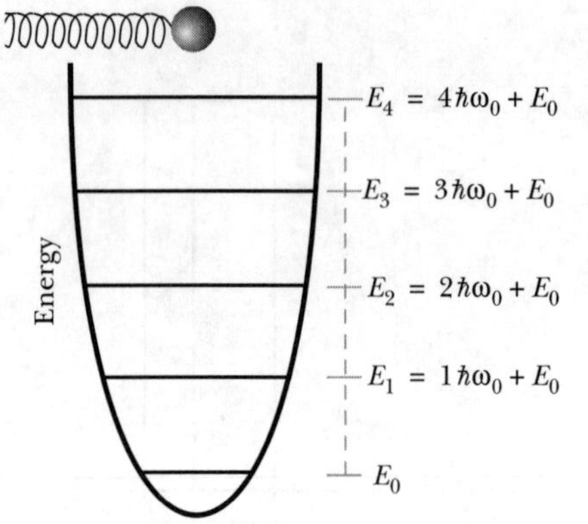

Figure 6.15 A quantum harmonic oscillator can have only certain energies, indicated by the horizontal lines on the graph. This idealized potential energy well represents the situation near the bottom of a real well (which is at a negative energy U_0).

Later, with the full development of quantum mechanics, it was recognized that one could model an atom in the wall of the furnace as a tiny spring-mass system whose energy levels are discrete (Figure 6.15). Classically, the energy of a mass on a spring is related to the amplitude of its oscillation ($E = \frac{1}{2}k_sA^2 + U_0$), so saying that a spring-mass oscillator can only have certain discrete amounts of energy is equivalent to saying that only certain amplitudes are allowed. (To see this, note that we have $K + U = \frac{1}{2}p^2/m + \frac{1}{2}k_ss^2 + U_0 = \frac{1}{2}k_sA^2 + U_0$, since when the stretch s is equal to the amplitude A, the momentum is zero.)

By solving the quantum mechanical Schrödinger equation, one can show that the spacing ΔE between the allowed, quantized energies has the following value:

$$\Delta E = \hbar\sqrt{\frac{k_s}{m}}, \text{ or } \Delta E = \hbar\omega_0, \text{ where } \omega_0 = \sqrt{\frac{k_s}{m}}$$

The quantity k_s is the interatomic spring stiffness, which we estimated in Chapter 3, m is the mass, and $\omega_0 = \sqrt{k_s/m}$ is the angular frequency of the oscillator (in radians per second), as we learned in our studies in Chapter 3. The quantity $\hbar$, pronounced "h-bar," is equal to h, "Planck's constant," divided by 2π. Planck's constant h has an astoundingly small value:

PLANCK'S CONSTANT

$$h = 6.6\times10^{-34} \text{ joule} \cdot \text{second}$$

$$\hbar = \frac{h}{2\pi} = 1.05\times10^{-34} \text{ joule} \cdot \text{second}$$

The quantized energies for an atomic harmonic oscillator are these:

QUANTIZED VIBRATIONAL ENERGY LEVELS

$$E_N = N\hbar\omega_0 + E_0,\ N = 0, 1, 2, \text{ etc., where } \omega_0 = \sqrt{\frac{k_s}{m}}$$

In other words, the energy can have only the following values:

$$E = E_0,\ 1\hbar\omega_0 + E_0,\ 2\hbar\omega_0 + E_0,\ 3\hbar\omega_0 + E_0, \text{ etc.}$$

It is an unusual and special feature of the quantized harmonic oscillator that the spacing between adjacent energy levels is the same for all levels ("uniform spacing"). This is quite different from the uneven spacing of energy levels in the hydrogen atom.

It is important to understand that in both the quantum and the classical picture, the angular frequency $\omega_0 = \sqrt{k_s/m}$ of a particular oscillator is fixed and is the same whether the energy of the oscillator is large or small. A useful way to think about the situation is to say that the frequency is determined by the parameters k_s and m, while the energy is determined by the initial conditions (or equivalently by the amplitude A). In the quantum world, it is as though the amplitude can have only certain specific values.

Ex. 6.6 Show that $\left(\hbar\sqrt{\frac{k_s}{m}}\right)$ has units of joules.

Ex. 6.7 Suppose a collection of quantum harmonic oscillators occupies the lowest 4 energy levels, and the spacing between levels is 0.4 eV. What is the complete emission spectrum for this system?

That is, what photon energies will appear in the emissions? Include all energies, whether or not they fall in the visible region of the electromagnetic spectrum.

6.5.3 Spring stiffness k_s

The energy quantum depends on the stiffness k_s of the "spring." A stiffer spring (larger value of k_s, Figure 6.16) means a steeper, narrower potential curve and larger spacing between the allowed energies. A smaller value of k_s (Figure 6.17) means a shallower, wider potential curve and smaller spacing between the allowed energies.

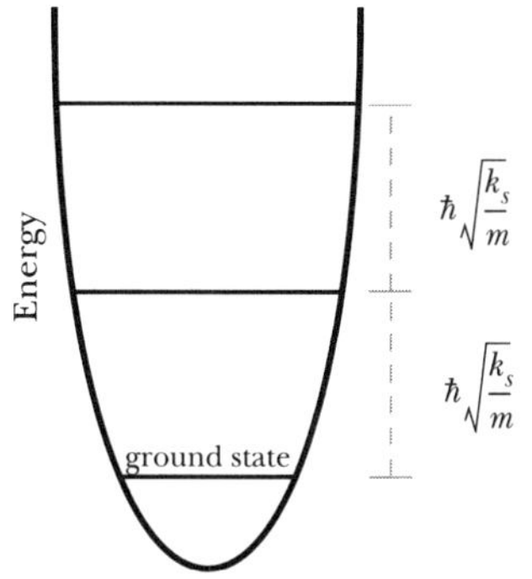

Figure 6.16 A quantum oscillator with a large k_s has widely spaced energy levels.

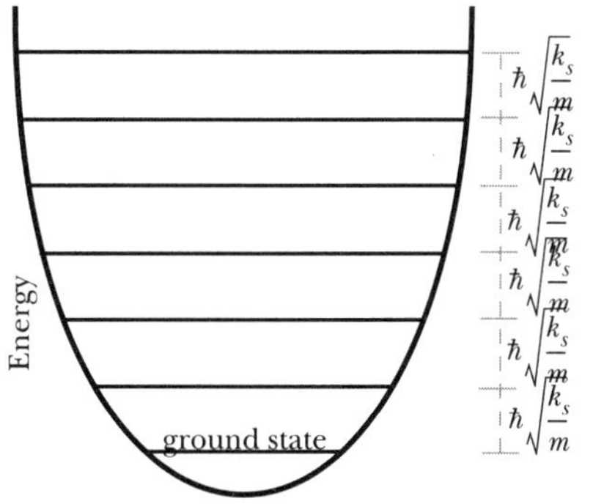

Figure 6.17 A quantum oscillator with a small k_s has closely spaced energy levels.

Ex. 6.8 You may have measured the properties of a simple spring-mass system in the lab. Suppose you found $k_s = 0.7$ N/m and $m = 0.02$ kg, and you observed an oscillation with an amplitude of 0.2 m. What is the approximate value of N, the "quantum number" for this oscillator? (That is, how many levels above the ground state is this oscillator?)

Ex. 6.9 Summarize the differences and similarities between different energy levels in a quantum oscillator. Specifically, for the first two levels in Figure 6.16, compare the angular frequency $\sqrt{k_s/m}$, the amplitude A, and the kinetic energy K at the same value of s. (In a full quantum-mechanical analysis the concepts of angular frequency and amplitude require reinterpretation. Nevertheless there remain elements of the classical picture. For example, larger amplitude corresponds to a higher probability of observing a large stretch.)

Ex. 6.10 What is the energy of the photon emitted when a harmonic oscillator with stiffness k_s and mass m loses 3 of its quanta of energy?

6.5.4 The ground state of a quantum oscillator

The lowest possible quantized energy E_0 is not zero relative to the bottom of the potential curve, but is $\frac{1}{2}(\hbar\sqrt{k_s/m})$ above the bottom of the well. This is closely related to the "Heisenberg uncertainty principle." When applied to the harmonic oscillator, this principle implies that if you knew that the mass on the spring were exactly at the center of the potential, its speed would be completely undetermined, whereas if you knew that the speed of the mass were exactly zero, the position of the mass could be literally anywhere! The minimum energy $\frac{1}{2}(\hbar\sqrt{k_s/m})$ above the bottom of the well can be viewed as a kind of a compromise, with neither the speed nor the position fixed at zero. The minimum energy is called the ground state of the system.

In most of our work with the quantized oscillator we will measure energy from the ground state rather than from the bottom of the potential well, so the offset from the bottom of the well won't come into our calculations.

6.5.5 Energy levels for the interatomic potential energy

If we model the atoms in a solid as simple harmonic oscillators, whose potential energy curve is a parabola, then the energy levels of these systems are uniformly spaced. However, if we use a more realistic interatomic potential energy such as is shown in Figure 6.18, the energy levels of the system follow

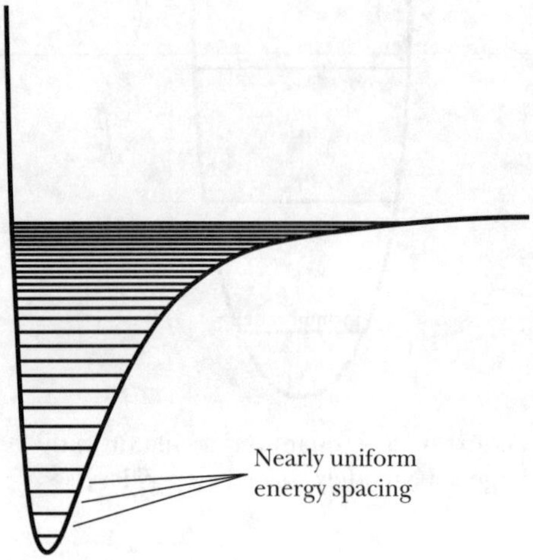

Figure 6.18 Energy levels for the interatomic potential energy, which describes the interaction of two neighboring atoms. This diagram is not to scale; the number of vibrational energy levels in the uniform region may be very large.

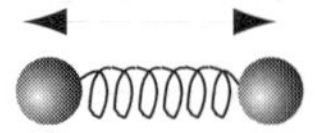

Figure 6.19 The vibrational energy of a diatomic molecule is quantized.

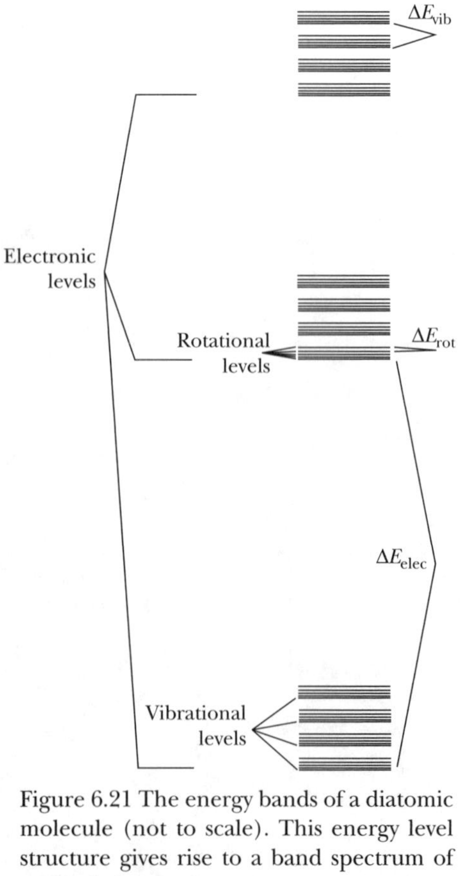

Figure 6.21 The energy bands of a diatomic molecule (not to scale). This energy level structure gives rise to a band spectrum of emitted photons.

a more complex pattern. The lowest energy levels are nearly evenly spaced, corresponding to the fact that for small amplitude the bottom of the potential energy curve can be fit rather well by a curve of the form $\frac{1}{2}k_s s^2 + U_0$. However, for large energies the energy spacing is quite small. And of course there are no bound states for energies larger than the binding energy. The spacing of the vibrational energy levels of a diatomic molecule (Figure 6.19) typically corresponds to photons in the infrared region of the spectrum (see Problem 6.9).

Energy spacing in various systems

Most quantum objects have energy level spacings that decrease as you go to higher levels. The harmonic oscillator potential energy is unusual in leading to uniform energy level spacing. Even less common is the situation where the energy level spacing actually increases with higher energy. One such exception is the so-called "square well" with very steep sides and a flat bottom. The energy levels in a square well actually get farther and farther apart for higher energies.

6.5.6 Rotational energy levels

A molecule in the gas phase has not only electronic energy and vibrational energy, but rotational energy as well. Like electronic and vibrational energy levels, rotational energy levels are also discrete—only certain rotational energies are permitted. As we will see in Chapter 9, the angular momentum of a molecule is quantized, and this quantization gives rise to discrete rotational energy levels. In this case all the energy of rotation is kinetic energy—there is no potential energy involved. Rotational energy levels for a diatomic molecule such as N_2 are shown in Figure 6.20. Note that the spacing between levels actually increases with increasing energy, which is the opposite of the trend seen in many atomic systems. Rotational energy levels are typically close together; transitions between these levels usually involve photons in the microwave region of the electromagnetic spectrum.

6.6 Case study: The diatomic molecule

Analyzing a diatomic molecule such as N_2 provides a good review of three important types of quantized energy levels: electronic, vibrational, and rotational.

Only certain configurations of the electron clouds in a diatomic molecule are possible, corresponding to discrete electronic energy levels. The spacings of the low-lying molecular electronic energy levels are typically in the range of 1 eV or more, like the spacings of the low-lying energy levels in the hydrogen atom.

For each major configuration of the electron clouds, many different vibrational energy states are possible, with energy spacings of the order of 10^{-2} eV (see Problem 6.9).

For each major configuration of the electron clouds, and each vibrational energy state, many different rotational kinetic energy states are possible, with energy spacings of the order of 10^{-4} eV, as we will see in Chapter 9.

These three different families of energy levels, with their very different energy level spacings, give rise to energy "bands" and a distinctive "band" spectrum of emitted photons. See Figure 6.21, which is not to scale due to the large differences in level spacing for the three kinds of quantized energy levels. An emitted photon can have the energy of an electronic transition plus a smaller-energy vibrational transition plus an even smaller-energy rotational transition. The various possible energy differences give rise to "bands" in the observed emission spectrum.

6.7 Nuclear energy levels

Just as atoms have discrete states corresponding to allowed configurations of the electron cloud, so nuclei have discrete states corresponding to allowed configurations of the nucleons (protons and neutrons). An excited nucleus can drop to a lower state with the emission of a photon. The energy spacing of nuclear levels is very large, of the order of 10^6 eV (1 MeV), so the emitted photons have energies in the range of millions of electron-volts. Such energetic photons are called "gamma rays."

Ex. 6.11 A gamma ray with energy greater than or equal to 2.2 MeV can dissociate a deuteron (the nucleus of deuterium, heavy hydrogen) into its constituents, a proton and a neutron. What is the nuclear binding energy of the deuteron?

6.8 Hadronic energy levels

Atoms have energy levels associated with the configurations of electrons in the atom, and nuclei have energy levels associated with the configurations of nucleons in the nucleus. Hadrons are particles made of quarks. In a simple model we represent baryons (particles like protons) as a combination of three quarks, and mesons (particles like pions) as a combination of a quark and an antiquark. Different configurations of quarks correspond to different energy levels of the multiquark system. These different configurations are the different hadrons we observe, and the corresponding energy levels are the observed rest energies of these particles. The energy spacing is of the order of hundreds of MeV, or about 10^8 eV. (Leptons, which include electrons, muons, and neutrinos, are not made of quarks.)

An example of a hadron is the Δ^+ particle, which has a rest energy of 1232 MeV. It can be considered an excited state of the three-quark system whose ground state is the proton, whose rest energy is 938 MeV. The Δ^+ particle can decay into a proton with the emission of a very high energy photon. However, this "electromagnetic" decay occurs in only about 0.5% of the decays. In most cases the Δ^+ decays into a proton plus a pion, a strong-interaction decay which has a much higher rate, corresponding to the much stronger interaction.

An excited electronic state of hydrogen is still called hydrogen, but for historical reasons we continue to give excited quark states distinctive names, such as Δ^+.

Ex. 6.12 What is the approximate energy of the photon emitted when the Δ^+ decays electromagnetically?

6.9 Comparison of energy level spacings

Here is a summary of typical energy scales for various quantum objects:

Type of state	Typical energy level spacing
hadronic	10^8 eV
nuclear	10^6 eV

Type of state	Typical energy level spacing
electronic (atoms, molecules)	1 eV
vibrational (molecules)	10^{-2} eV
rotational (molecules)	10^{-4} eV

It is important to distinguish among hadronic, nuclear, electronic, vibrational, rotational, and other kinds of quantized energy situations. We can generalize the main issues to any kind of generic quantum object. The key features of all these different kinds of objects are these:

- discrete energy levels
- discrete emissions whose energies equal differences in energy levels

Some of the homework problems at the end of the chapter ask you to analyze generic objects, independent of the details of what gave rise to the particular discrete energy levels.

6.10 *Case study: How a laser works

A laser offers an interesting case study of the interaction of light and atomic matter. We will give a basic introduction to lasers, but we can only scratch the surface of a vast topic which continues to be an active field of research. Laser action depends on a process called "stimulated emission" and on creating an "inverted population" of quantum states, which can lead to a chain reaction and produce a special kind of light, called "coherent" light.

Stimulated emission

If an atom can be placed into an excited state, that is, a state of higher energy than the ground state, normally it drops quickly to a lower energy state and emits a photon in the process. For example, in a neon sign high-energy electrons collide with ground-state neon atoms and raise them to an excited state. An excited neon atom quickly drops back to the ground state, emitting a photon whose energy is the difference in energy between the two quantized energy levels, and you see the characteristic orange color corresponding to these photons.

We call this emission process "spontaneous emission." Spontaneous emission happens at an unpredictable time. All that quantum mechanics can predict is the *probability* that the excited atom will drop to the ground state in the next small time interval. It is truly not possible to predict the exact time of the emission, just as it is impossible to predict exactly when a free neutron will decay into a proton, electron, and antineutrino.

A collection of excited atoms such as those in a neon sign emit spontaneously, at random times. In the full quantum-mechanical description of light, light has both the properties of a particle (the photon picture) and the properties of a wave; see Section 6.11. Emissions at random times have random phases in a wave description: sine, cosine, and all phases between a sine and a cosine. We say that such light is "incoherent."

Figure 6.22 Stimulated emission produces two identical photons from one.

"Stimulated emission" of light is a different process which produces coherent light. Consider an atom in an energy level E above the ground state. A photon of this same energy E can interact with the atom in a remarkable way, causing the atom to drop immediately to the ground state with the emission of a second, "clone" photon. Where there had been one photon of energy E, there are now two photons of exactly the same energy E, and in a wave description they have exactly the same phase (Figure 6.22). That is, the two waves are both sines, or both cosines, or both with exactly the same phase somewhere between a sine and a cosine. Such light is called coherent light, and it has valuable properties for such applications as making holograms.

Chain reaction

The phenomenon of stimulated emission can be exploited to create a chain reaction. Start with one photon of the right energy E to match the energy difference between the ground state and an excited state of the atoms in the system. If that photon causes stimulated emission from an excited atom, there are now two identical photons, both capable of triggering stimulated emission of other excited atoms, which yields four identical photons, then 8, 16, 32, 64, 128, 256, 512, 1024, etc.

However, the chain reaction won't proceed if all of the photons escape from the system, so some kind of containment mechanism is required. In a common type of gas laser the gas is in a long tube with mirrors at both ends (Figure 6.23). Photons that happen to be going in the direction of the tube are continually reflected back and forth through the tube, which increases the probability that they will interact with excited atoms and cause stimulated emission. Since the stimulated-emission photons are clones, they too are headed in the direction of the tube and they too are likely to interact. One of the mirrors is deliberately made to be an imperfect reflector, so a fraction of the photons leaks out and provides a coherent beam of light.

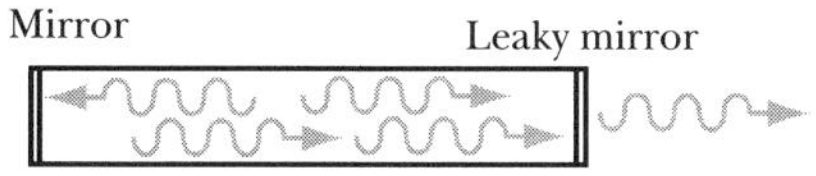

Figure 6.23 A gas laser with mirrors at the ends. Photons leak through one of the mirrors.

Unfortunately, the stimulated-emission photons also have just the right energy E to be absorbed by an atom in the ground state, in which case they can no longer contribute to the chain reaction. For this reason we need an "inverted population," with few atoms in the ground state to absorb the photons and stop the chain reaction. We discuss this issue next.

Inverted population

As we have discussed earlier in this chapter, in a normal collection of atoms more atoms are in the ground state than in any other state, and the number of atoms decreases in each successively higher state. In fact, for a collection of atoms in thermal equilibrium the fraction of atoms in a particular energy state decreases exponentially with the energy above the ground state, as we will see in Chapter 10.

By clever means it is possible to invert this population scheme, so that there are more atoms in an excited state than in the ground state (this is not a state of thermal equilibrium). An inverted population is crucial to laser action, so that stimulated emission with cloning of photons will dominate over simple absorption of photons by atoms in the ground state.

There are many different schemes for creating an inverted population. For example, the scheme used in the common helium-neon laser is to "pump" atoms to a high energy level (Figure 6.24). There are diverse energy input schemes, including electric discharges and powerful flashes of ordinary light. Atoms in this high energy level rapidly drop to an intermediate energy level, higher than the ground state. This intermediate state happens to have a rather long lifetime for spontaneous emission, so that it is possible to sustain an inverted population, with more atoms in the intermediate state than in the ground state. Stimulated emission drives these excited atoms down to the ground state, from which they are again pumped up to the high-energy state. Spontaneous emission from the high-energy state to the intermediate state maintains the inverted population.

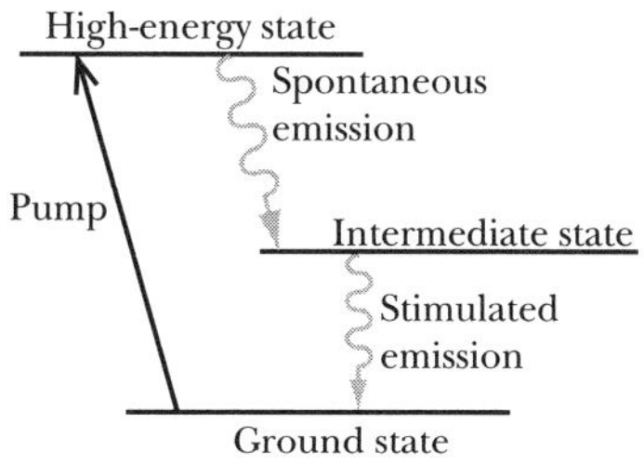

Figure 6.24 A three-state scheme for sustaining an inverted population.

6.11 *Wavelength of light

In this chapter we have emphasized the particle model of light, in terms of photons and their energy. For completeness we will mention a connection between the particle model and the wave model of light. Light consisting of photons whose energy is E can also be treated as a wave, and the energy E and wavelength λ are related as follows:

ENERGY AND WAVELENGTH OF LIGHT

$$E_{\text{photon}} = \frac{hc}{\lambda_{\text{light}}}$$

where h is Planck's constant. A model for light that incorporates both the features of a wave and a localized particle is a "wave packet," a wave whose magnitude is nonzero only in a small region of space (Figure 6.25). We will return to this topic in Chapter 24.

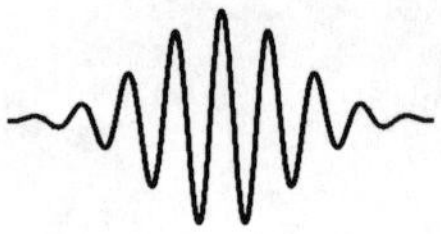

Figure 6.25 A wave packet is a travelling wave of finite spatial extent.

6.12 Summary

Fundamental Physical Principles

At the microscopic level the energy of bound states is discrete, not continuous. (Unbound states have a continuous range of energies.)

New concepts

Quantized systems can have only certain energies. They can emit or absorb photons when they change energy levels.

The temperature is a measure of the average energy of a collection of objects, so high temperature is associated with significant probability of finding particles in any of many states, from the ground state up to a high energy level. Very low temperature is associated with little probability of the particles being in any state other than the ground state.

Results

Discrete hydrogen atom energy levels:

$E_N = -(13.6\text{ eV})/N^2$, where N is a nonzero positive integer (1, 2, 3...)

Discrete harmonic oscillator energy levels:

$$E_N = N\hbar\omega_0 + E_0\,,\ N = 0, 1, 2, \text{etc., where } \omega_0 = \sqrt{\frac{k_s}{m}}$$

Planck's constant $h = 6.6\times10^{-34}$ joule · second

$$\hbar = \frac{h}{2\pi} = 1.05\times10^{-34}\text{ joule} \cdot \text{second}$$

$1\text{ eV} = 1.6\times10^{-19}\text{ J}$

Photon energy and wavelength: $E_{\text{photon}} = \dfrac{hc}{\lambda_{\text{light}}}$

6.13 Review questions

Energy quantization

RQ 6.1 If you double the amplitude, what happens to the frequency in a classical (non-quantum) harmonic oscillator? In a quantum harmonic oscillator?

RQ 6.2 If you try to increase the amplitude of a quantum harmonic oscillator by adding an amount of energy

$$\frac{1}{2}\hbar\sqrt{\frac{k_s}{m}}$$

the amplitude doesn't increase. Why not?

RQ 6.3 When starlight passes through a cold cloud of hydrogen gas, some hydrogen atoms absorb energy, then reradiate it in all directions. As a result, a spectrum of the star shows dark absorption lines at the energies for which less energy from the star reaches us. How does the spectrum of dark absorption lines for very cold hydrogen differ from the spectrum of bright emission lines from very hot hydrogen?

6.14 Problems

Problem 6.1 Determining energy levels
Suppose we have reason to suspect that a certain quantum object has only three quantum states. When we excite such an object we observe that it emits electromagnetic radiation of three different energies: 2.48 eV (green), 1.91 eV (orange), and 0.57 eV (infrared).

(a) Propose two possible energy-level schemes for this system.

(b) Explain how to use an absorption measurement to distinguish between the two proposed schemes.

Problem 6.2 Emission and absorption
A certain material is kept at very low temperature. It is observed that when photons with energies between 0.2 and 0.9 eV strike the material, only photons of 0.4 eV and 0.7 eV are absorbed. Next the material is warmed up so that it starts to emit photons. When it has been warmed up enough that 0.7 eV photons begin to be emitted, what other photon energies are also observed to be emitted by the material? Explain briefly.

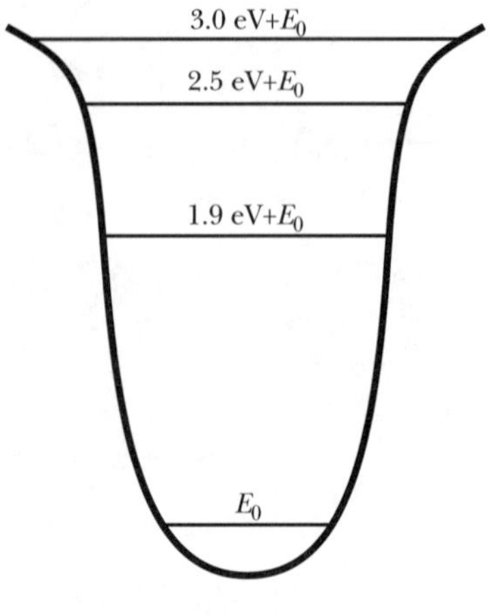

Figure 6.26 Potential energy diagram for Problem 6.3.

Problem 6.3 Predicting a spectrum
Suppose a hypothetical object has just four quantum states, with the energies shown on the potential energy diagram in Figure 6.26.

(a) Suppose that the temperature is high enough that in a material containing many such objects, at any instant some objects are found in all of these states. What are all the energies of photons that could be strongly emitted by the material? (In actual quantum objects there are often "selection rules" that forbid certain emissions even though there is enough energy; assume that there are no such restrictions here.)

(b) If the temperature is very low and electromagnetic radiation with a wide range of energies is passed through the material, what will be the energies of photons corresponding to missing ("dark") lines in the spectrum? (Assume that the detector is sensitive to a wide range of photon energies, not just energies in the visible region.)

Problem 6.4 Emission from hot metal

A hot bar of iron glows a dull red. Using our simple model of a solid, answer the following questions, explaining in detail the processes involved. The mass of one mole of iron is 56 grams.

(a) What is the energy of the lowest-energy spectral emission line? Give a numerical value.

(b) What is the approximate energy of the highest-energy spectral emission line? Give a numerical value.

(c) What is the quantum number of the highest-energy occupied state?

(d) Predict the energies of two other lines in the emission spectrum of the glowing iron bar.

(Note: our simple model is too simple—the actual spectrum is more complicated. However, this simple analysis gets at some important aspects of the phenomenon.)

Problem 6.5 Potential energy in a molecule

(a) Figure 6.27 is a graph of potential energy vs. interatomic distance for a particular molecule. What is the direction of the associated force at location *A*? At location *B*? At location *C*? Rank the magnitude of the force at locations *A*, *B*, and *C* (that is, which is greatest, which is smallest, are any of these equal to each other). For the energy level shown on the graph, draw a line whose height is the kinetic energy when the system is at location *D*.

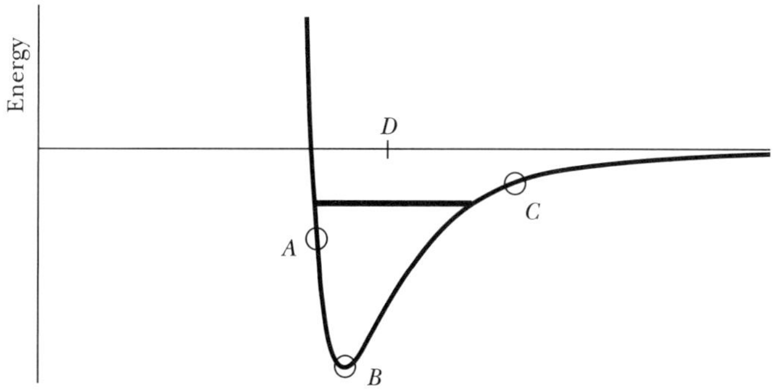

Figure 6.27 Potential energy vs. interatomic distance for a molecule (Problem 6.5a).

(b) Figure 6.28 shows all of the quantized energies (bound states) for one of these molecules. The energy for each state is given on the graph, in electron-volts ($1\ \text{eV} = 1.6\times10^{-19}$ J). How much energy is required to break a molecule apart, if it is initially in the ground state? (Note that the final state must be an unbound state; the unbound states are not quantized.)

(c) If the temperature is high enough, in a collection of these molecules there will be at all times some molecules in each of these states, and light will be emitted. What are the energies in electron-volts of the emitted light?

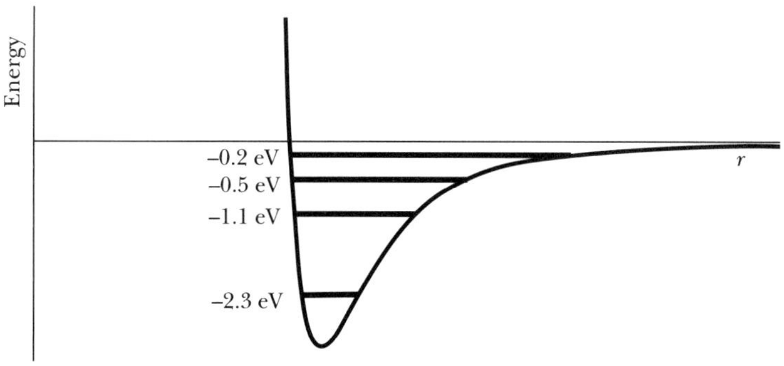

Figure 6.28 Quantized molecular vibrational energies (Problem 6.5b).

(d) The "inertial" mass of the molecule is the mass that appears in Newton's second law, and which determines how much acceleration will result from applying a given force. Compare the inertial mass of a molecule in the top energy state and the inertial mass of a molecule in the ground state. If there is a difference, briefly explain why and calculate the difference. If there isn't a difference, briefly explain why not.

Problem 6.6 A microscopic oscillator

Consider a microscopic spring-mass system whose spring stiffness is 50 N/m, and the mass is 4×10^{-26} kg.

(a) What is the smallest amount of vibrational energy that can be added to this system?

(b) What is the difference in mass (if any) of the microscopic oscillator between being in the ground state and being in the first excited state?

(c) In a collection of these microscopic oscillators, the temperature is high enough that the ground state and the first three excited states are occupied. What are possible energies of photons emitted by these oscillators?

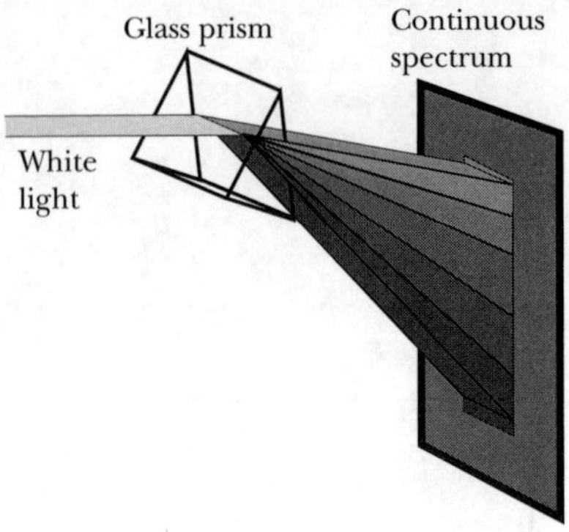

Figure 6.29 Spectrum of white light.

Problem 6.7 The spectrum of atomic hydrogen

The eye is sensitive to photons with energies in the range from about 1.8 eV, corresponding to red light, to about 3.1 eV, corresponding to violet light. White light is a mixture of all the energies in the visible region. If you shine white light through a slit onto a glass prism, you can produce a rainbow spectrum on a screen, because the prism bends different colors of light by different amounts (Figure 6.29).

If you replace the source of white light with an electric-discharge lamp containing excited atomic hydrogen, you will see only a few lines in the spectrum, rather than a continuous rainbow (Figure 6.30).

Predict how many lines will be seen in the visible spectrum of atomic hydrogen. Specify the atomic transitions that are responsible for these lines, given that the energies of the quantized states in atomic hydrogen are given by $E_N = -(13.6 \text{ eV})/N^2$, where N is a nonzero positive integer (1, 2, 3...).

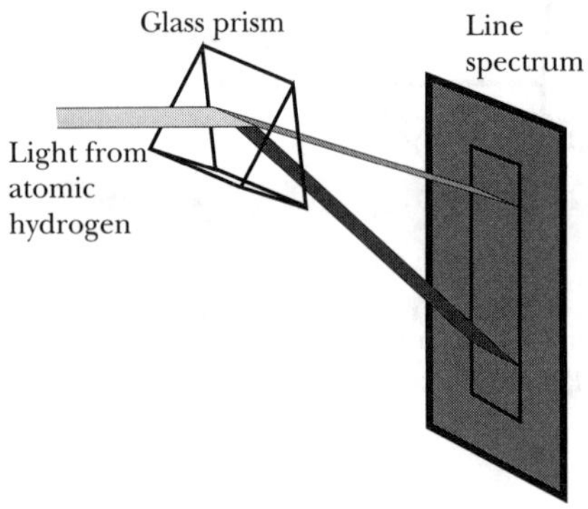

Figure 6.30 Atomic emission spectrum (Problem 6.7).

Problem 6.8 An experiment with quantum objects

Some material consisting of a collection of microscopic objects is kept at a high temperature. A photon detector capable of detecting photon energies from infrared through ultraviolet observes photons emitted with energies of 0.3 eV, 0.5 eV, 0.8 eV, 2.0 eV, 2.5 eV, and 2.8 eV. These are the only photon energies observed.

(a) Draw and label a possible energy-level diagram for one of the microscopic objects, which has 4 bound states. On the diagram, indicate the transitions corresponding to the emitted photons. Explain briefly.

(b) Would a spring-mass model be a good model for these microscopic objects? Why or why not?

(c) The material is now cooled down to a very low temperature, and the photon detector stops detecting photon emissions. Next a beam of light with a continuous range of energies from infrared through ultraviolet shines on the material, and the photon detector observes the beam of light after it passes through the material. What photon energies in this beam of light are observed to be significantly reduced in intensity ("dark absorption lines")? Explain briefly.

Problem 6.9 Quantized energy for a diatomic molecule

(a) A nitrogen molecule, N_2, can be considered to be a quantized harmonic oscillator, with quantized vibrational energy levels that are evenly spaced. Make a rough estimate of this uniform energy spacing in electron volts (where 1 eV = 1.6×10^{-19} J). You will need to make some rough estimates of atomic properties based on prior work. For comparison with the spacing of these vibrational energy states, note that the spacing between quantized energy levels for "electronic" states such as in atomic hydrogen is of the order of several electron volts.

(b) List several photon energies that would be emitted if a number of these vibrational energy levels were occupied due to electron excitation. To what region of the spectrum (x-ray, visible, microwave, etc.) do these photon belong? (See Figure 6.2 on page 272.)

6.15 Answers to exercises

6.1 (page 271) 13.6 eV

6.2 (page 272) About 2.5 eV

6.3 (page 272) About 3.5×10^{21} photons per square meter per second.

6.4 (page 273) Six different photon energies, corresponding to transitions from 4 to 3, 4 to 2, 4 to 1, 3 to 2, 3 to 1, and 2 to 1.

6.5 (page 274) Just one dark line at 2.4 eV

6.7 (page 278) 0.4 eV (three transitions), 0.8 eV (two transitions), and 1.2 eV (one transition)

6.8 (page 279) $N \approx 10^{31}$. A system in a very high quantum state behaves like a classical system.

6.9 (page 279) Upper level has same ω, larger A, larger K at same s.

6.10 (page 279) $3\hbar\sqrt{\frac{k_s}{m}}$

6.11 (page 281) 2.2 MeV

6.12 (page 281) 294 MeV

Chapter 7

Multiparticle Systems

Chapter 7

Multiparticle Systems

At times we have treated a block of metal as though it were a single particle. At other times we have modeled a block as being made up of a huge number of atoms connected by "springs" and in continual motion around their equilibrium positions.

When is it valid to treat a complex, multiparticle system as though it were one big particle? Why is this *ever* valid? In this chapter we will look in detail at multiparticle systems.

7.1 The motion of a multiparticle system

There are some major complications in applying the Momentum Principle to systems consisting of many particles. Suppose you pull on a thin wire that is attached to a metal block, and you accelerate the block across a low-friction surface (Figure 7.1).

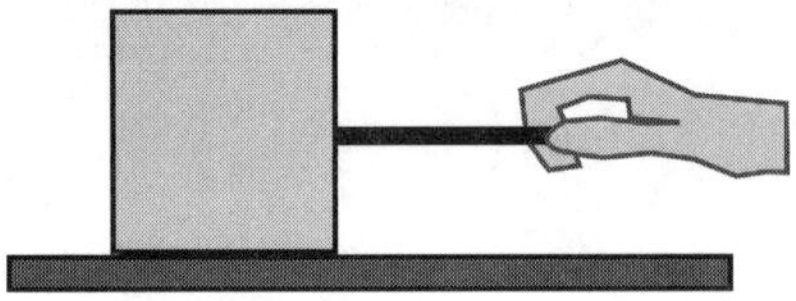

Figure 7.1 You pull on a wire attached to a block.

? Actually, you are not touching the block and are not exerting a force on the atoms in the block. How do these atoms know that they should accelerate?

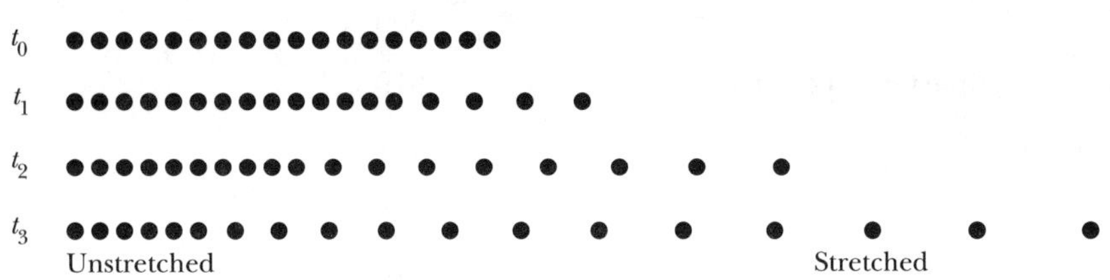

Figure 7.2 Strain in the wire increases with time.

When you first start pulling on your end of the wire, you very slightly lengthen the interatomic bonds between neighboring atoms in the nearby section of the wire (Figure 7.2). As a result of your pulling, these atoms stretch the bonds of their neighbors, and very quickly this new interatomic bond-stretching propagates all the way to the other end of the wire, and the whole wire is then in tension.

This process is very fast, but it is not instantaneous. These changes in the average lengths of the interatomic bonds propagate at the speed of sound in the material. (See page 94 in Chapter 3.)

Eventually the interatomic bonds at the left end of the wire will be stretched. The left-most atoms, which are attached to the metal block, will pull on atoms in the block. Now the block itself begins to stretch. As with the wire, initially, only those atoms of the block that are in direct contact with the wire are pulled to the right, but quickly (at the speed of sound in the material that the block is made of) these effects propagate throughout the block. Eventually it must be the case that all of the interatomic bonds in the block are stretched, but not equally.

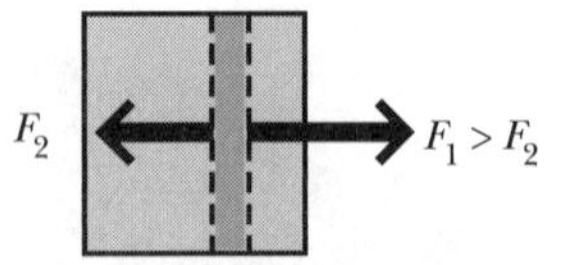

Figure 7.3 Forces acting on a slice of the block.

Consider a slice of the block, indicated by dashed lines in Figure 7.3. Since this slice is accelerating to the right, the interatomic forces acting on the right face of this slice (F_1) must be larger than those on the left face (F_2). This requires that the interatomic bonds on the right be stretched more than those on the left. While the block accelerates to the right, the pattern of strain in the block must look something like Figure 7.4 (strain is $\Delta L/L$, or fractional stretch). Actually, even this description is oversimplified, because there will be maximum strain near the point where the wire is attached, and the lines of strain will be somewhat curved.

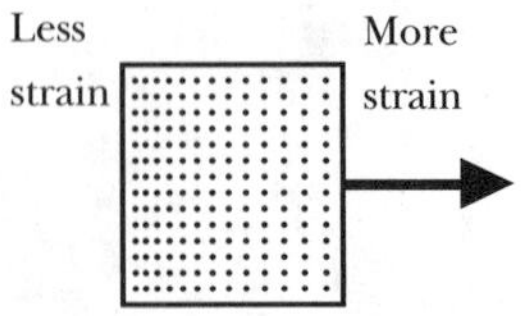

Figure 7.4 Pattern of strain in a block accelerating to the right.

For a metal block, the strain is normally too small to see with the unaided eye. Nevertheless, the situation is pretty complicated! And there is a similar gradient of strain in the wire. There is also random thermal motion of the atoms within the block. Given all this complexity, what hope is there of treating the block as though it were just one big, rigid particle?

Ex. 7.1 If you let go and allow the block to coast along at nearly constant speed, what must happen to the strain throughout the block? What is the mechanism for that happening at the atomic level?

Energy in multiparticle systems

Just as there are great complexities in dealing with forces in multiparticle systems, so also there are puzzling aspects of work and energy. For example, when you jump upwards the floor pushes up on your feet, yet this force does no work, because there is no displacement of the point of contact (Figure 7.5). However, it is clear that your kinetic energy has increased. Evidently there are some subtle issues here that we need to address.

Figure 7.5 The floor does no work on the jumper, yet K increases.

7.2 The Momentum Principle for multiparticle systems

In Chapter 2 (page 43) we derived an equation that predicts the overall motion of a complex multiparticle system, in terms of its total momentum $\vec{P}_{tot}$:

THE MOMENTUM PRINCIPLE FOR MULTIPARTICLE SYSTEMS

$$\frac{d\vec{P}_{tot}}{dt} = \vec{F}_{net,ext}$$

The basis for this result is that internal forces between pairs of particles cancel for gravitational and electric forces (reciprocity; Newton's third law). This result is what makes it possible for many purposes to treat a complex system as though it were a simple point particle. A concept that simplifies analyses of this kind is the "center of mass," which we discuss next.

7.2.1 Center of mass

For a system of particles moving slowly compared to the speed of light, with total mass M, we can show that $\vec{P}_{tot} = M\vec{v}_{cm}$, where $\vec{v}_{cm}$ is the velocity of the "center of mass" of the system. Since $d\vec{P}_{tot}/dt = \vec{F}_{net,ext}$, we can treat a complicated system as though it were a single particle of mass M, located at the center of mass of the system.

The center of mass can be thought of as a weighted average of the locations of all the particles in the system. In many cases it is simply the geometrical center of a system. For example, the center of mass of a disk or sphere or cube is the geometrical center of the object, if the object has uniform density. Let $\vec{r}_1$ be the location (relative to some arbitrary origin) of mass m_1, $\vec{r}_2$ be the location of mass m_2, etc., as shown in Figure 7.6. The center of mass is located at position $\vec{r}_{cm}$, defined like this:

CENTER OF MASS

$$\vec{r}_{cm} \equiv \frac{m_1\vec{r}_1 + m_2\vec{r}_2 + m_3\vec{r}_3 + \ldots}{m_1 + m_2 + m_3 + \ldots}$$

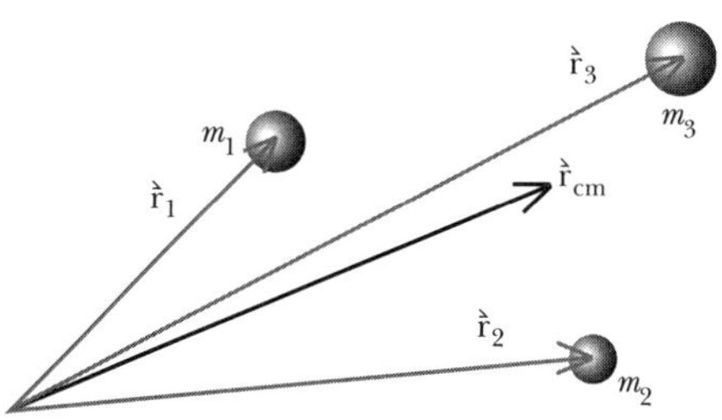

Figure 7.6 Location of the center of mass of the three-particle system.

Let's see whether this makes sense as a definition for the location of the center of mass. We'll consider several simple concrete situations and check that this definition yields results that are consistent with what we would intuitively think of as the center of mass.

? Consider the case where there are three masses very close to each other. Does the definition

$$\vec{r}_{cm} \equiv \frac{m_1\vec{r}_1 + m_2\vec{r}_2 + m_3\vec{r}_3 + \ldots}{m_1 + m_2 + m_3 + \ldots}$$

make sense in this case?

In this case, $\vec{r}_1 = \vec{r}_2 = \vec{r}_3$, and the definition of the location of the center of mass leads to the result that $\vec{r}_{cm} = \vec{r}_1$, which makes sense. If all three masses are located at the origin, then according to the definition the center of mass is also at the origin, which makes sense.

? Consider the case where there are two particles, with equal masses, located at <–5, 0, 0> and <+5, 0, 0>. Does the definition make sense in that case?

In this case we find that $\vec{r}_{cm} = 0$, which makes sense; the center of mass is equidistant between the two equal masses.

? What if there is a very small mass located at <–5, 0, 0> and a very large mass located at <+5, 0, 0>. Does the definition make sense in that case?

In this case the formula yields $\vec{r}_{cm} \approx \langle +5, 0, 0 \rangle$, corresponding to the fact that almost all of the total mass is concentrated at that location.

If M is the total mass of the system, we can write these two equivalent equations:

$$\vec{r}_{cm} = \frac{m_1\vec{r}_1 + m_2\vec{r}_2 + m_3\vec{r}_3 + \ldots}{M}$$

$$M\vec{r}_{cm} = m_1\vec{r}_1 + m_2\vec{r}_2 + m_3\vec{r}_3 + \ldots$$

Center of mass of several large objects

Often we need to know the location of the center of mass of a system consisting of several large objects, each of whose center of mass is known. In Figure 7.7, $\vec{R}_1$ is the location (relative to the origin) of the center of mass of the object whose total mass is M_1, etc.

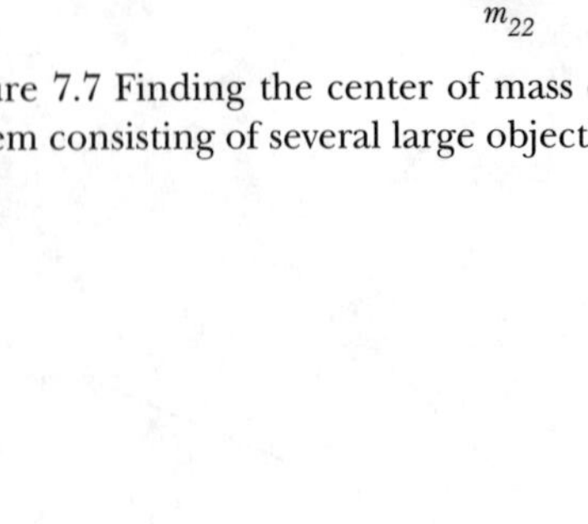

Figure 7.7 Finding the center of mass of a system consisting of several large objects.

For each of the large objects we have an equation for its center of mass of the form $M_1\vec{R}_1 = m_{11}\vec{r}_{11} + m_{12}\vec{r}_{12} + \ldots$, $M_2\vec{R}_2 = m_{21}\vec{r}_{21} + m_{22}\vec{r}_{22} + \ldots$, etc.

Group terms in the calculation of the location of the center of mass of the total system:

$$\vec{r}_{cm} \equiv \frac{(m_{11}\vec{r}_{11} + m_{12}\vec{r}_{12}) + \ldots + (m_{21}\vec{r}_{21} + m_{22}\vec{r}_{22}) + \ldots}{(m_{11} + m_{12} + \ldots) + (m_{21} + m_{22} + \ldots) + \ldots}$$

$$\vec{r}_{cm} = \frac{M_1\vec{R}_1 + M_2\vec{R}_2 + M_3\vec{R}_3 + \ldots}{M_1 + M_2 + M_3 + \ldots}$$

This looks just like the equation for finding the center of mass of a bunch of atoms, except that the $\vec{R}$ vectors refer to the centers of mass of large objects.

Ex. 7.2 Determine the location of the center of mass of a "T" whose thin vertical and horizontal members have the same length L and the same mass M. Use the formal definition to find the x and y

coordinates, and check your result by doing the calculation with respect to two different origins, one at the bottom of the vertical member and one at the left end of the horizontal member.

Ex. 7.3 Relative to an origin at the center of the Earth, where is the center of mass of the Earth-Moon system? The mass of the Earth is 6×10^{24} kg, the mass of the Moon is 7×10^{22} kg, and the distance from the center of the Earth to the center of the Moon is 4×10^{8} m. The radius of the Earth is 6400 km. One can show that the Earth and Moon orbit each other around their center of mass (for example, see Problem 7.4).

7.2.2 Total momentum and center of mass motion

Now that we have introduced the concept of center of mass, we can make a useful connection between the total momentum and the motion of the center of mass. The velocity of the center of mass of a system is obtained by taking the time derivative of the definition of the position of the center of mass, where $\vec{v}_1 = d\vec{r}_1/dt$, etc. Take the important case where all the individual masses are constants. In that case, since

$$M\vec{r}_{cm} = m_1\vec{r}_1 + m_2\vec{r}_2 + m_3\vec{r}_3 + \dots$$

we have this:

$$M\vec{v}_{cm} = M\frac{d\vec{r}_{cm}}{dt} = m_1\vec{v}_1 + m_2\vec{v}_2 + m_3\vec{v}_3 + \dots \text{ (constant masses)}$$

But $(m_1\vec{v}_1 + m_2\vec{v}_2 + m_3\vec{v}_3 + \dots)$ is the total (nonrelativistic) momentum of the system, so we have shown that

$$\vec{P}_{tot} = M\vec{v}_{cm} \text{ (if } v \ll c\text{, and constant mass)}$$

In summary, we restate the important results:

THE MOMENTUM PRINCIPLE FOR MULTIPARTICLE SYSTEMS

$$\frac{d\vec{P}_{tot}}{dt} = \vec{F}_{net,ext} \text{ (in general)}$$

$$\vec{P}_{tot} = M\vec{v}_{cm} \text{ (if } v \ll c\text{, and constant mass)}$$

For example, if you apply an external force F to one side of a block (through the tension in a wire, for example), the center of mass of the block moves as though that force were applied at the center of mass, producing an acceleration of the center of mass $dv_{cm}/dt = F/M$, even though the detailed motions of the many atoms in the block aren't simple (different amounts of strain in different parts of the block, thermal agitation of the atoms, etc.).

? A high diver jumps outward, then tucks into a tight ball and spins rapidly, then straightens out before hitting the water. What is the path of the diver's center of mass?

Although the diver's overall motion is very complex, the center of mass of the diver must move like that of a simple projectile! We do not yet have enough tools to analyze the diver's complicated rotation and stretching relative to the center of mass, yet we can easily analyze the motion of the center of mass.

7.2.3 Application: An ice skater pushes off from a wall

All of this may sound quite plausible, but it can lead to some rather odd consequences. Consider a woman on ice skates who pushes off from a wall with a nearly constant force F_N (normal to the wall) and moves backwards with increasing speed. Her center of mass is marked on each frame of the following sequence:

What would we predict for the motion of the skater's center of mass? If M is her total mass, and if we can assume that the frictional force of the ice on the skates is negligibly small, we predict this from the force diagram (Figure 7.8):

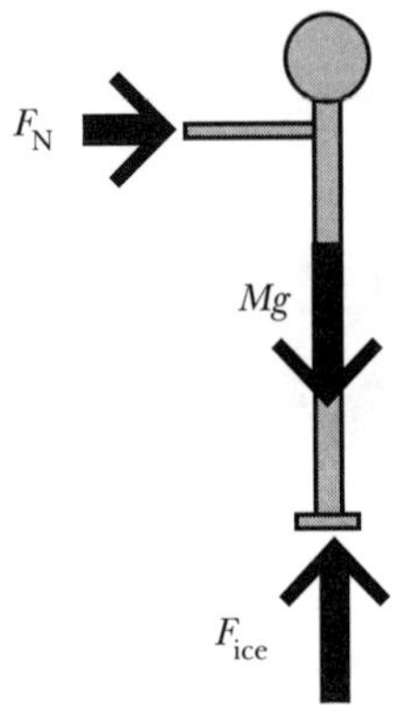

Figure 7.8 Force diagram for the skater.

$$\frac{dP_{tot}}{dt} = F_N$$

What's odd about this simple and intuitively appealing result is that during the time that the wall exerts a force F_N on the skater's hand, her hand does not move! (Note that the work done by the force exerted by the wall is zero; more about this later.) The place where the force F_N is applied is very far away from the mathematical point (the center of mass) whose rate of change of speed we predict. It is only because the electric interactions between neighboring atoms in the skater's body obey the principle of reciprocity (Newton's third law) that these internal forces cancel and lead to a simple prediction for the motion of the center of mass:

$$\frac{d\vec{P}}{dt} = \vec{F}_{net,ext}$$

Note that as the skater straightens her arms, the location in her body of the center of mass shifts toward her hands a little; it is not a point fixed at some place in her body. The motion that we predict is the motion of this mathematical center of mass point.

7.2.4 Application: Pull on two hockey pucks

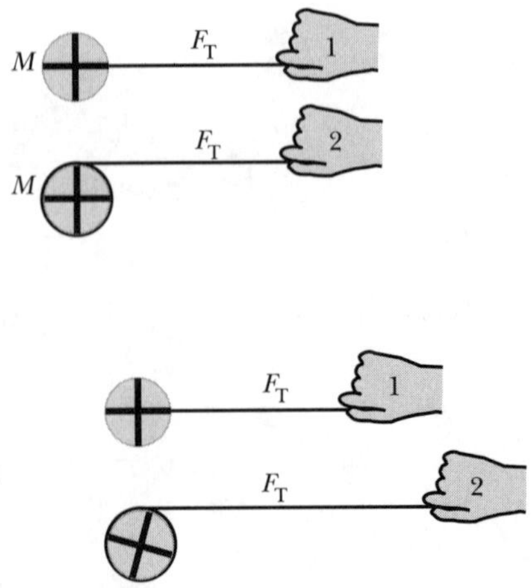

Figure 7.9 One puck is pulled by a string attached to its center. The other puck is pulled by a string wrapped around its edge, which unrolls as the puck is pulled.

We'll look at another odd consequence of this multiparticle version of the Momentum Principle. Tie a string to the center of a hockey puck, and wrap another string around the outside of a second hockey puck, as shown in Figure 7.9. Then two people pull on the two strings so that both strings have the same tension F_T. As shown in the figure, in a time Δt hand 2 pulling the bottom puck has traveled farther, because the string has unrolled from the puck a bit.

The pucks have the same mass M. Assume that friction with the ice rink is negligible. It would be reasonable to suppose that it would make a difference where you attach the string to the puck.

? What will be the velocity (direction and magnitude) of the center of the first puck after a short time Δt? Of the center of the second puck? How do you justify your predictions?

Both pucks are subjected to exactly the same external force, so their centers of mass must move in exactly the same way. After a short time Δt, the momentum of the center of mass must be $F_T\Delta t$, in the direction of the string, for both pucks. Despite the fact that one puck is pulled from the side, the change in momentum of its center of mass must be in the direction of the net external force, so it cannot acquire a momentum component at right angles to that force. Doesn't this seem a bit odd? For example, one might have expected the second puck to have some kind of sideways motion.

Ex. 7.4 Is there any difference at all in the motion of the two pucks? Think about the motion of your hands at the other ends of the strings. Which hand does more work? Which puck has more energy? In what forms? Does the Momentum Principle for the motion of the center of mass predict all the details of the motion of both pucks?

Ex. 7.5 A meter stick whose mass is 300 grams lies on ice (Figure 7.10). You pull at one end of the meter stick, at right angles to the stick, with a force of 6 newtons. The ensuing motion of the meter stick is quite complicated, but what are the initial magnitude and direction of the rate of change of the momentum of the stick, dP_{tot}/dt, when you first apply the force? What is the magnitude of the initial acceleration of the center of the stick?

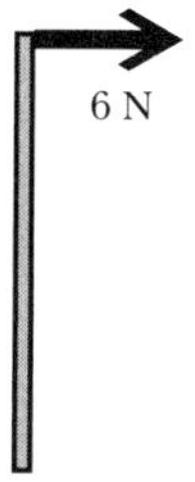

Figure 7.10 Pull one end of a meter stick that is lying on ice.

7.3 Energy in multiparticle systems

We have seen that we can predict the motion of a multiparticle object by treating it as if all its mass were concentrated at the center of mass, acted upon by a force that is the (vector) sum of all the external forces acting on the object. In a number of important situations we find that the energy of a multiparticle system can also be analyzed very simply.

7.3.1 Gravitational energy of a multiparticle system

Our first example is the gravitational energy of a block of mass M and the Earth, near the Earth's surface. The gravitational energy associated with the i-th atom in the block is $m_i g y_i$ (Figure 7.11).

We calculate the total gravitational energy by adding up the gravitational energy associated with the interaction of each of the N atoms with the Earth:

$$U_g = m_1 g y_1 + m_2 g y_2 + m_3 g y_3 + \ldots = g(m_1 y_1 + m_2 y_2 + m_3 y_3 + \ldots)$$

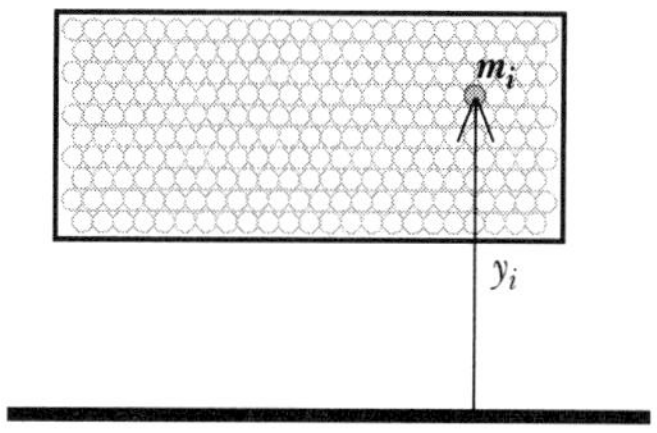

Figure 7.11 The i-th atom in the block is a distance y_i from the Earth's surface.

But the last expression in parentheses appears in the calculation of the y component of the location of the center of mass:

$$My_{cm} = m_1 y_1 + m_2 y_2 + m_3 y_3 + \ldots$$

Therefore for a block made up of many atoms, near the Earth's surface, we have this:

GRAVITATIONAL ENERGY U_g

$$U_g = Mgy_{cm} \text{ near the Earth's surface}$$

In other words, the gravitational energy of a block of material and the Earth, near the Earth's surface, can be evaluated as though the block were a tiny particle of mass M located at the center of mass of the block. This simple formula is not valid if the object is so large that the strength of the gravitational field is significantly different at different locations in the object, but in that case mg would not be a good approximation for the gravitational force, nor mgy a good approximation for the gravitational energy associated with each atom.

? In a calculation of gravitational energy in the voyage of a spacecraft to the Moon is it a valid approximation to treat the spacecraft as if all its mass were concentrated at one point?

This is a good approximation, because the spacecraft is very small compared to the distances to the Earth and Moon, so the gravitational force on each atom is nearly the same.

7.3.2 Kinetic energy of a multiparticle system

We can show that the total kinetic energy K_{tot} of a multiparticle system (that is, the sum of the kinetic energies of all the particles in the system) can usefully be split into two parts: the "translational" kinetic energy $K_{trans} = P_{tot}^2/(2M)$ of a fictitious particle of mass M located at the center of mass and moving with total momentum P_{tot} (which is equal to Mv_{cm}), plus the kinetic energy K_{rel} of the various atoms relative to the center of mass.

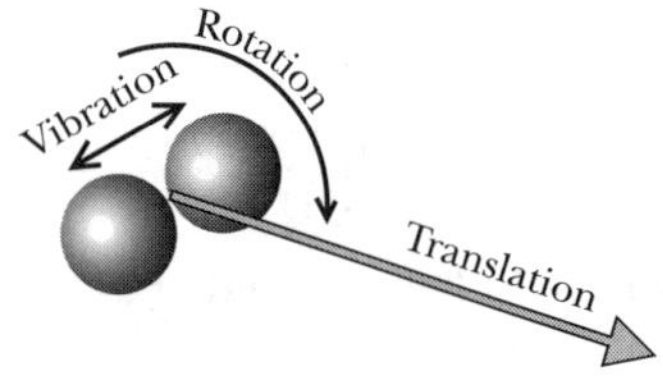

Figure 7.12 A diatomic molecule such as O_2 can vibrate and rotate while also translating.

MULTIPARTICLE KINETIC ENERGY

$$K_{tot} = K_{trans} + K_{rel}, \text{ where } K_{trans} = \frac{P_{tot}^2}{2M} = \frac{1}{2}Mv_{cm}^2$$

For example, the kinetic energy of a rotating, vibrating diatomic molecule such as oxygen (O_2) is the sum of the kinetic energy that the molecule would have if it were not rotating or vibrating (this simple motion is called "translational" motion), plus additional kinetic energy of rotation around the center of mass and kinetic energy of vibration relative to the center of mass (Figure 7.12).

This split of the total kinetic energy into different parts sounds reasonable. A formal derivation of this result is given in section 7.7 at the end of this chapter.

Rotation and translation

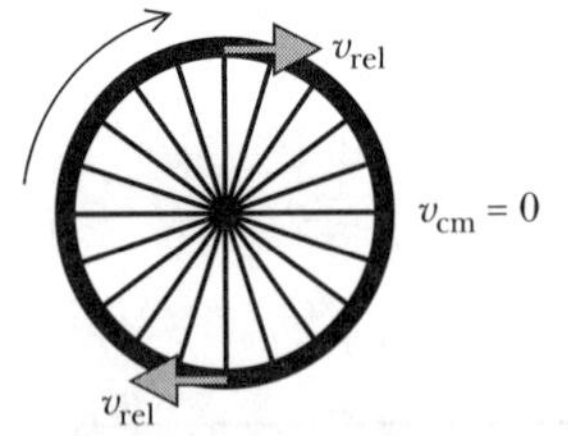

Figure 7.13 A bicycle wheel spinning on its axle, which is at rest. Atoms in the rim have a speed v_{rel} relative to the center of mass.

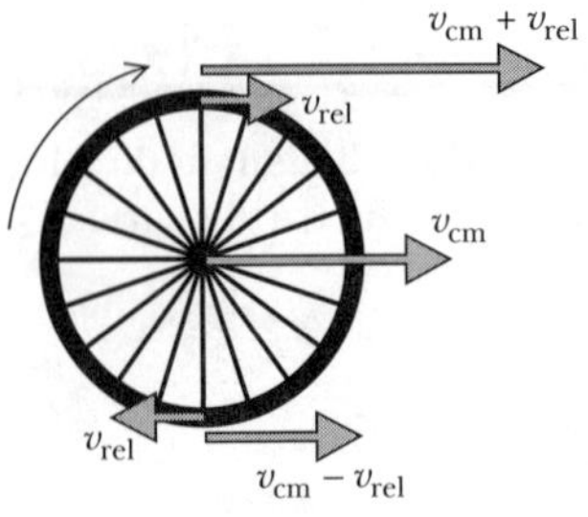

Figure 7.14 The velocity of an atom in the rim is the vector sum of the center-of-mass velocity and the velocity relative to the center of mass.

As an example, suppose you spin a bicycle wheel on its axis, and hold the axle stationary, as in Figure 7.13. The spinning wheel has some kinetic energy K_{rel}, which in this case we will call K_{rot}, because the atoms are moving in rotational motion relative to the center of mass. Almost all of the mass M of the wheel is in the rim, so if the rim is traveling at a (tangential) speed v_{rel}, the kinetic energy of the wheel is approximately $K_{rot} \approx \frac{1}{2}Mv_{rel}^2$. (This isn't exact, because there is some mass in the spokes, and the atoms in the spokes are moving at speeds smaller than v.)

With the wheel still spinning, now move the axle with some speed v_{cm}, without touching the wheel to the ground, as in Figure 7.14. When the center of mass moves, we say that there is *translational* motion. At the top of the wheel the speed of an atom is $v_{cm} + v_{rel}$, because it shares the overall translational motion of the wheel but is also moving forward relative to the axle. At the bottom of the wheel the speed of an atom is $v_{cm} - v_{rel}$, because it shares the overall translational motion of the wheel but is also moving backward relative to the axle. The motions of atoms at other locations around the rim are somewhat complicated. Nevertheless, when we add up all the kinetic energies of all the atoms, we find that $K_{tot} = \frac{1}{2}Mv_{cm}^2 + K_{rot}$. The wheel has kinetic energy corresponding to its overall motion (this is kinetic energy that it would have even if it weren't rotating) plus kinetic energy corresponding to its rotation (this is kinetic energy that it would have even if the center of mass weren't moving). It is appropriate to call the rotational kinetic energy a kind of "internal energy" of the system.

If you drop the spinning bicycle wheel, the Energy Principle can be expressed like this:

$$\left(\frac{P_{tot}^2}{2M} + K_{rot} + Mgy_{cm}\right)_f = \left(\frac{P_{tot}^2}{2M} + K_{rot} + Mgy_{cm}\right)_i$$

If you simply drop the wheel, nothing will make the wheel spin faster or slower around the axle, so K_{rot} doesn't change, and the energy equation for this very complicated motion, involving both translation and rotation, reduces to the Energy Principle for a simple point particle:

$$\left(\frac{P_{tot}^2}{2M} + Mgy_{cm}\right)_f = \left(\frac{P_{tot}^2}{2M} + Mgy_{cm}\right)_i$$

Vibration and translation

Another kind of energy that is internal to a system is vibrational energy, both elastic and kinetic. Consider an oxygen molecule (O_2) that has no translational motion (that is, the center of mass is stationary), but is vibrating (Figure 7.15), with elastic energy and kinetic energy continually interchanging, but with the sum of the two energies remaining constant, like the mass and spring you modeled in Chapter 3.

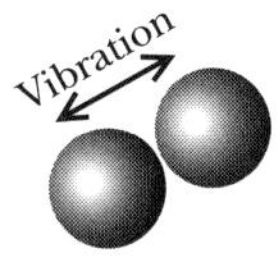

Figure 7.15 A vibrating oxygen molecule whose center of mass is at rest.

If in addition the oxygen molecule is translating (Figure 7.16), the total energy of the molecule is the sum of the translational kinetic energy, the vibrational kinetic energy (in terms of velocities of the two atoms relative to the center of mass), and the elastic energy of the "spring" holding them together, and the rest energies of the constituent atoms:

$$E_{tot} = \frac{P_{tot}^2}{2M} + K_{vib} + \frac{1}{2}k_s s^2 + U_0 + 2mc^2$$

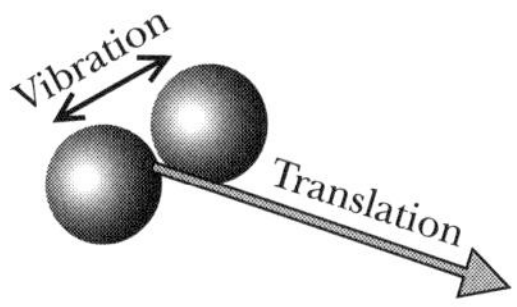

Figure 7.16 A translating, vibrating oxygen molecule.

Note that the elastic energy $\frac{1}{2}k_s s^2$ is clearly unaffected by the translational motion of the center of mass, because it depends only on the stretch of the distance between the two atoms.

A related example is the motion of a hot block of metal. In addition to its translational kinetic energy $P_{tot}^2/(2M)$ there is vibrational kinetic and elastic energy of the atoms around their equilibrium positions, which we call thermal energy (or internal energy E_{int}).

Rotation, vibration, and translation

The most general motion of an oxygen molecule involves rotation, vibration, and translation. It is easy to show that the internal (non-translational) kinetic energy separates cleanly into rotational and vibrational contributions. Consider one atom with momentum p relative to the center of mass, and resolve its momentum into "tangential" and "radial" components (Figure 7.17). By "radial" component we mean a component in the direction of the line from the center of mass to the particle. By "tangential" component we mean a component perpendicular to that line. If the particle is in circular motion around the center of mass, the radial component is zero, so tangential momentum is associated with rotation. Radial momentum is associated with vibration.

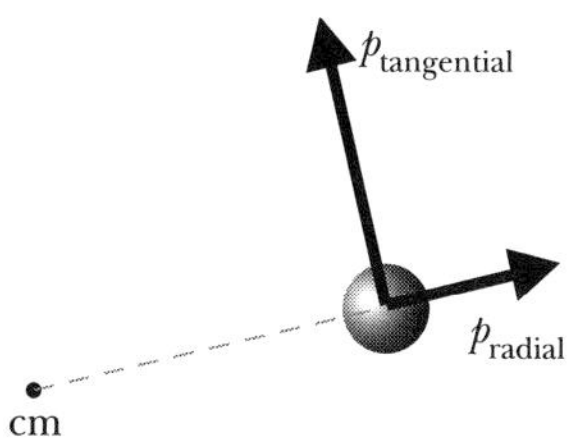

Figure 7.17 Radial and tangential components of momentum of a single atom, relative to the center of mass of an object.

We work out what the kinetic energy is in terms of these two momentum components:

$$\frac{(p^2)}{2m} = \frac{(p^2{}_{\text{tangential}} + p^2{}_{\text{radial}})}{2m} = K_{\text{rot}} + K_{\text{vib}}$$

Therefore the kinetic energy relative to the center of mass splits into two terms, rotational energy and vibrational energy:

ROTATIONAL + VIBRATIONAL ENERGY

$$K_{\text{rel}} = K_{\text{rot}} + K_{\text{vib}}$$

The total energy of a translating, rotating, vibrating oxygen molecule can be written like this:

$$E_{\text{tot}} = K_{\text{trans}} + K_{\text{rot}} + K_{\text{vib}} + \frac{1}{2}k_s s^2 + U_0 + 2mc^2\text{, where } K_{\text{trans}} = \frac{P_{\text{tot}}^2}{2M}$$

(If there are changes in height near the Earth's surface, there is also a term of the form Mgy_{cm} for the energy of the system of Earth plus molecule.)

If the object can fly apart into separate pieces, we might call the kinetic energy relative to the center of mass $K_{\text{explosion}}$! For example, a fireworks rocket has translational kinetic that doesn't change from just before to just after its explosion, and the pieces move at high speeds relative to the center of mass of the rocket. These large kinetic energies relative to the center of mass come from the chemical energy used up in the explosion.

7.4 The "point particle system"

In analyzing the motion and energy of deformable multiparticle systems (systems whose shape can change), we encounter situations in which it would be extremely convenient to be able to calculate only the change in the translational kinetic energy K_{trans} of the system. Perhaps this might be all we are interested in, or perhaps we might wish to subtract this energy change from the change of the total energy of the system to find out how much energy has gone into vibration or rotation. Fortunately, it is quite simple to calculate ΔK_{trans} by itself.

We have established that the center of mass of a multiparticle system moves exactly like a simple "point" particle whose mass is the mass of the entire system, under the influence of the net external force applied to the entire system. Pretend you could crush the real system down to a very tiny ball and apply to this "point particle" the same forces that acted on the real system. What would the motion of this point particle system look like? It would look exactly like the motion of the center of mass of the real system (Figure 7.18).

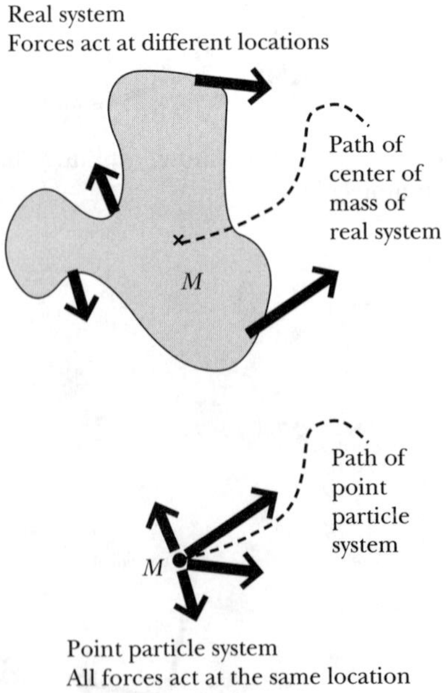

Figure 7.18 The path of the point particle system is exactly the same as the path of the center of mass of the real system.

These two different systems have the same total mass M, and both are acted on by the same net force, so the two paths are exactly the same. The difference is that the real system may rotate and stretch and vibrate due to the effects of the forces acting at different locations on the extended object. In contrast, the point particle system has no rotational motion, no vibrational motion, no internal energy of any kind. All of the forces act at the location of the point particle, and these forces don't stretch or rotate it. The only energy the point particle system can have is translational kinetic energy, and this is exactly the same as the translational kinetic energy of the real multiparticle system:

$$K_{\text{trans}} = \frac{P_{\text{tot}}^2}{2M} = \frac{1}{2}M_{\text{tot}}v_{\text{cm}}^2 \text{ for the real system or the point particle system}$$

An analysis of the point particle system gives the translational kinetic energy of the real system. This is why the point particle system is a useful concept. The force and energy equations for the fictitious point particle are these:

POINT PARTICLE SYSTEM (MASS *M* AT CM LOCATION)

$$\frac{d\vec{P}_{tot}}{dt} = \vec{F}_{net,ext}$$

$$\Delta K_{trans} = \Delta\left(\frac{P_{tot}^2}{2M}\right) = \int_i^f \vec{F}_{net,ext} \bullet d\vec{r}_{cm}$$

The integral is the work done on the point particle system by the net force, which acts at the location of the point itself. In Chapter 4 we showed that starting from the Momentum Principle for a single point particle we could derive the result that the change in the kinetic energy of the particle is equal to the work done on the particle by the net force. Here, we consider the net force to be applied to the fictitious point particle located at the center-of-mass point, doing work to increase the kinetic energy of the point particle. A more formal derivation of the energy equation for the point particle system may be found at the end of this chapter.

These equations tell us about the motion of the center of mass of a multiparticle system, but they tell us nothing about the vibration or rotation or thermal changes of the system. For a full treatment we need to combine analyses of the point particle system with analyses of the real system.

7.4.1 Application: Jumping up

As an example of how to exploit the point particle system, we'll analyze jumping (Figure 7.19). You crouch down and then jump straight upward. We want to calculate how fast your center of mass v_{cm} is moving at the instant that your feet leave the floor. Let h be the distance through which your center of mass rises in this process. Note that there is a change of shape of the system.

Let F_N be the electric contact force that the atoms in the floor exert on the bottom of your foot, and let M be your total mass. We don't know exactly how the floor force F_N varies with time, but to get an idea of what happens let's make the crude approximation that it is constant as long as you are in contact with the floor (of course it falls suddenly to zero as your foot leaves the floor). Consider the point particle system shown in Figure 7.20, which does not change shape.

Figure 7.19 As the jumper leaves the floor, the center of mass has risen h. The system changes shape.

Point Particle system

? What is the net force on you? What is the energy equation for the point particle system?

The net force in the upward direction is $(F_N - Mg)$ (Figure 7.20), and the energy equation for the point particle system is

$$\Delta K_{trans} = \frac{P_{tot}^2}{2M} - 0 = \int_i^f \vec{F}_{net,ext} \bullet d\vec{r}_{cm} = (F_N - Mg)h$$

As we will see, this is not the same as the energy equation for the real system consisting of you, the jumper! The reason F_N shows up in the point particle energy equation is that it contributes to the net force, which affects the center of mass motion. The work done on the point particle system involves the net force multiplied by the displacement of the center-of-mass point.

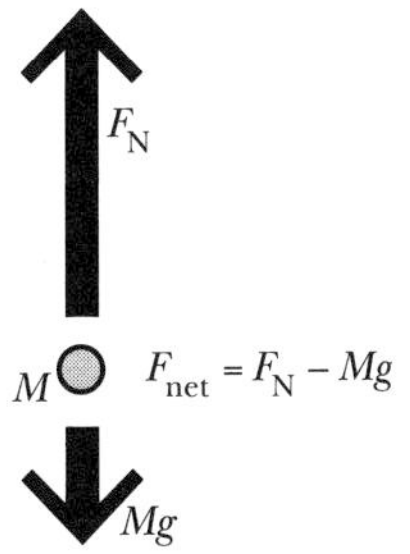

Figure 7.20 Force diagram for your body, considered as a point particle. This system does not change shape.

Real system

? What are some of the energy changes that occur during the jump that are not represented in this point particle energy equation?

There is change of kinetic energy of your moving arms and legs relative to the center of mass, decrease of chemical energy inside you, and increase of thermal energy inside you (your temperature rises somewhat).

? In the real system, how much work is done by the floor force F_N?

The floor force F_N is applied to the bottom of your foot, and the contact point does not move during the entire process leading to lift-off, so there is no work. The definition of the work done by a force involves (the parallel component of) the displacement of the point of contact, at the place where the force is applied to the object. No displacement, no work done on you.

The energy equation for the real system (you, the jumper) looks something like the following, ignoring thermal transfer of energy Q between you and the surrounding air:

$$\Delta E_{\text{sys}} = W = \sum_i \int \vec{F}_i \bullet d\vec{r}_i \quad (W \text{ is sum of work done by all external forces})$$

$$\Delta K_{\text{trans}} + \Delta K_{\text{other}} + \Delta E_{\text{thermal}} + \Delta E_{\text{chemical}} = -Mgh$$

The floor force F_N does not appear in the energy equation for the real system, because this force does no work (no displacement of the point of contact; $d\vec{r}_{\text{floor}} = 0$). (The floor force does however appear in the energy equation for the point particle system.) As before, the gravitational force does negative work on you.

ΔK_{other} includes the rotation of your upper and lower legs and the swinging of your arms. $\Delta E_{\text{thermal}}$ corresponds to the higher temperature that your body runs due to the exertion (which ultimately will lead to thermal transfer of energy Q out of your body into the surrounding, cooler air). $\Delta E_{\text{chemical}}$ represents the burning of chemical energy stored in your body, in order to support this activity. None of these terms appear in the energy equation for the point particle system, because the point particle equation deals just with the motion of the mathematical center of mass point, which doesn't rotate or stretch or get hot. A crucial difference is that you change shape (legs unbend, etc.), but the point particle system doesn't.

The point particle equation gives you $\Delta K_{\text{trans}} = (F_N - Mg)h$. Substitute this into the energy equation for the real system and solve for the other terms, $\Delta K_{\text{other}} + \Delta E_{\text{thermal}} + \Delta E_{\text{chemical}}$.

Comparison of the point particle and real energy equations

The important thing to notice is that the energy equation for the real system and for the point particle system are quite different. Most energy terms do not appear at all in the point particle equation, and the work that appears in the energy equation for the real system is not the same as the work that appears in the energy equation for the point particle system.

? Which of the energy changes in the energy equation shown above for the real system are positive? Which are negative?

All of these energy changes of your body are positive except for the change in chemical energy, which decreases to pay for all the other increases.

7.4.2 Application: Stretching a spring

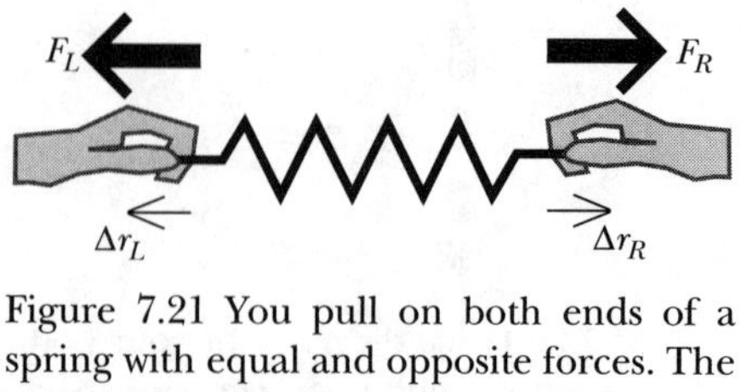

Figure 7.21 You pull on both ends of a spring with equal and opposite forces. The system (the spring) changes shape.

Let's check to make certain this kind of analysis gives us a sensible answer in a simple situation. You pull on one end of a spring (Figure 7.21) with a force F_L to the left, making a short displacement Δr_L to the left, and you pull on the other end with an equal but opposite force F_R and equal but opposite short displacement Δr_L. The system (the spring) changes shape. Our analysis should yield the expected result: the spring should not gain translational kinetic energy!

The real system

Consider the spring itself as the system.

> **?** Does your right hand do any work on the spring? Of what sign? Does your left hand do any work on the spring? Of what sign? Is there any change in the energy of the spring?

Your left hand exerts a force through a short displacement Δr_L and does some (positive) work $W_L = F_L \Delta r_L \cos(0°) = +F_L \Delta r_L$.

Your right hand also exerts a force through a short displacement Δr_R and does some (positive) work $W_R = F_R \Delta r_R \cos(0°) = +F_R \Delta r_R$.

The total work increases the energy of the spring, in the form of spring (elastic) energy corresponding to increased stretch of the spring. Note carefully that both hands do positive work. The energy equation for the real system undergoing this process (the spring) is this:

$$\Delta[\tfrac{1}{2}k_s s^2] = W_L + W_R$$

The point particle system

Next consider the "point particle system" consisting of a fictitious particle located at the center of mass of the spring (Figure 7.22). This point particle system does not change shape.

Figure 7.22 Force diagram for the "point particle system," which does not change shape.

> **?** What is the net force $\vec{F}_{net,ext}$? How far does the center of mass move? How much work $\sum \vec{F}_{net,ext} \bullet \Delta\vec{r}_{cm}$ is done? What is the change in $K_{trans} = P_{tot}^2/(2M)$?

The net force is zero ($F_R - F_L = 0$). Reassuringly, the center of mass does not move. The work done on the fictitious point particle is zero, so there is no change in the speed of the center of mass, so no change in the translational kinetic energy $K_{trans} = P_{tot}^2/(2M)$. The energy equation reduces to 0 = 0, which is certainly correct but provides no detailed information.

Comparison of the point particle and real energy equations

The key point here is that in the real system, work is calculated by adding up the amount of work done by *each* force: the work done by your right hand *plus* the work done by your left hand. The work done by each force is the force times the actual displacement of the point of application of that force, $F_L \Delta r_L$ and $F_R \Delta r_R$, both of which are positive.

But in the point particle system, work is calculated by first adding up all the forces to get the *net* force before calculating the work, and the net force is multiplied by the displacement of the center of mass. In the present example the net force is zero, and the displacement of the center of mass is zero, so the work done on the point particle system is zero.

Note also that the actual work done by an individual force involves the displacement of the point where the force is applied. In contrast, work done on the point particle system involves the displacement of the center of mass point, a point which may be very far away from any of the actual locations where the forces are applied, with a motion that is quite different from the motion of the point of application of an individual force.

Another way of looking at the situation is that the real multiparticle system (the spring) changes shape, and there is potential energy associated with that change of shape. The point particle system does not change shape, and there is no change in potential energy.

The energy equation for the real system and the energy equation for the point particle system are both valid and give correct results for any process, but they provide different kinds of information:

Real system: Info on change of total energy of system
Point Particle system: Info on change of translational kinetic energy

The differences are what make it useful to analyze certain kinds of phenomena both ways. Note that the two energy equations are different for this spring example because the system is deformable and changes shape, which leads to very different calculations of the work in the two systems.

Ex. 7.6 If a system such as a large block does not change shape (or rotate faster and faster due to tangential forces, and there are no changes of identity or temperature), the energy equations for the real system and for the point particle system are identical. Explain why.

Ex. 7.7 A runner whose mass is 50 kg accelerates from a stop to a speed of 10 m/s in 3 seconds. (A good sprinter can run 100 meters in about 10 seconds, with an average speed of 10 m/s.) What is the average horizontal component of the force that the ground exerts on the runner's shoes?

Ex. 7.8 How much displacement is there of the force that acts on the sole of the runner's shoes, assuming that there is no slipping? Therefore, how much work is done on the real system (the runner) by the force you calculated in the previous exercise? How much work is done on the point particle system by this force?

Ex. 7.9 The kinetic energy of the runner in the previous exercises increases—what kind of energy decreases? By how much?

Ex. 7.10 A very similar situation is the acceleration of a car on dry pavement, if there is no slipping. The axle moves at speed v, and the outside of the tire moves at speed v relative to the axle. The instantaneous velocity of the bottom of the tire is zero. How much work is done by the force exerted on the tire by the road? What is the source of the energy that increases the car's translational kinetic energy?

7.5 How to analyze point particle and real systems

Here is a procedure to use in analyzing the point particle and real systems.

- Imagine crushing the real system and its forces down to a point at the center of mass of the real system, including moving the tails of the force arrows to that point. Draw diagrams of the initial and final states, representing the point particle system as a dot, and showing all the forces that act on the real system.

- The tails of all the forces must be on the point particle itself.

- Write the Energy Principle for the point particle, paying attention to the fact that each force acts through the distance that the point particle moves, *not* the distance the tail of the force moves in the real system. The Energy Principle will include the term ΔK_{trans}.

- Draw diagrams of the initial and final states of the real system.

- Draw arrows representing the forces acting on the real system. It is very important to place the tail of each arrow at the location where that force acts on an object.

- Look carefully at how far the tails of the force arrows move, from the initial state to the final state. Use this information to calculate the work done by each force. Add up these work terms to find the work done on the real system.
- Write the Energy Principle for the real system, which will include the term ΔK_{trans}.
- Replace ΔK_{trans} in the equation for the real system with the ΔK_{trans} you found from the equation for the point particle.

7.5.1 Example: A box containing a spring

A thin box in outer space contains a large ball of clay of mass M, connected to an initially relaxed spring of stiffness k_s (Figure 7.23). The mass of the box and spring are negligible compared to M. The apparatus is initially at rest. Then a force of constant magnitude F is applied to the box. When the box has moved a distance b, the clay makes contact with the left side of the box and sticks there, with the spring stretched an amount s. See the diagram for distances.

(a) Immediately after the clay sticks to the box, how fast is the box moving?

(b) What is the increase in thermal energy of the clay?

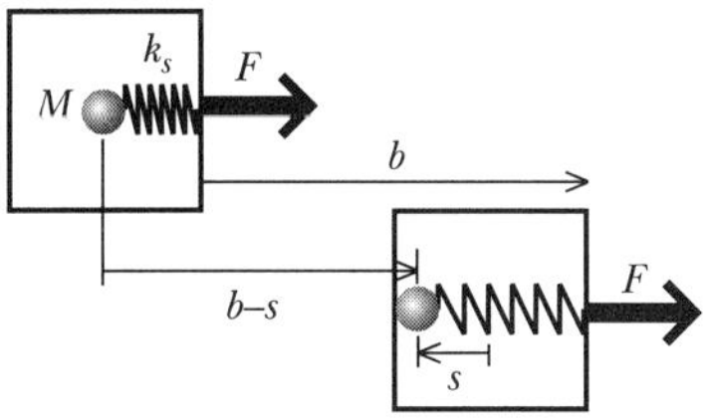

Figure 7.23 A thin box containing a spring and a ball of clay.

Solution

The fundamental principle here is the Energy Principle. We will apply it both to the real system and to the point particle system (a system where we imagine crushing the real system down to a small dense point, and apply all of the forces at that point).

(a) Since we're just interested in the translational speed, consider the point particle version of the system. Imagine you could crush the box and its contents down to a tiny ball with the total mass M of the real system. Apply the same forces to the point particle system that were applied to the real system, but with the tails of the force arrows attached to the point particle (Figure 7.24). The center of mass of the real system is at the center of the ball of clay, since essentially all of the mass is in the clay. The center of mass moves a distance $(b-s)$. The net force is F.

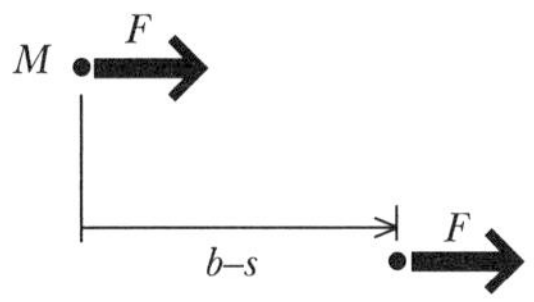

Figure 7.24 The point particle system, with forces attached to the point particle.

$$\Delta K_{trans} = F\Delta x_{cm} \text{ (Energy Principle applied to point particle system)}$$

$$\left(\tfrac{1}{2}Mv^2 - 0\right) = F(b-s)$$

$$v = \sqrt{\frac{2F(b-s)}{M}}$$

There is nothing about the spring or thermal energy in these equations, because they pertain to the point particle system, which doesn't stretch or get hot.

(b) Since we're interested in internal energy, we should apply the Energy Principle to the real system (the point particle system has no internal energy). The work done on the real system is Fb, because the point of application of the force (the tail of the force arrow) moves a distance b.

$$\Delta E_{system} = W + Q \text{ (Energy Principle applied to real system)}$$

Assume that the process takes place so quickly that there isn't time for any significant thermal transfer of energy Q to or from the surroundings:

$$\Delta(K_{trans} + U_{spring} + E_{thermal}) = Fb$$

Substitute the value of ΔK_{trans} obtained from the point particle analysis:

$$F(b-s)+\tfrac{1}{2}k_s s^2+\Delta E_{\text{thermal}} = Fb$$

$$\Delta E_{\text{thermal}} = Fs-\tfrac{1}{2}k_s s^2$$

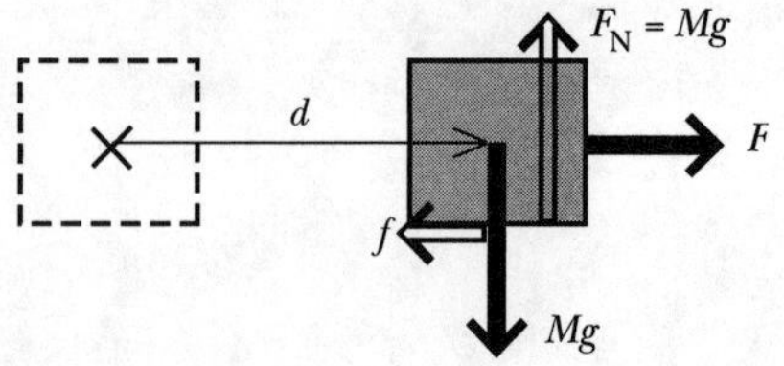

Figure 7.25 The center of mass of the block moves a distance d under the influence of a net force F–f.

7.6 *Modeling friction in detail

We can use what we've learned about complex systems to model the nature of sliding friction, which is a complicated phenomenon. In Chapter 5 we described a seeming paradox involving friction, that when you drag a block at constant speed across a table it seems that no work is done on the block, yet its temperature increases, indicating an increase in its thermal energy. With the new tools developed in this chapter we are in a position to analyze this phenomenon in more detail and resolve the paradox.

Point Particle system

A key issue that we will address is the question of how much work is done on the block by the friction force f exerted by the table. One of the new tools is the energy equation for the point particle system (Figure 7.25 and Figure 7.26). You apply a force F to the right and the table exerts a force f to the left. (Pretend for the moment that you don't know the magnitude of the friction force f.) The change in the kinetic energy of the fictitious center-of-mass point particle is given by the product of the net force $(F-f)$ times the displacement d of the center-of-mass point:

$$\Delta K_{\text{trans}} = (F-f)d$$

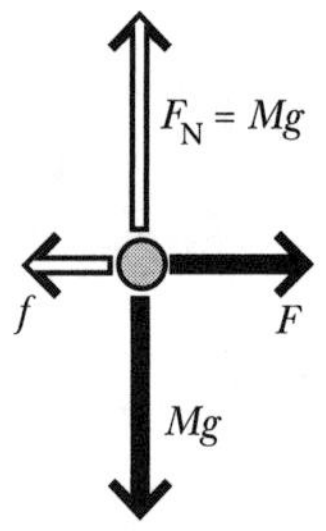

Figure 7.26 The block considered as a point particle.

Since the speed of the block is constant, $\Delta K_{\text{trans}} = 0$, so $F-f = 0$, which shows that the friction force is equal and opposite to the force you apply: $f = F$. The thermal energy does not appear in this energy equation for the point particle system, because this equation deals merely with the motion of the (mathematical) center of mass, not with the many kinds of energy in the real system.

We could also have obtained the result $f = F$ directly from the Momentum Principle: $dP_{\text{tot}}/dt = F-f = 0$. The Momentum Principle is closely related to the energy equation for the point particle system, because both involve the net force acting on a system.

Real system

The energy equation for the real system of the block can be written as follows, where Fd is the work done by you, and W_{fric} is the work done by the table:

$$\Delta K_{\text{trans}}+\Delta E_{\text{th}} = Fd+W_{\text{fric}}$$

where ΔE_{th} is the rise in the thermal energy of the block. We're assuming that the process takes a short enough time that there is negligible thermal transfer of energy Q between the system of the block and the surroundings (the table), because energy transfer due to a temperature difference between table and block is a relatively slow process. (To remove such energy transfer as a possible complication, we could consider a block sliding not on a table but on an identical block. Then the symmetry of the situation is such that there cannot be any net thermal transfer of energy into or out of the upper block. This tactic—reducing the complexity of a model—is a useful approach to complex problems.)

Since the speed of the center of mass v_{cm} does not change, $\Delta K_{\text{trans}} = 0$, and the energy equation reduces to

$$\Delta E_{th} = Fd + W_{fric}$$

If we conclude that the friction force does an amount of work $-Fd$, then we would have to conclude that the thermal energy of the block doesn't change, which is absurd. The block definitely gets hotter, indicating an increase in its thermal energy. We need to find a way around the entirely plausible but apparently incorrect conclusion that the friction force does an amount of work $-Fd$.

? Considering the sign of the change of thermal energy, must the magnitude of the work done by the friction force be greater or less than Fd?

Since the thermal energy change ΔE_{th} is surely positive (the block gets hotter, not colder), the magnitude of the work done by the friction force must be less than Fd. Since the friction force is definitely equal to F, the friction force must act through some effective distance d_{eff} *that must be less than d!*

Evidently the energy equation for the real system has the following form, with d_{eff} less than d:

$$\Delta E_{th} = Fd - Fd_{eff}$$

7.6.1 *A physical model for dry friction

How can the effective distance through which the friction force does work be less than the distance through which the block moves? We can understand this by looking at a microscopic picture of what happens on the surfaces in contact.

When a metal block slides on a metal surface, the block is supported by as few as three protruding "teeth," called "asperities" in the literature on friction (Figure 7.27). The very high load per unit area on these teeth makes the material partially melt and flow, and high local temperatures produced during sliding lead to adhesion (welding) in the contact regions. The frictional force divided by the tiny contact area corresponds to the large "shear" (sideways) stress required to break these welds. This shearing of contact welds is the dominant friction mechanism for a dry metal sliding on the same metal.

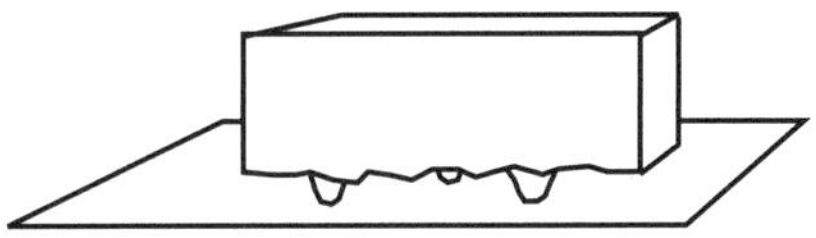

Figure 7.27 A block slides on a small number of protruding teeth (vertical scale exaggerated).

Because the tooth tips can become stronger than the bulk metal due to a process called "work-hardening" (which introduces dislocations in the otherwise regular geometrical arrangements of atoms), shearing often occurs in the weaker regions of the teeth, away from the tip. This is a major effect when the two objects are made of the same material, and chunks of metal can break off and embed in the other surface. Nevertheless, this wear will be ignored in the further discussion. It is in any case a symmetrical effect for identical blocks. If the metal surfaces have oxide coatings, this can reduce the shear stress required to break the temporary weld (which reduces the friction force) and can prevent the breaking off of chunks of metal, if the oxide contact area is the weakest section.

This is the model of dry friction developed in a classic treatise on friction, *The Friction and Lubrication of Solids*, Part 1 and Part II, F. P. Bowden and D. Tabor (Oxford University Press, 1950 and 1964). The physics and chemistry of friction continues to be an active field of research, because the effects of friction can be quite complex, and there is high practical interest in controlling friction. A recent textbook is *Friction, Wear, Lubrication*, K. C Ludema, (CRC Press, 1996).

Having briefly reviewed a basic model of dry friction, we proceed to use this model to calculate the work done by frictional forces exerted at the contact points. The key issue is that the surface is deformable, which leads to

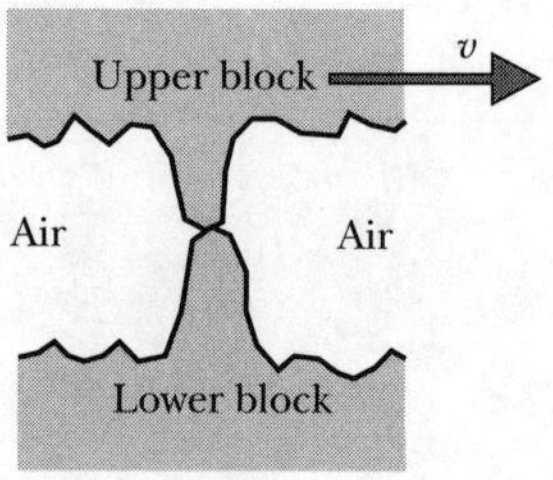

Figure 7.28 Two projecting teeth of identical blocks, which have temporarily adhered to each other. (Exaggerated vertical scale).

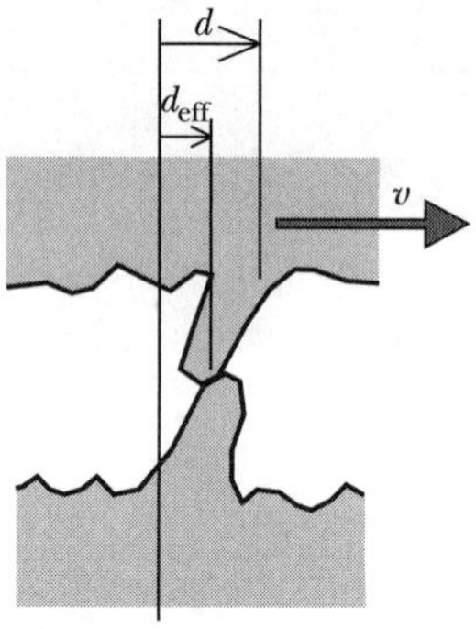

Figure 7.29 The top block has moved a distance d to the right, but the point of contact has moved only d_{eff}.

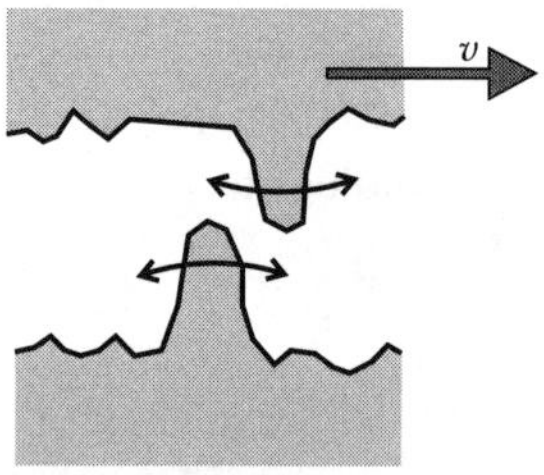

Figure 7.30 After the temporary weld breaks, the top tooth catches up to the rest of the block. Now no force acts on the teeth. Both teeth vibrate.

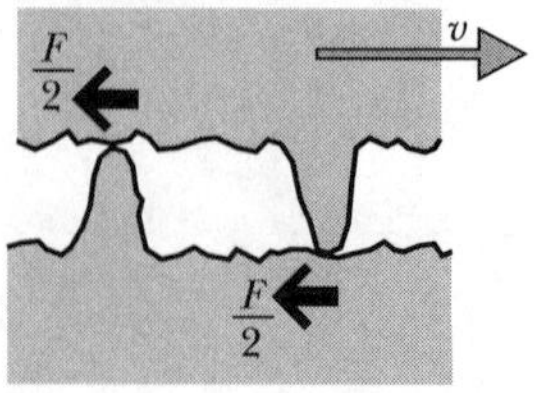

Figure 7.31 A more representative picture of contact between two surfaces, with projecting teeth in contact with flat areas.

differences in the energy equation for the real system compared with the energy equation for the point particle system.

Actual work done by friction forces

We'll consider a microscopic picture of the contact region between two identical blocks. Figure 7.28 shows in a schematic way two teeth that have temporarily adhered to each other. The vertical scale has been greatly exaggerated for clarity—machined surfaces have much gentler slopes

As the top block is dragged to the right, the teeth continue to stick together for a while. Both teeth must deform as a result; the top tooth is stretched backward, and the bottom tooth is stretched forward. In Figure 7.29 we see that when the top block has moved a distance d to the right, the point where the teeth touch each other has moved a distance of only d_{eff}, because the bottom block hasn't moved. For two identical blocks made of the same material, on average the two teeth will bend the same amounts, and d_{eff} will be equal to $d/2$.

> *This is the key issue:* The point of contact where the friction force is applied moves a shorter distance (d_{eff}) than the block itself moves (d). This means that the work done by the friction force is less than one might expect.

Eventually the weld breaks, and the top tooth snaps forward, so all the atoms in the tooth now catch up with the rest of the top block, but during this part of their journey, there is no external force acting on them (Figure 7.30).

The time average of the contact forces is indeed F (as determined by the fact that the net force must be zero, $F - F$), but the effective displacement d_{eff} at the point of contact of the frictional force is less than the displacement of the center of mass of the upper block. For two identical blocks we expect to find that $d_{eff} = d/2$.

Internal energy of the blocks

Once the weld has broken, the teeth can vibrate, as we saw in Figure 7.30. The vibration of the tooth belonging to the top block and thermal conduction upward from the hot tip into the main body of the block contribute to the increase in the thermal energy of the upper block. Similarly, vibration and thermal conduction in the tooth belonging to the bottom block end up as increase in the thermal energy of the bottom block.

A more physical picture

Having two long teeth in contact is rather unphysical. A better picture is the one shown in Figure 7.31 (again with a greatly exaggerated vertical scale), where we show one of the top block's longest teeth in contact with the average surface level of the bottom block, and one of the bottom block's longest teeth similarly in contact with the average surface level of the top block. The two teeth shown here are to be taken as representative of time and space averages of the sliding friction. The frictional force F exerted on the top block is divided on the average into two forces each of magnitude $F/2$ at the ends of the two sets of long teeth.

In Figure 7.32 we see that when the top block moves a distance d to the right the upper contact also moves a distance d, whereas the lower contact does not move at all. The frictional work is therefore $(-F/2)(0) + (-F/2)(d)$, which is $-Fd/2 = -Fd_{eff}$, so d_{eff} is equal to $d/2$.

Summary of dry friction

The fundamental reason why the friction force can act through a distance less than d is that the block is deformable. All atoms in the top block eventually move the same distance d, but because of the stick/slip contact between the blocks the friction force F acts only for a portion of the displacement ($d/2$ for identical blocks). The point of contact for the friction force does not move in the same way as the center of mass. Note again that the energy equation for the point particle system is not the same as the energy equation for the real system if the system changes shape..

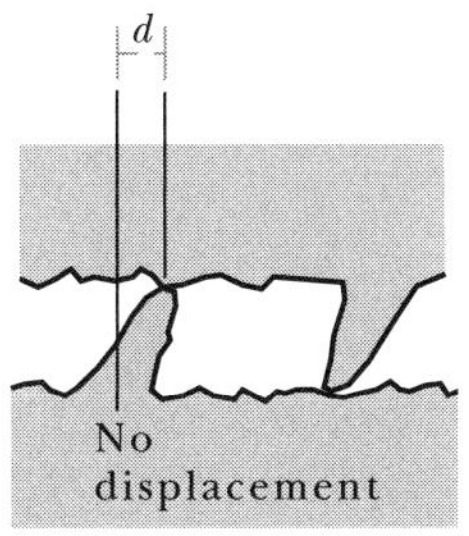

Figure 7.32 During this time interval the upper contact moves a distance d, while the lower contact does not move at all.

7.6.2 *A model-independent calculation of the effective distance

In the special case of a block sliding on an identical block we can calculate d_{eff}, independent of the particular model of the surfaces. Consider a system consisting of both blocks together (Figure 7.33). The bottom block is held stationary by applying a force to the left (to prevent it being dragged to the right). This constraining force acts through no distance (the bottom block stands still), so this force does no work. The only work done on the two-block system is done by you, of magnitude Fd. So the energy equation for the real two-block system is this:

$$\Delta E_{th,1} + \Delta E_{th,2} = Fd$$

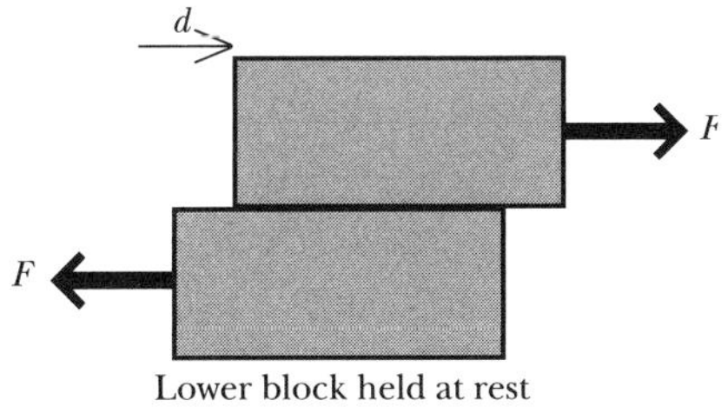

Figure 7.33 The lower block is held at rest, while the upper block slides a distance d. Forces in the vertical direction are not shown.

Since the two blocks are identical, half of this increased thermal energy shows up in the upper block ($Fd/2$), and half in the lower block ($Fd/2$). We can plug this into our earlier energy equation for the upper block:

$$\Delta E_{th,1} = \frac{Fd}{2} = Fd - Fd_{\text{eff}}$$

? Calculate d_{eff}.

We find that the effective distance through which the friction force acts is half the distance d that the block moves: $d_{\text{eff}} = d/2$. This is in agreement with the model of dry friction with bending teeth that we examined earlier, but our result is a general one that applies to any kind of model of friction surfaces for two symmetrical blocks.

7.6.3 *Lubricated friction

It is interesting to see that the model-independent result $d_{\text{eff}} = d/2$ is also consistent with the case of lubricated friction. We separate the two blocks with a film of viscous lubricating oil, so that the two blocks do not make direct contact, as shown in Figure 7.34.

It is a property of simple fluid flow that fluid layers immediately adjacent to the blocks are constrained to share the motion of the blocks. Also, for a common type of flow called "laminar" flow, the displacement profile in the oil is linear. In particular, at the midplane the fluid moves half as far as the top block moves. As a result, as the top block is pulled a distance d to the right, the top layer of the oil is dragged along and moves a distance d to the right, the bottom layer of oil doesn't move, and the layer in the midplane (halfway between the blocks) moves a distance $d/2$.

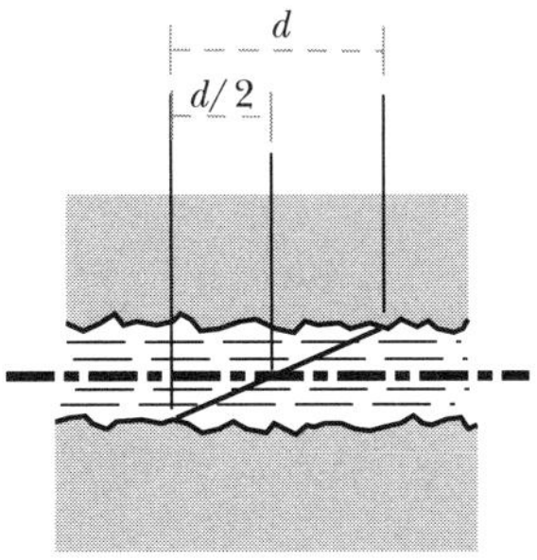

Figure 7.34 Two objects separated by a film of oil, which is modeled as a stack of fluid layers that can move relative to each other.

If we take as the symmetrical systems of interest the top block with the upper half of the oil, and the bottom block with the lower half of the oil, we see that the shear force between the two systems (at the midplane in the oil) acts through a distance which is again half the displacement of the top block: $d_{\text{eff}} = d/2$. (Of course the magnitude of the friction force is much reduced by the lubrication, and the applied force F must be much smaller if the velocity is to be constant.)

This discussion of friction is based on an article "Work and heat transfer in the presence of sliding friction," by B. Sherwood and W. Bernard, *American Journal of Physics* volume **52**, number 11, Nov. 1984, pages 1001-1007, which in turn draws on an earlier article "Real work and pseudowork," by B. Sherwood, *American Journal of Physics* volume **51**, number 7, July 1983, pages 597-602.

7.7 *Derivation: Kinetic energy of a multiparticle system

In this section we derive the important result that $K_{tot} = K_{trans} + K_{rel}$, where $K_{trans} = P_{tot}^2/(2M)$ and K_{rel} is the kinetic energy relative to the center of mass. This result sounds entirely plausible, but the formal proof is rather difficult.

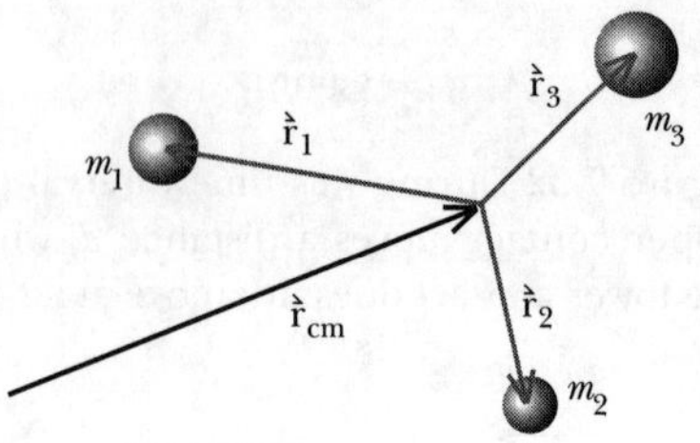

Figure 7.35 Center of mass of a multiparticle system.

As in the case of calculating the gravitational energy of a multiparticle object, the derivation of the kinetic energy of a multiparticle system hinges on the definition of the center of mass point of a collection of atoms. The (vector) location of the i-th atom of the object can be expressed as the sum of two vectors, one from the origin to the center of mass ($\vec{r}_{cm}$), plus another from the center of mass to the i-th atom ($\vec{r}_i$).

The kinetic energy $K_i = \frac{1}{2}mv_i^2$ of the i-th atom is this, where we write $\vec{v}_i = d\vec{r}_i/dt$, the velocity of the i-th particle relative to the center of mass:

$$K_i = \frac{1}{2}m_i\left|\frac{d}{dt}(\vec{r}_{cm} + \vec{r}_i)\right|^2$$
$$= \frac{1}{2}m_i|\vec{v}_{cm} + \vec{v}_i|^2$$

This can be expanded, using the fact that the magnitude squared of a vector can be written as a vector dot product, $|\vec{C}|^2 = \vec{C} \bullet \vec{C} = C^2\cos(0) = C^2$:

$$K_i = \frac{1}{2}m_i[\vec{v}_{cm} + \vec{v}_i] \bullet [\vec{v}_{cm} + \vec{v}_i]$$
$$= \frac{1}{2}m_iv_{cm}^2 + 2\frac{1}{2}m_i\vec{v}_{cm} \bullet \vec{v}_i + \frac{1}{2}m_iv_i^2$$

Now that we have the kinetic energy of the i-th atom in terms of its position $\vec{r}_i$ relative to the center of mass, we need to add up the total kinetic energy of all the atoms. The total kinetic energy K_{tot} can be written in the following compact way, where Σ (Greek capital sigma) means "sum," and the sum goes from $i = 1$ through $i = N$:

$$K_{tot} = \underbrace{\sum_{i=1}^{N}\frac{1}{2}m_iv_{cm}^2}_{\textbf{First term}} + \underbrace{\sum_{i=1}^{N}2(\tfrac{1}{2}m_i)\vec{v}_{cm} \bullet \vec{v}_i}_{\textbf{Second term}} + \underbrace{\sum_{i=1}^{N}\frac{1}{2}m_iv_i^2}_{\textbf{Third term}}$$

First term

The first term in this summation turns out to be the kinetic energy of a single particle of mass M, moving at the speed of the center of mass:

$$\sum_{i=1}^{N}\frac{1}{2}m_iv_{cm}^2 = \frac{1}{2}v_{cm}^2\left(\sum_{i=1}^{N}m_i\right) = \frac{1}{2}v_{cm}^2(m_1 + m_2 + m_3 + \ldots)$$
$$= \frac{1}{2}Mv_{cm}^2 = \frac{P_{tot}^2}{2M}$$

Second term

The second term can be shown to be zero:

$$\sum_{i=1}^{N}2(\tfrac{1}{2}m_i)\vec{v}_{cm} \bullet \vec{v}_i = \vec{v}_{cm} \bullet \sum_{i=1}^{N}m_i\vec{v}_i = \vec{v}_{cm} \bullet \frac{d}{dt}\left(\sum_{i=1}^{N}m_i\vec{r}_i\right) = 0$$

This result follows from the way that we calculate the location of the center of mass:

$$\vec{r}_{cm} = \frac{m_1\vec{r}_1 + m_2\vec{r}_2 + m_3\vec{r}_3 + \ldots}{m_1 + m_2 + m_3 + \ldots} = \frac{\sum_{i=1}^{N} m_i\vec{r}_i}{M}$$

We are measuring the location $\vec{r}_i$ of the i-th atom relative to the center of mass, and since the distance from the center of mass to the center of mass is of course zero, we have

$$\vec{r}_{cm} = 0 = \frac{\sum_{i=1}^{N} m_i\vec{r}_i}{M} \text{ and therefore } \sum_{i=1}^{N} m_i\vec{r}_i = 0 .$$

Third term

The third term by definition is the kinetic energy of the atoms relative to the (possibly moving) center of mass:

$$\sum_{i=1}^{N} \tfrac{1}{2}m_i v_i^2 = \tfrac{1}{2}m_1 v_1^2 + \tfrac{1}{2}m_2 v_2^2 + \tfrac{1}{2}m_3 v_3^2 + \ldots = K_{rel}$$

Putting the three pieces together, we find that the total kinetic energy splits into two parts: a term associated with the overall motion of the center of mass, plus the kinetic energy relative to the center of mass:

$$K_{tot} = K_{trans} + K_{rel}\text{, where } K_{trans} = \frac{P_{tot}^2}{2M} = \tfrac{1}{2}Mv_{cm}^2$$

7.8 *Derivation: The point particle energy equation

In this chapter we showed that the energy equation for the point particle version of the system follows from the fact that the motion of the center of mass is just like that of a point particle with the total mass of the real system and subjected to the *net* force acting on the real system. Here we give a more formal derivation of this important result.

Start from the x component of the Momentum Principle for a multiparticle system whose center of mass is moving at nonrelativistic speed:

$$\frac{dP_{tot,x}}{dt} = M_{tot}\frac{dv_{cm,x}}{dt} = F_{net,ext,x}$$

Integrate through the x displacement of the center of mass:

$$M_{tot}\int_i^f \frac{dv_{cm,x}}{dt}dx_{cm} = \int_i^f F_{net,ext,x}dx_{cm}$$

Switch dv and dx:

$$M_{tot}\int_i^f \frac{dx_{cm}}{dt}dv_{cm,x} = \int_i^f F_{net,ext,x}dx_{cm}$$

But dx_{cm}/dt is the x component of the center of mass velocity:

$$M_{tot}\int_i^f v_{cm,x}dv_{cm,x} = \int_i^f F_{net,ext,x}dx_{cm}$$

The integral on the left can be carried out:

$$M_{\mathrm{tot}}\left[\frac{1}{2}v_{\mathrm{cm,x}}^2\right]_i^f = \int_i^f F_{\mathrm{net,ext,x}}\,dx_{\mathrm{cm}}$$

$$\Delta\left[\frac{1}{2}M_{\mathrm{tot}}v_{\mathrm{cm,x}}^2\right] = \int_i^f F_{\mathrm{net,ext,x}}\,dx_{\mathrm{cm}}$$

We could repeat exactly the same argument for the *y* and *z* motions:

$$\Delta\left[\frac{1}{2}M_{\mathrm{tot}}v_{\mathrm{cm,y}}^2\right] = \int_i^f F_{\mathrm{net,ext,y}}\,dy_{\mathrm{cm}}$$

$$\Delta\left[\frac{1}{2}M_{\mathrm{tot}}v_{\mathrm{cm,z}}^2\right] = \int_i^f F_{\mathrm{net,ext,z}}\,dz_{\mathrm{cm}}$$

Note that $\frac{1}{2}M_{\mathrm{tot}}v_{\mathrm{cm,x}}^2 + \frac{1}{2}M_{\mathrm{tot}}v_{\mathrm{cm,y}}^2 + \frac{1}{2}M_{\mathrm{tot}}v_{\mathrm{cm,z}}^2 = \frac{1}{2}M_{\mathrm{tot}}v_{\mathrm{cm}}^2$.

Moreover, $F_{\mathrm{net,ext,x}}dx_{\mathrm{cm}} + F_{\mathrm{net,ext,y}}dy_{\mathrm{cm}} + F_{\mathrm{net,ext,z}}dz_{\mathrm{cm}} = \vec{F}_{\mathrm{net,ext}} \bullet d\vec{r}_{\mathrm{cm}}$.
Adding the three equations together (the sum of the left sides is equal to the sum of the right sides), we have the following equation for the translational kinetic energy of the system:

$$\Delta\left[\frac{1}{2}M_{\mathrm{tot}}v_{\mathrm{cm}}^2\right] = \int_i^f \vec{F}_{\mathrm{net,ext}} \bullet d\vec{r}_{\mathrm{cm}}$$

In words, the change in the translational kinetic energy of a system is equal to the integral of the *net* force acting through the displacement of the center of mass point.

The derivation that we have just carried out shows that although this equation looks like an energy equation, it is actually closely related to the Momentum Principle from which it was derived. The common element is the *net* force.

In contrast, the actual energy equation for the real system involves the work done by each individual force through the displacement of the point of application of that force. If the system deforms or rotates, these displacements of the individual forces need not be the same as the displacement of the center of mass.

7.9 Summary

Fundamental principles

No new fundamental principles.

New concepts

Center of Mass

$$\vec{r}_{cm} \equiv \frac{m_1\vec{r}_1 + m_2\vec{r}_2 + m_3\vec{r}_3 + \ldots}{m_1 + m_2 + m_3 + \ldots}$$

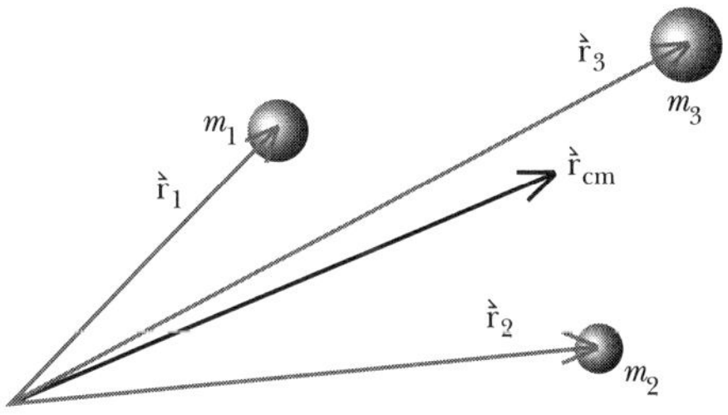

Center of mass

Problem Solving Techniques

Combined use of the energy equations for the real system and for the fictitious "point particle system" in analyzing complex phenomena involving multiparticle systems.
Point Particle system: Info on change of translational kinetic energy
Real system: Info on change of total energy

Results

The Momentum Principle for multiparticle systems, derived from Newton's second and third laws of motion, was extended by defining the center of mass:

THE MOMENTUM PRINCIPLE FOR MULTIPARTICLE SYSTEMS

$$\frac{d\vec{P}_{tot}}{dt} = \vec{F}_{net,ext} \text{ (in general)}$$

$$\vec{P}_{tot} = M\vec{v}_{cm} \text{ (if } v << c\text{, and constant mass)}$$

Gravitational energy of multiparticle systems plus Earth, near the Earth's surface:

$$U_g = Mgy_{cm}$$

Kinetic energy of multiparticle systems:

$$K_{tot} = K_{trans} + K_{rel}\text{, where } K_{trans} = \frac{P_{tot}^2}{2M} = \frac{1}{2}Mv_{cm}^2$$

Kinetic energy relative to center of mass can be split into two terms:

$$K_{rel} = K_{rot} + K_{vib}$$

POINT PARTICLE SYSTEM

$$\Delta K_{trans} = \Delta\left(\frac{P_{tot}^2}{2M}\right) = \int_i^f \vec{F}_{net,ext} \bullet d\vec{r}_{cm}$$

Sliding friction can deform the contact surfaces, with the result that the frictional force may act through a distance that is different from the distance through which the center-of-mass point moves. As with any deformable system, the energy equations for the real system and for the point particle system differ (though both are correct).

7.10 Review questions

The motion of a multiparticle system

RQ 7.1 Discuss qualitatively the motion of the atoms in a block of steel that falls onto another steel block. Why and how do large-scale vibrations damp out?

Center of mass

RQ 7.2 Can you give an example of a system that has no atoms located at its center of mass?

The Momentum Principle for multiparticle systems

RQ 7.3 Two people with different masses but equal speeds slide toward each other with little friction on ice with their arms extended straight out to the side (so each has the shape of a "t"). Her right hand meets his right hand, they hold hands and spin 90°, then release their holds and slide away. Make a rough sketch of the path of the center of mass of the system consisting of the two people, and explain briefly. (It helps to mark equal time intervals along the paths of the two people, and of their center of mass.)

RQ 7.4 The Momentum Principle for multiparticle systems would seem to say that the center of mass of a system moves just as though the system were a point particle. This led Chris to ask, "Wouldn't that mean that a piece of paper ought to fall in the same way as a small metal ball, if they have the same mass?" Explain carefully to Chris the resolution of this puzzle.

Gravitational energy of a multiparticle system

RQ 7.5 Consider the voyage to the Moon that you studied in Chapter 2. Would it make any difference, even a very tiny difference, whether the spacecraft is long or short, if the mass is the same? Explain briefly.

Kinetic energy of a multiparticle system

RQ 7.6 Outline how you would calculate the kinetic energy of the Earth as it moves through space.

The point particle system

RQ 7.7 Under what conditions does the energy equation for the point particle system differ from the energy equation for the real system? Give two examples of such a situation. Give one example of a situation where the two equations look exactly alike.

7.11 Problems

Problem 7.1 Jumping straight up—experiment

(a) Crouch down and jump straight up, as high as you can. Estimate the location of your center of mass, and measure its height at three stages in this process: in the crouch, at lift-off, and at the top of the jump. Report your measurements. You may need to have a friend help you make the measurements.

(b) Analyze this process as fully as possible, using all the theoretical tools now available to you, especially the concepts in this chapter. Include a calculation of the force of the floor on your feet, the change in your chemical energy, and the time of contact from the beginning of the jump to lift-off. Be sure to explain clearly what approximations and simplifying assumptions you made in modeling the process.

Problem 7.2 Ice skater pushes away from a wall

On page 296 we discussed a woman ice skater who pushes away from a wall.

(a) Estimate the speed she can achieve just after pushing away from the wall. Then estimate the average acceleration during this process. How many "g's" is this? (That is, what fraction or multiple of 9.8 m/s^2 is your estimate?) Be sure to explain clearly what approximations and simplifying assumptions you made in modeling the process.

(b) For this process, choose the woman as the system of interest and discuss the energy transfers, and the changes in the various forms of energy. Estimate the amount of each of these, including the correct signs.

Problem 7.3 Translation and rotation

A hoop of mass M and radius R rolls without slipping down a hill (Figure 7.36). The lack of slipping means that when the center of mass of the hoop has speed v, the tangential speed of the hoop relative to the center of mass is also equal to v, since in that case the instantaneous speed is zero for the part of the hoop that is in contact with the ground ($v - v = 0$).

(a) The initial speed of the hoop is v_i, and the hill has a height h. What is the speed v_f at the bottom of the hill?

(b) Replace the hoop with a bicycle wheel whose rim has mass M and whose hub has mass m (Figure 7.37). The spokes have negligible mass. What would the bicycle wheel's speed be at the bottom of the hill?

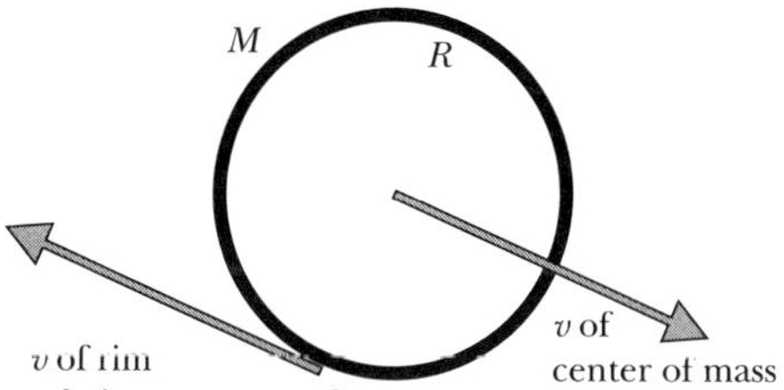

Figure 7.36 A hoop of mass M and radius R rolls down a hill (Problem 7.3a).

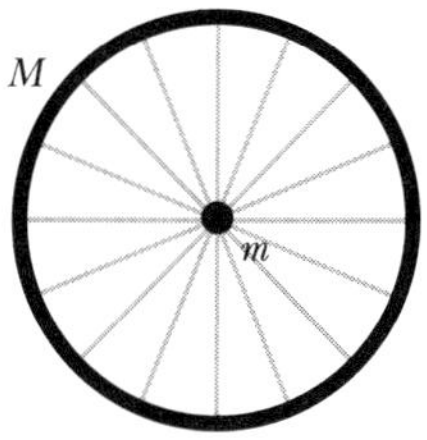

Figure 7.37 A bicycle wheel rolls down a hill (Problem 7.3b). The rim has mass M and the hub has mass m.

Problem 7.4 Moving in a canoe

A man whose mass is 80 kg and a woman whose mass is 50 kg sit at opposite ends of a canoe 5 m long, whose mass is 30 kg.

(a) Relative to the man, where is the center of mass of the system consisting of man, woman, and canoe? (Hint: Choose a specific coordinate system with a specific origin.)

(b) Suppose the man moves quickly to the center of the canoe and sits down there. How far does the canoe move in the water? Explain your work and your assumptions.

Problem 7.5 Pulling a disk from the side

A string is wrapped around a disk of mass M and radius R. Starting from rest, you pull the string with a constant force F along a nearly frictionless surface (Figure 7.38). At the instant when the center of the disk has moved a distance d, a length L of string has unwound off the disk.

(a) At this instant, what is the speed of the center of mass of the disk?

(b) At this instant, how much rotational kinetic energy does the disk have relative to its center of mass?

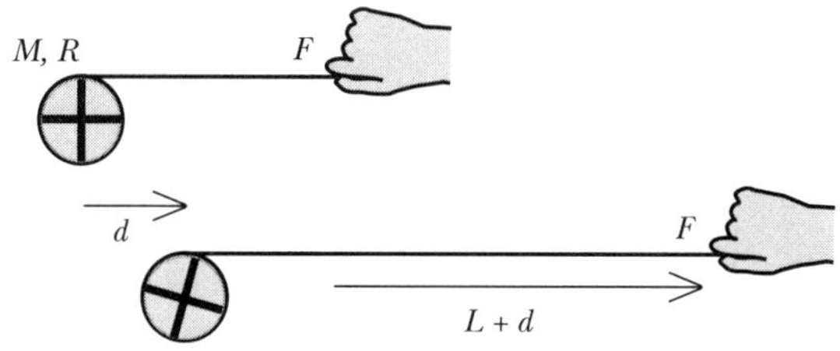

Figure 7.38 Pull a disk by a string wrapped around the disk (Problem 7.5).

Problem 7.6 Two disks colliding

Two disks are initially at rest, each of mass M, connected by a string between their centers (Figure 7.39). The disks slide on low-friction ice as the center of the string is pulled by a string with a constant force F through a distance d. The disks collide and stick together, having moved a distance b horizontally.

(a) What is the final speed of the stuck-together disks?

(b) When the disks collide and stick together, their temperature rises. Calculate the increase in thermal energy of the disks, assuming that the process is so fast that there is insufficient time for there to be much thermal transfer of energy to the ice. (Also, ignore the small amount of energy radiated away as sound produced in the collisions between the disks.)

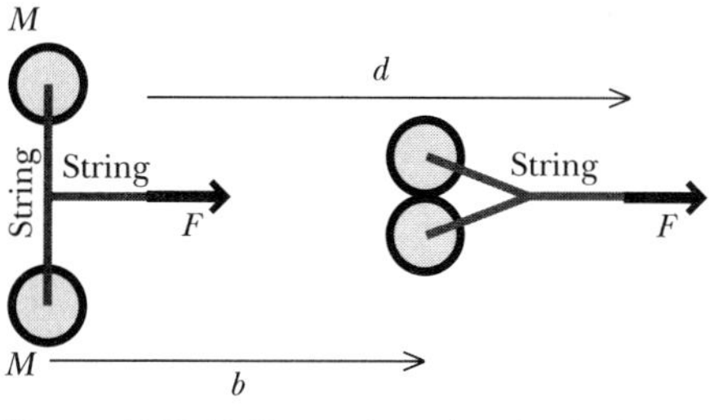

Figure 7.39 Pull a string that is tied to a string connecting two disks (Problem 7.6).

Problem 7.7 An uncoiling chain

A chain (mass M) of metal links is coiled up in a tight ball on a low-friction table. You pull on a link at one end of the chain with a constant force F. Eventually the chain straightens out to its full length L, and you keep pulling until you have pulled your end of the chain a total distance d.

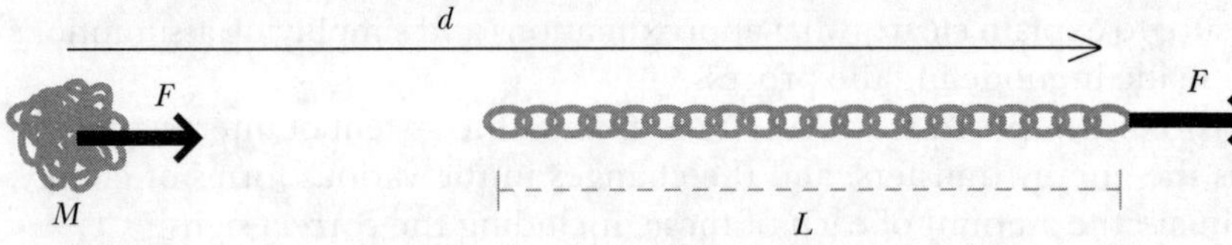

(a) What is the speed of the chain at this instant?

(b) In straightening out, the links of the chain bang against each other, and their temperature rises. Calculate the increase in thermal energy of the chain, assuming that the process is so fast that there is insufficient time for the chain there to be much thermal transfer of energy to the table. (Also, ignore the small amount of energy radiated away as sound produced in the collisions among the links.)

Problem 7.8 Friction

It is sometimes claimed that friction forces always slow an object down, but this is not true. If you place a box of mass M on a moving horizontal conveyor belt, the friction force of the belt acting on the bottom of the box speeds up the box. At first there is some slipping, until the speed of the box catches up to the speed v of the belt. The coefficient of friction between box and belt is μ.

(a) What is the distance d (relative to the floor) that the box moves before reaching the final speed v? Use energy arguments, and explain your reasoning carefully.

(b) How much time does it take for the box to reach its final speed?

(c) The belt and box of course get hot. Is the effective distance through which the friction force acts greater than or less than d? Give as quantitative an argument as possible. You can assume that the process is quick enough that you can neglect thermal transfer of energy Q between the belt and the box. Do not attempt to use the *results* of the friction analysis in this chapter; rather, apply the *methods* of that analysis to this different situation.

(d) Explain the result of part (c) qualitatively from a microscopic point of view, including physics diagrams.

Problem 7.9 Falling person

You hang by your hands from a tree limb that is a height L above the ground, with your center of mass a height h above the ground, and your feet a height d above the ground (Figure 7.40). You then let yourself fall. You absorb the shock by bending your knees, ending up momentarily at rest in a crouched position with your center of mass a height b above the ground. Your mass is M. You will need to draw labeled physics diagrams for the various stages in the process.

(a) What is the net internal energy change ΔE_{int} in your body (chemical plus thermal)?

(b) What is your speed v at the instant when your feet first touch the ground?

(c) What is the approximate average force F exerted by the ground on your feet during the time when your knees are bending?

(d) How much work is done by this force F?

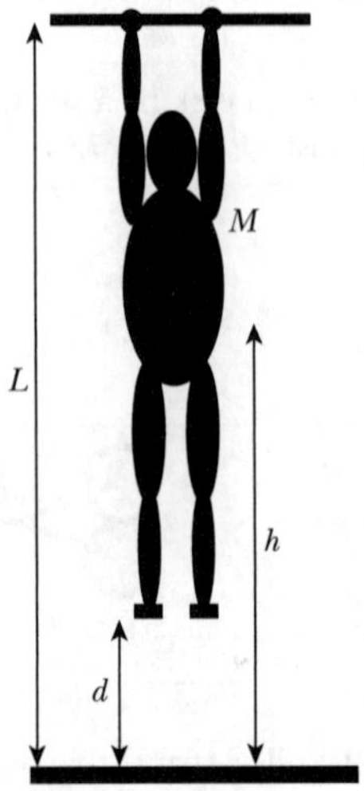

Figure 7.40 You hang from a tree limb and then drop to the ground (Problem 7.9).

Problem 7.10 Pulling a box containing an apparatus
A box and its contents have a total mass M. A string passes through a hole in the box, and you pull on the string with a constant force F (this is in outer space—there are no other forces acting).

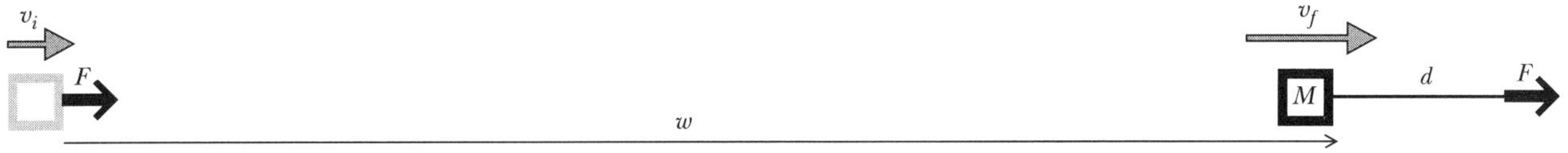

(a) Initially the speed of the box was v_i. After the box had moved a long distance w, your hand had moved an additional distance d (a total distance of $w + d$), because additional string of length d came out of the box. What is now the speed v_f of the box?

(b) If we could have looked inside the box, we would have seen that the string was wound around a hub that turns on an axle with negligible friction (Figure 7.41). Three masses, each of mass m, are attached to the hub at a distance r from the axle. Initially the angular speed of this apparatus was ω_i. In terms of the given quantities, what is the final angular speed ω_f of the apparatus?

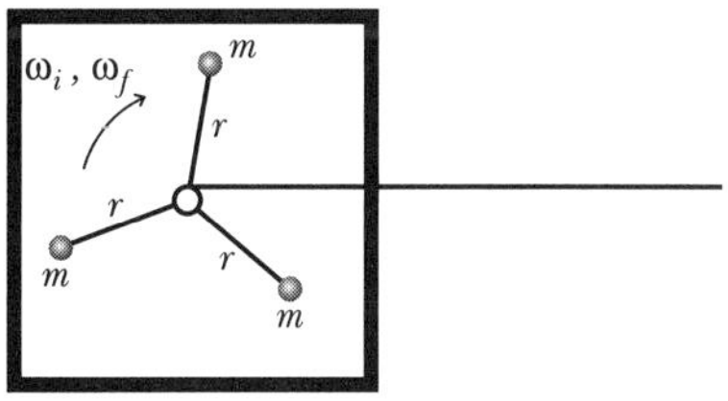

Figure 7.41 The mechanism inside the box (Problem 7.10).

Problem 7.11 Moving blocks connected by a spring
Two identical 0.1 kg blocks (labeled 1 and 2) were at rest on a nearly frictionless surface, connected by an unstretched spring whose stiffness is 100 N/m. Then a constant force of 5 N to the right was applied to block 2, and at a later time the blocks are in the new positions shown in the lower diagram.

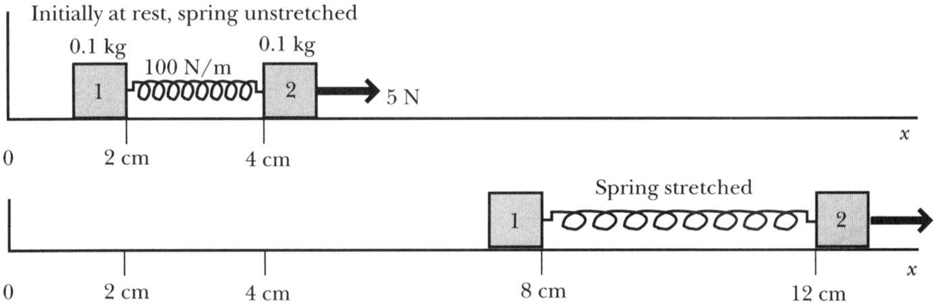

(a) At this later time, what is $K+U$ of the two-block system?
(b) What is the translational kinetic energy of the two-block system?
(c) What is the speed of the center of mass of the two-block system?
(d) What is the vibrational kinetic energy of the two-block system? Note that the spring is now stretched.

Problem 7.12 Spring and two masses
You exert an upward force of $2Mg$ to hold an object at rest that consists of two masses, each of mass M, connected by a low-mass spring whose stiffness is k_s (Figure 7.42). The initial stretch of the spring is s_i (equal to Mg/k_s). Then you suddenly start applying a larger force, of constant magnitude $F > 2Mg$. The diagram shows the situation some time later, when the blocks have moved upward, and the spring stretch has increased to s_f.

(a) What is now the speed of the center of mass of the two blocks? As part of your explanation, on a diagram show all of the forces that are acting. Also, be clear and explicit about what system you are analyzing.

(b) What is the vibrational kinetic energy of the two blocks (their kinetic energy relative to the center of mass)? Be clear and explicit about what system you are analyzing.

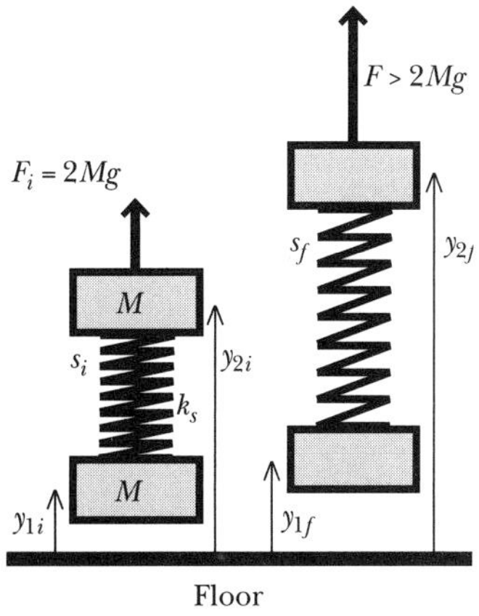

Figure 7.42 Pull up on a system consisting of two masses connected by a spring (Problem 7.12).

Problem 7.13 A binary star system

(a) About half of the visible "stars" are actually binary star systems, two stars that orbit each other with no other objects nearby. Describe the motion of the center of mass of a binary star system. Briefly explain your reasoning.

(b) For a particular binary star system, telescopic observations repeated over many years show that one of the stars (whose unknown mass we'll call M_1) has a circular orbit with radius $R_1 = 6\times10^{11}$ m, while the other star (whose unknown mass we'll call M_2) has a circular orbit of radius $R_2 = 9\times10^{11}$ m about the same point. Make a sketch of the orbits, and show the positions of the two stars on these orbits at some instant. Label the two stars as to which is which, and label their orbital radii. Indicate on your sketch the location of the center of mass of the system, and explain how you know its location, using the concepts and results of this chapter.

(c) This double star system is observed to complete one revolution in 40 years. What are the masses of the two stars? (For comparison, the distance from Sun to Earth is about 1.5×10^{11} m, and the mass of the Sun is about 2×10^{30} kg.) This method is often used to determine the masses of stars. The mass of a star largely determines many of the other properties of a star, which is why astrophysicists need a method for measuring the mass.

Problem 7.14 Kinetic energy of a two-particle system

By calculating numerical quantities for a multiparticle system, one can get a concrete sense of the meaning of the relationships $\vec{p}_{tot} = M_{tot}\vec{v}_{cm}$ and $K_{tot} = K_{trans} + K_{rel}$. Consider an object consisting of two balls connected by a spring, whose stiffness is 400 N/m. The object has been thrown through the air and is rotating and vibrating as it moves. At a particular instant the spring is stretched 0.3 m and the two balls at the ends of the spring have the following masses and velocities:

1: 5 kg, $\langle 8, 14, 0\rangle$ m/s
2: 3 kg, $\langle -5, 9, 0\rangle$ m/s

(a) For this system, calculate $\vec{p}_{tot}$.

(b) Calculate $\vec{v}_{cm}$.

(c) Calculate K_{tot}.

(d) Calculate K_{trans}.

(e) Calculate K_{rel}.

(f) Here is a way to check your result for K_{rel}. The velocity of a particle relative to the center of mass is calculated by subtracting $\vec{v}_{cm}$ from the particle's velocity. To take a simple example, if you're riding in a car that's moving with $v_{cm,x} = 20$ m/s, and you throw a ball with $v_{rel,x} = 35$ m/s, relative to the car, a bystander on the ground sees the ball moving with $v_x = 55$ m/s. So $\vec{v} = \vec{v}_{cm} + \vec{v}_{rel}$, and therefore we have $\vec{v}_{rel} = \vec{v} - \vec{v}_{cm}$. Calculate $\vec{v}_{rel} = \vec{v} - \vec{v}_{cm}$ for each mass and calculate the corresponding K_{rel}. Compare with the result you obtained in part (e).

7.12 Answers to exercises

7.1 (page 293) No acceleration, no force, no strain. Once you are no longer exerting a force on the right-most atoms in the block, atoms just to the left catch up and the interatomic force relaxes. This relaxation propagates to the left through the block, until finally there is no strain, and the block coasts along at a constant speed.

7.2 (page 294) In the vertical member, $3L/4$ up from its bottom.

7.3 (page 295) 4600 km (so inside the Earth)

7.4 (page 297) The puck with the string wound around it is rotating and so has additional kinetic energy. You have to pull farther, by the amount of string that unwinds, so you do more work, which corresponds to this puck having more kinetic energy. The Momentum Principle for the motion of the center of mass tells us nothing about the rotational aspects of the motion.

7.5 (page 297) 6 N to the right; 20 m/s^2

7.6 (page 304) All of the applied forces act through the same displacement, which is also the displacement of the center of mass. So it makes no difference whether you calculate the work done by each individual force, and then add them up, or first add up all the forces, and then calculate the work. Also, the only energy change in such a system is the overall kinetic energy.

7.7 (page 304) Average force = 167 newtons.

7.8 (page 304) Work done on real system = 0 joules. Work done on point particle system = 2500 joules.

7.9 (page 304) Change in chemical energy of runner = −2500 joules

7.10 (page 304) The instantaneous speed of the bottom of the tire is zero. The force by the road on the tire does no work. The increased translational kinetic energy of the car comes from a decrease in the chemical energy of the gasoline.

Chapter 8

Collisions: Exploring the Nucleus

Chapter 8

Collisions: Exploring the Nucleus

In high-energy particle accelerators, subatomic particles are hurled at each other. Changes of momentum and changes in identity of the particles involved in the collisions can reveal a great deal about the properties of the particles, and about their mutual interactions. We may not be able to observe what happens during a brief collision, but by applying conservation of momentum we can analyze the situation before and after the collision, and can obtain much useful information.

By "collision" we mean an interaction that takes place in a very short time, with little interaction before or after the collision. Neutral molecules in a gas experience collisions—the interatomic electric force is effectively short range, so there is little interaction until the molecules come nearly into contact. Similarly, collisions between macroscopic objects are quite common: before a ball hits a wall it has little interaction with the wall, and the contact between ball and wall lasts a very short time.

Although we have not yet talked very much about conservation of momentum, it is already familiar to us because it is simply a consequence of the momentum principle and the reciprocity of electric and gravitational forces. In this chapter we will learn how to apply conservation of momentum to the analysis of collisions. Our study of collisions will lead naturally into studying how processes look from different reference frames. We will see what we can learn about interactions by observing the probability distribution of deflection angles. We will investigate "elastic" collisions (no change of kinetic energy) and "inelastic" collisions (some energy is changed from kinetic to other forms of energy).

8.1 Momentum and impulse

In a collision an object receives a very short-duration impulse $\vec{F}\Delta t$ and its momentum changes by an amount $\Delta\vec{p} = \vec{F}\Delta t$. The following exercises review the relation between impulse and momentum change.

Example: A ball whose mass is 0.1 kg hits the floor with a speed of 6 m/s and rebounds upward with a speed of 5 m/s. What was the magnitude and direction of the impulse delivered to the ball by the floor? (Remember that momentum and impulse are vectors.)

Solution: With the y axis pointing upward,

$$\vec{p}_i = (0.1\text{ kg})(\langle 0, -6, 0\rangle\text{ m/s})$$

$$\vec{p}_f = (0.1\text{ kg})(\langle 0, 5, 0\rangle\text{ m/s})$$

$$\vec{p}_f - \vec{p}_i = \langle 0, +1.1, 0\rangle\text{ kg·m/s} = \text{the impulse}$$

Ex. 8.1 In the preceding exercise, if the ball was in contact with the floor for 1 ms (1×10^{-3} s), what was the average magnitude of the force exerted on the ball by the floor? Compare with the weight of the ball mg.

This illustrates the class of interactions we're talking about. We can't neglect the gravitational mg force before or after the collision, but *during* the brief collision the force exerted by the floor is so large that the other forces are negligible.

8.2 Conservation of momentum

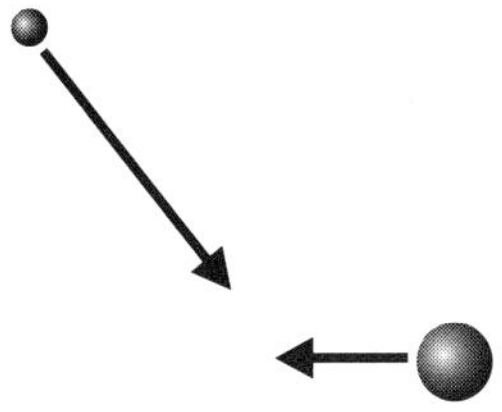

Figure 8.1 Two particles moving toward each other, free from external forces.

Figure 8.2 During the collision the two particles exert large forces on each other.

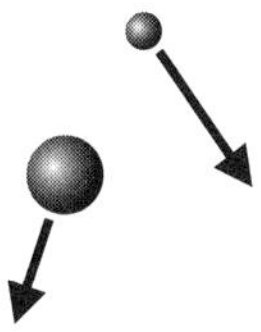

Figure 8.3 After the collision each particle's momentum has changed.

Momentum conservation is particularly useful in analyzing collisions between two particles. As an example, consider two particles in outer space (Figure 8.1), free of external forces, which move toward each other and eventually collide. During the short time of the collision, the particles exert large forces on each other (Figure 8.2), which according to the momentum principle will cause changes in each particle's momentum (Figure 8.3).

Consider the system consisting of both particles. The net external force on this combined system is zero. Therefore there is no change in the total momentum of the system:

$$\Delta\vec{P}_{tot} = \Delta(\vec{p}_1 + \vec{p}_2) = 0$$

Another way to predict the same result is to note that by the principle of reciprocity (Newton's third law), the force (and impulse) exerted on particle 1 by particle 2 is equal and opposite to the force (and impulse) exerted on particle 2 by particle 1, so the change in the momentum of particle 1 is equal and opposite to the change of the momentum of particle 2. Therefore the net change in the total momentum is zero. (Note that it was the principle of reciprocity that allowed us to derive the multiparticle version of the momentum principle.)

Here is a statement of this very important principle:

CONSERVATION OF MOMENTUM

$$\Delta\vec{P}_{system} + \Delta\vec{P}_{surroundings} = 0$$

A system that is free of external forces exerted by the surroundings will have a total momentum that does not change.

? Are there any systems of particles (smaller than the Universe itself) that are completely free of external forces?

No object is completely free of gravitational interactions, so the momentum for systems of interest may change slightly, yet a constant momentum can be an excellent approximation to the real situation, especially in collisions. Why? A collision takes a very short time Δt, from just before to just after the initial contact. The nonzero external forces are small compared to the large forces the colliding objects exert on each other, and therefore have only a small effect during the short duration of the collision, since $\Delta\vec{P}_{tot} = \vec{F}_{net,ext}\Delta t$.

As long as Δt is very short, and the net external force is of ordinary magnitude, the external impulse will be small and the change it causes in the total momentum of the two-object system will usually be negligible. Moreover, the change in the total momentum is likely to be negligible compared to the change in the momentum of one of the colliding particles, which receives a large impulse due to a very large force. An example is a collision of two baseballs in midair (Figure 8.4).

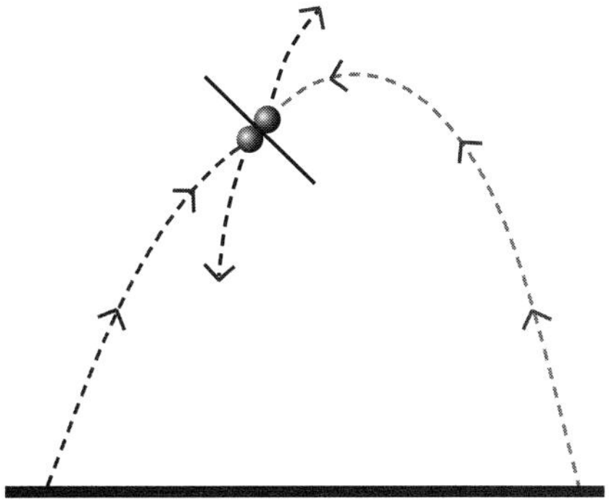

Figure 8.4 Midair collision between two baseballs.

? Look at the curving trajectories before the collision. Does the total momentum of the two baseballs change during the long period before the collision? What is the cause?

The total momentum of the two-ball system changes, because gravitational and air-resistance forces act on the system.

The collisional contact lasts a very short time, and involves very large contact forces, as is evident from the abrupt changes of directions of both balls.

> **?** How does the external impulse $\vec{F}_{\text{net,ext}}\Delta t$ (gravitational force, air resistance) acting on just one of the baseballs compare with the impulse due to the other baseball, during contact? Is it a good approximation to say that from just before contact to just after contact, the total momentum of the two balls is (nearly) constant?

The external impulse involves an ordinary force for a very short time, while the impulse due to the other ball involves a huge force acting for that same short time. Therefore the momentum change of one of the balls is very nearly the opposite of the momentum change of the other ball. Approximately, during the short contact time, the total momentum of the two-ball system is constant.

These considerations justify an approach to the analysis of collisions that proves fruitful. Up until the collision, and after the collision, we do need to take into account external forces, which affect the total momentum. But during the brief time interval of the collision itself, we can use the approximation that the external impulse is small, and the change in total momentum of the interacting objects is negligible.

"Collisions" without contact

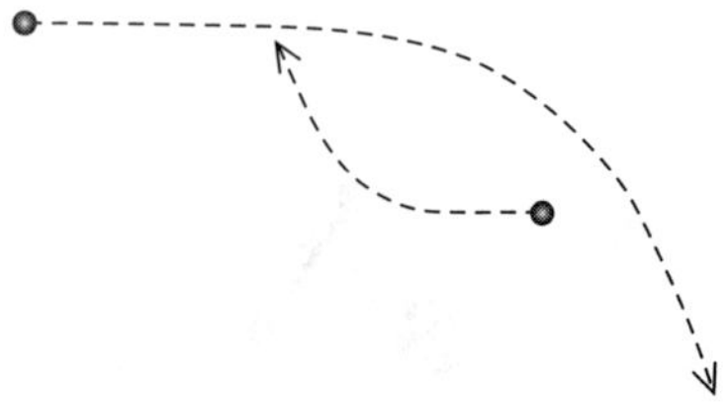

Figure 8.5 Electric or gravitational interactions without contact; these too are considered to be "collisions."

Consider the electric interaction of a proton and electron, or the gravitational interaction of two asteroids, as shown in Figure 8.5. There is no contact—no touching—yet we call such an interaction a "collision." A collision is any process in which there is little interaction before and after a short time interval, and large interactions during that short time interval.

8.2.1 A simple example

In the following sections we'll apply the principle of conservation of momentum to several different kinds of processes. In this section we consider a simple but instructive example.

A ball of mass m is thrown with speed v against a wall and rebounds with nearly the same speed v (Figure 8.6).

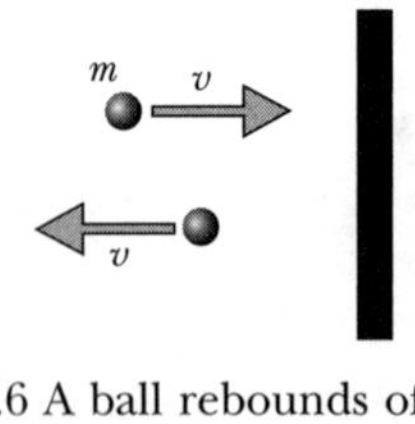

Figure 8.6 A ball rebounds off a wall with nearly the same speed v.

> **?** Is the momentum of the ball constant?

No. Even though the magnitude of the momentum doesn't change, the x component of the momentum vector does change, from positive (going to the right) to negative (going to the left).

> **?** What caused the momentum of the ball to change?

The wall exerted a large force to the left on the ball for a short time, and this impulse changed the momentum of the ball.

> **?** What was the magnitude and direction of the momentum change, $\Delta\vec{p}$?

The initial momentum in the x direction was $+p$ and the final momentum in the x direction was $-p$ (Figure 8.7), so there was a change of x component of momentum of the following amount:

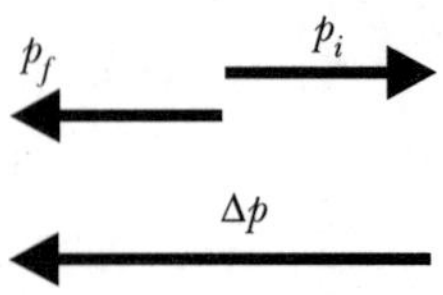

Figure 8.7 Change in momentum of the ball.

$$\Delta p_x = (p_f - p_i) = -2p$$

> **?** What was the magnitude and direction of the impulse exerted by the wall on the ball?

Since change of momentum is equal to the impulse, evidently the impulse must have had a magnitude of $2p$, and a direction to the left; the x component of the impulse was $-2p$.

? But isn't something constant when the ball bounced off the wall? After all, the speed hardly changed.

Yes, the energy of the ball hardly changed—specifically, the kinetic energy before and after was nearly the same. When the ball compressed against the wall, momentarily the kinetic energy went to zero, but elastic energy in the ball restored the kinetic energy on the rebound. Note that energy is a scalar represented by a single number, whereas momentum is a vector represented by three numbers: $\langle p_x, p_y, p_z \rangle$.

? But isn't momentum always unchanged in a process?

No, the momentum of a system changes if an impulse (a force acting for a time) is applied to the system by the surroundings, in which case the momentum of the surroundings has an equal and opposite change (momentum conservation). Here the wall exerted a force on the ball, so the momentum of the ball was not constant. However, if you consider a larger system consisting of the Earth (including the wall) plus the ball, the momentum of the combined system *is* constant in the collision. When the ball bounces to the left, the wall (and the Earth to which it is connected) recoil to the right! In the following sections we will develop a method of analysis that can be used to determine the recoil speed of the Earth.

8.2.2 A head-on collision

Later in this chapter we will study a famous experiment that led to the discovery of the nucleus inside the atom. The experiment was carried out in 1911 by a group led by Ernest Rutherford. In that experiment high-speed alpha particles (now known to be helium nuclei, consisting of two protons and two neutrons) were shot at a thin gold foil.

In the original Rutherford experiment, the velocity of the alpha particles was not high enough to allow them to make actual contact with the nucleus. It is nonetheless appropriate to think of the process as a collision, because there are sizable electric forces acting between the alpha particle and the gold nucleus for a very short time. We call any process of this kind a collision, even if there is no direct contact but only gravitational, electric, or magnetic interactions at a distance.

As an example of the use of the momentum concept, we will analyze the head-on collision of an alpha particle and a gold nucleus (consisting of 79 protons and 118 neutrons). We treat a head-on collision first because it is a simple but important case, with only x components of momentum.

We make "before" and "after" diagrams of the relevant momenta in a head-on collision (Figure 8.8 and Figure 8.9). Let m be the mass of the alpha particle and M be the mass of the gold nucleus. The incoming alpha particle has a known momentum $\vec{p}_1$, the gold nucleus initially has a known momentum $\vec{p}_2 = 0$, the outgoing alpha particle has a momentum $\vec{p}_3$, and the gold nucleus picks up a momentum $\vec{p}_4$. Our task is to calculate p_3 and p_4.

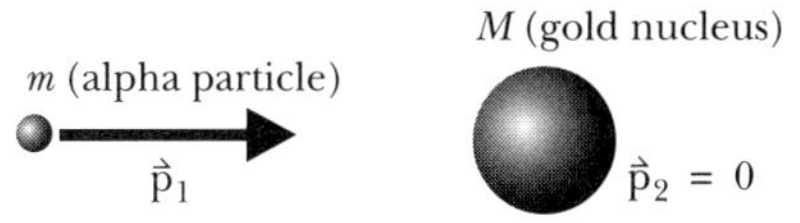

Figure 8.8 Before a head-on collision between an alpha particle and a gold nucleus.

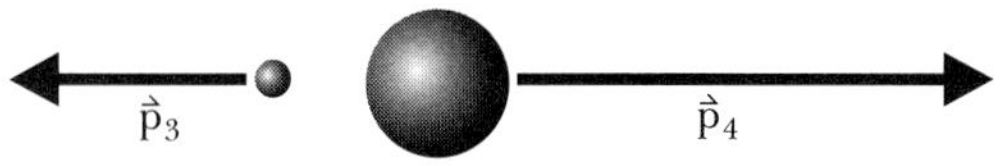

Figure 8.9 After the collision.

In the Rutherford experiment the speeds of all of the particles were small compared to the speed of light, so we expect to be able to use the low-speed approximations for momentum and energy to analyze the collision. The outgoing speed of the gold nucleus is quite small despite the large momentum vector, because it has a large mass: $v_4 = p_4/M$.

? What can you say about the relation of the total momentum of the alpha particle plus gold nucleus *before* the collision to their total momentum *after* the collision? Why?

During the collision external forces are negligible, so the total momentum of the two-particle system will not change. Therefore we can write the following momentum conservation equation relating the total momentum before the collision to the total momentum after the collision:

$$\vec{p}_1 + \vec{p}_2 = \vec{p}_3 + \vec{p}_4$$

$$p_{1x} + 0 = p_{3x} + p_{4x}$$

We expect to find that p_{3x} is negative, corresponding to the alpha particle bouncing straight back when it hits the massive gold nucleus head-on. Note that if the alpha particle's momentum essentially turns around, $p_{3x} \approx -p_{1x}$. the gold nucleus picks up a lot of momentum:

$$p_{4x} = p_{1x} - p_{3x} \approx p_{1x} - (-p_{1x}) = 2p_{1x}$$

The outgoing gold nucleus has about twice the momentum of the incoming alpha particle, as suggested in Figure 8.8 and Figure 8.9. However, the speed of the gold nucleus, about $2p_{1x}/M$, is small because its mass is very large.

The conservation of momentum has given us one equation in two unknowns, p_{3x} and p_{4x} (the incoming momentum p_{1x} is known for these alpha particles). We don't have enough information to solve for the unknowns and check our qualitative conclusions. Can we use the energy principle to get a second equation involving the two unknown momenta?

8.2.3 Elastic collisions

Since conservation of energy is an important principle in its own right, and separate from conservation of momentum, it may be fruitful to apply energy conservation to the analysis of a head-on collision between an alpha particle and a gold nucleus in the Rutherford experiment. The conditions of this experiment were such that there was no change in the internal energy of either the alpha particle or the gold nucleus.

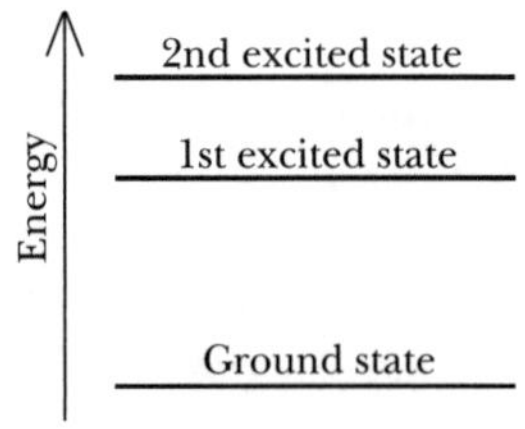

Figure 8.10 Energy levels in a nucleus. The potential energy curve is not shown.

There do exist "excited" states of nuclei, in which there is a change in the configuration of the nucleons corresponding to a higher energy than the ground state, the normal configuration of the nucleus. The internal energy of the nucleus is quantized, and there are no states with energy intermediate between these quantized energies (Figure 8.10). In the Rutherford experiment, the interaction did not raise the gold nucleus nor the alpha particle to an excited state; both nuclei remained in their ground states. This happy accident made it much easier for Rutherford to analyze the experiment. Such a collision, in which there is no change in the internal energy of the colliding particles, is called an "elastic" collision.

ELASTIC COLLISION

In an "elastic" collision, the internal energy of the objects in the system does not change: $\Delta E_{int} = 0$

We want to write an energy equation for the collision of the alpha particle and the gold nucleus. This is an elastic collision; the internal energy of the system does not change.

What about the electric potential energy associated with the electric forces? Long before the collision the two nuclei were far from each other, and long after the collision they are again far from each other. Like gravitational energy, electric energy is proportional to $1/r$, where r is the distance between the two charged objects. Therefore from long before the collision to long after the collision the change in electric energy is very nearly zero; that is, $\Delta U_{el} \approx 0$.

Can we in fact claim that in the initial and final states the particles are far enough apart to justify the approximation $\Delta U_{el} \approx 0$? Because the electrons of the atom are much less massive than the alpha particle they do not affect its path very much, so we can start our analysis when the alpha particle is deep inside the atom, with the electrons of the gold atom mostly far outside the region of interest. An atomic diameter is about 10^{-10} m, whereas the diameter of a gold nucleus is much smaller, less than 10^{-14} m, so we might start our analysis of the collision when the particles are about 10^{-12} m apart, a distance that is very small compared to an atom and very large compared to a nucleus. At this distance the electric potential energy is

$$U_{el} = \frac{1}{4\pi\varepsilon_0}\frac{(2e)(79e)}{r} = \left(9\times10^9 \frac{\text{N}\cdot\text{m}^2}{\text{C}^2}\right)\frac{(2)(79)(1.6\times10^{-19}\text{ C})^2}{(10^{-12}\text{ m})}$$

$$U_{el} = (3.6\times10^{-14}\text{ J})\frac{(1\text{ eV})}{(1.6\times10^{-19}\text{ J})} = 0.2\times10^6\text{ eV} = 0.2\text{ MeV}$$

The radioactive source used by the Rutherford group provided alpha particles whose kinetic energy was 10 MeV, so if we take as the initial state a separation of 10^{-12} m the electric energy is only 2% of the kinetic energy. And if we also take as the final state a separation of 10^{-12} m we have $\Delta U_{el} = 0$. Evidently the only significant energy changes in this experiment are changes in the kinetic energies of the alpha particle and the gold nucleus.

Here then is the additional equation we get from energy conservation, involving only kinetic energies:

$$K_1 + 0 = K_3 + K_4$$

Rewrite in terms of the momentum components:

$$\frac{p_{1x}^2}{2m} = \frac{p_{3x}^2}{2m} + \frac{p_{4x}^2}{2M}$$

Remember that momentum conservation gave us this equation:

$$p_{1x} = p_{3x} + p_{4x}$$

Now we have two equations in two unknowns, p_{3x} and p_{4x}.

Ex. 8.2 Try to solve the two equations to determine the final momenta, in terms of the known initial momentum p_{1x}. A good way to start is to solve the momentum equation for p_{4x} (the final momentum of the gold nucleus) and plug this into the energy equation, then solve for the single unknown p_{3x} (the final momentum of the alpha particle).

8.2.4 Light projectile, heavy target

In the preceding exercise, you probably used the quadratic formula to find the following result for the final momentum p_{3x} of the alpha particle:

$$p_{3x} = \left[\frac{m \pm M}{m + M}\right]p_{1x}$$

This equation has two solutions. If you take the "+" sign, you get $p_{3x} = p_{1x}$, implying that instead of bouncing back the alpha particle kept going in the forward direction, with its original momentum p_{1x}. This solution simply corresponds to passing right through the gold nucleus without interacting! It is interesting that this is a possible solution, since it would be valid if a neutrino instead of an alpha particle were the projectile, since typically a neutrino would pass right through a gold nucleus without interacting. It also corresponds to the case where the alpha particle passes very far from the gold nucleus and so is hardly deflected by the tiny electric force.

The other solution, with the "–" sign, is the interesting one for the alpha particle final momentum:

$$p_{3x} = \left[\frac{m-M}{m+M}\right]p_{1x}$$

? Should the magnitude of the final momentum p_3 of the alpha particle be greater or smaller than the initial momentum p_1? Why? Is this solution consistent with that prediction?

Since the gold nucleus acquires some nonzero kinetic energy, the kinetic energy of the alpha particle must decrease, and that means that the magnitude of the alpha particle momentum must decrease. Since the magnitude of $(m-M)/(m+M)$ is less than one, the result is consistent with the loss of kinetic energy to the gold nucleus.

? We assumed that the low-mass alpha particle would bounce backward upon colliding with the high-mass gold nucleus, and that p_{3x} should be negative. Is it?

Yes, $(m-M)$ is negative, since $m<M$.

Approximate final momentum of the light projectile

? What approximation can we make to get a simpler expression for the final momentum of the alpha particle?

The alpha particle contains 4 nucleons (2 protons and 2 neutrons), while the gold nucleus contains 197 nucleons (79 protons and 118 neutrons). Since protons and neutrons have nearly the same mass, the mass of a nucleus containing N nucleons is approximately equal to the mass of N nucleons (neglecting the small mass difference associated with the binding energy of the nucleus).

Since the mass of the alpha particle is very small compared to the mass of the gold nucleus ($m << M$), $(m-M)/(M+m)$ is nearly equal to –1, and

$$p_{3x} \approx -p_{1x}$$

? Is this consistent with what happens when you throw a (low-mass) ping pong ball at a (high-mass) bowling ball?

Our approximate formula predicts that a fast lightweight object hitting a heavy object at rest will bounce back with nearly the initial momentum (and speed), which is what we observe when a ping pong ball bounces off a bowling ball (which is a nearly elastic collision).

Rutherford, who was familiar with the result we have just derived, was extremely surprised when his group observed some alpha particles bounce straight back from a gold foil at high speed.

? Why might Rutherford have been surprised? Given what you know (and Rutherford didn't) about the structure of atoms, is his observation surprising to you?

Rutherford didn't know about the nucleus: he was about to discover it! He didn't know that there was a very small, massive core at the heart of the atom. Knowing that a gold nucleus is much more massive than an alpha particle, it isn't surprising to us that the alpha particle could bounce straight back.

It is important to note that we didn't need to know what kind of force was involved in the collision. Our results would be valid for nonelectric forces.

Approximate final momentum of the heavy target

If you solve the momentum and energy conservation equations for the momentum of the recoiling gold nucleus you get the following:

$$p_{4x} = \frac{2M}{M+m}p_{1x}$$

Since the mass M of the gold nucleus is much larger than the mass m of the alpha particle, $2M/(M+m)$ is approximately equal to 2, and the gold nucleus recoils with about twice the momentum of the alpha particle:

$$p_{4x} \approx 2p_{1x}$$

This justifies our original qualitative analysis. If the alpha particle simply turns around, its momentum change is about $-2p_{1x}$, and the gold nucleus acquires a momentum of about $2p_{1x}$.

Ex. 8.3 We can use our results for head-on elastic collisions to analyze the recoil of the Earth when a ball bounces off a wall embedded in the Earth. Suppose a professional baseball pitcher hurls a baseball ($m = 155$ grams) with a speed of 100 miles per hour ($v_1 = 44$ m/s) at a wall, and the ball bounces back with little loss of kinetic energy. What is the recoil speed of the Earth ($M = 6\times10^{24}$ kg)?

Ex. 8.4 Calculate the recoil kinetic energy of the Earth and compare to the kinetic energy of the baseball. The Earth gets lots of momentum (twice the momentum of the baseball) but very little kinetic energy.

8.2.5 Heavy projectile, light target

We have seen what happens when a light object collides with a stationary massive one. For completeness, it is interesting to consider the case of a gold nucleus hitting head-on an alpha particle that is initially at rest (Figure 8.11). We can determine the final speeds and directions without completely repeating our analysis. Merely switch M and m in the final results! In that case we predict that after the collision the momentum of the gold nucleus is

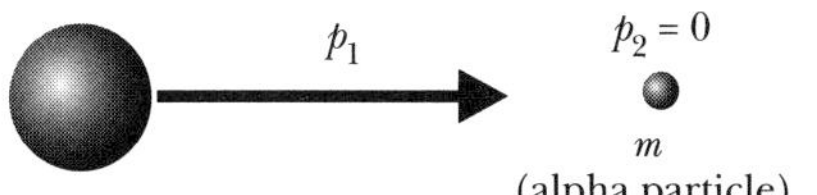

Figure 8.11 A gold nucleus collides head on with an alpha particle that is initially at rest.

$$p_{3x} = \left[\frac{M \pm m}{M+m}\right]p_{1x} \approx p_{1x} \text{ (since } M \text{ is much larger than } m\text{)}$$

Taking the "+" sign, we find exactly $p_{3x} = p_{1x}$, which would be the case if there were no interaction (or the gold nucleus passes far from the alpha particle). Taking the "–" sign, $(M-m)/(M+m)$ is only very slightly less than +1, which means that the gold nucleus keeps going with nearly its initial momentum. This makes sense: if you throw a heavy bowling ball at a light ping pong ball, you expect the bowling ball to keep going, and not to lose much speed.

What about the low-mass target—the alpha particle (or ping pong ball)? We use our previous result for the final target momentum, but interchange M and m:

$$p_{4x} = \frac{2m}{m+M}p_{1x} \approx \frac{2m}{M}p_{1x}$$

In terms of the final speed v_4 of the alpha particle (ping pong ball) and the initial speed v_1 of the gold nucleus (bowling ball), this yields the following:

$$mv_4 \approx \frac{2m}{M}Mv_1 \text{ or } v_4 \approx 2v_1$$

This is a surprisingly simple result. The low-mass target moves at *twice* the speed of the incoming high-mass projectile, independent of the precise ratio of the masses. If you throw a bowling ball with speed v_1 at a ping-pong ball, the bowling ball keeps going with a speed almost equal to v_1, and the ping pong ball moves out ahead of the bowling ball with a speed that is approximately $2v_1$.

Viewed from a different reference frame

It is possible to understand this simple result, $v_4 \approx 2v_1$, by considering the original case of an alpha particle hitting a stationary gold nucleus, in which case we know that the alpha particle rebounds with nearly unchanged speed v_1. Run along with the velocity of the incoming alpha particle, and you will see it at rest, with an oncoming gold nucleus. The outgoing alpha particle heads to the left with a speed of approximately $2v_1$ from your point of view, because it is headed to the left with a speed of nearly v_1 while you are headed to the right with a speed v_1.

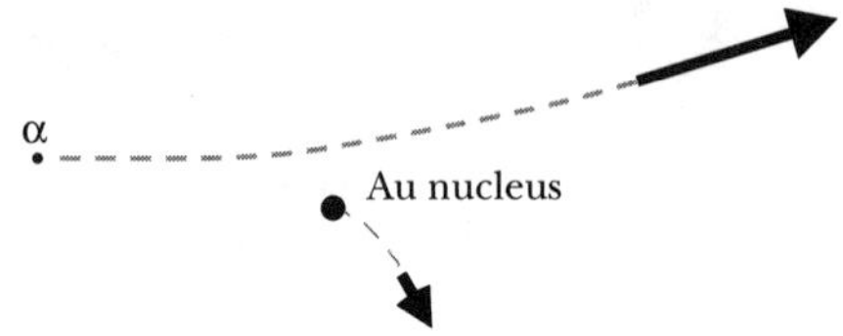

Figure 8.12 A collision that is not head-on.

8.3 Scattering

We considered the simplest case of collisions between an alpha particle and a gold nucleus, a head-on collision. What happens if the particles don't collide head-on? Figure 8.12 shows what the trajectories look like in this case.

In atomic or nuclear collisions, we can't observe in detail the curving trajectories inside the tiny interaction region. We only observe the trajectories before and after the collision, when the particles are far apart and their mutual interaction is very weak, so they are traveling in nearly straight lines.

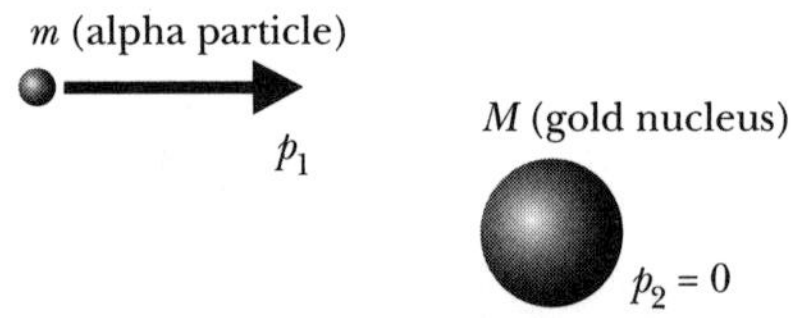

Figure 8.13 Before the collision, the alpha particle (mass m) approaches the gold nucleus (mass M).

The alpha particle is deflected through some final angle θ, and the gold nucleus recoils at some other final angle ϕ. The final angles are measured relative to the initial direction of the alpha particle (Figure 8.13 and Figure 8.14). The process is called "scattering," and we say that the alpha particle "scattered" through an angle θ.

We'll again assume that there is no internal energy change inside either the alpha particle or the gold nucleus, so that the initial kinetic energy (before the collision) should be equal to the final kinetic energy (after the collision). We'll take the "before" time early enough and the "after" time late enough that the two particles are sufficiently far away from each other that we can neglect their electric potential energy. As for conservation of momentum, we must write two equations, for the x and y components of momentum conservation.

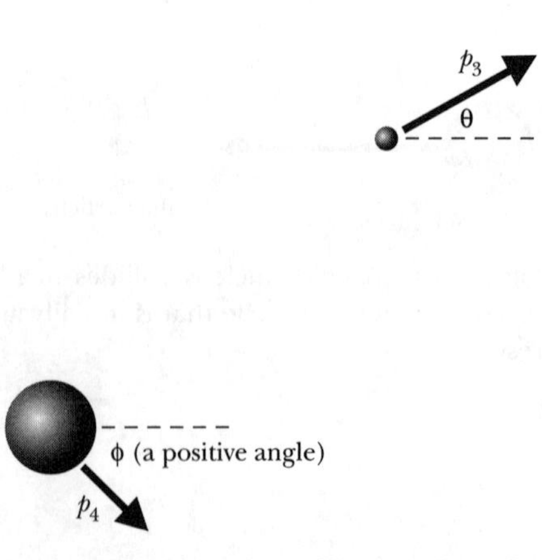

Figure 8.14 After the collision, when the particles are far apart and moving with nearly uniform velocities. Final angles are measured relative to the initial direction of the alpha particle.

? Try to write the two momentum-conservation equations, and the energy-conservation equation. Choose x in the direction of the incoming alpha particle, and y at right angles to it. Consider ϕ to be a positive angle.

We write the three equations in terms of the magnitudes of the momenta:

$$p_1 = p_3\cos\theta + p_4\cos\phi \quad \text{momentum conservation; } x \text{ components}$$

$$0 = p_3\sin\theta - p_4\sin\phi \quad \text{momentum conservation; } y \text{ components}$$

$$\frac{p_1^2}{2m} = \frac{p_3^2}{2m} + \frac{p_4^2}{2M} \quad \text{energy conservation; scalar}$$

Make absolutely sure that you understand every detail of these three equations! In particular, pay attention to the fact that since momentum is a vector, the momentum equations are written in terms of *x* and *y* components of the momentum.

> **?** It is fruitful to reflect on which of the quantities in these three equations are known and which are unknown. We know the masses of the alpha particle and the gold nucleus, and we know the initial speed of the alpha particle. Which quantities in these three equations are unknown?

You should have identified the two final momenta and the two final angles as unknown quantities (p_3, p_4, θ, and ϕ, all of which represent positive numbers). But if there are four unknowns and only three equations, how can we solve for the unknowns? The simple answer is, we can't!

Nevertheless, there are many situations where these three equations can be exploited. For example, if we measure the final direction of motion of the alpha particle, the angle θ, we can solve for the other three unknown quantities and thereby predict the final momenta p_3 and p_4 of the two particles and the direction of motion ϕ of the gold nucleus, even without measuring all of these quantities.

It is worth mentioning that in experimental particle physics the momentum is often measured in a very direct way. Charged particles curve in a magnetic field due to magnetic forces that act on moving charges, and the radius of curvature is a direct measure of the momentum of the particle, if you know how much charge it has, which you do know if you know what kind of particle it is. Usually the momentum of a particle can be measured more accurately than the velocity, though once you have measured the momentum you can deduce the velocity if you know what kind of particle it is, since in that case you know its mass.

Example: Elastic collision between identical particles, one initially at rest

An interesting example of an elastic collision is one involving particles of equal mass, such as an elastic collision between two alpha particles, or two gold nuclei, or two billiard balls (whose collision is nearly elastic), in which one of the two objects (the "target") is initially at rest (Figure 8.15 and Figure 8.16).

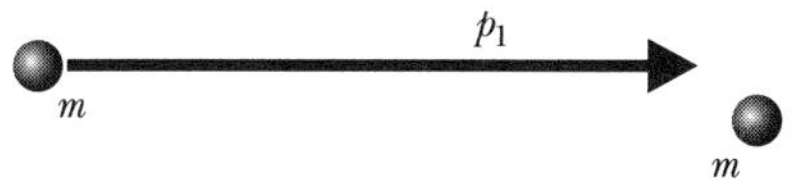

Figure 8.15 Before a collision of two identical particles, one initially at rest.

Because the vector momentum is conserved, we can write $\vec{p}_1 = \vec{p}_3 + \vec{p}_4$, then calculate p_1^2 by expanding the dot product:

$$p_1^2 = \vec{p}_1 \bullet \vec{p}_1 = (\vec{p}_3 + \vec{p}_4) \bullet (\vec{p}_3 + \vec{p}_4) = p_3^2 + p_4^2 + 2p_3p_4\cos A$$

Divide by $2m$:

$$\frac{p_1^2}{2m} = \frac{p_3^2}{2m} + \frac{p_4^2}{2m} + \frac{2p_3p_4\cos A}{2m}$$

Since the kinetic energy K can be written as $p^2/2m$, we have

$$K_1 = K_3 + K_4 + \frac{2p_3p_4\cos A}{2m}$$

But energy conservation for this elastic collision says that $K_1 = K_3 + K_4$.

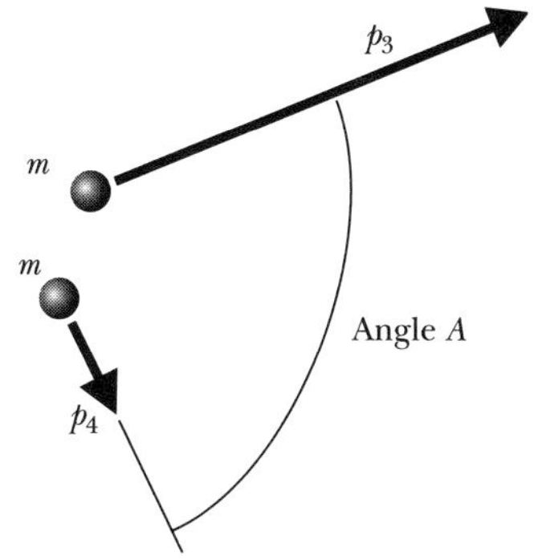

Figure 8.16 After the collision of two identical particles, one initially at rest.

> **?** What can you conclude about $\cos A$? What is the angle A?

In this special situation, an elastic collision between equal particles one of which is initially at rest, the angle between the final velocities must be 90°, as long as neither p_3 nor p_4 is zero (in that case the equation reduces to $0\cos A = 0$, in which case $\cos A$ is indeterminate). If one of the final momenta is indeed zero, the other particle must have all the momentum of the initial particle. This can happen either because the incoming particle missed the target entirely and kept going in the original direction, or because the incoming particle hit the target dead center, in which case the only consistent solution of the equations is for the incoming particle to come to rest and the target particle to move forward with the initial momentum.

8.3.1 Impact parameter

The distance between centers perpendicular to the incoming velocity is called the "impact parameter" and is usually denoted by b. A head-on collision has an impact parameter of zero. Here are possible elastic collisions between two billiard balls, for various impact parameters:

If the impact parameter, b, is too large, there is no interaction at all between two billiard balls.

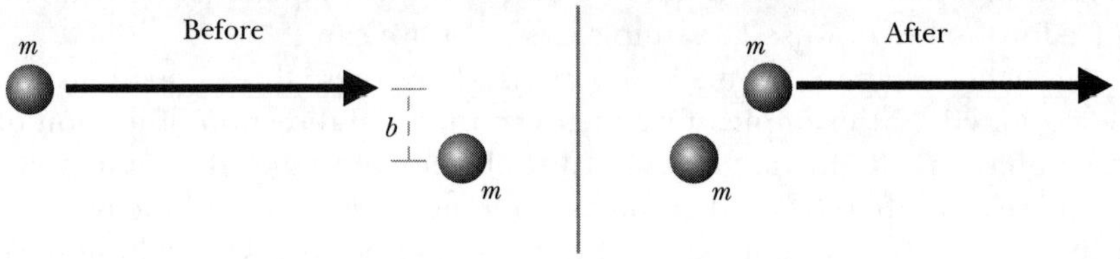

A grazing impact results in small angle scattering.

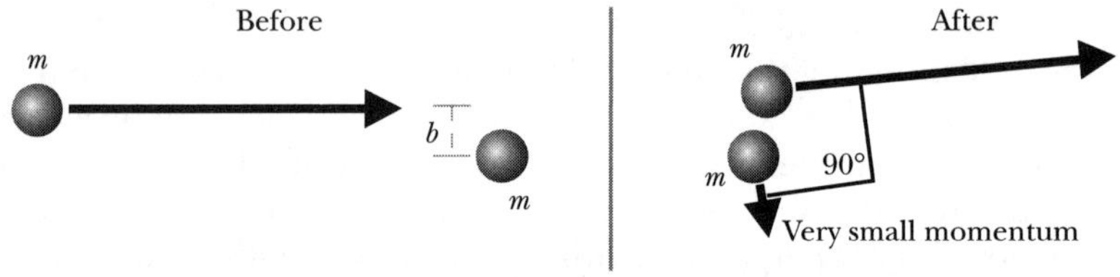

A medium impact parameter results in symmetric scattering.

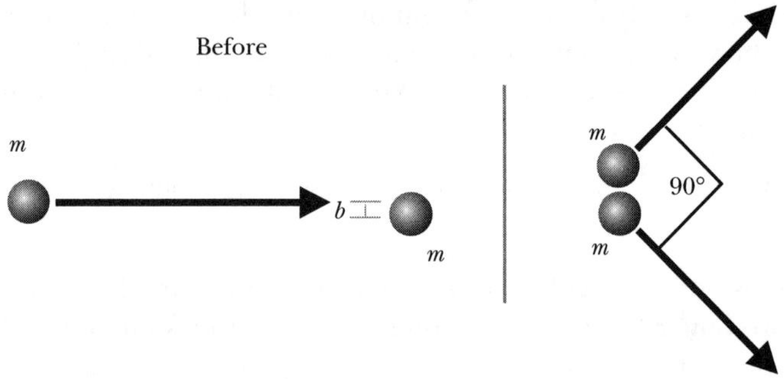

In a head-on collision the impact parameter $b = 0$. In this case all the momentum is transferred to the target.

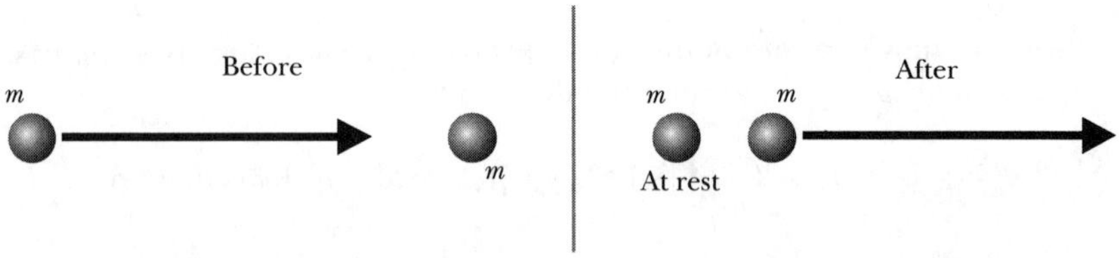

As you can see, the smaller the impact parameter, the more severe is the collision, and the larger the deflection angle of the incoming particle (larger "scattering"), except for a head-on collision, where the incoming particle stops dead and the target particle gets the entire momentum.

8.4 Discovering the nucleus inside atoms

In 1911 Ernest Rutherford and his coworkers decided to perform a scattering experiment in order to probe the internal structure of matter. They were familiar with the relationships of momenta in collisions that we have just studied, and they reasoned that by shooting microscopic particles at a thin layer of metal, they might be able to learn something about the microscopic structure of the metal by looking at patterns of scattering.

8.4.1 The plum pudding model of an atom

For about a decade, the prevailing model of an atom had been what was called a "plum pudding" model proposed by J. J. Thomson. Electrons were discovered in 1897 by Thomson, using a cathode ray tube, a device similar to a television picture tube. Thomson was able to show that the "cathode rays" that he and others had observed in electric gas discharges were in fact particles that were negatively charged, by the already existing convention for labeling electrical charges. This was the first conclusive demonstration of the existence of what we now call "electrons," one of the major constituents of all atoms.

Since these "electrons" were extracted from ordinary matter, which is normally not electrically charged, Thomson's discovery implied that matter also contained something having a positive charge. Because the electrons apparently had little mass, it seemed likely that most of the mass in ordinary matter was charged positively. Thomson proposed what was called a "plum pudding" model of atoms: a positively charged substance of uniform, rather low density, with tiny electrons distributed like raisins throughout it. Rutherford's group expected to gather evidence supporting this model from their collision experiments. Rutherford was extremely surprised by the results of these experiments, which led to a radical revision of the model.

8.4.2 The Rutherford experiment

A new source of fast-moving particles was provided through the discoveries of natural radioactivity by Henri Becquerel and by Marie and Pierre Curie. These newly discovered particles included high-speed, massive, positively charged "alpha particles," which we now know to be the nuclei of helium atoms, consisting of two protons and two neutrons bound together. Alpha particles are emitted in the radioactive decay of heavy elements. Alpha particles were convenient microscopic projectiles, and Rutherford's group decided to shoot alpha particles at a very thin piece of gold foil and observe where they reappeared.

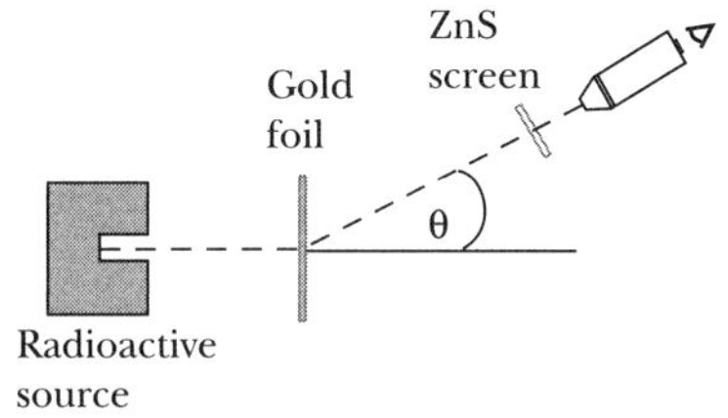

Figure 8.17 Schematic diagram of the Rutherford experiment. A radioactive source emits alpha particles, which are directed at a thin piece of gold foil. Particles scattered by gold nuclei in the gold foil hit a screen coated with ZnS, which fluoresces, giving off light, when hit by an energetic particle. The light was observed by the experimenter.

? Based on the plum pudding model, what would Rutherford and his colleagues have expected to see in their collision experiments?

The researchers expected the alpha particles to be deflected only very slightly by interactions with the low-mass electrons and with the low-density "pudding" of positively charged matter in the gold atoms. They could detect a deflected ("scattered") alpha particle by a pulse of light that was emitted when the particle struck a fluorescent material (zinc sulfide). They moved their detector around at various angles to the incoming beam of alpha particles and measured what fraction of the alpha particles were scattered through various angles (Figure 8.17).

Rutherford and his coworkers were astounded to observe that an alpha particle, which is much more massive than an electron, sometimes bounced straight backwards, as though it had encountered an even more massive yet very tiny positive particle inside the gold foil, rather than being slightly deflected in its passage through the "pudding" of positive charge.

"It was quite the most incredible event that has ever happened to me in my life," Rutherford said. "It was almost as incredible as if you fired a 15-inch [artillery] shell at a piece of tissue paper and it came back and hit you."

He continued, "On consideration, I realized that this scattering backward must be the result of a single collision, and when I made the calculations I saw that it was impossible to get anything of that order of magnitude unless you took a system in which the greater part of the mass of the atom was concentrated in a minute nucleus. It was then that I had the idea of an atom with a minute massive center, carrying a charge." (See Figure 8.18.)

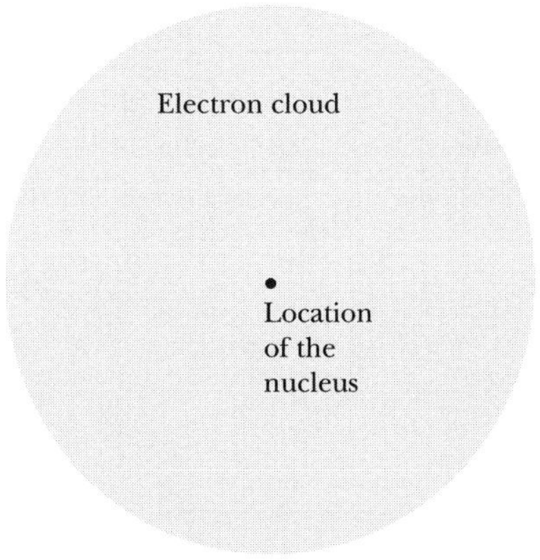

Figure 8.18 This diagram of an atom is not to scale; the radius of the nucleus is only about 1/100000 the radius of the electron cloud, so the nucleus would not be visible at all on this scale.

8.4.3 Computer modeling of the Rutherford experiment

Since the details of the electric interaction are now known, we can model the Rutherford experiment in detail; see Problem 8.1. Remember that Coulomb's law for electric forces is very similar to Newton's law of gravitation. The electric force F_e between two small charged particles carrying amounts of charge q_1 and q_2, and located a distance r apart is this:

$$F_e = \frac{1}{4\pi\varepsilon_0}\frac{q_1 q_2}{r^2}$$

The charges q_1 and q_2 are measured in units of coulombs, and the proportionality factor is measured to be $1/(4\pi\varepsilon_0) = 9\times10^9 \text{ N}\cdot\text{m}^2/\text{coulomb}^2$. The electric charge of the proton is $e = +1.6\times10^{-19}$ coulomb, and the electric charge of the electron is $-e = -1.6\times10^{-19}$ coulomb.

8.4.4 Distribution of scattering angles

In general, we have three equations describing an elastic collision—two momentum equations (in the x and y directions) and one energy equation, but there are four unknown quantities (the outgoing speeds and directions for each particle). The concept of impact parameter provides an additional quantity, and you have seen on page 332 how the impact parameter determines the scattering angle θ of the incoming particle, from which you can predict the other final values (p_3, p_4, and ϕ).

In a macroscopic situation, such as shooting billiard balls, you can choose an impact parameter simply by carefully aiming at a point a distance b from the center of the target. Unfortunately, in experiments where you fire alpha particles or other subatomic projectiles at nuclei, you can't aim at a specific impact parameter because the target is so incredibly small. All you can do is fire lots of projectiles at the subatomic target and expect a probabilistic distribution of impact parameters

When we fire a projectile at this microscopic target, the probability of hitting one particular ring (whose radius corresponds to the impact parameter) is proportional to the area of that ring (Figure 8.19). If you calculate the range of scattering angles that corresponds to hitting somewhere within a particular ring, you can predict the angular distribution of scattering angles that will be observed.

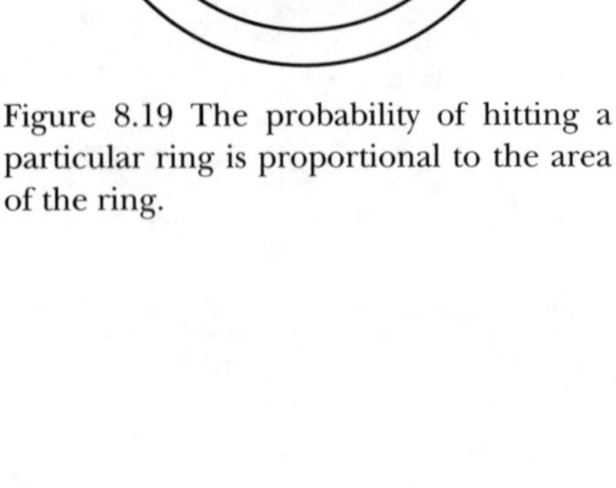

Figure 8.19 The probability of hitting a particular ring is proportional to the area of the ring.

Cross section

If a target particle such as a gold nucleus has a radius R, we say that its "geometrical cross section" is πR^2. A high-speed alpha particle that hits within this cross-sectional area will interact strongly with the gold nucleus. Quantum-mechanical effects can make the effective target size bigger or smaller than the geometrical cross section, and the effective target size is called the "cross section" for interactions of this kind. Cross sections are measured experimentally by observing how often projectiles are deflected by an interaction, instead of passing straight through the material without interacting. Cross sections have units of square meters.

A "differential" cross section for scattering within a particular range of angles is the area in square meters of a ring bounded by impact parameters corresponding to the given angular range.

8.4.5 Conservation laws *vs.* details of the interaction

We used general principles to determine the final momenta of two colliding particles (conservation of momentum, conservation of energy). We didn't have to say what kind of force was involved. But doesn't it make a difference whether the interaction is electric or nuclear? Yes, the type of interaction determines the *distribution* of scattering angles. For example, in what fraction of the collisions is the scattering angle between 0 and 10 degrees? Between 10 and 20 degrees? Between 20 and 30 degrees? These fractions are different for different forces.

To put it another way, if the scattering angle is 20 degrees, the momenta are determined. The question is, how often is the scattering angle in fact near 20 degrees? Measurement of the distribution of scattering angles in collisions at modern particle accelerators is one of the ways used to study the nature of the strong and weak interactions.

Rutherford used Coulomb's law of electric force to make a detailed prediction of the distribution of scattering angles, assuming that the positive nucleus was point-like and very massive. That is, he tried to predict what fraction of the alpha particles would be scattered by less than 10 degrees, by an angle between 10 and 20 degrees, by an angle between 20 and 30 degrees, etc. The observed distribution of scattering angles agreed with the predicted distribution (Figure 8.20 shows a portion of this distribution).

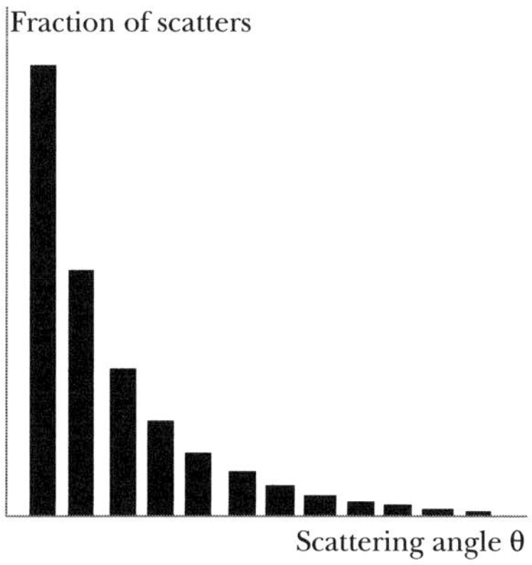

Figure 8.20 For an electric interaction, the fraction of scatters at different scattering angles falls rapidly with angle.

The calculation that Rutherford did involves a lot of complicated geometry, algebra, and probability (with a random distribution of impact parameters), so we won't go through it. But the main features of Figure 8.20 are easily understood. Because the gold nucleus is very small, most of the time the alpha particle will miss the nucleus by a large distance. With a large impact parameter the electric force is very small, and the deflection (scattering angle) is very small. For that reason small scattering angles are common. Large scattering angles occur only for nearly head-on collisions (small impact parameter), and these events are rare.

Since Rutherford's prediction assumed a point-like nucleus, the observations were consistent with the nucleus being very small. Also, the good fit between prediction and observation strengthened the belief that the interaction was purely electric in nature (the strong interaction played no role because it is a very short-range interaction, and the alpha particle was not going fast enough to overcome the electric forces and contact the gold nucleus).

If the alpha particle has high enough energy, it can overcome the electric repulsion and make contact with the gold nucleus. In that case the strong interaction comes into play, and the angular distribution changes. The radius of the gold nucleus can be determined by finding out how much energy an alpha particle has to have in order to obtain an angular distribution that deviates from what is expected due just to the electric interaction.

? For a particular scattering angle, can we use momentum and energy conservation to determine the unknown angle and momenta, even if the alpha particle makes contact with the gold nucleus?

Yes. These conservation principles are completely general and apply to all types of interactions. But they cannot by themselves predict the angular distribution—that is, how often a particular range of scattering angles will occur. To do that you have to have a mathematical description for the specific interaction, such as the $1/r^2$ force law for electric interactions.

8.5 Relativistic momentum and energy

Rutherford's original experiments were carried out with alpha particles whose kinetic energy was about 10 MeV (10 million electron-volts), with a corresponding speed that was small compared to the speed of light. In order to probe the nucleus more deeply, physicists later built special machines, called "particle accelerators," which use electric forces to accelerate charged particles to very much higher energies, with speeds that may be as much as 99.9999% of the speed of light. For such high speeds we cannot use the low-speed approximate formulas for momentum and energy.

Physicists analyzing collisions produced by high-energy accelerators routinely do calculations based on the relativistic energy and momentum formulas. Consider a collision in which a particle of mass m_1 and momentum p_1 strikes a stationary particle of mass m_2 (Figure 8.21). There is a change of identity, with the production of two new particles of mass m_3 and m_4, with momenta p_3 and p_4, at angles θ and ϕ to the horizontal (Figure 8.22).

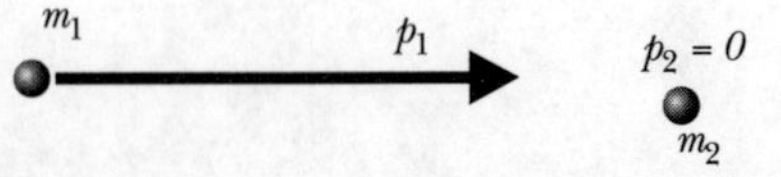

Figure 8.21 Before a high-energy collision.

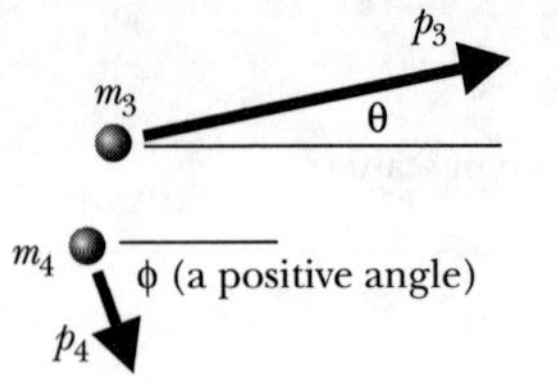

Figure 8.22 After the collision.

We will write down relativistically correct equations for momentum conservation and energy conservation, in terms of momentum p and energy E, knowing that we can determine E from p, since $E^2 - (pc)^2 = (mc^2)^2$:

$$p_1 = p_3\cos\theta + p_4\cos\phi \quad \text{momentum conservation; } x \text{ components}$$

$$0 = p_3\sin\theta - p_4\sin\phi \quad \text{momentum conservation; } y \text{ components}$$

$$E_1 + m_2c^2 = E_3 + E_4 \quad \text{energy conservation; scalar}$$

The energy equation is equivalent to

$$\sqrt{(p_1c)^2 + (m_1c^2)^2} + m_2c^2 = \sqrt{(p_3c)^2 + (m_3c^2)^2} + \sqrt{(p_4c)^2 + (m_4c^2)^2}$$

? Suppose the initial momentum p_1 of the projectile is known, and all four masses are known (because we were able to identify the particles). How many unknown quantities are there, considering that we can write the energy equation in terms of p?

There are four unknown quantities, the magnitudes of the momenta and the directions of the two outgoing particles (p_3, p_4, θ, and ϕ), but only three equations. A typical situation in a high-energy particle experiment is that one of these unknown quantities is measured, and momentum and energy conservation are used to calculate the other three quantities.

As in the Rutherford experiment, measurements of the outgoing particle momenta do not tell us anything about the detailed nature of the interaction involved in the collision, because they are determined by general momentum and energy conservation laws. The identity changes of the particles tell us a lot, and the distribution of scattering angles also tells us a lot.

Photons and neutrinos

If a particle has zero mass (the photon and possibly the neutrino), the relationship $E^2 - (pc)^2 = (mc^2)^2$ reduces simply to $E = pc$. For the photon we can't write its energy as $mc^2/\sqrt{1 - v^2/c^2}$, because both the numerator and denominator are zero, and $0/0$ is indeterminate. Nevertheless, $E = pc$ is valid: a photon of energy E carries momentum $p = E/c$.

Identifying a particle

A variation on the relativistic collision analysis can be used to identify one of the outgoing particles. Suppose we measure p_3 and θ, and we know m_3 but not m_4. We can use our three equations in the three unknowns (m_4, p_4, and ϕ) to solve for the unknown mass m_4. This is a very powerful technique when we are unable to observe the particle directly. For example, neutrinos don't have electromagnetic or strong interactions and therefore typically fail to register in ordinary detectors. But if we measure the other outgoing particle we can determine the properties of the unseen particle.

The discovery of the neutrino

A specific example of such analyses was the original motivation for proposing the existence of the neutrino. The neutron (either free or when bound into some nuclei) was observed to decay with the creation of a proton and an electron, both charged particles that are easy to detect. But the amount of energy and momentum carried by the proton and electron varied from one decay to the next, which made no sense:

$$n \rightarrow p^+ + e^-\text{, violating momentum and energy conservation??}$$

It was proposed that there was an unseen third particle, called the neutrino (or antineutrino in neutron decay), which carried energy and momentum that would salvage the conservation laws:

$$n \rightarrow p^+ + e^- + \bar{\nu}\text{, conserving momentum and energy}$$

It was possible to solve for the properties of the unseen particle by the techniques outlined above, and it was found that the neutrino must have zero mass and must travel at the speed of light. (Recent experimental evidence suggests that the neutrino may have a very small nonzero mass. If so, it must travel at slightly less than the speed of light.)

For many years the neutrino was just a theoretical construct which balanced the accounts, but eventually the neutrino was detected directly. The neutrino is very hard to detect because it does not have electromagnetic or strong interactions, only weak interactions, so the most probable thing for it to do is to pass right through any detector you place in its way. In fact, neutrinos emitted in nuclear reactions in the Sun mostly pass right through the entire Earth! However, with extremely intense neutrino beams produced in nuclear reactors or accelerators, or with gigantic underground detectors used to identify neutrinos coming from the Sun, it is possible to have a measurable rate of neutrino reactions. (A current puzzle is that many fewer neutrinos are detected coming from the Sun than expected.)

8.6 Inelastic collisions

Collisions in which the initial and final internal energies of the particles are the same are called "elastic" collisions. The Rutherford experiment involved elastic collisions, because the internal energies of the gold nucleus and the alpha particle didn't change during the interaction. When there is an internal energy change, or the production of additional particles or the emission of light, we call the collision "inelastic."

If we shoot alpha particles at a gold nucleus with higher energy than the energy of alpha particles in the Rutherford experiment, we may be able to excite the gold nucleus to a higher quantum energy level, an amount ΔE above the ground state of the gold nucleus (Figure 8.23). Such a state corresponds to a different configuration of the nucleons inside the gold nucleus.

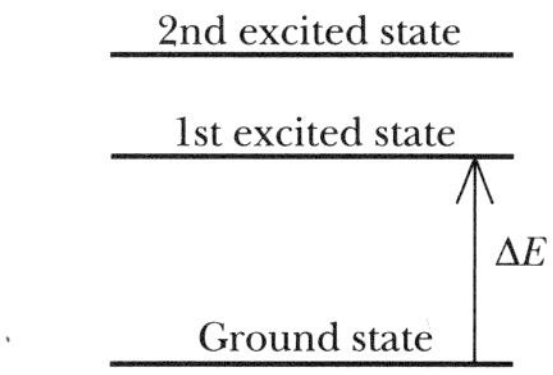

Figure 8.23 Energy levels for a gold nucleus. A sufficiently energetic alpha particle may have enough energy to raise the nucleus above its ground state.

Suppose that you know the incoming momentum p_1 of the alpha particle, and you are able to measure its outgoing momentum p_3 and scattering angle θ (Figure 8.24).

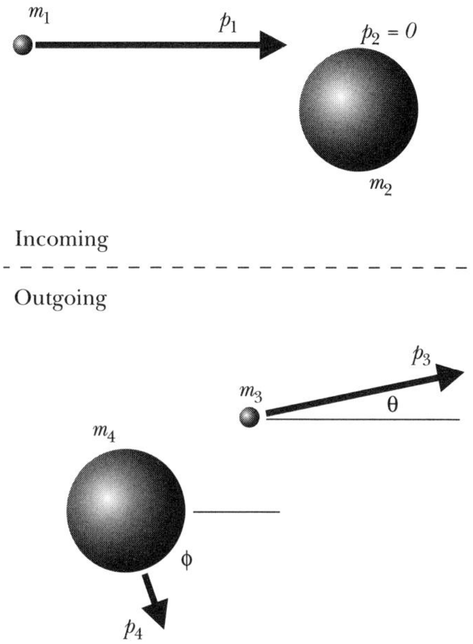

Figure 8.24 You measure the incoming and outgoing momenta of the alpha particle and deduce something about the energy levels in the gold nucleus.

? What is the form of the energy-conservation and momentum-conservation equations in this case? Don't assume that the speeds are small compared to the speed of light. Note that the rest energy of the recoiling gold nucleus is $M_4c^2 + \Delta E$ because it is in an excited state.

Here are the relevant equations:

$$p_1 = p_3\cos\theta + p_4\cos\phi\text{,}$$

where p_4 and ϕ refer to the gold nucleus, which is not observed in the experiment.

$$0 = p_3\sin\theta - p_4\sin\phi$$

$$E_1 + Mc^2 = E_3 + E_4\,,\ \text{where } E_4 = \sqrt{(p_4c)^2 + (M_4c^2 + \Delta E)^2}$$

? How many unknown quantities are there in the energy and momentum equations? Is there enough information to be able to determine the internal energy difference ΔE? (This is one of the methods used to study the energy levels of nuclei. This energy level determination can be verified by measuring the energy of a gamma-ray photon emitted when the nucleus drops back to the ground state.)

There are three unknowns (p_4, ϕ, and ΔE), and there are three equations (energy, x and y momentum), so we could determine ΔE.

8.6.1 Change of identity

Instead of merely exciting the target particle to a higher internal energy state, suppose that there is a change of identity in which the target particle turns into some other particle. An example is the following particle reaction, which was used in the 1950's to discover the Δ^+ particle of mass m_Δ, which is a particle more massive than a proton:

$$\pi^- + p^+ \rightarrow \pi^- + \Delta^+$$

A beam of negative pions (π^-) of known mass m_π and known kinetic energy, produced by a particle accelerator, is directed at a container of hydrogen gas, and the targets for the collisions are the protons of mass m_p that are the nuclei of the hydrogen atoms. Sometimes an incoming pion interacts through the strong interaction with a proton to produce a Δ^+ particle. The momentum of the scattered pion can be measured by measuring the radius of curvature of its curving trajectory in a magnetic field.

? What is the form of the energy-conservation and momentum-conservation equations in this case? (Don't assume that the speeds are small compared to the speed of light.)

Here are the relativistically correct equations:

$$p_1 = p_3\cos\theta + p_4\cos\phi$$

$$0 = p_3\sin\theta - p_4\sin\phi$$

$$E_1 + m_pc^2 = E_\pi + E_\Delta\,,\ \text{where } E^2 - (pc)^2 = (mc^2)^2$$

? How many unknown quantities are there in the energy and momentum equations? Is there enough information to be able to determine the mass of the Δ^+ particle?

There are three unknowns (p_4, ϕ, and E_Δ) and three equations, so it is possible to determine the mass of the Δ^+ particle.

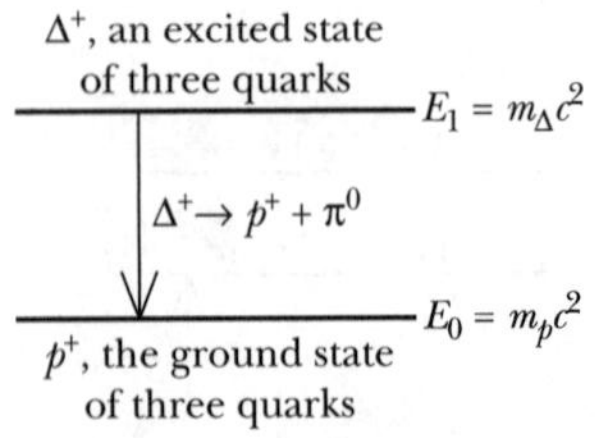

Figure 8.25 The Δ^+ particle can be considered to be an excited state of three quarks, with the proton representing the ground state. A decay to the ground state can occur with the emission of a neutral pion (π^0).

8.6.2 What we learn from inelastic collisions

The analysis for finding the mass change in the preceding exercises is essentially the same as the analysis for determining the internal energy change of an excited gold nucleus. In fact, we can say that the Δ^+ particle is an excited state of three quarks, with the proton representing the ground state of three quarks (Figure 8.25). The excited state (the Δ^+ particle) can decay to the ground state (the proton) with the emission of a neutral pion (π^0), which is a quark-antiquark pair. The extra kinetic energy is equal to $(m_\Delta c^2 - m_pc^2)$. Scattering experiments have been an important tool for discovering new particles (or new excited states of quark systems).

8.6.3 A sticking collision

To illustrate the wide applicability of the momentum principle, we'll consider an inelastic collision in which two objects stick together when they make contact. A macroscopic example is the collision of a truck and a car in an icy intersection (Figure 8.26). In the figure we have shown velocity vectors rather than momentum vectors; a typical truck would have much more momentum than the car, given the truck's much greater mass.

Let M be the mass of the truck, m the mass of the car, and θ the angle of their mutual final velocity, as shown in the diagram. Also let v_1 be the initial speed of the truck, v_2 the initial speed of the car, and v_3 the final speed when they're stuck together and sliding on the ice.

? How does the fact that the intersection is icy affect our ability to apply conservation of momentum to the truck-car collision?

Friction with the road is small if the intersection is icy, so we can presumably ignore road friction and air resistance during the short time of the actual collision. Therefore the total momentum of the truck+car system should be (approximately) unchanged. Momentum conservation is really three equations, one each for the x, y, and z directions. In the z direction (vertically upward) the ice simply supports the weights of the vehicles, so the net force is zero, and the zero z component of momentum remains zero throughout the collision.

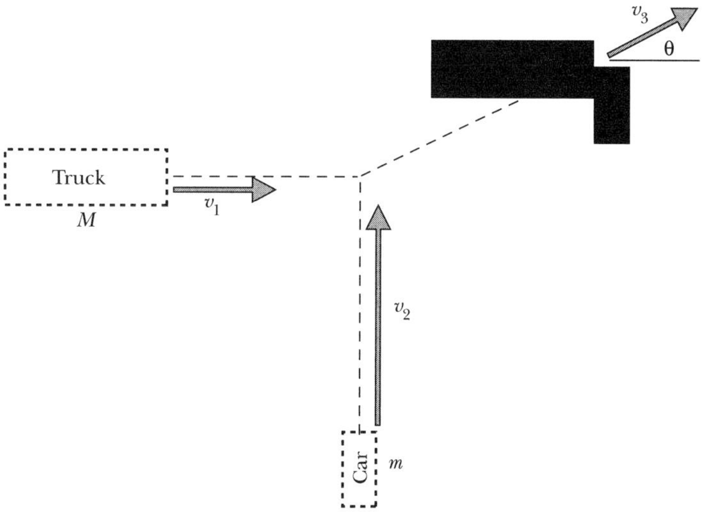

Figure 8.26 A sticking collision between a car and a truck on an icy road.

? In terms of the named quantities, write two equations for the x and y components of the momentum conservation equation, equating momentum components before the collision to momentum components after the vehicles stick together.

Here are the x and y components of the momentum conservation:

$$Mv_1 = (M+m)v_3\cos\theta$$
$$mv_2 = (M+m)v_3\sin\theta$$

Assume that we know the initial speeds of the truck and car (v_1 and v_2). Then we have two equations in two unknowns (v_3 and θ), and we can solve for the final speed and angle. First, divide the y equation by the x equation:

$$\frac{(M+m)v_3\sin\theta}{(M+m)v_3\cos\theta} = \frac{mv_2}{Mv_1} = \tan\theta\text{, so } \theta = \arctan\left(\frac{mv_2}{Mv_1}\right)$$

Next square the equations and add them:

$$(Mv_1)^2 + (mv_2)^2 = [(M+m)v_3]^2[(\sin\theta)^2 + (\cos\theta)^2]$$

$$\sqrt{(Mv_1)^2 + (mv_2)^2} = (M+m)v_3$$

$$v_3 = \frac{\sqrt{M^2v_1^2 + m^2v_2^2}}{M+m}$$

We used the trig identity $\sin^2\theta + \cos^2\theta = 1$, which corresponds to the Pythagorean theorem applied to a triangle whose hypotenuse has length 1.

Energy in a sticking collision

A sticking collision is unusual in that momentum conservation alone gives us enough equations (two) to be able to solve for the unknown final speed and unknown direction. If the car and truck bounce off each other instead of sticking, there are four unknowns instead of two, because there are two final velocities, each with a speed and a direction to be calculated. In that case momentum conservation alone doesn't provide enough equations to be able to analyze the situation fully.

There is an important issue associated with energy in a sticking collision such as this one. Using our result for the final speed, we can write the kinetic energy after the collision:

$$\frac{1}{2}(M+m)v_3^2 = \frac{1}{2}(M+m)\left[\frac{(Mv_1)^2+(mv_2)^2}{(M+m)^2}\right]$$

After minor rearrangements we get the following:

$$\frac{1}{2}(M+m)v_3^2 = \left(\frac{M}{M+m}\right)\left(\frac{1}{2}Mv_1^2\right)+\left(\frac{m}{M+m}\right)\left(\frac{1}{2}mv_2^2\right)$$

Since $M/(M+m)$ is less than 1, and $m/(M+m)$ is also less than 1, the final kinetic energy is less than the initial kinetic energy.

? Where is the "missing" kinetic energy?

Part of the energy was radiated away as sound waves in the air, produced when the vehicles collided with a bang. Some of the energy went into raising the temperature of the vehicles, so that their thermal energy is now higher than before (larger random motions of the atoms).

Some of the energy went into deforming the metal bodies of the truck and car. A bent fender has a higher internal energy, associated with a change in the configuration of the atoms. Sometimes you can get such energy back (as when you compress a spring, changing the configuration of the atoms in the spring, and later let it push back, returning to its original configuration), and we speak of "elastic" energy. But sometimes you can't get the energy back, because the configuration change gets "locked in."

A sticking collision is an example of an "inelastic" collision, referring to the effective "loss" of kinetic energy into forms that are inaccessible.

Ex. 8.5 Two cars collide, one headed north and one headed west. They stick together and leave skid marks on the pavement, which show that car 1 was deflected 30° (so car 2 was deflected 60° . What can you conclude about the cars before the collision?

8.7 Example: Photodissociation of the deuteron

A photon of energy E is absorbed by a stationary deuteron. As a result, the deuteron breaks up into a proton and a neutron. The photon energy is sufficiently low that the proton and neutron have speeds small compared to the speed of light. The proton is observed to move in a direction at an angle θ to the direction of the incoming photon.

Write equations that could be solved to determine the unknown magnitudes of the proton momentum p_p and neutron momentum p_n, and the positive angle α of the neutron to the direction of the incoming photon. Let the precise masses of the particles be M_d, M_p, and M_n for the deuteron, proton, and neutron. Explain clearly on what principles your equations are based. Provide a physics diagram of the reaction and label the angles. Do not attempt to solve the equations.

Solution

Fundamental principles: momentum and energy conservation.

Momentum conservation (Figure 8.27):

$$\vec{p}_{\text{photon}} = \vec{p}_p + \vec{p}_n$$

A photon with energy E has momentum $p = E/c$:

$$\left\langle \left(\frac{E}{c}\right), 0, 0 \right\rangle = \langle (p_p \cos\theta + p_n \cos\alpha), (p_p \sin\theta - p_n \sin\alpha), 0 \rangle$$

This is equivalent to two component equations:

$$\frac{E}{c} = (p_p \cos\theta + p_n \cos\alpha)$$

$$0 = (p_p \sin\theta - p_n \sin\alpha)$$

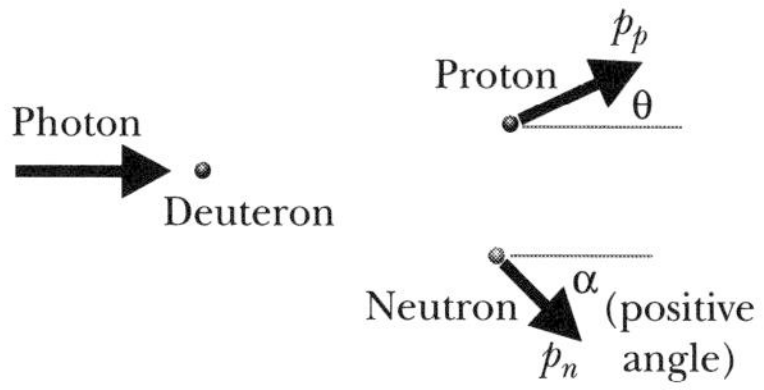

Figure 8.27 Physics diagram for photodissociation of deuteron.

Energy conservation:

$$E + M_d c^2 = \left(M_p c^2 + \frac{p_p^2}{2M_p}\right) + \left(M_n c^2 + \frac{p_n^2}{2M_n}\right)$$

We have three equations (two momentum component equations and one energy equation) in three unknowns: p_p, p_n, and α. So in principle we can solve for the unknown quantities, though the algebra may be messy.

We could also have written the energy equation relativistically, using the relationship $E^2 - (pc)^2 = (mc^2)^2$ which gives $E = \sqrt{(pc)^2 + (mc^2)^2}$:

$$E + M_d c^2 = \sqrt{(p_p c)^2 + (M_p c^2)^2} + \sqrt{(p_n c)^2 + (M_n c^2)^2}$$

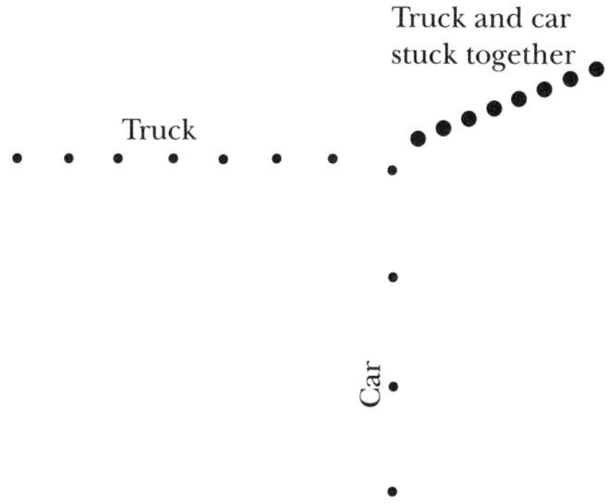

Figure 8.28 View of collision from hovering (stationary) helicopter.

8.8 Viewing collisions from other reference frames

We can show that the truck-car collision process looks a lot simpler from the vantage point of a helicopter flying over the scene with the final velocity of the truck+car combined system. This will lead to new insights. Suppose the accident takes place at night, and from the helicopter all you can see are lights mounted on top of the two vehicles. Figure 8.28 shows a multiple exposure of the scene that you would see from a helicopter that is hovering over the intersection.

Suppose instead that your helicopter flies over the intersection with the same velocity that the stuck-together truck and car share after their collision, as shown in Figure 8.29. The view thus obtained is in an important sense "simpler," as shown in Figure 8.30. From the moving observation platform you see the truck and car run into each other and then stand still (since you and they now have the same velocity). In particular, in this (moving) reference frame it is obvious that there is a loss of kinetic energy into other forms of energy, because after the collision the kinetic energy is clearly zero. In contrast, we had to go through a fair amount of algebra in the non-moving reference frame in order to establish the loss of kinetic energy.

The reason that we imagine making these observations at night is so that you won't be distracted by the moving background. If you actually view this collision from a moving helicopter in daylight, there is a strong tendency to compensate mentally for your motion and to put yourself (mentally) into the reference frame of the road, in which case you could miss the important fact that *relative to you*, the car and truck are stationary after the collision.

You may have had an experience related to this situation. You're sitting in a bus or train, waiting to leave. You think you've started forward, but then you realize that you were fooled by a neighboring bus or train moving backwards. For a moment you were mentally in the reference frame of the moving vehicle.

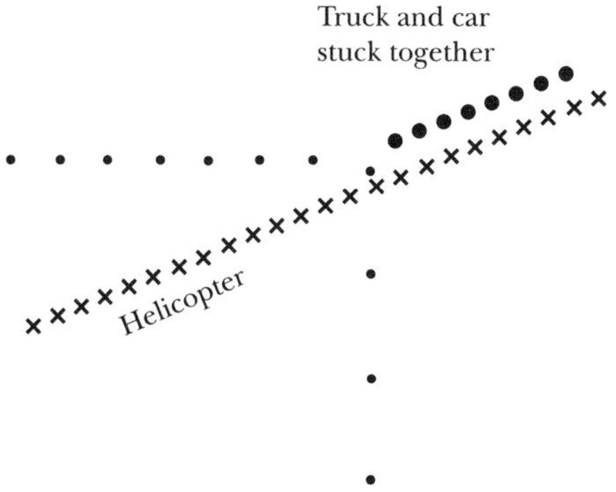

Figure 8.29 Path of moving helicopter that produces the simple view seen in Figure 8.30.

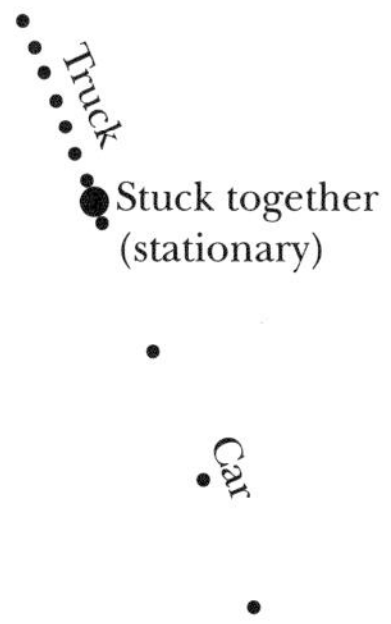

Figure 8.30 View of collision from helicopter moving with velocity of stuck-together wreck.

8.8.1 Total momentum is zero in the center-of-mass frame

Viewing a collision from a moving reference frame can provide insights into all kinds of collisions, not just sticking collisions. The key point that leads to a useful generalization is that the total momentum of the truck+car system after their collision is zero when the vehicles are motionless relative to your helicopter.

? In this moving reference frame, if the total momentum is zero after the collision, what was the total momentum before the collision? Why?

The helicopter is moving at constant velocity (constant speed and direction), and we know that the momentum principle is valid in this uniformly moving reference frame. So the principle of momentum conservation still holds, and the initial total momentum of the car+truck system must be zero if the final total momentum is zero. That's what makes the process look so simple: the total momentum is zero before and after the collision in the moving reference frame.

In viewing the scene from the moving helicopter, we have subtracted from each of the velocities an unchanging, constant vector velocity, the velocity $\vec{v}_h$ of the helicopter, and this subtraction makes no change in the momentum principle for the car (or truck):

$$\frac{d[m(\vec{v}-\vec{v}_h)]}{dt} = \frac{d[m\vec{v}]}{dt} = \vec{F}_{\text{net}}$$

The net force is the same in both reference frames because the contact electric interaction forces depend on interatomic distances, which are the same in both reference frames.

Evidently we gain insight into a collision by viewing the collision in a moving reference frame where the total momentum is zero. How do we determine the appropriate velocity for such a reference frame? Let this velocity be $\vec{v}_{\text{fr}}$, which is to be subtracted from all the original velocities to obtain the velocities as viewed in the moving reference frame. We want the total momentum to add up to zero in the moving frame:

$$\vec{P}_{\text{tot,new frame}} = m_1(\vec{v}_1 - \vec{v}_{\text{fr}}) + m_2(\vec{v}_2 - \vec{v}_{\text{fr}}) = 0$$

$$(m_1 + m_2)\vec{v}_{\text{fr}} = m_1\vec{v}_1 + m_2\vec{v}_2$$

$$\vec{v}_{\text{fr}} = \frac{m_1\vec{v}_1 + m_2\vec{v}_2}{m_1 + m_2}$$

We have solved for a particular velocity for the moving reference frame that makes the total momentum in that frame be zero. But look at the form of this equation.

? Does this equation look familiar?

Evidently $\vec{v}_{\text{fr}} = \vec{v}_{\text{cm}}$, the velocity of the center of mass of the system, since

$$\vec{r}_{\text{cm}} = \frac{m_1\vec{r}_1 + m_2\vec{r}_2}{m_1 + m_2}$$

We have obtained an important result. In a reference frame moving with the center of mass of a system, the total momentum of the system is zero. If no external forces act, the total momentum is unchanged, and the total momentum is zero before and after a collision.

We can also express the frame velocity directly in terms of the total momentum:

$$\vec{v}_{cm} = \frac{m_1\vec{v}_1 + m_2\vec{v}_2}{m_1 + m_2} = \frac{\vec{P}_{tot}}{M_{tot}}$$

This provides another way to calculate the frame velocity and shows why this is often called the "center-of-momentum" velocity.

Let's reanalyze the car-truck collision from this new point of view. Again let M be the mass of the truck, m the mass of the car, and θ the angle of their mutual final velocity, as shown in Figure 8.31. Also let v_1 be the initial speed of the truck, v_2 the initial speed of the car, and v_3 the final speed when they're stuck together and sliding on the ice.

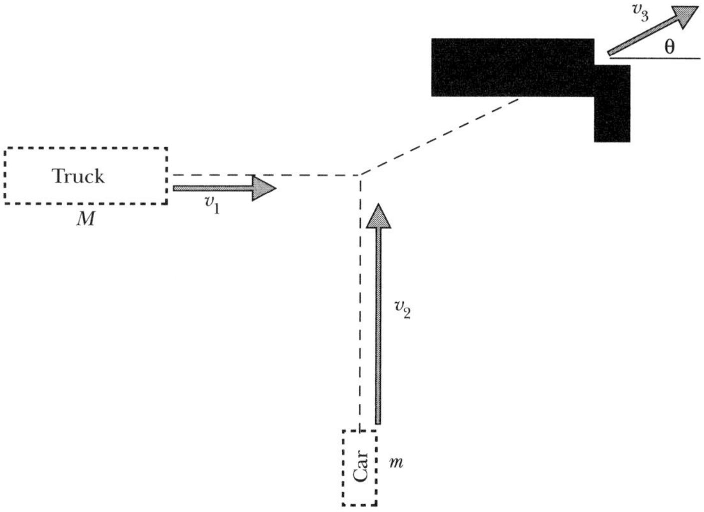

Figure 8.31 Collision of car and truck.

? What are the x and y components of $\vec{v}_{cm}$ in terms of v_1 and v_2?

$$\vec{v}_{cm} = \langle \frac{Mv_1}{M+m}, \frac{mv_2}{M+m}, 0 \rangle$$

? In the moving center-of-momentum reference frame, what are the x and y components of the car's velocity before the collision? Remember that you find the car's velocity (or its components) in the center-of-momentum frame by subtracting the center-of-momentum velocity (or its components) from that of the car. (That is, the view from the helicopter is obtained by subtracting the helicopter's velocity from the velocities of the car and truck.)

$$\langle -\frac{Mv_1}{M+m}, v_2 - \frac{mv_2}{M+m}, 0 \rangle = \langle -\frac{Mv_1}{M+m}, \frac{Mv_2}{M+m}, 0 \rangle$$

? In the moving center-of-momentum reference frame, what are the x and y components of the truck's velocity before the collision? Show that this velocity is in a direction opposite to the velocity of the car, and that the total momentum is zero.

$$\langle v_1 - \frac{Mv_1}{M+m}, -\frac{mv_2}{M+m}, 0'''' \rangle = \langle \frac{mv_1}{M+m}, -\frac{mv_2}{M+m}, 0 \rangle$$

The x and y components of velocity are in the opposite direction to those of the car (multiply the car's velocity by $-m/M$ and you have the truck's velocity). Also, the total momentum is zero in the center-of-momentum frame:

$$\langle \left[m\left(-\frac{Mv_1}{M+m}\right) + M\left(\frac{mv_1}{M+m}\right)\right], \left[m\left(\frac{Mv_2}{M+m}\right) + M\left(-\frac{mv_2}{M+m}\right)\right], 0 \rangle = 0$$

? In the moving center-of-momentum reference frame, what is the speed of the car after the collision? Of the truck? Why?

Both are zero; in the center-of-mass frame the total momentum is zero, and with the car and truck stuck together the only way for this to be true is for them to be at rest. Because of this, the kinetic energy loss to other kinds of energy is equal to the total kinetic energy in the center-of-momentum frame before the collision.

? Now, after the collision, transform back to the (non-moving) reference frame of the intersection, by adding back in the $\vec{v}_{cm}$ that had been subtracted. What are the speed and direction of the car after the collision? Of the truck? Compare with the results obtained in our earlier analysis of this collision.

$$\vec{v}_3 = \left\langle \frac{Mv_1}{M+m}, \frac{mv_2}{M+m}, 0 \right\rangle, \text{ which yields}$$

$$v_3 = \frac{\sqrt{M^2 v_1^2 + m^2 v_2^2}}{M+m}; \ \theta = \arctan\left(\frac{mv_2}{Mv_1}\right)$$

These are the same results we obtained in our earlier analysis.

8.8.2 Energy in the two reference frames

As far as energy is concerned, note that the kinetic energy of the car+truck system is as usual

$$K_{tot} = K_{trans} + K_{rel}, \text{ where } K_{trans} = \frac{P_{tot}^2}{2M_{tot}}$$

In the original reference frame, $K_{trans} > 0$ but is the same before and after the collision, since P_{tot} doesn't change. In the center-of-momentum reference frame, $K_{trans} = 0$ both before and after the collision, because $P_{tot} = 0$ in the center-of-momentum reference frame.

In either reference frame, the kinetic energy relative to the center of mass, K_{rel}, decreases to zero in the collision, because the car and truck after the collision are traveling with the center of mass. It is ΔK_{rel} that represents the loss of kinetic energy in this inelastic collision, and this quantity is the same in both reference frames.

Also note that distances between atoms are the same in both reference frames, so potential energy associated with bending of metal parts is the same in both frames.

Ex. 8.6 A 1000 kg car moving east at 30 m/s runs head-on into a 3000 kg truck moving west at 20 m/s. The vehicles stick together. Use the concept of the center-of-momentum frame to determine how much kinetic energy is lost.

8.9 Summary

Fundamental principles

CONSERVATION OF MOMENTUM

$$\Delta\vec{P}_{\text{system}} + \Delta\vec{P}_{\text{surroundings}} = 0$$

New concepts

The notion of a "collision"
Scattering, and scattering distributions
The center-of-mass reference frame

$$\vec{v}_{\text{cm}} = \frac{m_1\vec{v}_1 + m_2\vec{v}_2}{m_1 + m_2} = \frac{\vec{P}_{\text{tot}}}{M_{\text{tot}}}$$

Results

Evidence for the nuclear model of an atom

Problem solving techniques

Conservation of momentum applied to collisions
Simplification of calculations in the center-of-mass reference frame

8.10 Review questions

Momentum and impulse

RQ 8.1 A force <30, 0, 0> N acts for 0.2 seconds on an object of mass 1.2 kg whose initial velocity was <0, 20, 0> m/s. What is the new velocity?

Conservation of momentum

RQ 8.2 Under what conditions is the momentum of a system constant? Can the x component of momentum be constant even if the y component is changing? In what circumstances? Give an example of such behavior.

Collisions

RQ 8.3 What happens when a high-mass object hits a low-mass object head-on? What happens when a low-mass object hits a high-mass object head-on?

RQ 8.4 What properties of the alpha particle and the gold nucleus in the original Rutherford experiment were responsible for the collisions being elastic collisions?

RQ 8.5 Give an example of what we can learn about matter through the use of momentum and energy conservation applied to scattering experiments. Explain what it is that we cannot learn this way, for which we need to measure the distribution of scattering angles.

Sticking collisions

RQ 8.6 In an elastic collision involving known masses and initial momenta, how many unknown quantities are there after the collision? How many equations are there? In a sticking collision involving known masses and initial momenta, how many unknown quantities are there after the collision? Explain how you can determine the amount of kinetic energy change.

RQ 8.7 In order to close a door, you throw an object at the door. Which would be more effective in closing the door, a 50 gram tennis ball or a 50 gram lump of sticky clay? Explain clearly what physics principles you used to draw your conclusion.

RQ 8.8 A bullet of mass m traveling horizontally at a very high speed v embeds itself in a block of mass M that is sitting at rest on a nearly frictionless surface.

(a) What is the speed of the block after the bullet embeds itself in the block?

(b) Calculate the total translational kinetic energy before and after the collision. Compare the two results and explain why there is a difference.

Changing reference frame

RQ 8.9 What is it about analyzing collisions in the center-of-mass frame that simplifies the calculations?

8.11 Homework problems

Problem 8.1 The Rutherford experiment

Use numerical integration techniques to explore possible trajectories of an alpha particle when it collides with a gold nucleus at different impact parameters. Your program should display the paths of the alpha particle and of the gold nucleus, and also plot the x and y components of momentum of both particles as a function of time. We will consider motion only in two dimensions (a slice through a 3-D beam of alpha particles).

An alpha particle is a helium nucleus (two protons and two neutrons, charge +2e); in the Rutherford experiment its initial kinetic energy was about 10 MeV (remember that 1 eV = 1.6×10^{-19} J). The gold nucleus is inside a piece of gold foil and is initially stationary; its charge is +79e and it contains 197 nucleons. Initially we will ignore the interactions of the alpha particle with the cloud of electrons surrounding the nucleus, which are in fact small.

(a) In this model it is important to pick reasonable initial conditions and display scales—otherwise you may not see anything meaningful. Before beginning your program, calculate or estimate reasonable values for the following and report your results:

- *Closest approach.* For a head-on collision, calculate approximately the distance of closest approach, given that the alpha particle and the gold nucleus repel each other. Assume that the gold atom's electrons are scattered away by the incoming alpha particle.
- *Initial position of the alpha particle.* Given the distance of closest approach, what would be a reasonable initial x for the alpha particle? What would be a good range to explore for the impact parameter b (initial y coordinate of the alpha particle)? Use these estimates to decide on a scale for the coordinates used to display the trajectories.
- *Value of Δt.* Trial and error alone will not be very helpful in estimating Δt. You can calculate the initial speed of the alpha particle, and you know the approximate size scale for the collision. Choose a value of Δt such that the alpha particle doesn't move very far in one step.
- *Time scale.* To plot p_x and p_y vs. t during the interaction, you must select a scale for the time axis that allows you to observe the entire interaction. About how long should the time axis be?
- *Momentum scale.* What should the maximum value of the p_x axis be? You should scale the p_y axis to the same value, to make comparisons easy.
- *Exit criterion.* You will need to exit from your computational loop when the particles have moved far apart. How will you test for this?

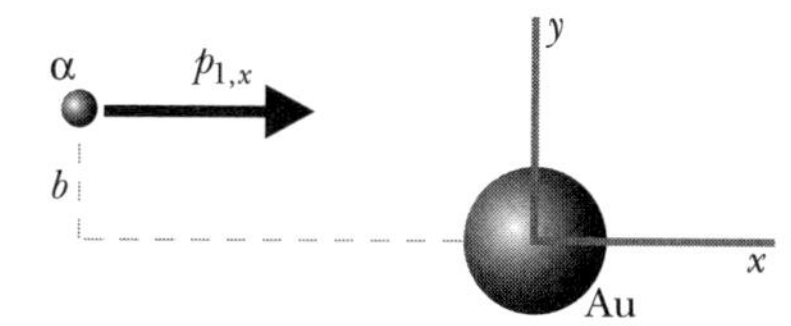

Coordinate system for Rutherford scattering calculations.

(b) Write a program that displays:

- The paths of the alpha particle and the gold nucleus.
- The value of the impact parameter b, and the value of the corresponding scattering angle (angle between the initial path of the alpha particle and its final path). The angle can be calculated by taking the arctangent of p_y/p_x, using the final values.
- A plot of p_x for each particle and total p_x, and a similar plot for p_y. (Consider what these plots should look like if the program is working properly.)

Report what value of Δt gives adequate accuracy. What is your criterion?

Report values of the impact parameter b that produce the following:

- A very small scattering angle
- An angle around 90 degrees
- A large deflection angle (back-scattering)

(c) What other physical quantity could you plot that would provide a check on the accuracy of your calculation? Carry out this check and report what value of Δt provides good accuracy according to this check.

Continued on next page.

(d) To see what a different model of matter would have predicted, replace the gold nucleus in your model by a proton. Describe briefly what you observe. (Note that if the protons were uniformly spread out throughout the volume of the atom, they would be far enough apart that on average an alpha particle would interact with only one of them.)

Problem 8.2 Deviations from Rutherford scattering

(a) A gold nucleus contains 197 nucleons packed tightly against each other. A single nucleon (proton or neutron) has a radius of about 10^{-15} m. Calculate the approximate radius of the gold nucleus.

(b) Rutherford correctly predicted the angular distribution for 10-MeV (kinetic energy) alpha particles colliding with gold nuclei. What kinetic energy must alpha particles have in order to make contact with a gold nucleus? (In this case the angular distribution will deviate from that predicted by Rutherford, which was based solely on electric interactions.)

Problem 8.3 The Rutherford experiment in the center-of-mass frame

Redo the analysis of the Rutherford experiment, this time using the concept of the center-of-momentum reference frame. Let m = the mass of the alpha particle and M = the mass of the gold nucleus. Consider the specific case of the alpha particle rebounding straight back. The incoming alpha particle has a momentum p_1, the outgoing alpha particle has a momentum p_3, and the gold nucleus picks up a momentum p_4.

(a) Determine the velocity of the center of momentum of the system.

(b) Transform the initial momenta to that frame (by subtracting the center-of-momentum velocity from the original velocities).

(c) Show that if the momenta in the center-of-momentum frame simply turn around (180°), with no change in their magnitudes, both momentum and energy conservation are satisfied, whereas no other possibility satisfies both conservation principles. (Try drawing some other momentum diagrams.)

(d) After the collision, transform back to the original reference frame (by adding the center-of-momentum velocity to the velocities of the particles in the center-of-mass frame). Show that the momenta of the alpha particle and gold nucleus are consistent with the results obtained earlier, on page 327 and page 329. Note that although using the center-of-momentum frame may be conceptually more difficult, the algebra for solving for the final speeds is much simpler.

Problem 8.4 Gamma-ray emission

A ^{57}Fe nucleus is at rest and in its first excited state, 14.4 keV above the ground state (14.4×10^3 eV, where 1 eV = 1.6×10^{-19} J). The nucleus then decays to the ground state with the emission of a gamma ray (a high-energy photon).

(a) What is the recoil speed of the nucleus?

(b) Calculate the slight difference in eV between the gamma-ray energy and the 14.4 keV difference between the initial and final nuclear states.

(c) The "Mössbauer effect" is the name given to a related phenomenon discovered by Rudolf Mössbauer in 1957, for which he received the 1961 Nobel prize for physics. If the ^{57}Fe nucleus is in a solid block of iron, occasionally when the nucleus emits a gamma ray the entire solid recoils as one object. This can happen due to the fact that neighboring atoms and nuclei are connected by the electric interatomic force. In this case, repeat the calculation of part (b) and compare with your previous result. Explain briefly.

Problem 8.5 Collisions and springs

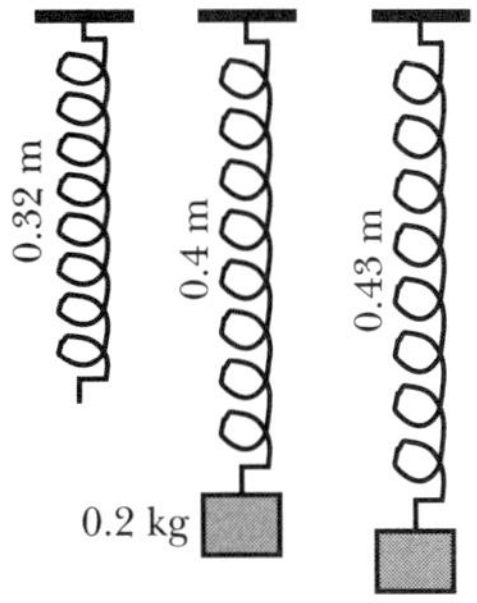

(a) A spring has an unstretched length of 0.32 m. A block with mass 0.2 kg is hung at rest from the spring, and the spring becomes 0.4 m long. Next the spring is stretched to a length of 0.43 m and the block is released from rest. Air resistance is negligible. How long does it take for the block to return to where it was released?

(b) Next the block is again positioned at rest, hanging from the spring (0.4 m long). A bullet of mass 0.003 kg traveling at a speed of 200 m/s straight upward buries itself in the block, which then reaches a maximum height h above its original position. What is the speed of the block immediately after the bullet hits?

(c) Now write an equation that could be used to determine how high the block goes after being hit by the bullet (a height h), but you need not actually solve for h.

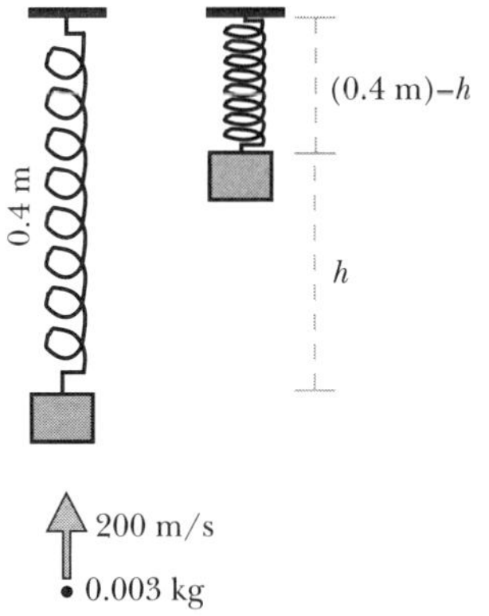

Problem 8.6 Solar sailing

It has been proposed to propel spacecraft through the Solar System with a large sail that is struck by photons from the Sun.

(a) Which would be more effective, a black sail that absorbs photons or a shiny sail that reflects photons back toward the Sun? Explain briefly.

(b) Suppose N photons hit a shiny sail per second, perpendicular to the sail. Each photon has energy E. What is the force on the sail? Explain briefly.

Problem 8.7 Slowing down neutrons

In a nuclear fission reactor, each fission of a uranium nucleus is accompanied by the emission of one or more high-speed neutrons which travel through the surrounding material. If one of these neutrons is captured in another uranium nucleus, it can trigger fission, which produces more fast neutrons, which could make possible a chain reaction.

However, fast neutrons have low probability of capture and usually scatter off uranium nuclei without triggering fission. In order to sustain a chain reaction, the fast neutrons must be slowed down in some material, called a "moderator" (Figure 8.32). For reasons having to do with the details of nuclear physics, slow neutrons have a high probability of being captured by uranium nuclei.

In the following analyses, remember that neutrons have almost no interaction with electrons. Neutrons do however interact strongly with nuclei, either by scattering or by being captured and made part of the nucleus. Therefore you should think about neutrons interacting with nuclei, not atoms.

(a) Based on what you now know about collisions, explain why fast neutrons moving through a block of uranium experience little change in speed.

(b) Explain why carbon should be a much better moderator of fast neutrons than uranium.

(c) Should water be a better or worse moderator of fast neutrons than carbon? Explain briefly.

Background: The first fission reactor was constructed in 1941 in a squash court under the stands of Stagg Field at the University of Chicago by a team

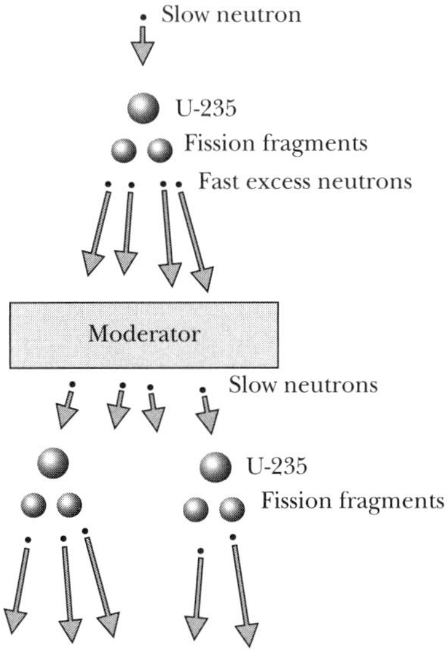

Figure 8.32 A slow neutron induces fission of U-235, with the emission of additional (fast) neutrons. The moderator is some material that slows down the fast neutrons, enabling a chain reaction.

led by the physicist Enrico Fermi. The moderator consisted of blocks of graphite, a form of carbon. The graphite had to be exceptionally pure because certain kinds of impurities have nuclei that capture neutrons with high probability, removing them from contributing to the chain reaction. Many reactors use ordinary "light" water as a moderator, though sometimes a proton captures a neutron and forms a stable deuterium nucleus, in which case the neutron is lost to contributing to the chain reaction. Heavy water, D_2O, in which the hydrogen atoms are replaced by deuterium atoms, actually works better as a moderator than light water, because the probability of a deuterium nucleus capturing a neutron to form tritium is quite small.

Problem 8.8 Fusion reaction revisited

Here is a modified version of a problem from Chapter 4, before we had discussed collisions. One of the thermonuclear or fusion reactions that takes place inside a star such as our Sun is the production of helium-3 (^{3}He, with two protons and one neutron) and a gamma ray (high-energy photon) in a collision between a proton (^{1}H) and a deuteron (^{2}H, the nucleus of "heavy" hydrogen, consisting of a proton and a neutron):

$$^1\text{H} + {}^2\text{H} \rightarrow {}^3\text{He} + \gamma$$

The rest mass of the proton is 1.0073 u (unified atomic mass unit, 1.66×10^{-27} kg), the rest mass of the deuteron is 2.0136 u, the rest mass of the helium-3 nucleus is 3.015 u, and the gamma ray is a massless photon.

The strong interaction has a very short range and is essentially a contact interaction. For this fusion reaction to take place, the proton and deuteron have to come close enough together to touch. The approximate radius of a proton or neutron is about 10^{-15} m.

(Continued on next page.)

(a) In the center-of-momentum frame (that is, the frame in which the total momentum is zero), draw a diagram showing the momentum vectors before the reaction and after the reaction. Clearly label the momentum vectors.

(b) In the center-of-momentum frame, what minimum kinetic energy must the proton have, and what minimum kinetic energy must the deuteron have, in order for the reaction to take place? Express your results in eV. (In the lab frame where the deuteron was at rest, the proton must have more kinetic energy, because the deuteron moves away due to electric repulsion and the proton has to catch up to it.)

(c) In this case, still in the center-of-momentum frame, what will be the kinetic energy of the helium-3 nucleus and the energy of the gamma ray in eV? Hint: You will find that the speed of the helium-3 nucleus is very small compared to the speed of light. You may find it useful to use the relationship $E^2 - (pc)^2 = (mc^2)^2$. Remember that the fundamental definition of the center-of-momentum frame is that it is the frame in which the total momentum is zero. This is the definition that must be used when dealing with zero-mass particles such as photons.

Problem 8.9 Particle decay

There is an unstable particle called the "sigma-minus" (Σ^-), which can decay into a neutron and a negative pion (π^-): $\Sigma^- \rightarrow n + \pi^-$. The mass of the Σ^- is 1196 MeV/c^2, the mass of the neutron is 939 MeV/c^2, and the mass of the π^- is 140 MeV/c^2. Write equations that could be used to calculate the momentum and energy of the neutron and the pion. You do not need to solve the equations, which would involve some messy algebra. But be clear in showing that you have enough equations that you could in principle solve for the unknown quantities in your equations.

It is advantageous to write the equations not in terms of v but rather in terms of E and p; remember that $E^2 - (pc)^2 = (mc^2)^2$.

Problem 8.10 Pion production of a particle

A beam of high energy π^- (negative pions) is shot at a flask of liquid hydrogen, and sometimes a pion interacts through the strong interaction with a proton in the hydrogen, in the reaction $\pi^- + p^+ \rightarrow \pi^- + X^+$, where X^+ is a positively-charged particle of unknown mass.

The incoming pion momentum is 3 GeV/c (1 GeV = 1000 MeV = 10^9 electron-volts). The pion is scattered through 40°, and its momentum is measured to be 1510 MeV/c (this is done by observing the radius of curvature of its circular trajectory in a magnetic field). A pion has a rest energy of 140 Mev, and a proton has a rest energy of 938 Mev.

What is the rest mass of the unknown X^+ particle, in MeV/c^2? Explain your work carefully.

It is advantageous to write the equations not in terms of v but rather in terms of E and p; remember that $E^2 - (pc)^2 = (mc^2)^2$.

There are additional collision problems in Chapter 2.

8.12 Answers to exercises

8.1 (page 322) 1.1×10^{3} N; $mg = 0.98$ N

8.3 (page 329) 2.3×10^{-24} m/s (!) So it is an excellent approximation to say that the Earth doesn't recoil.

8.4 (page 329) $K_{\text{Earth}} = 1.6\times10^{-23}$ J; $K_{\text{baseball}} = 150$ J

8.5 (page 340) Car 1 must have had more momentum than car 2

8.6 (page 344) We outline the entire solution. First find $v_{\text{cm}, x}$:

$$v_{\text{cm}, x} = \frac{(1000 \text{ kg})(30 \text{ m/s}) + (3000 \text{ kg})(-20 \text{ m/s})}{(4000 \text{ kg})} = -7.5 \text{ m/s}$$

Transform to center-of-momentum frame:

$$v_{\text{car}, x} = (30 \text{ m/s}) - (-7.5 \text{ m/s}) = 37.5 \text{ m/s}$$

$$v_{\text{truck}, x} = (-20 \text{ m/s}) - (-7.5 \text{ m/s}) = -12.5 \text{ m/s}$$

Check to make sure that the momenta are equal and opposite in this frame:

$$p_{\text{car}, x} = (1000 \text{ kg})(37.5 \text{ m/s}) = 37500 \text{ kg}\cdot\text{m/s}$$

$$p_{\text{truck}, x} = (3000 \text{ kg})(-12.5 \text{ m/s}) = -37500 \text{ kg}\cdot\text{m/s}$$

In this frame the kinetic energy before the collision is this;

$$K = \frac{(37500 \text{ kg}\cdot\text{m/s})^2}{2(1000 \text{ kg})} + \frac{(37500 \text{ kg}\cdot\text{m/s})^2}{2(3000 \text{ kg})} = 9.4\times10^{5} \text{ J}$$

After the collision, the velocities are zero in this frame, so kinetic energy in this frame goes to zero, and there is an increase in the internal energy of the mangled car and truck of amount 9.4×10^{5} J. There must be the same internal energy increase in the original reference frame, so the kinetic energy lost in the original reference frame was 9.4×10^{5} J.

Chapter 9

Angular Momentum

Chapter 9

Angular Momentum

Rotational motion is such an important type of motion that special techniques have been developed to describe and predict it. A particularly important concept that arises from the study of rotational motion is "angular momentum," the rotational analogue of the ordinary "linear" momentum with which we are already familiar. Angular momentum is conserved, just as energy and linear momentum are. In the atomic world angular momentum is quantized, which has far-reaching consequences for the structure of atoms and nuclei.

9.1 Angular momentum

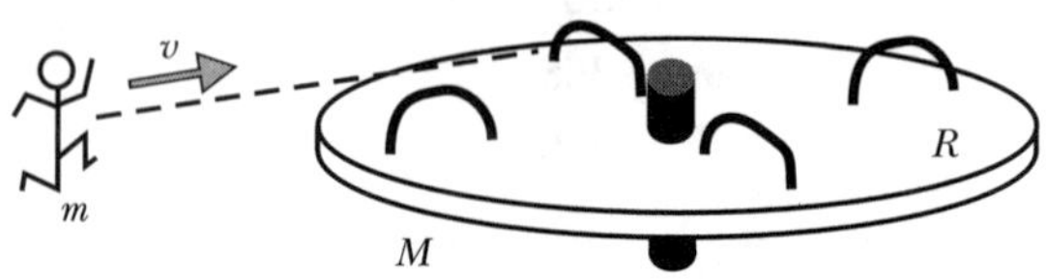

Figure 9.1 A child jumps onto a playground ride, and the disk spins. What is the rate of spin?

A playground ride consists of a disk of mass M and radius R mounted on a low-friction axle (Figure 9.1). A child of mass m runs at speed v on a line tangential to the disk and jumps onto the outer edge of the disk. The disk spins, but at what rate?

The Momentum Principle can be used to determine the impulse exerted by the axle on the disk, which prevents the disk from translating sideways, but it tells us nothing about the spin rate. The Energy Principle can be used to determine how much thermal energy is produced in the inelastic sticking collision of the child and the disk, but it tells us nothing about the spin rate. We need a new principle: the Angular Momentum Principle.

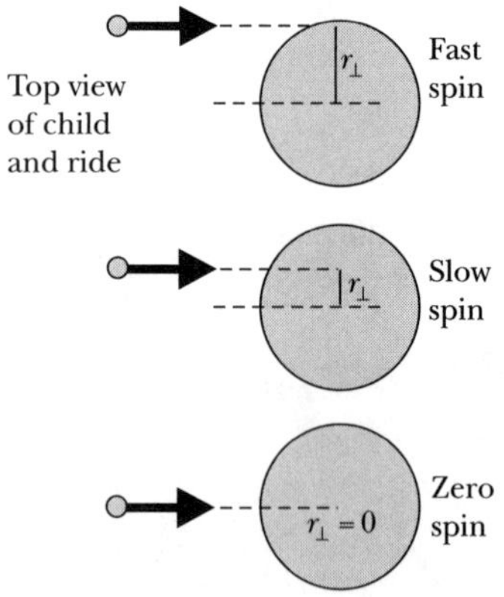

Figure 9.2 How much the disk spins depends on the impact parameter $r_\perp$.

In Figure 9.2 we see a top view of the child heading toward the disk along several different paths, with different effects. Evidently the child's effectiveness in spinning the disk depends on the impact parameter $r_\perp$, which measures how far off-center is the child's initial path. Presumably if the child has more mass m and/or more speed v, these too will affect the spin rate.

Putting these influences together, the spin rate of the disk should be proportional to $r_\perp mv$, and we call this the magnitude of the "angular momentum" of the child relative to the center of the disk. The term "angular momentum" is used to reflect the fact that it is often associated with something rotating through an angle (in this case, the disk), and it is proportional to the familiar momentum mv. When necessary to emphasize the difference between angular momentum and ordinary momentum, the latter is called "linear momentum."

More generally, we define L_A, the magnitude of the angular momentum of a particle relative to a location A like this, as shown in Figure 9.3:

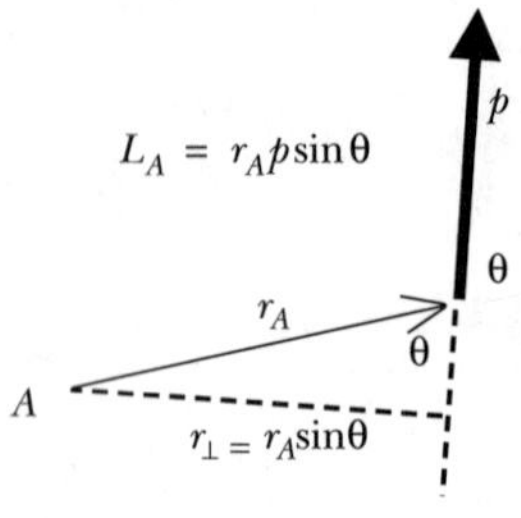

Figure 9.3 Magnitude of the angular momentum of a particle relative to location A.

MAGNITUDE OF ANGULAR MOMENTUM OF A PARTICLE RELATIVE TO LOCATION *A*

$$L_A = r_\perp p = r_A p \sin\theta$$

Here $\vec{r}_A$ is a position vector of the particle measured from location A to the particle, $\vec{p}$ is the ordinary "linear" momentum of the particle, and θ is the angle between the position vector and the momentum vector. $r_\perp$ is the perpendicular distance from location A to the line of motion.

We will find that the sum of the angular momentum of the child plus that of the disk is constant, which is what makes angular momentum useful.

9.1.1 The cross product

The magnitude of angular momentum isn't everything: the disk may rotate clockwise or counterclockwise, depending on the child's path (Figure 9.4). How do we take this into account in a definition of the child's angular momentum? The angular momentum of a particle is defined as a vector, using the "cross product," to be explained in a moment:

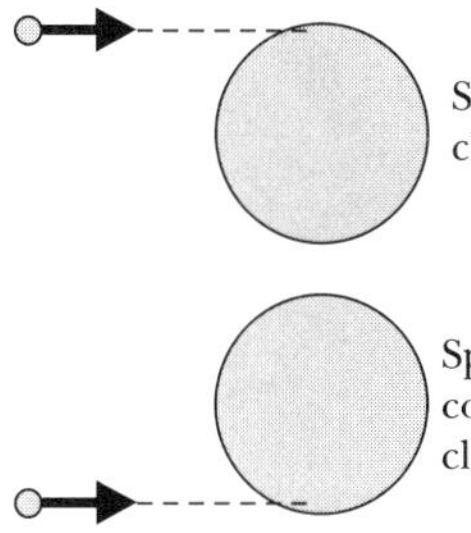

Figure 9.4 The disk will rotate clockwise or counterclockwise, depending on the child's path.

ANGULAR MOMENTUM OF A PARTICLE RELATIVE TO LOCATION *A*

$$\vec{L}_A = \vec{r}_A \times \vec{p}$$

The cross product $\vec{A} \times \vec{B}$ is a vector in a direction perpendicular to the plane defined by $\vec{A}$ and $\vec{B}$. In Figure 9.5, both $\vec{A}$ and $\vec{B}$ lie in the plane, and the vector $\vec{A} \times \vec{B}$ is perpendicular to the plane.

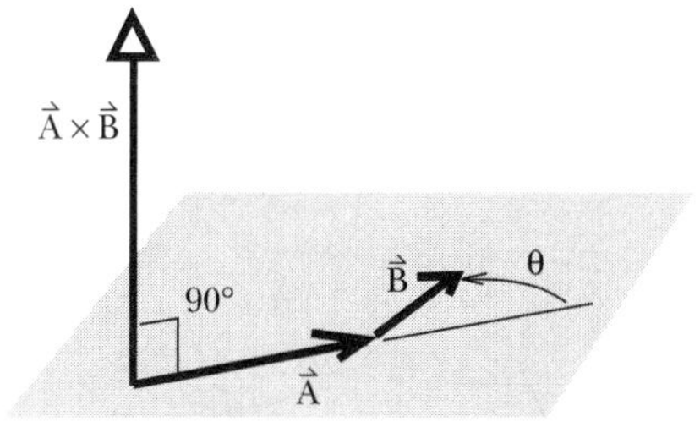

Figure 9.5 The cross product produces a vector that is perpendicular to the two original vectors.

CROSS PRODUCT

$\vec{A} \times \vec{B}$ is a vector with

magnitude $|\vec{A} \times \vec{B}| = AB\sin\theta$ and

direction given by a "right-hand rule"

$$\vec{A} \times \vec{B} = \langle (A_yB_z - A_zB_y), (A_zB_x - A_xB_z), (A_xB_y - A_yB_x) \rangle \text{ (see Section 9.1.3)}$$

In the particular use of the cross product to describe angular momentum, angular momentum is a vector $\vec{L}_A = \vec{r}_A \times \vec{p}$ that is perpendicular to $\vec{r}_A$ and perpendicular to $\vec{p}$. The angular momentum vector lies along the axis about which there would be rotation if the particle struck a disk whose axle is at location *A*, and which way it points along this rotation axis is related to whether the disk would rotate clockwise or counterclockwise (more on this in a moment). The magnitude of the angular momentum vector is proportional to how much spin the particle would produce in the disk.

9.1.2 Right-hand rule for cross products

There are several different schemes for evaluating cross products. We will describe a scheme that has significant advantages for our purposes. We will illustrate a general "right-hand rule" for the particular case of angular momentum, $\vec{L}_A = \vec{r}_A \times \vec{p}$.

1) With your right hand open and flat (fingers and thumb in the same plane), with your thumb perpendicular to your fingers, point your fingers in the direction of the first vector ($\vec{r}_A$). See Figure 9.6.
2) Bend your fingers toward the second vector ($\vec{p}$), noting the angle of bend θ, which is always less than 180° (Figure 9.7). You may have to rotate your wrist to get your hand into a position that lets you bend your fingers this way; this wrist rotation is an essential part of the right-hand rule. You must get your right hand into a position that permits bending your fingers from pointing along $\vec{r}_A$ to pointing along $\vec{p}$.
3) Your thumb now points in the direction of the cross-product vector $\vec{r}_A \times \vec{p}$. This direction is perpendicular to the plane defined by $\vec{r}_A$ and $\vec{p}$. The fingers of your right hand curl in the direction that a disk with center at *A* would rotate if struck by the particle. It may seem odd to have the angular momentum vector perpendicular to the motion, but the point is that it lies along the (possible) axis of rotation.
4) The magnitude of the cross-product vector is the magnitude of the first vector, times the magnitude of the second vector, times the sine of the angle of bend of your fingers: $L_A = r_A p\sin\theta$. One of the advantages of this particular way of dealing with cross products is that you have a visual representation of the angle θ whose sine goes into the magnitude of the cross product.

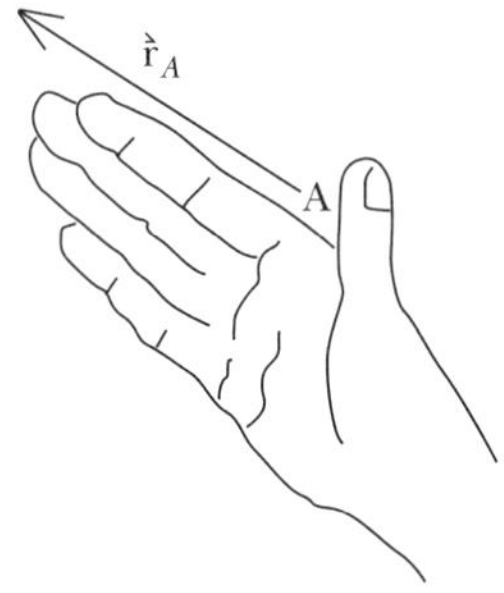

Figure 9.6 Open right hand, fingers point in direction of $\vec{r}_A$.

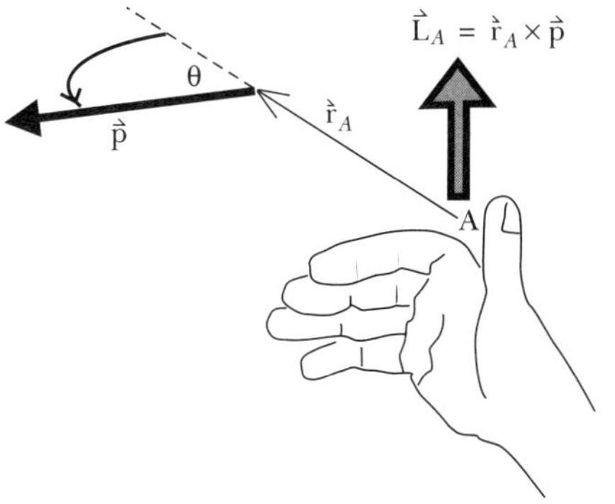

Figure 9.7 Bend fingers toward lining up with $\vec{p}$. Thumb points in direction of $\vec{r}_A \times \vec{p}$.

Things to watch out for

The rotation of the wrist is an important part of the right-hand rule. Consider the situation in Figure 9.8. With your right hand in this position, you can't bend your fingers backwards from the first vector $\vec{r}_A$ toward the second vector $\vec{p}$—it is a physical impossibility.

You need to rotate your wrist into a position from which it is possible to bend the fingers, as shown in Figure 9.9. The direction of your thumb is down, which is the correct direction for the cross product in this situation. The right-hand rule forces you to rotate your wrist into an orientation that permits bending your fingers from the first vector toward the second vector, in which case your thumb will point in the correct direction of the cross product.

Important: Pay attention to the size of the angle through which you bend your fingers. This is the angle whose sine is part of the definition of the magnitude of the cross product. Also note that this angle should never be more than 180°. If it is, you have made a mistake in orienting your hand; probably you need to rotate your wrist.

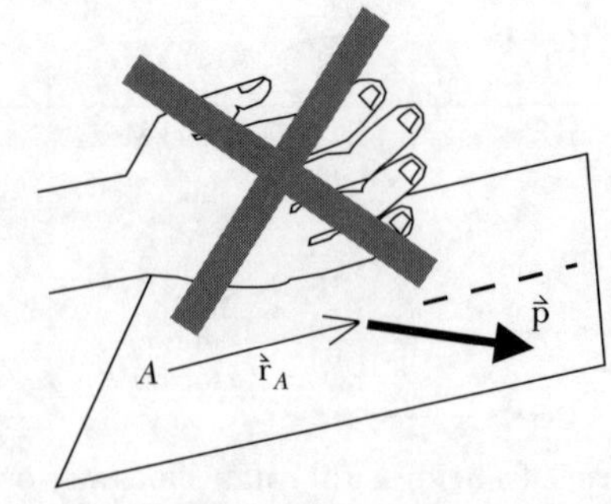

Figure 9.8 You can't bend your fingers backward. You must rotate the wrist into a position that lets you bend the fingers.

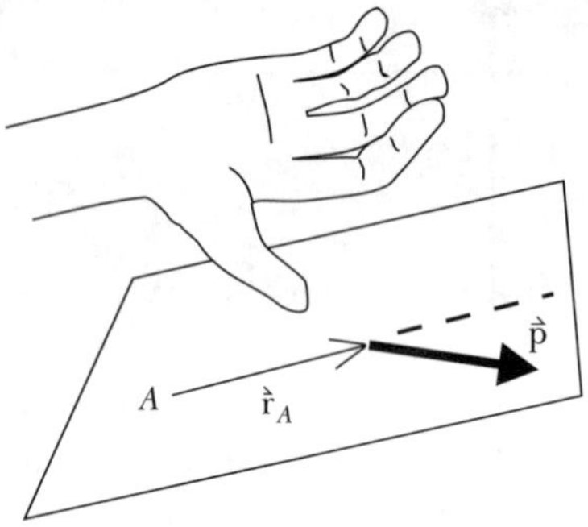

Figure 9.9 After rotating the wrist, it is possible to bend the fingers.

Two-dimensional projections

Because it is difficult to sketch a situation in three dimensions, whenever possible we will work with two-dimensional projections onto the *x-y* plane. If $\vec{r}_A$ and $\vec{p}$ lie in the *x-y* plane, the angular momentum $\vec{r}_A \times \vec{p}$ points in the $+z$ direction (out of the page) or in the $-z$ direction (into the page). A special notation is used to indicate "out of the page" ($\odot$) and "into the page" ($\otimes$); see Figure 9.10. The symbol $\odot$ can be thought of as showing the tip of an arrow (a vector) pointing out at you, while the symbol $\otimes$ represents the feathers of an arrow (a vector) pointing away from you.

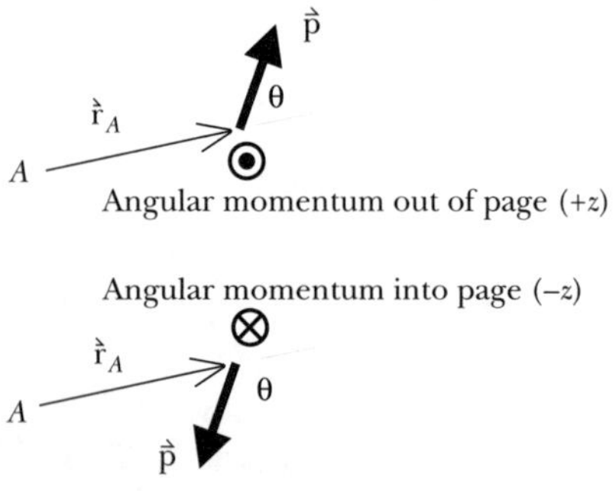

Figure 9.10 2-D projections: symbols for angular momentum out of and into page.

Calculating the magnitude

It is often easiest to calculate the magnitude $r_A p \sin\theta$ of the angular momentum by associating the sine of the angle with r_A or with p. In Figure 9.11, note that the perpendicular component $r_\perp$ of the "lever arm" is $r_A \sin\theta$, while the perpendicular component $p_\perp$ of the momentum is $p\sin\theta$. Study Figure 9.11 to learn two efficient methods for calculating the magnitude of the angular momentum of a particle.

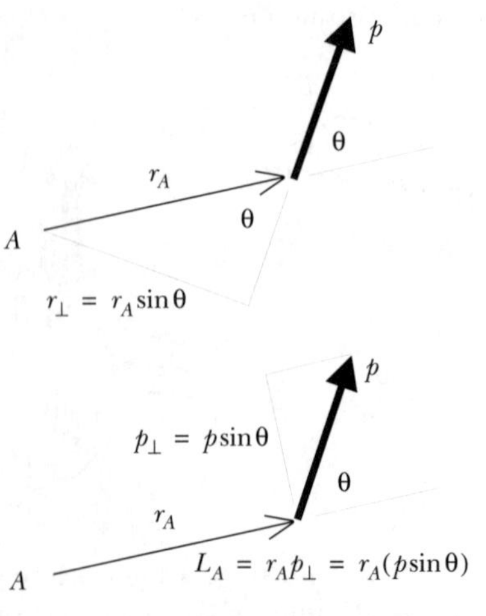

Figure 9.11 Two different ways to calculate the magnitude of the cross product.

Practice

Be sure that you do and check the following exercises before going on, because you need to be able to use the concept of cross product quickly and easily in later discussions.

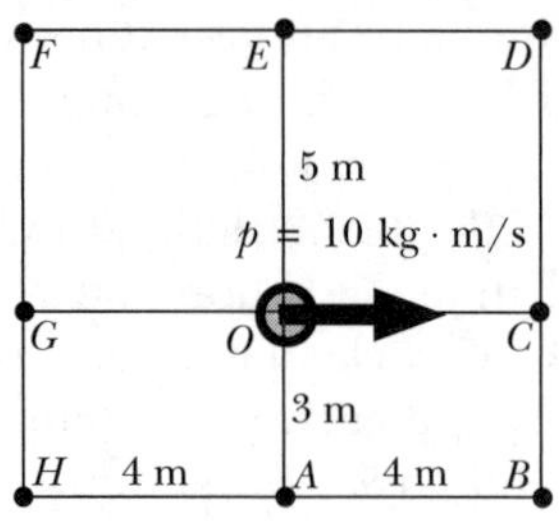

Ex. 9.1 Determine both the direction and magnitude of the angular momentum of the particle at location *O* relative to each point: *A*, *B*, *C*, *D*, *E*, *F*, *G*, and *H*.

Ex. 9.2 What are the magnitude and direction of the angular momentum about location *K* below?

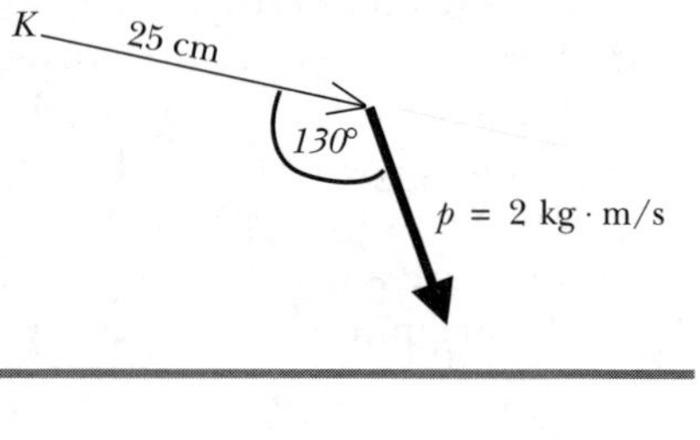

9.1.3 The cross product and unit vectors

It is also possible to evaluate the cross product in terms of unit vectors. First, note that $\hat{\imath}\times\hat{\imath} = 0$, $\hat{\jmath}\times\hat{\jmath} = 0$, and $\hat{k}\times\hat{k} = 0$, since when we cross a vector with itself the angle between the two vectors is zero, and $\sin 0° = 0$.

Second, $\hat{\imath}\times\hat{\jmath} = \hat{k}$, since the angle is 90° and the right-hand rule gives a result in the $+z$ direction (out of the page; Figure 9.12). On the other hand, $\hat{\jmath}\times\hat{\imath} = -\hat{k}$, because the right-hand rule gives a result in the $-z$ direction (into the page). Similarly, $\hat{\jmath}\times\hat{k} = \hat{\imath}$, $\hat{k}\times\hat{\jmath} = -\hat{\imath}$, $\hat{k}\times\hat{\imath} = \hat{\jmath}$, and $\hat{\imath}\times\hat{k} = -\hat{\jmath}$. Putting this all together, we obtain the following general result:

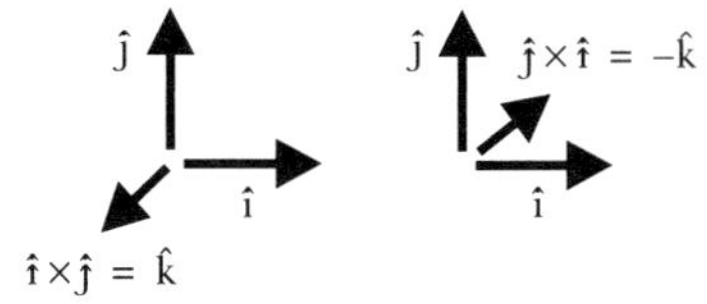

Figure 9.12 Cross products of unit vectors.

$$\vec{A}\times\vec{B} = (A_yB_z - A_zB_y)\hat{\imath} + (A_zB_x - A_xB_z)\hat{\jmath} + (A_xB_y - A_yB_x)\hat{k} \text{ or}$$

$$\vec{A}\times\vec{B} = \langle(A_yB_z - A_zB_y), (A_zB_x - A_xB_z), (A_xB_y - A_yB_x)\rangle$$

This approach to calculating a cross product is particularly useful in computer calculations. Note the cyclic nature of the subscripts: *xyz, yzx, zxy*.

Ex. 9.3 Given that $\vec{r}_A = \langle x, y, z\rangle$ and $\vec{p} = \langle p_x, p_y, p_z\rangle$, show that angular momentum can be calculated in the following way:

COMPONENT CALCULATION OF ANGULAR MOMENTUM

$$\vec{L}_A = \langle(yp_z - zp_y), (zp_x - xp_z), (xp_y - yp_x)\rangle$$

9.2 Angular momentum of multiparticle systems

Later we will show that when the child in Figure 9.1 jumps on the disk-shaped playground ride, the initial angular momentum of the child is equal to the sum of the final angular momenta of the child and disk. In order to calculate the angular momentum of a multiparticle system such as the disk, we need additional tools.

Recall from Chapter 7 that the kinetic energy of a multiparticle system can be split into overall translational kinetic energy plus rotational and vibrational kinetic energy relative to the center of mass:

$$K_{\text{tot}} = K_{\text{trans}} + K_{\text{rel}} = K_{\text{trans}} + K_{\text{rot}} + K_{\text{vib}}$$

In an analogous fashion, angular momentum can be split into a term associated with the motion of the center of mass (which we will call translational angular momentum, sometimes called "orbital" angular momentum) plus a term associated with rotation relative to the center of mass (which we will call rotational angular momentum, sometimes called "spin" angular momentum):

$$\vec{L}_A = \vec{L}_{\text{trans},A} + \vec{L}_{\text{rot}}$$

Splitting the angular momentum into these two terms is very useful in analyzing multiparticle systems such as the Earth, which orbits the Sun but also rotates on its axis. You may also know that electrons orbiting the nucleus of an atom have both translational and rotational angular momentum.

We will calculate the angular momentum relative to location A of a 3-mass system (Figure 9.13). We specify the positions of each mass relative to location A by going from A to the center of mass (vector $\vec{r}_{\text{cm}}$) and then to the individual masses. (The location A need not be in the same plane as the masses, nor are the momentum vectors necessarily in the plane of the masses.)

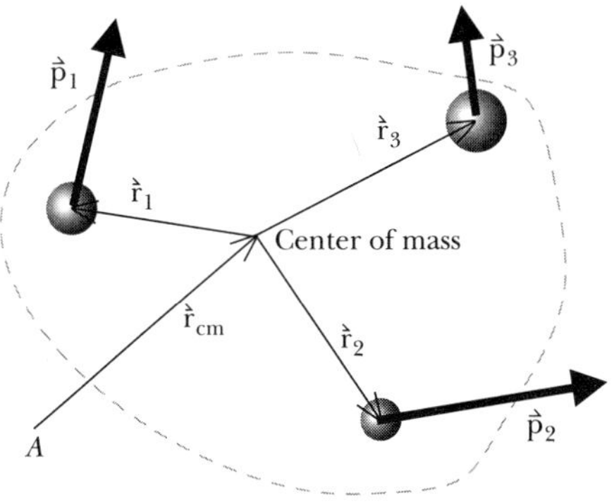

Figure 9.13 Position vectors relative to the center of mass.

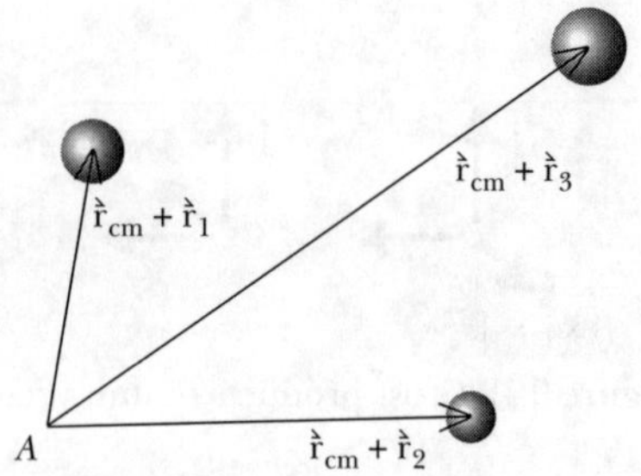

Figure 9.14 Position vectors relative to location A.

The position of mass m_1 relative to location A is $\vec{r}_{cm} + \vec{r}_1$, the position of mass m_2 is $\vec{r}_{cm} + \vec{r}_2$, and the position of mass m_3 is $\vec{r}_{cm} + \vec{r}_3$ (Figure 9.14). Using these vectors we calculate the total angular momentum relative to location A:

$$\vec{L}_A = (\vec{r}_{cm} + \vec{r}_1) \times \vec{p}_1 + (\vec{r}_{cm} + \vec{r}_2) \times \vec{p}_2 + (\vec{r}_{cm} + \vec{r}_3) \times \vec{p}_3$$

$$\vec{L}_A = [\vec{r}_{cm} \times (\vec{p}_1 + \vec{p}_2 + \vec{p}_3)] + [\vec{r}_1 \times \vec{p}_1 + \vec{r}_2 \times \vec{p}_2 + \vec{r}_3 \times \vec{p}_3]$$

Since $(\vec{p}_1 + \vec{p}_2 + \vec{p}_3)$ is the total momentum $\vec{P}_{tot}$, we have the following:

$$\vec{L}_A = \underbrace{[\vec{r}_{cm} \times \vec{P}_{tot}]}_{\text{"translational"}} + \underbrace{[\vec{r}_1 \times \vec{p}_1 + \vec{r}_2 \times \vec{p}_2 + \vec{r}_3 \times \vec{p}_3]}_{\text{"rotational"}}$$

The first term in the total angular momentum,

$$\vec{r}_{cm} \times \vec{P}_{tot}$$

is the angular momentum the system would have if all the mass were concentrated at the center of mass. We call this term the "translational" angular momentum $\vec{L}_{trans,A}$. This term is also called the "orbital" angular momentum.

The second term in the total angular momentum,

$$\vec{r}_1 \times \vec{p}_1 + \vec{r}_2 \times \vec{p}_2 + \vec{r}_3 \times \vec{p}_3$$

is the angular momentum of the system relative to the center of mass of the system. We call this term the "rotational" angular momentum $\vec{L}_{rot}$. An example of rotational angular momentum is the angular momentum due to the rotation of the Earth on its axis. This term is also called the "spin" angular momentum. In Figure 9.15 we show both the translational and rotational angular momentum vectors for the Earth relative to the Sun.

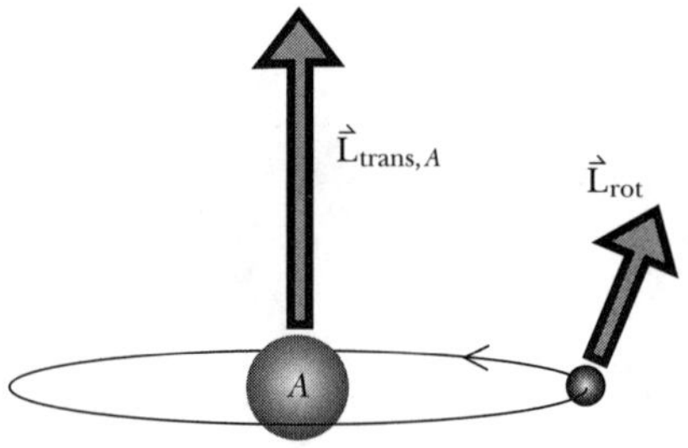

Figure 9.15 Translational angular momentum of the Earth relative to the Sun, with rotational angular momentum relative to the Earth's center of mass. At this instant the linear momentum of the Earth is into the page.

TRANSLATIONAL PLUS ROTATIONAL ANGULAR MOMENTUM

$$\vec{L}_A = \vec{L}_{trans,A} + \vec{L}_{rot}$$

$$\text{Translational: } \vec{L}_{trans,A} = \vec{r}_{cm,A} \times \vec{P}_{tot}$$

$$\text{Rotational: } \vec{L}_{rot} = \vec{r}_1 \times \vec{p}_1 + \vec{r}_2 \times \vec{p}_2 + \vec{r}_3 \times \vec{p}_3$$

The translational angular momentum differs for different choices of the location A, so we need the "A" subscript on the translational angular momentum and the total angular momentum to remind us of this dependence. The rotational angular momentum doesn't need a subscript because it is calculated relative to the center of mass, and this calculation is unaffected by our choice of the point A, and by the motion of the center of mass, which may even be accelerating.

For a point particle, or the "collapsed" point-particle version of a multiparticle system, the translational angular momentum is the only kind of angular momentum there is, just as translational kinetic energy is the only kind of kinetic energy a point particle can have.

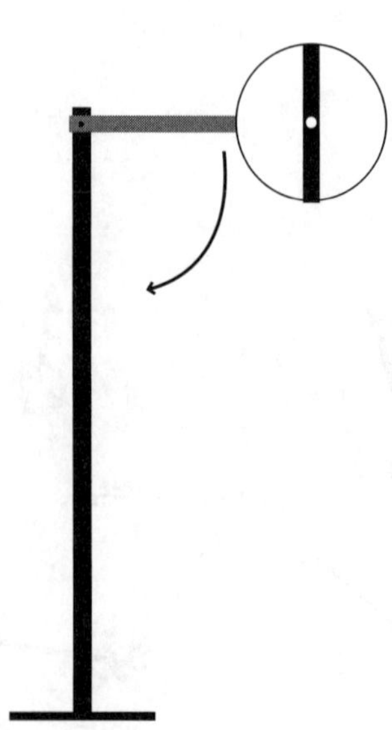

Figure 9.16 A rod rotating in the vertical plane with a wheel attached.

Ex. 9.4 A rod rotates in the vertical plane around a horizontal axle. A wheel is free to rotate on the rod, as shown in Figure 9.16. A vertical stripe is painted on the wheel. As the rod rotates clockwise, the vertical stripe on the wheel remains vertical. Is the translational angular momentum of the wheel zero or nonzero? If nonzero, what is its direction? Is the rotational angular momentum of the wheel zero or nonzero? If nonzero, what is its direction?

Ex. 9.5 Consider a system similar to that in the previous exercise, but with the wheel welded to the rod (not free to turn). As the rod rotates clockwise, does the stripe on the wheel remain vertical? Is the translational angular momentum of the wheel zero or nonzero? If nonzero, what is its direction? Is the rotational angular momentum of the wheel zero or nonzero? If nonzero, what is its direction?

Ex. 9.6 Pinocchio rides a horse on a merry-go-round turning counterclockwise as viewed from above, with his long nose always pointing forward, in the direction of his velocity. Is Pinocchio's translational angular momentum zero or nonzero? If nonzero, what is its direction? Is his rotational angular momentum zero or nonzero? If nonzero, what is its direction?

9.2.1 Rotational angular momentum: moment of inertia

An important special case of rotational motion is one where a system is rotating on an axis, with all the atoms in the system sharing the same "angular speed" in radians per second but with different linear speeds in meters per second, depending on their distances from the axis (Figure 9.17). This is the situation for a rigid object.

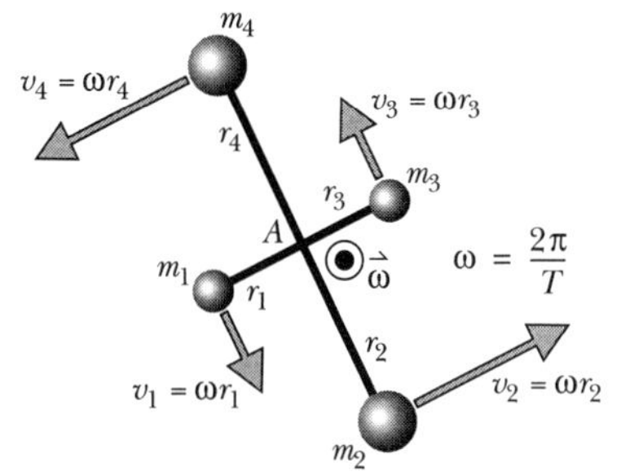

Figure 9.17 A case of rigid rotation about an axis with angular speed ω. Note the different speeds, with $v_i = \omega r_i$.

As we saw in Chapter 2, angular speed, normally denoted by ω (lowercase Greek omega), is a measure of how fast something is rotating. If an object makes one complete turn of 360 degrees (2π radians) in a time T, we say that its angular speed is $\omega = 2\pi/T$ radians per second. The time T is called the period.

We showed in Chapter 2 that the magnitude of the time rate of change of a rotating vector $\vec{X}$ is $|d\vec{X}/dt| = \omega X$. One application of this general result is that the speed v of an atom rotating around an axis at a distance r from the axis is $|d\vec{r}/dt| = v = \omega r$. We measure angular speed ω in the "natural" units of radians per second rather than degrees per second, because the fundamental geometrical relationship between angle and arc length (arc length = $r\theta$) is valid only if the angle θ is measured in radians.

An advantage of using angular speed can be seen in calculating the rotational angular momentum of a rotating rigid object, all of whose parts share a common angular speed ω, but have different linear speeds. We define an angular velocity vector:

ANGULAR VELOCITY VECTOR $\vec{\omega}$

magnitude: radians/second

direction: see Figure 9.18

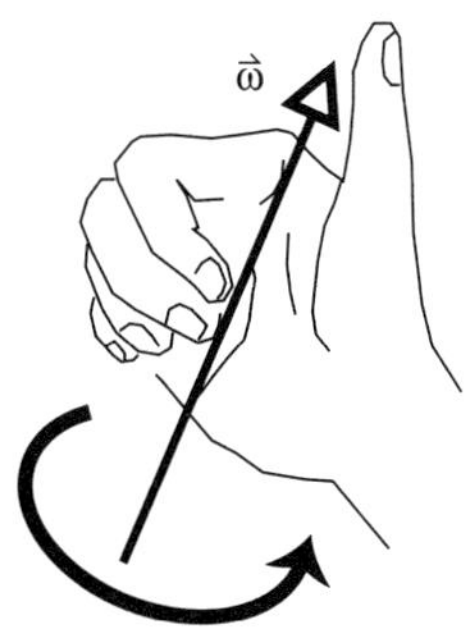

Figure 9.18 Finding the direction of the angular velocity vector.

Start with your fingers headed radially outward, bend them in the direction of the rotation, and your thumb points in the direction of the angular velocity, along the axis of rotation (angular velocity is also shown in Figure 9.17).

The magnitude of the angular momentum of each piece is this, where $r_\perp$ is the perpendicular distance of a piece from the axis:

$$|\vec{r} \times \vec{p}| = r_\perp mv = r_\perp m(\omega r_\perp) = mr_\perp^2\omega$$

So the magnitude of the rotational angular momentum is this:

$$|\vec{L}_{\text{rot}}| = [m_1 r_{\perp 1}^2 + m_2 r_{\perp 2}^2 + m_3 r_{\perp 3}^2 + m_4 r_{\perp 4}^2]\omega$$

The quantity in brackets is called the "moment of inertia" and is usually denoted by the letter I:

MOMENT OF INERTIA

$$I = m_1 r_{\perp 1}^2 + m_2 r_{\perp 2}^2 + m_3 r_{\perp 3}^2 + m_4 r_{\perp 4}^2 + \ldots$$

Since the rotational angular momentum points in the direction of $\vec{\omega}$, we have the following for the angular momentum relative to the center of mass, even if the center of mass is moving in some complicated way:

ROTATIONAL ANGULAR MOMENTUM

$$\vec{L}_{rot} = I\vec{\omega}$$

This way of evaluating rotational angular momentum is often easier to use than the more basic expression in terms of cross products $\vec{r} \times \vec{p}$, especially because the moments of inertia for common objects can be looked up in reference books. For example, the moment of inertia about an axis passing through the center of a uniform solid disk of radius R is $\frac{1}{2}MR^2$ (see Problem 9.3 (page 388)), and the moment of inertia about an axis passing through the center of a uniform solid sphere of radius R is $\frac{2}{5}MR^2$.

Ex. 9.7 What is the moment of inertia of a nitrogen molecule around its center of mass if the mass of each atom is M and the distance between nuclei is d? (Note that the electrons hardly contribute to the moment of inertia because their mass is so small.)

Ex. 9.8 A bicycle wheel has almost all its mass M located in the outer rim at radius R. If it rotates on a stationary axle with angular speed ω, what is the angular momentum of the wheel? (Hint: It's helpful to think of dividing the wheel into the atoms it is made of.)

9.2.2 Rotational kinetic energy

The kinetic energy of the system shown in Figure 9.19, which is a rotating rigid object, is $\frac{1}{2}m_1v_1^2 + \frac{1}{2}m_2v_2^2 + \frac{1}{2}m_3v_3^2 + \ldots$. Since the speed of each mass is $v = \omega r_\perp$, we can write the rotational kinetic energy like this:

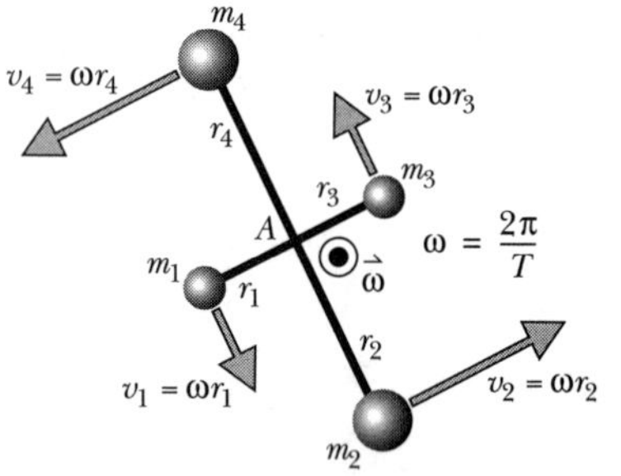

Figure 9.19 A case of rigid rotation about an axis with angular speed ω. Note the different speeds, with $v_i = \omega r_i$.

$$K_{rot} = \frac{1}{2}[m_1(r_{\perp 1}\omega)^2 + m_2(r_{\perp 2}\omega)^2 + m_3(r_{\perp 3}\omega)^2 + \ldots]$$

$$= \frac{1}{2}[m_1 r_{\perp 1}^2 + m_2 r_{\perp 2}^2 + m_3 r_{\perp 3}^2 + \ldots]\omega^2$$

We again see the moment of inertia showing up in the bracket, so we have the following calculation of rotational kinetic energy:

ROTATIONAL KINETIC ENERGY

$$K_{rot} = \frac{1}{2}I\omega^2 = \frac{1}{2}\frac{(I\omega)^2}{I} = \frac{L_{rot}^2}{2I}$$

Ex. 9.9 A uniform-density disk whose mass is 10 kg and radius is 0.4 m makes one complete rotation every 0.2 s. What is its kinetic energy? (The moment of inertia of a uniform-density disk about its center is $\frac{1}{2}MR^2$.)

Complications

We will deal with situations where the axis of rotation of a rigid body passes through the center of mass, and the object is symmetric or nearly so, in which case we can calculate the rotational angular momentum and rotational kinetic energy using the formulas $\vec{L}_{rot} = I\vec{\omega}$ and $K_{rot} = \frac{1}{2}I\omega^2$. In Section 9.11 on page 384 we discuss some complications that apply in the more general case. We will avoid these more complex rotational situations, which are treated in more advanced courses.

9.2.3 More exercises on angular momentum

The following exercises constitute an important worksheet that gives you practice in calculating angular momentum. Be sure to do all of these exercises and check your answers at the end of the chapter.

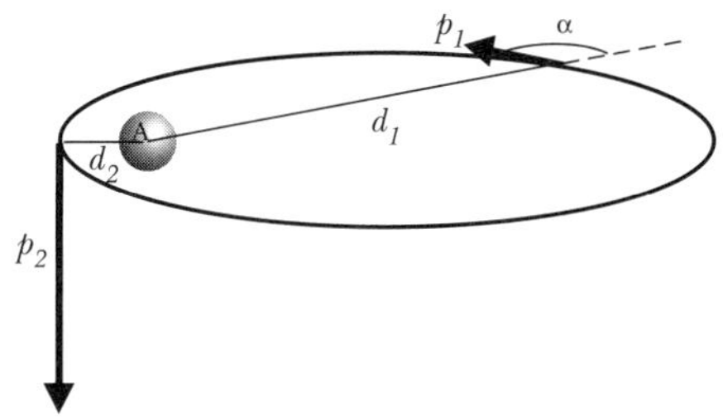

Figure 9.20 A comet orbits the Sun. Location A is at the center of the Sun.

Ex. 9.10 A comet orbits the Sun (Figure 9.20). When it is at location 1 it is a distance d_1 from the Sun, and has momentum $\vec{p}_1$. Location A is at the center of the Sun.

(a) When the comet is at location 1, what is the direction of $\vec{L}_A$?

(b) When the comet is at location 1, what is the magnitude of $\vec{L}_A$?

When the comet is at location 2, it is a distance d_2 from the Sun, and has momentum $\vec{p}_2$.

(c) When the comet is at location 2, what is the direction of $\vec{L}_A$?

(d) When the comet is at location 2, what is the magnitude of $\vec{L}_A$?

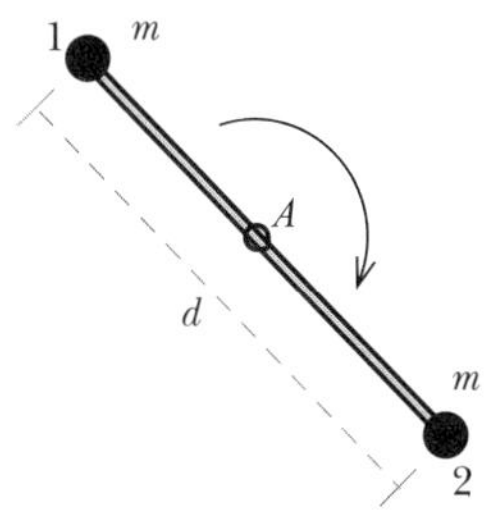

Figure 9.21 A rotating barbell.

Ex. 9.11 A barbell spins around a pivot at its center at A (Figure 9.21). The barbell consists of two small balls, each with mass m, at the ends of a very low mass rod of length d. The barbell spins clockwise with angular speed ω_0 radians/s.

(a) Calculate $\vec{L}_{trans,1,A}$ (both direction and magnitude).

(b) Calculate $\vec{L}_{tot,A}$ (both direction and magnitude).

(c) Calculate I for the barbell.

(d) What is the direction of $\vec{\omega}_0$?

(e) Calculate $\vec{L}_{rot}$ (both direction and magnitude).

(f) How does $\vec{L}_{rot}$ compare to $\vec{L}_{tot,A}$?

(g) Calculate K_{rot}.

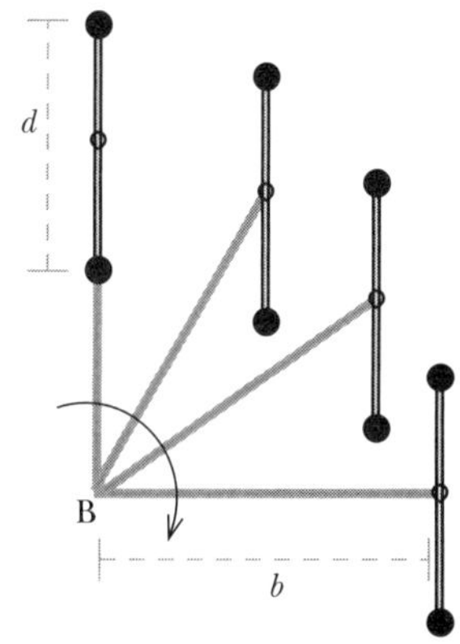

Figure 9.22 A barbell pivoted on a low-mass rotating rod. The barbell does not rotate.

Ex. 9.12 The barbell in the previous exercise is mounted on the end of a low mass rigid rod of length b (Figure 9.22). The apparatus is started in such a way that although the rod rotates clockwise with angular speed ω_1 rad/s, the barbell maintains its vertical orientation.

(a) Calculate $\vec{L}_{rot}$ (both direction and magnitude).

(b) Calculate $\vec{L}_{trans,B}$ (both direction and magnitude).

(c) Calculate $\vec{L}_{tot,B}$ (both direction and magnitude).

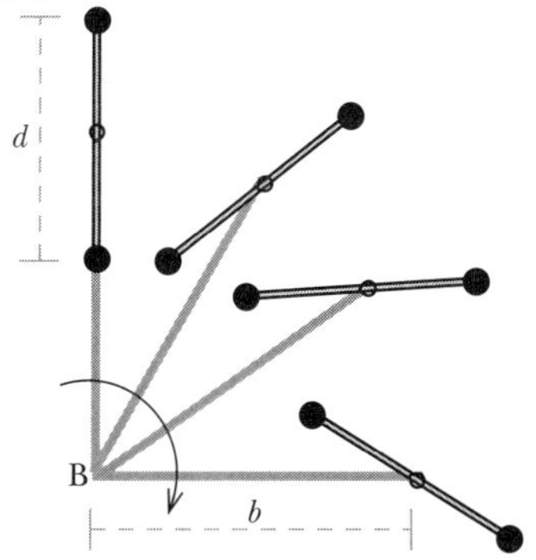

Figure 9.23 A barbell pivoted on a low-mass rotating rod. The barbell rotates.

Ex. 9.13 The apparatus in the previous exercise is restarted in such a way that it again rotates clockwise with angular speed ω_1 rad/s, but in addition, the barbell rotates clockwise about its center, with an angular speed ω_2 rad/s (Figure 9.23).

(a) Calculate $\vec{L}_{rot}$ (both direction and magnitude).

(b) Calculate $\vec{L}_{trans,B}$ (both direction and magnitude).

(c) Calculate $\vec{L}_{tot,B}$ (both direction and magnitude).

9.3 The Angular Momentum Principle

So far we have concentrated on defining and calculating angular momentum. Next we will show that an off-center force that applies a twist to a system can change the angular momentum of the system.

For a single particle, the Momentum Principle says that the time derivative of the momentum is the net force: $d\vec{p}/dt = \vec{F}_{net}$. Let's see what the time derivative of angular momentum turns out to be. By applying the product rule for derivatives, we have

$$\frac{d(\vec{r}_A \times \vec{p})}{dt} = \frac{d\vec{r}_A}{dt} \times \vec{p} + \vec{r}_A \times \frac{d\vec{p}}{dt}$$

Since $\dfrac{d\vec{r}_A}{dt} = \vec{v}$ and $\vec{p} = \dfrac{m\vec{v}}{\sqrt{1-(v^2/c^2)}}$, we have

$$\frac{d\vec{r}_A}{dt} \times \vec{p} = \frac{\vec{v} \times m\vec{v}}{\sqrt{1-(v^2/c^2)}} = 0$$

since $|\vec{v} \times \vec{v}| = v^2 \sin(0°) = 0$. Therefore we have proved that the following is true for a single particle:

$$\frac{d(\vec{r}_A \times \vec{p})}{dt} = \vec{r}_A \times \vec{F}_{net}$$

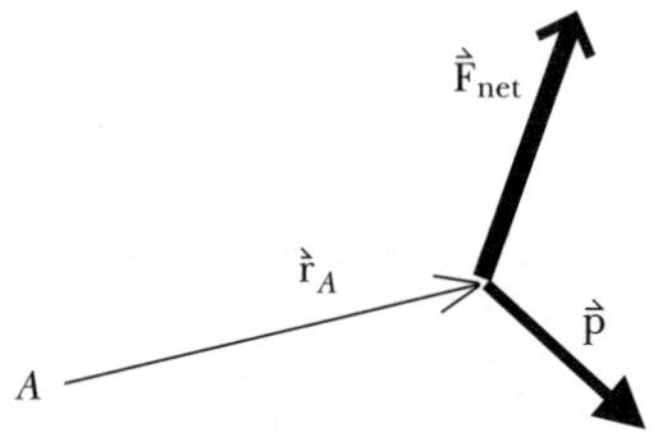

Figure 9.24 Torque relative to location A is defined as $\vec{\tau}_A = \vec{r}_A \times \vec{F}_{net}$.

We have previously defined the quantity $\vec{L}_A = \vec{r}_A \times \vec{p}$ to be the angular momentum of a particle relative to location A. We call the quantity $\vec{\tau}_A = \vec{r}_A \times \vec{F}_{net}$ the "torque" exerted about location A (Figure 9.24). Torque is usually written as τ, Greek lowercase letter tau. Torque is a word of Latin origin and simply means "twist." Twisting a system changes its angular momentum.

We can read the equation as saying "the rate of change of the angular momentum of a particle relative to location A is equal to the torque applied to the particle about location A."

THE ANGULAR MOMENTUM PRINCIPLE FOR A POINT PARTICLE

$$\frac{d\vec{L}_A}{dt} = \vec{r}_A \times \vec{F}_{net} = \vec{\tau}_A \text{ or } \Delta\vec{L}_A = (\vec{r}_A \times \vec{F}_{net})\Delta t = \vec{\tau}_A \Delta t$$

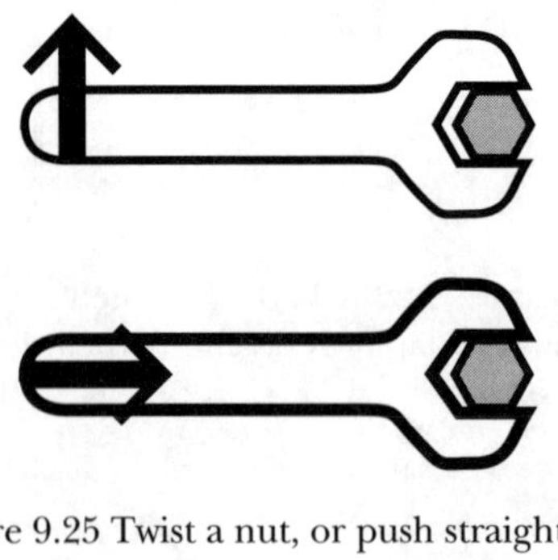

Figure 9.25 Twist a nut, or push straight at it?

The Angular Momentum Principle is similar to the Momentum Principle $d\vec{p}/dt = \vec{F}_{net}$ or $\Delta\vec{p} = \vec{F}_{net}\Delta t$. The term $\vec{\tau}_A \Delta t$ is the "angular impulse." We will apply this new principle after practicing how to calculate torques.

9.3.1 Calculating torque

Torque and angular momentum are both calculated by using the cross product. In this section we offer some additional ways to think about cross products, in the context of practice in calculating torques.

? In Figure 9.25, which way of pushing on a wrench is more effective in twisting the nut?

You can see that a force parallel to the handle is completely ineffective in twisting the nut. Evidently it is only the perpendicular component of a force that is effective in twisting an object.

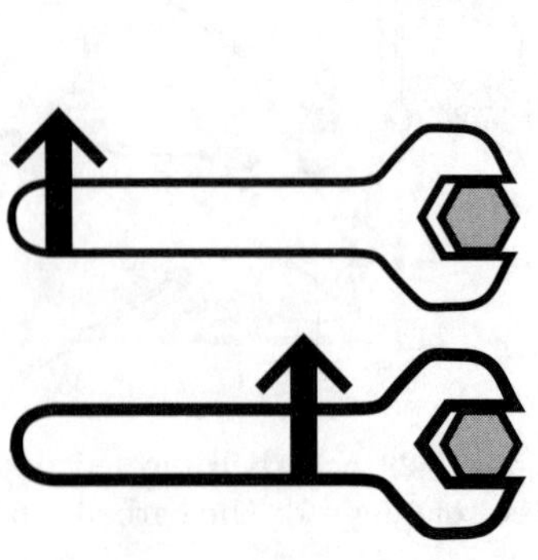

Figure 9.26 Push at the end, or near the nut?

? In Figure 9.26, where should you push on the wrench—at the end of the wrench or near the nut?

The farther away from the nut we push, the more effective we are in twisting the nut. The advantage of a long wrench is to provide more leverage in twisting.

Suppose we apply a force at an angle θ to the radius (Figure 9.27). Evidently the component of the force that is effective in twisting is $F\sin\theta$, the perpendicular component. Moreover, the bigger the lever arm (r_A, the distance from the nut at location A), the more effective we'll be. To capture both effects, we define the magnitude of "torque" (which means "twist") exerted by a force $\vec{F}$ relative to a location A as follows:

MAGNITUDE OF TORQUE RELATIVE TO LOCATION *A*

$$\text{torque} = r_A F\sin\theta$$

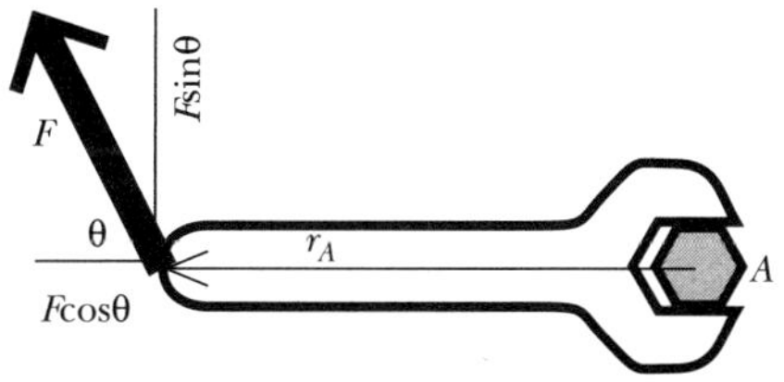

Figure 9.27 Push on the wrench at an angle.

For a purely perpendicular force, $\theta = 90°$, $\sin\theta = 1$, and the torque is r_AF. For a force that is parallel to the lever arm, $\theta = 0°$, $\sin\theta = 0$, and the torque is zero. See Figure 9.25 for these extreme examples.

Direction of torque

The direction of the torque is given by the cross product and associated right-hand rule.

? In Figure 9.27, is the torque vector into the page or out of the page? Use the right-hand rule.

With your right hand at the location of the nut, point your fingers toward the point of application of the force (at the end of the wrench). Next curl your fingers toward the direction of the force vector. The only way you can do this results in your extended right thumb pointing into the page. If you try to keep your extended right thumb pointing out of the page, you will find that you can't curl your fingers toward the direction of the force.

Most nuts and bolts have what are called "right-handed threads." In Figure 9.27 a right-handed nut will advance in the direction of the torque vector, into the page. If you reverse the direction of the force, so the torque vector points out of the page, a right-handed nut will advance out of the page. (Exceptions include left-handed threads on gas cylinders containing explosive gases; the unexpected behavior alerts users to danger.)

The fact that a physical right-handed nut advances in the direction of the torque vector is another justification for drawing torque cross product vectors along an axis of rotation, perpendicular to the plane of the motion.

Ex. 9.14 If $r_A = 3$ m, $F = 4$ N, and $\theta = 30°$, what is the magnitude of the torque about location A, including units?

Ex. 9.15 If the force in Figure 9.27 were perpendicular to $\vec{r}_A$, but gave the same torque as in the preceding exercise, what would its magnitude be?

Ex. 9.16 At $t = 15$ s, a particle has angular momentum $\langle 3, 5, -2\rangle$ $\text{kg}\cdot\text{m}^2/\text{s}$ relative to location A. A constant torque $\langle 10, -12, 20\rangle$ N·m relative to location A acts on the particle. At $t = 15.1$ s, what is the angular momentum of the particle?

9.3.2 Example: Falling object

Let's compare the Momentum Principle and the Angular Momentum Principle in a simple situation. Consider a mass m falling near the Earth (Figure 9.28). We choose a location A off to the side, on the ground.

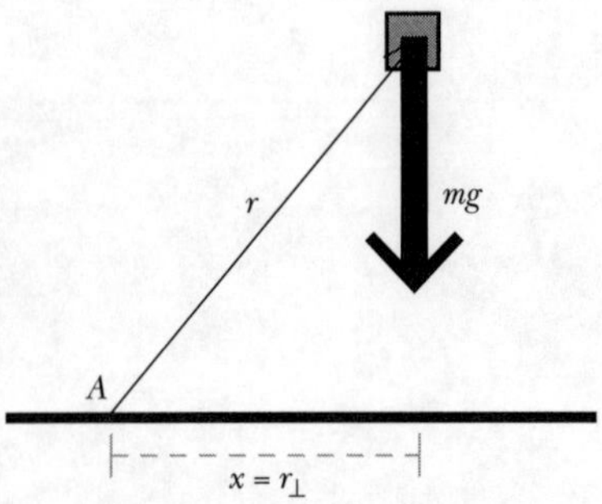

Figure 9.28 Analyze a falling mass using force and momentum, and using torque and angular momentum.

Neglecting air resistance, the Momentum Principle gives $dp/dt = mg$, yielding $dv/dt = g$ (nonrelativistic). Next we see what the Angular Momentum Principle gives.

? What is the direction of the torque on the mass relative to location A? What is its magnitude?

On the diagram you can see that $x = r_\perp$, so the torque on the mass relative to point A is $r_\perp mg = xmg$. The right-hand rule shows that the direction of the torque is into the page (⊗).

? What is the direction of the angular momentum of the mass relative to point A? What is its magnitude?

The angular momentum of the mass relative to point A is $r_\perp mv = xmv$, since $x = r_\perp$. The right-hand rule shows that the direction of the angular momentum is into the page (⊗). The Angular Momentum Principle has components into the page of $d(xp)/dt = xmg$. Since x is a fixed value, this yields $dp/dt = mg$ as expected. So the Angular Momentum Principle for a point particle is equivalent to the Momentum Principle. This isn't a big surprise, since we derived the new equation starting from the Momentum Principle.

9.4 Conservation of angular momentum

Since we just found that the Angular Momentum Principle didn't tell us anything new about a falling object, what's the point of all this? An important point is that the angular momentum is constant under certain circumstances. If the torque around a location A is zero, the angular momentum about that location does not change. There may be forces acting, and causing changes in the linear momentum, but if these forces don't exert any torque, the rate of change of angular momentum is zero. A major reason for defining torque is to be able to identify situations in which the torque about a location A is zero, in which case angular momentum is constant.

Angular momentum is a conserved quantity:

CONSERVATION OF ANGULAR MOMENTUM

$$\Delta\vec{L}_{A,\,\text{system}} + \Delta\vec{L}_{A,\,\text{surroundings}} = 0$$

If a system gains angular momentum, the surroundings lose that amount. An important special case that merits separate attention is the case of no torque acting on the system, in which case the angular momentum does not change. Comet orbits provide a nice example. Most comets have very long elliptical orbits around the Sun (Figure 9.29). We can treat the comet as a point particle, because it is tiny compared to the distances involved.

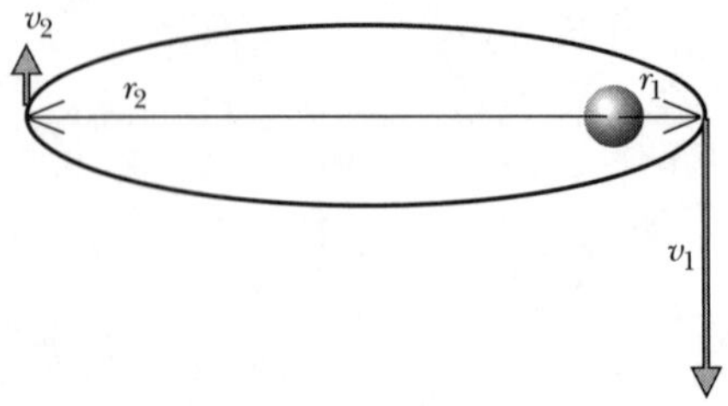

Figure 9.29 A long elliptical orbit of a comet around the Sun.

Ex. 9.17 Relative to the center of the Sun, explain why the torque exerted by the Sun's gravitational force on the comet is zero at every point along the orbit. What does that say about the angular momentum of the comet relative to the center of the Sun?

Ex. 9.18 In terms of the comet's mass m and speed v_1 when nearest the Sun (distance r_1 from the Sun), what is the comet's angular momentum relative to the center of the Sun (direction and magnitude)? What is the comet's angular momentum relative to the center of the Sun (direction and magnitude) when it is farthest from the Sun (distance r_2 from the Sun) and traveling at speed v_2? What can you conclude about the relationship between v_1 and v_2?

Comments about comets

Some comets go far beyond Pluto and spend most of their time there, because they're traveling very slowly, as you have just shown. We see them when they come near the Sun, but only for a few months, because they're now traveling fast.

The result $r_1 v_1 = r_2 v_2$ at the closest and farthest points in the orbit is correct no matter what kind of "central" force is involved. For any force that acts along a line connecting two objects, there is no torque about a point at the center of one of the objects, so angular momentum is constant about that location. In the particular case of the gravitational force, we also have an energy equation relating v_1 and v_2:

$$\frac{1}{2}mv_1^2 - GMm/r_1 = \frac{1}{2}mv_2^2 - GMm/r_2$$

Historical note: Kepler and elliptical orbits

Based on careful, accurate naked-eye measurements made by Tycho Brahe before the invention of the telescope, Johannes Kepler in 1609 announced his discovery that the planets follow elliptical orbits around the Sun.

Kepler also stated that he had found that "a radius vector joining any planet to the Sun sweeps out equal areas in equal lengths of time." This is equivalent to conservation of angular momentum, as can be seen with the aid of Figure 9.30. The area swept out in a time Δt is the area of a triangle whose base is $v\Delta t$ and whose altitude is $r\sin\theta$. This area is $\frac{1}{2}(rv\sin\theta)\Delta t$, which is proportional to the angular momentum $rmv\sin\theta$.

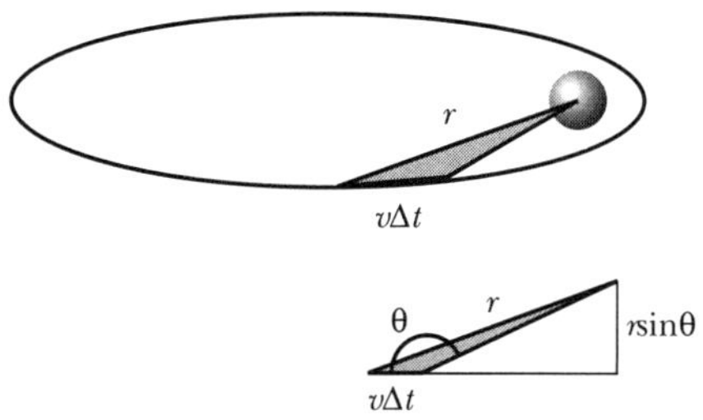

Figure 9.30 "Equal area in equal time."

In 1618 Kepler announced his discovery that the square of the time it takes a planet to go around the Sun is proportional to the cube of the mean distance from the Sun (a result you derived in Chapter 2 for the simpler case of circular orbits).

Later in the 1600's Newton explained all three of these discoveries as derivable from his second law of motion plus his universal law of gravitation. Kepler's insights provided important tests for Newton's theories.

9.5 Multiparticle systems

The Angular Momentum Principle has already given us new insight into the motion of a single particle acted on by a torque due to a single force. The equation becomes a really powerful tool when we extend it to apply to multiparticle systems acted on by multiple torques. For example, we will be able to understand the counterintuitive behavior of spinning tops and gyroscopes.

We will now derive a multiparticle version of the Angular Momentum Principle, following closely the derivation in Chapter 2 of a multiparticle version of the Momentum Principle.

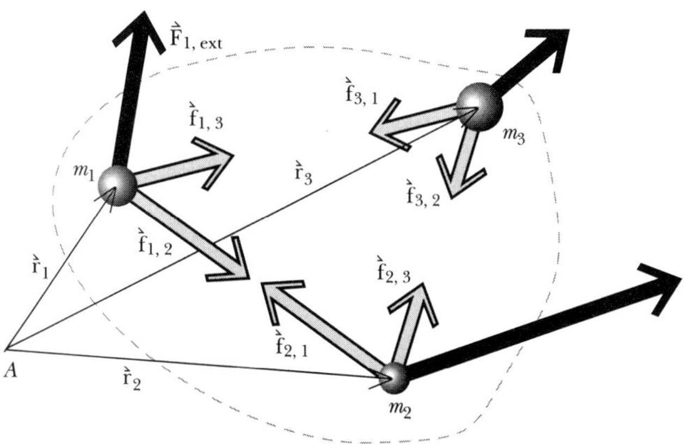

Figure 9.31 The Angular Momentum Principle in a multiparticle system.

To be as concrete as possible, we'll consider a system consisting of just three particles. It is easy to see how this generalizes to larger systems, including a block consisting of an astronomically huge number of atoms.

In Figure 9.31 we show all of the forces acting on each particle, where the lower case $\vec{f}$'s are internal forces, and the upper case $\vec{F}$'s are external forces, such as the gravitational attraction of the Earth. We write the Angular Momentum Principle applied to each of the three particles. We measure angular momenta and torques relative to location A, but to avoid clutter we don't write the subscript A in any of these equations until the end. ($\vec{L}_1$ is the angular momentum of m_1 relative to A.)

$$\frac{d\vec{L}_1}{dt} = \vec{r}_1 \times (\vec{F}_{ext,1} + \vec{f}_{1,2} + \vec{f}_{1,3})$$

$$\frac{d\vec{L}_2}{dt} = \vec{r}_2 \times (\vec{F}_{ext,2} + \vec{f}_{2,1} + \vec{f}_{2,3})$$

$$\frac{d\vec{L}_3}{dt} = \vec{r}_3 \times (\vec{F}_{ext,3} + \vec{f}_{3,1} + \vec{f}_{3,2})$$

Nothing new so far. But now we add up these three equations. That is, we create a new equation by adding up all the terms on the left sides of the three equations, and adding up all the terms on the right sides, and setting them equal to each other. In doing so, we take into account that many of these terms cancel. By the reciprocity principle (Newton's third law), we have

$$\vec{f}_{1,2} = -\vec{f}_{2,1}$$

$$\vec{f}_{1,3} = -\vec{f}_{3,1}$$

$$\vec{f}_{2,3} = -\vec{f}_{3,2}$$

Therefore the torques of the internal forces cancel. For example, consider this piece of the sum:

$$\vec{r}_1 \times \vec{f}_{1,2} + \vec{r}_2 \times \vec{f}_{2,1} = \vec{r}_1 \times \vec{f}_{1,2} - \vec{r}_2 \times \vec{f}_{1,2} = (\vec{r}_1 - \vec{r}_2) \times \vec{f}_{1,2}$$

The vector $\vec{r}_1 - \vec{r}_2$ points from m_2 toward m_1, so the angle between it and $\vec{f}_{1,2}$ is either 0° or 180°, and $\sin\theta$ is zero. Consequently all of the torques associated with the internal forces cancel, and all that remains is this:

$$\frac{d(\vec{L}_1 + \vec{L}_2 + \vec{L}_3)}{dt} = \vec{r}_1 \times \vec{F}_{ext,1} + \vec{r}_2 \times \vec{F}_{ext,2} + \vec{r}_3 \times \vec{F}_{ext,3}$$

The right side of this equation represents the net torque due to external forces, $\vec{\tau}_{net,ext,A}$. The great importance of this equation is that the reciprocity of gravitational and electric forces has allowed us to get rid of all the torques due to internal forces, the torques that the particles in the system exert on each other. We rewrite our result like this:

THE ANGULAR MOMENTUM PRINCIPLE FOR A MULTIPARTICLE SYSTEM

$$\frac{d\vec{L}_{tot,A}}{dt} = \vec{\tau}_{net,ext,A} \text{ or } \Delta\vec{L}_{tot,A} = \vec{\tau}_{net,ext,A}\Delta t$$

In words, the rate of change of the "total angular momentum" of a system relative to a location A, $\vec{L}_{tot,A} = \vec{L}_{1,A} + \vec{L}_{2,A} + \vec{L}_{3,A} + \ldots$, is equal to the net torque due to external forces exerted on that system, relative to location A. (Or, the change in the angular momentum is equal to the angular impulse.) The internal forces and torques do not appear in this multiparticle version of the Angular Momentum Principle.

Often it pays to choose location A to be at the place where the center of mass happens to be at that instant. In that case, the Angular Momentum Principle simplifies:

$$\frac{d\vec{L}_{cm}}{dt} = \frac{d}{dt}[(\vec{r}_{cm,cm} \times \vec{P}_{tot}) + \vec{L}_{rot}] = \frac{d\vec{L}_{rot}}{dt}, \text{ since } \vec{r}_{cm,cm} = 0$$

That is, the position of the center of mass, relative to the center of mass itself, is of course a zero vector. Therefore we have the important and useful special case for the motion relative to the center of mass

THE ANGULAR MOMENTUM PRINCIPLE RELATIVE TO THE CENTER OF MASS

$$\frac{d\vec{L}_{rot}}{dt} = \vec{\tau}_{net,cm} \text{ or } \Delta\vec{L}_{rot} = \vec{\tau}_{net,cm}\Delta t$$

The child and the playground ride

We began the chapter with a child running and jumping onto a playground ride. The axle about which the disk turns may exert a force on the disk, but this force has a very small lever arm measured from the center of the axle. If we consider the combined multiparticle system of child plus disk, we expect that the angular momentum to be (nearly) constant. The initial angular momentum of the child should be equal to the sum of the final angular momentum of the child (on the now rotating disk) and the angular momentum of the disk.

Ex. 9.19 Consider a rotating star far from other objects. Its rate of spin stays constant, and its axis of rotation keeps pointing in the same direction. Why?

Ex. 9.20 Because the Earth is nearly perfectly spherical, gravitational forces act on it effectively through its center. Explain why the Earth's axis points at the North Star all year long. Also explain why the Earth's rotation speed stays the same throughout the year (one rotation per 24 hours). In your analysis, does it matter that the Earth is going around the Sun?

In actual fact, the Earth is not perfectly spherical. It bulges out a bit at the equator, and tides tend to pile up water at one side of the ocean. As a result, there are small torques exerted on the Earth by other bodies, mainly the Sun and the Moon. Over many thousands of years there are changes in what portion of sky the Earth's axis points toward (change of direction of rotational angular momentum), and changes in the length of a day (change of magnitude of rotational angular momentum). See page 383.

9.5.1 Summary: Angular momentum and torque

At last we have all the tools we need to investigate a variety of situations involving angular momentum. Let's summarize these new tools and concepts:

- A force exerts a torque relative to a chosen location A: $\vec{\tau}_A = \vec{r}_A \times \vec{F}$.
- Torque causes changes in angular momentum; the rate of change of the angular momentum is equal to the net torque:

$$\frac{d\vec{L}_A}{dt} = \vec{\tau}_{net,ext,A}$$

- If the net torque relative to A is zero, the angular momentum relative to A does not change. (And if we know the angular momentum isn't changing about some location, as in static equilibrium, we know that the net torque must be zero about that location.)

- The angular momentum can be divided into two pieces: translational angular momentum of the system as a whole (as though all the mass were concentrated at the center-of-mass point), plus rotational angular momentum relative to the center of mass: $\vec{L}_A = \vec{L}_{trans,A} + \vec{L}_{rot}$.
- The rate of change of the rotational angular momentum is equal to the torque relative to the center of mass:

$$\frac{d\vec{L}_{rot}}{dt} = \vec{\tau}_{net,ext,cm}$$

- Rotational angular momentum is given by moment of inertia times angular velocity: $\vec{L}_{rot} = I\vec{\omega}$, where $I = m_1 r_{\perp 1}^2 + m_2 r_{\perp 2}^2 + m_3 r_{\perp 3}^2 + ...$
- Rotational kinetic energy about the center of mass can be calculated as

$$K_{rot} = \frac{1}{2}I\omega^2 = \frac{1}{2}\frac{(I\omega)^2}{I} = \frac{L_{rot}^2}{2I}$$

9.6 The three fundamental principles of mechanics

We now have three fundamental principles that relate a change in some property of a system to the interactions of the system with external objects. In the special case in which there are no external interactions, these principles indicate that a property of the system doesn't change (momentum or energy or angular momentum).

Momentum	Angular momentum	Energy
$\frac{d\vec{P}}{dt} = \vec{F}_{net,ext}$	$\frac{d\vec{L}_A}{dt} = \vec{\tau}_{net,ext,A}$	$\Delta E = W + Q$
If external forces: momentum changes.	If external torques: angular momentum changes.	If energy inputs: energy of system changes.
If no external forces: the momentum of a system is constant.	If no external torques: the angular momentum of a system is constant.	If no energy inputs: the energy of a system is constant.
Location of object does not matter.	Location of object relative to point A is important.	Location of object does not matter.

9.6.1 More exercises on the Angular Momentum Principle

The following exercises constitute an important worksheet that gives you practice in using the Angular Momentum Principle. Be sure to do all of these exercises and check your answers at the end of the chapter.

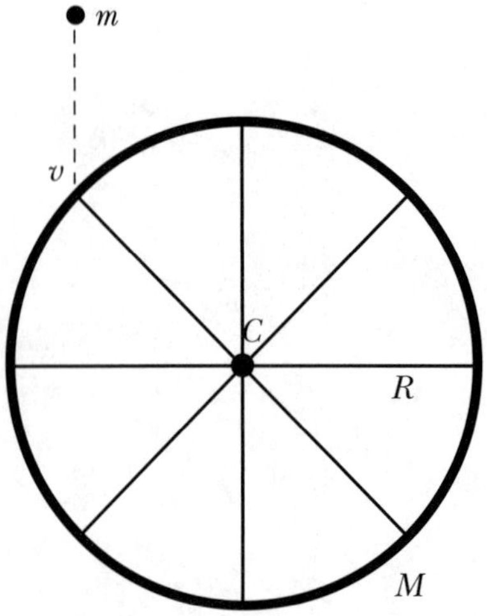

Figure 9.32 A lump of clay falls onto a wheel mounted on a frictionless axle.

Ex. 9.21 A stationary bicycle wheel of radius R is mounted in the vertical plane on a horizontal frictionless axle (Figure 9.32). The wheel has mass M, all concentrated in the rim (the spokes have negligible mass). A lump of clay with mass m falls and sticks to the outer edge of the wheel at the location shown. Just before the impact the clay has a speed v.

(a) Just before the impact, what is the angular momentum of the combined system of wheel plus clay about the center C?

(b) Just after the impact, what is the angular momentum of the combined system of wheel plus clay about the center *C*?

(c) Just after the impact, what are the magnitude and direction of the angular velocity of the wheel?

(d) Qualitatively, what happens to the linear momentum of the combined system? Why?

Ex. 9.22 A disk of radius 8 cm is pulled along a frictionless surface with a force of 10 N by a string wrapped around the edge (Figure 9.33). 24 cm of string has unwound off the disk.

What are the magnitude and direction of the torque exerted about the center of the disk at this instant?

Ex. 9.23 In Figure 9.33, the uniform solid disk has mass 0.4 kg (moment of inertia $I = \frac{1}{2}MR^2$). At the instant shown, the angular velocity is 20 radians/s into the page.

(a) At this instant, what are the magnitude and direction of the angular momentum about the center of the disk?

(b) At a time 0.2 s later, what are the magnitude and direction of the angular momentum about the center of the disk?

(c) At this later time, what are the magnitude and direction of the angular velocity?

Ex. 9.24 A barbell is mounted on a nearly frictionless axle through its center (Figure 9.34). At this instant, there are two forces of equal magnitude applied to the system as shown, with the directions indicated, and at this instant the angular velocity is 60 radians/s, counterclockwise. In the next 0.001 s, the angular momentum relative to the center increases by an amount 2.5×10^{-4} kg·m^2/s.

What is the magnitude of each force?

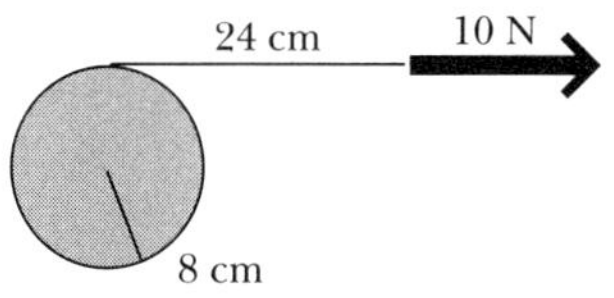

Figure 9.33 Pull a disk on a frictionless surface, unwinding the string.

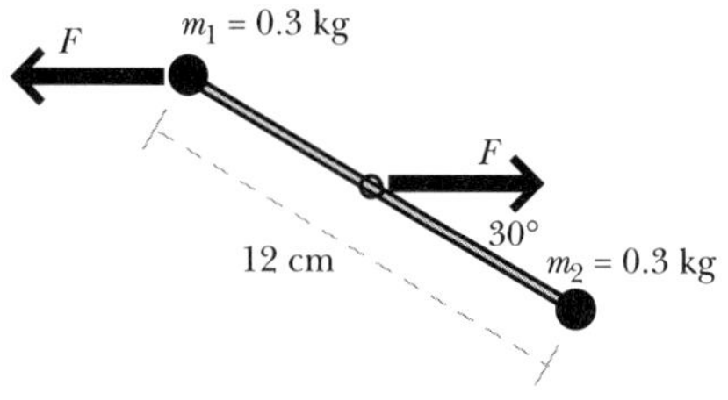

Figure 9.34 Forces applied to a barbell.

9.7 Applications to systems with no torques

We will apply our tools to two situations where no torques are applied.

You may have seen an ice skater spin vertically on the tip of one skate, with her arms and one leg outstretched, then pull her leg in and bring her arms to a vertical position above her head. She then spins much faster.

What's going on? There is some frictional force of the ice on the tip of the skate, but this force is applied so close to the axis of rotation that the torque is small.

? During the short time when the skater quickly changes her configuration, what can you say about the rotational angular momentum? Why?

Small torque implies small change in angular momentum per unit time, so in a short time the angular momentum of the skater will hardly change.

? But if the rotational angular momentum hardly changes, how can the skater spin faster?

When she changes her configuration, the moment of inertia decreases, because some parts of her body are now closer to the axis of rotation. For the rotational angular momentum not to change, the angular velocity must increase to compensate for the decreased moment of inertia: $I_1\omega_1 = I_2\omega_2$, and therefore $\omega_2 = I_1\omega_1/I_2$.

Now you may quite legitimately be puzzled about how this actually works! *Why* does moving her arms and leg closer to the spin axis increase her angular speed? At one level of discussion, you can close your eyes (and maybe hold your nose) and say, "Well, we did all that general analysis of the effect of torques on the angular momentum of multiparticle systems, so if that's what we get when we apply these general principles, I guess that's that." But at another level it would be very nice to get a better sense of the detailed mechanisms involved in this odd phenomenon.

One approach is to analyze the changes in energy involved in this skating maneuver. The angular momentum L_{rot} of the skater doesn't change, so as her moment of inertia decreases, the kinetic energy of the skater $L_{rot}^2/(2I)$ actually increases.

? Where does this increased energy come from?

Evidently the skater has to expend chemical energy in order to increase her kinetic energy. In fact, at high spin rates it takes a noticeable effort to pull her arms and leg toward the spin axis. This is even more dramatic when holding heavy weights.

A popular physics demonstration is to sit on a rotating stool holding a dumbbell in each hand. Start spinning slowly with the arms held out, then pull your hands in toward your chest. It requires considerable effort to pull the dumbbells in. After all, the dumbbells would tend to move in straight lines, and you must exert a radial force just to keep turning them in a circle (though in that case you do no mechanical work, because the motion of the dumbbells has no radial component in the direction of the force you apply). To move the dumbbells into a smaller radius requires applying an even larger force which does work on the dumbbells, because there is now a radial component of the motion, in the direction of the radial force you exert.

9.7.1 A high dive

You may have seen a skilled diver leap off a high board, tuck himself into a tight ball (holding onto his ankles), and rotate quite fast for a few turns. Then he straightens out and enters the water like a knife, hardly ruffling the surface.

? What can you say about the diver's rotational angular momentum while he is in the air? Why?

We can probably neglect air resistance, because the diver is very far from reaching terminal speed. Then the only significant force acting is the Earth's gravitational force, which effectively acts through the center of mass. Therefore there is negligible torque about the center of mass, and the rotational angular momentum does not change.

? Early in the dive the diver is spinning rapidly. If rotational angular momentum is constant during the dive, where did that rapid spin come from?

When the diver jumps off the board, he must thrust with his feet in such a way that the force of the board on his feet has a sizable lever arm about his center of mass, so that by the time his feet lose contact with the board, the diver already has acquired a sizable rotational angular momentum.

? When the diver pulls out of the tucked position, why does he stop spinning rapidly?

There is a large increase in the moment of inertia, because many atoms are now much farther from the center of mass than they were when the diver was in the tucked position. Larger moment of inertia implies smaller angular speed, since the rotational angular momentum $I\omega$ does not change.

? Can his body go straight from then on?

His body can't really go completely straight, because he still has rotational angular momentum. But his angular speed may be so small that you hardly notice it, especially in comparison with the very rapid spin in the preceding tucked position. Moreover, his body could approximately follow the curving path of his center of mass, with the body rotating just enough to stay tangent to the trajectory. This enhances the illusion of straight motion. The most important aspect for good form is to arrange that the body rotate into the vertical position at the time of entering the water, so as to make little splash.

See Problem 9.24, which asks you to analyze a diver's motion.

9.7.2 Example: A satellite with solar panels

A satellite has four low-mass solar panels sticking out as shown. The satellite can be considered to be approximately a uniform solid sphere. Originally it is traveling to the right with speed v and rotating clockwise with angular speed ω_i.

A tiny meteor traveling at high speed v_1 rips through one of the solar panels and continues in the same direction but at reduced speed v_2. Afterwards, calculate the v_x and v_y components of the center-of-mass velocity of the satellite, and its angular velocity ω_f (magnitude and direction). Additional data are provided on the diagram.

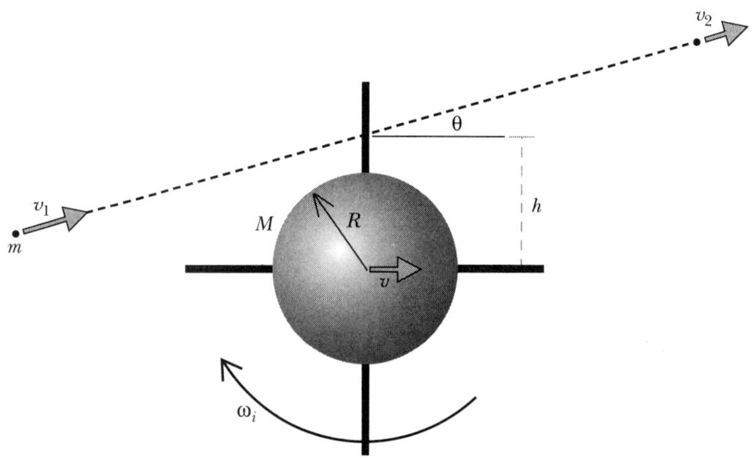

Solution

What fundamental principles are useful starting points? The Momentum Principle and the Angular Momentum Principle. The Energy Principle would not be a useful starting point, because we don't know how much thermal energy increase is associated with the meteor ripping through the panel. Take the satellite and meteor together as the system of interest, in which case the momentum and the angular momentum will remain constant, there being no external forces or torques.

Momentum principle, from just before to just after collision:

$$\langle (Mv + mv_1\cos\theta),\ m_1 v_1 \sin\theta,\ 0\rangle =$$

$$\langle (Mv_x + mv_2\cos\theta),\ (Mv_y + mv_2\sin\theta),\ 0\rangle$$

Angular Momentum Principle, about the original center of the satellite, component into the page, from just before to just after collision:

$$I\omega_i + hmv_1\cos\theta = I\omega_f + hmv_2\cos\theta\text{, where } I = \tfrac{2}{5}MR^2$$

Calculating just before and just after the collision makes it particularly easy to see how to calculate the translational angular momentum of the meteor, since it is directly above the center of the satellite at those instants. Note that we need $\cos\theta$, not $\sin\theta$, in calculating the translational angular momentum of the meteor. In the right-hand-rule, your fingers rotate through an angle that is $(90° - \theta)$.

Solving for the unknown final quantities:

$$v_x = v + \frac{m}{M}(v_1 - v_2)\cos\theta$$

$$v_y = \frac{m}{M}(v_1 - v_2)\sin\theta$$

$$\omega_f = \omega_i + \frac{hm}{(\frac{2}{5}MR^2)}(v_1 - v_2)\cos\theta\text{, clockwise, into the page}$$

9.8 Applications to systems with nonzero torques

In the remainder of this chapter we will analyze systems where there are external torques applied to the system, which can change the angular momentum of the system.

9.8.1 A meter stick on the ice

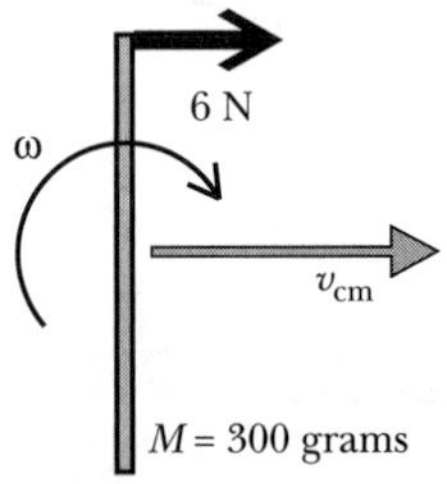

Figure 9.35 Pull one end of a meter stick that is lying on ice (negligible friction).

Consider a meter stick whose mass is 300 grams and which lies on ice (in Figure 9.35 we're looking down on the meter stick). You pull at one end of the meter stick, at right angles to the stick, with a force of 6 newtons. Assume that friction with the ice is negligible.

> **?** The ensuing motion of the meter stick is quite complicated, but what is the initial rate of change dv_{cm}/dt of the center-of-mass velocity v_{cm} of the stick, at the instant when you first apply the force?

Using the Momentum Principle, $d\vec{P}/dt = \vec{F}_{net}$, we find that $dv_{cm}/dt = (6\text{ N})/(0.3\text{ kg}) = 20\text{ m/s}^2$.

Next we'll determine the initial rate of change $d\omega/dt$ of the angular speed ω about the center of mass. We'll use the Angular Momentum Principle about the center of mass, $d\vec{L}_{rot}/dt = \vec{\tau}_{net,ext,cm}$, with $L_{rot} = I\omega$.

> **?** What is the torque about the center of mass? State the direction as well as the magnitude.

By the right-hand rule, the torque points into the page. The magnitude is $(0.5\text{ m})(6\text{ N})\sin 90° = 3\text{ N}\cdot\text{m}$.

> **?** The moment of inertia I around the center of mass of a uniform rod of mass M and length L can be shown to be $ML^2/12$ (or looked up in a handbook). What is the initial rate of change $d\omega/dt$ of the angular speed ω?

We find that $d\omega/dt = (3\text{ N}\cdot\text{m})/[(0.3\text{ kg}\cdot\text{m}^2)/12] = 120\text{ radians/s}^2$. In vector terms, $d\vec{\omega}/dt$ points into the page, corresponding to the fact that the angular velocity points into the page and is increasing.

Alternative analysis—taking torques around the end of the stick

It is interesting to analyze the motion of the meter stick by calculating torque and angular momentum about the end of the stick where the force is applied, rather than about the center of mass. To be cautious and correct, we should say that we are taking torques about a location fixed in the ice next to the place where the end of the stick is momentarily located. This is a fixed location, not tied to the moving stick.

> **?** About that location (by the end of the stick), what is the magnitude of the net external torque?

The torque is zero, because the distance is zero from that location to the point where the force is exerted.

? Therefore, what do you know about the angular momentum measured about that same location?

The angular momentum about that location is constant, since there is no torque.

? After a very short time interval the stick has a center-of-mass speed v_{cm} and some angular speed ω. At that instant, what is the angular momentum about the end of the stick where the force is applied?

The angular momentum is the sum of the translational angular momentum and the rotational angular momentum. The translational angular momentum is $(L/2)Mv_{cm}$, where L is one meter, and its direction is out of the page or up out of the ice (using the right-hand rule). The rotational angular momentum is $I\omega = (ML^2/12)\omega$, and its direction is into the page or down into the ice. Taking the net direction to be into the page, we have this:

$$\frac{d}{dt}\left[\frac{ML^2}{12}\omega - \frac{L}{2}Mv_{cm}\right] = 0$$

$$\frac{d\omega}{dt} = \frac{6}{L}\frac{dv_{cm}}{dt}\text{, where } L = 1\text{ m.}$$

Using the Momentum Principle we determined that $dv_{cm}/dt = 20\text{ m/s}^2$. Evidently $d\omega/dt$ should equal $(6/1\text{ m})(20\text{ m/s}^2) = 120\text{ radians/s}^2$. This agrees with our analysis where we took torques about the center of mass of the stick.

9.8.2 A puck with string wound around it

Wrap a string around the outside of a hockey puck. Then pull on the string with a constant tension F_T (Figure 9.36). The puck has mass M and radius R. Assume that friction with the ice rink is negligible. Evidently $dv_{cm}/dt = F_T/M$.

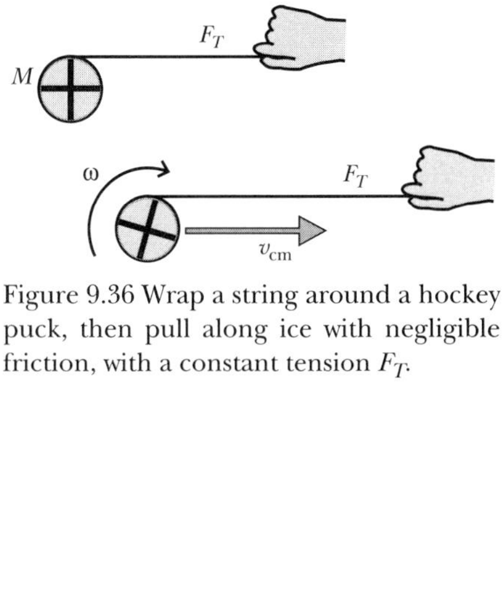

Figure 9.36 Wrap a string around a hockey puck, then pull along ice with negligible friction, with a constant tension F_T.

? The moment of inertia of a uniform solid puck of mass M and radius R can be shown to be $MR^2/2$ (see Problem 9.3; or this result can be looked up in a handbook). What is the initial rate of change $d\omega/dt$ of the angular speed ω?

The torque about the center of mass is RF_T, into the page (down into the ice). The rotational angular momentum is $(MR^2/2)\omega$, into the page (down into the ice). Therefore $(MR^2/2)d\omega/dt = RF_T$, and we have $d\omega/dt = (2F_T)/(MR)$.

Ex. 9.25 Redo the analysis, calculating torque and angular momentum relative to a fixed location in the ice anywhere underneath the string (similar to the analysis of the meter stick around one end). Show that the two analyses of the puck are consistent with each other.

9.8.3 Static equilibrium

If the net torque on a system is zero, its angular momentum is constant. Conversely, if we know that the angular momentum is not changing, we can conclude that the net torque must be zero. For example, if a system is in static equilibrium, not only must the net force on the system be zero; the net torque must also be zero. This allows us to make conclusions about the individual torques whose vector sum is zero.

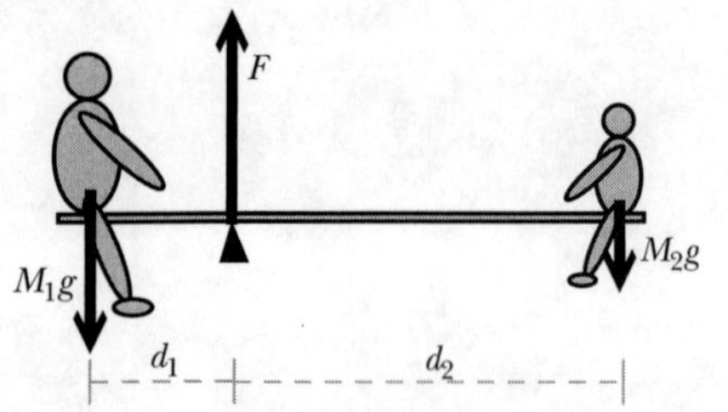

Figure 9.37 A seesaw in static equilibrium.

Ex. 9.26 As a simple application, consider a seesaw that is not rotating around its frictionless support, so that the angular momentum is not changing and is in fact zero (Figure 9.37). Write a force equation, and a torque equation around the support, assuming that the board itself has negligible mass compared to the people sitting on it.

Ex. 9.27 Write another torque equation around a location fixed in space where the person on the left is sitting, and show that it is in fact equivalent to the torque equation around the axle, when you take the force equation into consideration.

It is often convenient to choose your fixed location so that some of the forces create no torque around that location and therefore don't appear in the Angular Momentum Principle.

9.9 Atomic and nuclear angular momentum

Many elementary particles have rotational angular momentum (or, if you wish to be very cautious, these particles behave as though they have rotational angular momentum, which comes down to essentially the same thing). Electrons bound in an atom may have translational angular momentum relative to the nucleus as well as their own rotational angular momentum. Atoms as a whole may have angular momentum, as do many nuclei. The total angular momentum is as expected the sum of the translational angular momenta plus the rotational angular momenta.

The surprising thing about angular momentum at the atomic and subatomic level is that it is quantized. The basic quantum of angular momentum is this:

$$\text{The angular momentum quantum} = \hbar = \frac{h}{2\pi} = 1.05\times10^{-34}\ \text{joule}\cdot\text{second}$$

where h is "Planck's constant," 6.63×10^{-34} joule-second. Whenever you measure a vector component of the angular momentum (that is, along the x or y or z axis), you get either a half-integer or integer multiple of this quantum of angular momentum.

Ex. 9.28 Show that $\hbar$ and angular momentum have the same units.

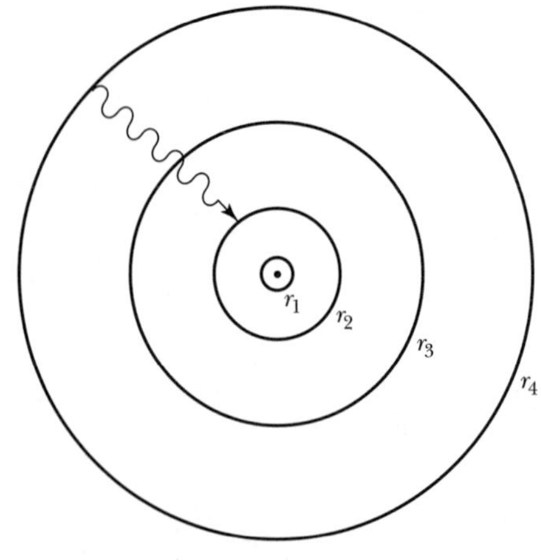

Figure 9.38 In the Bohr model of the hydrogen atom, electrons can be in only those circular orbits for which angular momentum is a multiple of $\hbar$. In a transition between allowed orbits, a photon is emitted.

9.9.1 The Bohr model of the hydrogen atom

As discussed in the previous chapter, in 1911 Rutherford and his group discovered the nucleus in atoms. Stimulated by this discovery, in 1913 the Danish physicist Niels Bohr made a bold conjecture that a hydrogen atom could be modeled as an electron going around a proton in circular orbits, but only in those orbits whose translational angular momentum is an integer multiple of $\hbar$. Consequently, these orbits would have only certain radii, and only certain values of energy. The differences in energies between these quantized orbits match the observed energies of photons emitted by atomic hydrogen (Figure 9.38).

Bohr's basic hypothesis—that the angular momentum of electrons in a atom is quantized—has proven to be an important insight, and has been retained as a fundamental tenet of the quantum mechanical model of an atom. As this model has been refined, a probabilistic view of the motion of electrons has been adopted. In more sophisticated models electrons do not have precise trajectories, but only a "probability density"—a probability of being found at a particular location. These more complex quantum mechanical models explain a much wider range of atomic and molecular phenomena than does Bohr's original model.

The simple Bohr model does, however, predict correctly the allowed electronic energy levels for atomic hydrogen (as determined from the emission and absorption spectra of hydrogen atoms). We will work with this model to see how quantization of angular momentum leads to the prediction of electronic energy levels in a hydrogen atom.

Allowed radii of electron orbits

Because the proton has much more mass than an electron (about 2000 times as much), we'll make the approximation that the proton is at rest, with the electron in a circular orbit around the proton.

? If the electron momentum is p, and the radius of the circular orbit is r, what is the translational angular momentum?

The translational angular momentum of the electron relative to the location of the proton is $L_{\text{trans},\,C} = rp\sin(90°) = rp$ (Figure 9.39). Bohr proposed that the only possible states of the hydrogen atom are those where the electron is in a circular orbit whose translational angular momentum is an integer multiple of $\hbar$:

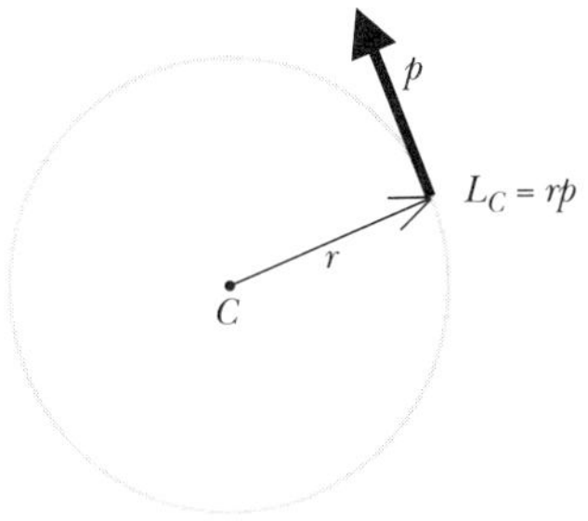

Figure 9.39 The angular momentum of the electron in a circular Bohr orbit relative to the proton. L_C is out of the page.

BOHR: ANGULAR MOMENTUM IS QUANTIZED

$$|\vec{L}_{\text{trans},C}| = rp = N\hbar, \text{ where } N \text{ is an integer } (1, 2, 3, ...)$$

The electric force and a circular orbit

? What is the magnitude of the electric force that the proton exerts on the electron?

$$F_{\text{el}} = \frac{1}{4\pi\varepsilon_0}\frac{e^2}{r^2}$$

where $+e$ is the electric charge on the proton ($-e$ for the electron).

? Use the Momentum Principle to relate this force to the circular motion of the electron.

In a circular orbit at constant speed we know that the rate of change of the momentum is $(v/r)p$, so we have in the approximation that $v << c$, where m is the mass of the electron,

$$\left(\frac{v}{r}\right)p = \frac{mv^2}{r} = F = \frac{1}{4\pi\varepsilon_0}\frac{e^2}{r^2}$$

Solving for *r*

We are looking for the allowed values of the orbit radius r, so we look for ways to express v in terms of r. Bohr's angular momentum condition gives us v in terms of r:

$$|\vec{L}_{\text{trans},C}| = rp = N\hbar \text{ leads to } v = \frac{p}{m} = \frac{N\hbar}{mr}$$

Putting this result for r into the result obtained from the Momentum Principle, $mv^2/r = F$, we have

$$\frac{m}{r}\left(\frac{N\hbar}{r}\right)^2 = \frac{1}{4\pi\varepsilon_0}\frac{e^2}{r^2}$$

Solving the latter relation for r, we obtain the following result (Figure 9.40):

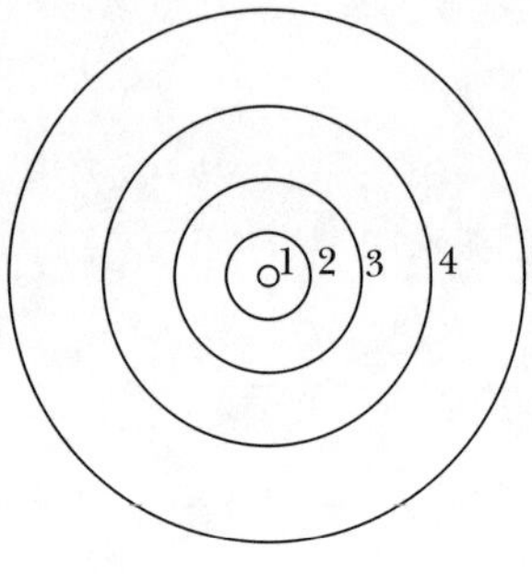

Figure 9.40 Bohr radii for N=1 through 5. The nucleus is not shown.

ALLOWED BOHR RADII FOR ELECTRON ORBITS

$$r = N^2\frac{\hbar^2}{\left(\frac{1}{4\pi\varepsilon_0}\right)e^2 m}$$

Let's evaluate this result numerically, so that later we can compare with measurements of the spectrum of light emitted by excited atomic hydrogen. Using

$\hbar = 1.05\times10^{-34}$ J·s,

$1/4\pi\varepsilon_0 = 8.99\times10^{9}$ N·m^2/coulomb^2,

$e = 1.60\times10^{-19}$ coulomb,

and the mass of the electron $m = 9.11\times10^{-31}$, we find this:

$$r = N^2(0.53\times10^{-10}\text{ m})$$

This is a striking result. The simple Bohr model predicts that the smallest permissible electron radius ($n = 1$) is 0.53×10^{-10} m , and atoms are in fact observed to have radii of approximately this size.

Energy for a circular orbit

We cannot observe directly the radius of the orbit of an electron in an atom. What we can observe are photons emitted by excited atoms, which tell us the differences in energies between various energy levels of the atom. Using our model we can predict the energies of these energy levels, and see if the differences between these levels match the energies of photons observed in an atomic spectrum.

Given a set of possible values for r, we can calculate the possible values for energy of the electron, starting from this observation:

$$\text{The result } mv^2/r = F \text{ leads to } mv^2 = \frac{1}{4\pi\varepsilon_0}\frac{e^2}{r}.$$

? Now write a formula for the kinetic energy plus electric potential energy of the hydrogen atom, using the Bohr model.

Kinetic energy:

$$K = \frac{1}{2}mv^2 = \frac{1}{2}\left(\frac{1}{4\pi\varepsilon_0}\frac{e^2}{r}\right) \text{ (from previous equation)}$$

Electric potential energy:

$$U_{\text{el}} = -\frac{1}{4\pi\varepsilon_0}\frac{e^2}{r}$$

So the energy is this (omitting the rest energies):

$$E = K + U_{\text{el}} = -\frac{1}{2}\left(\frac{1}{4\pi\varepsilon_0}\frac{e^2}{r}\right)$$

According to the Bohr model only certain radii will actually occur. Inserting our previous expression for r, we get:

BOHR MODEL ENERGY LEVELS

$$E = -\frac{\left(\frac{1}{4\pi\varepsilon_0}\right)^2 e^4 m}{2N^2\hbar^2}$$

Evaluating this expression we get

$$E = -\frac{2.17\times10^{-18}\ \text{J}}{N^2}$$

Converting to electron-volts (1 eV = 1.6×10^{-19} J) we have (Figure 9.41)

$$E = -\frac{13.6\ \text{eV}}{N^2},\ N = 1, 2, 3,...$$

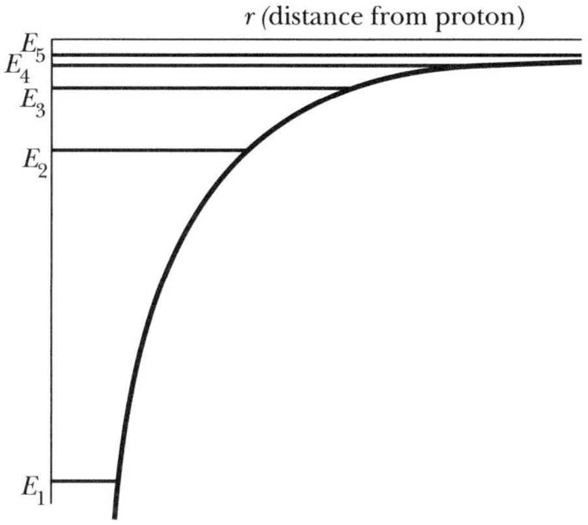

Figure 9.41 Bohr model prediction of electronic energy levels for a hydrogen atom. The potential energy curve for the electron–proton system is also shown.

This prediction for the quantized energies of atomic hydrogen agrees well with the observed electronic spectrum of hydrogen. We quoted this result for the quantized energy levels in Chapter 6. The energies are negative, corresponding to bound states.

Photons emitted by atomic hydrogen

Bohr proposed that electromagnetic radiation would be emitted when there was a sudden change from a higher energy level to a lower one. We saw in Chapter 6 that the Bohr energy formula does correctly predict the energies of photons emitted by excited atomic hydrogen.

? In the Bohr model, does photon emission correspond to the radius of the circular orbit getting larger or smaller? Does the quantum number N increase or decrease?

The kinetic plus potential energy is negative, corresponding to a bound state, so a higher energy is one that is less negative and therefore corresponds to a larger radius (since the energy is proportional to $-1/r$). Also, higher energy corresponds to a larger value of N, which is also associated with a larger radius. Therefore photon emission is associated with a decrease in r and a decrease in N (Figure 9.42). Photon absorption involves an increase in r and N.

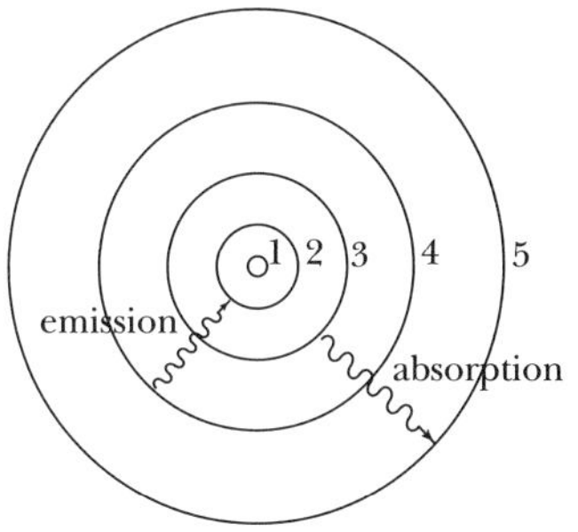

Figure 9.42 A photon is emitted in a transition from a higher to a lower energy level. A photon is absorbed in a transition from a lower to a higher energy level.

Refinements of the Bohr model

The Bohr model was an important predecessor to current quantum mechanical models of the atom. It predicts the main aspects of the hydrogen energy levels correctly (neglecting various small effects). The idea that photon emission is associated with a drop from a higher energy level to a lower energy level, with the photon energy equal to the difference in the hydrogen energy levels, is fundamental to the extended quantum-mechanical treatment.

The notion of deterministic circular orbits has been abandoned and replaced with a picture of a probabilistic electron "cloud," but even in more sophisticated models the average radius of this electron cloud is larger for higher energies, just as in the Bohr model. From your study of chemistry you are probably familiar with the shapes of the "probability density" functions or "orbitals" describing the electron's location in the atom in different energy states, some of which are spherical, others dumbbell shaped.

The idea that angular momentum is quantized has far-reaching consequences and is a major element in modern quantum mechanics. However, it has become clear that the actual quantization rules in a hydrogen atom are more complex than those that are assumed in the Bohr model. For example, the translational angular momentum in the ground state ($N = 1$) is actually zero, not $\hbar$, and for the next higher state ($N = 2$), the z component of translational angular momentum can be either 0 or $\hbar$.

9.9.2 Particle spin

An important example of angular momentum quantization is the observation that many elementary particles have rotational angular momentum, and it is quantized. We know this in part because of observations of the interaction of these particles with other particles having nonzero angular momentum; in these interactions total angular momentum is constant.

For example, the electron, muon, and neutrino are said to have spin $1/2$, because measurement of a component of their rotational angular momentum always yields $\pm(1/2)\hbar$. Every quark has spin $1/2$, and particles built out of three quarks, such as protons and neutrons, necessarily have half-integral spin. The spin of the proton and neutron is $1/2$, as though two of the quarks have their spins opposed ($\Uparrow\Downarrow\Uparrow$), with no translational angular momentum of the quarks. Some short-lived three-quark particles have spin $3/2$. One way to get this is for all the quarks to have their spins aligned ($\Uparrow\Uparrow\Uparrow$), or there can be some translational angular momentum of the quarks. While rotational angular momentum can be half-integral, translational angular momentum is always integral (0, 1, 2, etc.).

Particles made out of one quark and one antiquark, called mesons, have integral spin. For example, the pion spin is zero (quark and antiquark spins opposed, $\Uparrow\Downarrow$), and the spin of the "rho" meson is 1 (quark and antiquark spins aligned, $\Uparrow\Uparrow$).

Presumably the angular momentum of a macroscopic object such as a baseball is also quantized, but the angular momentum of a baseball is huge compared to Planck's constant, and you don't notice the quantization, which is on an exceedingly fine scale (10^{-34} J·s!). Only at the atomic level are the effects of quantization really evident.

9.9.3 Consequences of angular momentum quantization

The quantization of angular momentum has far-reaching consequences. Angular momentum quantization plays a major role in determining the structure of atoms, and the nature of the chemical periodic table. The two lowest-energy-state electrons in an atom always have zero translational angular momentum and zero rotational angular momentum (because the two rotational angular momenta are always oppositely aligned and add up exactly to zero).

There is a very deep connection between angular momentum and the statistical behavior of particles. Particles such as the electron and the proton that have half-integral spin are called "fermions" and exhibit the peculiar behavior that two fermions cannot be in the exact same quantum state. This is why only two electrons can be in the lowest energy state of an atom, with their spins opposed. Their spin directions are forbidden to be the same, because then their energy and angular momentum would be exactly the same, which is forbidden for fermions. This prohibition is called the "Pauli exclusion principle."

On the other hand, particles with integral spin, called "bosons," are not subject to the Pauli exclusion principle, and there is no limit to the number of bosons that can be in the same energy state. In fact, there is a special state of matter called a "Bose-Einstein condensation" in which very large numbers of particles end up in exactly the same quantum state. This state of matter was predicted long before it was actually created and observed in 1995.

In a diatomic gas molecule such as oxygen (O_2), the kinetic energy associated with rotation of the molecule is of course $L_{rot}^2/(2I)$. But the rotational angular momentum L_{rot} is quantized, so the rotational energies of oxygen are quantized. The phenomenon of angular momentum quantization affects the specific heat of diatomic gases at low temperatures, as will be discussed in a later chapter. (You can of course refer to the angular momentum of a oxygen molecule as translational angular momentum of the two atoms instead of rotational angular momentum of the molecule.)

It is an important prediction of quantum mechanics that the x, y, or z component of the angular momentum (L_x, L_y, or L_z) can only be an integer or half-integer multiple of $\hbar$, whereas the square magnitude of the angular momentum has the quantized values $l(l+1)\hbar^2$, where l has integer or half-integer values, depending on whether a component has integer or half-integer values:

QUANTIZED VALUES OF L^2

$$L^2 = l(l+1)\hbar^2, \text{ where } l \text{ is integer or half-integer}$$

All nuclei that have an even number of protons and an even number of neutrons ("even-even" nuclei) have a total angular momentum of zero. Examples include carbon-12 (6 protons and 6 neutrons) and oxygen-16 (8 protons and 8 neutrons). The spins of protons and neutrons in even-even nuclei are paired up in the lowest-energy state of a nucleus with each other in such a way as to produce zero net angular momentum. This would be an exceedingly unlikely arrangement if angular momentum weren't quantized.

9.10 *Gyroscopes

A gyroscope is a fascinating device, and its unusual properties have been exploited to stabilize ships and spacecraft. Its behavior provides a good model and analogy for some important aspects of the quantum mechanical behavior of atoms and nuclei. Magnetic resonance imaging (MRI) is based on the gyroscope-like behavior of spinning nuclei.

A gyroscope has a spinning disk mounted on an axle. If you place one end of the axle on a vertical support, you may observe a very complex motion (Figure 9.43). The gyroscope rises and falls ("nutation") as it revolves around the support ("precession").

Figure 9.43 A gyroscope can exhibit both "nutation" and "precession."

One fruitful approach in modeling a complex physical system is to try to analyze the simplest motion that the system is capable of. For example, we can observe a pure precession of a gyroscope, with no nutation. It is possible with care to start the gyroscope with special initial conditions so that it merely precesses, without bobbing up and down. We'll try to analyze this special kind of gyroscope motion. In fact, we'll start with a particularly simple form of precession, with the rotational axis horizontal (Figure 9.44).

We define some variables to make it easier to describe and discuss the situation. The gyroscope disk rotates on its axis with some spin angular speed that we'll call ω, and the disk has a moment of inertia I. The rotational angular momentum $\vec{L}_{rot}$ of the disk always points horizontally but is continually changing direction as the gyroscope precesses (if we neglect friction, the angular speed ω is constant). The gyroscope revolves around the support with angular speed Ω (Greek uppercase omega), constant in magnitude and direction (if we neglect friction). Note that $v_{cm} = \Omega r$ and $\Omega = (v_{cm}/r)$, where r is the distance from the support to the center of mass. The rate of change of the momentum is ΩP, and the same argument can be used to show that the rate of change of the angular momentum is ΩL.

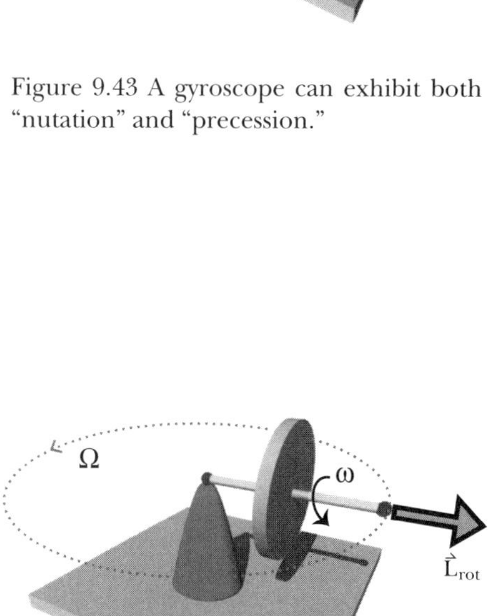

Figure 9.44 A gyroscope which is precessing horizontally, with no nutation.

? What forces act on the gyroscope? What relationships are there among these forces?

The support pushes up with a force we'll call F_N, and the Earth pulls down with a force Mg through the center of mass (Figure 9.45). Since the center of mass stays at the same height all the time in the case of simple precession, it must be that the vertical component of the net force must be zero. Therefore $F_N = Mg$.

This isn't quite the whole story, though. Note that the center of mass of the gyroscope is moving in a circle. That requires that there be a radially inward force to make the momentum vector rotate (with angular speed Ω). The only object that can exert this force is the support. There must be a small horizontal frictional force f such that $|d\vec{P}/dt| = \Omega P = f$, where $P = Mv_{cm} = M(\Omega R)$; R is the radius of the circle traveled by the center of mass. This force, $f = MR\Omega^2$, is small if the precession rate Ω is small.

It is easy to observe with a toy gyroscope that as the spin of the disk slows down due to friction (smaller ω), the precession actually speeds up (larger Ω). If we could predict the precession speed for a given spin, we would largely "understand" this simple example of gyroscope motion. What can we use to attempt a prediction?

We've already used the Momentum Principle to conclude something about the forces F_N and f. What about energy? If the gyroscope precesses in the horizontal plane, there's no change in gravitational energy, because the center of mass remains always at the same height. There's no change in kinetic energy, either. So energy considerations don't seem to tell us anything interesting about precession (though they would be useful in analyzing the more complex rise and fall of nutation). The remaining principle to apply is the Angular Momentum Principle (no wonder we left gyroscopes until this chapter).

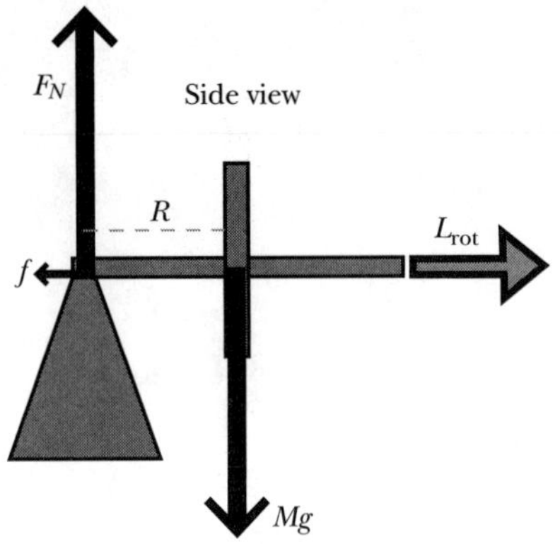

Figure 9.45 Side view showing external forces acting on the gyroscope, and the rotational angular momentum vector.

? What is the magnitude of the rotational angular momentum of the gyroscope? What is the magnitude of the translational angular momentum of the gyroscope around the support?

We said that the rotating disk has moment of inertia I and angular speed ω, so the rotational angular momentum is simply $L_{rot} = I\omega$. The magnitude of the translational angular momentum is this:

$$L_{support} = |\vec{R} \times \vec{P}| = RP$$

Evidently the magnitudes of both the rotational and translational angular momenta are constant, not changing with time. But what about the directions of these angular momenta? Do they change with time?

? How does the direction of the translational angular momentum change with time? How does the direction of the rotational angular momentum change with time?

Neither the magnitude nor the direction of the translational angular momentum changes (neglecting friction): it has constant magnitude and points vertically upward at all times. But the direction of the rotational angular momentum is constantly changing as the gyroscope precesses. We need to calculate the rate at which the rotational angular momentum vector is changing, $d\vec{L}_{rot}/dt$. Since the rotational angular momentum vector is a rotating vector, which rotates with the precession angular speed Ω, we have this:

$$\left|\frac{d\vec{L}_{rot}}{dt}\right| = \Omega L_{rot}$$

? Therefore, what is the Angular Momentum Principle for the gyroscope?

The remaining element we need for the Angular Momentum Principle is the torque that acts around the center of mass, at the center of the spinning disk. Look again at the force diagram (Figure 9.46), and you see that the magnitude of the torque about the center of mass is $RF_N = RMg$, and the direction of the torque is into the page. Seen from above, the torque points at a right angle to the rotational angular momentum (Figure 9.47). Therefore the Angular Momentum Principle yields

$$\left|\frac{d\vec{L}_{rot}}{dt}\right| = \Omega L_{rot} = \tau_{cm}\text{, where } L_{rot} = I\omega\text{ , and } \tau_{cm} = RMg$$

Solving for the precession angular speed Ω, we have

$$\Omega = \frac{\tau_{cm}}{L_{rot}} = \frac{RMg}{I\omega}$$

This is a surprisingly simple result for such a complicated system, and it agrees at least qualitatively with observations. A smaller spin ω is associated with a larger precession rate Ω. Conversely, a gyroscope which spins very fast precesses very slowly. If possible, your instructor will provide an opportunity to test the theory quantitatively by observing an actual gyroscope whose spin and precession rates can be measured.

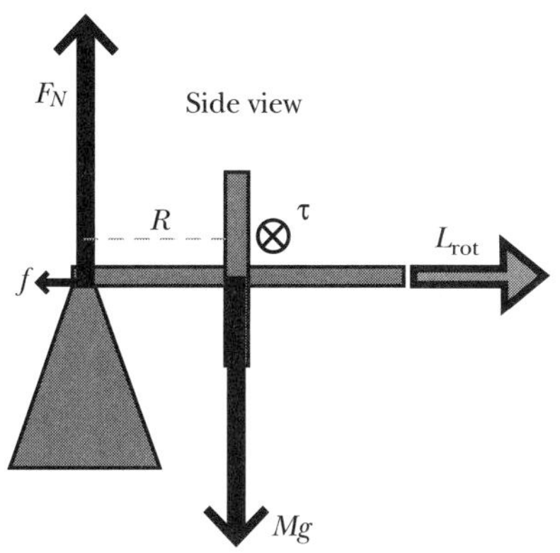

Figure 9.46 Side view showing external forces acting on the gyroscope, and the rotational angular momentum vector. There is a torque around the center of mass, into the page, due to the F_N force.

Reflection

Take some time to reflect on how we arrived at this result. Look over the steps involved. The key point is that the rotational angular momentum varies in direction without varying in magnitude, and the magnitude of the rate of change of the rotational angular momentum is simply Ω times L_{rot}. This rate of change of the rotational angular momentum is equal to the net torque acting on the system (around the center of mass).

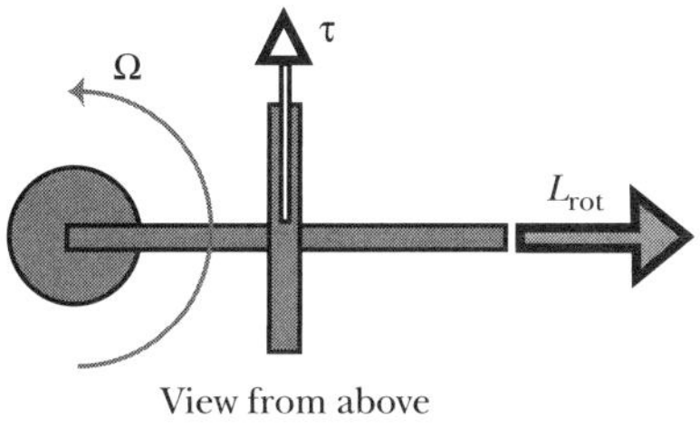

Figure 9.47 View from above, showing the torque vector and the rotational angular momentum vector.

Ex. 9.29 To complete this reflection, determine the relationship between ω and Ω for the case of pure precession, but with the spin axis at an arbitrary angle θ to the vertical (Figure 9.48; $\theta = 90°$ is the case we treated of horizontal precession). If you have the opportunity, see whether this relationship holds for a real gyroscope.

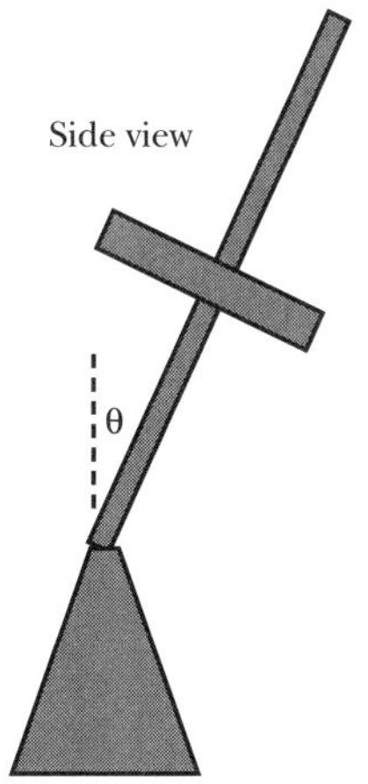

Figure 9.48 What is the precession rate when the spin axis is at an angle θ to the vertical?

Uses of gyroscopes

The precession we calculated is the result of a nonzero torque acting around the center of mass. Through clever mounting of the gyroscope it is possible to make this torque vanishingly small, in which case the axis does not precess but always maintains the same direction, which has been useful in navigation. The satellite-based Global Positioning System (GPS) can tell you where your airplane is; gyroscopically based instruments can tell you the orientation and attitude of your airplane at this known location.

Large gyroscopes have been used to stabilize ships against rolling in the sea. A gyroscope with very high rotational angular momentum (large moment of inertia and large angular speed) is hard to turn quickly, and this can provide mechanical stability. Gyroscopes are also used to stabilize the orientation of spacecraft and satellites.

Friction

We deliberately neglected friction in the calculation. Friction has two main effects on a gyroscope. Friction in the spin axis makes ω decrease with time, which leads to an increased precession rate Ω. Friction on the top of the support slows down the precession, which means that Ω is no longer equal to RMg/L_{rot} as it should be for pure precession. As a result of these effects, what you observe with a gyroscope that starts out in pure precession is that the spin axis eventually starts to tip lower, accompanied by faster precession. The gyroscope starts out with a stately, dignified, slow precession, but toward the end gives the impression of motion that is more and more frantic.

9.10.1 *Magnetic Resonance Imaging (MRI)

Electrically charged atomic and subatomic objects that have angular momentum act like little bar magnets with north and south poles along the angular momentum axis (we say that they have "magnetic moments"). It is as though the spinning charged particle constituted little loops of current, and current loops produce a magnetic field. Even some electrically neutral particles such as the neutron have magnetic moments, because they are built out of electrically charged quarks that have magnetic moments.

In the presence of an applied magnetic field (typically produced by other current-carrying coils of wire), a bar magnet tends to twist to line up with the applied magnetic field. That is, a magnetic field exerts a torque on a bar magnet. A well-known example is that a compass needle aligns with the Earth's magnetic field, which is useful for navigation. (The compass needle is magnetized and is itself a little bar magnet.)

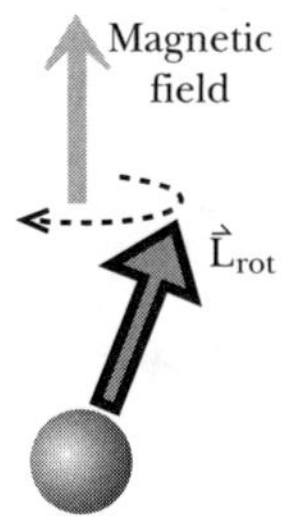

Figure 9.49 Precession of proton spin in an external magnetic field.

Since a magnetic field applies a torque to a bar magnet, or to an atomic or subatomic particles that has angular momentum, such particles precess in the presence of an applied magnetic field (Figure 9.49). This phenomenon is exploited in a technique important in physics, chemistry, and biology, called NMR—nuclear magnetic resonance. A particularly useful application of NMR is in magnetic resonance imaging (MRI).

In MRI, a large, steady, uniform magnetic field created by large direct-current-carrying coils of wire surrounding the patient makes protons in the person's body precess. Many of the most common nuclei in the body are even-even nuclei (in particular, the most common isotopes of carbon and oxygen) and have zero angular momentum and therefore no magnetic moment, and these do not precess. But hydrogen nuclei (protons) are very common in the body, have rotational angular momentum, and do precess. The stronger the magnetic field, the larger the torque acting on the proton nuclear magnets, and the faster their precession. (Note that for a gyroscope, $\Omega = \tau/L_{rot}$; the precession rate is proportional to the applied torque τ.)

With the protons precessing in the presence of the large steady magnetic field, a small high-frequency (time-varying) magnetic field tuned to match exactly the precession frequency of the protons will flip the spins upside down. This is called a "resonance" phenomenon. The high-frequency signal is turned off, and the protons revert back to being aligned with the steady magnetic field. The act of flipping back emits radiation (at the precession frequency) that can be detected by coils connected to a receiver, and the strength of the signal indicates how many protons were affected.

That's the basic physical mechanism, but this by itself would not yield spatial detail about the interior of the body. The trick is to superimpose on the large steady magnetic field a small nonuniform magnetic field, which has the effect of establishing slightly different torques on protons at different locations in the body. As a result, when the protons flip back into alignment they radiate signals whose frequencies indicate their locations. A computer algorithm calculates how much of each frequency is present in the signal, and this indicates how many protons are at each location. In this way a very detailed image is built up of the interior of the body.

9.10.2 *Precession of spin axes in astronomy

As was mentioned earlier, the Earth is subject to small torques due to gravitational forces of the Sun and Moon acting on its nonspherical, "oblate" shape. There is a torque on the Earth's equatorial bulge, perpendicular to the spin rotation axis, and this causes the Earth's axis to precess very slowly, once around about every 26000 years, so that the "North Star" hasn't always been the star we call Polaris. This effect is called the "precession of the equinoxes" because it leads to a change in what month the spring and fall equinoxes occur. To see why there is such an effect, consider Figure 9.50.

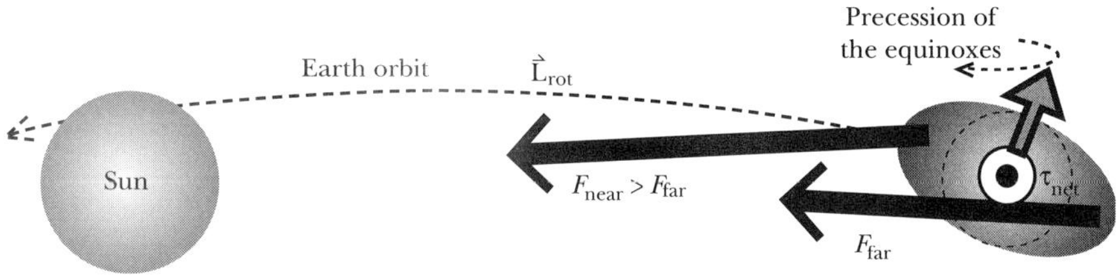

Figure 9.50 Forces on the Earth due to the Sun. The diagram is not to scale; the asymmetry of the Earth is greatly exaggerated. The net torque about the center of mass of the Earth is out of the page.

The Sun (or Moon) exerts a slightly larger force on the closer bulge, due to the $1/r^2$ dependence of the force, so there is a nonzero net torque around the center of mass of the Earth, pointing toward you, out of the page. This makes the rotational angular momentum $\vec{L}_{rot}$ of the Earth precess in a "retrograde" way—that is, in the opposite direction to most rotations in our Solar System. Slowly the direction of the axis of the Earth changes due to the torque. (The size of the equatorial bulge, and the differences in F_{near} and F_{far}, have been greatly exaggerated in Figure 9.50).

Since the Earth's axis is tipped 23° away from perpendicular to the Earth-Sun orbit, the precession of the axis makes a big change in the location of the "North Star." 13000 years from now the "North Star" will be a star that is 46° away from Polaris in the night sky.

The Moon going around the Earth is a kind of gyroscope. The Earth-Moon orbit is inclined a few degrees to the plane of the Earth-Sun orbit, and the Sun exerts a nonzero torque perpendicular to the angular momentum of the Earth-Moon system. To see why, consider a simpler case where the Earth and Moon have equal mass, and you see the same effect as with the equatorial bulge of the Earth (Figure 9.51).

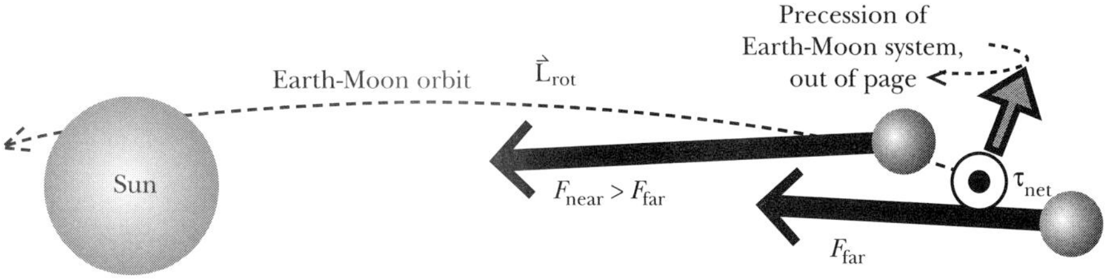

Figure 9.51 Forces on the Earth-Moon system due to the Sun. The diagram is not to scale; the Earth and Moon are shown as if their masses were equal. The net torque on the Earth-Moon system about its center of mass is out of the page.

This torque varies in magnitude during a month, but the averaged effect is that the Earth-Moon system precesses once around in about 18 years (this precession is also "retrograde"). This has an effect on the timing of eclipses, which can occur only when the Moon in its orbit is passing through the plane of the Earth-Sun orbit.

Tidal torques

If the Earth did not rotate, the Moon would create tides in the oceans that would pile up in line with the Moon. However, the rotating Earth drags these tidal bulges so that they are no longer in line with the Moon, and the Moon exerts a small torque (Figure 9.52). This torque is directed along the axis and acts to slow down the spin rate, so that the day is getting longer than 24 hours. This effect is called "tidal friction." This interaction with the Moon has the effect that as the rotational angular momentum of the Earth decreases, the translational angular momentum of the Earth-Moon system increases (conservation of total angular momentum), with the result that the Earth and Moon are getting farther apart. (These tidal forces have a small net tangential component that acts on both Earth and Moon to increase their translational angular momenta.)

Something like this presumably happened in the past to the Moon, when it was molten. The rotational angular momentum of the Moon decreased due to tidal torques exerted by the Earth, which are much larger than the tidal torques that the low-mass Moon exerts on the Earth. When the Moon's spin angular speed had decreased to be the same as its translational angular speed (currently about 2π radians or 360° per month), the process terminated. That is why the Moon now always displays nearly the same face to the Earth.

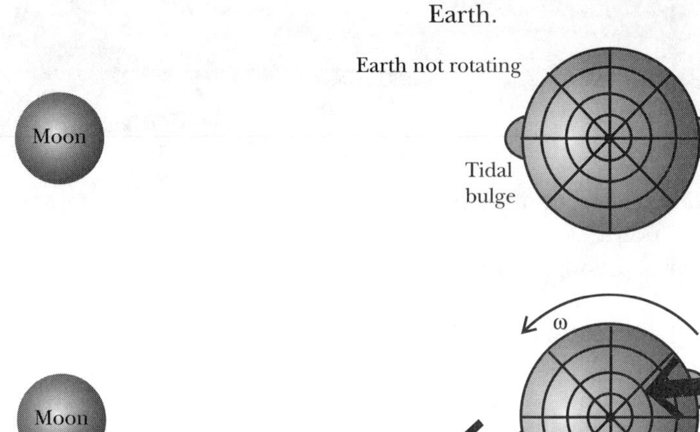

Figure 9.52 Tidal torques due to the Moon, viewed from a point above the North Pole. In the top image the Earth is not rotating; in the bottom image it is rotating. Not to scale; tidal effects greatly exaggerated.

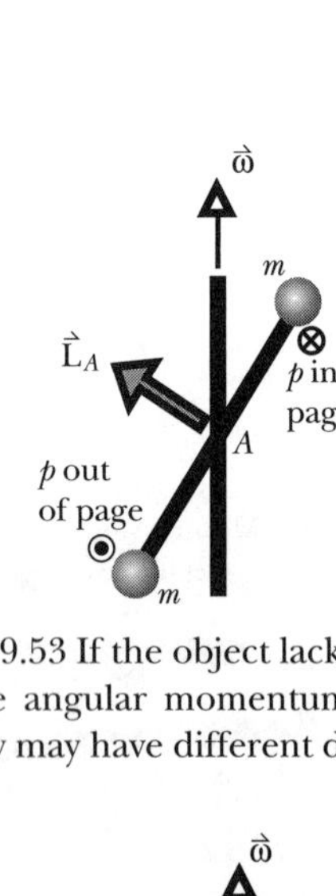

Figure 9.53 If the object lacks axial symmetry, the angular momentum and angular velocity may have different directions!

9.11 *More complex rotational situations

In more advanced courses you may study nonsymmetric rotational situations like that in Figure 9.53. The angular velocity points along the axis of rotation, but the (translational) angular momentum of each mass is $\vec{r}_A \times \vec{p}$, so the total angular momentum $\vec{L}_A$ does *not* point along the axis! Moreover, the angular momentum vector continually changes direction.

A symmetric object like that shown in Figure 9.54 has an angular momentum that does point along the axis, because the perpendicular components of the angular momenta cancel each other. This is the simpler kind of situation we have dealt with in this chapter. In this symmetric case it is true that $I = m_1 r_{\perp 1}^2 + m_2 r_{\perp 2}^2 + m_3 r_{\perp 3}^2 + m_4 r_{\perp 4}^2 + \ldots$ about the axis of rotation, and the simple formulas $\vec{L}_{rot} = I\vec{\omega}$ and $K_{rot} = \frac{1}{2}I\omega^2$ are valid.

We calculated moment of inertia as a sum (integral) over $r_\perp^2$, the perpendicular distance to an axis. To deal with rotational motion in general requires expressing the moment of inertia as a "tensor," a 3 by 3 array of numbers representing integrals over x^2, xy, xz, y^2, yx, yz, z^2, zx, and zy. Matrix multiplication of this tensor times the angular velocity vector $\vec{\omega}$ yields an angular momentum vector that need not point in the same direction as $\vec{\omega}$. This is beyond the scope of this introductory course.

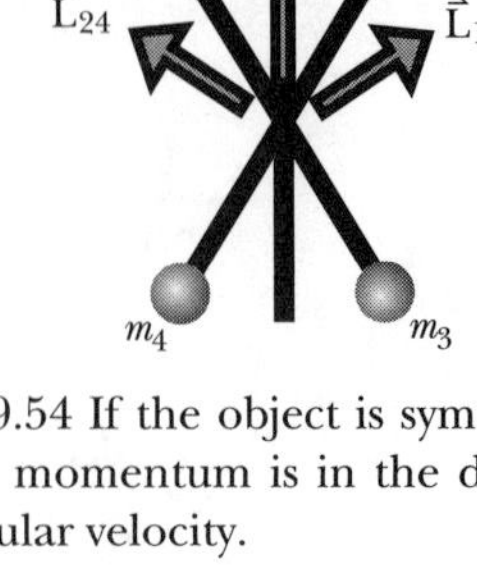

Figure 9.54 If the object is symmetric, the angular momentum is in the direction of the angular velocity.

9.12 *Vector rate of change of a rotating vector

A minor comment: The magnitude of the time rate of change of a rotating vector $\vec{X}$ is $|d\vec{X}/dt| = \omega X$. If you look at a diagram involving the angular velocity vector $\vec{\omega}$ you can see that the direction of $d\vec{X}/dt$ is given by the cross product $\vec{\omega} \times \vec{X}$. It can be shown rather easily that

$$\frac{d\vec{X}}{dt} = \vec{\omega} \times \vec{X}$$

gives both the magnitude and direction of a rotating vector (Figure 9.55). This is another illustration of the usefulness of cross products.

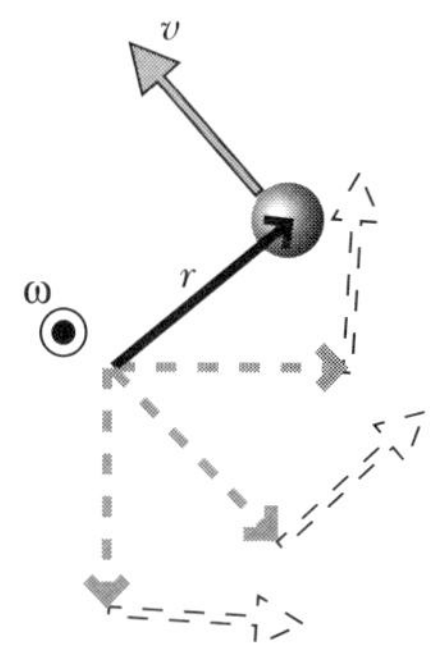

Figure 9.55 In circular motion, both $\vec{r}$ and $\vec{v}$ are rotating vectors. $\vec{\omega}$ is out of the page. At any instant, $\vec{\omega} \times \vec{r}$ is in the direction of $\vec{v}$, while $\vec{\omega} \times \vec{v}$ is toward the center, in the direction of $\vec{a}$, as predicted by the relation $d\vec{X}/dt = \vec{\omega} \times \vec{X}$.

9.13 Summary

Fundamental Physical Principles

The Angular Momentum Principle for a system:

$$\frac{d\vec{L}_{tot,A}}{dt} = \vec{\tau}_{net,ext,A} \quad \text{or} \quad \Delta\vec{L}_{tot,A} = \vec{\tau}_{net,ext,A}\Delta t$$

New concepts

Angular momentum of a particle about location A (Figure 9.56):

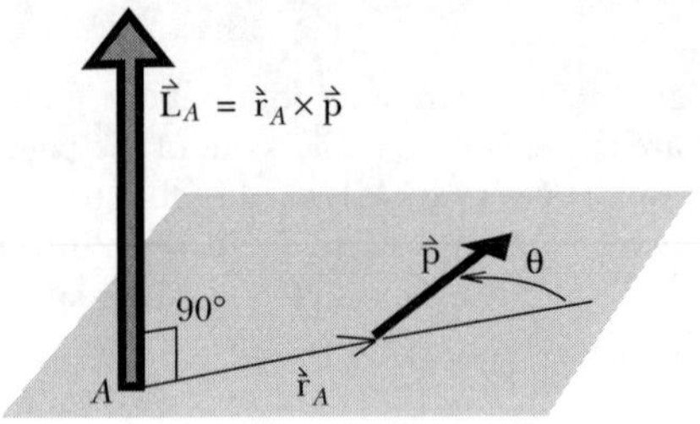

Figure 9.56 Angular momentum of a particle relative to location A.

$$\vec{L}_A = \vec{r}_A \times \vec{p}$$

$$L_A = r_\perp p = r_A p\sin\theta \text{ , direction given by right-hand rule}$$

$$\vec{L}_A = \langle (yp_z - zp_y), (zp_x - xp_z), (xp_y - yp_x) \rangle$$

Torque is a measure of the twist imparted by a force about location A:

$$\vec{\tau}_A = \vec{r}_A \times \vec{F}$$

Moment of inertia about an axis of rotation: $I = m_1 r_{\perp 1}^2 + m_2 r_{\perp 2}^2 + m_3 r_{\perp 3}^2 + \ldots$

Angular momentum is quantized in microscopic systems:

A component of angular momentum is an integer or half-integer multiple of $\hbar$.

The square magnitude of angular momentum has quantized values $L^2 = l(l+1)\hbar^2$, where l has integer or half-integer values.

Results

Angular momentum for a multiparticle system about location A can be divided into translational and rotational angular momentum (Figure 9.57):

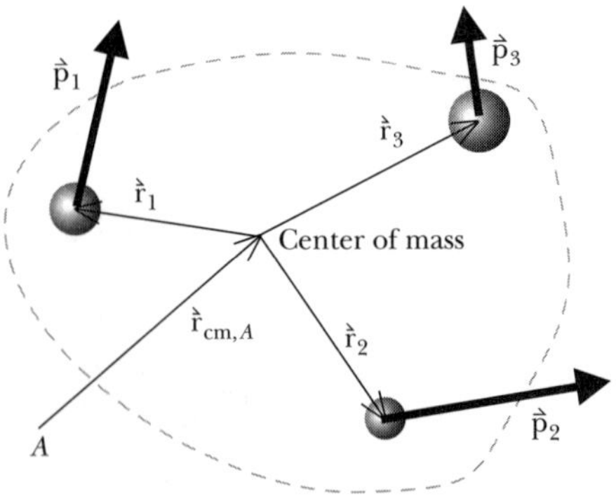

Figure 9.57 Calculating translational and rotational angular momentum.

$$\vec{L}_A = \vec{L}_{trans,A} + \vec{L}_{rot}$$

$$\vec{L}_{trans,A} = \vec{r}_{cm,A} \times \vec{P}_{tot} \text{ , } \quad \vec{L}_{rot} = \vec{r}_1 \times \vec{p}_1 + \vec{r}_2 \times \vec{p}_2 + \vec{r}_3 \times \vec{p}_3$$

Rotational angular momentum in terms of moment of inertia: $\vec{L}_{rot} = I\vec{\omega}$

Rate of change of rotational angular momentum:

$$\frac{d\vec{L}_{rot}}{dt} = \vec{\tau}_{net,ext,cm}$$

Kinetic energy relative to center of mass in terms of moment of inertia and rotational angular momentum:

$$K_{rot} = \frac{1}{2}I\omega^2 = \frac{1}{2}\frac{(I\omega)^2}{I} = \frac{L_{rot}^2}{2I}$$

Angular speed of precession for a gyroscope: $\Omega = \dfrac{\tau}{L_{rot}} = \dfrac{RMg}{I\omega}$

Bohr model of hydrogen: $E = -\dfrac{13.6 \text{ eV}}{N^2}$, $N = 1, 2, 3, \ldots$

$I_{disk} = I_{cylinder} = \frac{1}{2}MR^2$, $I_{sphere} = \frac{2}{5}MR^2$ for axis passing through center

Uniform solid cylinder of length L, radius R, about axis perpendicular to cylinder, through center of cylinder: $I = \frac{1}{12}ML^2 + \frac{1}{4}MR^2$

9.14 Review questions

In addition to these review questions, note that the exercises on page 361 and page 368 constitute important worksheets that gives you practice in calculating angular momentum and torque.

Angular momentum

RQ 9.1 Give an example of a situation where an object is traveling in a straight line, yet has nonzero angular momentum.

RQ 9.2 Under what circumstances is angular momentum constant? Give an example of a situation where the x component of angular momentum is constant, but the y component isn't.

Translational and rotational angular momentum

RQ 9.3 Give examples of translational angular momentum and rotational angular momentum in our Solar System.

RQ 9.4 Give an example of a physical situation in which the angular momentum is zero yet the translational and rotational angular momenta are both nonzero.

The Bohr model of hydrogen

RQ 9.5 What features of the Bohr model of hydrogen are consistent with the later, full quantum mechanical analysis? What features of the Bohr model had to be abandoned?

Particle spin

RQ 9.6 Give some examples of atomic or nuclear phenomena associated with particle spin.

Torque

RQ 9.7 Under what conditions is the torque about some location equal to zero?

RQ 9.8 Make a sketch showing a situation where the torque due to a single force about some location is 20 N·m in the positive z direction, whereas about another location the torque is 10 N·m in the negative z direction.

Static equilibrium

RQ 9.9 What must be true for a system to be in static equilibrium (that is, not moving)?

Gyroscopes

RQ 9.10 Two gyroscopes are made exactly alike except that the spinning disk in one is made of low-density aluminum, whereas the disk in the other is made of high-density lead. If they have the same spin angular speeds and the same torque is applied to both, which gyroscope precesses faster?

9.15 Homework problems

Problem 9.1 Angular momentum in an elliptical orbit

In Chapter 2 you wrote a program to model the motion of a planet going around a fixed star. In this problem you will build on that program.

(a) Use initial conditions that produce an elliptical orbit. At each step calculate the translational angular momentum $\vec{L}_{trans,A}$ of the planet with respect to a location A chosen to be in the orbital plane but outside the orbit. Display this in two ways (i and ii below), and briefly describe in words what you observe:

(i) Display $\vec{L}_{trans,A}$ as an arrow with its tail at location A, throughout the orbit. Since the magnitude of $\vec{L}_{trans,A}$ is quite different from the magnitudes of the distances involved, you will need to scale the arrow by some factor to fit it on the screen.

(ii) Graph the component of $\vec{L}_{trans,A}$ perpendicular to the orbital plane as a function of time.

(b) Repeat part (a), but this time choose a different location B at the center of the fixed star, and calculate and display $\vec{L}_{trans,B}$ relative to that location B. As in part (a), display $\vec{L}_{trans,B}$ as an arrow (scaled appropriately), and also graph the component of $\vec{L}_{trans,B}$ perpendicular to the orbital plane, as a function of time, and briefly describe in words what you observe.

(c) Choose a location C which is not in the orbital plane, and calculate and display $\vec{L}_{trans,C}$ as an arrow throughout the orbit. (You do not need to make a graph.) Briefly describe in words what you observe.

Problem 9.2 Playground ride

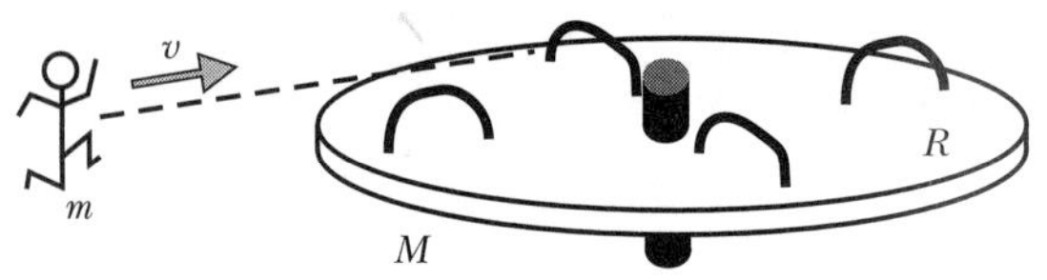

Figure 9.58 A child runs and jumps on a playground ride (Problem 9.2).

A playground ride consists of a disk of mass M and radius R mounted on a low-friction axle (Figure 9.58). A child of mass m runs at speed v on a line tangential to the disk and jumps onto the outer edge of the disk.

(a) If the disk was initially at rest, now how fast is it rotating? (The moment of inertia of a uniform disk is $\frac{1}{2}MR^2$.)

(b) If you were to do a lot of algebra to calculate the kinetic energies in part (a), you would find that $K_f < K_i$. Where has this energy gone?

(c) Calculate the change in *linear* momentum of the system consisting of the child plus the disk (but not including the axle), from just before to just after impact. What caused this change in the linear momentum?

(d) The child on the disk walks inward on the disk and ends up standing at a new location a distance $R/2$ from the axle. Now what is the angular speed?

(e) If you were to do a lot of algebra to calculate the kinetic energies in part (d), you would find that $K_f > K_i$. Where has this energy come from?

(f) In part (a), estimate numerical values for all of the quantities and determine a value for the spin rate ω. Is your result reasonable?

Problem 9.3 Moment of inertia of a disk

Show that the moment of inertia of a disk of mass M and radius R is $\frac{1}{2}MR^2$. Divide the disk into narrow rings each of radius r and width dr. The contribution to I by one of these rings is simply $r^2 dm$, where dm is the amount of mass contained in that particular ring. The mass of any ring is the mass per unit area times the area of the ring. The area of this ring is approximately $2\pi r dr$. Use integral calculus to add up all the contributions.

Problem 9.4 Rotating disk

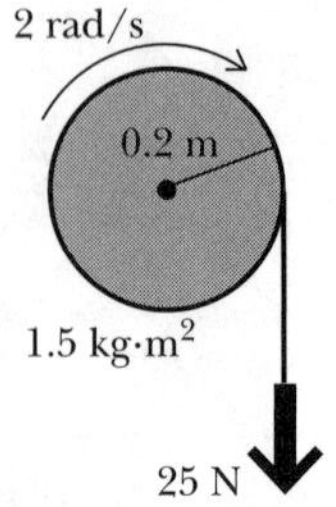

Figure 9.59 A disk rotates on a nearly frictionless axle (Problem 9.4).

A disk of radius 0.2 m and moment of inertia 1.5 kg·m^2 is mounted on a nearly frictionless axle (Figure 9.59). A string is wrapped tightly around the disk, and you pull on the string with a constant force of 25 N. After a while the disk has reached an angular speed of 2 radians/s. What is its angular speed 0.1 seconds later? Explain briefly.

Problem 9.5 Momentum and angular momentum

Two small objects each of mass m are connected by a lightweight rod of length d (Figure 9.60). At a particular instant they have velocities as shown and are subjected to external forces as shown. The system is moving in outer space.

In the following questions involving vectors, give components along the axes shown, and state which axis you're using for each component.

(a) What is the total (linear) momentum $\vec{P}_{\text{total}}$ of this system?

(b) What is the velocity $\vec{v}_{\text{cm}}$ of the center of mass?

(c) What is the total angular momentum $\vec{L}_A$ of the system relative to point A?

(d) What is the rotational angular momentum $\vec{L}_{\text{rot}}$ of the system?

(e) What is the translational angular momentum $\vec{L}_{\text{trans}}$ of the system relative to point A?

After a short time interval Δt,

(f) What is the total (linear) momentum $\vec{P}_{\text{total}}$ of the system?

(g) What is the rotational angular momentum $\vec{L}_{\text{rot}}$ of the system?

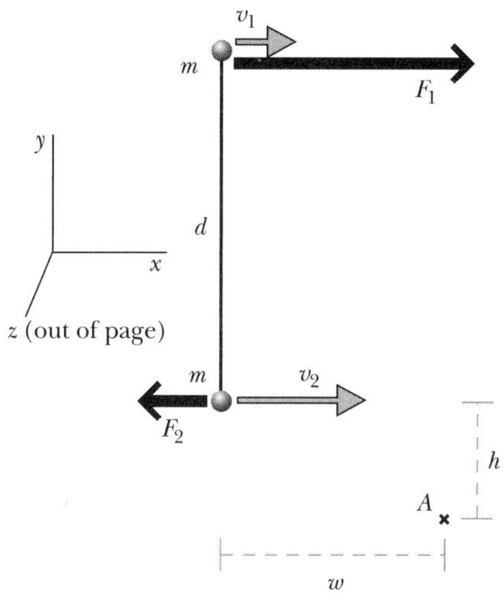

Figure 9.60 Two masses connected by a lightweight rod (Problem 9.5).

Problem 9.6 Ice skater

An ice skater whirls with her arms and one leg stuck out as shown, making one complete turn in one second (Figure 9.61). Then she quickly moves her arms up above her head and pulls her leg in as shown.

(a) Estimate how long it now takes for her to make one complete turn. Explain your calculations, and state clearly what approximations and estimates you make.

(b) Estimate the minimum amount of chemical energy she must expend to change her configuration.

Figure 9.61 A spinning skater pulls in one leg (Problem 9.6).

Problem 9.7 Rotating stool and barbells

You sit on a rotating stool and hold barbells in both hands with your arms fully extended horizontally. You make one complete turn in 2 seconds. You then pull the barbells in close to your body.

(a) Estimate how long it now takes you to make one complete turn. Be clear and explicit about the principles you apply and about your assumptions and approximations.

(b) About how much energy did you expend?

Problem 9.8 An asteroid collision

A spherical non-spinning asteroid of mass M and radius R moving with speed v_1 to the right collides with a similar non-spinning asteroid moving with speed v_2, to the left, and they stick together (Figure 9.62). The impact parameter is d. Note that $I_{\text{sphere}} = \frac{2}{5}MR^2$.

After the collision, what is the velocity v_{cm} of the center of mass and the angular velocity ω about the center of mass? (Note that each asteroid rotates about its own center with this same ω.)

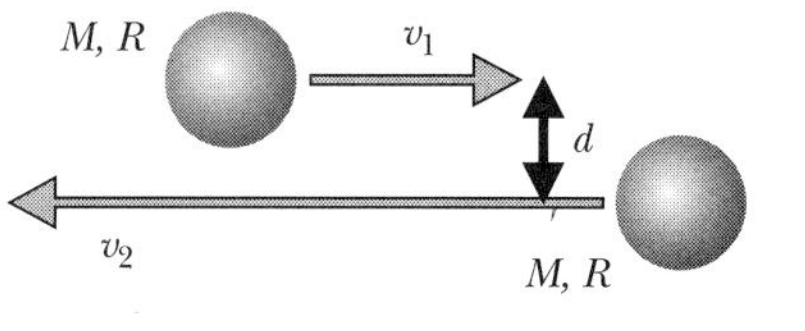

Figure 9.62 Two asteroids collide and stick together (Problem 9.8).

Problem 9.9 A collision with a rod with masses on the ends

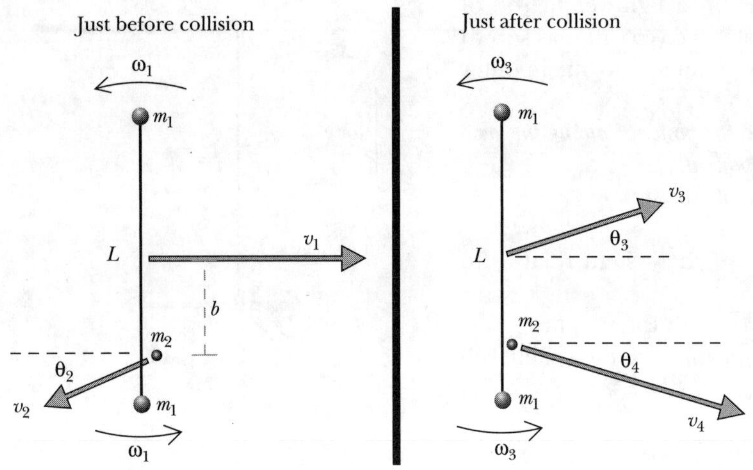

Figure 9.63 A spinning rod is struck by a small object (Problem 9.9).

Two small objects each of mass m_1 are connected by a lightweight rod of length L (Figure 9.63). At a particular instant the center of mass speed is v_1 as shown, and the object is rotating counterclockwise with angular speed ω_1. A small object of mass m_2 traveling with speed v_2 collides with the rod at an angle θ_2 as shown, at a distance b from the center of the rod. After being struck, the mass m_2 is observed to move with speed v_4, at angle θ_4. All the quantities are positive magnitudes. This all takes place in outer space.

For the object consisting of the rod with the two masses, write equations that, in principle, could be solved for the center of mass speed v_3, direction θ_3, and angular speed ω_3 in terms of the given quantities. State clearly what physical principles you use to obtain your equations.

Don't attempt to solve the equations; just set them up.

Problem 9.10 Space junk

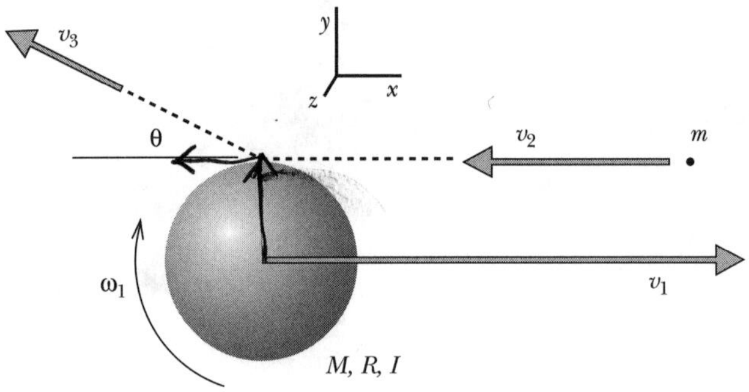

Figure 9.64 Space junk strikes a satellite (Problem 9.10).

A tiny piece of space junk of mass m strikes a glancing blow to a spherical satellite (Figure 9.64). After the collision the space junk is traveling in a new direction and moving more slowly. The space junk had negligible rotation both before and after the collision. The velocities of the space junk before and after the collision are shown in the diagram. The center of mass of the satellite is at its geometrical center. The satellite has mass M, radius R, and moment of inertia I about its center. Before the collision the satellite was moving and rotating as shown in the diagram.

(a) Just after the collision, what are the components of the center-of-mass velocity of the satellite (v_x and v_y) and its rotational speed ω?

(b) Calculate the rise in the thermal energy of the satellite and space junk combined. You do *not* need to substitute in the values for quantities you already calculated in part (a).

Problem 9.11 Space station

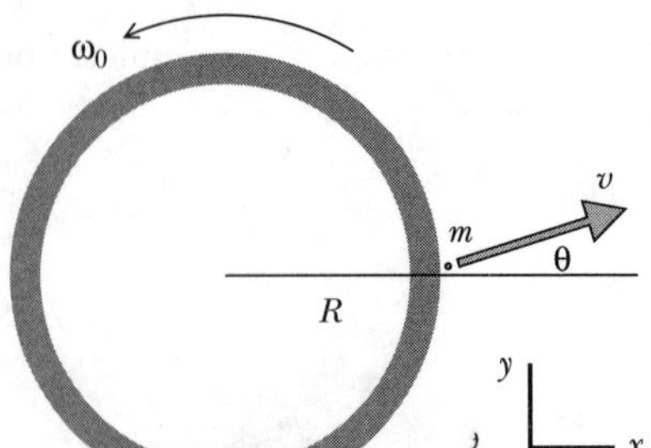

Figure 9.65 Launch a package from a space station (Problem 9.11).

A space station has the form of a hoop of radius R, with mass M (Figure 9.65). Initially its center of mass is not moving, but it is spinning with angular speed ω_0. Then a small package of mass m is thrown by a spring-loaded gun toward a nearby spacecraft as shown; the package has a speed v after launch. Calculate the center-of-mass velocity of the space station (v_x and v_y) and its rotational speed ω after the launch.

Problem 9.12 Rotational spectrum of a diatomic molecule

(a) Calculate the energies of the quantized rotational energy levels for O_2. The most common oxygen nucleus contains 8 protons and 8 neutrons. Estimate any quantities you need. See discussion of diatomic molecules on page 378; the parameter l has values 0, 1, 2, 3...

(b) Describe the emission spectrum for electromagnetic radiation emitted in transitions among the rotational O_2 energy levels. Include a calculation of the lowest-energy emission in electron volts (1.6×10^{-19} J).

(c) It is transitions among "electronic" states of atoms that produce visible light, with photon energies on the order of a couple of electron-volts. Each electronic energy level has quantized rotational and vibrational (harmonic oscillator) energy sublevels. Explain why this leads to a visible spectrum that contains "bands" rather than individual energies.

Problem 9.13 The Bohr model

The Bohr model correctly predicts the main energy levels not only for atomic hydrogen but also for other "one-electron" atoms where all but one of the atomic electrons has been removed, such as in He^+ (one electron removed) or Li^{++} (two electrons removed).

(a) Predict the energy levels in eV for a system consisting of a nucleus containing Z protons and just one electron. You need not recapitulate the entire derivation for the Bohr model, but do explain the changes you have to make to take into account the factor Z.

(b) The negative muon (μ^-) behaves like a heavy electron, with the same charge as the electron but with a mass 207 times as large as the electron mass. As a moving μ^- comes to rest in matter, it tends to knock electrons out of atoms and settle down onto a nucleus to form a "one-muon" atom. For a system consisting of a lead nucleus (Pb^{208} has 82 protons and 126 neutrons) and just one negative muon, predict the energy in eV of a photon emitted in a transition from the first excited state to the ground state. The high-energy photons emitted by transitions between energy levels in such "muonic atoms" are easily observed in experiments with muons.

(c) Calculate the radius of the smallest Bohr orbit for a μ^- bound to a lead nucleus (Pb^{208} has 82 protons and 126 neutrons). Compare with the approximate radius of the lead nucleus (remember that the radius of a proton or neutron is about 10^{-15} m, and the nucleons are packed closely together in the nucleus).

Comments: This analysis in terms of the simple Bohr model hints at the result of a full quantum-mechanical analysis, which shows that in the ground state of the lead-muon system there is a rather high probability for finding the muon *inside* the lead nucleus. Nothing in quantum mechanics forbids this penetration, especially since the muon does not participate in the strong interaction.

The eventual fate of the μ^- in a muonic atom is that it either decays into an electron, neutrino, and antineutrino, or it reacts through the weak interaction with a proton in the nucleus to produce a neutron and a neutrino. This "muon capture" reaction is more likely if the probability is high for the muon to be found inside the nucleus, as is the case with heavy nuclei such as lead.

Problem 9.14 Nuclear gamma ray

The nucleus dysprosium-160 (containing 160 nucleons) acts like a spinning object with quantized angular momentum, $L^2 = l(l+1)\hbar^2$, and for this nucleus it turns out that l must be an even integer (0, 2, 4...). When a Dy-160 nucleus drops from the $l = 2$ state to the $l = 0$ state, it emits an 87 keV photon (87×10^3 eV).

(a) What is the moment of inertia of the Dy-160 nucleus?

(b) Given your result from part (a), find the approximate radius of the Dy-160 nucleus, assuming it is spherical. (In fact, these and similar experimental observations have shown that some nuclei are not quite spherical.)

(c) The radius of a (spherical) nucleus is given approximately by $(1.3\times10^{-15}\text{ m})A^{1/3}$, where A is the total number of protons and neutrons. Compare this prediction with your result in part (b).

Problem 9.15 A comet

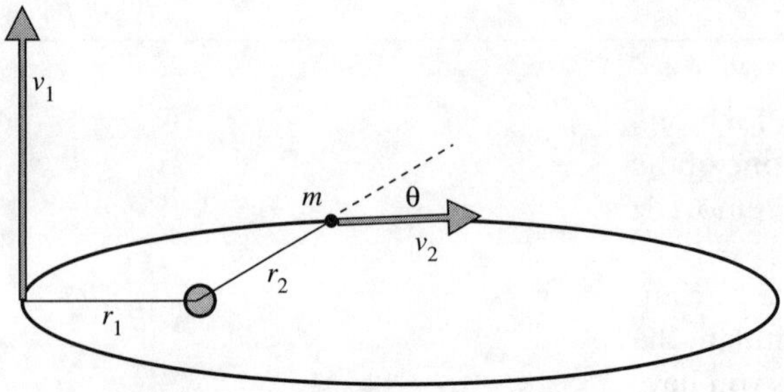

A certain comet of mass m at its closest approach to the Sun is observed to be at a distance r_1 from the center of the Sun, moving with speed v_1. At a later time the comet is observed to be at a distance r_2 from the center of the Sun, and the angle between $\hat{r}_2$ and the velocity vector is measured to be θ. What is v_2? Explain briefly.

Problem 9.16 Yo-yo

A yo-yo is constructed of three disks (Figure 9.66): two outer disks of mass M, radius R, and thickness d, and an inner disk (around which the string is wrapped) of mass m, radius r, and thickness d. The yo-yo is suspended from the ceiling and then released with the string vertical.

Calculate the tension in the string as the yo-yo falls. Note that when the center of the yo-yo moves down a distance y, the yo-yo turns through an angle y/r, which in turn means that the angular speed ω is equal to v_{cm}/r. The moment of inertia of a uniform disk is $\frac{1}{2}MR^2$.

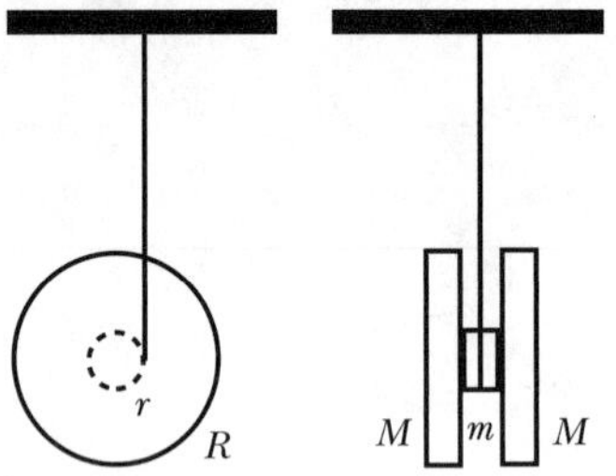

Figure 9.66 A yo-yo is released and moves downward (Problem 9.15).

Problem 9.17 A hovering yo-yo

String is wrapped around an object of mass M and moment of inertia I. You pull the string with your hand straight up with some constant force F such that the center of the object does not move up or down, but the object spins faster and faster (Figure 9.67). This is like a yo-yo; nothing but the vertical string touches the object.

When your hand is a height y_0 above the floor, the object has an angular speed ω_0. When your hand has risen to a height y above the floor, what is the angular speed ω of the object?

Your result should not contain F nor the (unknown) radius of the object. Explain what physics principles you are using.

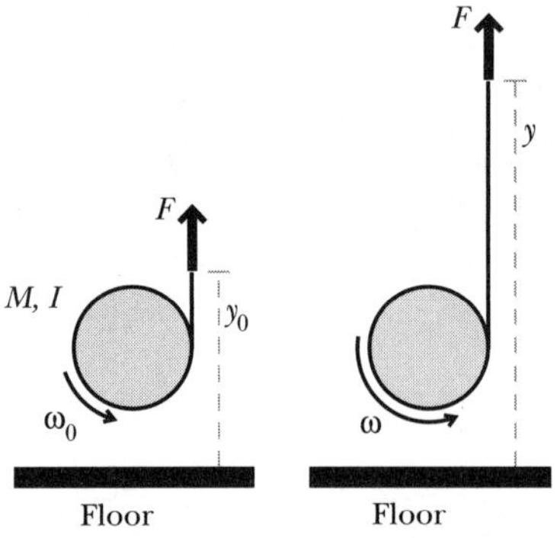

Figure 9.67 The center of this yo-yo doesn't move up or down (Problem 9.17).

Problem 9.18 Bullet and stick

A stick of length L and mass M hangs from a low-friction axle (Figure 9.68). A bullet of mass m traveling at a high speed v strikes near the bottom of the stick and quickly buries itself in the stick.

(a) During the brief impact, is the linear momentum of the stick+bullet system constant? Explain why or why not. Include in your explanation a sketch of how the stick shifts on the axle during the impact.

(b) During the brief impact, around what point does the angular momentum of the stick+bullet system remain constant?

(c) Just after the impact, what is the angular speed ω of the stick (with the bullet embedded in it)? (Note that the center of mass of the stick has a speed $\omega L/2$. The moment of inertia of a uniform rod about its center of mass is $\frac{1}{12}ML^2$.)

(d) Calculate the change in kinetic energy from just before to just after the impact. Where has this energy gone?

(e) The stick (with the bullet embedded in it) swings through a maximum angle θ_{max} after the impact, then swings back. Calculate θ_{max}.

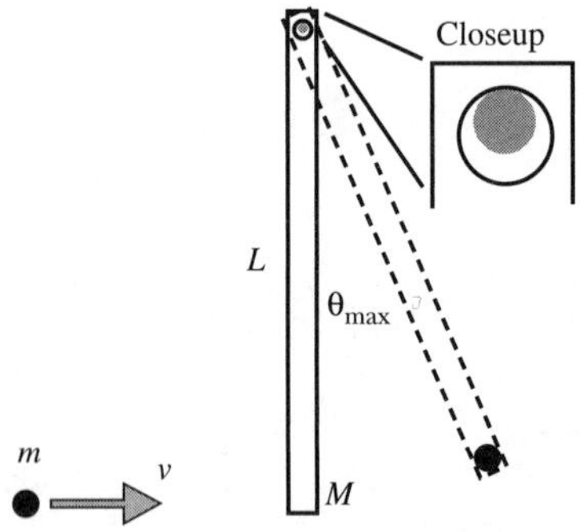

Figure 9.68 A bullet strikes a stick that is suspended from an axle (Problem 9.15).

Problem 9.19 An object rolls down a ramp

A solid object of uniform density with mass M, radius R, and moment of inertia I rolls without slipping down a ramp at an angle θ to the horizontal. The object could be a hoop, a disk, a sphere, etc.

(a) Carefully follow the complete analysis procedure explained in Chapter 3, but with the addition of the Angular Momentum Principle about the center of mass. Note that in your force diagram you must include a small frictional force f that points *up* the ramp. Without that force the object will slip. Also note that the condition of nonslipping implies that the instantaneous velocity of the atoms of the object that are momentarily in contact with the ramp is zero, so $f < \mu F_N$ (no slipping). This zero-velocity condition also implies that $v_{cm} = \omega R$, where ω is the angular speed of the object, since the instantaneous speed of the contact point is $v_{cm} - \omega R$.

(b) The moment of inertia about the center of mass of a uniform hoop is MR^2, for a uniform disk it is $(1/2)MR^2$, and for a uniform sphere it is $(2/5)MR^2$. Calculate the acceleration dv_{cm}/dt for each of these objects.

(c) If two hoops of different mass are started from rest at the same time and the same height on a ramp, which will reach the bottom first? If a hoop, a disk, and a sphere of the same mass are started from rest at the same time and the same height on a ramp, which will reach the bottom first?

(d) Write the energy equation for the object rolling down the ramp, and for the point-particle system. Show that the time derivatives of these equations are compatible with the force and torque analyses.

Problem 9.20 Pulling a rotating device

A string is wrapped around a uniform disk of mass M and radius R. Attached to the disk are four low-mass rods of radius b, each with a small mass m at the end (Figure 9.69).

The apparatus is initially at rest on a nearly frictionless surface. Then you pull the string with a constant force F. At the instant when the center of the disk has moved a distance d, a length w of string has unwound off the disk.

(a) At this instant, what is the speed of the center of the apparatus? Explain your approach.

(b) At this instant, what is the angular speed of the apparatus? Explain your approach.

(c) You keep pulling with constant force F for an additional time Δt. By how much ($\Delta\omega$) does the angular speed of the apparatus increase in this time interval Δt?

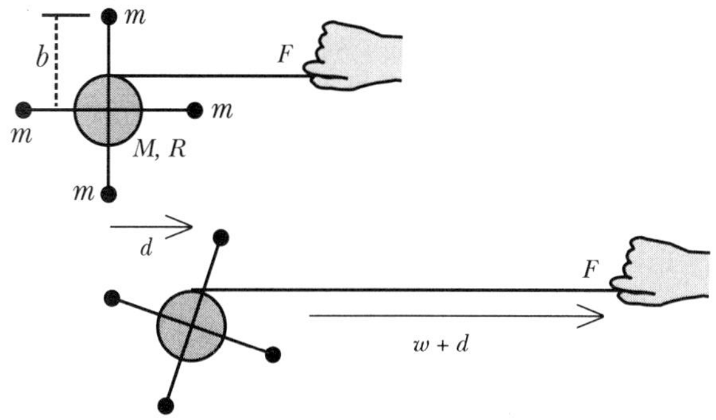

Figure 9.69 A rotating disk with four masses attached (Problem 9.20).

Problem 9.21 A rotating disk with masses sliding on a rod

A rod of length L and negligible mass is attached to a uniform disk of mass M and radius R (Figure 9.70). A string is wrapped around the disk, and you pull on the string with a constant force F. Two small balls each of mass m slide along the rod with negligible friction. The apparatus starts from rest, and when the center of the disk has moved a distance d, a length of string s has come off the disk, and the balls have collided with the ends of the rod and stuck there. The apparatus slides on a nearly frictionless table. Here is a view from above:

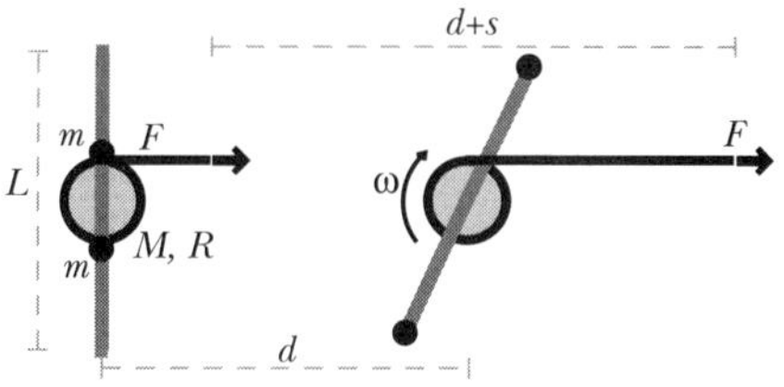

Figure 9.70 The masses slide on the rod (Problem 9.21).

(a) At this instant, what is the speed v of the center of the disk?

(b) At this instant the angular speed of the disk is ω. How much thermal energy has been produced?

(c) In the next short amount of time Δt, by how much will the angular speed change ($\Delta\omega$)?

Problem 9.22 Asteroid collision

Suppose an asteroid of mass 2×10^{21} kg is nearly at rest outside the solar system, far beyond Pluto. It falls toward the Sun and crashes into the Earth at the equator, coming in at an angle of 30 degrees to the vertical as shown, against the direction of rotation of the Earth (Figure 9.71). It is so large that its motion is barely affected by the atmosphere.

(a) Calculate the impact speed.

(b) Calculate the change in the length of a day due to the impact.

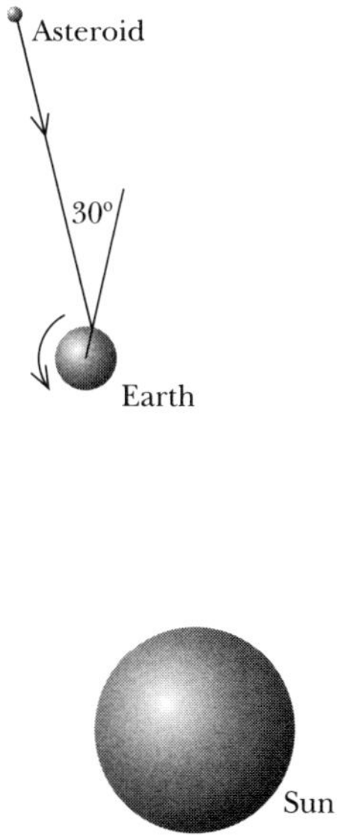

Figure 9.71 An asteroid crashes into the Earth (Problem 9.22; not to scale).

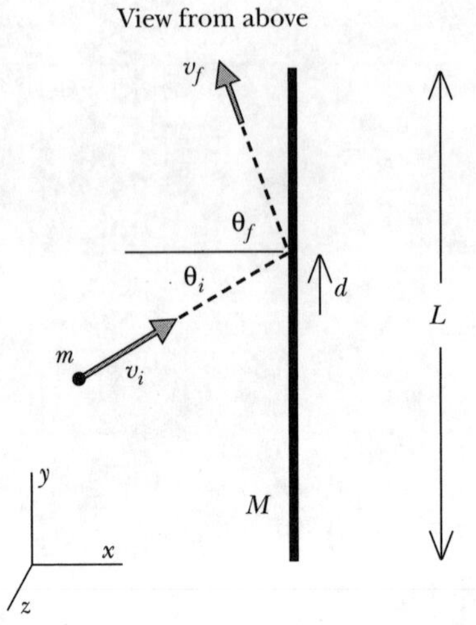

Figure 9.72 A stick lying on ice is struck by a small mass (Problem 9.23).

Figure 9.73 Tucked position (side view).

Figure 9.74 Entering water (front view).

Problem 9.23 Stick on ice

A stick of mass M and length L is lying on ice (Figure 9.72). A small mass m traveling at high speed v_i strikes the stick a distance d from the center and bounces off with speed v_f as shown in the diagram, which is a top view of the situation. The magnitudes of the initial and final angles to the x axis of the small mass's velocity are θ_i and θ_f. All of the symbols in the diagram represent positive numbers.

(a) Afterwards, what are the velocity components v_x and v_y of the center of the stick? Explain briefly.

(b) Afterwards, what is the magnitude and direction of the angular velocity ω of the stick?

(c) What is the increase in thermal energy of the objects? You can leave your expression in terms of the initial quantities and v_x, v_y, and ω.

Problem 9.24 Diver

A diver dives from a high platform. When he leaves the platform, he tucks tightly (Figure 9.73) and performs three complete revolutions in the air, then straightens out with his body fully extended before entering the water (Figure 9.74). He is in the air for a total time of 1.4 seconds. What is his angular speed ω just as he enters the water? Give a numerical answer.

Be explicit about the details of your model, and include (brief) explanations. You will need to estimate some quantities.

Problem 9.25 Gyroscope experiment—qualitative

This problem requires that you have a toy gyroscope available. The purpose of this problem is to make as concrete as possible the unusual motions of a gyroscope and their analysis in terms of fundamental principles. In all of the following studies, the effects are most dramatic if you give the gyroscope as large a spin angular speed as possible.

(a) Hold the spinning gyroscope firmly in your hand, and try to rotate the spin axis quickly to point in a new direction. Explain qualitatively why this feels "funny." Also explain why you *don't* feel anything odd when you move the spinning gyroscope in any direction *without* changing the direction of the spin axis.

(b) Support one end of the spinning gyroscope (on a pedestal or in an open loop of the string) so that the gyroscope precesses *counterclockwise* as seen from above. Explain this counterclockwise precession direction; include sketches of top and side views of the gyroscope.

(c) Again support one end of the spinning gyroscope so that the gyroscope precesses *clockwise* as seen from above. Explain this clockwise precession direction; include sketches of top and side views of the gyroscope.

Problem 9.26 Gyroscope experiment—quantitative

This problem requires that you have a toy gyroscope available. The purpose of this problem is to make as concrete as possible the unusual motions of a gyroscope and their analysis in terms of fundamental principles. In all of the following studies, the effects are most dramatic if you give the gyroscope as large a spin angular speed as possible.

(a) If you knew the spin angular speed of your gyroscope, you could predict the precession rate. Invent an appropriate experimental technique and determine the spin angular speed approximately. Explain your experimental method and your calculations. Then predict the corresponding precession rate, and compare with your measurement of the precession rate. You will have to measure and estimate some properties of the gyroscope and how it is constructed.

(b) Make a *quick* measurement of the precession rate with the spin axis horizontal, then make another quick measurement of the precession rate with the spin axis nearly vertical. (It you make quick measurements, friction on the spin axis doesn't have much time to change the spin angular speed.) Repeat, this time with the spin axis initially nearly vertical, then horizontal. Making all four of these measurements gives you some indication of how much the spin unavoidably changes due to friction while you are quickly changing the angle. What do you conclude about the dependence of the precession rate on the angle, assuming the same spin rate at these different angles? What is the theoretical prediction for the dependence of the precession rate on angle (for the same spin rate)?

Problem 9.27 Wood or steel top

(a) A solid wood top spins at high speed on the floor, with a spin direction shown in Figure 9.75. Using appropriately labeled diagrams, explain the direction of motion of the top (you do not need to explain the magnitude).

(b) How would the motion change if the top had a higher spin rate? Explain briefly.

(c) If the top were made of solid steel instead of wood, explain how this would affect the motion (for the same spin rate).

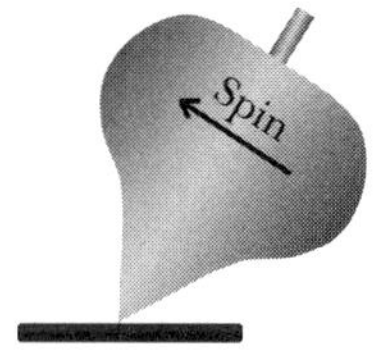

Figure 9.75 A wood or steel top (Problem 9.27).

Problem 9.28 Bicycle wheel

A bicycle wheel with a heavy rim is mounted on a lightweight axle, and one end of the axle rests on top of a post. The wheel is observed to precess in the horizontal plane. With the spin direction shown, does the wheel precess clockwise or counterclockwise? Explain in detail, including appropriate diagrams.

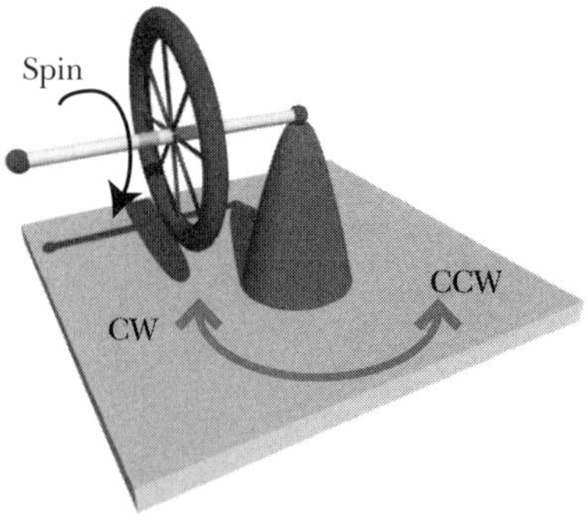

Figure 9.76 A bicycle wheel on a pivot (Problem 9.28).

Problem 9.29 Gyroscope

The axis of a gyroscope is tilted at an angle of 30° to the vertical. The rotor has a radius of 15 cm, mass 3 kg, moment of inertia 0.06 kg·m^2, and spins on its axis at 30 radians/s. It is supported in a cage (not shown) in such a way that without an added weight it does not precess. Then a mass of 0.2 kg is hung from the axis at a distance of 18 cm from the center of the rotor.

(a) Viewed from above, does the gyroscope precess in a 1) clockwise or a 2) counterclockwise direction? That is, does the top end of the axis move 1) out of the page or 2) into the page in the next instant? Explain your reasoning.

(b) How long does it take for the gyroscope to make one complete precession?

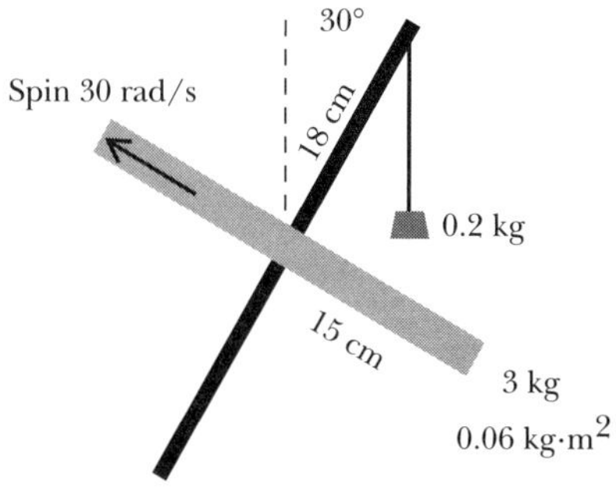

Figure 9.77 A gyroscope (Problem 9.29).

Problem 9.30 Angular momentum of the Earth

(a) Calculate the magnitude of the translational angular momentum of the Earth relative to the center of the Sun. See data on inside back cover.

(b) Calculate the magnitude of the rotational angular momentum of the Earth. How does this compare to your result in part (a)?

(The angular momentum of the Earth relative to the center of the Sun is the sum of the translational and rotational angular momenta. The rotational axis of the Earth is tipped 23.5° away from a perpendicular to the plane of its orbit.)

9.16 Answers to exercises

9.1 (page 356) A: 30 kg·m^2/s, into page ($\otimes$)
B: 30 kg·m^2/s, into page ($\otimes$)
C: 0
D: 50 kg·m^2/s, out of page ($\odot$)
E: 50 kg·m^2/s, out of page ($\odot$)
F: 50 kg·m^2/s, out of page ($\odot$)
G: 0
H: 30 kg·m^2/s, into page ($\otimes$)

9.2 (page 356) 0.38 kg·m^2/s, into page ($\otimes$)

9.4 (page 358) Nonzero; into the page; zero; (no direction)

9.5 (page 358) No; nonzero; into the page; nonzero; into the page

9.6 (page 359) Nonzero; up (toward the sky); nonzero; up

9.7 (page 360) $2M\left(\frac{d}{2}\right)^2 = \frac{1}{2}Md^2$

9.8 (page 360) $MR^2\omega$

9.9 (page 360) 395 J

9.10 (page 361) (a) up; (b) $d_1 p_1 \sin\alpha$; (c) up; (d) $d_2 p_2$

9.11 (page 361) (a) $(d/2)p = (d/2)[m(d/2)\omega_0]$ into page
(b) $2m(d/2)^2\omega_0$ into page ($\otimes$)
(c) $2m(d/2)^2$
(d) into page
(e) $2m(d/2)^2\omega_0$ into page ($\otimes$)
(f) they are equal
(g) $m(d/2)^2\omega_0^2$

9.12 (page 361) (a) 0
(b) $b(2m)(b\omega_1)$ into page ($\otimes$)
(c) $b(2m)(b\omega_1)$ into page ($\otimes$)

9.13 (page 361) (a) $2m(d/2)^2\omega_2$ into page ($\otimes$)
(b) $b(2m)(b\omega_1)$ into page ($\otimes$)
(c) $2m[b^2\omega_1 + (d/2)^2\omega_2]$ into page ($\otimes$)

9.14 (page 363) 6 N·m

9.15 (page 363) 2 N

9.16 (page 363) (4,3.8,0) kg·m^2/s

9.17 (page 364) Force directed toward center of Sun, so zero torque about the location of the Sun.
Angular momentum about center of Sun *constant.*

9.18 (page 364) $r_1 m v_1$, into page; $r_2 m v_2$, into page; $r_1 v_1 = r_2 v_2$

9.19 (page 367) Negligible external forces means negligible external torques.
Therefore magnitude and direction of angular momentum constant.

9.20 (page 367) No torques around center of mass means no change in rotational angular momentum, so rotational angular momentum stays constant in magnitude (which determines length of day) and direction (which determines what "North Star" the axis points at). Doesn't matter that Earth is going around Sun; rotational angular momentum affected solely by torque around center of mass.

9.21 (page 368) (a) $R\cos(45°)mv$ out of page ($\odot$)

(b) $R\cos(45°)mv$ out of page ($\odot$)

(c) $\dfrac{R\cos(45°)mv}{(M+m)R^2}$ out of page ($\odot$)

(d) There is a change of linear momentum of the clay (but not of the wheel), caused by a force applied by the axle to the combined system.

9.22 (page 369) 0.8 N·m, into the page, or <0,0,–0.8> N·m

9.23 (page 369) (a) 0.0256 kg·m²/s, into page ($\otimes$), or <0,0,–0.0256> kg·m²/s

(b) 0.1856 kg·m²/s, into page ($\otimes$), or <0,0,–0.1856> kg·m²/s

(c) 145 rad/s, into page ($\otimes$), or <0,0,–145> rad/s

9.24 (page 369) 8.33 N

9.25 (page 373) $\dfrac{dL_{\text{string}, z}}{dt} = \dfrac{d}{dt}\left[RMv_{cm} - \dfrac{1}{2}MR^2\omega\right] = 0$, since no torque here. Differentiating, we get $\dfrac{d\omega}{dt} = \dfrac{2}{R}\dfrac{dv_{cm}}{dt} = \dfrac{2F_T}{MR}$, as before.

9.26 (page 374) $\dfrac{dP_z}{dt} = F_N - M_1g - M_2g = 0$

$\dfrac{dL_{\text{support}, z}}{dt} = d_1M_1g - d_2M_2g = 0$ (+z out of page)

9.27 (page 374) $\dfrac{dL_{\text{left person}, z}}{dt} = d_1F_N - (d_1 + d_2)M_2g = 0$

Substitute $F_N = (M_1 + M_2)g$ and get

$d_1M_1g - d_2M_2g = 0$

9.29 (page 381) Same as in horizontal case: $\Omega = \dfrac{RMg}{I\omega}$

Chapter 10

Entropy: Limits on the Possible

Chapter 10

Entropy: Limits on the Possible

We have now studied in some detail the three fundamental principles of classical mechanics: the Energy Principle, the Momentum Principle, and the Angular Momentum Principle. These three powerful principles, together with the statistical concepts to be introduced in this chapter, are sufficient to allow us to model quite large and complex systems.

In this chapter we go deeply into the statistical nature of the behavior of systems that include a huge number of atoms. Our goal is to explain some of the macroscopic thermal properties of matter by reasoning about the behavior of individual atoms. A theme that will run through all of the topics in this chapter is the question of how the total energy of a system is divided up among the various components of the system.

10.1 Statistical issues

Temperature

We have repeatedly encountered a connection between the temperature of an object and the average motion of the atoms that make up the object, but our understanding of the meaning and role of temperature is incomplete. What is the quantitative relationship between temperature and microscopic energy?

Direction of thermal energy flow

When two blocks of different temperatures are placed in contact with each other, we observe that energy flows from the hotter block to the colder block (Figure 10.1). We can understand this in a rough way. The average energy of atoms in the hotter block is greater than the average energy of atoms in the colder block. In the interface region, where atoms of the two blocks are in contact with each other, it seems more likely that an atom in the hotter block will lose energy rather than gain energy when it collides with an atom in the colder block.

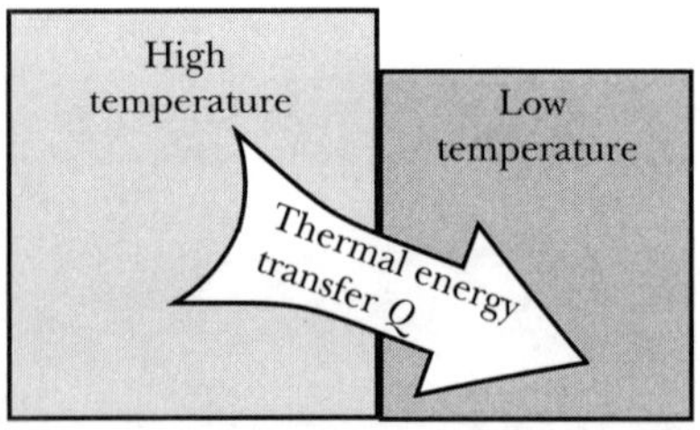

Figure 10.1 We observe that energy flows from a hotter object to a cooler one.

But couldn't energy flow the other way? After all, sometimes an individual atom in the hotter block might happen to have a lot less than its average energy when it collides with an individual atom in the colder block that happens to have a lot more than its average energy. In that case energy would be transferred from the colder block to the hotter block (Figure 10.2).

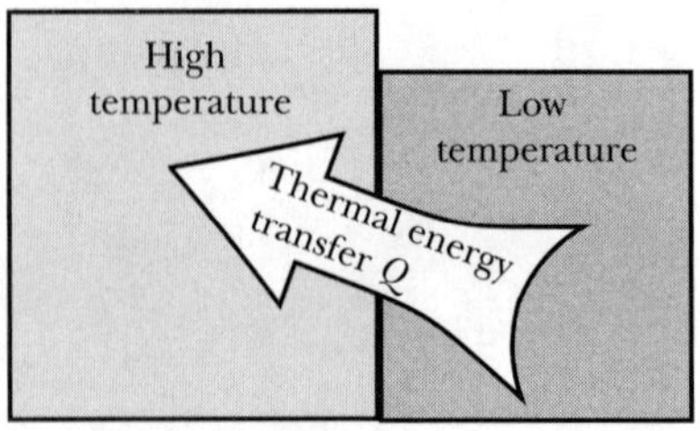

Figure 10.2 Could energy be transferred from the cooler object to the hotter one?

This "uphill" movement of energy seems unlikely, but how unlikely is it? Should we occasionally observe a hot block get hotter when we put it in contact with a cold block? If you put an ice cube in your drink, could your drink get warmer and the ice cube get colder? This would not violate conservation of energy, because the total energy of the two objects would still be the same. So what physical principle would be violated?

Reversibility

A related question, and a very deep one, has to do with the "reversibility" of processes. The fundamental physical interactions seem to be completely "reversible" in the following technical, physics sense: Make a movie of an alpha particle scattering off a gold nucleus, then run the movie forward or reversed. Both views look entirely reasonable. In contrast, make a movie of a ball bouncing on the floor with decreasing height on each bounce, then run the movie forward or reversed. The reversed movie looks silly: you see a ball bouncing higher and higher with each bounce! Would such a motion violate conservation of energy? Not necessarily. It could be that the atoms in the floor happen on average to give energy to the ball, and the floor is getting colder as a result of this continuing loss of energy. So what physical principle is violated by a process represented by the reversed movie?

Statistical models

This chapter addresses these questions by applying probabilistic and statistical ideas to the behavior of systems. This subject is called "statistical mechanics." We will first apply statistical mechanics to a simple model of a solid, and we'll find that we can go surprisingly far with just a few new concepts. The main concepts that we will develop for the specific case of a solid apply in general to a wide variety of systems.

How will we know whether our statistical model explains anything? Our criterion for understanding is whether the predictions of our microscopic model agree with measurements of macroscopic systems, such as measurements of heat capacity (the amount of energy required to increase the temperature of an object by one degree).

10.2 A statistical model of solids

We have noted in previous chapters that the interactions between atoms in a solid are electric in nature, but that a detailed model of them involves quantum mechanics. The interatomic potential energy function encountered in Chapter 4 provides a reasonably accurate description of these interactions. Because of the similarity of this function to the potential energy curve for a harmonic oscillator (a mass on a spring), in previous chapters we have modeled a solid as a large number of tiny masses (the atoms) connected to their neighbors by springs (the interatomic bonds), as shown in Figure 10.3. This model has allowed us to understand qualitatively how solids interact with other objects. We would now like to use this model to ask detailed quantitative questions about the distribution of energy in a solid.

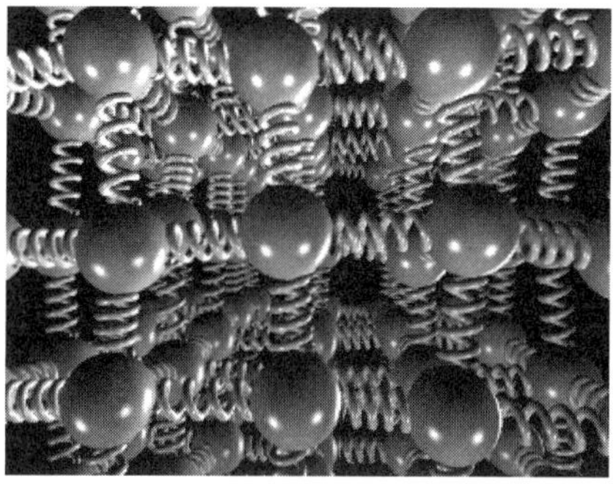

Figure 10.3 Our familiar model of a solid as massive balls connected by springs.

In this chapter we focus on calculating how probable a particular distribution of speeds or energies in a solid would be. We do this for a solid because it happens to be easier to calculate probabilities for a simple model of a solid than it is for a gas or a liquid. The atoms in a solid are nearly fixed in position, so we don't have to consider how likely different spatial arrangements might be (unlike the situation for a gas). Reasoning about the energy distribution in a solid will also enable us to draw conclusions about the transfer of thermal energy from one solid object to another.

10.2.1 The Einstein model of a solid

Because our goal is to calculate the probability of particular distributions of energy among atoms, we can simplify our model of a solid even further. We can think of each atom as moving independently of its neighbors, as though it were connected to rigid walls rather than to other atoms (Figure 10.4). Of course in a real solid energy is exchanged with neighboring atoms. However, to address our current questions we do not need to worry about the mechanism of energy exchange between atoms, since we will focus on calculating probabilities for various distributions of energy among the atoms of the solid. This is a different kind of model from those we have so far constructed, because this very simple model would not allow us to predict in detail the dynamic motion of each atom in the solid, nor to ask how long it would take for energy to flow from one end of a solid to the other. It will not

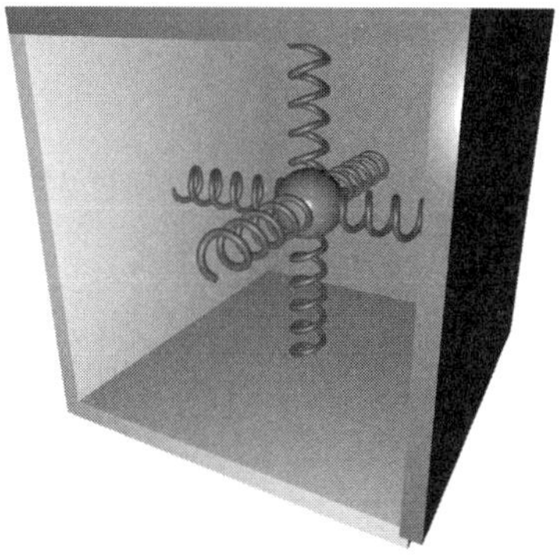

Figure 10.4 A single atom can oscillate in three dimensions. We simplify our model by assuming that each atom moves independently of the surrounding atoms—in effect, that it is connected to rigid walls instead of moving atoms.

shed light on the details of a process, but it does allow us to ask questions about initial and final states. Since at the moment the questions we want to address involve initial and final states, this model is useful, and it is mathematically much simpler than the connected-atoms model.

Einstein proposed this simple model in 1907, and he found that some basic properties of solids such as heat capacity could in fact be understood using this model. This model also allows us to understand in detail the statistical nature of energy transfer between a hot object and a cold object, and why two objects come to "thermal equilibrium" (the same final temperature). This simple model of a solid will help us gain a more sophisticated and powerful understanding of the meaning of temperature.

A three-dimensional oscillator—three one-dimensional oscillators

We will consider each atom in a solid to be connected by springs to immovable walls. Each isolated atom is a three-dimensional spring-mass system, with $\vec{s}$ representing the three-dimensional vector displacement away from a fixed equilibrium position (because we are ignoring collective motions of groups of atoms). Since $p^2 = p_x^2 + p_y^2 + p_z^2$ and $s^2 = s_x^2 + s_y^2 + s_z^2$, we can write the energy of a three-dimensional classical oscillator as consisting of three parts, corresponding to the x, y, and z oscillations:

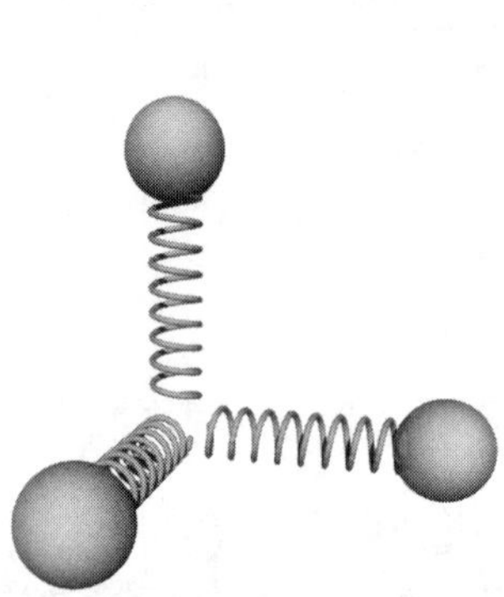

Figure 10.5 Since we are considering the atoms to be independent, in our model we can replace a single three-dimensional oscillator (one atom) by three independent one-dimensional oscillators.

$$K_{\text{vib}} + U_s = \left(\frac{p_x^2}{2m} + \frac{1}{2}k_s s_x^2 + U_0\right) + \left(\frac{p_y^2}{2m} + \frac{1}{2}k_s s_y^2 + U_0\right) + \left(\frac{p_z^2}{2m} + \frac{1}{2}k_s s_z^2 + U_0\right)$$

Recall from Chapter 6 that a quantized oscillator (a quantum mechanical "ball and spring") has evenly spaced energy levels. A complete quantum mechanical analysis of an oscillator that is free to oscillate in three dimensions rather than just one dimension leads to the conclusion that the motion of a 3-D oscillator can be separated into x, y, and z components, each of which has the same energy level structure as the familiar one-dimensional oscillator. This is mathematically equivalent to replacing each three-dimensional oscillator (an atom) with three ordinary one-dimensional oscillators, which we will do to simplify our model. We will think of a block as containing N one-dimensional oscillators, corresponding to $N/3$ atoms (Figure 10.5).

In Chapter 6 we discussed energy quantization in atomic spring-mass systems. As predicted by quantum mechanics and abundantly confirmed by experiments, energy can be added to a one-dimensional atomic oscillator only in multiples of one "quantum" of energy $\hbar\omega_0$, where $\omega_0 = \sqrt{k_s/m}$ (Figure 10.6):

$$\text{Added energy can be } 0,\ 1\hbar\omega_0,\ 2\hbar\omega_0,\ 3\hbar\omega_0,\ \text{etc.}$$

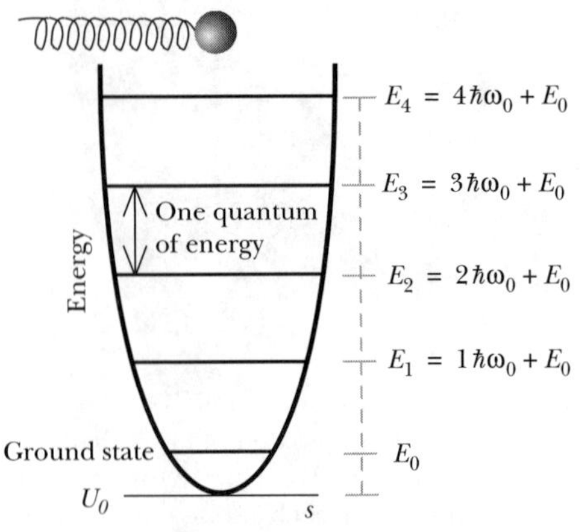

Figure 10.6 The quantized energy levels of a one-dimensional atomic oscillator.

Here $\hbar = h/(2\pi) = 1.05\times10^{-34}$ joule · second, k_s is the spring stiffness, and m is the mass of the atom. The "ground state" (lowest energy level) of the quantum oscillator is not U_0 but $\frac{1}{2}\hbar\omega_0 + U_0$. This offset doesn't matter for what we are going to do; it just sets the baseline energy level, and we will measure energies starting from the ground state.

10.2.2 Distributing energy among objects

When two identical blocks are brought into contact, their total energy is shared among all the oscillators (atoms) in both blocks. It seems plausible that the most likely outcome would be that the total thermal energy be shared equally between identical blocks. However, might there be some probability that the first block would have more energy than the other, or even have all of it? With a large number of atoms the probability of even a small deviation from the most probable distribution turns out to be hugely unlikely. In order to look at this question in detail, we need to find a way to calculate the probability of various possible distributions of the energy. If we can find the most probable energy distribution, we can predict the eventual equilibrium distribution of energy between two objects brought into thermal contact with each other.

Distributing energy: A single atom inside a solid

Consider a single atom inside a solid, whose energy we model in terms of the energy of three one-dimensional oscillators (x, y, z), neglecting interactions with the neighbors. Each of these three oscillators can have 0, 1, 2, etc. number of "quanta" of vibrational energy ($\hbar\omega_0$) added to its ground state.

Suppose the total vibrational energy added to the atom is 4 quanta, and we ask how we might distribute this energy among the three oscillators. We will soon see that enumerating all the possible ways of distributing the energy among the oscillators leads to a deeper understanding of the statistical behavior of a solid. We could give all the energy to the first one-dimensional oscillator and none to the others, or all to the second, or all to the third, as shown in Figure 10.7.

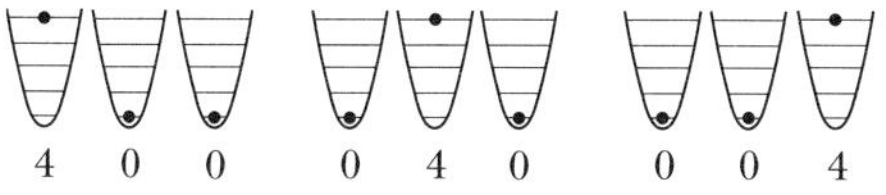

Figure 10.7 Three ways of distributing four quanta of vibrational energy among three one-dimensional oscillators.

Or we could give three quanta to one oscillator, and give the remaining quantum to one of the others, as shown in Figure 10.8.

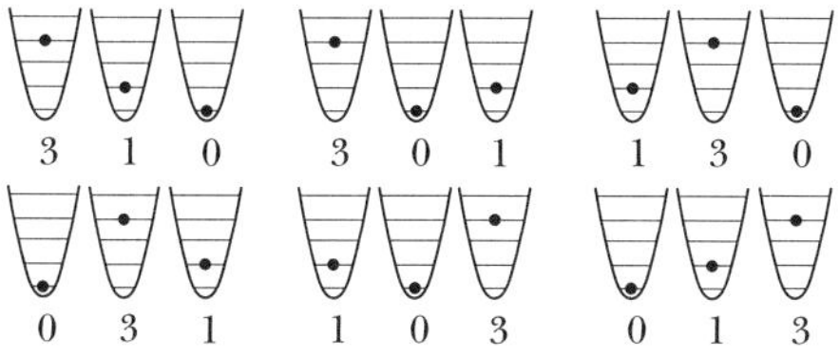

Figure 10.8 Six more ways of distributing four quanta of vibrational energy among three one-dimensional oscillators.

Or we could give two quanta to one oscillator, and distribute the other two to the others, as shown in Figure 10.9.

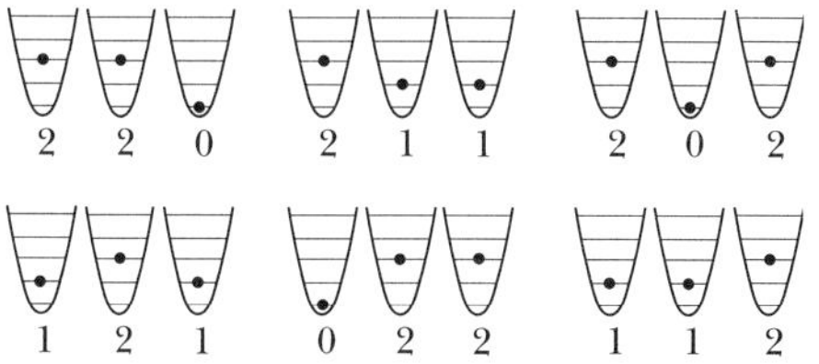

Figure 10.9 Still more ways of distributing four quanta of vibrational energy among three one-dimensional oscillators.

That's it—there aren't any more ways to distribute the energy. (Check to make sure that you can't think of any other arrangements.) By explicitly listing all the possible arrangements, we see that there are 15 different ways that the four energy quanta could be distributed among the three one-dimensional oscillators.

In each of these 15 cases the total energy of the three-oscillator atom is exactly the same. When we make macroscopic measurements of the energy of a block, we don't know and we usually don't care exactly how the energy is distributed among the atoms that make up the block, because the internal energy of the block is the same for all of these distributions. However, the number of different arrangements affects the probability of certain processes occurring, as we shall see.

"Microstate" and "macrostate"

Some terminology: We say that there are 15 "microstates," such as the microstate with 3 quanta in the first oscillator, 0 in the second, and 1 in the third (the second microstate in Figure 10.8). The 15 microstates correspond to one "macrostate," which is characterized by the total energy being equal to 4 quanta of energy, no matter how distributed.

THE FUNDAMENTAL ASSUMPTION OF STATISTICAL MECHANICS

> A fundamental assumption of statistical mechanics is that in our state of microscopic ignorance, each microstate (microscopic distribution of energy) corresponding to a given macrostate (total energy) is equally probable.

That is, we assume that if we could observe the detailed arrangement whenever these three oscillators (corresponding to one atom) have a total energy of 4 quanta, we would find on the average that each of these arrangements would occur 1/15th of the times that we looked. This fundamental assumption is plausible, but ultimately it is justified by the fact that deductions based on this assumption do agree with experimental observations.

Two interacting atoms

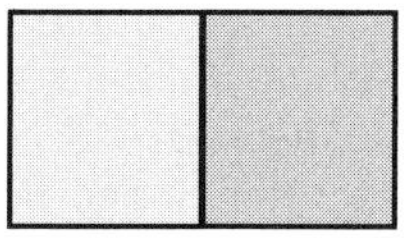

Figure 10.10 Two small blocks in thermal contact.

We can now consider the case of two blocks in thermal contact (Figure 10.10). The importance of counting arrangements ("microstates") can be seen when we have systems of oscillators interacting with each other. Consider the smallest possible "blocks," two neighboring atoms (a total of 6 one-dimensional oscillators), and suppose 4 quanta of vibrational energy, $4\hbar\omega_0$, are distributed among these 6 oscillators. We already know that there are 15 ways to distribute the 4 quanta among the three one-dimensional oscillators of the first atom, and in that case there is only one way to distribute zero quanta to the second atom. Similarly, in Figure 10.11 we see that there are 15 ways to distribute the 4 quanta among the three one-dimensional oscillators of the second atom, and in that case there is only one way to distribute zero quanta to the first atom.

atom 1	atom 2	# ways
4	0	15·1
0	4	1·15

Figure 10.11 If we give all four quanta to one atom or the other, there are 30 ways of distributing four quanta of energy between the two atoms (i.e. among six independent oscillators).

But since the two atoms are in contact with each other, the 4 quanta could also be shared between the atoms. For example, the first atom might have only 3 quanta, with the second atom having the other quantum.

? How many ways are there to distribute 3 quanta among 3 one-dimensional oscillators? Try to list all the possibilities, as we did for the case of 4 quanta.

You should have found ten ways, which we could list as 300 (that is, 3 quanta on the first oscillator, 0 on the second, and 0 on the third), 030, 003, 201, 210, 021, 120, 012, 102, and 111.

atom 1	atom 2	# ways
3	1	10·3
1	3	3·10

Figure 10.12 We find 60 more ways to distribute four quanta among two atoms (six oscillators) if we give three quanta to one atom and one quantum to the other.

? For each of these arrangements, how many ways are there to distribute the one remaining quantum among the other 3 one-dimensional oscillators?

There are just three ways to arrange one quantum among the other 3 oscillators: 100, 010, and 001.

The product of these two numbers is the total number of ways of distributing the 4 quanta in such a way that there are 3 quanta on the first atom and 1 on the other. This product is $10 \cdot 3 = 30$ ways. These results are summarized in Figure 10.12.

Now let's give just 2 quanta to the first atom (leaving 2 quanta for the other atom).

atom 1	atom 2	# ways
2	2	6·6

Figure 10.13 We find 36 ways of distributing four quanta of energy by giving two quanta to each atom.

? How many ways are there to distribute 2 quanta among 3 one-dimensional oscillators?

You should find that there are six ways: 200, 020, 002, 101, 110, and 011.

? For each of these arrangements, how many ways are there to distribute the 2 remaining quanta among the other 3 one-dimensional oscillators?

Clearly there are also six ways to arrange the remaining 2 quanta among the other 3 oscillators. The total number of ways of distributing the 4 quanta in such a way that there are 2 quanta on the first atom and 2 on the other is $6 \cdot 6 = 36$ (Figure 10.13)

Now we have carried out enough calculations to be able to make a table of all the ways of distributing 4 quanta between two atoms (that is, among 6 one-dimensional oscillators). In this table, q_1 is the number of quanta (0 to 4) assigned to the first atom, and q_2 is the number of quanta assigned to the second (with q_1+q_2 necessarily equal to 4). In Figure 10.14 we show the number of ways to arrange the quanta among the first three one-dimensional oscillators, among the second three oscillators, and the product, which is the total number of ways of distributing q_1 and q_2 quanta.

q_1	q_2	# Ways$_1$	# Ways$_2$	# Ways$_1$ · # Ways$_2$
0	4	1	15	15
1	3	3	10	30
2	2	6	6	36
3	1	10	3	30
4	0	15	1	15

Figure 10.14 Summary of the 126 different ways to distribute four quanta of vibrational energy between two atoms (each consisting of three independent oscillators).

Figure 10.15 is a graph of the total number of ways of distributing the quanta vs. q_1. Remember that the basic assumption of statistical mechanics is that all of these 126 different arrangements (15 + 30 + 36 + 30 + 15 = 126 microstates) with a total energy of 4 quanta are equally probable.

? Therefore, if two atoms share 4 quanta, what is the most likely division of the energy between the two? What is the probability at some instant that the energy is equally divided? What is the probability at some instant that the first atom has all the energy?

Evidently the most probable division is 2 and 2, since there are 36 ways for this to be done, whereas all other divisions have fewer ways for this to happen. The probability for a 2-2 division is 36/126, or 29%, so even though this is the most probable division, it happens less than one-third of the time. The probability that the first atom has all the energy is the number of ways for this to happen (15) divided by 126: 15/126 = 12%.

This is a microcosm of what happens when two identical blocks are brought into contact: the most probable division of the thermal energy is that it is shared equally. In order to understand in detail how this works in a real macroscopic system, we need to consider large numbers of atoms and large numbers of quanta. For numbers of atoms or quanta even slightly larger than the few we've been considering, it becomes practically impossible to figure out the number of arrangements of quanta by explicitly listing all the possibilities as we did up to now. We need a formula for calculating the number of arrangements.

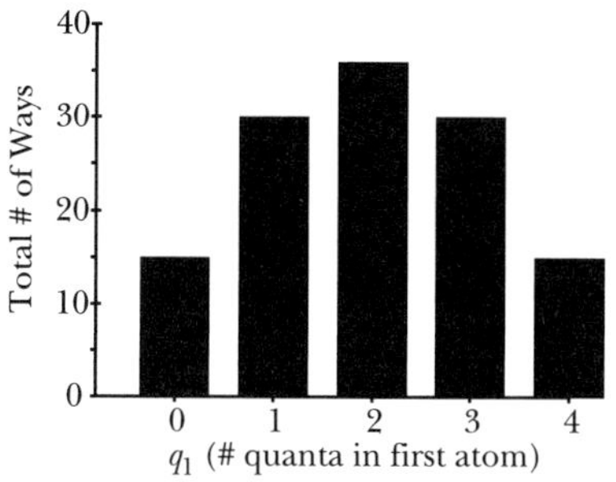

Figure 10.15 Bar graph showing the total number of ways of distributing four quanta of vibrational energy between two atoms (six one-dimensional oscillators).

Ex. 10.1 For practice in counting microstates, determine how many ways there are to arrange 3 quanta among 4 one-dimensional oscillators. (This would represent one-and-a-third atoms, so this doesn't make physical sense.)

10.2.3 One system or many?

The results for distributing 4 quanta among 2 atoms (6 oscillators) can be thought of in two complementary ways. You can say, "I will make frequent observations of my two-atom system, and I expect that in 29% of these observations I'll find the energy split evenly (2-2)." Alternatively, you can set up 100 of these two-atom systems and say, "Whenever I look at my 100 systems, I expect that 29 of them will have the energy split evenly (2-2)." Sometimes the "one system, many observations" view is particularly helpful, and sometimes the "many systems, one observation" view is the more useful way to think about the statistical nature of a phenomenon.

10.2.4 A formula for the number of arrangements of quanta

Clearly it would be not only tedious but impractical to keep on counting states as we did above for systems involving very large numbers of atoms. We need a formula for calculating how many different ways a set of objects can be arranged. We can develop this formula in a general way, using concepts of probability, and then apply it to solids made up of atomic oscillators.

Suppose you have five billiard balls in a bag, numbered 1 through 5. You draw them out of the bag, one at a time, and record the sequence of numbers, such as 34152 (Figure 10.16). Then you put them back in the bag and repeat. How many different number sequences are possible?

- The first ball might be any one of the five balls.
- For each of these 5 possibilities, there are four possibilities for the second ball, or 5·4 = 20 choices so far.
- For each of the 5·4 = 20 choices of the first two balls, there are three different possible choices for the third ball, or 5·4·3 choices so far.

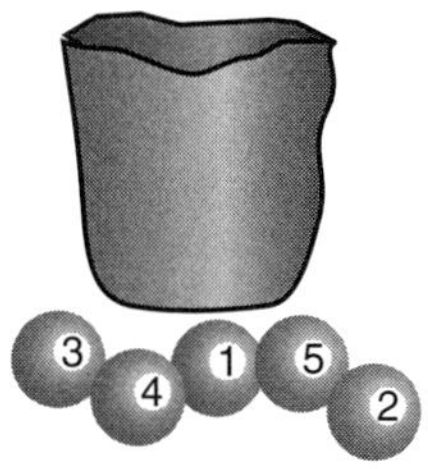

Figure 10.16 Five numbered billiard balls are taken one at a time from a bag. How many different number sequences are possible?

- For each of the 5·4·3 = 60 choices of the first three balls, there are 2 possible choices for the fourth ball, for a total of 5·4·3·2·1 = 120 possibilities, since there is only one remaining ball to choose.

Evidently there are 120 different "permutations" of the five integers. It would be exceedingly tedious to list all these different arrangements, but we have a simple formula to calculate how many there are: 5!, which is the standard notation for "5 factorial," meaning 5·4·3·2·1. Fortunately, for analyzing a solid all we care about is how many arrangements (microstates) there are for a particular total energy, not the details about how much of the energy is assigned to which oscillators. So all we need from a formula is the number of arrangements.

? We can easily check this factorial formula for the case of three balls. Explicitly list all possible permutations of the numbers 1, 2, and 3, and verify that there are indeed 3! = 3·2·1 = 6 possible arrangements.

Making fewer distinctions

Now suppose that of the five balls, three are red and two are green, and we ask how many different arrangements of the colors are possible, such as RGRRG or GGRRR. We know that there are 5! = 120 different arrangements of the numbered balls, but if we're only interested in the color sequence, the numerical order of the red balls is irrelevant, as is the numerical order of the green balls.

There are 3! = 6 permutations of the red balls among each other, and 2! = 2 permutations of the green balls among each other. Therefore there are many fewer than 120 distinctively different color sequences, and we need to correct for this by dividing by the extra permutations:

$$\text{\# of color sequences} = \frac{5!}{3!2!} = \frac{120}{(6)(2)} = 10$$

? Check this result by listing all the different ways of ordering 3 R's and 2 G's.

Arranging quanta among oscillators

By an appropriate choice of visual representation, we can convert our problem of calculating the number of arrangements of q quanta among N one-dimensional oscillators ($N/3$ atoms) into the problem we just solved, for which we have a formula. Consider again the specific case of $q = 4$ quanta distributed among $N = 3$ one-dimensional oscillators. We'll represent a quantum of energy by the symbol •, and a (fictitious) boundary between oscillators by a vertical bar **|**. Here is a picture of a particular situation where the first oscillator has 2 quanta, the second oscillator has 1 quantum, and the third oscillator has 1 quantum; together with the boundaries between oscillators we have 6 objects arranged in a particular sequence:

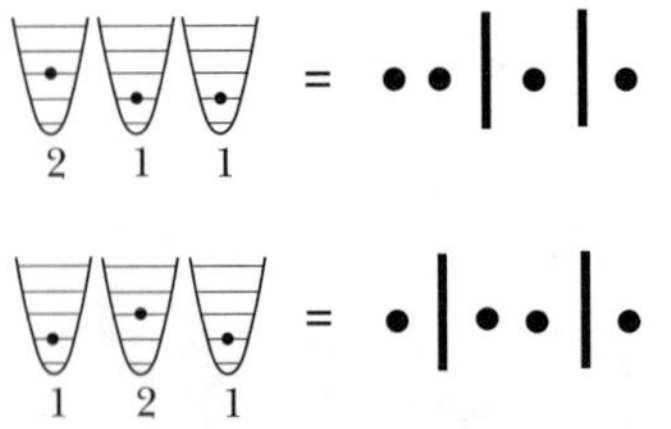

Figure 10.17 Two different representations of number of quanta in three oscillators.

• • **|** • **|** •

This sequence represents a total energy of 4 quanta and 2 boundaries (a total of 6 things, if we consider a boundary to be a "thing"). Figure 10.17 illustrates two such sequences. Notice that we need $N - 1 = 2$ vertical bars to be able to indicate which oscillators have how much energy. How many such sequences like this are there? Rearranging quanta and boundaries is equivalent to moving quanta between oscillators. In this pictorial form, the problem is like having 4 red balls and 2 green ones. Therefore we can write a formula for the number of different arrangements (number of different microstates):

$$\frac{6!}{4!2!} = 15 \text{ ways of arranging 4 quanta among 3 oscillators}$$

This agrees with our earlier calculations. Generalizing to arbitrary numbers of quanta distributed among arbitrary numbers of oscillators, we have this important result for the number of microstates, written as Ω (Greek capital omega):

 # WAYS Ω TO ARRANGE *q* QUANTA AMONG *N* ONE-DIMENSIONAL OSCILLATORS

$$\Omega = \frac{(q+N-1)!}{q!(N-1)!}$$

Ex. 10.2 Verify that this formula gives the correct number of ways to arrange 0, 1, 2, 3, or 4 quanta among 3 one-dimensional oscillators, given in the table on page 404.

Very big numbers

As you increase q and/or N, this formula gets very big very fast. For example, the number of ways to distribute 100 quanta among 300 oscillators (100 atoms) is about 1.7×10^{96}, which is 17 followed by 95 zeros!

In striking contrast, there is only one way to arrange to have all 100 energy quanta be placed on just one particular oscillator out of all the 300 oscillators. Although this arrangement would satisfy the requirement that the total energy of the system be 100 quanta, this is extremely improbable. The fundamental assumption of statistical mechanics is that all microstates are equally probable, so the odds that the actual microstate is the one with all the energy given to just one particular oscillator of your choice is 1 in 1.7×10^{96}, which makes this unlikely event essentially impossible.

A typical macroscopic object such as a block of ordinary size contains 10^{23} or more atoms, not a mere hundred, and the number of quanta is even larger. How likely is it that all the energy will be found concentrated on one particular atom? The mind boggles at the astronomically huge odds against this ever happening. Could it happen? Yes. Is any human ever likely to observe it? No.

Ex. 10.3 Suppose you look once every second at a system with 300 oscillators and 100 energy quanta, to see whether your favorite oscillator happens to have all the energy (all 100 quanta) at the instant when you look. You expect that just once out of 1.7×10^{96} times you will find all of the energy concentrated on your favorite oscillator. On the average, about how many years will you have to wait? Compare this to the age of the Universe, which is thought to be about 10^{10} years. (1 year $\approx \pi\times10^7$ seconds.)

10.3 Thermal equilibrium of two blocks in contact

We now have the tools necessary for analyzing in some detail what will happen when two blocks are brought into contact and approach thermal equilibrium. We'll choose two blocks made of the same material, so a quantum of energy is the same for the atomic oscillators in both blocks (same atomic mass m; same interatomic forces, so same "spring stiffness" k_s). We'll make the analysis somewhat general by choosing blocks of different sizes. One block contains N_1 one-dimensional oscillators ($N_1/3$ atoms) and initially contains q_1 quanta of energy, and the other block contains N_2 one-dimensional oscillators ($N_2/3$ atoms) and initially contains q_2 quanta of energy.

? We want to treat the simple case where the total energy $q_1 + q_2$ of the two-block system remains fixed at all times. What simplifying assumption should we make about the situation?

During the entire process we assume that there is little energy transferred into or out of the surrounding air or the supports for the blocks, so that the total energy of the two blocks doesn't change. However, the number of quanta in each block, q_1 or q_2, need not stay fixed, since energy can flow back and forth between the two blocks.

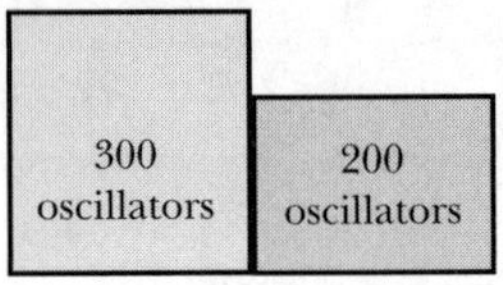

Figure 10.18 A total of 100 quanta of vibrational energy are available to be distributed between two systems, one consisting of 300 oscillators (100 atoms), the other consisting of 200 oscillators (about 67 atoms).

Consider a very concrete example. Suppose $N_1 = 300$ (100 atoms), $N_2 =$ 200 (about 67 atoms; or choose 201 oscillators if you wish to be exact), and there is a total energy distributed throughout the two blocks corresponding to $q_1 + q_2 = 100$ quanta (Figure 10.18). We use a computer to calculate the number of ways that q_1 quanta can be distributed among the $N_1 = 300$ oscillators of the first block (using the formula we developed earlier), and we multiply this number times the number of ways that $q_2 = (100 - q_1)$ quanta can be distributed among the $N_2 = 200$ oscillators of the second block.

The product of these two calculations is the number of ways that we can arrange the 100 quanta so that the first block has q_1 quanta and the second block has $q_2 = (100 - q_1)$ quanta, for a total energy shared between the two blocks of 100 quanta. We have the computer do this calculation for $q_1 = 0$, 1, 2, 3,...99, 100 quanta, which corresponds to $q_2 = 100, 99, 98,...1, 0$ quanta. The following table shows the first few results, where the number of microstates is denoted by Ω (Greek uppercase omega):

q_1	$q_2 = (100 - q_1)$	$\Omega_1 = \frac{(q_1 + 300 - 1)!}{q_1!(300-1)!}$	$\Omega_2 = \frac{(q_2 + 200 - 1)!}{q_2!(200-1)!}$	Total # of Ways $\Omega_1\Omega_2$
0	100	1	2.772 E+81	2.772 E+81
1	99	300	9.271 E+80	2.781 E+83
2	98	4.515 E+04	3.080 E+80	1.391 E+85
3	97	4.545 E+06	1.016 E+80	4.619 E+86
4	96	3.443 E+08	3.331 E+79	1.147 E+88
	...	...	...	...

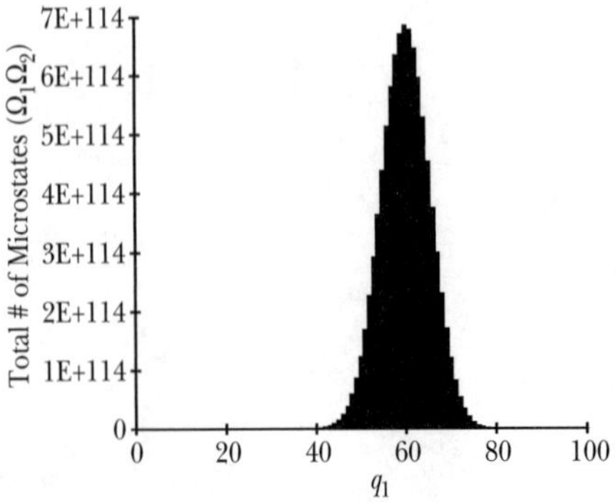

Figure 10.19 The number of ways of distributing 100 quanta of vibrational energy between two blocks having 300 and 200 oscillators, respectively. q_1 is the number of quanta in the first (larger) block.

We see in the last column that there is an enormous number of ways of arranging 100 quanta among 500 oscillators, and that this number is growing rapidly in the table. In Figure 10.19 the possible number of arrangements of quanta is plotted on the y axis, and q_1, the number of quanta assigned to the first block, is plotted on the x axis.

The most probable arrangement (indicated by the highest point of the peak on the graph) is that 60 of the 100 quanta will be found in the first block, which contains 300 oscillators out of the total of 500 oscillators. This seems reasonable, since it does seem most probable that the energy would be distributed uniformly throughout the two blocks (since they're made of the same material), and that would give 3/5 (60%) of the energy to the first block. It is gratifying that our statistical analysis leads to this plausible result.

It appears from the width of the peak shown in the graph that we shouldn't be too surprised if occasionally we would find that the first block contains anywhere between 40 to 80 of the 100 quanta, but it appears that it is very unlikely to find fewer than 40 or more than 80 of the quanta in the first block.

Relatively speaking, how likely is it for *none* of the energy to be in the first block? In that case, $q_1 = 0$, and from the table we see that the number of ways to arrange the 100 quanta this way is 2.772 E+81 (2.772×10^{81}). That's an awfully big number, but how big is it compared to the most probable arrangement, where $q_1 = 60$, for which there are about 7×10^{114} ways according to the graph? Evidently it is less likely by about a factor of 10^{33} (!) to find the energy split 0-100 rather than 60-40. Is it possible according to the laws of physics for none of the energy to be in the first block? Yes. Is it likely that we would ever observe such an unusual distribution? Most emphatically not!

Note that the number of ways to arrange the quanta 0-100 (2.772×10^{81}) isn't actually visible on the graph, because it is 10^{33} times smaller than the peak of the graph. Values of q_1 outside the 40 to 80 range are invisible on the graph, because they are relatively so very small compared to the peak.

Now do Problem 10.1, in which these ideas are made very concrete.

Width of the distribution

Compared to the graph on page 405 of our earliest calculation (6 oscillators and 4 quanta), the peak shown in Figure 10.19 occupies a much narrower range of the graph, reflecting the sharply decreased relative probability of observing extreme distributions of the energy.

For further comparison, Figure 10.20 is a graph of the number of ways of distributing 1000 quanta among two blocks containing 3000 and 2000 oscillators. This peak is much narrower than the peak in the previous graph (where there were only 100 quanta distributed among only 300 and 200 oscillators). This is a general trend. The larger the number of quanta and oscillators, the narrower is the peak around the most probable distribution.

The fractional width of the peak is the width of the peak at half height divided by the value of q_1 that gives the maximum probability (in the present case, the width divided by 600). It can be shown that the fractional width of the peak is proportional to $1/\sqrt{q}$ or to $1/\sqrt{N}$, whichever is larger. That is, if you quadruple the number of quanta or the number of oscillators, the fractional width of the peak decreases by a factor of 2.

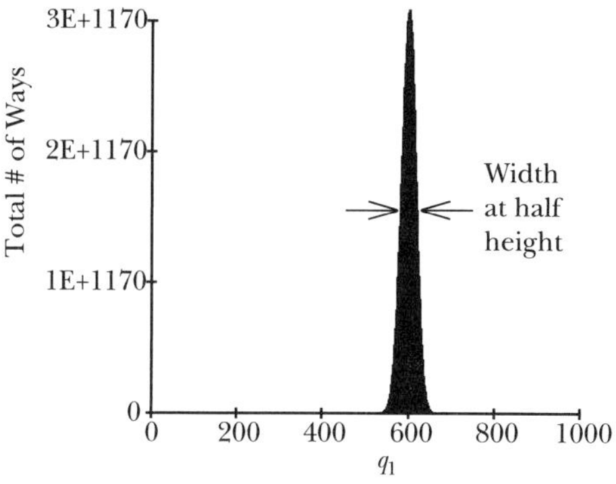

Figure 10.20 The number of ways of distributing 1000 quanta between two blocks containing 3000 and 2000 oscillators, respectively.

? Consider two blocks that are of ordinary macroscopic size, containing 3×10^{23} atoms and 2×10^{23} atoms, and many quanta per atom. Qualitatively, what would you expect about the width of the peak, if you could calculate it? How likely is it that you would ever observe a significant fluctuation away from the most probable 60-40 split?

You would expect the peak to be extremely narrow, and hence the probability of significant fluctuations would be very low.

The most probable is the only real possibility

These considerations show that in the world of macroscopic objects such as ordinary-sized blocks, the most probable arrangement is essentially the *only* arrangement that is ever observed. That is why you do not see a block suddenly leap up from the table when all of the thermal energy of the table floods into the block. On the other hand, at the microscopic level we should not be surprised if the energy of one of the atoms in a block varies a lot, since the most probable distribution includes many different arrangements of the quanta, with varying numbers of quanta on one particular atom.

10.3.1 Entropy and equilibrium

We now have a good statistical description of the thermal equilibrium of two blocks in contact. At equilibrium, energy is distributed between the two blocks in the most probable manner, based on having the largest number of ways of achieving this distribution (largest number of microstates for the macrostate of given total energy). We have been studying very small systems consisting of a hundred or more atoms, and we found that in small systems there is some significant probability of finding the energy distributed somewhat differently than the most probable 60-40 division. (Objects consisting of only a few hundred atoms are called "nanoparticles" and are currently the subject of intense research, because many of their properties and behaviors are intermediate between those of atoms and those of large-scale objects.) For large macroscopic objects containing 10^{20} atoms or more, the most probable distribution of the energy is essentially the only energy distribution we will ever observe, because the probability of distributions that are only slightly different is very small.

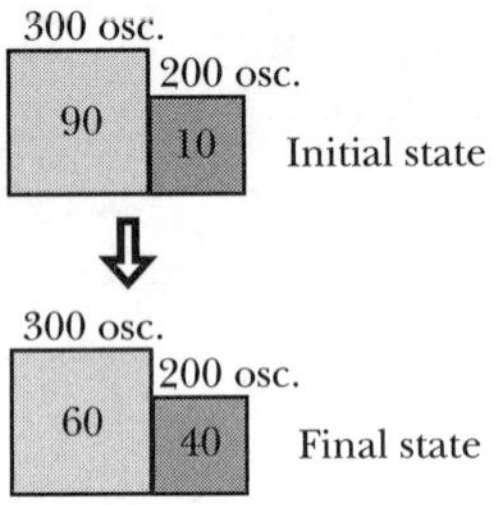

Figure 10.21 If the initial energy distribution between two systems in thermal contact is not the most probable energy distribution, energy will be exchanged until the most probable distribution is reached.

Next we will study more deeply the details of why a particular thermal equilibrium becomes established. Suppose for example that the first block (a nanoparticle with 300 oscillators, or 100 atoms) starts out with 90 quanta and the second (200 oscillators, or about 67 atoms) with only 10 quanta (Figure 10.21). When we put them together, we expect this to shift toward a 60-40 distribution (Figure 10.22). Studying the details of why this shift occurs will lead us to a deep understanding of the concepts of temperature and something called "entropy."

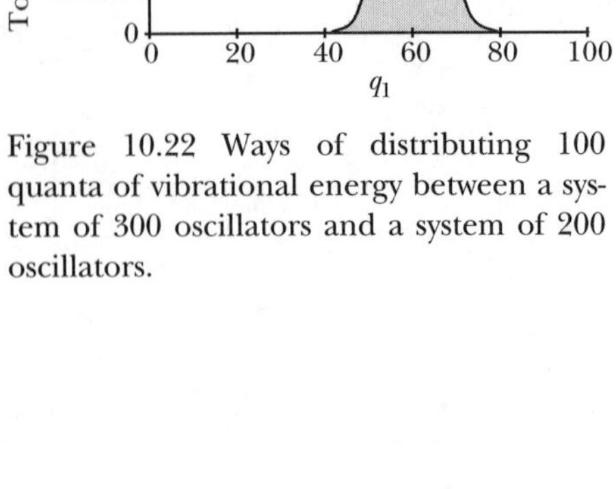

Figure 10.22 Ways of distributing 100 quanta of vibrational energy between a system of 300 oscillators and a system of 200 oscillators.

Since we start from a non-equilibrium energy distribution, it would be nice if we could make some kind of graph where we could see the total number of ways to arrange the quanta even when we are far from equilibrium. As we have seen, the graph of the total number of microstates is so strongly peaked that we don't see anything outside the peak. For example, on the graph in Figure 10.22 where we can see 10^{114} we can't see the relatively much smaller value of 10^{81}. A way to get around this problem is to plot the logarithm of the data, which makes the data visible across the whole range of energy distributions.

In statistical mechanics it is standard practice to use the base-*e* or "natural" logarithm ("ln"). For the case of distributing 100 quanta among 300 and 200 one-dimensional oscillators, which we studied before, Figure 10.23 shows a plot of the natural logarithms of the number of ways Ω_1 to arrange q_1 quanta in block 1, the number of ways Ω_2 to arrange $q_2 = (100 - q_1)$ quanta in block 2, and the total number of ways $\Omega_1\Omega_2$ to arrange the 100 quanta among the 500 oscillators.

Because we are considering discrete microstates, each of these curves is really a set of 100 dots, but we connect the dots and make continuous curves. This is particularly appropriate when we deal with macroscopic objects, where an increment of one quantum along the *x* axis is practically an infinitesimal fraction of the axis.

By taking (natural) logarithms, we have converted a highly-peaked graph into a slowly varying one. To make sure you understand how this graph is related to the peaked graph, do the following:

? For this situation, we calculated on page 408 that there is just 1 way to arrange 0 quanta in block 1, and 2.772×10^{81} ways to arrange 100 quanta in block 2. Take the natural logarithm of this number and verify that all three curves on the graph make sense to you at $q_1 = 0$ ($q_2 = 100$). Note that at the right end of the graph, at $q_1 = 100$, the curve for $\ln(\Omega_1)$ goes higher than $\ln(2.772\times10^{81})$, because block 1 has a larger number of oscillators among which to distribute the 100 quanta (300 oscillators compared with 200).

Definition of entropy

We will deal repeatedly with the natural logarithm of the number of ways to arrange energy among a group of atoms (the number of microstates corresponding to a particular macrostate of specified energy). This quantity, $\ln(\Omega)$, when multiplied by the Boltzmann constant k, is called the "entropy" of the object and is denoted by the letter "S" (the triple equal sign means "is defined as"):

DEFINITION OF ENTROPY *S*

$$S \equiv k \ln \Omega$$

The Boltzmann constant is $k = 1.4\times10^{-23}$ J/K, so entropy has units of joules per kelvin. The Boltzmann constant is included in the definition for consistency with an older, macroscopically based definition of entropy. Figure 10.23 is a plot of S/k, or you can think of it as a plot of S measured in units of k, just as we have been measuring energy in units (quanta) of $\hbar\sqrt{k_s/m}$, where k_s is the effective spring stiffness. Alternatively, you could make a graph of S (= $k\ln\Omega$) instead of $\ln\Omega$.

Since $k\ln(\Omega_1\Omega_2) = k\ln(\Omega_1) + k\ln(\Omega_2)$ (this is a property of logarithms), the entropy S of the two-block system is equal to the entropy S_1 of the first block plus the entropy S_2 of the second block. A consequence of defining entropy in terms of a logarithm is that we can consider entropy as describing a property of a system, and when there is more than one object in a system we get the total entropy simply by adding up the individual entropies, $S_{tot} = S_1+S_2$, as is the case with energy ($E_{tot} = E_1 + E_2$).

Since both $k\ln(\Omega_1\Omega_2)$ and $\Omega_1\Omega_2$ go through a maximum at the same value of q_1, we can state the condition for thermal equilibrium of the two blocks, in terms of entropy:

In equilibrium the most probable energy distribution is that which maximizes the total entropy $S_{tot} = S_1+S_2$ of the two blocks.

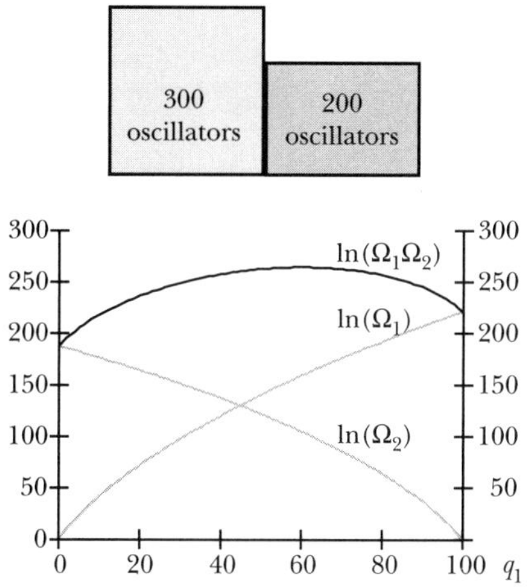

Figure 10.23 A logarithmic plot of the ways of distributing 100 quanta of vibrational energy between two systems. The number of quanta in system 1 is plotted on the x axis, and the natural log of the number of arrangements on the y axis.

? In Figure 10.23, what is the physical significance of the fact that the top curve goes through a maximum for $q_1 = 60$?

The most probably energy distribution is a 60/40 sharing, corresponding to the 300/200 sizes of the two objects.

? In Figure 10.23, what is the physical significance of the point where the two lower curves cross each other?

Nothing! This point isn't anything special. It is the point where the entropy in one object is equal to the entropy in the other object (and equal to half the entropy of the combined system). It turns out that this doesn't have any real physical significance.

Now do Problem 10.2 to see in detail how to produce Figure 10.23.

10.4 The second law of thermodynamics

The "first law of thermodynamics" is another name for the Energy Principle $\Delta E = W + Q$, which by now should be very familiar to you. The "second law of thermodynamics" is however something new. It can be stated in a number of equivalent forms, but there is a particularly useful formulation in terms of entropy:

THE SECOND LAW OF THERMODYNAMICS

If a closed system is not in equilibrium, the most probable consequence is that the entropy will increase.

To say it differently, a closed system will tend toward maximum entropy.

As a specific example of the second law of thermodynamics, consider our two blocks. The most probable energy distribution is the one for which the total entropy is a maximum, and if initially the energy distribution is something else, it is highly likely that the entropy will increase. For nanoparticles there can be significant fluctuations away from this state of maximum entropy (with accompanying decrease in total entropy), but for ordinary-sized systems these fluctuations are extremely small, and for practical purposes the entropy of a closed macroscopic system never decreases.

Loosely, one can restate the second law of thermodynamics like this: "A closed system tends toward increasing disorder." This is overly vague without precise definitions of "order" and "disorder," and it is the definition of entropy as $k\ln\Omega$ that provides the needed precision. A closed system tends toward increasing entropy, which is a measure of the number of microstates corresponding to a particular macrostate of given energy. An extreme example of high "order" is a situation in which all the energy is concentrated on a single oscillator, for there is only one way to arrange the energy like this. In this extreme case, not only do we correctly perceive the situation to be one of high order and low disorder, but the quantitative measurement of disorder (the entropy) is $k\ln(1)$, which equals 0, the lowest possible value.

Irreversibility

Any process in which the entropy of the Universe increases is "irreversible" in the technical, physics sense: a reversed movie of the process looks odd. But any process in which the entropy of the Universe doesn't change is in principle "reversible," and there do exist processes that are approximately or nearly reversible. For example, a steel ball bearing that bounces vertically on a steel plate rebounds almost to the same height from which it was dropped, and this represents a (nearly) reversible process, because the system returns (nearly) to its original state.

However, if we watch for a while, the ball bearing returns to lower and lower heights and eventually settles down to rest on the plate. If the ball bearing kept returning to its original height, this would not be a violation of energy conservation. It is the second law of thermodynamics that says that we cannot expect the process to be completely reversible. There is a larger number of ways to share the energy between the ball and the plate than the number of ways for the ball to keep all the energy.

Suppose we make a movie of the ball bearing bouncing lower and lower and coming to rest on the plate. Then we run it backwards, and our friends see the ball starting to bounce, and bouncing higher and higher (Figure 10.24). This needn't violate energy conservation: energy in the plate could be flowing into the ball. But this reversed movie certainly looks odd, presumably because we have an instinctive sense, based on lots of experience, that the entropy of the Universe increases rather than decreases.

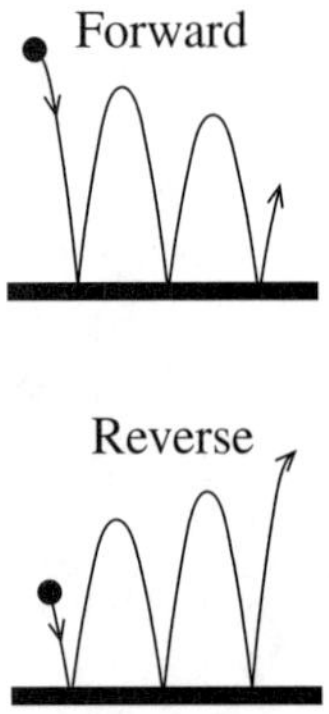

Figure 10.24 If the height of a bouncing ball increased with time instead of decreasing with time, this would not necessarily violate conservation of energy.

REVERSIBLE PROCESS: $\Delta S_{sys} + \Delta S_{surroundings} = 0$

IRREVERSIBLE PROCESS: $\Delta S_{sys} + \Delta S_{surroundings} > 0$

Ex. 10.4 You see a movie in which a shallow puddle of water coalesces into a perfectly cubical ice cube. How do you know the movie is being played backwards? Otherwise, what physical principle would be violated?

10.5 What is temperature?

Until now, we have associated temperature with the average energy of a molecule. The concept of entropy makes possible a deeper connection between our macroscopic measurements of temperature and a fundamental, atomic, statistical view of matter and energy. We will develop a statistically based definition of temperature.

Consider again our two blocks. On the graph in Figure 10.25 we plot entropy $S = k\ln\Omega$ rather than $\ln\Omega$. Remember that we are approximating a set of dots by a smooth curve, since the dots are very close together (tiny energy spacing). At the maximum of the total entropy $S = S_1+S_2$, the total entropy curve is horizontal (slope = 0), and the following relationship is true at equilibrium:

$$\frac{dS}{dq_1} = \frac{dS_1}{dq_1} + \frac{dS_2}{dq_1} = 0$$

Since $q_2 = (100 - q_1)$, this leads to

$$\frac{dS_1}{dq_1} - \frac{dS_2}{dq_2} = 0 \text{ (at equilibrium)}$$

dS_2/dq_2 is the slope moving from right to left and is positive, since q_2 increases from right to left. Therefore we obtain the following result:

$$\frac{dS_1}{dq_1} = \frac{dS_2}{dq_2} \text{ (at equilibrium)}$$

dS_1/dq_1 is a measure of the state of the first block, and dS_2/dq_2 is a measure of the state of the second block.

? What physical property can these derivatives represent? What physical property of the blocks is the same when the two blocks reach thermal equilibrium?

The temperature. It must be that the derivative of the entropy with respect to the energy is somehow related to temperature.

To see what this relationship is, suppose that when the two blocks were initially brought into contact, the first block contained $q_1 = 90$ quanta and the second block only $q_2 = 100 - 90 = 10$ quanta, as illustrated in Figure 10.26. Note that in Figure 10.26 the initial slope of the entropy curve for block 2 is steeper than the initial slope of the entropy curve for block 1. (Remember: dS_2/dq_2 is the slope moving from right to left and is positive, since q_2 increases from right to left.) Therefore, if we remove one quantum of energy from block 1 and give it to block 2, we'll increase the entropy in block 2 more than we'll decrease the entropy in block 1.

? Will this result in a net increase or a net decrease in the total entropy S? (See Figure 10.27, a copy of Figure 10.26.)

Since the entropy of block 2 increases more than the entropy of block 1 decreases, there is a net increase of the entropy of the two blocks together. We already established that the state of the two blocks will evolve toward greater total entropy, not less total entropy.

? Consider which block gives up energy, and which block takes it up. Which block is initially at a higher temperature?

Evidently block 1 must be at the higher temperature, because on average it will give energy to block 2 in order that the total entropy increase.

? Does a steeper or a less steep slope of entropy vs. energy correspond to higher temperature?

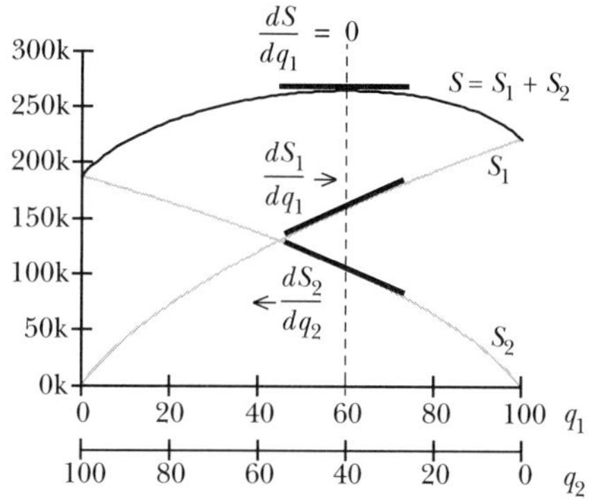

Figure 10.25 Entropy *vs.* number of quanta of energy in system 1.

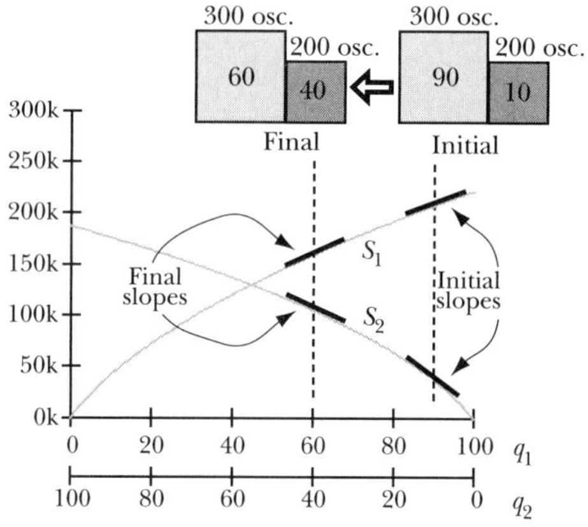

Figure 10.26 Location of the initial and final states of the two-block system on a plot of entropy *vs.* number of energy quanta in system 1.

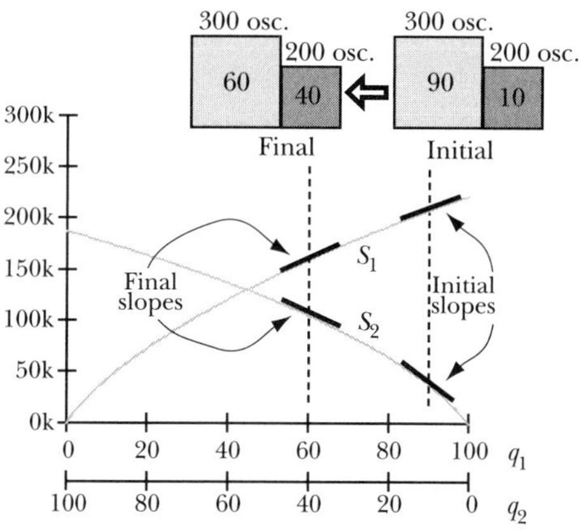

Figure 10.27 Location of the initial and final states of the two-block system on a plot of entropy *vs.* number of energy quanta in system 1.

Initially, block 1 has the higher temperature, and S_1 vs. q_1 has the smaller slope. Initially, block 2 has the lower temperature, and S_2 vs. q_2 has the larger slope.

? Therefore, which of the following looks like a better guess for relating temperature to entropy: $T_1 \propto dS_1/dq_1$? Or $1/T_1 \propto dS_1/dq_1$?

Since block 1 has the higher temperature and the smaller slope (smaller dS/dq), it seems possible that $1/T_1 \propto dS_1/dq_1$.

10.5.1 Definition of temperature

For these reasons, we *define* temperature in terms of dS/dq. But the magnitude of one quantum of energy varies for different systems, so we can't compare different systems if we use q. So instead of q we use E_{int}, the internal energy above the ground state of a system measured in joules. For a group of oscillators, E_{int} is the number of quanta q times the energy per quantum: $E_{int} = q\hbar\sqrt{k_s/m}$ for a system of harmonic oscillators, where k_s is the spring stiffness.

DEFINITION OF TEMPERATURE *T*

$$\frac{1}{T} \equiv \frac{dS}{dE_{int}}$$

The Boltzmann constant k in the definition $S \equiv k\ln\Omega$ makes the units come out right. With the internal energy E_{int} of a system measured in joules (J), and entropy S measured in joules per kelvin (J/K), temperature T is measured in kelvins (K). (Remember that we're approximating a set of closely-spaced dots by a curve, so taking a derivative makes sense.)

This is a highly sophisticated and abstract way of defining temperature. How does this relate to the temperature that is measured by an ordinary thermometer? We will see in Chapter 11 that the temperature defined by $1/T = dS/dE_{int}$ is the same as the "absolute" temperature, which appears for example in the ideal gas law $PV = N_{moles}RT$, which you have probably encountered in previous chemistry or physics courses. On this temperature scale, ice melts at a temperature of +273 K (0° C), and water boils at +373 K (100° C). To put it another way, absolute zero Kelvin is at −273° Celsius.

Review of temperature definition

To recapitulate, the greater the dependence of the entropy on the energy for an object (the steeper the slope of S *vs.* E_{int}), the more "eager" the object is to take in energy to contribute to increasing the total entropy of the total system. It will do this if it can take energy from another object that has a smaller dependence of entropy on energy (smaller slope), since the second object's entropy will decrease less than the first object's entropy will increase. An object with a large dS/dE_{int} has a low temperature, since it is low-temperature objects that take in energy from high-temperature objects.

Conversely, the smaller the dependence of the entropy on the energy, the less reluctant the object is to give up energy and decrease its own entropy, if it can give the energy to an object with a greater dependence of entropy on energy. An object with a small dS/dE_{int} has a high temperature, since it is high-temperature objects that donate energy to low-temperature objects.

Now do Problem 10.3 to calculate the temperature of an object as a function of the amount of energy in the object.

Work, thermal transfer of energy, and entropy

We should really write $1/T = \partial S/\partial E_{\text{int}}$, involving a "partial derivative," which means that we hold everything but S and E constant when we take the derivative. In particular, we hold the volume constant, which means that we do no work on the system, so we should write $1/T = (\partial S/\partial E_{\text{int}})_V$, where the subscript V means "take the (partial) derivative holding the volume constant."

One might ask whether there is any entropy change of the surroundings associated with energy exchange in the form of work. The answer turns out to be no for processes that are reversible (there do exist irreversible processes in which work can be associated with an increase in entropy). Thermal transfer of energy (associated with a temperature difference between objects in contact) is "disorganized" energy transfer and is associated with entropy change. Work is "organized" energy transfer and in many situations doesn't affect the entropy. It is beyond the scope of this course to prove this rigorously, but we can illustrate the basic issues in terms of an Einstein solid.

If you mechanically squeeze or compress a block, doing work on it to raise its energy, it can be shown that you shift the energy levels upward without changing which state an oscillator is in, as shown in Figure 10.28. The "spring" stiffnesses don't change, but there is more energy stored in the "springs." On the other hand, if there is thermal transfer of energy into the system (due to a temperature difference), there is no change in the energy level, but an oscillator will jump to a higher level. This jump from one level to another affects the probability calculations and the entropy, which is the natural log of the number of ways to arrange the energy quanta.

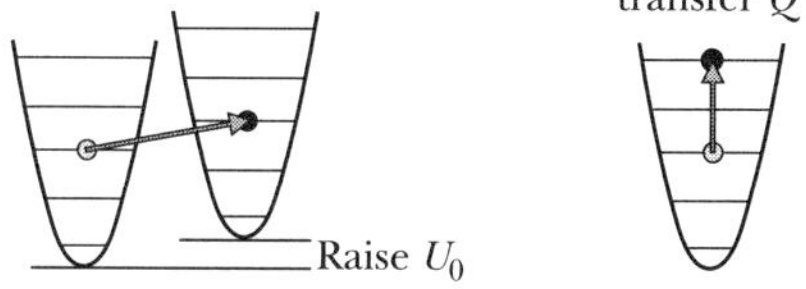

Figure 10.28 Mechanical work (compression or stretching of an oscillator system) compared to thermal transfer of energy to the same system.

Roughly speaking, thermal transfer of energy Q (due to a temperature difference) alters which state the system is in and affects the entropy. Work W alters the energy levels without changing which level the system is in, and doesn't affect the entropy. A small amount of thermal transfer of energy into a system of amount Q, with $\Delta E_{\text{int}} = Q$, leads to an entropy change of that system.

Since $\frac{1}{T} = \frac{\Delta S}{\Delta E_{\text{int}}} = \frac{\Delta S}{Q}$, we conclude the following:

ENTROPY CHANGE ASSOCIATED WITH SMALL *Q*

$$\Delta S = \frac{Q}{T} \text{ (small } Q\text{, associated with nearly constant } T\text{)}$$

If there is a great deal of thermal transfer of energy, causing a large temperature change, we must add up (integrate) the contributions to the entropy change.

Ex. 10.5 There was thermal transfer of energy of 5000 J into a system, and the entropy increased by 10 J/K. What was the approximate temperature of the system?

Ex. 10.6 It takes about 335 joules to melt one gram of ice. During the melting, the temperature stays constant. Which has higher entropy, a gram of liquid water at 0° C or a gram of ice at 0° C? Does this make sense? How large is the entropy difference?

10.5.2 Can the entropy of an object decrease?

? We analyzed two blocks that initially had a 90-10 energy distribution. In the process of changing to a 60-40 distribution, was it possible for the entropy of *one* of the blocks to decrease significantly? Is this a violation of the second law of thermodynamics?

The entropy of block 1 decreased, but the entropy of block 2 increased even more. We believe that the Universe is a closed system, and that the total entropy of the Universe continually increases and never decreases. However, some portions of the Universe may experience a decrease of entropy, as long as there is at least as much increase elsewhere.

Irreversibility again

We started with a 90-10 distribution of the energy but found the most probable final distribution to be 60-40. This change is effectively irreversible. The 60-40 distribution is so enormously more probable than the 90-10 arrangement that an observer will essentially *never* see a 90-10 distribution spontaneously recur. Closed macroscopic systems (such as two blocks inside an insulating box) always evolve toward the most probable arrangement, and stay there (or so close that you can hardly tell the difference).

10.6 Heat capacity of a solid

In Problem 10.3 you calculated the relationship between the temperature of a block and its energy. How can we compare these of calculations with experiment, when we don't have a good way to measure the total energy in a solid block? It is, however, possible to measure *changes* in energy and temperature. Suppose that we add a known amount of energy ΔE to a system, and we measure the resulting rise in temperature ΔT. The ratio of the energy input to the temperature rise on a per-atom basis is called the heat capacity on a per-atom basis, C:

$$C = \frac{\Delta E_{\text{atom}}}{\Delta T}, \text{ where } \Delta E_{\text{atom}} = \frac{\Delta E_{\text{system}}}{N_{\text{atoms}}}$$

The heat capacity can also be defined on a per-gram or per-kilogram basis, or on a per-mole basis ("molar heat capacity").

Different materials have different heat capacities. For example, at room temperature the heat capacity of water (on a per-gram basis) is 4.2 J/K/gram, while the heat capacity of iron at room temperature is 0.84 J/K/gram. Heat capacity is an important property of a material for two reasons. It has practical consequences in science and engineering because it, along with the thermal conductivity, determines the thermal interactions that the material has with other objects. Also, heat capacity is an experimentally measurable quantity that we can compare with theoretical calculations, to test the validity of our models.

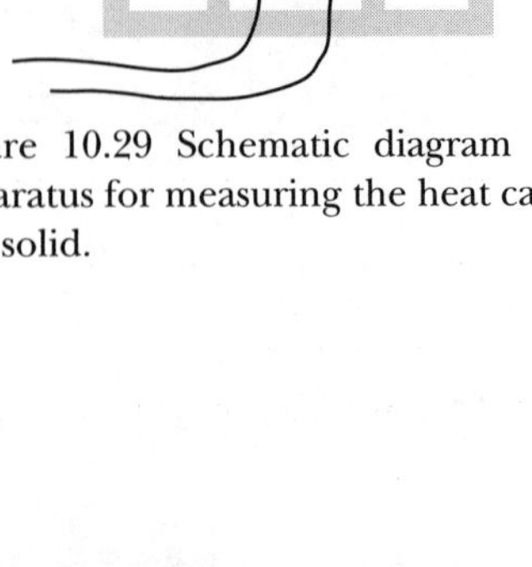

Figure 10.29 Schematic diagram of an apparatus for measuring the heat capacity of a solid.

How would we measure the heat capacity of a block? One way is to enclose the block inside an insulating box and use an electric heater inside the box to raise the temperature of the block (Figure 10.28). The electric power (in watts) times the amount of time the current runs (in seconds) gives us the input energy ΔE (in joules), and we can use a thermometer to measure the temperature rise ΔT of the block. The heat capacity (per atom) is $\Delta E_{\text{atom}}/\Delta T = (\Delta E_{\text{block}}/\Delta T)/N_{\text{atoms}}$. The heater itself should have a small mass compared to the sample of material whose heat capacity we're measuring, so that the heater isn't a significant part of the material that is being warmed up.

The following exercise illustrates another way to measure heat capacity:

Ex. 10.7 A 100 gram block of metal at a temperature of 20° C is placed into an insulated container with 400 grams of water at a temperature of 0° C. The temperature of the metal and water ends up at 2° C. What is the heat capacity of this metal, per gram? (The heat capacity of water is 4.2 joules/K/gram.)

Experimental results for heat capacity

Measurements of the heat capacity of aluminum (circles) and lead (squares) are displayed in Figure 10.30 (the numerical data are shown in a table below). They are plotted as heat capacity (J/K) per atom to facilitate comparison. Note that at low temperatures the heat capacity of both substances changes dramatically with temperature. It is not initially obvious why this should be the case, and in fact this temperature dependence was not predicted by the classical theory, developed in the 1800's. Note also that at room temperature and above, the heat capacity (per atom) of both materials is about the same.

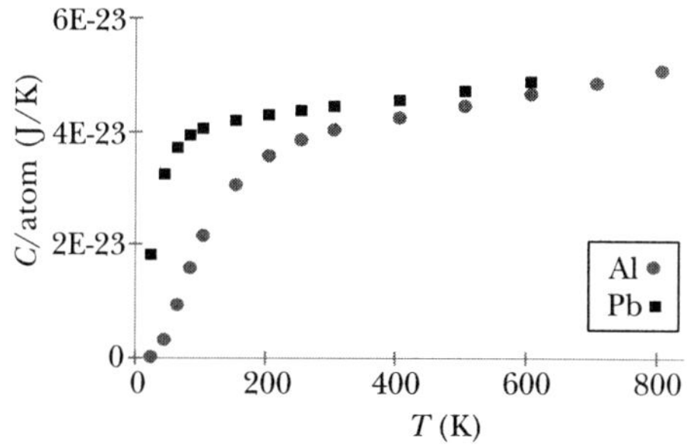

Figure 10.30 Measured heat capacities of aluminum and lead, shown on a per-atom basis.

Will our simple statistical model of a solid be able to predict the temperature dependence of the heat capacity of a solid? In Problem 10.3 you determined the relationship between energy and temperature for a block (in fact, you did the calculation for two different blocks). This calculation can easily be extended to predict the heat capacity per oscillator, which is $\Delta E_{osc}/\Delta T$, where E_{osc} is the average energy per oscillator. We can compare this prediction with experimental data by noting that the heat capacity per atom is 3 times the heat capacity per oscillator.

Effective spring stiffness

In Chapter 3 (page 81) we determined the interatomic spring stiffness k_s of aluminum and lead by relating it to the macroscopic stress-strain relationship (Young's modulus). In the following computation (Problem 10.4) you will use the Einstein independent-oscillator model of a solid to predict the heat capacity of aluminum and lead. We need to use a value for k_s in this model, in order to convert the number of quanta to a value for energy in joules. Presumably the spring stiffness k_s that we use in $\hbar\sqrt{k_s/m}$ should be related to the spring stiffness k_s determined from Young's modulus. However, we shouldn't expect exact agreement, because our model ignores the fact that the atoms are not actually isolated from each other, which is a significant simplification.

Moreover, the effective spring stiffness for oscillations in the *x*, *y*, or *z* directions can be expected to be larger than the value of k_s estimated in Chapter 3 for two reasons. When an atomic core is displaced a distance s_x from its equilibrium position, the spring to the left and the spring to the right each exert a force of $k_s s_x$, so the combined force is $2k_s s_x$. This implies that the effective spring stiffness for oscillations would be $2k_s$.

In addition, in the Einstein model we divide the metal into cubes surrounding each independent atomic oscillator, so each of the two springs to the left and right of an atomic core is effectively half the length of a full interatomic spring, and each half-spring would have a spring stiffness of $2k_s$ (see Ex. 3.5 on page 78).

These considerations suggest that the effective spring stiffness for thermal oscillations in our model might be a factor of 4 larger than the spring stiffness we determined in Chapter 3 from stretching a wire.

T (K)	C, Al (J/K/mole)	C, Pb (J/K/mole)
20	0.23	11.01
40	2.09	19.57
60	5.77	22.43
80	9.65	23.69
100	13.04	24.43
150	18.52	25.27
200	21.58	25.87
250	23.25	26.36
300	24.32	26.82
400	25.61	27.45

Problem 10.4 Heat capacity data.

Now do Problem 10.4 to calculate the heat capacity of aluminum and lead and compare with the actual experimental data shown here.

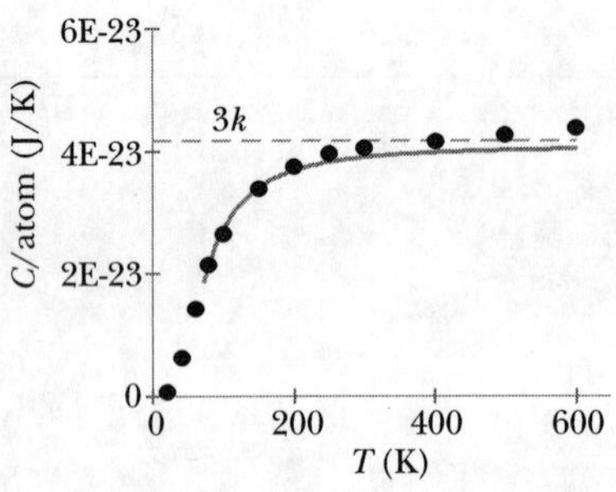

Figure 10.31 Computed heat capacity of copper, compared with experimental values.

Figure 10.31 displays the result of a similar calculation, shown with heat capacity data for copper. Our simple model of a solid as a collection of independent harmonic oscillators does a surprisingly good job of fitting experimental data over a wide range of temperatures. If we were able to include more oscillators in our calculations (by using double precision arithmetic to handle larger numbers, or by using mathematical techniques such as Stirling's approximation for factorials), we could extend our predictions to lower temperatures.

The deviation of the experimental data from the prediction at very high temperatures suggests that at these temperatures a simple harmonic oscillator is not a good model of the atoms in this solid. At high quantum levels, the harmonic oscillator potential energy curve is a poor approximation to the actual potential energy curve describing interatomic interactions (see See Figure 6.18 on page 280).

Note an interesting aspect of the graph: at room temperatures, the predicted heat capacity per atom approaches a constant value of $3k$ (three times the Boltzmann constant, 1.4×10^{-23}). This value agrees quite well with the measured heat capacity of a variety of substances at ordinary temperatures.

Energy quantization and heat capacity

The key difference between our model of solids (the Einstein model) and earlier classical models that did not predict a temperature dependence of heat capacity lies in the quantization of energy in atomic oscillators. Statistical mechanics was originally developed in the 1800's, before the beginning of quantum theory. The classical theory did predict that at high temperatures the energy of each atom in a solid would be $3kT$ (kT per non-quantized one-dimensional oscillator), so that the heat capacity per atom would be $3k$.

If you examine your calculations, you'll find that the high-temperature limit (heat capacity per atom = $3k$) corresponds to temperatures high enough that kT is significantly larger than one quantum of energy. In a situation where the average energy per one-dimensional oscillator is about kT and is large compared to one quantum of energy, the quantization of the energy doesn't make much difference in the analysis. That is, if energy quanta are small compared to the energies of interest, mathematical analysis can be carried out adequately by considering energy to be continuous rather than discrete. In such cases pre-quantum and quantum calculations will give the same results, as is the case here at high temperatures.

On the other hand, at low temperatures the average energy per one-dimensional oscillator is comparable to or smaller than one quantum of energy, and the continuous, nonquantum calculations are not valid. The classical theory provided no explanation for the discovery that as materials were cooled down to very low temperatures, the heat capacity decreased with decreasing temperature. In 1907 Einstein carried through the analysis we have just done and predicted the curve we have just plotted. The good agreement with both low-temperature *and* high-temperature measurements of heat capacity was strong additional evidence for the hypothesis that the energy of oscillators is indeed quantized. (At extremely low temperatures the model of a solid must be refined, taking into account the electrons in the metal, among other things, to achieve full agreement between theory and experiment.)

The fact that the heat capacity for all materials decreases at low temperatures has practical consequences. For example, it makes it difficult to cool a sample to a very low temperature. Cooling a sample depends essentially on putting the sample in contact with a "sink," a large object that is already at a lower temperature, so that there is thermal transfer of energy (Q) out of the sample into the sink, lowering the temperature of the sample and not raising the temperature of the sink very much. But at very low temperatures the sink has a low heat capacity, so this is difficult to achieve.

Ex. 10.8 In an insulated container a 100-watt electric heating element of small mass warms up a 300 gram sample of copper for 6 seconds. The initial temperature of the copper was 20° C (room temperature). Predict the final temperature of the copper.

Ex. 10.9 Since $\Delta T = \Delta E/C$, what will happen at low temperatures to the temperature of the sink when some energy ΔE is transferred to it from the sample? Why is this unfortunate?

10.6.1 Which of our results are general?

We have analyzed simple models of solid matter. Nevertheless the basic conclusions are quite general. For example, if our two model blocks were made of different materials, so that the energy quanta were of different size in the two blocks, this would complicate the procedures for evaluating the number of ways Ω to arrange the energy, but the basic conclusion would remain, that the entropy will increase to a maximum.

10.7 Example: A lead nanoparticle

In Chapter 3, from Young's modulus for lead we found that the effective interatomic "spring" stiffness was about 5 N/m.

(a) For a nanoparticle consisting of 3 lead atoms, what is the approximate temperature when there are 5 quanta of energy in the nanoparticle?

(b) What is the approximate specific heat per lead atom at this temperature? Compare with the approximate high-temperature specific heat per atom for lead.

Solution

We'll use the Einstein model of a solid, though with only three atoms our conclusions will be very approximate.

(a) We model the three atoms as 9 independent quantized oscillators. One quantum of energy in one of these oscillators is this many joules:

$$\hbar\omega_0 = \hbar\sqrt{k_s/m} = (1.05\times10^{-34}\ \text{J·s})\sqrt{\frac{4(5\ \text{N/m})}{(207\times1.7\times10^{-27}\ \text{kg})}} = 7.92\times10^{-22}\ \text{J}$$

We need to calculate the number of ways to arrange q quanta in the neighborhood of $q = 5$, and the associated entropy (Figure 10.32).

Since $1/T = \partial S/\partial E$, $T \approx \Delta E/\Delta S$. Take differences from $q = 4$ to $q = 6$, in order to approximate the slope at $q = 5$; energy increase is 2 quanta:

$$T \approx \frac{N\hbar\omega_0}{k\Delta(\ln\Omega)} = \frac{2(7.92\times10^{-22}\ \text{J})}{(1.4\times10^{-23}\ \text{J/K})(8.01-6.20)} \approx 62.5\ \text{K}$$

(b) Find T_1 at midpoint of 4 to 5 interval, T_2 at midpoint of 5 to 6 interval, energy increase from T_1 to T_2 is one quantum:

$$T_1 \approx \frac{(7.92\times10^{-22}\ \text{J})}{(1.4\times10^{-23}\ \text{J/K})(7.16-6.20)} \approx 58.9\ \text{K}$$

$$T_2 \approx \frac{(7.92\times10^{-22}\ \text{J})}{(1.4\times10^{-23}\ \text{J/K})(8.01-7.16)} \approx 66.6\ \text{K}$$

q	Ω	$\ln\Omega$
4	$\frac{12!}{4!8!} = 495$	6.20
5	$\frac{13!}{5!8!} = 1287$	7.16
6	$\frac{14!}{6!8!} = 3003$	8.01

Figure 10.32 Calculation of entropy ($S = k\ln\Omega$).

As expected, these two temperatures bracket the temperature of 62.5 K at $q = 5$. We want the heat capacity on a per-atom basis, and there are 3 atoms:

$$C_{\text{per atom}} = \frac{1}{3}\frac{\Delta E}{\Delta T} \approx \frac{1}{3}\frac{(7.92\times10^{-22}\ \text{J})}{(66.6-58.9)\ \text{K}} \approx 3.4\times10^{-23}\ \text{J/K}$$

The approximate high-temperature heat capacity for a solid is $3k$ per atom, which is $3(1.4\times10^{-23}\ \text{J/K}) = 4.2\times10^{-23}\ \text{J/K}$. This suggests that at a temperature of about 62.5 K, lead is not quite at the high-temperature limit.

The actual experimental value for lead at 60 K (page 418) is 22.43 J/K/mole, which on a per-atom basis is

$$(22.43\ \text{J/K/mole})\left(\frac{1\ \text{mole}}{6\times10^{23}\ \text{atoms}}\right) = 3.7\times10^{-23}\ \text{J/K}$$

Our 3-atom calculation is not a very accurate model of a macroscopic block of lead, in part because with only three atoms it is not a very good model to say that each atom is connected by spring-like bonds to six neighboring atoms. Despite the failings of our simple model, our result is rather close to the actual value.

10.8 The Boltzmann distribution

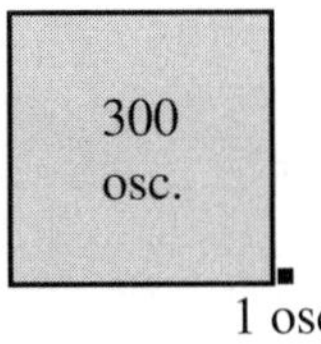

Figure 10.33 A single oscillator in contact with a system of 300 oscillators (100 atoms).

So far we have mainly been concerned with the thermal equilibrium of two blocks, and how thermal equilibrium arises as that particular distribution of energy between the two blocks that has (by far) the largest number of ways to arrange the quanta. What can we say about the probability of observing a particular amount of energy associated with one particular atom (or one particular one-dimensional oscillator)? Addressing this question will lead us to the "Boltzmann distribution," which provides insight into the behavior of a very wide variety of physical, chemical, and biological phenomena.

Consider a collection of 300 one-dimensional oscillators (100 atoms; object 1) in contact with just a single one-dimensional oscillator (object 2), as shown in Figure 10.33. There are 100 quanta of energy distributed between the two objects.

No matter how many quanta are in object 2, there is only one way to distribute them in that single oscillator. Therefore the number of ways of arranging q_2 quanta in the one oscillator (object 2) is just $\Omega_2 = 1$, so in our calculations of $\Omega = \Omega_1\Omega_2$ we can focus on Ω_1, which is equal to Ω.

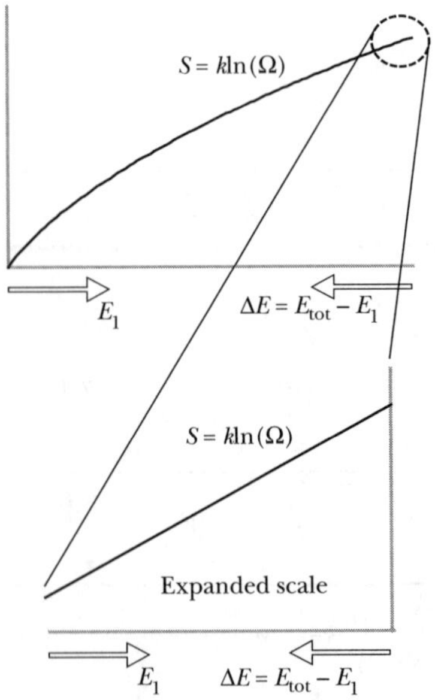

Figure 10.34 Entropy (in units of k) as a function of E_1, for the larger object.

In Figure 10.34 we show the familiar calculation of the entropy $S = k\ln\Omega$ for the larger object 1, as a function of the energy E_1 (q_1 times one quantum of energy in joules). On an expanded scale we show a portion of a plot of the entropy *vs.* E_1 near the maximum possible value of E_1, which is the total energy E_{tot} of the combined system (object 1 plus object 2).

As usual, the total entropy of the two systems does tend to maximize, but in this case the maximum is at the right end of the scale, and the slope is nonzero there. We see that $\ln(\Omega)$ is falling approximately *linearly* with increasing energy in the single oscillator (decreasing energy in the large object 1), so the number of microstates Ω must be falling *exponentially* with increasing energy $\Delta E = (E_{\text{tot}} - E_1)$ in the single oscillator. We can be quantitative about this.

What we would like to know is the probability for the single oscillator to have a certain amount of the total energy, so we need to evaluate Ω for small values of $\Delta E = (E_{\text{tot}} - E_1)$. We can obtain this probability by the following argument. The straight line on the graph has a slope dS/dE_1, so in this region, S can be represented by

$$S = A - \frac{dS}{dE_1}\Delta E$$

where A is some constant, so that the entropy decreases with increasing energy ΔE in the small system.

But $\frac{dS}{dE_1} = \frac{1}{T}$, so we have

$$S = k\ln\Omega = A - \frac{\Delta E}{T}, \text{ and } \ln\Omega = \frac{A}{k} - \frac{\Delta E}{kT}$$

Take the exponential of this equation—that is, use both sides of the equation as powers of *e:*

$$e^{\ln\Omega} = e^{\frac{A}{k} - \frac{\Delta E}{kT}}$$

Since *e* raised to the natural logarithm of some quantity is that quantity, we have the following, where B is some other constant:

$$\Omega = Be^{-\frac{\Delta E}{kT}}$$

This is what we were looking for—the "Boltzmann distribution" of energy to be found in a microscopic system. We have calculated the number of ways (number of microstates) associated with finding an amount ΔE of energy in a single oscillator. This is proportional to the probability of finding this energy in the single oscillator. By ΔE we mean the amount of energy in a small object in thermal equilibrium with a much larger object. We're measuring ΔE from the ground state (minimum-energy state) of the small object.

Taking ratios of probabilities, we have this:

$$\frac{\text{Prob. oscillator has energy } \Delta E \text{ above ground state}}{\text{Prob. oscillator is in ground state}} = \frac{Be^{-\frac{\Delta E}{kT}}}{Be^{-\frac{0}{kT}}} = e^{-\frac{\Delta E}{kT}}$$

Here is a memorable way to state our result, where again ΔE is the energy above the ground state:

THE BOLTZMANN DISTRIBUTION

The probability of finding a microscopic system to be in a state with energy ΔE above the ground state is proportional to

$$e^{-\frac{\Delta E}{kT}}$$

Although we have derived the Boltzmann distribution by considering a particular situation (a single quantum oscillator in thermal equilibrium with a large number of oscillators), the result is actually very general and applies to a very wide variety of phenomena.

? What is the most likely value of ΔE, the energy to be found in a microscopic system that is in thermal equilibrium with a large system? Are you likely to find an energy that is much larger than kT?

The exponential is largest for $\Delta E = 0$, so the most likely energy for the microscopic system is zero. That is, you're most likely to find the system in its ground state. Since e^{-1} is 0.37, the probability of the energy being much larger than $\Delta E = kT$ is rather small.

If the temperature is so low that kT is small compared to the energy E_1 of the first excited state, you are unlikely ever to find the system in one of its excited states. The system is thermally inert, because the probability of taking in *any* amount of energy is very small. In contrast, if the temperature is

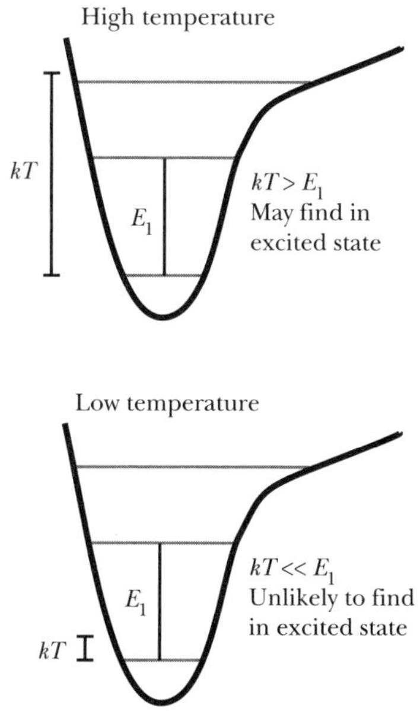

Figure 10.35 At a low temperature it is unlikely to find a system in an excited state, but at a high temperature the likelihood increases.

high enough that $kT > E_1$, then you will sometimes find the system in one of its excited states, although it is still true that the *most* likely situation is that it will be in its ground state (Figure 10.35).

The Boltzmann distribution has far-reaching consequences. For example, chemical and biochemical reaction rates typically depend strongly on temperature because with higher temperature the reactants are moving faster and may be found in excited states. Physical reaction rates are also affected. At the very high temperatures found in the interior of our Sun, kinetic energies are high enough to overcome the electric repulsion between nuclei and to allow the nuclei to come in contact, so that they can undergo thermonuclear fusion reactions. A gas becomes a plasma if the temperature is so high that kT is comparable to the ionization energy.

Ex. 10.10 A certain atom has its first and second excited states E_1 and E_2 above the ground state energy. Calculate the relative probability of finding the atom in the second excited state compared to finding it in the first excited state.

Ex. 10.11 At room temperature, show that $kT \approx \frac{1}{40}$ eV. It is useful to memorize this result, because it tells a lot about what phenomena are likely to occur at room temperature.

10.8.1 One microscopic system *vs.* many

We have found that one microscopic system in contact with a large system has an exponential energy distribution. The meaning of this result is this: You make repeated observations of the microscopic system, and you count the number of times you find the system in its ground state, how many times you find it in the first excited state, etc. You will find that the probabilities for finding the system in each of these states are proportional to the Boltzmann distribution.

But we have also determined the energy distribution in a macroscopic system! Take an energy "snapshot" of a macroscopic system. That is, suppose you can make a simultaneous measurement of the energy contained in each of the microscopic components of this large system. What fraction of the microscopic systems will be in the ground state, or the first excited state? The fraction given by the Boltzmann distribution! This follows because we could analyze any one of these microscopic systems as being a small system in thermal contact with a large system.

These two views of the Boltzmann distribution, one microscopic system observed repeatedly, or a large number of microscopic systems observed once, complement each other. Sometimes one view is more helpful, sometimes the other.

10.9 Application: The Boltzmann distribution in a gas

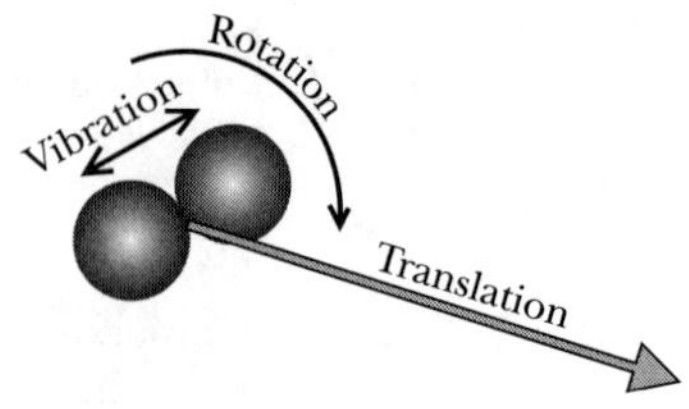

Figure 10.36 Energy of an oxygen molecule.

The Boltzmann distribution applies to any kind of system—not just a solid. As a major application of the Boltzmann distribution, we will study a gas consisting of molecules which don't interact much with each other. Examples are the so-called "ideal gas" (with no interactions at all), and any real gas at sufficiently low density that the molecules seldom come near each other.

In order to apply the Boltzmann distribution, we need a formula for the energy of a molecule in the gas. We will omit rest energy, and we will also omit nuclear energy and electronic energy from our total, because at ordinary temperatures there is not enough energy available in the surroundings to raise the molecule to an excited nuclear state or an excited electronic state. Also, for simplicity, instead of writing ΔE_{vib} to represent an amount of vibrational energy above the ground vibrational state, we will simply write E_{vib}.

The energy of a single gas molecule in the gravitational field near the Earth's surface (excluding rest energy, nuclear energy, and electronic energy) is this:

$$K_{\text{trans}} + E_{\text{vib}} + E_{\text{rot}} + Mgy_{\text{cm}}$$

where E_{vib} and E_{rot} are the vibrational and rotational energies relative to the center of mass (if the molecule contains more than one atom, as in Figure 10.36) and y_{cm} is the height of the molecule's center of mass above the Earth's surface. The mass of the molecule is M. For brevity we will speak of "the energy of the gas molecule" but of course the gravitational potential energy really applies to the system of gas molecule plus Earth.

? For this expression for the energy to be accurate, why must the gas be an "ideal" gas (or a real gas at low density)?

In a real gas at somewhat high density we may not be able to neglect the potential energy associated with the intermolecular forces, in which case this expression for the energy would not be adequate.

If the temperature T is the same everywhere in the (ideal) gas, the probability that a particular molecule will have a certain amount of energy is proportional to this, where P stands for probability:

$$P(E) \propto e^{-\frac{[K_{\text{trans}} + E_{\text{vib}} + E_{\text{rot}} + Mgy_{\text{cm}}]}{kT}}$$

Again, we mentally divide the gas into two systems: one particular molecule of interest, and all the rest of the molecules. These two systems are in thermal equilibrium with each other, because the gas molecules are continually colliding with each other and can share energy. The energy for the one particular molecule is then expected to follow the Boltzmann distribution.

We know that vibration and rotation are quantized. When the gas is confined inside a finite container the momentum (or velocity) and height are also quantized, but under almost all conditions the size of the energy quantum associated with momentum (or velocity) and position is so small compared to kT that it is appropriate to take a nonquantum approach for these variables. We speak of the probability that the gas molecule has an x component of velocity within the range between some v_x and $v_x + dv_x$, and similarly for v_y and v_z. Here dv_x is considered to be a very small amount, small compared to the average speed of the molecules. We also speak of the probability that the x component of the gas molecule's position lies between some x and $x+dx$, and similarly for y and z. Here dx is considered to be a very short distance, small compared to the size of the container, but large compared to the size of a molecule.

(Note: To avoid excess subscripts, in the following discussion we will simply write v for the center-of-mass speed v_{cm}, and y for the center-of-mass height y_{cm}.)

Given these considerations, the formula

$$P(E) \propto e^{-\frac{[K_{\text{trans}} + E_{\text{vib}} + E_{\text{rot}} + Mgy]}{kT}} dxdydzdv_xdv_ydv_z$$

is proportional to the probability that one particular molecule of interest will have the specified energy, with its position and velocity components within dx, dv_x, etc. We will use this formula to explore various properties of a gas.

Separating the various factors

It is useful to group the terms of the distribution formula like this, where the velocity and position distributions are treated in a classical (non-quantum) way:

$$P(E) \propto \left[e^{-\frac{K_{\text{trans}}}{kT}} dv_x dv_y dv_z \right] \left[e^{-\frac{Mgy}{kT}} dxdydz \right] \left[e^{-\frac{E_{\text{vib}}}{kT}} \right] \left[e^{-\frac{E_{\text{rot}}}{kT}} \right]$$

The first bracket gives us the distribution of velocities, the second the distribution of positions, the third the distribution of vibrational energy, and the fourth the distribution of rotational energy. We will discuss each of these individually before putting it all together.

10.9.1 Height distribution in a gas

A striking property of Earth's atmosphere is that in high mountains the air density and pressure are significantly lower than at sea level. The air density is so low on top of Mount Everest that most climbers carry oxygen tanks. Can we explain this? In order to get at the main issue, the variation of density with height, we make the rough approximation that the temperature is constant—the same in the mountains as at sea level.

? How bad an approximation is it to consider the temperature to be the same at all altitudes?

Even in high mountains the temperature is typically above –29° C (–20° F), which is 244 K, and this is only 17% lower than room temperature of 293 K (20° C or 68° F). So maybe this approximation isn't too bad. Focus just on that part of the Boltzmann distribution that deals with position:

$$P(y) \propto e^{-\frac{Mgy}{kT}} dxdydz$$

This is proportional to the probability of finding one particular molecule between x and $x+dx$, y and $y+dy$, and z and $z+dz$. Evidently there's nothing very interesting about x and z (directions parallel to the ground). But in the vertical direction there is an exponential fall-off with increasing height for the probability of finding a particular molecule at height y (Figure 10.37).

Looked at another way, our exponential formula tells us how the number density of the atmosphere depends on height, because in telling us about the behavior of one representative molecule, the formula also tells us something about all the molecules.

We shifted gears from thinking about one molecule to thinking about many. Think again about one single air molecule. Suppose you place it on a table, which is at room temperature. It is not a very unusual event for a thermally agitated atom in the table to hit our air molecule hard enough to send it kilometers high into the air! Actually, our particular molecule will very soon run into another air molecule and not make it very high in one great leap, but on average we do find lots of air molecules very high above sea level rather than finding them all lying on the ground.

Notice again that kT is the important factor in understanding the statistical behavior of matter. Here it sets the scale for the variation with height of the atmosphere's number density.

At a height where the gravitational potential energy $U_g = mgy$ is equal to kT, the number density has dropped by a factor of $e^{-1} = 1/e = 0.37$. Dry air at sea level is 78% nitrogen (one mole N_2 = 28 grams), 21% oxygen (one mole O_2 = 32 grams), about 1% argon, and 0.03% CO_2. An average mass of 29 grams per mole is good enough for most calculational purposes.

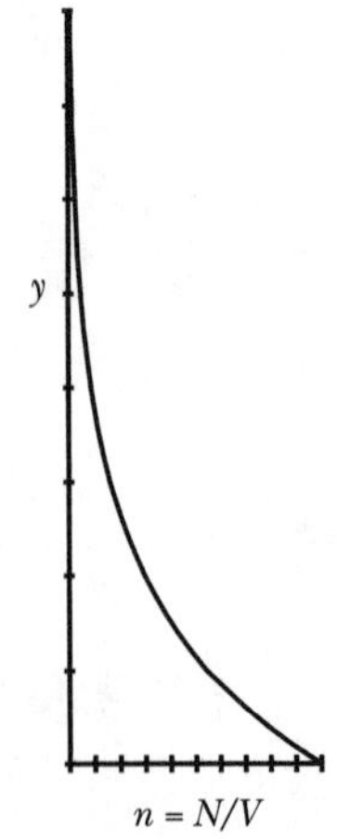

Figure 10.37 Number density vs. height in a constant-temperature atmosphere.

Ex. 10.12 Suppose you put one air molecule on your desk, so that it is in thermal equilibrium with the desk at room temperature. Suppose that there is no atmosphere to get in the way of this one molecule bouncing up and down on the desk. Calculate the typical height that the air molecule will be above your desk, so that $Mgy \approx kT$.

Ex. 10.13 Marbles of mass $M = 10$ grams are lying on the floor. They are of course in thermal equilibrium with their surroundings. What is a typical height above the floor for one of these marbles? That is, for what value of y is $Mgy \approx kT$?

Ex. 10.14 Approximately what fraction of the sea-level air density is found at the top of Mount Everest, a height of 8848 meters above sea level?

10.9.2 Distribution of velocities in a gas

Next we look at the distribution of molecular speeds. Since we're explicitly interested in v_{cm}, it will be useful to express translational kinetic energy as $\frac{1}{2}Mv_{cm}^2$ rather than $p^2/(2M)$ in this discussion. Since $v^2 = v_x^2 + v_y^2 + v_z^2$, we can rewrite the formula for the distribution of velocity in a gas (remember that we are simply writing v for the center-of-mass speed v_{cm}):

$$P(v_x, v_y, v_z) \propto \left[e^{-\frac{\frac{1}{2}Mv_x^2}{kT}}dv_x\right]\left[e^{-\frac{\frac{1}{2}Mv_y^2}{kT}}dv_y\right]\left[e^{-\frac{\frac{1}{2}Mv_z^2}{kT}}dv_z\right]$$

The distribution for each velocity component is a bell-shaped curve, called a "Gaussian." In Figure 10.38 we show the distribution of the x component of velocity for helium at room temperature.

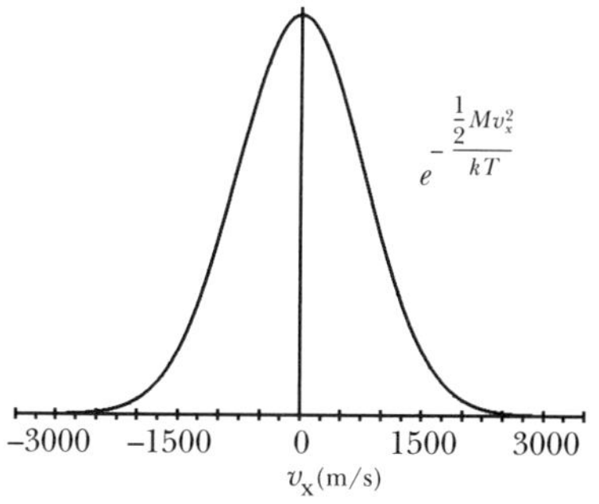

Figure 10.38 Distribution of the x component of velocity for helium at room temperature.

? Judging from this graph, what is the average value for v_x? What are the average values of v_y and v_z? Why are these results plausible?

Evidently the average value for the x component of the velocity is zero. This is reassuring, because a molecule is as likely to be headed to the right as it is to be headed to the left. Similarly, the average values for the y and z components of velocity are zero.

It would not be particularly surprising to find a gas molecule with an x component of velocity such that $\frac{1}{2}Mv_x^2 \approx kT$. On the other hand, it would be very surprising to find a gas molecule with a value of v_x many times larger. As usual, kT sets the scale for thermal phenomena.

Distribution of speeds in a gas

By converting from rectangular coordinates to "spherical" coordinates, the velocity-component distribution can be converted into a speed distribution. We won't go into the details, but the main idea is that the volume element in rectangular coordinates in "velocity space" $dv_x dv_y dv_z$ turns into the volume of a spherical shell with surface area $4\pi v^2$, and thickness dv, as shown in Figure 10.39.

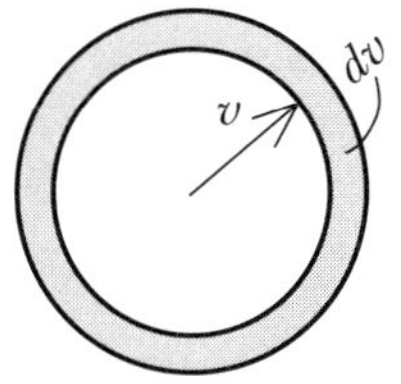

Figure 10.39 The volume of a shell with radius v and thickness dv is $4\pi v^2 dv$.

Transferring to spherical coordinates, we have the following:

$$e^{-\frac{\frac{1}{2}Mv^2}{kT}}dv_x dv_y dv_z = e^{-\frac{\frac{1}{2}Mv^2}{kT}}4\pi v^2 dv$$

The only thing missing is a "normalization" factor in front to make the integral over all speeds from 0 to infinity be equal to 1.0 (since our one molecule must have a speed somewhere in that range). Here is the "Maxwell-Boltzmann" distribution for a low-density gas:

MAXWELL-BOLTZMANN SPEED DISTRIBUTION (LOW-DENSITY GAS)

$$P(v) = 4\pi\left(\frac{M}{2\pi kT}\right)^{\frac{3}{2}} v^2 e^{-\frac{\frac{1}{2}Mv^2}{kT}}$$

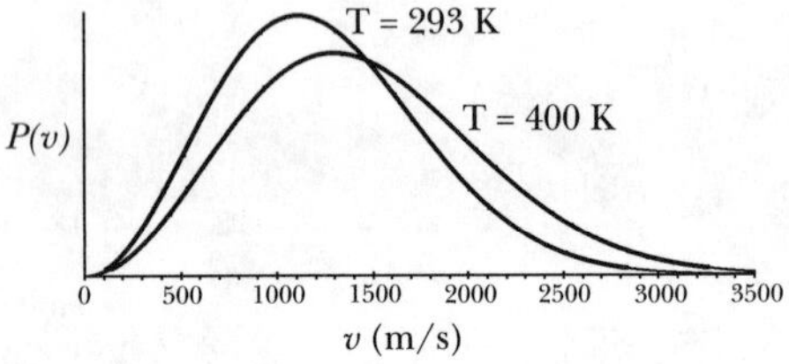

Figure 10.40 Distribution of speeds of helium atoms at two temperatures.

The probability that a molecule of a gas have a center-of-mass speed within the range v to $v+dv$ is given by $P(v)dv$. The distribution function for helium is shown for two different temperatures in Figure 10.40. You can see that the average speed of a helium atom is predicted to be about 1200 m/s at room temperature (20° Celsius, which is 293 Kelvin). The prediction was first made by Maxwell, in the mid-1800's, long before the development of quantum mechanics. In retrospect, this worked because the quantum levels of the translational kinetic energy are so close together as to be almost a continuum, so the classical model is a good approximation.

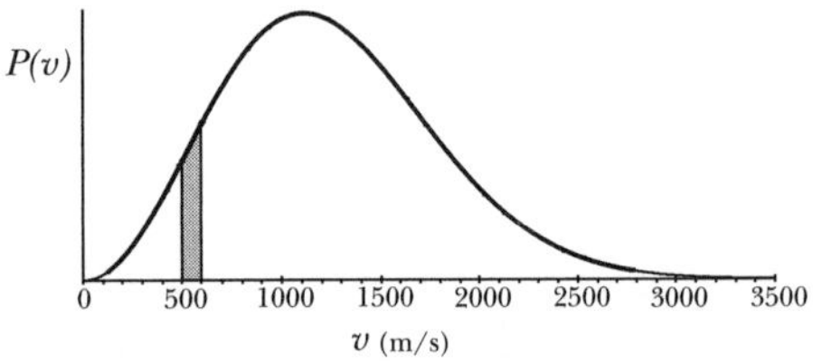

Figure 10.41 Fraction of helium atoms at 293 K with speeds between 500 and 600 m/s.

The average number of helium atoms in a container that have speeds in the range of 415.4 m/s to 415.7 m/s can be calculated by evaluating Maxwell's formula with v = 415.4 m/s and multiplying by dv = 0.3 m/s, which gives the probability of finding one molecule in this speed range. If there are N atoms in the container, multiplying by N gives the total number of molecules likely to be in that speed range at any given instant. The area of the vertical slice shown in Figure 10.41 represents the fraction of helium atoms that are likely to have speeds between 500 m/s and 600 m/s. At higher temperatures the distribution shifts to higher speeds (Figure 10.40).

10.9.3 Measuring the distribution of speeds in a gas

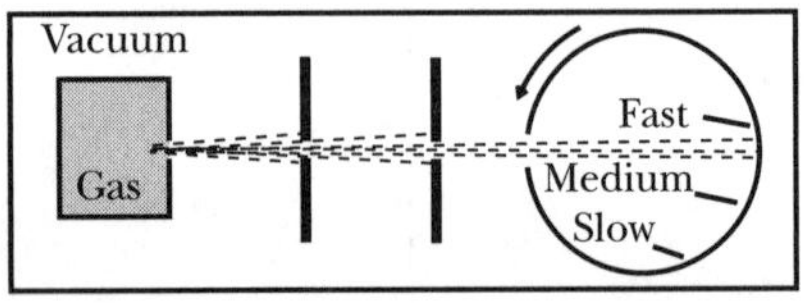

Figure 10.42 Apparatus for measuring the distribution of speeds of gas molecules.

The actual distribution of speeds of molecules in any gas can be measured by an ingenious experiment. Make a tiny hole in a container of gas and let the molecules leak out into a vacuum (that is, a region from which the air is continually pumped out). The moving molecules run through collimating holes and then through a slot in a rapidly rotating drum and strike a row of devices at the other side of the drum that can detect gas molecules (Figure 10.42). Such measurements confirm the Maxwell prediction.

Fast molecules strike soonest, followed by medium-speed molecules, and finally slow molecules, so molecules with different speeds strike at different locations along the rotating drum. The number of molecules striking various locations along the drum is a direct indication of the distribution of speeds of molecules in the gas. (A correction has to be applied to these data to obtain the speed distribution inside the gas, because high-speed molecules emerge through the hole more frequently than low-speed molecules do, even if their numbers are the same in the interior of the gas.)

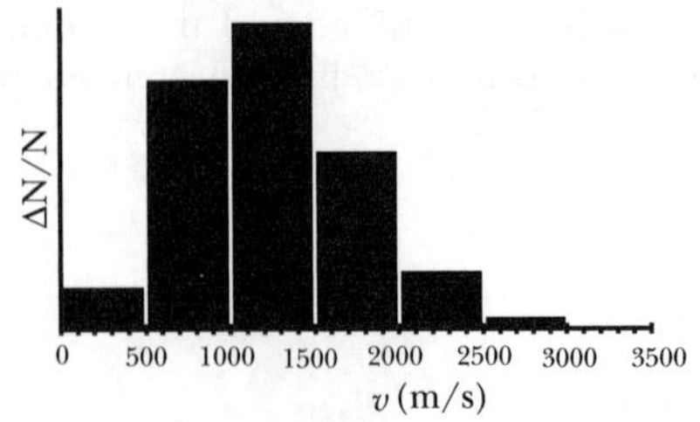

Figure 10.43 Distribution showing fraction of atoms with speeds in a given range.

How might we display the results of this experiment? We could present the data as a graph (Figure 10.43) of the fraction $\Delta N/N$ of the N helium atoms that have speeds between 0 and 500 m/s, between 500 and 1000 m/s, between 1000 and 1500 m/s, etc.

10.9.4 Average translational kinetic energy in a gas

For a gas confined inside a container, quantum mechanics predicts that the kinetic energy $\frac{1}{2}Mv^2$ is quantized, but if the container is of ordinary macroscopic size, the energy quanta are extremely small compared to the average kinetic energy, even at very low temperatures. Most gases liquefy before the temperature drops so low as to invalidate the pre-quantum analysis for the velocity distribution.

Whenever the energy has a term containing a square of a position or momentum component, such as x^2 or p_x^2, pre-quantum theory predicts an associated average energy of $\frac{1}{2}kT$. This result is valid in quantum mechanics for high temperatures, where kT is large compared to the quantum energy spacing. This result follows from using an integral to take the average value of a quadratic term when the distribution is a Gaussian involving that quadratic term. That is, using integral tables (or integrating by parts), you can show that the average value of a quadratic term is $\frac{1}{2}kT$:

$$\overline{w^2} = \frac{\int_0^\infty w^2 e^{-\frac{w^2}{kT}} dw}{\int_0^\infty e^{-\frac{w^2}{kT}} dw} = \tfrac{1}{2}kT$$

The average value is obtained by weighting the values of w^2 by the probability of finding that value; the denominator takes care of normalizing the distribution properly.

As an example, when we did quantum-based calculations for the Einstein solid, we found that at high temperatures the heat capacity per atom was $3k$, which implies an average energy of $3kT$. We modeled an atomic oscillator as three one-dimensional oscillators, each having an average energy of $\frac{1}{2}kT$ corresponding to kinetic energy $p^2/(2M)$ and another $\frac{1}{2}kT$ for the potential energy $\frac{1}{2}k_s x^2$. This is a total of six quadratic terms, implying an average energy per atom of $6(\frac{1}{2}kT)$, or $3kT$. We therefore expect the heat capacity per atom in a solid at high temperature to be $3k$, which is indeed what is observed.

HIGH-TEMPERATURE AVERAGE ENERGY

If $kT >>$ the energy quantum, the average energy associated with a quadratic energy term is $\frac{1}{2}kT$.

The number of quadratic terms in the expression for the energy is often called the "degrees of freedom."

? What is the average translational kinetic energy $\frac{1}{2}M\overline{v^2}$ in terms of kT? (A bar over a quantity means "average value.") What is the associated contribution to the heat capacity of the gas?

Since there are three quadratic terms in the translational kinetic energy, $K_{trans} = \frac{1}{2}mv_x^2 + \frac{1}{2}mv_y^2 + \frac{1}{2}mv_z^2$, the average value is three times $\frac{1}{2}kT$. We have the following important result:

AVERAGE K_{trans} FOR AN IDEAL GAS

$$\overline{K}_{trans} = \frac{1}{2}M\overline{v^2} = \frac{3}{2}kT$$

The contribution to the heat capacity is $\frac{3}{2}k$.

Root-mean-square speed

The square root of $\overline{v^2}$ is called the "root-mean-square" or "rms" speed:

$$v_{rms} \equiv \sqrt{\overline{v^2}}$$

With this definition of v_{rms}, we write $\frac{1}{2}m\overline{v^2} = \frac{1}{2}mv_{rms}^2 = \frac{3}{2}kT$.

Ex. 10.15 Calculate v_{rms} for a helium atom in the room you're in (whose temperature is probably about 293 K).

Ex. 10.16 Calculate v_{rms} for a nitrogen molecule (N_2; molecular mass 28) in the room you're in (whose temperature is probably about 293 K). Air is about 80% nitrogen.

10.9.5 Average speed vs. rms speed

The way you find the average speed of molecules in a gas is to weight the speed by the number of molecules that have that speed, and divide by the total number of molecules:

$$\bar{v} = \frac{N_1 v_1 + N_2 v_2 + \text{etc.}}{N} = \frac{N_1}{N} v_1 + \frac{N_2}{N} v_2 + \text{etc.}$$

For a continuous distribution of speeds this turns into an integral, where the weighting factors are given by the Maxwell speed distribution, which gives the probability that a molecule have a center-of-mass speed in the range v to $v+dv$. Using integral tables one finds this:

$$\bar{v} = \int_0^\infty 4\pi\left(\frac{m}{2\pi kT}\right)^{\frac{3}{2}} e^{-\frac{\frac{1}{2}mv^2}{kT}} (v) v^2 \, dv = \sqrt{\frac{8}{3\pi}}\sqrt{\frac{3kT}{m}} = 0.92\, v_{rms}$$

This shows that the average speed is smaller than the rms speed ($0.92 v_{rms}$). This same calculational scheme can be used to determine any average. For example, if you calculate the average value of v^2 by this method, you find $v^2 = 3kT/m$, as expected.

On page 426 we commented that the average speed $\bar{v}$ of helium atoms at room temperature is about 1200 m/s. In Ex. 10.15 on page 428 you found that v_{rms} at room temperature is 1360 m/s. This is an example of the fact that the rms speed is higher than the average speed.

Ex. 10.17 The rms speed is somewhat higher than the average speed due to the averaging of squared speeds. Calculate the average of the numbers 1, 2, 3, and 4, then calculate the rms average (the square root of the average of their squares), and show that the rms average is larger than the simple average.

10.9.6 Application: Retaining a gas in the atmosphere

Since $v_{rms} = \sqrt{3k(T/m)}$, helium atoms typically travel much faster than nitrogen molecules in our atmosphere, due to the small mass of the helium atoms. Some few helium atoms will be going much faster than the average and may attain a high enough speed to escape from the Earth entirely (escape velocity from the Earth is about 1.1×10^4 m/s). This leads to a continuous leakage of high-speed helium atoms and explains why there is essentially no helium in our atmosphere, and no hydrogen either (Figure 10.44).

Figure 10.44 Low mass atoms or molecules in our atmosphere typically have speeds high enough to escape from the Earth entirely.

? Why doesn't the Moon have any atmosphere at all?

The Moon's gravitational field is so weak that escape speed is quite small, and all common gases can escape, not just hydrogen and helium.

Ex. 10.18 Calculate the escape speed from the Moon and compare with typical speeds of gas molecules. The mass of the Moon is 7×10^{22} kg, and its radius is 1.75×10^{6} m.

10.9.7 Application: Speed of sound

Sound waves in a gas consist of propagation of variations in density, and the fundamental mechanism for this kind of wave propagation involves collisions between neighboring molecules, whose speeds are proportional to v_{rms} (and roughly comparable to the speed of sound). For example, compare the v_{rms} for nitrogen, 496 m/s, which you calculated in Ex. 10.16 (page 428) with the speed of sound in air (which is mostly nitrogen) at 293 K, which is measured to be 344 m/s. (You may know the approximate rule that a 1-second delay between lightning and thunder indicates a distance of about 1000 feet, which is about 300 m.)

Figure 10.45 Periodic variations in gas density make a sound wave, which travels through a gas with speed v (the speed of sound).

Ex. 10.19 Should the speed of sound in air increase or decrease with increasing temperature? What percentage change would result from doubling the absolute temperature? (This effect is readily observed by measuring the speed of sound in a gas as a function of absolute temperature. The excellent agreement between theory and experiment provides additional evidence for our understanding of gases.)

10.9.8 Vibrational energy in a diatomic gas molecule

We repeat that the distribution of velocity, position, vibrational energy, and rotational energy is this:

$$P(E) \propto \left[e^{-\frac{K_{\text{trans}}}{kT}} dv_x dv_y dv_z\right]\left[e^{-\frac{Mgy}{kT}} dxdydz\right]\left[e^{-\frac{E_{\text{vib}}}{kT}}\right]\left[e^{-\frac{E_{\text{rot}}}{kT}}\right]$$

We have treated the distribution of velocity and position. Next we discuss the distribution of vibrational energy:

$$P(E_{vib}) \propto e^{-\frac{E_{\text{vib}}}{kT}}$$

For a monatomic gas such as helium or argon, there is no vibrational energy term. But for a diatomic molecule such as N_2 or O_2 or HCl, the vibrational energy is

$$E_{vib} = \frac{p_1^2}{2m_1} + \frac{p_2^2}{2m_2} + \frac{1}{2}k_s s^2$$

where k_s is the effective spring stiffness corresponding to the interatomic electric force (not to be confused with k, the Boltzmann constant), and s is

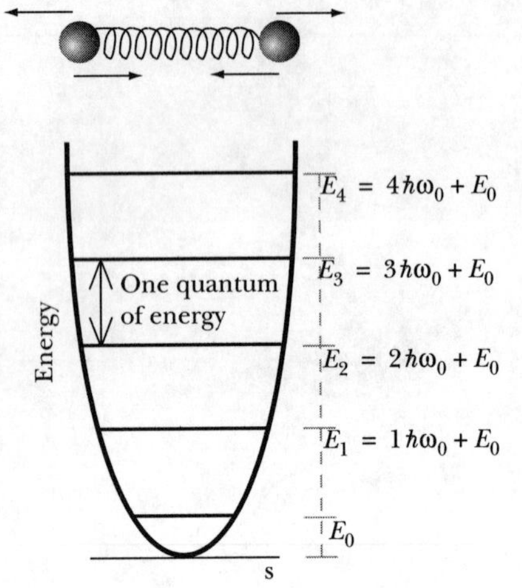

Figure 10.46 The quantized vibrational energy levels of a diatomic oscillator.

the stretch of the interatomic bond. This formula for the vibrational energy is essentially the same as the formula for a one-dimensional spring-mass oscillator, which we have studied in detail (Figure 10.46).

Actually, since $p_1 = p_2$ for momenta p_1 and p_2 relative to the center of mass, we can write

$$\frac{p_2^2}{2m_2} = \frac{p_1^2}{2m_2} = \left(\frac{m_1}{m_2}\right)\left(\frac{p_1^2}{2m_1}\right)$$

$$\frac{p_1^2}{2m_1} + \left(\frac{m_1}{m_2}\right)\frac{p_1^2}{2m_1} + \frac{1}{2}k_s s^2 = \left(1 + \frac{m_1}{m_2}\right)\frac{p_1^2}{2m_1} + \frac{1}{2}k_s s^2$$

Note that this equation has a form exactly like that for a spring-mass oscillator, with a different mass.

Earlier we modeled a solid as a large number of isolated three-dimensional atomic oscillators (each corresponding to three one-dimensional oscillators, because there are spring-like interatomic forces on an atom from neighboring atoms in all directions). This is an overly simplified model of a solid, because in a solid the atoms interact with each other. For example, if an atom moves to the left this affects the atoms to the right and to the left.

In a low-density gas however the vibrational oscillators really are independent of each other, because the gas molecules aren't even in contact with each other except when they happen to collide. So the analysis we carried out for the Einstein model of a solid applies even better to the vibrational portion of the energy in a real gas than it does to a real solid.

Among the results that apply immediately are that the heat capacity associated with the vibrational energy of one oscillator is k at high temperatures. The heat capacity decreases at very low temperatures where kT is small compared to the energy spacing of the quantized oscillator energies. Just as it was a surprise when the heat capacity of metals was found to decrease at low temperatures, there was a similar surprise in the measurements of the heat capacity of diatomic gases at low temperatures, because the contribution of the vibrational motion vanished.

? There are two quadratic terms in the vibrational energy (kinetic and spring). Therefore, at high temperatures what is the average vibrational energy in terms of kT? What is the associated contribution to the heat capacity of the gas?

At high temperatures the vibrational motion of a diatomic molecule (a one-dimensional oscillator) is expected to have an average energy that is approximately $2(\frac{1}{2}kT) = kT$, and contributes k to the heat capacity.

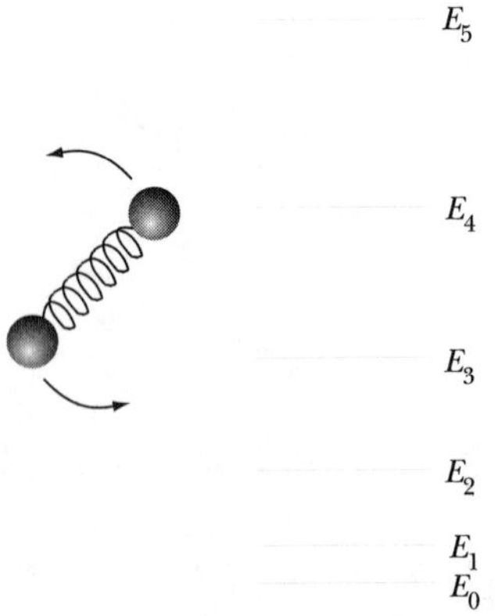

Figure 10.47 Rotational kinetic energy levels of a diatomic molecule.

10.9.9 Rotational energy in a diatomic gas molecule

Finally we consider the rotational-energy portion of the distribution formula:

$$P(E_{rot}) \propto e^{-\frac{E_{rot}}{kT}}$$

For a monatomic gas there is no rotational term. But for a diatomic gas there is rotational kinetic energy associated with rotation of the "dumbbell" consisting of the two nuclei. The rotational kinetic energy is this (Figure 10.47):

$$\frac{1}{2}I\omega_x^2 + \frac{1}{2}I\omega_y^2 = \frac{L_{rot,x}^2}{2I} + \frac{L_{rot,y}^2}{2I}$$

This corresponds to rotational angular momenta $L_{rot,x}$ and $L_{rot,y}$ about the x axis and about the y axis. Note that only the nuclei contribute significantly to the rotational kinetic energy, because the electrons have much less mass.

Rotation about the z axis connecting the two nuclei is irrelevant, for a somewhat subtle reason. The angular momentum is quantized, which leads to energy quantization. For rotations around the z axis the energy is $L^2_{rot,z}/(2I)$, where $L_{rot,z}$ is the z component of the rotational angular momentum. Since I for the tiny nuclei is extremely small compared with the moment of inertia about the x and y axes of the diatomic molecule, the rotational energy of the associated first excited state is enormous, and this state is not excited at ordinary temperatures (Figure 10.48).

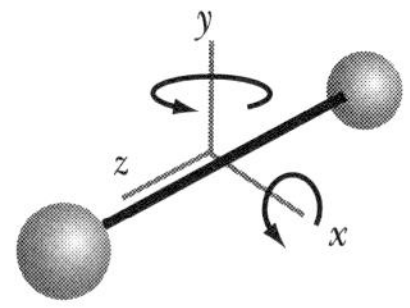

Figure 10.48 A diatomic molecule can rotate around the x or the y axis. We say there are two rotational degrees of freedom.

Since there are two quadratic energy terms associated with rotation, we conclude that at high temperatures the rotational motion of a diatomic molecule has an average energy that is approximately $2(\frac{1}{2}kT) = kT$, and contributes k to the heat capacity.

The spacing between quantized rotational energies for a diatomic molecule is even smaller than the vibrational energy quantum. As a result, the gas must be cooled to a very low temperature before the pre-quantum results become invalid. Many gases liquefy before this low-temperature regime is reached.

10.9.10 Heat capacity of a gas

We are now in a position to discuss the heat capacity of a gas as a function of temperature. This property is important in calculating the thermal interactions of a gas. Historically, measurements of the heat capacity of gases were important in testing theories of statistical mechanics. We'll concentrate on calculating c_v, the heat capacity at constant volume (meaning no mechanical work is done on the gas). The associated experiment would be to add a known amount of energy to a gas in a rigid container and measure the temperature rise of the gas.

Consider the average energy of a diatomic molecule such as N_2 or HCl, consisting of the translational kinetic energy associated with the motion of the center of mass, plus the vibrational energy, plus the rotational energy, plus the gravitational energy of the molecule plus the Earth:

$$\underbrace{\left\{\frac{\overline{p_x^2}}{2m} + \frac{\overline{p_y^2}}{2m} + \frac{\overline{p_z^2}}{2m}\right\}}_{\text{Translation}} + \underbrace{\left\{\left(1 + \frac{m_1}{m_2}\right)\frac{\overline{p_1^2}}{2m_1} + \frac{1}{2}k_s\overline{s^2}\right\}}_{\text{Vibration}} + \underbrace{\left\{\frac{\overline{L^2_{rot,x}}}{2I} + \frac{\overline{L^2_{rot,y}}}{2I}\right\}}_{\text{Rotation}} + \underbrace{Mg\bar{y}_{cm}}_{\text{Gravitation}}$$

? At high temperature, what should the heat capacity at constant volume be?

There are seven quadratic terms in the expression for the energy, so if the temperature is very high, we expect an average energy of $7(\frac{1}{2}kT)$ and a heat capacity at constant volume $c_v = \frac{7}{2}k$.

The discrete energy levels for a diatomic molecule include electronic, vibrational, and rotational energy levels (Figure 10.49). The electronic energy levels correspond to particular configurations of the electron clouds, and the spacing between these levels is typically 1 eV or more. Since at room temperature kT is about $\frac{1}{40}$ eV, the electronic levels do not contribute to the heat capacity at ordinary temperatures. The vibrational energy spacing is much smaller, and the rotational energy spacing is smaller still. A diatomic gas at high temperature has a band-like spectrum, since transitions between electronic levels may be accompanied by one or more vibrational or rotational quanta.

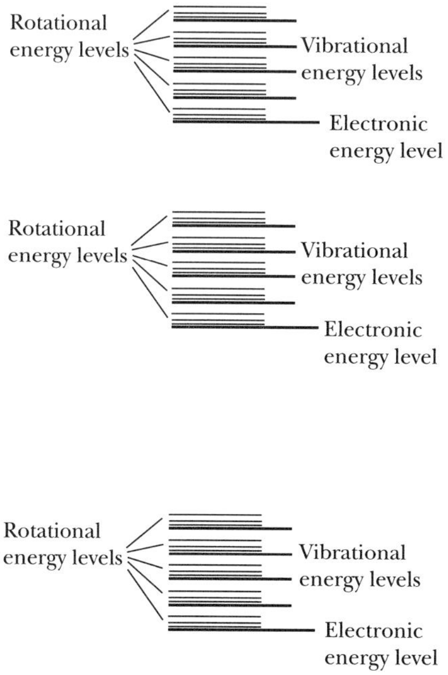

Figure 10.49 Electronic, vibrational, and rotational energy levels for a diatomic molecule.

What about *Mgy*?

What about the Mgy_{cm} term in the energy? Remember that we are trying to predict the heat capacity of a gas at constant volume (so that no mechanical work is done on the system). We could measure c_v by having a known amount of thermal transfer of energy Q into a closed container of gas and observing the temperature rise of the gas. During this process the height y_{cm} of the center of mass of the gas in the container hardly changes, so the Mgy term doesn't contribute to the heat capacity.

Actually there is an extremely small increase in y_{cm} of the gas in the container due to the temperature dependence of the height distribution (page 424), but the associated contribution to the heat capacity is negligible.

Heat capacity *vs.* temperature

Figure 10.50 is a graph of the heat capacity (at constant volume) of a diatomic gas as a function of temperature. In the following exercises, see whether you can explain these values of the heat capacity in terms of our analyses of the various contributions to the energy of a diatomic gas molecule.

Remember that kT is about $\frac{1}{40}$ eV at room temperature.

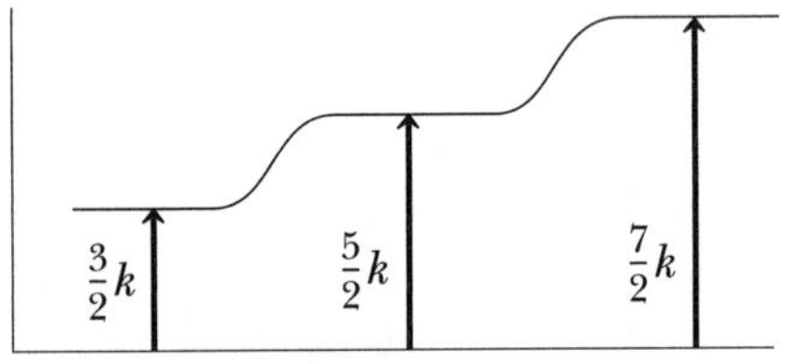

Figure 10.50 Heat capacity c_v of a diatomic gas.

Ex. 10.20 What is the heat capacity per molecule (at constant volume) of a monatomic gas such as helium or neon? Why doesn't the heat capacity depend on temperature?

Ex. 10.21 At high temperatures, what is the heat capacity (at constant volume) per molecule of a diatomic gas such as oxygen or nitrogen?

Ex. 10.22 Suppose we lower the temperature of a diatomic gas to a point where kT is small compared to the energy of the first excited vibrational state, but still large compared to the energy of the first excited rotational state. Now what is the heat capacity (at constant volume)?

Ex. 10.23 Lower the temperature even more, so that kT is small compared to the energy of the first excited rotational state (some gases liquefy before this low temperature is reached). What is the heat capacity (at constant volume) at this low temperature?

Ex. 10.24 As the temperature is decreased, the heat capacity (at constant volume) for H_2 eventually decreases to $\frac{3}{2}k$, before the gas liquefies. Does this transition to $c_v = \frac{3}{2}k$ happen at a higher or lower temperature for D_2 (deuterium, whose nuclei each contain a proton plus a neutron)? Why? (Hint: Remember that it is the rotational angular momentum that is quantized.)

10.10 Summary

Fundamental physical principles

Fundamental assumption of statistical mechanics: Each arrangement of the same total energy within a system is equally probable.

Two blocks in thermal contact evolve to that division of energy that has associated with it the largest number of ways $\Omega = \Omega_1\Omega_2$ of arranging the total energy among the atoms (largest number of microstates for a macrostate of given energy).

Second law of thermodynamics: The entropy of a closed system never decreases. Only in a reversible process does the entropy of a closed system stay constant.

New concepts

Entropy: $S \equiv k\ln\Omega$, where $k = 1.4\times10^{-23}$ J/K

Temperature: $\frac{1}{T} = \frac{\partial S}{\partial E}$

Adding small Q to a system raises entropy by $\Delta S = \frac{Q}{T}$

Heat capacity per atom: $C = \frac{\Delta E_{\text{atom}}}{\Delta T}$, where $\Delta E_{\text{atom}} = \frac{\Delta E_{\text{system}}}{N_{\text{atoms}}}$

Results

 # WAYS TO ARRANGE *q* QUANTA AMONG *N* ONE-DIMENSIONAL OSCILLATORS

$$\Omega = \frac{(q+N-1)!}{q!(N-1)!}$$

The Boltzmann distribution for the energy of a microscopic system in thermal contact with a macroscopic system:

$$e^{-\frac{\Delta E}{kT}}$$

For each energy term involving a quadratic term such as x^2 or p_x^2 ("degree of freedom"), if the average energy is large compared to kT the average energy is $\frac{1}{2}kT$ and the contribution to the heat capacity is $\frac{1}{2}k$. This contribution decreases at low temperatures. At high temperatures, heat capacity per atom in a solid $\approx 3k$.

Distribution of energy for a diatomic gas:

$$P(E) \propto \left[e^{-\frac{K_{\text{trans}}}{kT}}dv_x dv_y dv_z\right]\left[e^{-\frac{Mgy}{kT}}dxdydz\right]\left[e^{-\frac{E_{\text{vib}}}{kT}}\right]\left[e^{-\frac{E_{\text{rot}}}{kT}}\right]$$

Speed distribution for a gas molecule:

$$P(v) = 4\pi\left(\frac{M}{2\pi kT}\right)^{\frac{3}{2}} e^{-\frac{\frac{1}{2}Mv^2}{kT}} v^2$$

10.11 Review questions

Arranging things

RQ 10.1 List explicitly all the ways of arranging 2 quanta among 4 one-dimensional oscillators.

A formula for the number of arrangements

RQ 10.2 How many different ways are there to get 5 heads in 10 throws of a true coin? How many different ways are there to get no heads in 10 throws of a true coin?

RQ 10.3 How many different ways are there to arrange 4 quanta among 3 atoms in a solid?

Thermal equilibrium

RQ 10.4 Energy conservation for two blocks in contact with each other is satisfied if all the energy is in one block and none in the other. Would you expect to observe this distribution in practice? Why or why not?

RQ 10.5 What is the advantage of plotting the (natural) logarithm of the number of ways of arranging the energy among the many atoms (natural logarithm of the number of microstates)?

RQ 10.6 Which has a higher temperature, a system whose entropy changes rapidly with increasing energy, or one whose entropy changes little with increasing energy?

Heat capacity

RQ 10.7 Explain why it is a disadvantage for some purposes that the heat capacity of all materials decreases at low temperatures.

RQ 10.8 Sketch and label graphs of heat capacity *vs.* temperature for hydrogen gas (H_2) and oxygen gas (O_2), using the same temperature scale. Explain briefly.

Second law of thermodynamics

RQ 10.9 Two blocks with different temperatures had entropies of 10 J/K and 35 J/K before they were brought in contact. What can you say about the entropy of the combined system after the two came in contact with each other?

The Boltzmann distribution

RQ 10.10 Explain qualitatively the basis for the Boltzmann distribution. Never mind the details of the math for the moment. Focus on the trade-offs involved with giving energy to a single oscillator *vs.* giving that energy to a large object.

RQ 10.11 Many chemical reactions proceed at rates that depend on the temperature. Discuss this from the point of view of the Boltzmann distribution.

Heat capacity

RQ 10.12 Which has more internal energy at room temperature—a mole of helium or a mole of air?

Distributions

RQ 10.13 Figure 10.51 shows the distribution of speeds of atoms in a particular gas at a particular temperature. Approximately what is the average speed? Is the rms (root-mean-square) speed bigger or smaller than this? Approximately what fraction of the molecules have speeds greater than 1000 m/s?

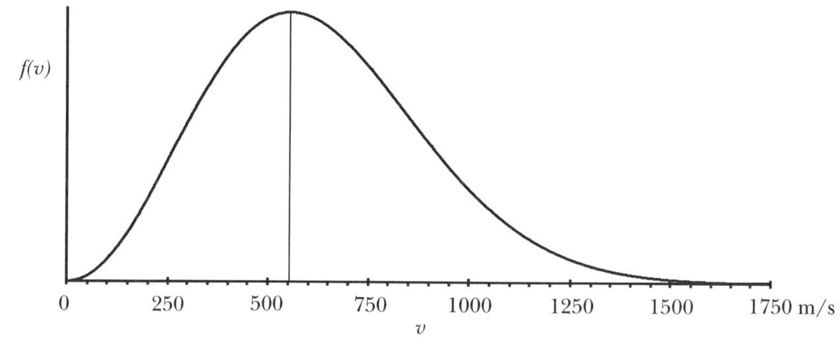

Figure 10.51 Distribution of speeds of atoms in a gas at a particular temperature (RQ 10.13).

10.12 Homework problems

It is important to work through the first four problems in detail in order to make the ideas concrete.

Problem 10.1 Probability distribution

(a) Model a system consisting of two atoms (three oscillators each), among which 4 quanta of energy are to be distributed. Write a program to display a histogram showing the total number of possible microstates of the two-atom system *vs.* the number of quanta assigned to atom 1. Compare your calculations and histogram to the one on page 405 (you should get the same distribution).

(b) Model a system consisting of two solid blocks, block 1 containing 300 oscillators and block 2 containing 200 oscillators. Find the possible distributions of 100 quanta among these blocks, and plot number of microstates *vs.* number of quanta assigned to block 1. Compare your calculations and histogram to those on page 408. Determine the distribution of quanta for which the probability is half as large as the most probable 60-40 distribution.

(c) Do a series of calculations distributing 100 quanta between two blocks whose total number of oscillators is 500, but whose relative number of atoms varies. For example, consider equal numbers of oscillators, and ratios of 2:1, 5:1, etc. Describe your observations.

Problem 10.2 Natural logarithm (ln) of ways to arrange energy

Start with your solution to Problem 10.1 (b). For the same system of two blocks, with $N_1 = 300$ oscillators and $N_2 = 200$ oscillators, plot $\ln(\Omega_1)$, $\ln(\Omega_2)$, and $\ln(\Omega_1\Omega_2)$, for q_1 running from 0 to 100 quanta. Your graph should look like the one in Figure 10.23. Determine the maximum value of $\ln(\Omega_1\Omega_2)$ and the value of q_1 where this maximum occurs. What is the significance of this value of q_1?

Problem 10.3 Temperature

Modify your calculations from Problem 10.2 to plot the temperature in kelvins of block 1 as a function of the number of quanta q_1 present in the first block. On the same graph plot the temperature in kelvins of block 2 as a function of q_1 (of course, $q_2 = q_{tot} - q_1$).

In order to plot the temperature in kelvins, you must determine the values of ΔE and ΔS that correspond to a one-quantum change in energy. Consider the model we are using. The energy of one quantum, in joules, is $\Delta E = \hbar\sqrt{k_s/m}$. The increment in entropy corresponding to this increment in energy is $\Delta S = k\Delta(\ln\Omega)$. Assume that the blocks are made of aluminum. In Problem 3.2 (page 82) you made an estimate of the interatomic spring constant k_s for aluminum. We show later, on page 417, that the effective k_s for oscillations in the Einstein solid is expected to be about 4 times the value obtained from measuring Young's modulus.

What is the significance of the value of q_1 (and of q_2) where the temperature curves for the two blocks cross (the temperatures are equal)?

Problem 10.4 Heat capacity

Modify your analysis of Problem 10.3 to determine the heat capacity as a function of temperature for a single block of metal. In order to see all of the important effects, consider a single block of 35 atoms (105 oscillators) with up to 300 quanta of energy. Note that in this analysis you are calculating quantities for a *single* block, not two blocks in contact.

To make specific comparison with experimental data, consider the cases of aluminum (Al) and lead (Pb). For each metal, plot the theoretical heat capacity C per atom *vs.* T *(K)*, with dots showing the actual experimental data (converted to the same basis, J/K per atom).

Adjust the interatomic spring stiffness k_s until your calculations approximately fit the experimental data given in the adjoining table. (You will need to convert to a per-atom value.) In trying to fit the data, ignore the lowest-temperature data, where electrons in the metal make a significant contribution that is not included in the Einstein model.

What value of k_s gives a good fit? Compare with the estimated values of k_s obtained in Chapter 3, but remember that those estimated values are probably 4 times smaller than the effective spring constant for oscillations.

Show from your graph that the high-temperature limit of the heat capacity is about $3k$ per atom. The rise above this limit at high temperatures is probably due to the fact that the assumed uniform spacing of the quantized oscillator energy levels isn't a good approximation for highly excited states. See Figure 6.18 on page 280.

T (K)	C, Al (J/K/mole)	C, Pb (J/K/mole)
20	0.23	11.01
40	2.09	19.57
60	5.77	22.43
80	9.65	23.69
100	13.04	24.43
150	18.52	25.27
200	21.58	25.87
250	23.25	26.36
300	24.32	26.82
400	25.61	27.45

Problem 10.4 Heat capacity data.

Problem 10.5 Square root of N

For some examples of your choice, demonstrate by carrying out actual computer calculations that the "square-root" rule holds true for the fractional width of the peak representing the most probable arrangements of the energy. Warning: Check to see what is the largest number you can use in your computations; some programs or programming environments won't handle numbers bigger than 10^{307}, for example, and larger numbers are treated as "infinite."

Problem 10.6 Copper and aluminum blocks

A 50-gram block of copper (one mole has a mass of 63.5 grams) at a temperature of 35° C is put in contact with a 100-gram block of aluminum (molar mass 27 grams) at a temperature of 20° C. The blocks are inside an insulated enclosure, with little contact with the walls. At these temperatures, the high-temperature limit is valid for the heat capacity. Calculate the final temperature of the two blocks. Do NOT look up the heat capacities of aluminum and copper; you should be able to figure them out on your own.

Problem 10.7 Low-temperature heat capacity of copper

Young's modulus for copper is measured by stretching a copper wire to be about 1.2×10^{11} N/m^2. The density of copper is about 9 grams/cm^3, and the mass of a mole is 63.5 grams. Starting from a very low temperature, use these data to estimate roughly the temperature T at which we expect the heat capacity for copper to approach $3k$. Compare your estimate with the data shown on page 418.

Problem 10.8 Experiment: Measurement of the heat capacity of water
The goal of this experiment is to understand, in a concrete way, what heat capacity is and how it can be measured. You will need a microwave oven, a styrofoam coffee cup, and a clock or watch.

In the range of temperature where water is a liquid (0 C to 100 C), it is approximately true that it takes 4.2 J of energy (1 calorie) to raise the temperature of 1 gram of water through 1 degree Kelvin. To measure this heat capacity of water, we need some way to raise a known mass of water from a known initial temperature to a final temperature which can also be measured, while we keep track of the energy supplied to the water. One way to do this, as discussed in the text, is to warm up water in a well-insulated container within which a heater, whose power output is known, warms up the water. In this experiment, instead of a well-insulated box with a heater, we will use microwave power which preferentially warms up water by exciting rotational modes of the water molecules, as opposed to burners or heaters which warm up water in a pan by first warming up the pan.

Use a styrofoam coffee cup of known volume in which water can be warmed up. The density of water is 1 gram/cm^3.

> It is a good idea not to fill the cup completely full, because this makes it more likely to spill.

One method of recording the initial temperature of the water is to get water from the faucet and wait for it to equilibrate with room temperature (which can either be read off a thermostat or estimated based on past experience). After waiting about a half hour for this to happen, place the cup in the microwave oven and turn on the oven at maximum power. The cup needs to be watched as it warms up, so that when the water starts to boil, the elapsed time can be noted accurately.

> BE CAREFUL! A styrofoam cup full of hot liquid can buckle if you hold it near the rim. Hold the cup near the bottom. If the cup is full, do not attempt to move the cup while the water is hot. A spill can cause a painful burn.

On the back of the microwave oven (or inside the front door), there is usually a sticker with specifications which says "Output Power = ...Watts" which can be used to calculate the energy supplied. If there is no indication, use a typical value of 600 watts for a standard microwave oven. Using all the quantities measured above and knowing the temperature interval over which you have warmed up the water, you can calculate the heat capacity of water.

(a) Show and explain all your data and calculations, and compare with the accepted value for water (4.2 J/K per gram).

(b) Discuss why your result might be expected to differ from the accepted value. For each effect that you consider, state whether this effect would lead to a result that is larger or smaller than the accepted value.

Problem 10.9 Atmosphere at very high altitudes
In studying a voyage to the Moon in Chapter 2 we somewhat arbitrarily started at a height of 50 kilometers above the surface of the Earth.

(a) At this altitude, what is the density of the air as a fraction of the density at sea-level?

(b) Approximately how many air molecules are there in one cubic centimeter at this altitude?

(c) At what altitude is air density one-millionth (10^{-6}) that at sea level?

Problem 10.10 Two nanoparticles of iron
A nanoparticle consisting of four iron atoms (object 1) initially has 1 quantum of energy. It is brought into contact with a nanoparticle consisting of two iron atoms (object 2) which initially has 2 quanta of energy. The mass of one mole of iron is 56 grams. (Continued on next page.)

(a) Using the Einstein model of a solid, calculate and plot ln Ω_1 *vs.* q_1 (the number of quanta in object 1), ln Ω_2 *vs.* q_1, and ln Ω_{total} *vs.* q_1 (put all three plots on the same graph). Show your work and explain briefly.

(b) Calculate the approximate temperature of the objects at equilibrium. State what assumptions or approximations you made.

Problem 10.11 Creating a plasma

At sufficiently high temperatures, the thermal speeds of gas molecules may be high enough that collisions may ionize a molecule (that is, remove an outer electron). An ionized gas in which each molecule has lost an electron is called a "plasma." Determine approximately the temperature at which air becomes a plasma.

Problem 10.12 Heat capacity

100 joules of thermal transfer of energy are given to air in a 50-liter rigid container and to helium in a 50-liter rigid container, both initially at STP (standard temperature and pressure).

(a) Which gas experiences a greater temperature rise?

(b) What is the temperature rise of the helium gas?

Problem 10.13 Heat capacity of hydrogen

(a) Below about 80 K the heat capacity at constant volume for hydrogen gas (H_2) is $\frac{3}{2}k$ per molecule, but at higher temperatures the heat capacity increases to $\frac{5}{2}k$ per molecule due to contributions from rotational energy states. Use these observations to estimate the distance between the hydrogen nuclei in an H_2 molecule.

(b) At about 2000 K the heat capacity at constant volume for hydrogen gas (H_2) increases to $\frac{7}{2}k$ per molecule due to contributions from vibrational energy states. Use these observations to estimate the stiffness of the "spring" that approximately represents the interatomic force.

Problem 10.14 Five microscopic objects

Figure 10.52 shows a one-dimensional row of 5 microscopic objects each of mass 4×10^{-26} kg, connected by forces that can be modeled by springs of stiffness 15 N/m. These objects can move only along the x axis.

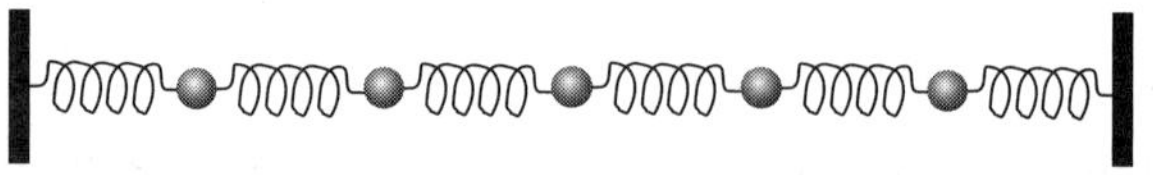

Figure 10.52 A one-dimensional row of microscopic masses and springs (Problem 10.14).

(a) Using the Einstein model, calculate the approximate entropy of this system for total energy of 0, 1, 2, 3, 4, and 5 quanta.

(b) Calculate the approximate temperature of the system when the total energy is 4 quanta.

(c) Calculate the approximate heat capacity per object when the total energy is 4 quanta.

(d) If the temperature is raised very high, what is the approximate heat capacity per object? Give a numerical value and compare with your result in part (c).

Problem 10.15 Pluto's atmosphere

In 1988 telescopes viewed Pluto as it crossed in front of a distant star. As the star emerged from behind the planet, light from the star was slightly dimmed as it went through Pluto's atmosphere. The observations indicated that the atmospheric density at a height of 50 km above the surface of Pluto is about one-third the density at the surface. The mass of Pluto is known to be about 1.5×10^{22} kg, and its radius is about 1200 km. Spectroscopic data indicate that the atmosphere is mostly nitrogen (N_2). Estimate the temperature of Pluto's atmosphere. State what approximations and/or simplifying assumptions you made.

Problem 10.16 Buckminsterfullerene

Buckminsterfullerene, C_{60}, is a large molecule consisting of sixty carbon atoms connected to form a hollow sphere. The diameter of a C_{60} molecule is about 7×10^{-10} m. It has been hypothesized that C_{60} molecules might be found in clouds of interstellar dust, which often contain interesting chemical compounds. The temperature of an interstellar dust cloud may be very low, around 3 K. Suppose you are planning to try to detect the presence of C_{60} in such a cold dust cloud by detecting photons emitted when molecules undergo transitions from one rotational energy state to another.

Approximately, what is the highest-numbered rotational level from which you would expect to observe emissions? Rotational levels are $l = 0, 1, 2, 3, ...$

Problem 10.17 A disk and brake

A box contains a uniform disk of mass M and radius R which is pivoted on a low-friction axle through its center. A block of mass m is pressed against the disk by a spring, so that the block acts like a brake, making the disk hard to turn. The box and the spring have negligible mass. A string is wrapped around the disk (out of the way of the brake) and passes through a hole in the box. A force of constant magnitude F acts on the end of the string. The motion takes place in outer space. At time t_i the speed of the box is v_i, and the rotational speed of the disk is ω_i. At time t_f the box has moved a distance x, and the end of the string has moved a longer distance d, as shown in Figure 10.53.

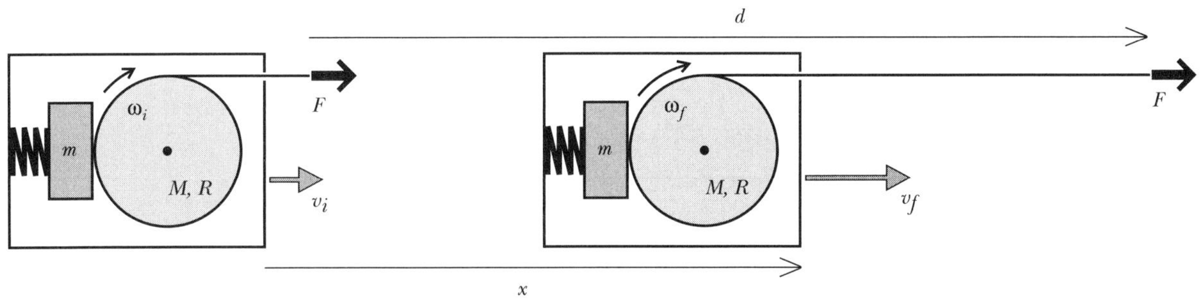

Figure 10.53 The box in Problem 10.17 shown both at its initial position, and after it has moved a distance x.

(a) At time t_f, what is the speed v_f of the box?

(b) During this process, the brake exerts a tangential friction force of magnitude f. At time t_f, what is the angular speed ω_f of the disk?

(c) At time t_f, assume you know (from part b) the rotational speed ω_f of the disk. From time t_i to time t_f, what is the increase in thermal energy of the apparatus?

(d) Suppose the increase in thermal energy in part c is 8×10^4 J. The disk and brake are made of iron, and their total mass is 1.2 kg. At time t_i their temperature was 350 K. At time t_f, what is their approximate temperature?

Problem 10.18 Coin tosses

The reasoning developed for counting microstates applies to many other situations involving probability. For example, if you flip a coin 5 times, how many different sequences of 3 heads and 2 tails are possible? Answer: 10 different sequences, such as HTHHT or TTHHH. In contrast, how many different sequences of 5 heads and 0 tails are possible? Obviously only one, HHHHH, and our formula gives $5!/[5!0!] = 1$, using the standard definition that 0! is defined to equal 1.

If the coin is equally likely on a single throw to come up heads or tails, any specific sequence like HTHHT or HHHHH is equally likely. But there is only one way to get HHHHH, while there are 10 ways to get 3 heads and 2 tails, so this is 10 times more probable than getting all heads.

Use the formula $5!/[N!(5\text{-}N)!]$ to calculate the number of ways to get 0 heads, 1 head, 2 heads, 3 heads, 4 heads, or 5 heads in a sequence of 5 coin tosses. Make a graph of the number of ways *vs.* the number of heads.

10.13 Answers to exercises

10.1 (page 405)	20
10.3 (page 407)	5.4×10^{88} years; 5.4×10^{78} times the lifetime of the Universe!
10.4 (page 412)	Looks *very* improbable; second law of thermodynamics being violated.
10.5 (page 415)	500 K
10.6 (page 415)	Liquid; yes, liquid more disordered; 1.23 J/K (note that 0° C is 273 K)
10.7 (page 417)	1.9 J/K/gram
10.8 (page 419)	25° C
10.9 (page 419)	Large rise in temperature of sink; no longer as cold as before
10.10 (page 422)	$e^{-(E_2 - E_1)/kT}$
10.12 (page 425)	About 8.4 kilometers! (about 5 miles)
10.13 (page 425)	About 4×10^{-20} m !
10.14 (page 425)	About 1/3
10.15 (page 428)	1360 m/s
10.16 (page 428)	496 m/s
10.17 (page 428)	2.50; 2.74
10.18 (page 429)	2300 m/s; not very large compared with 500 m/s
10.19 (page 429)	Increase; factor of $\sqrt{2}$, so increase about 40%
10.20 (page 432)	$\frac{3}{2}k$; translational energy quanta are very small.
10.21 (page 432)	$\frac{7}{2}k$
10.22 (page 432)	$\frac{5}{2}k$
10.23 (page 432)	$\frac{3}{2}k$
10.24 (page 432)	Lower temperature. Rotational energy is $\frac{S_x^2}{2I}$, so energy of rotational first excited state for D_2 is smaller (larger nuclear mass means larger moment of inertia, since internuclear distance about the same).

Chapter 11

The Kinetic Theory of Gases

Chapter 11

The Kinetic Theory of Gases

In this chapter we will extend our model of gases. The emphasis is on description and analysis at the atomic level, and how the atomic picture relates to large-scale, macroscopic observations of the behavior of gases. Building on our prior studies, we will address the following issues:

- The rate at which gas molecules escape through a small hole.
- How far gas molecules go between collisions.
- The dependence of pressure on temperature, and the ideal gas law.
- Energy relations in processes involving gases.

This variety of applications illustrates the usefulness of an atomic-level approach, which is known as "the kinetic theory of gases."

11.1 Gases, solids, and liquids

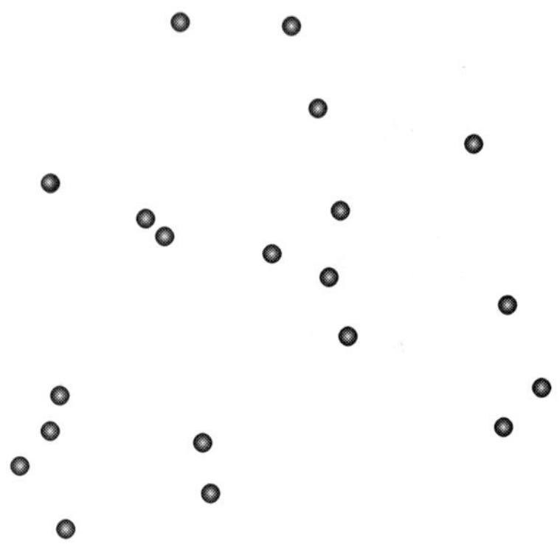

Figure 11.1 Molecules in a gas.

Remember that a gas has no fixed structure, unlike a solid. The molecules are not bound to each other but move around very freely, which is why a gas does not have a well-defined shape of its own; it fills whatever container you put it in (Figure 11.1). Think for example of the constantly shifting shape of a cloud, or the deformability of a balloon, in contrast with the rigidity of a block of aluminum. On average, gas molecules are sufficiently far apart that most of the time they hardly interact with each other. This low level of interaction is what makes it feasible to model a gas in some detail, using relatively simple concepts.

The molecular motion must be sufficiently violent that molecules can't stay stuck together. At high enough temperatures, any molecules that do manage to bind to each other temporarily soon get knocked apart again by high-speed collisions with other molecules. At a low enough temperature however, molecules move sufficiently slowly that collisions are no longer violent enough to break intermolecular bonds. Rather, more and more molecules stick to each other in a growing mass as the gas turns into a liquid or, at still lower temperatures, a solid.

Why we don't study liquids

Figure 11.2 Molecules in a liquid.

For completeness, we should say why we don't study liquids in this course. A liquid is intermediate between a solid and a gas. The molecules are sufficiently attracted to each other that the liquid doesn't fly apart like a gas (Figure 11.2), yet the attraction is not strong enough to keep each molecule near a fixed equilibrium position. The molecules can slide past each other, giving liquids their special property of fluid flow (unlike solids) with fixed volume (unlike gases).

The analysis of liquids in terms of atomic, microscopic models is quite difficult compared with gases, where the molecules only rarely come in contact with other atoms, or compared with solids, where the atoms never move very far away from their equilibrium positions. For this reason, in this introductory course with its emphasis on atomic-level description and analysis we concentrate mostly on understanding gases and solids.

11.2 Gas leaks through a hole

In Chapter 10 we used statistical mechanics to determine the average speed of a gas molecule. Here we'll see some interesting phenomena where the average speed plays a role. One example is the leakage rate of a gas through a small hole in a container filled with the gas. First we'll consider a simplified one-directional example, in order to understand the basic issues before stating the results for a real three-dimensional gas.

One-directional gas

The chain of reasoning that we follow is basically geometric. Consider a situation where many gas molecules are all traveling to the right inside a tube. For the moment, temporarily assume that they all have the same speed v (Figure 11.3). The cross-sectional area of the tube is A (Figure 11.4). There are N gas molecules in the tube of volume V, and we will frequently use the symbol $n = N/V$ to stand for the number of molecules per cubic meter:

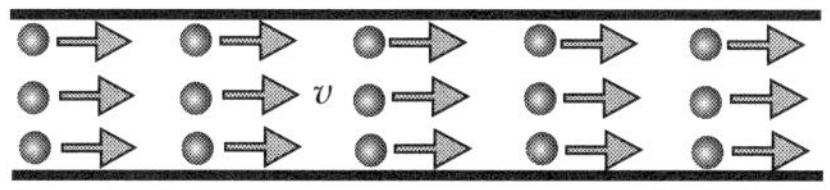

Figure 11.3 Side view of molecules all traveling to the right with speed v inside a tube. There are n molecules per cubic meter inside the tube.

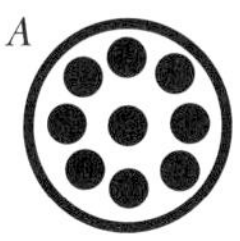

Figure 11.4 End view of molecules all traveling the same direction in a tube of cross-sectional area A.

NUMBER DENSITY: NUMBER PER CUBIC METER

$$\text{Definition: } n \equiv \frac{N}{V} \; \frac{\text{number of molecules}}{\text{m}^3}$$

Because we eventually want to be able to calculate how fast a gas will leak through a hole in a container, we will calculate how many molecules leave this tube in a short time interval Δt. For a molecule that is traveling at speed v to be able to reach the right end of the tube in this time interval Δt, it must be within a distance $v\Delta t$ of the end(Figure 11.5).

Since the cross-sectional area of the tube is A, the volume V of the tube that contains just those molecules that will leave the tube during the time interval Δt is simply $A(v\Delta t)$; see Figure 11.6.

There are n molecules per m^3 inside the tube, and $N = nV$ molecules inside the volume V. So the number N of molecules that will leave in the time interval Δt is

$$N = nV = n(Av\Delta t)$$

Dividing by the time interval Δt, we find the following important result:

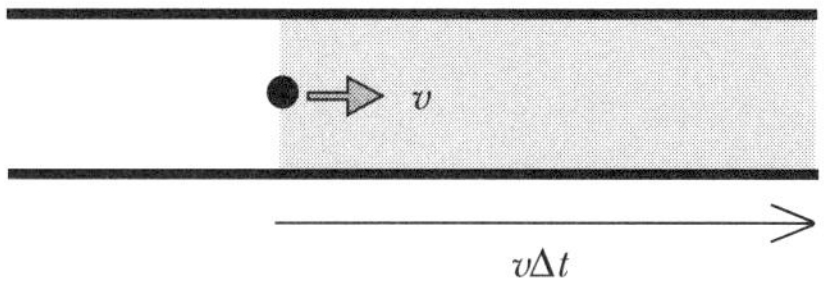

Figure 11.5 A molecule that will leave the tube in a time Δt must be within a distance $v\Delta t$ from the end.

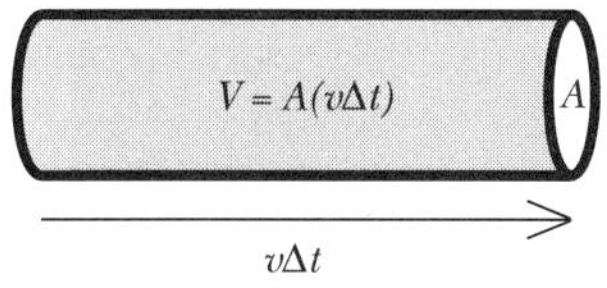

Figure 11.6 Volume containing the molecules that will leave the tube in time Δt.

 # OF MOLECULES CROSSING AREA *A* PER SECOND

nAv (one-directional case; all molecules have same speed)

? Does this formula make sense? What would you expect if you increased the number of molecules per cubic meter, or the cross-sectional area, or the speed?

The formula does make sense. The more molecules per unit volume ($n = N/V$), the more molecules will reach the right end of the tube per unit time. The bigger the cross-sectional area (A), the more molecules that will pass through that area per unit time. The faster the molecules are moving (v), the more molecules from farther away that can reach the end of the tube in a given time interval. The units are right: $(\#/\text{m}^3)(\text{m}^2)(\text{m/s}) = \#/\text{s}$.

Effect of different speeds

We need to account for the fact that the gas molecules don't all have the same speed v. Suppose n_1 molecules per unit volume have speeds of approximately v_1, n_2 molecules per unit volume have speeds of approximately v_2, etc. The number of molecules crossing an area A per second is

$$n_1Av_1 + n_2Av_2 + \text{etc.}$$

The average speed of all the molecules is by definition the following, where we weight each different speed by the number of molecules per unit volume that have that approximate speed:

$$\bar{v} = \frac{n_1v_1 + n_2v_2 + \text{etc.}}{n}$$

Therefore,

$$n_1 A v_1 + n_2 A v_2 + \text{etc.} = nA\bar{v}$$

A horizontal bar over a symbol is a standard notation for "average," and $\bar{v}$ means average speed.

Finally we have a valid formula for the number of molecules in a one-directional flow leaving the right end of the tube, even in the situation where they have a distribution of different speeds:

 # OF MOLECULES CROSSING AREA *A* PER SECOND

$nA\bar{v}$ (one-directional case; molecules have various speeds)

A "one-directional" gas may sound a bit silly, but this formula does apply to real one-directional flows such as the flow of water or gas through a pipe or the flow of electrons through a copper wire in an electric circuit.

A three-dimensional gas

A more realistic model of a gas would have approximately equal numbers of molecules heading to the left as well as to the right in the tube, in which case the number of molecules leaving the right end of the tube would be only $\frac{1}{2}nA\bar{v}$. In a real three-dimensional gas, molecules are moving in all directions. Only those molecules that are headed in the $+x$ direction can pass through a hole located to the right, not those moving in the $-x$ direction nor in the $\pm y$ or $\pm z$ directions. The molecules are headed randomly in all six directions, so we might expect our formula would have a factor of 1/6.

However, the actual factor is 1/4, which comes from detailed averaging over all directions and is related to our use of the average speed (magnitude of velocity), rather than averages of velocity components v_x or v_y or v_z. We don't want to get bogged down in the rather heavyweight mathematics required to prove this, so we just state that the factor is 1/4 rather than 1/6.

 # OF MOLECULES CROSSING AREA *A* PER SECOND

$\frac{1}{4}nA\bar{v}$ (three-dimensional case; molecules have various speeds)

$$\text{Remember that } n = \frac{N}{V}\ \frac{\text{\# of molecules}}{\text{m}^3}$$

Let's see what this formula predicts for the rate at which helium will escape from a balloon through a hole. You probably learned in chemistry that at "standard temperature and pressure" (STP, which means a temperature of 0° Celsius or 273 Kelvin, and a pressure of 1 atmosphere), one mole of a gas occupies 22.4 liters (a liter is 1000 cubic centimeters). As we saw in Chapter 10, the average speed $\bar{v}$ of helium atoms at ordinary temperatures is about 1200 m/s.

Ex. 11.1 Suppose we make a circular hole 1 millimeter in diameter in a balloon. Calculate the initial rate at which helium escapes through the hole (at 0° C), in number of helium atoms leaving the balloon per second.

11.2.1 Cooling of the gas

There is an interesting and important effect of gas escaping through a hole. If you look back over the derivations of the formulas, you can see that faster molecules escape disproportionately to their numbers. Faster molecules can be farther away from the hole than is true for slower molecules and still escape through the hole in the next short time interval. As a result, the distribution of speeds of molecules inside the container becomes somewhat depleted of high speeds.

Ex. 11.2 What can you say about the temperature of the gas inside the container as the gas escapes? Why?

11.2.2 Example: Leakage from a balloon

A party balloon is about one foot in diameter (about 30 centimeters). The volume of a sphere is $\frac{4}{3}\pi r^3$, where r is the radius of the sphere. Refer to the calculation you made in Exercise 11.1 on page 444 for the initial leak rate through a circular hole 1 millimeter in diameter.

(a) If the helium were to escape at a constant rate equal to the initial rate, about how long would it take for all the helium to leak out?

(b) As the helium escapes, the balloon shrinks, and the number of helium atoms per cubic meter n will stay roughly constant, in which case our analysis is pretty good. However, remember that the temperature of the gas drops due to a preferential loss of high-speed atoms. Would this effect make the amount of time to empty be more or less than the value you calculated in part (a)?

Solution

(a) The leak rate was 6.3×10^{21} atoms per second. Calculate number of helium atoms in the balloon originally:

$$\frac{4}{3}\pi(0.3\text{ m})^3\left(\frac{6\times10^{23}\text{ atoms}}{22.4\times10^3\text{ cm}^3}\right)\left(\frac{10^6\text{ cm}^3}{\text{m}^3}\right) = 3\times10^{24}\text{ atoms}$$

Assuming constant rate:

$$\frac{3\times10^{24}\text{ atoms}}{6.3\times10^{21}\text{ atoms/s}} = 500\text{ s}$$

Note that when you blow up an ordinary rubber balloon the pressure is higher than one atmosphere, which means higher number density n but also a correspondingly higher leak rate, which is proportional to n. To a first approximation the density doesn't matter in this estimate of the time to empty. Also note that often when you puncture a balloon the balloon rips, creating a large opening; this is not the case we are considering.

(b) With the preferential loss of high-speed atoms, the speed distribution inside the balloon shifts to lower speeds (corresponding to a lower temperature). If the average speed is lower, the leak rate is lower, and it should take longer for the balloon to empty than we calculated in part (a), where we assumed a constant leak rate.

We've implicitly done the analysis in vacuum. If the balloon is in air, air molecules enter the balloon through the same hole.

11.3 Mean free path

There is another important property of a gas that we can understand at this point in terms of our atomic model. Suppose we place a special molecule somewhere in the helium gas, one whose movements we can trace. For example, it might be a molecule of perfume. On average, how far does this molecule go before it runs into a molecule of the gas? The average distance between collisions is called the "mean free path." It plays an important role in many phenomena, including the creation of electric sparks in air (discussed in Volume II of this textbook).

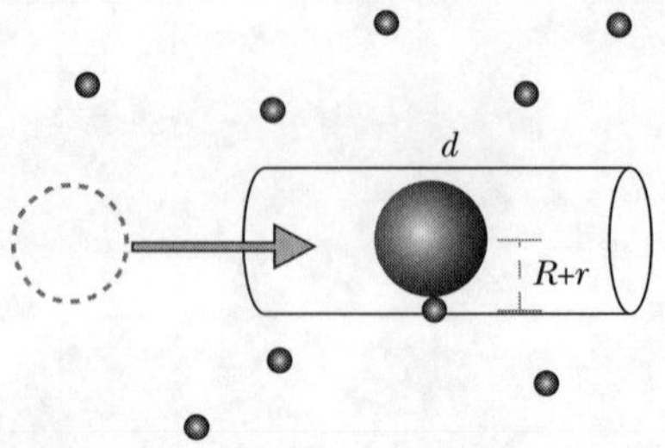

Figure 11.7 If a "special molecule" (dashed circle) enters a cylinder of length d and radius $R+r$, it will collide with a gas molecule (dark).

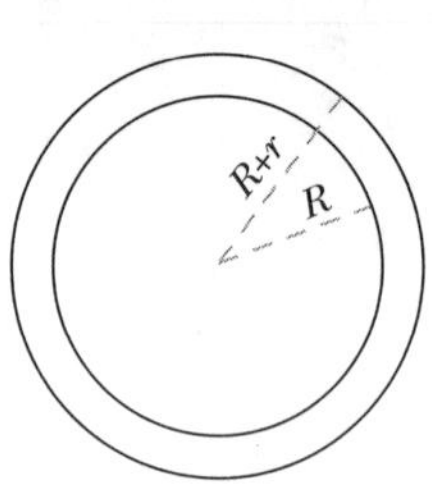

Figure 11.8 End view of cylinder.

The calculation of the mean free path depends on a simple geometrical argument. Draw a cylinder along the direction of motion of the special molecule, with length d and radius $R+r$, where R is the radius of the molecule, and r is the radius of a gas molecule, as shown in Figure 11.7 and Figure 11.8.

The geometrical significance of this cylinder is that if the path of the special molecule comes within one molecular radius of a gas molecule, there will be a collision, so any gas molecule whose center is inside the cylinder will be hit. How long should the cylinder be for there to be a collision?

We define d to be the average distance the special molecule will travel before colliding with another molecule, so the cylinder drawn in Figure 11.7 should contain on average about one gas molecule. The cross-sectional area of the cylinder is $A \approx \pi(R+r)^2$ and the volume of the cylinder is Ad. If n stands for the number of gas molecules per cubic meter, we can write a formula involving the mean free path:

MEAN FREE PATH

$$nAd \approx 1 \text{ where } A \approx \pi(R+r)^2 \text{ (and } n = N/V)$$

We are ignoring some subtle effects in this calculation, but this analysis gives us the main picture: the mean free path is shorter for higher density or larger molecular size.

To get an idea of the order of magnitude of a typical mean free path, assume that an N_2 molecule in the air has a radius of approximately 2×10^{-10} m (the radius of one of the N atoms being about 10^{-10} m), and let's calculate approximately the mean free path d of an N_2 molecule moving through air. At "standard temperature and pressure" (STP, 0° Celsius, 1 atmosphere pressure) one mole of air (6×10^{23} molecules) has a volume of 22.4 liters (22.4×10^3 cm^3).

$$n = \frac{6\times10^{23} \text{ molecules}}{(22.4\times10^3 \text{ cm}^3)(10^{-6} \text{ m}^3/\text{cm}^3)} = 2.7\times10^{25} \frac{\text{molecules}}{\text{m}^3}$$

$$A \approx \pi(2\times2\times10^{-10} \text{ m})^2 = 5\times10^{-19} \text{ m}^2$$

$$d \approx \frac{1}{nA} \approx \frac{1}{\left(2.7\times10^{25}\frac{\text{molecules}}{\text{m}^3}\right)(5\times10^{-19} \text{ m}^2)} = 7\times10^{-8} \text{ m}$$

Ex. 11.3 It is interesting to compare the mean free path of about 7×10^{-8} m to the average spacing L between air molecules, which is the cube root of the volume occupied on average by one molecule (Figure 11.9). Calculate L. You may be surprised to find that the mean free path d is much larger than the average molecular spacing L. The molecules represent rather small targets.

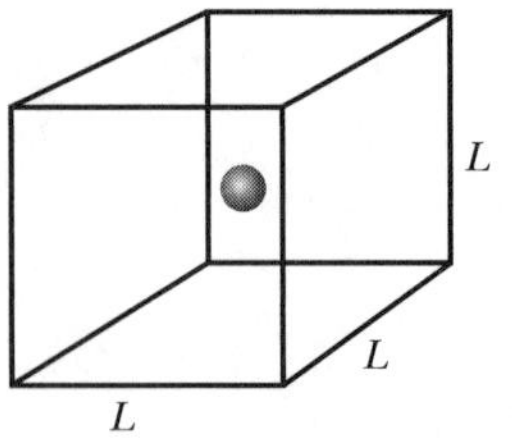

Figure 11.9 The volume of air occupied on average by one molecule.

11.4 Pressure and temperature

Next we will use statistical ideas to relate what we know about the motion of molecules in a gas to the pressure that the gas exerts on its container.

In a party balloon, helium atoms are continually hitting the rubber walls of the balloon. In Ex. 11.1 on page 444 you found that on average, at STP (standard temperature and pressure) about 6×10^{21} helium atoms hit a 1-mm-diameter section of the balloon every second. The time-and space-averaged effect of this bombardment is an average force exerted on every square millimeter of the balloon. This average force per unit area is called the "pressure" P and is measured in N/m^2, also called a "pascal." We are going to calculate how big the pressure is in terms of the average speed of the helium atoms, thus building a link between the microscopic behavior of the helium atoms and the macroscopic time- and space-averaged pressure.

When the velocity distribution is stable, on average a helium atom bounces off the wall with no change of kinetic energy. We emphasize that this is the average behavior. Any individual atom may happen to gain or lose energy in the collision with a vibrating atom in a rubber molecule, but if the balloon is not being warmed up or cooled down, the velocity distribution in the gas is stable, and on average the helium atoms rebound from the wall with the same kinetic energy they had just before hitting the wall (Figure 11.10).

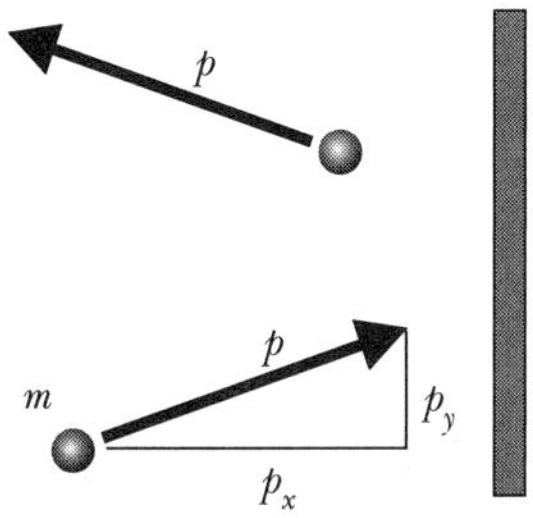

Figure 11.10 On average an atom bounces off the wall without changing speed.

? If the kinetic energy doesn't change when an atom bounces off the wall, is there any change in the atom's momentum?

Unlike speed or kinetic energy, momentum is a vector quantity, and there is a large change in the x component of momentum, from $+p_x$ to $-p_x$. (There is no change in the y component.) Therefore the momentum change of the helium atom is $\Delta p_{x,\,\text{helium}} = -2p_x$.

? What caused this change in the momentum of the helium atom?

A force is required to change the momentum of an object. In this case the force was applied by the wall of the container (or in more detail, by an atom in a rubber molecule in the balloon). By the principle of reciprocity (Newton's third law), the helium atom must have applied an equal and opposite force to the wall (or more precisely, to that atom in the rubber molecule). Therefore the wall must have acquired an amount of momentum $\Delta p_{x,\,\text{wall}} = 2p_x$.

If we could calculate the average time Δt between collisions of helium atoms with a small area A of the wall, we could express the pressure (force per unit area) as follows, since $d\vec{p}/dt = \vec{F}$:

$$P = \frac{F}{A} = \frac{1}{A}\frac{\Delta p_x}{\Delta t}$$

Assuming one direction and one speed

We already know that if there are n_{right} helium atoms per unit volume that have an x component of velocity equal to $+v_x$ (that is, moving to the right toward the wall), the number of such atoms that hit an area A to their right in a time Δt is $nAv_x\Delta t$. Therefore the average time between collisions is

$$\Delta t = 1/(n_{\text{right}}Av_x)$$

and we find

$$P = \frac{F}{A} = \frac{1}{A}\frac{\Delta p_x}{\left(\frac{1}{n_{\text{right}}Av_x}\right)} = (n_{\text{right}}v_x)(2p_x) = 2n_{\text{right}}\left(\frac{p_x}{m}\right)p_x = 2n_{\text{right}}\frac{p_x^2}{m}$$

An alternative argument

Alternatively, we can say that $n_{\text{right}}Av_x$ helium atoms with this x component of velocity hit the area A every second, and each of them delivers $2p_x$ of momentum to the wall, so that the total momentum transfer to the wall per second is

(# hits per second) (momentum transfer per hit)

$$(n_{\text{right}}Av_x)(2p_x)$$

Divide by A to get the force per unit area, and you again get

$$P = (n_{\text{right}}v_x)(2p_x) = 2n_{\text{right}}\frac{p_x^2}{m}$$

Taking the speed distribution into account

Actually, this is just the contribution to the pressure made by those atoms that happened to have this particular value of v_x to the right. On average only half the atoms are headed to the right, so we replace n_{right} by $n/2$, where n is the number of atoms per unit volume (going in either direction).

$$P = 2n_{\text{right}}\frac{p_x^2}{m} = n\frac{p_x^2}{m}$$

We also need to average over the slow and fast atoms:

$$P = n\frac{\overline{p_x^2}}{m}$$

We'll explain what we mean by this average. Suppose n_1 atoms per unit volume have x components of momentum of approximately p_{x1}, n_2 atoms per unit volume have x components of momentum of approximately p_{x2}, etc. The pressure is

$$P = n_1\left(\frac{p_{x1}^2}{m}\right) + n_2\left(\frac{p_{x2}^2}{m}\right) + \ldots$$

The average value of p_x^2 for all the atoms is by definition the following, where we weight each different value of p_x^2 by the number of atoms per unit volume that have that approximate value:

$$\overline{p_x^2} = \frac{n_1p_{x1}^2 + n_2p_{x2}^2 + \text{etc.}}{n}\text{, so we have } n_1p_{x1}^2 + n_2p_{x2}^2 + \text{etc.} = n\overline{p_x^2}$$

Therefore we can write this:

$$P = n_1\left(\frac{p_{x1}^2}{m}\right) + n_2\left(\frac{p_{x2}^2}{m}\right) + \ldots = n\frac{\overline{p_x^2}}{m}$$

Taking direction into account

We can re-express the pressure in terms of momentum p rather than p_x by the following steps:

First, note that $p = \sqrt{p_x^2 + p_y^2 + p_z^2}$, or $p^2 = p_x^2 + p_y^2 + p_z^2$.

Second, since the atoms are flying around in random directions, there should be no difference in averaging in the x, y, or z direction, so $\overline{p_x^2} = \overline{p_y^2} = \overline{p_z^2}$, which implies that $\overline{p^2} = 3\overline{p_x^2}$.

Taking these factors into account, instead of $P = n\frac{\overline{p_x^2}}{m}$ we can write the following important result:

GAS PRESSURE IN TERMS OF ATOMIC QUANTITIES

$$P = \frac{1}{3}n\frac{\overline{p^2}}{m}\text{, where } n = N/V$$

? Look back over the line of reasoning that led us to this result, and reflect on the nature of the argument. Stripped of the details, try to summarize the major steps leading to this result.

We reached this result by combining two effects: the number of molecules hitting an area per second is proportional to v, and the momentum transfer is also proportional to v. Hence the force per unit area is proportional to v^2.

11.4.1 The ideal gas law

It is useful to rewrite our result for the pressure, factoring out the term $\overline{p^2}/(2m)$, which is the average translational kinetic energy K_{trans} of a molecule of mass m:

$$P = \frac{2}{3}n\left(\frac{\overline{p^2}}{2m}\right) = \frac{2}{3}n\overline{K}_{trans}$$

The pressure of a gas is proportional to the number density (number of molecules per cubic meter, N/V), and proportional to the average translational kinetic energy of the gas molecules. This gives us a connection to temperature, because we found in Chapter 10 that $\overline{K}_{trans} = \frac{3}{2}kT$.

Substituting into the pressure formula, $P = \frac{2}{3}n\overline{K}_{trans} = \frac{2}{3}n(\frac{3}{2}kT) = nkT$, which relates pressure to temperature and is called the "ideal gas law":

MOLECULAR VERSION OF THE IDEAL GAS LAW

$$P = nkT$$

$n = N/V$ = number of molecules per m^3

$k = R/6\times10^{23}$ is the "Boltzmann constant" (1.4×10^{-23} J/K)

Ex. 11.4 A mole of a gas (6×10^{23} molecules) occupies 22.4 liters under standard conditions (temperature of 0° Celsius and one atmosphere pressure). Calculate the value of one atmosphere pressure in units of N/m^2.

11.4.2 The macroscopic ideal gas law

We can compare our molecular version of the gas law with experiments that measure the macroscopic properties of gases. For gases with fairly low densities, measurements of the pressure for all gases (helium, oxygen, nitrogen, carbon dioxide, etc.) are well summarized by the macroscopic "ideal gas law" which is probably familiar to you from chemistry, and which describes the observed behavior of any low-density gas:

$$P = \frac{(\#\text{ of moles})RT}{V} \quad \text{(mole version of the gas law)}$$

Here R is the "gas constant" (8.3 J/K/mole), T is the absolute temperature in Kelvins, and V is the volume in cubic meters. To compare with our microscopic prediction, we can convert this macroscopic version of the ideal gas law to a version involving microscopic quantities. Note the following:

$$\text{\# of moles} = \frac{(\text{\# of molecules})}{(6\times10^{23}\ \text{molecules/mole})} = \frac{N}{6\times10^{23}}$$

where N is the number of molecules in the gas. Therefore we have

$$P = \left(\frac{N}{6\times10^{23}}\right)\frac{RT}{V} = \left(\frac{N}{V}\right)\left(\frac{R}{6\times10^{23}}\right)T$$

But this is simply $P = nkT$, where $n = N/V$ and $R/6\times10^{23}$ is the Boltzmann constant k:

$$k = \frac{R}{6\times10^{23}} = \frac{8.3\ \text{J/K/mole}}{6\times10^{23}\ \text{molecules/mole}} = 1.4\times10^{-23}\ \text{J/K}$$

Warning about the meaning of "*n*"

You may be familiar with the gas law written in the form $PV = nRT$, where n is the number of moles, whereas we mean something else by "n"—the number of molecules per unit volume, N/V. On the rare occasions when we need to refer to the number of moles, we'll write it out as "# of moles."

11.4.3 Temperature from entropy or from the ideal gas law

In Chapter 10 we found the relationship $\overline{K}_{\text{trans}} = \frac{3}{2}kT$, based on the Boltzmann distribution as applied to a low-density gas. The Boltzmann distribution in turn was based on the statistical mechanics definition of temperature in terms of entropy as $1/T = \partial E/\partial S$.

When we inserted $\overline{K}_{\text{trans}} = \frac{3}{2}kT$ into the kinetic theory result for pressure, $P = \frac{1}{3}n(\overline{p^2/m})$, we obtained the molecular version of the ideal gas law, $P = nkT$, which we showed was equivalent to the macroscopic version of the ideal gas law, $P = (\text{\# of moles})RT/V$.

Low-density gases are described well by the ideal gas law, and for that reason gases are used to make accurate thermometers. You measure the pressure P and volume V of a known number of moles of a low-density gas and determine the temperature from the macroscopic gas law:

$$T = \frac{PV}{(\text{\# of moles})R} \quad \text{(gas thermometer)}$$

The fact that we could start from $1/T = \partial E/\partial S$ and derive the ideal gas law proves that the temperature measured by a gas thermometer is exactly the same as the "thermodynamic temperature" defined in terms of entropy.

11.4.4 Real gases

Our analysis works well for low-density gases. For high-density gases, there are two major complications. First, the molecules themselves take up a considerable fraction of the space, so the effective volume is less than the geometrical volume, and V in the gas law must be replaced by a smaller value.

Second, at high densities the short-range electric forces between molecules have some effect. In this context these intermolecular forces are called "van der Waals" forces, which are the gradient of the interatomic potential energy discussed in Chapter 4. In a low-density gas almost all of the energy is kinetic energy, but in a high-density gas some of the energy goes into configurational energy associated with the interatomic potential energy. The effect is that in the gas law P must be replaced by a smaller value.

When these two effects are taken into effect, the resulting "van der Waals" equation fits the experimental data quite well for all densities of a gas, although the corrections are different for different gases due to differences in molecular sizes and intermolecular forces.

11.4.5 Energy of a diatomic gas

The formula for the average translational kinetic energy $\bar{K}_{\text{trans}} = \frac{3}{2}kT$ is valid even for a gas with multiatom molecules such as nitrogen (N_2) or oxygen (O_2). But, as you will recall from Chapter 7, this translational kinetic energy is only a part of the total energy of the molecule. In addition to translational energy, a diatomic molecule can have rotational and vibrational energy relative to the center of mass (Figure 11.11), so the energy contains additional terms. We write the energy of a diatomic molecule in the following way:

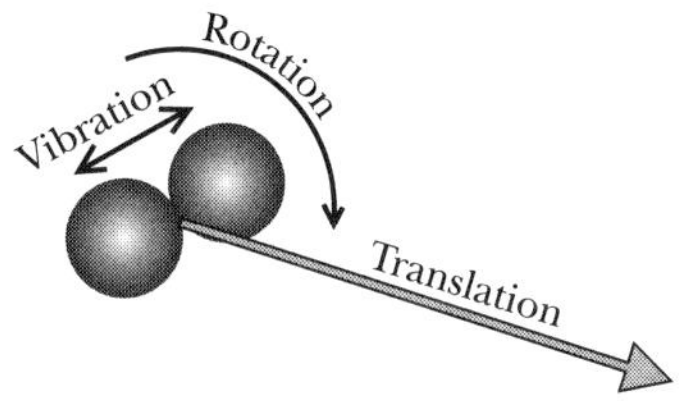

Figure 11.11 Energy of an oxygen molecule.

$$K_{\text{trans}} + E_{\text{vib}} + E_{\text{rot}} + Mgy_{\text{cm}}$$

The key point, as we discussed in Chapter 10, is that the average energy of a diatomic molecule is greater than the average energy of an atom in a monatomic gas at the same temperature. For a monatomic gas such as helium the average total energy per molecule is just $\bar{K}_{\text{trans}} = \frac{3}{2}kT$.

11.4.6 Application: Weight of a gas in a box

The exponential dependence of number density ($n \propto e^{-\frac{mgy}{kT}}$) and therefore of pressure on height makes it possible to understand something that otherwise might be rather odd about weighing a box that contains a gas (Figure 11.12). Let's do the weighing in a vacuum so we don't have to worry about buoyancy forces due to surrounding air. The box definitely weighs more on the scales if there is gas in it than if there is a vacuum in it. In fact, if the mass of the gas is M, the additional weight is Mg, because the momentum principle for multiparticle systems refers to the net force on a system, and the gravitational contribution to the net force is the sum of the gravitational forces on each individual gas molecule in the box, Mg.

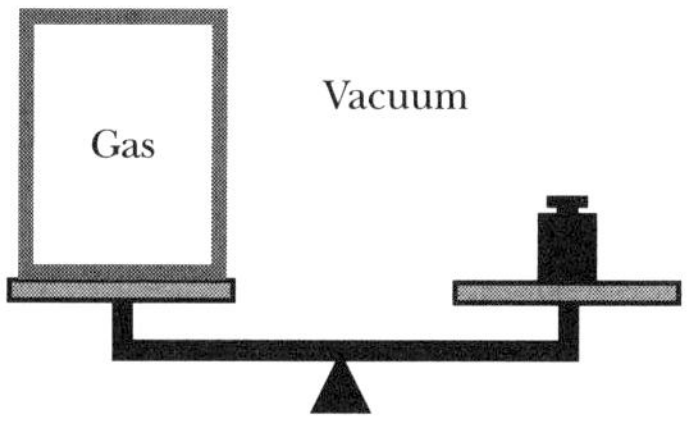

Figure 11.12 Weighing a box containing a gas.

That seems reasonable until you think about the details of what is going on inside the box. At any given instant, the vast majority of the gas molecules are not touching the box! And some of the gas molecules are colliding with the top of the box, exerting an upward force on the box. How can these molecules possibly contribute to the weight measured by the scales? What is the mechanism for the momentum principle working out correctly for this multiparticle system?

The number density n at the bottom of the box is slightly larger than the number density at the top of the box, if all the gas is at the same temperature, which is a good assumption in most situations. Therefore the pressure, $P = nkT$, is larger at the bottom than at the top. Let's calculate the y components of the forces associated with these slightly different pressures, where A is the area of the top (and bottom) of the box, and h is the height of the box (Figure 11.13).

$$F_{\text{net},y} = P_{\text{top}}A - P_{\text{bottom}}A = A\Delta P$$

where ΔP is the pressure difference from bottom to top inside the box. We can calculate ΔP directly, by starting from the fact that for a small height change Δy, we have this:

$$\frac{\Delta P}{\Delta y} \approx \frac{dP}{dy}, \text{ so } \Delta P \approx \frac{dP}{dy}\Delta y$$

If we can evaluate dP/dy, we can determine the small pressure difference ΔP. Since $P = nkT$, the height dependence of the pressure is the same as for the number density (for constant temperature), and we can write this:

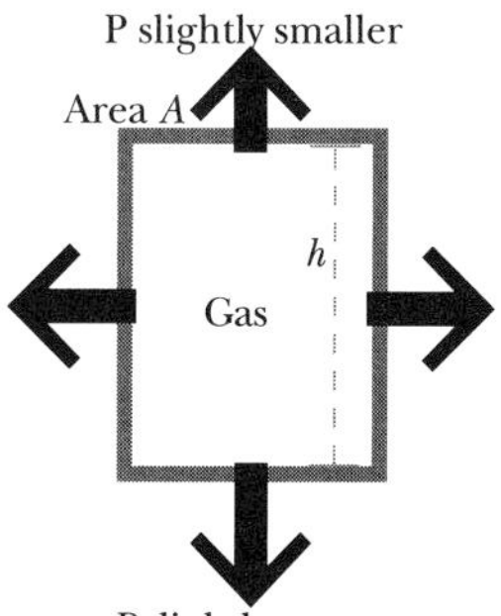

Figure 11.13 The force on the top of the box differs slightly from the force on the bottom of the box.

$$\frac{dP}{dy} = \frac{d}{dy}\left[P_{\text{bottom}} e^{-\frac{mgy}{kT}}\right]$$

$$\frac{dP}{dy} = P_{\text{bottom}} e^{-\frac{mgy}{kT}}\left[-\frac{mg}{kT}\right] \approx -P_{\text{bottom}}\left[\frac{mg}{kT}\right]$$

because the factor $e^{-\frac{mgy}{kT}}$ is very close to 1. The result is negative because the force the gas exerts on the top of the box is smaller than the force on the bottom. Choosing $\Delta y = h$, and writing $P_{\text{bottom}} = nkT$, we have this result for ΔP:

$$\Delta P = \frac{dP}{dy}\Delta y \approx -nkT\left[\frac{mg}{kT}\right]h = -nmgh$$

Now that we know the pressure difference, we can calculate the y component of the net force:

$$F_{\text{net},y} = A\Delta P = -nmg(Ah)$$

? But Ah is the volume V of the box, and $n = N/V$ is the number of molecules per m^3 in the box, so what does this formula reduce to?

We have the striking result that the air inside the box pushes down on the box with a force equal to the combined weight Mg of all N molecules in the box:

$$F_y = -Nmg = -Mg \text{ !!}$$

We have shown that the difference in the time- and space-averaged momentum transfers by molecular collisions to the top and bottom of the box is equal to the weight of the gas in the box, as predicted by the momentum principle for a multiparticle system. The pressure difference is very slight, but then the weight of the gas is very small, for that matter. What is surprising is that any particular instant, relatively few of the molecules are actually in contact with the box, yet the effects of these relatively few molecules is the same as though they were all sitting on the bottom of the box.

You could think of a box full of water in the same way. The water pressure is larger at the bottom than at the top, but with water this difference is quite large, corresponding to the much higher density of water. In fact, a column of water only 10 meters high makes a pressure equal to that produced by the many kilometers of atmosphere (atmospheric pressure at sea level is about 10^5 N/m^2).

Ex. 11.5 In the preceding discussion the box was in vacuum, but under normal conditions the box we were weighing would have been surrounded by air. Suppose the box has thin walls and is initially open to the air. Then we close the lid, trapping air inside it. Consider all forces including buoyancy forces, and determine what the scales will read: the weight of the box alone or the box plus air?

11.4.7 Application: Weight of a bouncing molecule

There is a related calculation that is amusing. If you did a related homework problem in Chapter 8 on collisions, you already went through a similar calculation. Consider a single molecule (otherwise in vacuum) that bounces up and down on a scale without losing significant energy. Figure 11.14 shows a snapshot at the instant that the molecule is just bouncing up after hitting the scale.

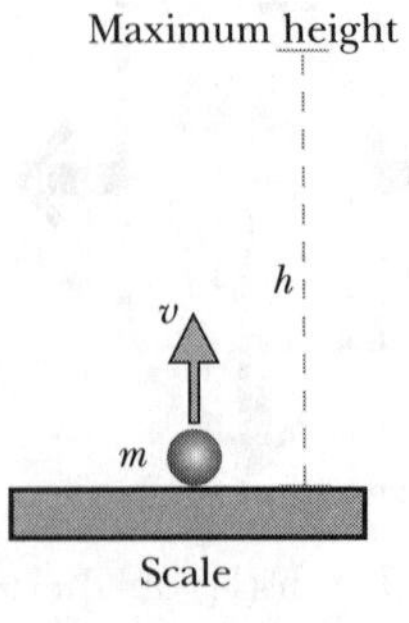

Figure 11.14 The ball has just hit the scale and has rebounded upward.

What do the scales read? If the scales can respond quickly, we will see brief spikes each time the molecule strikes it (Figure 11.15). Let's determine what the time-averaged force is.

? How long does it take for the molecule to reach the top of its trajectory, h, starting with a speed v? What then is the time Δt between impacts? How much momentum transfer is there to the scale on each impact?

It takes a time interval v/g to go up (for the speed to decrease from v to 0 with acceleration $-g$) and another time interval v/g to come down, so the time between impacts is $\Delta t = 2v/g$. Each impact transfers an amount of momentum $\Delta P = 2mv$. Therefore the time-averaged force is

$$\frac{\Delta P}{\Delta t} = \frac{2mv}{2v/g} = mg \text{ !!}$$

If the scale is sluggish, and can't respond in a time as short as $\Delta t = 2v/g$, the scale will simply register the value mg, just as though the molecule were sitting quietly on the scale.

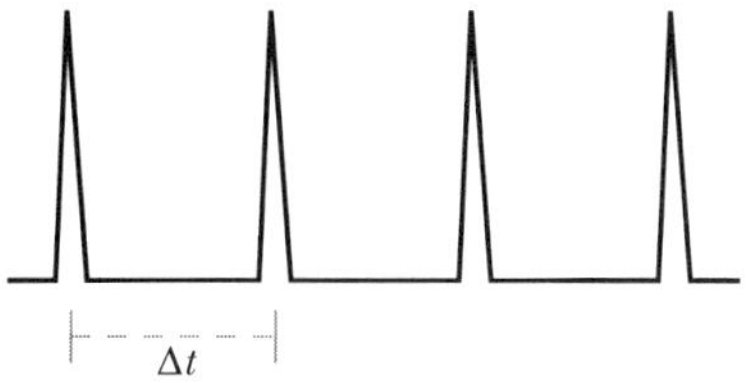

Figure 11.15 Impacts of the ball on the scale.

11.5 Energy transfers with the surroundings

The final topic in this chapter on gases is an analysis of energy transfers between a gas and its surroundings. We will address such questions as these: How much thermal energy transfer Q is required to raise the temperature of a gas by one degree (the heat capacity of a gas)? How does the temperature of a gas change when you compress it quickly? Do the answers to these questions depend on what kind of gas is involved?

We need a device that lets us control the flow of energy into and out of a gas, in the form of work or in the form of thermal energy transfer. We will use a system consisting of a cylindrical container containing gas that is enclosed by a piston that can move in and out of the cylinder with little friction but which fits tightly enough to keep the gas from leaking out (Figure 11.16). This is similar to a cylinder in an automobile engines, into which is sprayed a mixture of gasoline vapor and air. The mixture is ignited by a spark, and the chemical reactions raise the temperature and pressure very high very quickly. The piston is pushed outward, which turns a shaft which ultimately drives the wheels.

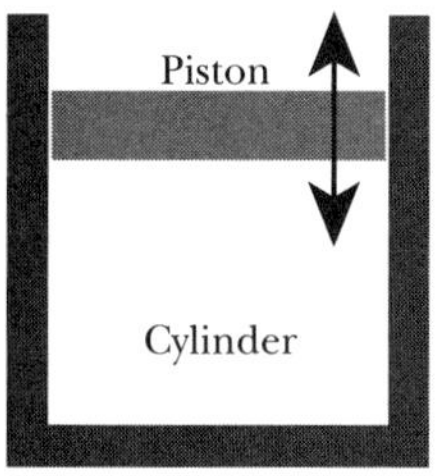

Figure 11.16 A cylinder with a piston that can move vertically.

Force and pressure

Figure 11.17 shows a cylinder with a vertical-running piston on which we can load varying amounts of sand, in order to be able to control the pressure of the contained gas, and to be able to do controlled amounts of work on the gas. We will also put the cylinder in contact with hot or cold objects and allow thermal energy transfer into or out of the cylinder.

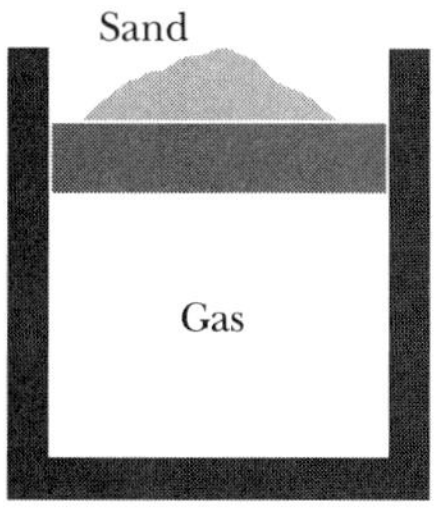

Figure 11.17 The piston can be loaded with varying amounts of sand.

? Consider the piston plus sand as the system of interest for a moment and think about what forces act on this system.

A physics diagram for the piston+sand system (Figure 11.18) includes the downward gravitational forces on the piston (Mg) and on the sand (mg), an upward time- and space-averaged force due to the pressure of the enclosed gas on the lower surface of the piston, and a downward time- and space-averaged force due to the pressure of the outside atmosphere on the upper surface of the system.

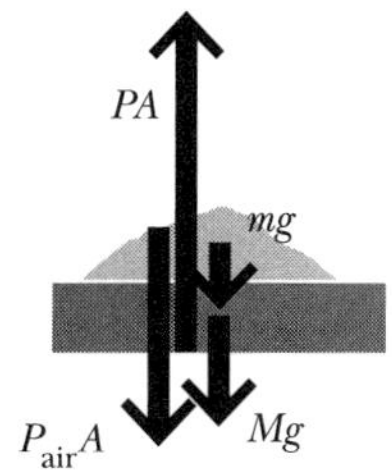

Figure 11.18 Forces on the piston.

Since pressure is force per unit area, the upward force is PA, where A is the surface area of the bottom of the piston. Similarly, the downward force contributed by the outside atmosphere is $P_{air}A$. These pressure-related forces are not actually continuous, since they are the result of collisions of individual gas molecules with the piston and sand. However, the rate of collisions is so extremely high over the area of the piston that the force seems essentially constant. For example, in Ex. 11.1 on page 444 you found that on average, at STP (standard temperature and pressure) about 6×10^{21} helium atoms strike a tiny 1-mm-diameter section every second.

? In mechanical equilibrium (velocity of piston not changing), solve for the gas pressure inside the cylinder.

The momentum principle tells us that in equilibrium the net force on the piston+sand system must be zero, from which we are able to deduce that the pressure of the gas inside the cylinder is

$$P = P_{air} + \frac{Mg + mg}{A}$$

A sudden change

By varying the amount of sand (m) we can vary the pressure of the gas under study.

? If you suddenly add a lot of sand, what happens?

If you suddenly add a lot of sand to the piston, there is suddenly a sizable nonzero net downward force on the piston+sand system: $F_{net} = AP_{air} + Mg + mg - AP \neq 0$. The piston starts to pick up speed downwards. As it does so, it runs into gas molecules and tends to increase their speeds.

? What happens to the temperature of the gas in the cylinder?

Since higher average speed means higher temperature, the gas temperature starts to increase, at the same time that the volume of the gas is decreasing. This is a double whammy: both increased temperature and decreased volume contribute to increased pressure, since $P = (N/V)kT$, where N is the number of gas molecules in the cylinder. Therefore the pressure in the gas quickly rises, and eventually there will be a new equilibrium with a lower piston (supporting more sand) and a higher gas pressure in the cylinder.

However, getting to that new equilibrium is pretty complicated. If there is no friction or other energy dissipation, the piston will oscillate down and up, with the gas pressure going up and down. It is even possible to determine an effective "spring constant" for the gas and calculate the frequency of the oscillation. However, in any real system there will be some friction, so we know that the system will eventually settle down to a new equilibrium configuration.

Quasistatic processes

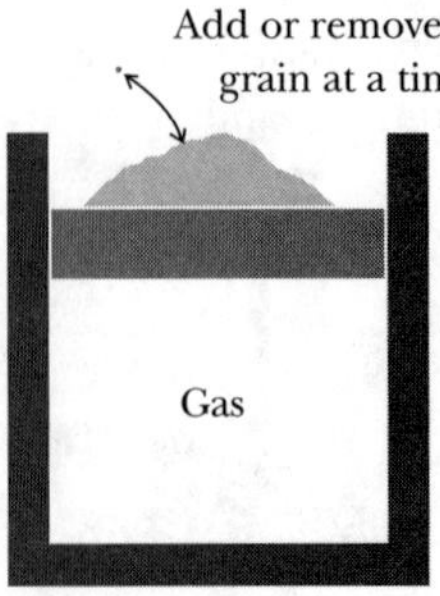

Figure 11.19 By adding or removing one grain of sand at a time we effect a "quasistatic" compression or expansion.

To avoid these complicated (though interesting) transient effects, we will study what happens when we add sand very carefully and very slowly, one grain at a time, and we assume that the new equilibrium is established almost immediately. This is called a "quasistatic compression" because the system is at all times very nearly in static equilibrium (Figure 11.19). Similarly, if we slowly remove one grain at a time, we can carry out a "quasistatic expansion." Note in particular that at no time does the piston have any significant amount of kinetic energy, and macroscopic kinetic energy is essentially zero at all times. Of course there is plenty of microscopic kinetic energy in the gas molecules and the outside air molecules, and in the thermal motion of atoms in the cylinder walls, piston, and sand.

Suppose we carry out a lengthy, time-consuming quasistatic compression by adding lots of sand, one grain at a time, very slowly. The piston goes down quite a ways, the pressure in the gas is a lot higher, and the volume of the gas is a lot smaller. If we know the pressure and volume, we can calculate the temperature by using the ideal gas law $P = (N/V)kT$ (assuming the gas density isn't too high to make this invalid).

But can we predict how low the piston will go for a given amount of added sand? Oddly enough, no—not without knowing something more about this device. There are two extreme cases that are both important in practice and calculable for an ideal gas. If the apparatus is made of metal (a very good thermal conductor) and is in good thermal contact with a large object at temperature T, the process will proceed at nearly constant temperature, with thermal energy transfer from the gas into the surroundings. If the apparatus is made of glass (a very poor thermal conductor), the temperature of the gas inside the cylinder will rise in a predictable way, with negligible energy transfer to the surroundings in the form of thermal energy transfer Q. We will analyze both kinds of processes—constant temperature processes, and no-Q processes.

11.5.1 Constant-temperature (isothermal) compression

Suppose the cylinder is made of metal (which is a very good thermal conductor) and is sitting in a very big tub of water whose temperature is T, as shown in Figure 11.20.

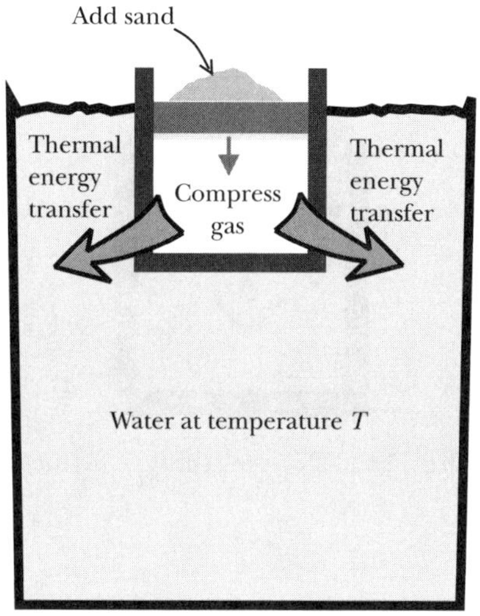

Figure 11.20 A cylinder of gas immersed in a large volume of water.

As we compress the gas, the temperature in the gas starts to increase. However, this will lead to energy flowing out of the gas into the water, because whenever the temperatures differ in two objects that are in thermal contact with each other, we have seen that there is a transfer of energy from the hotter object into the colder object. In fact, for many materials the rate of energy transfer is proportional to the temperature difference—double the temperature difference, double the rate at which energy transfers from the hotter object into the colder one.

The mechanism for thermal energy transfer is that atoms in the hotter object are on average moving faster than atoms in the colder object, so in collisions between atoms at the boundary between the two systems it is likely that energy will be gained by the colder object and lost by the hotter object.

Energy transfer out of the gas will lower the temperature of the gas, since the total energy of the gas is proportional to the temperature. Quickly the temperature of the gas will fall back to the temperature of the water. The temperature of the big tub of water on the other hand will hardly change as a result of the energy added to it from the gas, because there is such a large mass of water to warm up.

Therefore the entire quasistatic compression takes place essentially at the temperature of the water, and the final temperature of the gas is nearly the same as the initial temperature of the gas. This is a constant-temperature compression (also called an "isothermal" compression, which just means constant temperature). Similarly, we can slowly remove sand from the piston and carry out a quasistatic constant-temperature expansion.

? Suppose the original gas pressure and volume were P_1 and V_1, and as the result of a constant-temperature process the final volume is V_2. What is the final pressure P_2?

Since the temperature hasn't changed, from the ideal gas law $P = (N/V)kT$ we can deduce that $P_1V_1 = P_2V_2$, and therefore $P_2 = P_1(V_1/V_2)$.

Work and thermal energy transfer in a constant-temperature compression

A more difficult question we can ask (and answer!) is this: How much thermal energy was added to the water in the constant-temperature compression?

? What energy inputs and outputs were made to the gas? What energy change occurred in the gas?

The piston did work on the gas, and there was thermal energy transfer out of the gas (and into the water). The Earth's gravitational force did work on the gas (since the center of mass of the gas went down), but this is negligibly small compared to the work done by the piston (the lowering of the heavy piston involves much more gravitational energy than the lowering of the low-mass gas).

? Did the total energy of the gas change (ignoring the small gravitational energy change)?

Since the total energy of an ideal gas is proportional to temperature (including rotational and vibrational energy if the gas is not monatomic), and we made sure that we kept the temperature constant, the total energy of the gas did not change. Therefore we have this energy equation for the open system that is the gas:

? Δ(energy of gas) = (W by piston) – ($|Q|$ that flowed into water) = 0

Symbolically, $\Delta E_{gas} = W + Q = 0$, where Q is negative (thermal energy transfer out of system). In thermal processes of this kind, the energy equation is called "the first law of thermodynamics."

If we can calculate the work done by the piston, we can equate that result to the amount of thermal energy transfer into the water. The piston exerts a (time- and space-averaged) force PA on the gas, where P is the pressure in the gas and A is the cross-sectional area of the piston (Figure 11.21).

Figure 11.21 Force on the gas by the piston.

? If the piston drops a distance h, is the work it does PAh?

As the piston drops, the pressure in the gas increases, so the force is not constant. We have to integrate the variable force through the distance h in order to determine the work. If we measure x downward from the initial piston position, at each step dx of the way the increment of work done is $PAdx$. But Adx is an increment of volume (base area A time altitude dx), and the change in the volume is negative. Putting this all together, we have this:

WORK DONE BY A PISTON ON A GAS

$$W = -\int_{V_1}^{V_2} PdV$$

Check the sign: If the volume decreases, the integral will be negative, and the minus sign in front of the integral makes the work done on the gas be positive, which is correct. Conversely, in an expansion of a gas, the integral is positive, and the work is negative, because the gas is doing work on the piston rather than the other way round.

? In a constant-temperature (isothermal) compression, work is done on the gas. Where does this energy flow go? How does the temperature stay constant?

As the piston moves down, it increases the average energy of the gas molecules that run into it. Therefore the input work starts to raise the temperature of the gas (higher speeds), but this higher temperature causes thermal energy transfer out of the cylinder, into the (very slightly) lower-temperature water. The net effect is for energy to flow into the gas in the form of mechanical work, and out of the gas into the water in the form of thermal energy transfer. There is no change in the total energy of the gas, because the temperature of the gas didn't change. The outward thermal energy transfer brings the gas temperature back to what it was before the falling piston tried to increase the temperature.

Now we can calculate quantitatively the work done by the piston in the constant-temperature compression, which is equal to the thermal energy transfer from the gas to the surrounding water. We add lots of sand, one grain at a time (to maintain quasi-static equilibrium), and we compress the gas from an initial volume V_1 to a final volume V_2 (Figure 11.22).

Replace P by NkT/V, where T is a constant:

$$W = -\int_{V_1}^{V_2} PdV = -\int_{V_1}^{V_2} \frac{NkT}{V} dV$$

$$W = -NkT\int_{V_1}^{V_2} \frac{dV}{V} = -NkT[\ln V]_{V_1}^{V_2} = -NkT\ln\left(\frac{V_2}{V_1}\right)$$

Since the process was a compression, V_2 is smaller than V_1. Since $-\ln(V_2/V_1) = \ln(V_1/V_2)$, we have

$$W_{\text{by piston}} = |Q_{\text{into water}}| = NkT\ln\left(\frac{V_1}{V_2}\right)$$

Just as the symbol W is normally used for "work," so the symbol Q is normally used for "thermal energy transfer." Here Q is negative (energy output).

Figure 11.22 The gas is compressed at a constant temperature T (the cylinder is immersed in a large tub of water at temperature T).

Graphical representation

There is a useful graphical representation of this constant-temperature process. We plot the pressure P vs. the volume V, as shown in Figure 11.23, and the process is represented by a curve of constant T (which is also a curve of constant PV, since $PV = NkT$). The area under the process curve represents the work W, which is also the thermal energy transfer Q that flows to the surroundings.

$$W = -\int_{V_1}^{V_2} PdV$$

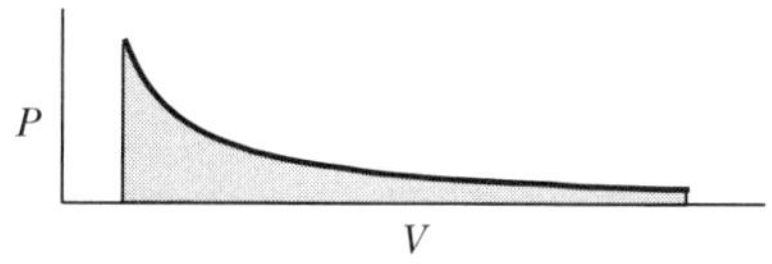

Figure 11.23 If T is kept constant, then the product PV is also constant.

The first law of thermodynamics

For historical reasons, in thermal processes the energy principle is called "the first law of thermodynamics."

THE FIRST LAW OF THERMODYNAMICS

$$\Delta E_{\text{sys}} = W + Q$$

In the particular case of constant-temperature (isothermal) compression of an ideal gas, we have

$$\Delta E_{\text{int}} = W + Q = 0$$

where E_{int} is the "internal" energy of the gas (the sum of the translational, rotational, vibrational, and other energy terms of all the molecules).

Ex. 11.6 A mole of nitrogen is compressed (by piling lots of sand on the piston) to a volume of 12 liters at room temperature (293 K). The cylinder is placed on an electric heating element whose temperature is maintained at 293.001 K. A quasistatic expansion is carried out at constant temperature by very slowly removing grains of sand from the top of the piston, with the temperature of the gas staying constant at 293 K. When the volume is 18 liters, how much thermal energy transfer Q has gone from the heating element into the gas? How much work W has been done on the piston by the gas? How much has the energy of the gas changed?

(You must assume that there is no thermal energy transfer from the gas to the surrounding air, and no friction in the motion of the piston, all of which is pretty unrealistic in the real world! Nevertheless there are processes that can be approximated by a constant-temperature expansion. This problem is an idealization of a real process.)

11.5.2 Heat capacity

We were able to calculate the important properties of a constant-temperature compression, where the apparatus was in good thermal contact with its surroundings. Before analyzing the opposite extreme, where the apparatus is thermally insulated from its surroundings and no thermal energy transfer is involved, we will review the concept of heat capacity in this context.

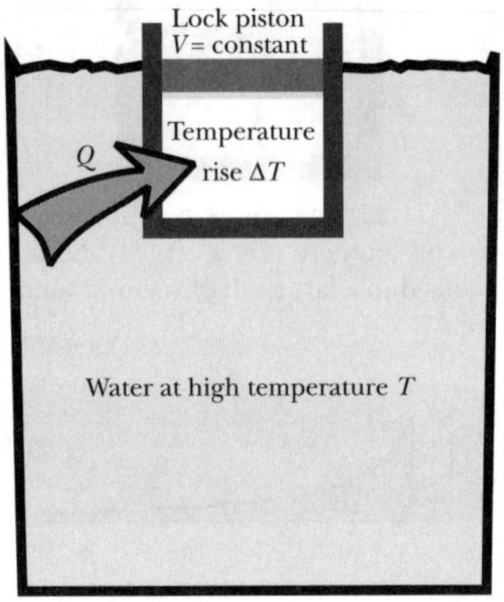

Figure 11.24 If the piston is locked so the volume of the gas cannot change, we can measure c_v of the gas.

Lock the piston in position so that the volume of the gas won't change (Figure 11.24). Put the apparatus in a big tub of water whose temperature is higher than the gas temperature, so that thermal energy transfer will go from the water to the gas. Allow an amount of thermal energy transfer Q to enter the gas and observe how much temperature rise ΔT has occurred in the gas. We define the "molecular heat capacity C_V at constant volume" in the following way, where N as usual is the total number of molecules in the gas:

$$Q = NC_V\Delta T \text{ (constant volume)}$$

The larger the heat capacity, the smaller the temperature rise ΔT for a given energy input Q.

? What is C_V for a monatomic gas such as helium?

Since the total energy in a monatomic gas is $N\overline{K}_{\text{trans}} = \frac{3}{2}NkT$, thermal energy transfer Q will increase the total energy of the gas by an amount $\frac{3}{2}Nk\Delta T$, so $C_V = \frac{3}{2}k$. Experimental measurements of the heat capacity of monatomic gases agree with this prediction.

For gases that are not monatomic, such as nitrogen (N_2), c_V can be larger than $\frac{3}{2}k$ because the total energy can be larger than $\frac{3}{2}NkT$ due to rotational and vibrational energy relative to the center of mass of the molecule. Before quantum mechanics, theoretical predictions for the contribution of the rotational and vibrational energies to the heat capacity of gases did not agree with experimental measurements. This was one of the puzzles that was eventually resolved by quantum mechanics, as we found in Chapter 10.

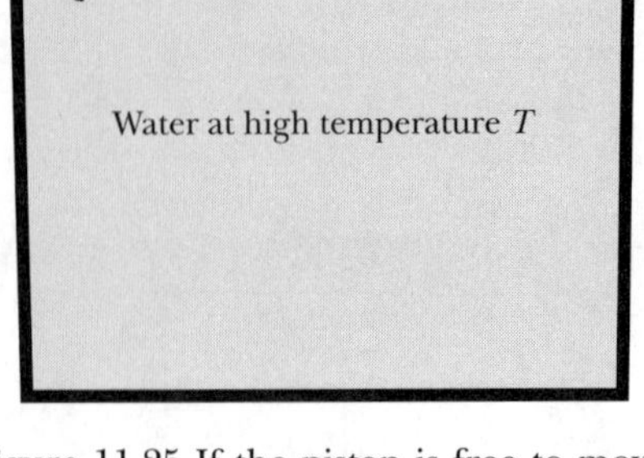

Figure 11.25 If the piston is free to move, the pressure inside the cylinder remains the same, and we can measure c_p of the gas.

HEAT CAPACITY AT CONSTANT VOLUME, PER MOLECULE

$$C_V = \frac{3}{2}k \text{ for a monatomic gas (He, etc.)}$$

$$C_V \geq \frac{3}{2}k \text{ for other gases (N}_2\text{, etc.)}$$

Heat capacity at constant pressure

If we don't lock the piston but let the gas expand at constant pressure, the incoming thermal energy transfer Q not only raises the energy of the gas by an amount $NC_V\Delta T$ but also raises the piston, which involves an amount of work W (Figure 11.25). We define the "molecular heat capacity C_P at constant pressure" as follows:

$$Q = NC_P\Delta T \text{ (constant pressure)}$$

Evidently C_P is bigger than C_V, because $Q = NC_P\Delta T = NC_V\Delta T + W$. We can calculate the work done on the piston by the gas as the gas expands, which is the negative of the work done by the piston on the gas:

$$W = \int_{V_1}^{V_2} PdV = P\int_{V_1}^{V_2} dV = PV_2 - PV_1 = NkT_2 - NkT_1 = Nk\Delta T$$

Therefore for an ideal gas there is a simple relationship between C_V and C_P:

$$Q = NC_P\Delta T = NC_V\Delta T + Nk\Delta T$$

$$C_P = C_V + k$$

Ex. 11.7 What is the molecular heat capacity at constant pressure C_P for helium?

Molar heat capacity

Heat capacity is often given on a per-mole basis rather than a per-molecule basis. The molar heat capacity at constant volume for an ideal gas is $C_V = \frac{3}{2}R$, where R is the molar gas constant (8.3 J/K), which is $6\times10^{23}k$. The molar heat capacity at constant pressure is $C_P = C_V + R$ for an ideal gas.

11.5.3 No-Q (adiabatic) compression

With the apparatus made of metal and sitting in a big tub of water, the temperature during the compression didn't change. Now we analyze the opposite extreme. We make the cylinder and piston out of glass or ceramic (which are poor conductors for thermal energy transfer) to minimize thermal energy transfer out of the gas as the gas gets hotter during the compression. In fact, we make the approximation that there is no thermal energy transfer at all (Figure 11.26). Such a no-Q process is also called "adiabatic" (which means "no thermal energy transfer").

How realistic is such a no-Q approximation? For many real situations this can be a rather good approximation if the compression or expansion is fast, so that there isn't enough time for significant thermal energy transfer. The flow of thermal energy from one object to another is a rather slow process. For example, a cup of coffee sitting on a table may stay quite hot for ten minutes or more.

But didn't we say that we were going to carry out compressions and expansions very slowly, "quasistatically"? Yes, so there is a contradiction. However, it is often the case that motion may be slow enough to be a good approximation to a quasistatic process and nevertheless may also be fast compared to the time required for significant thermal energy transfer. Again, think of how long it takes a cup of coffee to cool off.

Figure 11.26 If the cylinder and piston are made of insulating material, no thermal energy transfer can occur between the gas and its surroundings. This is called an adiabatic compression.

We again consider the gas in the cylinder as the system of interest, and the work done by the piston is equal to the increase in energy of the gas:

$$W = -\int_{V_1}^{V_2} PdV = NC_V\Delta T$$

? Why did we use C_V in this equation? The volume isn't constant in this compression!

An ideal gas is unique among materials in that its total energy is entirely determined by the temperature, independent of volume or pressure. So the change of energy of the gas itself is just $NC_V\Delta T$, even when the volume is not constant. As a practical matter, real gases behave this way as long as the density is not too high. At high densities the energy of the gas includes a significant term associated with the interatomic forces.

We can work through the integration, starting from a "differential" form of the equation shown above:

$$-PdV = NC_V dT$$

Use the gas law $P = (N/V)kT$ to substitute for the pressure:

$$-NkT\frac{dV}{V} = NC_V dT$$

Divide through by kT and rearrange:

$$\left(\frac{C_V}{k}\right)\frac{dT}{T} + \frac{dV}{V} = 0$$

$$\left(\frac{C_V}{k}\right)\int\frac{dT}{T} + \int\frac{dV}{V} = 0$$

Now integrate:

$$\left(\frac{C_V}{k}\right)\ln T + \ln V = \text{constant}$$

Rewrite using the properties of logarithms:

$$\ln T^{\left(\frac{C_V}{k}\right)} + \ln V = \ln\left[T^{\left(\frac{C_V}{k}\right)}V\right] = \text{constant}$$

And therefore we have

$$T^{\left(\frac{C_V}{k}\right)}V = \text{constant}$$

Ex. 11.8 Use the ideal gas law to eliminate T from this expression, and show that $PV^{\gamma} = \text{constant}$ in a no-Q process, where the parameter γ is defined as the ratio of the constant-pressure heat capacity to the constant-volume heat capacity, $\gamma \equiv C_P/C_V$. (Since C_p is greater than C_v, γ is greater than 1.)

In particular, if the initial pressure and volume are P_1 and V_1, and the final pressure and volume are P_2 and V_2, then $P_2V_2^{\gamma} = P_1V_1^{\gamma}$. Alternatively, we also have $T_2^{(C_V/k)}V_2 = T_1^{(C_V/k)}V_1$.

Graphical representation

Again, there is a useful graphical representation of this no-Q process. In Figure 11.27 we plot the pressure P vs. the volume V, and the process is represented by a curve along which $PV^{\gamma} = \text{constant}$. The area under the process curve represents the work W. For comparison we also show the curve for a constant-temperature process (T and PV constant). The no-Q curve is much steeper than the constant-temperature curve.

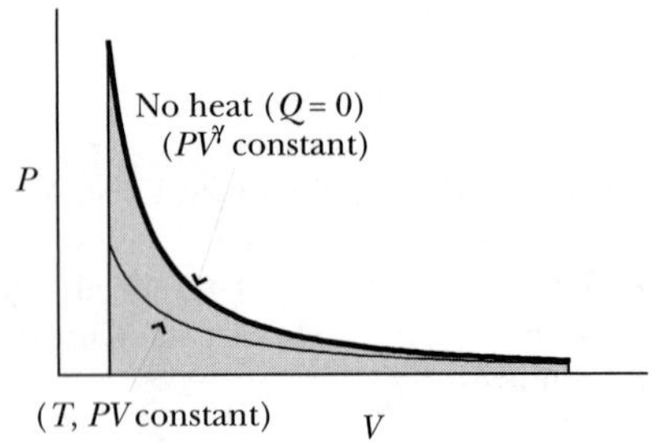

Figure 11.27 *PV* curves for a no-*Q* (adiabatic) process and a constant temperature (isothermal) process.

11.5.4 Work vs. thermal energy transfer

We have studied the response of a gas to energy inputs and outputs, both mechanical (work) and thermal (thermal energy transfer). Work represents organized, orderly, macroscopic energy input. Thermal energy transfer represents disorganized, disorderly, microscopic energy input. There is randomness at the atomic level in the collisions of atoms, which is the basic mechanism for thermal energy transfer. In Chapter 10 we saw deep consequences of the distinction between work W and thermal energy transfer Q.

11.6 *Application: A random walk

If you would find it useful to study one more application of the basic statistical concepts underlying our analysis of a gas, here is an interesting one. If we could watch one special molecule wandering around in the air, colliding frequently with air molecules, its path would look something like Figure 11.28. This kind of motion is called a "random walk." It may surprise you to find that despite the random nature of this motion we can calculate something significant about the motion, using simple statistical reasoning.

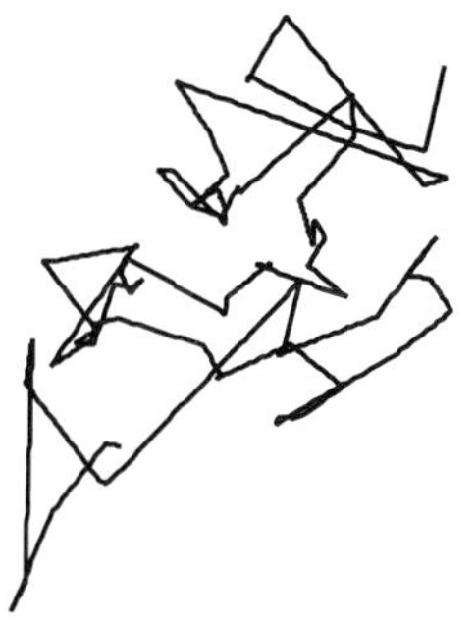

Figure 11.28 A random walk, generated by a computer program, using random numbers.

One dimension

For simplicity, first we'll consider just the x component of the motion. Pick the origin of the x axis to be at the original position of the special molecule. The first thing that happens is that the special molecule moves until it collides with an air molecule. We call this first x component of the displacement Δx_1. This component of the displacement may be to the right ($+x$ direction) or to the left ($-x$ direction). As a result of the collision, the special molecule may change direction, and change speed. The next displacement before another collision we call Δx_2, etc. After N collisions, the net displacement Δx away from the origin is

$$\Delta x = \Delta x_1 + \Delta x_2 + \Delta x_3 + \Delta x_4 + \ldots + \Delta x_N$$

After N collisions, what is the average (most probable) position of the special molecule? On average, it is just as likely that it has moved to the right as moved to the left, so the average x component of displacement ought to be zero. It is easy to see that this will be the case, by taking the average value:

$$\overline{\Delta x} = \overline{\Delta x_1} + \overline{\Delta x_2} + \overline{\Delta x_3} + \overline{\Delta x_4} + \ldots + \overline{\Delta x_N}$$

Each of the individual x displacements are equally likely to be to the left or the right, so the average value of the nth x displacement (for $n = 1, 2, 3, \ldots N$) is zero. Therefore the average value of the net x component of displacement is also zero.

On average, after N collisions the special molecule ends up to the left of the origin as often as it ends up to the right. But we can ask the question, "On average, how far away from the origin (along the x axis) does the special molecule get, no matter whether it ends up to the left or the right?" A good indicator of this distance is the average value of the square of the net x-displacement, $(\Delta x)^2$, because that's always a positive number. We can calculate this average.

Since $\Delta x = \Delta x_1 + \Delta x_2 + \Delta x_3 + \Delta x_4 + \ldots + \Delta x_N$, we have

$$(\Delta x)^2 = (\Delta x_1)^2 + (\Delta x_2)^2 + \ldots + (\Delta x_N)^2 + 2\Delta x_1 \Delta x_2 + 2\Delta x_1 \Delta x_3 + 2\Delta x_1 \Delta x_4 + \ldots$$

In this square of the net x displacement, there are two kinds of terms: squares of individual x displacements such as $(\Delta x_2)^2$ and "cross terms" like $2\Delta x_1 \Delta x_3$. Given the random nature of the process, we expect that one fourth of the time both Δx_1 and Δx_3 are positive (product is positive), one-fourth of the time they're both negative (product is positive), one-fourth of the time Δx_1 is positive with Δx_3 negative (product is negative), and one-fourth of the time Δx_1 is negative with Δx_3 positive (product is negative). Therefore, the average value of each cross term $\overline{2\Delta x_i \Delta x_j}$ $(i \neq j)$ is zero.

As for the other terms, the squares of individual x-displacements such as $(\Delta x_2)^2$, the average value of each of these terms is some number d_x^2, related to d^2, the square of the mean free path that we discussed in Section 11.2.2 on page 445.

? Count up how many of these square terms there are for N collisions and calculate the average value of the square of the net displacement:

$$\overline{(\Delta x)^2} = ?$$

There are N of these terms, and the square root of this quantity is the "root-mean-square" or "rms" value of the net displacement:

$$\Delta x_{\text{rms}} = \sqrt{\overline{(\Delta x)^2}} = (\sqrt{N})\, d_x$$

We can also write this formula in terms of time. If we let v be the average speed of the special molecule between collisions, and we let T be the average time between collisions, we have $v = d/T$. Also, the total time t for the N collisions is $t = NT$.

? Rewrite the formula for the rms displacement in terms just of d_x, v, and t (that is, eliminate N and T): $\Delta x_{\text{rms}} = ?$

We find that $\Delta x_{\text{rms}} = \sqrt{N d_x d_x} = \sqrt{N(vT) d_x} = \sqrt{v d_x}\sqrt{t}$. This is a somewhat curious result. For ordinary motion at constant speed, displacement increases proportional to time: double the time, double the displacement. In contrast, the rms displacement in a random walk grows with the square root of the time: on average, the rms displacement doubles when the time quadruples.

This calculation was done for the x component of the motion, but we can generalize our result to real three-dimensional motion in a gas. In Figure 11.29 is a three-dimensional displacement involving Δx, Δy, and Δz.

You can see in Figure 11.29 a three-dimensional version of the Pythagorean theorem for triangles. We can assume that the motion in each dimension is independent of the motions in the other two dimensions, and we have the following result:

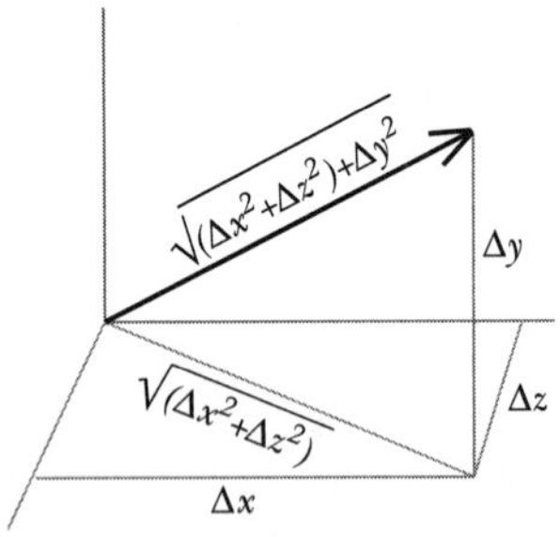

Figure 11.29 Pythagorean theorem in three dimensions.

$$\Delta r_{\text{rms}} = \sqrt{\overline{(\Delta x)^2} + \overline{(\Delta y)^2} + \overline{(\Delta z)^2}} = \sqrt{N d_x^2 + N d_y^2 + N d_z^2} = (\sqrt{N})\, d$$

(Here, d_x is the component in the x direction of the three-dimensional mean free path d.)

Writing the rms displacement in terms of the average speed v, the time t, and the mean-free path d, we have

$$\Delta r_{\text{rms}} = \sqrt{vd}\sqrt{t}$$

This result is somewhat unusual, because it predicts (correctly, it turns out) that for this random process the displacement from the starting location is proportional to the square root of the time rather than to the time.

Ex. 11.9 As you have calculated, the average speed of an air molecule at room temperature is about 500 m/s. What is Δr_{rms} for an air molecule after one second?

11.7 Summary

Fundamental physical principles

No new fundamental physical principles; application of older principles plus statistical reasoning.

New concepts

Statistical reasoning about systems containing large number of atoms.

Mean free path d: $n[\pi(R+r)^2(d)] \approx 1$

Root-mean-square speed: $v_{rms} \equiv \sqrt{\overline{v^2}}$

Work done on a gas: $W = -\int_{V_1}^{V_2} P dV$

First law of thermodynamics: $\Delta E_{\text{sys}} = W + Q$

Molecular heat capacity at constant volume C_v: $Q = NC_V \Delta T$

Molecular heat capacity at constant pressure C_p: $Q = NC_P \Delta T$

Random walk root-mean-square displacement:

$$\Delta r_{\text{rms}} = (\sqrt{N})d = \sqrt{vd}\sqrt{t}$$

Results

\# of gas molecules hitting an area A per second = $\frac{1}{4} nA\bar{v}$

$P = \frac{2}{3} n\left(\frac{\overline{p^2}}{2m}\right) = nkT$, where $n = N/V$ (# of molecules per cubic meter)

For a multiatom gas molecule, $\Delta \bar{E}_{\text{tot}} = \Delta(\frac{3}{2}kT + \bar{E}_{\text{rot}} + \bar{E}_{\text{vib}})$

$C_V = \frac{3}{2}k$ for a monatomic gas (He, etc.)

$C_V \geq \frac{3}{2}k$ for other gases (N_2, etc.)

$C_P = C_V + k$ (molecular heat capacity);

$C_P = C_V + R$ (molar heat capacity)

In a constant-temperature (isothermal) compression,

$$W_{\text{by piston}} = Q_{\text{into surroundings}} = NkT\ln\left(\frac{V_1}{V_2}\right)$$

In a no-Q (adiabatic) compression,

$$PV^{\gamma} = \text{constant}, \text{ where } \gamma = C_p/C_v, \text{ and also } T^{\left(\frac{C_V}{k}\right)}V = \text{constant}$$

$k = 1.4\times10^{-23}$ J/K $\qquad R = (6\times10^{23})k = 8.3$ J/K

At "standard temperature and pressure" (STP, which means a temperature of 0° Celsius or 273 Kelvin, and a pressure of 1 atmosphere), one mole of a gas occupies 22.4 liters (a liter is 1000 cubic centimeters).

Sea-level pressure (1 atm) is 10^5 N/m^2

11.8 Review questions

Gas leaks

RQ 11.1 Gas leaks at a rate L (in molecules per second) through a small circular hole. If the density of the gas is doubled, and the Kelvin temperature is doubled, and the radius of the hole is doubled, what is the new leak rate?

Mean free path

RQ 11.2 How does the mean free path of an atom in a gas change if the temperature is increased, with the volume kept constant?

RQ 11.3 What is the approximate time between collisions for one particular air molecule?

Speed of sound

RQ 11.4 How does the speed of sound in a gas change when you raise the temperature from 0° C to 20° C? Explain briefly.

Compressions and expansions

RQ 11.5 If you expand the volume of a gas containing N molecules to twice its original volume, while maintaining a constant temperature, how much thermal energy transfer is there from the surroundings?

RQ 11.6 If you compress a volume of helium containing N atoms to half the original volume in a well-insulated cylinder, what is the ratio of the final pressure to the original pressure?

11.9 Homework problems

Problem 11.1 Leakage from a tank

In the example problem (page 445) we considered leakage from a flexible party balloon. If the leakage is from a rigid container (a metal storage tank, for example), the number of atoms per unit volume, n, will decrease with time t. If the total volume of the tank is V, there are $N = nV$ atoms in the tank at any instant, so we can write the following "differential" equation (that is, an equation that involves derivatives):

$$\frac{d}{dt}(nV) = -\tfrac{1}{4}nA\bar{v}$$

This says "the rate of change of the number of atoms in the tank is equal to the (negative) of the rate at which atoms are leaving the tank."

(a) Show that the differential equation is satisfied if $n = n_{\text{initial}}e^{-\frac{A\bar{v}}{4V}t}$, where t is the time elapsed since the hole was made. Just plug this function of n, and its derivative, into the equation and show that the two sides of the equation are equal for all values of the time t. Also show that the initial particle density n is equal to n_{initial}.

(b) Despite the fancy math, this solution is really only approximate, because the average speed isn't a constant but is decreasing. Suppose however that we use a heater to keep the container and the gas at a nearly constant temperature, so that the average speed does remain nearly constant. Suppose the (rigid) container is again a sphere 30 cm in diameter, with a circular hole 1 millimeter in diameter. About how long would it take for most of the helium to leak out? Explain your choice of what you mean by "most."

Problem 11.2 A helium leak

A rigid, thermally insulated container with a volume of 22.4 liters is filled with one mole of helium gas (4 grams per mole) at a temperature of 0 Celsius (273 K). The container is sitting in a room, surrounded by air at STP.

(a) Calculate the pressure inside the container in N/m^2.

(b) Calculate the root-mean-square average speed of the helium atoms.

(c) Now open a tiny square hole in the container, with area 10^{-8} m^2 (the hole is 0.1 mm on a side). After 5 seconds, how many helium atoms have left the container?

(d) Air molecules from the room enter the container through the hole during these 5 seconds. Which is greater, the number of air molecules that enter the container or the number of helium atoms that leave the container? Explain briefly.

(e) Does the pressure inside the container increase slightly, stay the same, or decrease slightly during these 5 seconds? Explain briefly. If you have to make any simplifying assumptions, state them clearly.

Problem 11.3 Gaseous diffusion

Natural uranium ore consists mostly of the isotope U-238 (92 protons and 146 neutrons), but 0.7% of the ore consists of the isotope U-235 (92 protons and 143 neutrons). Because only U-235 fissions in a reactor, industrial processes are used to enrich the uranium by enhancing the U-235 content.

One of the enrichment methods is "gaseous diffusion." The gas UF_6, uranium hexafluoride, is manufactured from supplies of natural uranium and fluorine (each of the six fluorine atoms has 9 protons and 10 neutrons). A container is filled with UF_6 gas. There are tiny holes in the container, and gas molecules leak through these holes into an adjoining container, where pumps sweep out the leaked gas.

(a) Explain why the gas that initially leaks into the second container has a slightly higher fraction of U-235 than is found in natural uranium.

(b) Estimate roughly the practical change in the concentration of U-235 that can be achieved in this single-stage separation process. Explain what approximations or simplifying assumptions you have made to obtain your estimate.

(c) A typical nuclear reactor requires uranium that has been enriched to the point where about 3% of the uranium is U-235. Estimate roughly the number of stages of gaseous diffusion required (that is, the number of times the gas must be allowed to leak from one container into another). Note that the effects are multiplicative.

This is why a practical gaseous diffusion plant has a large number of stages, each operating at high pressure, which makes this an expensive process. The first large gaseous diffusion plant was constructed during World War II at Oak Ridge, Tennessee, and used cheap Tennessee Valley Authority electricity.

Problem 11.4 Spacecraft emergency

You are on a spacecraft measuring 8 m by 3 m by 3 m when it is struck by a piece of space junk, leaving a circular hole of radius 4 mm, unfortunately in a place that can be reached only by making a time-consuming spacewalk. About how much time do you have to patch the leak? Explain what approximations you make in assessing the seriousness of the situation.

Problem 11.5 Meteor in air

A roughly spherical meteor made mainly of iron (density about 8 grams per cubic centimeter) is hurtling downward through the air at low altitude. At an instant when its speed is 10^4 m/s, calculate the approximate rate of change of the meteor's speed. Do the analysis for two different meteors—one with a radius of 10 meters and one with a radius of 100 meters.

Start from fundamental principles. Do not try to use some existing formula that applies to a very different situation. Follow the kind of *reasoning* used in this chapter, applied to the new situation, rather than trying to use the *results* of this chapter.

A major difference from our earlier analyses is that the meteor is traveling much faster than the average thermal speed of the air molecules, so it is a good approximation to consider the air molecules below the meteor to be essentially at rest, and to assume that no air molecules manage to catch up with the meteor and hit it from behind. The meteor drills a temporary hole in the atmosphere, a vacuum, that gets filled explosively by air rushing in after the meteor has passed.

There is good evidence that a very large meteor, perhaps 10 kilometers in diameter, hit the Earth near the Yucatan Peninsula in southern Mexico 65 million years ago and caused so much damage that the dinosaurs became extinct. See the excellent account in *T. rex and the Crater of Doom*, by Walter Alvarez (Princeton University Press, 1997). Alvarez is the geologist who made the first discoveries leading to our current understanding of this cataclysmic event.

Problem 11.6 Manipulating a gas

A horizontal cylinder 10 cm in diameter contains helium gas at room temperature and atmospheric pressure. A piston keeps the gas inside a region of the cylinder 20 cm long.

(a) If you *quickly* pull the piston outward a distance of 12 cm, what is the approximate temperature of the helium immediately afterwards? What approximations did you make?

(b) How much work did you do, including sign? (Note that you need to consider the outside air as well as the inside helium in calculating the amount of work *you* do.)

(c) Immediately after the pull, what force must you exert on the piston to hold it in position (with the helium enclosed in a volume that is 20+12 = 32 cm long)?

(d) You wait while the helium slowly returns to room temperature, maintaining the piston at its current location. After this wait, what force must you exert to hold the piston in position?

(e) Next, you very *slowly* allow the piston to move back into the cylinder, stopping when the region of helium gas is 25 cm long. What force must you exert to hold the piston at this position? What approximations did you make?

Problem 11.7 Pumping up a bicycle tire

Atmospheric pressure at sea level is about 10^5 N/m^2, which is about 15 pounds per square inch (psi). A bicycle tire typically is pumped up to 50 psi above atmospheric pressure (psi "gauge"), for an actual pressure of about 65 psi. In rapidly pumping up a bicycle tire, starting from atmospheric pressure, about how high does the temperature of the air rise? Explain what approximations and simplifying assumptions you make.

11.10 Answers to exercises

11.1 (page 444) 6.3×10^{21} atoms per second

11.2 (page 445) T decreases because average v decreases.

11.3 (page 446) $L \approx 0.3\times10^{-10}$ m

11.4 (page 449) 10^5 N/m^2

11.5 (page 452) Box alone; buoyancy force is Mg upward.

11.6 (page 457) 1000 J; 1000 J; 0

11.7 (page 459) $\frac{3}{2}k + k = \frac{5}{2}k$

11.8 (page 460) $\left(\frac{PV}{Nk}\right)^{c_v/k} V = \text{constant}$

$P^{c_v/k} V^{c_v/k+1} = \text{new constant}$

Take (c_v/k) root:

$PV^{1+k/c_v} = PV^{(c_v+k)/c_v} = PV^{c_p/c_v}$

11.9 (page 462) 6 millimeters

Chapter 12

The Efficiency of Engines

Chapter 12

The Efficiency of Engines

12.1 Fundamental limitations on efficiency

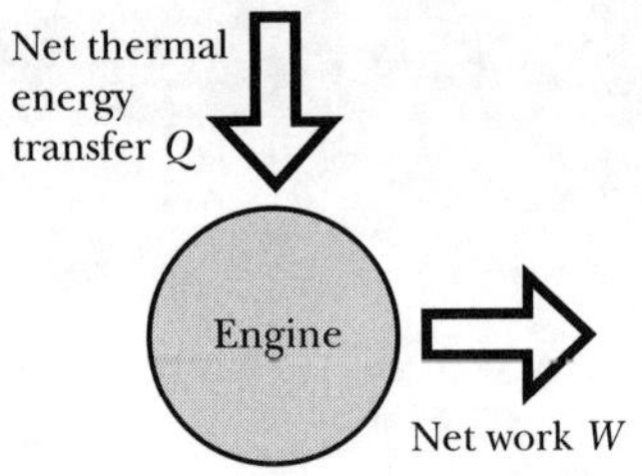

Figure 12.1 An engine converts net thermal energy transfer Q into net work W.

An important technology involves the conversion of thermal energy transfer Q into useful work W (Figure 12.1). For example, in a steam-powered electricity generating plant, coal is burned to warm up water that drives a steam engine to turn a generator, which converts the work done by the steam engine into electric energy. Energy conservation of course puts limits on how much useful work you can get from the burning of the coal. But there is a further limitation due to the second law of thermodynamics. It turns out that in a practical generating plant only about one-third of the energy of the coal can be turned into useful work! The fundamental problem is that thermal energy transfer is a disorderly energy transfer, and the second law of thermodynamics takes that into account.

We will find that the most efficient processes are "reversible" processes, that is, processes that produce no change in the entropy of the Universe (a reversed movie of such a process looks possible). In the next sections we discuss two processes, mechanical friction and finite-temperature-difference thermal energy transfer, which are major contributors to entropy production and whose effects must be minimized in order to obtain the most efficient performance in a mechanical system. Where we are headed is to establish limits on how efficiently thermal energy can be turned into mechanical energy in an engine, as a consequence of the second law of thermodynamics. This is a revealing example of the power of the second law of thermodynamics to set limits on phenomena.

12.1.1 Friction and entropy production

The second law of thermodynamics says that the entropy of the Universe never decreases. Portions of the Universe may experience a decrease of entropy, but only if the entropy of other portions increases at least as much (and typically more). An example of a process that increases the total entropy of the Universe is sliding friction. A block sliding across a table slows down as kinetic energy associated with the overall motion of the block turns into thermal energy inside the block, with an increase in entropy.

This friction process is irreversible. We would be astonished if after coming to rest the block suddenly started picking up speed back toward its starting position, although this would not violate conservation of energy. But the probability is vanishingly small that the thermal energy in the block could convert back into an orderly motion of the block.

12.1.2 Thermal energy transfer and entropy production

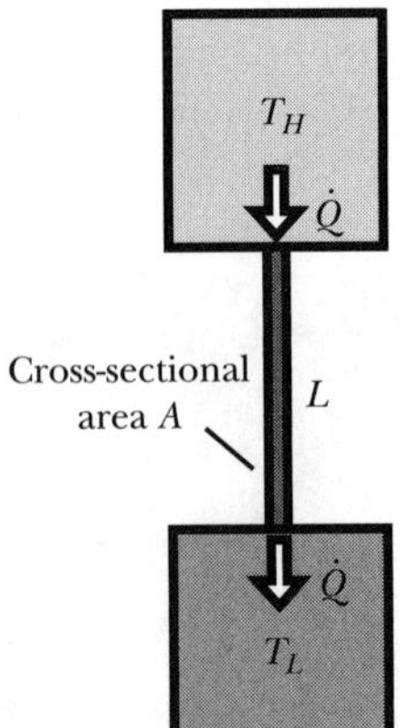

Figure 12.2 Energy flows as thermal energy transfer at a rate $\dot{Q}$ from high temperature to low.

Mechanical friction contributes to increasing the entropy of the Universe. We will show that thermal energy transfer between two objects that have different temperatures also makes the total entropy of the Universe increase. This is then a process to avoid if possible in an efficient engine.

Connect a metal bar between a large block at a high temperature T_H and another large block at a low temperature T_L (Figure 12.2). We write the rate of thermal energy transfer from the high-temperature block (the "source") to the lower-temperature block (the "sink") as $\dot{Q}$. The dot over the letter Q means "rate" or "amount per second." This thermal energy transfer rate is

- proportional to the "thermal conductivity" σ of the bar (metals have higher thermal conductivity than glass or plastic),
- proportional to the temperature difference (twice the temperature difference, twice the rate of energy transfer),
- proportional to the cross-sectional area A of the bar (twice the cross-sectional area is like having two bars), and
- inversely proportional to the length L of the bar (twice the length of the bar, half the energy transfer rate):

RATE OF THERMAL ENERGY TRANSFER

$$\dot{Q} = \sigma A \frac{(T_H - T_L)}{L} \quad \text{(J/s)}$$

The quantity $(T_H - T_L)/L$ is called the "temperature gradient." The larger the gradient (the more rapidly the temperature changes with distance along the bar), the larger the number of joules per second of thermal energy transfer.

We want to show you that thermal energy transfer between different temperatures increases the total entropy of the Universe. Suppose the source and sink are so large that for a while the thermal energy transfer doesn't change the temperatures much. Since $1/T = (dS)/(dE)$, it follows that a small thermal energy input (dE) into a system, of amount Q, leads to an entropy change of that system:

$$\Delta S = \frac{Q}{T}$$

Suppose the thermal energy transfer *rate* is $\dot{Q}$. In a short time interval Δt, the high-temperature source at temperature T_H loses an amount of energy $\dot{Q}\Delta t$.

? In this short time interval Δt, does the entropy of the source increase or decrease? By how much?

Since $\Delta S_H = (\Delta E)/T_H$, and the high-temperature source loses energy, we have $\Delta S_H = -(\dot{Q}\Delta t)/T_H$.

In this same time interval, the low-temperature sink at temperature T_L gains the same amount of energy, $\dot{Q}\Delta t$.

? In this short time interval Δt, does the entropy of the sink increase or decrease? By how much?

Since the low-temperature sink gains energy, we have $\Delta S_L = +(\dot{Q}\Delta t)/T_L$

Here is a crucial and perhaps not entirely obvious point: Does the entropy of the bar change during this time interval? No. The blocks are so large that their temperatures don't change much in a short time interval, so the temperatures at the ends of the bar, and all along the bar, are not changing. Every second, energy enters the bar, but the same amount of energy leaves the bar, so the energy of the bar isn't changing. Nothing about the bar is changing. It is merely a conduit for the thermal energy transfer, but its state is not changing.

Therefore the change in the total entropy of the Universe in a time interval Δt is

$$\Delta S_{\text{source}} + \Delta S_{\text{sink}} + \Delta S_{\text{bar}} = -\frac{\dot{Q}\Delta t}{T_H} + \frac{\dot{Q}\Delta t}{T_L} + 0 = \dot{Q}\Delta t\left(\frac{1}{T_L} - \frac{1}{T_H}\right)$$

? Is this quantity positive or negative? Is this consistent with the second law of thermodynamics?

Since $T_L < T_H$, $1/T_L > 1/T_H$, and the entropy change of the Universe in this process is positive. This is consistent with the second law of thermodynamics, which states that the entropy of the Universe should never decrease.

Reversible and irreversible processes

An increase in the total entropy of the Universe is associated with an irreversible process, because to return to the earlier state would require that the total entropy of the Universe actually decrease, which won't happen with macroscopic systems. We would be astonished if the thermal energy transfer suddenly reversed and ran from the colder block thermally back "uphill" to the hotter block!

There is only one way to carry out thermal energy transfer reversibly or nearly so (that is, with little or no change in the total entropy of the Universe)—do it between systems whose temperatures are very nearly equal to each other ($T_H \approx T_L$). But there's a practical problem.

? If the two temperatures are nearly equal (to avoid increasing the entropy of the Universe), at what rate is energy transferred from the (slightly) hotter source to the (slightly) cooler sink?

Thermal energy transfer will flow extremely slowly, because $\dot{Q}$ is proportional to the temperature difference (or more specifically to the temperature gradient). If there is hardly any temperature difference, the energy transfer rate will be very small. So we can carry out thermal energy transfers nearly reversibly, hardly changing the entropy of the Universe, but only if we do it so slowly as to be of little practical use in driving some kind of mechanical engine that converts thermal energy transfer into work.

Ex. 12.1 The thermal conductivity of copper (a good thermal conductor) is 400 watt/K/m (for comparison, the thermal conductivity of iron is about 70 watt/K/m, and that of glass is only about 1 watt/K/m). One end of a copper bar 1 cm on a side and 30 cm long is immersed in a large pot of boiling water (100° C), and the other end is embedded in a large block of ice (0° C). It takes about 335 joules to melt one gram of ice. How long does it take to melt one gram of ice? Does the entropy of the Universe increase, decrease, or stay the same?

12.2 A maximally efficient process

Despite the serious practical problem that reversible thermal energy transfer proceeds infinitesimally slowly, we will analyze reversible "engines" in which (nearly) reversible thermal energy input with tiny temperature differences is used (very slowly!) to do something mechanically useful, such as lift a weight or turn an electric generator. We will also assume that we can nearly eliminate sliding friction. The idea that we will pursue is to see how efficiently we can convert disorderly thermal energy into orderly mechanical energy (a lifted weight). We expect that reversible processes, which don't increase the total entropy of the Universe, should be the most efficient processes, though we recognize that these most efficient processes must proceed excruciatingly slowly.

Since we will use nearly reversible processes to do work, we could run the engine backwards, and such a backwards-running engine turns out to be a refrigerator—an engine in which work input to the engine can lead to extracting thermal energy from something to make it colder or to keep it cold.

The conception of a theoretically most efficient possible engine, and the recognition that this ideal engine would have to be reversible and not increase what we now call the total entropy of the Universe, was due to a young French engineer, Sadi Carnot, in 1824. His achievement is all the more remarkable because it came before the principle of energy conservation was established!

After using the second law of thermodynamics to determine the maximum possible efficiency of infinitely slow engines, in the last part of the chapter we will analyze engines that run at useful speeds. We will find that they are even less efficient than the infinitely slow engines.

12.2.1 A cyclic process of a reversible engine

Some of the first practical engines drove pumps to pump water out of deep mines in England, repeatedly lifting large amounts of water large distances. The energy came from burning coal, which boiled water to make steam in a cylinder which pushed up on a piston, and operated the pump. People started wondering how efficient such an engine could be. What limits the amount of useful work you can get from burning a ton of coal?

We describe a scheme for running an engine in a reversible way, which ought to be as efficient as possible. For concreteness, our engine is a device consisting of a cylinder containing an ideal gas, with a piston and some sand on the piston to adjust the pressure on the gas (Figure 12.3). This is a simple and familiar device which permits energy exchanges in the form of work and thermal energy transfer. As we'll see, the actual construction details of the engine don't matter for the theoretical question regarding the maximum possible efficiency, though they matter very much in the actual construction of a useful engine.

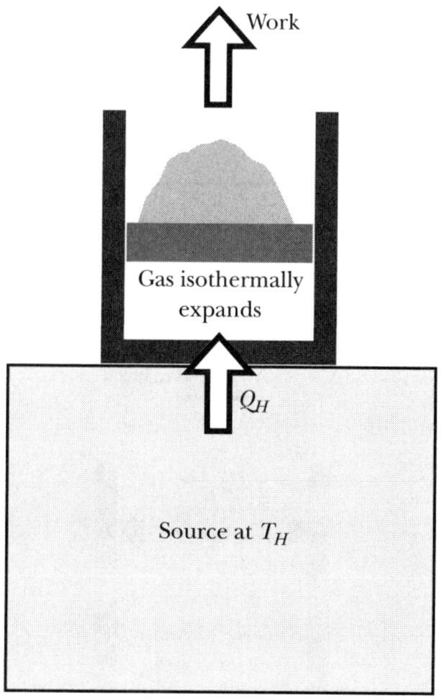

Figure 12.3 Constant-temperature expansion; gas temperature just slightly lower than T_H.

A constant-temperature (isothermal) expansion

In addition to the engine itself, we need a large high-temperature source, large enough that extracting some energy from it will cause only a negligible decrease in its temperature. Start by arranging that the engine (the gas cylinder with its piston) has a temperature very slightly lower than the temperature T_H of the high-temperature source. We connect the engine to the high-temperature source and perform a reversible constant-temperature (isothermal) expansion of the engine, doing some work on the surroundings in the process (Figure 12.3). We lift the piston by slowly shifting some sand sideways from the piston onto nearby shelves. This requires almost no work on our part but results in the lifting of some weight. We make sure that the temperature of the gas doesn't change during this process. There is thermal energy transfer into the engine because it is very slightly cooler, but the temperature difference is so slight that there is negligible change in the entropy of the Universe.

Entropy and energy in lifting the piston

Remember that the entropy change of a system when there is thermal energy transfer Q is $\Delta S = Q/T$. In lifting the piston the entropy of the high-temperature source has changed by an amount $\Delta S_H = -Q_H/T_H$, where Q_H is a positive quantity. The entropy of the gas has changed by an amount $\Delta S_{\text{gas}} = +Q_H/T_H$. The entropy of the Universe hasn't changed at all. (The increased entropy of the gas is associated with the fact that there are more ways to arrange the molecules in a large volume than in a small volume.)

Because the temperature hasn't changed, there is no change in the energy of the ideal gas. Recall that the energy of an ideal gas, or a low-density real gas, is a function solely of the temperature, not the volume. (A dense real gas is more complicated, because the electric potential energy for pairs of molecules depends on distance, and therefore the energy of the gas depends on volume as well as temperature.)

We have succeeded in converting all of the thermal energy transfer from the high-temperature source into useful work (lifting the piston), because none of the input energy went into changing the energy of the gas. This is 100% efficiency in converting thermal energy into useful work on the surroundings. What's the problem?

The need for a cycle

The problem is this: We need to be able to do this again, and again, and again. For example, we want to keep pumping water out of the deep mine, over and over. But to repeat the lifting process, we have to return to the original state, with the gas compressed. We could simply reverse the process, doing work on the gas (constant-temperature compression), with thermal energy transfer from the gas into the high-temperature source. But the net effect would be that there was zero net thermal energy transfer to the gas, and zero net work done on the surroundings.

We need to run our engine in a non-trivial repetitive "cycle," where we can lift the piston repeatedly. Each time we lift the piston and do useful work, we need to get the piston back down again without having to do the same amount of work to push it down. If we can get the gas back to its original state with net work done on the surroundings, we will have a useful device.

One possibility for bringing the piston down would be to let the gas cool down. But to lower the temperature we would have to place the gas cylinder in contact with a large object at some low temperature T_L, called the "sink" because, as we'll see, we will dump some thermal energy into it. We can't place the gas cylinder in contact with the sink immediately, because the gas is at a high temperature T_H, and placing the hot gas in contact with the cold sink would mean that there would be a large temperature difference, and there would be a large increase in the entropy of the Universe associated with the thermal energy transfer from the hot gas to the cold sink.

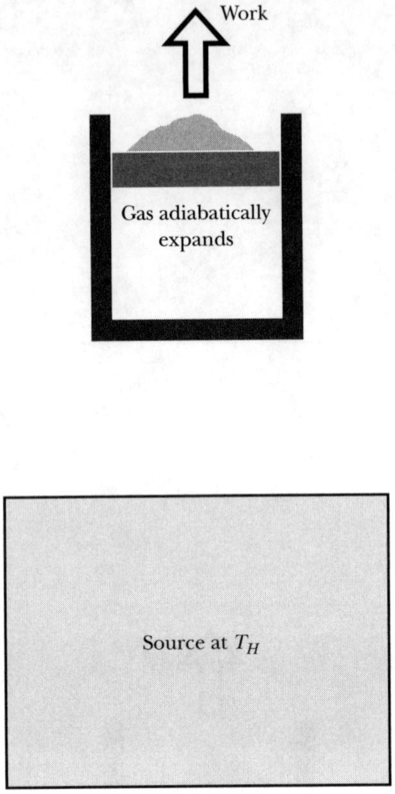

Figure 12.4 No-Q expansion; gas temperature drops to T_L.

A no-Q (adiabatic) expansion

To avoid this large production of entropy, we need to bring the temperature of the gas down almost to T_L *before* making contact with the sink. To do this, we disconnect the gas cylinder from the high-temperature source and perform a reversible no-Q (adiabatic) expansion, which does some more work on the surroundings and lowers the temperature of the engine (Figure 12.4). This is accomplished by slowly removing some weight from the piston, allowing the gas to expand and the temperature to fall. We stop the expansion when the temperature of the gas is just slightly higher than the low temperature T_L of the sink.

A constant-temperature (isothermal) compression

We can now safely place the gas cylinder in contact with the low-temperature sink. This is okay as far as entropy production is concerned, because we made sure that the temperature of the gas is only very slightly higher than the sink temperature T_L. The piston is now quite high, and we need to bring it down. We do this by slowly adding some weight to the piston, performing a constant-temperature (isothermal) compression (Figure 12.5). There is thermal energy transfer into the sink. Because the temperature of the ideal gas is nearly T_L at all times, there is no energy change in the gas. Therefore the work that we do is equal to the thermal energy transfer Q_L into the sink.

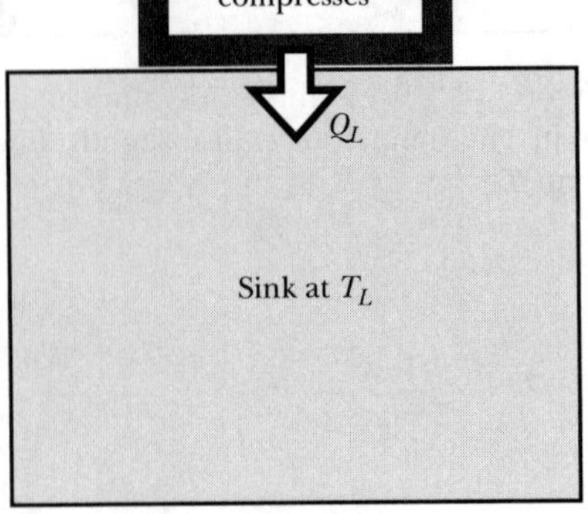

Figure 12.5 Constant-temperature compression; gas temperature just slightly higher than T_L.

There is an increase in the entropy of the sink $\Delta S_L = +Q_L/T_L$, and there is an entropy decrease in the entropy of the gas $\Delta S_{\text{gas}} = -Q_L/T_L$. The entropy of the Universe doesn't change. (The decreased entropy of the gas is associated with the fact that there are fewer ways to arrange the molecules in a small volume than in a large volume.

A no-Q (adiabatic) compression

We stop the compression at a particular state of the gas chosen with care so that we can do the following: We disconnect the engine from the low-temperature sink and perform a no-Q compression that raises the temperature of the gas to a temperature just barely less than T_H, the temperature of the high-temperature source. We do some work to perform this compression (Figure 12.6).

This compression brings the system back to its original state (density, pressure, temperature), so we can repeat the cycle of four processes all over again. The possible usefulness of this engine is that we can run it repeatedly, over and over. What remains to be analyzed is how much work is done on the surroundings in one cycle, and how much thermal energy input we need to supply. Of course if we find that the net work is zero, the engine won't be useful. However, explicit calculations for an ideal gas show that at least in that case there is net work done on the surroundings. We will find that this is true for any reversible engine run in such a cycle.

This reversible cycle of two constant-temperature (isothermal) processes and two no-Q (adiabatic) processes running between a high-temperature source and a low-temperature sink is called a "Carnot cycle."

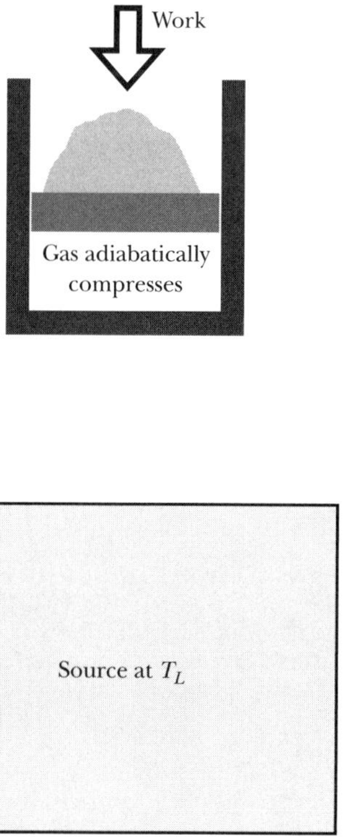

Figure 12.6 No-Q compression; gas temperature rises to T_H. Engine is now back to its original state, ready to begin another cycle.

Summary of the cycle

Here is a summary of the four processes of the cycle:

- Isothermal expansion in contact with T_H
- Adiabatic expansion (no contact); temperature falls to T_L
- Isothermal compression in contact with T_L
- Adiabatic compression (no contact); temperature rises to T_H

The four processes return the gas to its original temperature and volume. The question is, did we get any net work out of this cycle?

12.2.2 Entropy in a cycle of a reversible engine

Let Q_H be the (absolute value of the) thermal energy extracted from the source at temperature T_H during the constant-temperature expansion, and let Q_L be the (absolute value of the) thermal energy dumped into the sink at temperature T_L during the constant-temperature compression. Remember that the entropy change of a system when there is thermal energy transfer Q is $\Delta S = Q/T$.

$$\text{The entropy change of the high-temperature source in one cycle} = -\frac{Q_H}{T_H}$$

$$\text{The entropy change of the low-temperature sink in one cycle} = +\frac{Q_L}{T_L}$$

? Is there any entropy change *in the engine* in one cycle?

The entropy change of the engine in one cycle is zero, because the state of the gas is brought back to its original state.

? Therefore, in one cycle, how much entropy change is there in the surroundings as a result of the work done on and by the surroundings?

Only the thermal energy transfer affects the entropy during one cycle (work doesn't contribute), and we calculated these through the formula $\Delta S = Q/T$. We were careful to run the engine in a reversible way, avoiding making any increase in the total entropy of the Universe, so in one cycle we have

$$-\frac{Q_H}{T_H}+\frac{Q_L}{T_L}+0 = 0$$

Therefore we can write this important relationship for one cycle of the engine:

ONE CYCLE OF A REVERSIBLE ENGINE

$$\frac{Q_H}{T_H}=\frac{Q_L}{T_L}$$

This is a surprisingly simple result for such a complicated process. What does this simple result depend on? Solely on the second law of thermodynamics, that statistically the total entropy of the Universe will essentially never decrease.

We illustrated the processes in the cycle with a cylinder filled with an ideal gas, but the result above doesn't actually depend at all on the details of how the engine is constructed. In particular, it doesn't matter whether the engine contains helium gas, or a block of rubber, or a liter of alcohol. There would be an easily visible difference between using a solid or liquid rather than a gas in the engine, because the distance through which the piston would move would be much smaller than with a gas. But the simple result shown above would remain the same.

12.2.3 Energy in a cycle of a reversible engine

Let's review the energy changes in the engine in one cycle. The source inputs an amount of energy Q_H. The sink removes an amount of energy Q_L. The working substance returns to its original state, so it undergoes no change of energy. Is there anything else? Yes, it may be that there has been net work W done on the surroundings. How can we tell? We can use conservation of energy of the engine for one cycle:

$$\Delta E_{\text{engine}} = Q_H - Q_L + W = 0$$

? Use the result that $Q_H/T_H = Q_L/T_L$ (because the total entropy of the Universe doesn't change), to determine whether W is positive or negative.

Because $Q_H = (T_H/T_L)Q_L > Q_L$, we have $W = Q_L - Q_H < 0$. The fact that W is negative means that our engine does net work on the surroundings in one cycle. Figure 12.7 outlines the basic scheme. In one cycle the net effect is that the engine takes in energy Q_H from a hot source, does some work on something (for example, turns an electric generator), and exhausts the remaining energy Q_L into a cold sink. The magnitude of the work done is $|W| = Q_H - Q_L$.

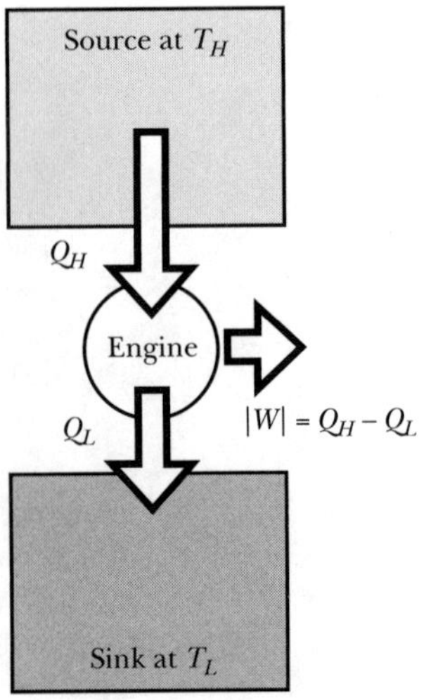

Figure 12.7 Conservation of energy lets us determine the net work output.

The crucial issue is that the engine will not run in repeatable cycles without exhausting some energy to the low-temperature sink. You *cannot* convert all of the thermal energy Q_H into useful work. We need the low-temperature block to allow us to begin the constant-temperature compression phase of the cycle in a way that avoids any thermal energy transfer with significant temperature difference, which would increase the total entropy of the Universe.

We have to pay for the high-temperature energy Q_H which we use, and the loss of some of the input energy in the form of Q_L is an unfortunate fact of life. We pay for coal or fuel oil or electricity to warm something up to the high temperature T_H, from which we can draw energy to do work for us.

For example, in an old railroad steam engine a fire maintained water at a high temperature to constitute the high-temperature source. In an automobile engine we burn gasoline to create a high temperature and push the pistons. In an electricity generating station we burn coal or fuel oil, or use nuclear fission reactions, to create a high-temperature source from which to drive the generators. We'd like to get our full money's worth (Q_H), but we don't. We only get $|W| = Q_H - Q_L$, having "lost" some of the energy to a low-temperature sink, which typically is our ordinary surroundings at around 20° C (293 K).

Efficiency

This leads to the question, what fraction of the energy Q_H we pay for turns into useful work W? This fraction is called the "efficiency" of the engine:

$$\text{efficiency} \equiv \frac{|W|}{Q_H}$$

? Given this definition of efficiency, prove the following result:

THE EFFICIENCY OF A REVERSIBLE ENGINE

$$\text{Efficiency} = 1 - \frac{T_L}{T_H}$$

We have $|W|/Q_H = (Q_H - Q_L)/Q_H = 1 - Q_L/Q_H = 1 - T_L/T_H$. This may seem surprising. The efficiency of a reversible engine depends solely on the ratio of the absolute temperatures of the source and sink. We emphasize that it doesn't matter how the engine is designed, or what kind of materials it is made of. This is the highest efficiency we can ever achieve for converting thermal energy into useful work.

Ex. 12.2 What is the efficiency of a reversible engine if the source is a large container of boiling water and the sink is a large block of melting ice?

Ex. 12.3 An actual electricity generating plant is powered by a nuclear reactor, with a high temperature of 300° C and low temperature of 25° C (near room temperature). What is the efficiency, assuming that we can treat the processes as being reversible?

12.2.4 No other engine can be more efficient

We can show that no engine running between temperatures T_H and T_L can be more efficient than a reversible engine.

The proof is a "proof by contradiction." Suppose that an inventor claims to have invented some cleverly designed new kind of (cyclic) engine, with a higher efficiency than $(1 - T_L/T_H)$ when run between these same two temperatures. In that case for a given Q_H the new engine would exhaust less Q_L than is exhausted from a reversible engine, and the entropy change of the Universe with this new engine would be negative instead of zero:

$$-\frac{Q_H}{T_H} + \frac{Q_L}{T_L} < 0 \quad \text{(Impossible!)}$$

This would be a violation of the second law of thermodynamics, so it is impossible. We can be quite sure that the inventor's claims are not valid. Note carefully that the inventor's claims don't violate energy conservation. Nothing about energy conservation forbids converting 100% of the input energy Q_H into work W in a cycle. The impossibility lies rather in the massive improbability of seeing the total entropy of the Universe decrease.

We took great pains to make a reversible engine, for which the total entropy change of the Universe was zero. No other engine can be more efficient. In fact, any real engine will have some friction and some thermal energy transfer across differing temperatures, in which case the entropy change of the Universe will be positive. So the second law of thermodynamics leads to the following, where "= 0" applies only to ideal, reversible engines, and "> 0" applies to real engines:

ENTROPY CHANGE OF THE UNIVERSE FOR REAL ENGINES

$$\Delta S_{\text{Universe}} = -\frac{Q_H}{T_H} + \frac{Q_L}{T_L} \geq 0$$

We should not underestimate the need for the ingenuity of inventors, however. The reversible engine made with an ideal gas cylinder is pretty useless for practical purposes. The challenge of ingenuity is to design engines that are practical, and this involves many kinds of engineering design decisions and trade-offs. But no matter how ingenious the design, the second law of thermodynamics puts a rigid limitation on how efficient *any* engine can be.

Ex. 12.4 If $\Delta S_{\text{Universe}} > 0$ in one cycle of an engine, show that the efficiency is less than the ideal efficiency $(1 - T_L/T_H)$ obtained with a reversible engine.

12.2.5 Running the engine backwards: A refrigerator or heat pump

We could run our engine backwards, since we took care to make all aspects of the cycle reversible. Starting from our original starting point, we would disconnect from the hot block, perform a no-Q expansion to lower the temperature to that of the cold block, connect to the cold block, do a constant-temperature expansion, then disconnect from the cold block. Next we do a no-Q compression to raise the temperature to that of the hot block, connect to the hot block, and do a constant-temperature compression back to the original state.

The net effect in one cycle is that the low-temperature block gives some energy Q_L to the engine, the high-temperature block absorbs some energy Q_H from the engine, and we do some (positive) work on the system (instead of the engine doing work on the surroundings). Here is the formula for conservation of energy of the engine when we run this reversed engine:

$$\Delta E_{\text{engine}} = Q_L - Q_H + W = 0$$

? Use the result that $Q_H/T_H = Q_L/T_L$ (which is still valid for our reversed engine, as you can convince yourself) to determine whether W is positive or negative.

Since $Q_H = (T_H/T_L)Q_L > Q_L$, we have $W = Q_H - Q_L > 0$. This seems an odd kind of "engine": it doesn't do any work for us—we have to do work on it. Is that of any use? Yes! At the cost of doing some work, we extract energy from a low-temperature block and exhaust it into a high-temperature block. This is a refrigerator (Figure 12.8).

Consider how we keep food cold in an ordinary home refrigerator. A reversed engine maintains the inside of the refrigerator at a low temperature

T_L, while the room is at a higher temperature T_H. Although the door and walls are heavily insulated, some thermal energy transfer does leak through from the room, and this energy, Q_L, must be removed to keep the food at a constant low temperature T_L. We exhaust an amount of energy Q_H into the room at temperature T_H. It takes an amount of work W to achieve this effect of moving energy out of the cold region and into the warm region.

In the most favorable case (no increase in total entropy) we have the following two conditions stemming from entropy and energy considerations (the high-temperature exhaust energy must equal the sum of the low-temperature energy plus the work we put in to drive the cycle):

$$\frac{Q_H}{T_H} = \frac{Q_L}{T_L} \text{ and } Q_H = Q_L + W$$

Here our view of "efficiency" shifts a bit. What we care about is how much work W we have to do to remove an amount of leakage energy Q_L from the inside of the refrigerator.

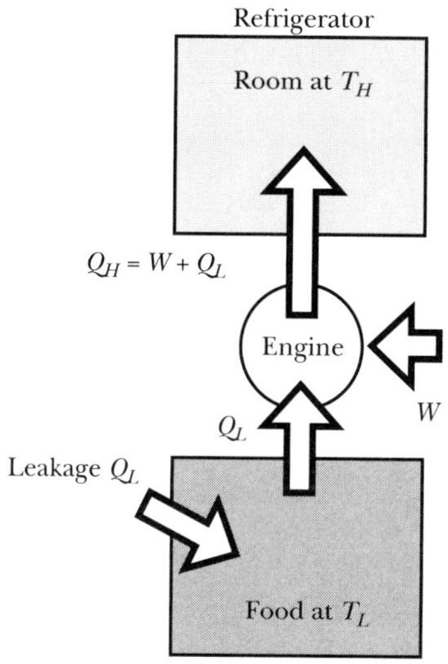

Figure 12.8 Run the engine backwards and you have a refrigerator.

? Show that the energy removal we get per amount of work done is as follows:

$$\frac{Q_L}{W} = \frac{1}{(T_H/T_L) - 1} > 1$$

Isn't this result somewhat surprising? Are we getting something for nothing when Q_L is bigger than the work W that we do? No, there's no violation of energy conservation. It's just that Q_L plus W equals the output energy Q_H, and we only have to add a modest amount of work to drive the energy "uphill."

Heat pump

A related device is the "heat pump" used in some areas to warm homes by driving low-temperature energy "uphill" into a higher-temperature house. Figure 12.9 shows a house whose interior is maintained at a temperature T_H despite continual leakage Q_H into the outside air. Energy Q_L at low temperature T_L is pumped out of the ground into the house, by the addition of some work W that we do in an engine called a heat pump. This is somewhat similar to the refrigerator situation, except that now what we care about is how much high-temperature energy Q_H we get per amount of work W done by the heat pump.

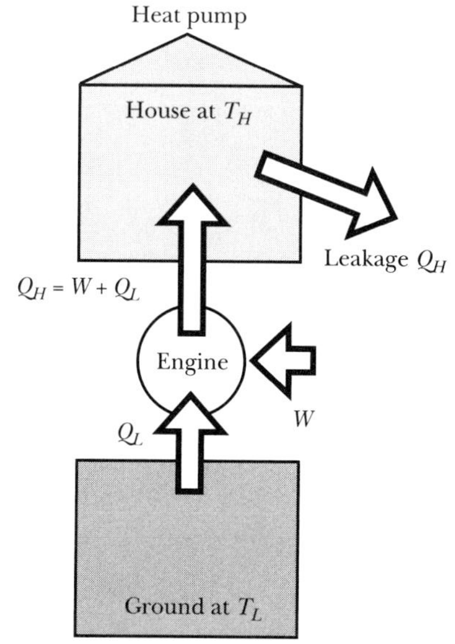

Figure 12.9 Run the engine backwards and you have a heat pump.

? Show that this ratio is given by the following formula for the heat pump:

$$\frac{Q_H}{W} = \frac{1}{1 - (T_L/T_H)} > 1$$

Again, this feels like we're getting something for nothing, since Q_H is greater than the work W that we do. After all, if you warm the house directly with gas or oil or electric energy, you have to pay for every joule of leakage, not some small fraction of the leakage. So why aren't heat pumps much more commonly used than they are? Basically because we've been calculating the best deal we can possibly obtain (the case of the reversible engine—that is, one whose operation leads to no increase in the total entropy of the Universe). Real engines running either forward or reversed don't attain the theoretical maximum performance, as we will see in the next sections.

Ex. 12.5 Suppose there is a leakage rate of 50 watts through the insulation into a refrigerator, which we maintain at 3° C. What is the minimum electric power required to continually remove this leakage energy, to maintain the low temperature? In that case, what is the rate at which thermal energy is exhausted into the room? Room temperature is about 20° C.

Ex. 12.6 Suppose the temperature underground from where we draw low-temperature energy for a heat pump is about 5° C, and we want to keep the house at a temperature of 20° C. How many joules of work must our heat pump supply for every joule of leakage of energy there is out of the house?

12.3 *Why don't we attain the theoretical efficiency?

In Ex. 12.3 on page 477 you calculated the reversible-engine efficiency of an actual nuclear-powered electricity generating plant to be 48%, but the observed efficiency of this plant is only 30%. Real engines do not attain the efficiency predicted by the second law of thermodynamics for ideal reversible engines. In fact, the real efficiency is often only about half of the theoretical efficiency. The reason for less than optimum performance in real engines is partly due to mechanical friction, but this effect can be minimized by good design and proper lubrication. The main limitation on performance comes from the practical necessity of incorporating thermal energy transfers between parts of the system that are at significantly different temperatures, which leads to sizable increases in the total entropy of the Universe, and to much reduced efficiency.

The problem is speed. As we saw earlier in the chapter, the rate of thermal energy transfer $\dot{Q}$ in joules per second between two objects (such as the hot or cold block and the engine) at temperatures T_H and T_L connected by some conducting material of length L and cross-sectional area A, with thermal conductivity σ is this:

$$\dot{Q} = \sigma A \frac{(T_H - T_L)}{L}$$

? So what is the rate of thermal energy transfer in a reversible engine when in contact with the hot block or the cold block?

Alas, a reversible engine is totally impractical when it comes to getting anything done in a finite amount of time, because the rate of thermal energy transfer is zero. If an engine is perfectly reversible, a cycle takes an infinite amount of time!

Consider the portion of an engine's cycle where the working substance is in thermal contact with the high-temperature source. In order to carry out the expansion quickly, there must be a high rate of thermal energy transfer from the source into the working substance. That means we need a large contact area A, a short distance L, and a high thermal conductivity σ. Most significant of all, the temperature of the source (T_H) must be considerably higher than the temperature of the engine, leading to irreversibility and increase of the entropy of the Universe.

The efficiency of a nonreversible engine

It is actually possible to calculate the effect of such nonreversible thermal energy transfer, as was first pointed out in an article titled "Efficiency of a

Carnot engine at maximum power output," by F. L. Curzon and B. Ahlborn, *American Journal of Physics* volume **43**, pages 22-24 (January 1975).

Suppose that when the engine is in contact with the high-temperature block at temperature T_H, the engine is at a lower temperature T_1, so that there is a finite thermal energy transfer rate into the engine of $\dot{Q}_H = b(T_H - T_1)$ joules per second, where b is a constant that lumps together the factors of thermal conductivity, cross-sectional area, and length $(\sigma A/L)$. Similarly, assume that when the engine is in contact with the low-temperature block at temperature T_L, the engine is at a higher temperature T_2, so that there is a finite thermal energy transfer rate into the working substance of $\dot{Q}_L = b(T_2 - T_L)$ joules per second. We're making the factor b be the same for both the high-temperature and the low-temperature contacts. This simplifies the algebra, and it turns out that using different b factors doesn't affect the final result anyway.

Now the energy flow diagram looks like Figure 12.10, with an ideal reversible engine running between the temperatures T_1 and T_2, but with irreversible finite-rate thermal energy transfer between this reversible engine and the source and sink at temperatures T_H and T_L.

Energy enters the engine from the high-temperature source at a rate $\dot{Q}_H$, work is done at a rate $\dot{W}$, and energy is dumped into the low-temperature sink at a rate $\dot{Q}_L$. All of these quantities are measured in watts (joules per second). From energy conservation we have

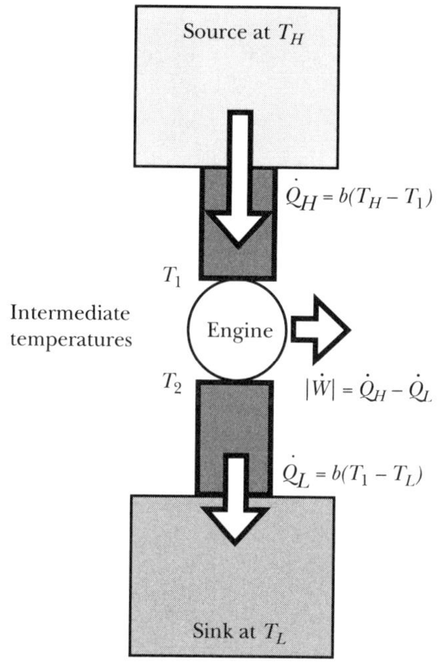

Figure 12.10 An engine that runs at a nonzero rate.

$$\dot{Q}_H = \dot{W} + \dot{Q}_L$$

Curzon and Ahlborn pointed out that although this engine will necessarily be less efficient than a reversible engine (due to the irreversible thermal energy transfers), it might be useful to ask the question, "What is the *maximum* output power $\dot{W}$ per input power $\dot{Q}_H$?" This question comes down to the question of what value of b will maximize the output power. This is essentially an engineering design question, yet there is an unusually simple answer to this question about the maximum output power.

There must be a maximum power output

By considering two extreme designs (Figure 12.11) it is easy to see that there must be a maximum possible power output (per power input). One extreme set of operating conditions is the reversible engine we've already considered, in which all the energy flow rates are zero, including the output power $\dot{W}$.

Another extreme is to arrange conditions in such a way that $T_1 \approx T_2$, so that we have essentially connected the source and sink by a conducting bar. In that case the input power $\dot{Q}_H$ is equal to $\dot{Q}_L$, and the useful output power $\dot{W}$ is again zero.

Somewhere between these two extremes, for both of which there is zero output power, we expect to find a maximum output power. We now look for the conditions that maximize the output power. The search will be carried out by expressing the output power in terms of T_1, the higher of the two temperatures to which the engine is directly exposed in Figure 12.10. Then we will vary T_1 (by adjusting b in Figure 12.10) until we maximize the power output. The details are given in a derivation at the end of the chapter. We obtain the following surprisingly simple result:

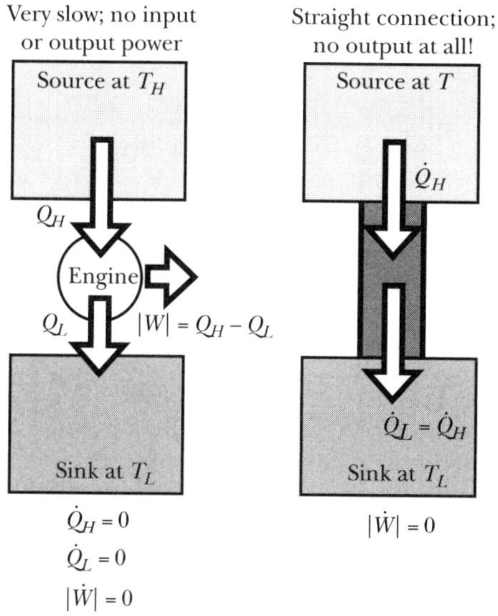

Figure 12.11 Two extreme designs where there is zero power output.

THE EFFICIENCY OF A MAXIMUM-POWER ENGINE

$$\frac{\dot{W}}{\dot{Q}_H} \le 1 - \sqrt{\frac{T_L}{T_H}}$$

This gives us the upper bound on the fraction of the thermal input power that can end up as output mechanical power if we maximize the output power. The formula looks quite similar to the reversible-engine zero-power efficiency, but the square root makes a big difference.

Comparison with the real world

Curzon and Ahlborn give an example of a plant powered by a nuclear reactor, with high temperature 300° C and low temperature 25° C. Using the formula given above, the calculated efficiency is 28%, and the actual observed efficiency is 30% (see exercise at the end of this section). This confirms the estimate that Curzon and Ahlborn made that

- the main source of irreversibility is in finite-temperature-difference thermal energy transfer, and
- power plants will normally be run in such a way as to maximize output power.

There is an economic issue here. Suppose that you are responsible for building and operating an electricity generating plant, and you must supply one megawatt of power (one million joules per second). If the operating expense of buying fuel is extremely high but capital investment in generators is inexpensive, it makes sense to build lots of generators and run them very slowly, nearly reversibly, to make the required megawatt of electric power. If the operating expense of buying fuel is relatively low but capital investment in generators is expensive, it makes sense to build few generators and go for maximum power, even though the fuel is not used very efficiently in producing the megawatt of electric power.

Similar issues apply to refrigerators and heat pumps. Thermal energy leaks into a refrigerator at some rate (joules per second) and must be removed by "pushing it thermally uphill" to a higher temperature, the temperature of the room. But in order to have a nonzero rate of thermal energy flow from the food to the engine, the engine must reach an even lower temperature than the food, with entropy-producing thermal energy transfer from the food into the engine. Also, in order for there to be a nonzero rate of thermal energy flow from the engine to the room air, the engine must reach an even higher temperature than the air, with entropy-producing thermal energy transfer from the engine into the air. Schematically the situation looks like Figure 12.12.

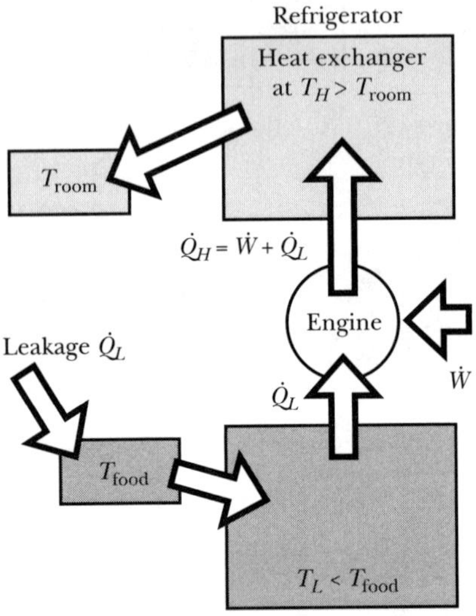

Figure 12.12 A refrigerator with nonzero cooling rate.

Often a refrigerator or freezer has exposed coils (a "heat exchanger") where the thermal energy transfer occurs between the working substance circulating in the coils and the air. If you touch the coils, you find that they are indeed much hotter than room temperature, in order to drive a sufficiently high rate of thermal energy transfer into the air. Notice not only that the heat exchanger must be hotter than the air, but that it is a rather large and costly device because it has to have a large surface area to get a large rate of energy transfer.

In the case of heat pumps, where the heat pump picks up low-temperature energy underground, the heat pump must reach an even lower temperature in order to get a nonzero rate of thermal energy flow from the ground into the engine. Also, in order for there to be a nonzero rate of thermal energy flow into the house, the engine must reach an even higher temperature than the air, with entropy-producing thermal energy transfer from the engine into the air. The radiators (heat exchangers) in the house must be considerably hotter than the air. There is considerable expense in all the metal in the ground and in the house that enables adequate thermal energy transfer rate. Schematically the situation looks like Figure 12.13.

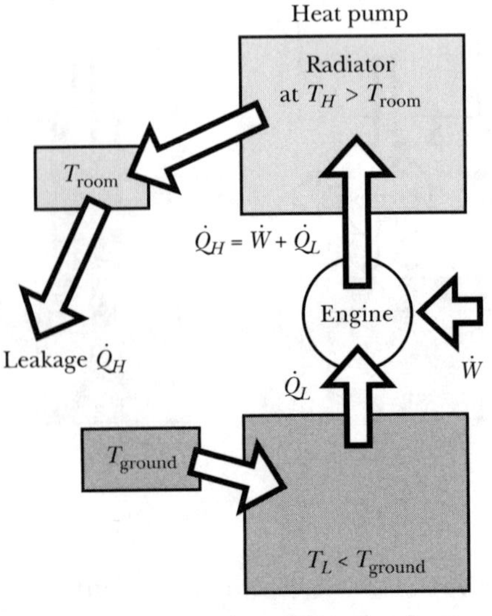

Figure 12.13 A heat pump with nonzero thermal energy transfer rate.

So although a heat pump does have a theoretical advantage in warming a house in part from energy in the cold ground, the expense of the heat exchangers and the problems of going to rather low temperatures in the ground are practical limitations, especially in very cold climates. Heat pumps are more useful in climates where the winters are not too severe.

Ex. 12.7 Curzon and Ahlborn give an example of a plant powered by a nuclear reactor, with high temperature 300° C and low temperature 25° C. In Ex. 12.3 on page 477 you calculated the reversible-engine zero-power efficiency to be 48%. Now calculate the non-reversible-engine maximum-power efficiency for this power plant and compare with the observed efficiency, which is 30%.

12.4 *Derivation: Maximum-power efficiency

In this appendix we find the conditions under which the power output is maximized for an engine that connects to high- and low-temperature reservoirs through finite-temperature-difference connections.

Expressing the power in terms of T_1

We will express the power output in terms of T_1, the higher temperature to which the reversible engine is subjected. We assume that the engine, the part that runs between the temperatures T_1 and T_2, can be considered to be a reversible engine because there are no different-temperature thermal energy transfers within it, and we're assuming that we can make mechanical friction negligibly small.

? In terms of the temperatures T_1 and T_2, what does the second law of thermodynamics say about the relationship between $\dot{Q}_H$ and $\dot{Q}_L$?

Since we have a reversible engine, we know that $\dot{Q}_H/T_1 = \dot{Q}_L/T_2$. Using this relationship, we can write a formula for the thermal energy transfer into the low-temperature sink in two different ways:

$$\dot{Q}_L = b(T_2 - T_L)\text{, and also } \dot{Q}_L = \frac{T_2}{T_1}\dot{Q}_H = \frac{T_2}{T_1}b(T_H - T_1)$$

Therefore we have

$$b(T_2 - T_L) = \frac{T_2}{T_1}b(T_H - T_1)$$

With a bit of manipulation we can solve this equation for T_2 in terms of T_1:

$$T_2 = \frac{T_L T_1}{2T_1 - T_H}$$

? For the inner, reversible engine, the input power is of course $\dot{Q}_H = b(T_H - T_1)$, and using the result for the efficiency of a reversible engine, show that

$$\dot{W} = b(T_H - T_1)\left(1 - \frac{T_2}{T_1}\right)$$

Using our solution for T_2 in terms of T_1, this becomes

$$\dot{W} = b(T_H - T_1)\left(1 - \frac{T_L}{2T_1 - T_H}\right)$$

Varying the thermal conduction to maximize the power

We consider the source and sink temperatures T_H and T_L to be fixed, so the only temperature that is unspecified in this equation is T_1, which we will vary until we maximize the output power $\dot{W}$. Physically, this corresponds to varying the parameter b that determines the thermal energy transfer rate. As usual, we can maximize $\dot{W}$ by differentiating it with respect to T_1, then set this derivative equal to zero (corresponding to a maximum, where the slope is zero):

$$\frac{d\dot{W}}{dT_1} = b\left[-1 + \frac{T_L}{2T_1 - T_H} + (T_H - T_1)\frac{2T_L}{(2T_1 - T_H)^2}\right] = 0$$

Solving for T_1 and simplifying, we get a quadratic equation:

$$4T_1^2 - (4T_H)T_1 + (T_H^2 - T_L T_H) = 0$$

There are two solutions to the quadratic equation:

$$T_1 = \frac{T_H \pm \sqrt{T_L T_H}}{2}$$

Only the "+" sign makes physical sense in this situation, because if T_L were nearly as large as T_H, the solution with the "–" sign would make T_1 nearly 0 instead of lying between T_L and T_H.

We conclude that if $T_1 = (T_H + \sqrt{T_L T_H})/2$, the output power $\dot{W}$ will be maximized. Since $\dot{Q}_H = b(T_H - T_1)$, our value for T_1 determines the value of b (= $\sigma A/L$) that will make T_1 come out right to maximize the power output for a given $\dot{Q}_H$.

Finally we can plug our value for T_1 back into the expression for $\dot{W}$ to find out what the efficiency at maximum power turns out to be:

$$\frac{\dot{W}}{\dot{Q}_H} = \frac{\dot{W}}{b(T_H - T_1)} = 1 - \frac{T_L}{2T_1 - T_H} = 1 - \frac{T_L}{\sqrt{T_L T_H}}$$

Here is the final result, where we write "≤" because we have calculated an upper bound on the efficiency (having taken into account issues of thermal energy transfer rate but not mechanical friction):

THE EFFICIENCY OF A MAXIMUM-POWER ENGINE

$$\frac{\dot{W}}{\dot{Q}_H} \le 1 - \sqrt{\frac{T_L}{T_H}}$$

This gives us the fraction of the thermal input power that ends up as output mechanical power if we maximize the output power. The formula looks quite similar to the reversible-engine zero-power efficiency, but the square root makes a big difference.

12.5 Summary

Fundamental physical principles

No new physical principles.

New concepts

Efficiency of an engine.

Results

RATE OF THERMAL ENERGY TRANSFER

$$\dot{Q} = \sigma A \frac{(T_H - T_L)}{L} \quad \text{(J/s)}$$

For a cyclic engine running between high temperature T_H and low temperature T_L we have

ENTROPY CHANGE OF THE UNIVERSE FOR REAL ENGINES

$$\Delta S_{\text{Universe}} = -\frac{Q_H}{T_H} + \frac{Q_L}{T_L} \geq 0$$

THE EFFICIENCY OF A REAL ENGINE

$$\text{Efficiency} = \frac{W}{Q_H} \leq 1 - \frac{T_L}{T_H}$$

The "=" sign applies only if the engine is reversible (extremely slow processes, no friction). A heat engine run in reverse is a refrigerator (or a heat pump).

If we maximize power output, we have

THE EFFICIENCY OF A MAXIMUM-POWER ENGINE

$$\frac{\dot{W}}{\dot{Q}_H} \leq 1 - \sqrt{\frac{T_L}{T_H}}$$

12.6 Review questions

Thermal conduction

RQ 12.1 An aluminum bar 30 cm long and 3 cm by 4 cm on its sides is connected between two large metal blocks at temperatures of 135° C and 20° C, and energy is transferred from the hotter block to the cooler block at a rate $\dot{Q}_1$. If instead the two blocks were connected by an aluminum bar that is 20 cm long and 3 cm by 2 cm on its sides, what would be the thermal energy transfer rate?

Engines

RQ 12.2 A (nearly) reversible engine is used to melt ice as well as do some useful work. If the engine does 1000 joules of work and dumps 400 joules into the ice, what is the temperature of the high-temperature source?

RQ 12.3 For engines where the high and low temperatures are not very different from each other, show that the efficiency of a maximum-power engine is about half of the efficiency of a reversible engine.

12.7 Homework problems

Problem 12.1 An engine containing an ideal gas
In one cycle of a reversible engine running between a high-temperature source and a low-temperature sink, a consequence of the second law of thermodynamics is that $Q_H/T_H = Q_L/T_L$. This result is independent of what kind of material the engine contains. It is instructive to check this general result for a specific model where we can calculate everything explicitly. Consider a reversible cycle of an ideal gas of N molecules, starting at high temperature T_H and volume V_1.

(1) Perform a constant-temperature (isothermal) expansion to volume V_2, and calculate the associated thermal energy transfer Q_H into the gas.

(2) Next perform a no-Q (adiabatic) expansion to volume V_3 and temperature T_L.

(3) Next perform a constant-temperature (isothermal) compression to volume V_4, and calculate the associated thermal energy transfer Q_L out of the gas.

(4) Finally perform a no-Q (adiabatic) compression back to the starting state, temperature T_H and volume V_1.

Show that $Q_H/T_H = Q_L/T_L$ for this cycle. (Hint: Determine the ratio of the final volume to the initial volume for each of the four processes.)

12.8 Answers to exercises

12.1 (page 472) 25 seconds; entropy of Universe increases

12.2 (page 477) 0.27

12.3 (page 477) 0.48

12.4 (page 478) $\Delta S_{\text{Universe}} > 0$ implies $\frac{Q_L}{T_L} > \frac{Q_H}{T_H}$, which implies

$\frac{Q_L}{Q_H} > \frac{T_L}{T_H}$, so

$$W = Q_H - Q_L = Q_H\left(1 - \frac{Q_L}{Q_H}\right) < Q_H\left(1 - \frac{T_L}{T_H}\right)$$

12.5 (page 480) 3 watts; 53 watts

12.6 (page 480) 0.05 J

12.7 (page 483) 28%, which is close to but less than the observed 30%. This implies that the power plant is not run at maximum power.

Appendix A

Basic Derivatives

A.1 Derivatives

In case you have not already studied calculus, or it has been a while since you last used calculus, we offer a calculation of some simple derivatives that are particularly useful in physics. You are probably familiar with the rate of change of velocity as the ratio of the velocity change $\Delta\vec{v}$ to the time interval Δt, $\Delta\vec{v}/\Delta t$. If we let the time interval Δt get very small, this ratio approaches what we call the "derivative" of the velocity with respect to the time:

$$\text{As } \Delta t \rightarrow 0,\ \frac{\Delta\vec{v}}{\Delta t} \rightarrow \frac{d\vec{v}}{dt}$$

More generally, we can approximate the rate of change of any physical quantity Q, as some other physical quantity R is varied, by the ratio $\Delta Q/\Delta R$, and if we consider a very small change ΔR, we approach the derivative of Q with respect to R:

$$\frac{\Delta Q}{\Delta R} \rightarrow \frac{dQ}{dR}$$

For concreteness, we'll take derivatives with respect to the time t, but the same principle applies if you take a derivative with respect to some other varying quantity, such as position x or y.

Derivative of a constant with respect to t

$$\frac{d(\text{constant})}{dt} = 0 \text{ because the change } \Delta(\text{constant}) = 0.$$

Derivative of t with respect to t

$$\frac{\Delta(t)}{\Delta t} = 1, \text{ no matter whether } \Delta t \text{ is large or small, so } \frac{d(t)}{dt} = 1$$

Derivative of t^2 with respect to t

The rate of change of t^2 with respect to t can be obtained starting from the fact that when t increases by an amount Δt, the quantity t^2 increases to $(t+\Delta t)^2$:

$$\begin{aligned}\frac{\Delta(t^2)}{\Delta t} &= \frac{(t+\Delta t)^2 - t^2}{\Delta t}\\ &= \frac{t^2 + 2t\Delta t + (\Delta t)^2 - t^2}{\Delta t}\\ &= \frac{2t\Delta t + (\Delta t)^2}{\Delta t}\\ &= 2t + \Delta t\end{aligned}$$

This is nearly equal to $2t$ if Δt is very small compared to $2t$. Therefore we conclude this:

$$\frac{d(t^2)}{dt} = 2t$$

In a similar fashion it is possible to show the following:

$$\frac{d(t^n)}{dt} = nt^{n-1} \text{ for any constant value of } n \text{ (positive } or \text{ negative)}$$

Derivative with a constant multiplier

The derivative of a quantity that contains a constant multiplier C is simply C times the basic derivative:

$$\frac{\Delta(CQ)}{\Delta t} = \frac{C\Delta Q}{\Delta t} = C\frac{\Delta Q}{\Delta t}$$

$$\text{so } \frac{d(CQ)}{dt} = C\frac{dQ}{dt}$$

Derivative of sine and cosine

We can calculate the rate of change (the derivative) of the sine and cosine functions by using the trigonometric identities for the sines and cosines of sums of angles:

$$\sin(A+B) = \sin A \cos B + \cos A \sin B$$

$$\cos(A+B) = \cos A \cos B - \sin A \sin B$$

Here is the derivative of the sine:

$$\frac{\Delta(\sin t)}{\Delta t} = \frac{\sin(t+\Delta t) - \sin t}{\Delta t}$$

$$\frac{\Delta(\sin t)}{\Delta t} = \frac{\sin t \cos \Delta t + \cos t \sin \Delta t - \sin t}{\Delta t}$$

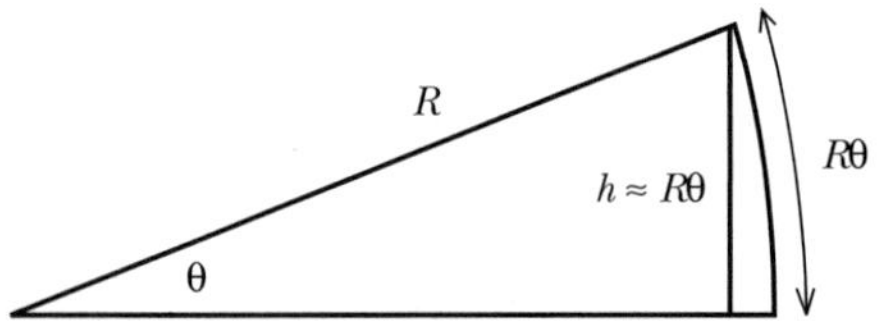

Figure B1.1 The sine of a small angle is approximately equal to the angle, measured in radians.

The cosine of a very small angle is approximately equal to 1, and the sine of a very small angle is approximately equal to the angle, measured in radians. You can see from Figure B1.1 that if the angle θ is small, the height h is approximately equal to the arc length $R\theta$ (with θ measured in radians), in which case $\sin(\theta) = h/R \approx \theta$. As an example, $30° = \pi/6$ radians $= 0.524$ radians, which is very close to $\sin(30°) = 0.5$. The approximation gets even better for smaller angles.

Therefore if Δt is very small, we have approximately

$$\frac{\Delta(\sin t)}{\Delta t} \approx \frac{\sin t + (\cos t)\Delta t - \sin t}{\Delta t} = \cos t$$

Therefore the derivative of the sine is a cosine. You can go through the same kind of reasoning to find that the derivative of the cosine is a negative sine. In summary:

$$\frac{d(\sin t)}{dt} = \cos t$$

$$\frac{d(\cos t)}{dt} = -\sin t$$

Do these derivatives make sense? Note that at $t = 0$, the function $\sin(t)$ is growing, and its derivative $\cos(0) = +1$, which makes sense. At $t = \pi/2$ (90°), the sine function reaches a maximum and is momentarily neither increasing nor decreasing, and its derivative $\cos(\pi/2) = 0$, which makes sense.

The chain rule

Often we need to calculate the rate of change of a function of some variable with respect to some *other* variable, such as differentiating a function of x with respect to t:

$$\frac{d[f(x)]}{dt}$$

In this example, suppose x changes by an amount Δx when t changes by an amount Δt. Then we have the following:

$$\frac{\Delta[f(x)]}{\Delta t} = \left\{\frac{\Delta[f(x)]}{\Delta x}\right\}\left\{\frac{\Delta x}{\Delta t}\right\}$$

$$\text{But } \frac{\Delta[f(x)]}{\Delta x} \to \frac{d[f(x)]}{dx}$$

$$\text{and } \frac{\Delta x}{\Delta t} \to \frac{dx}{dt}$$

As we let Δt get very small, we obtain the chain rule:

$$\frac{d[f(x)]}{dt} = \left\{\frac{d[f(x)]}{dx}\right\}\left\{\frac{dx}{dt}\right\}$$

Here is an example of the use of the chain rule in physics (ω is a constant):

$$\frac{d[\cos(\omega t)]}{dt} = \left\{\frac{d[\cos(\omega t)]}{d(\omega t)}\right\}\left\{\frac{d(\omega t)}{dt}\right\} = \{-\sin(\omega t)\}\{\omega\}$$

A.2 Self-test

Here are some short questions on derivatives that you can use as a self-test. Try not to look back through this appendix as you work on these questions. If you find that you do need to look up something in order to answer a question, or if you find that your answer disagrees with the answers given at the end of this appendix, be sure to make a note of which question this was, to guide your further study and review.

Ex. A.1 What is the derivative of $4t^2$ with respect to t?

Ex. A.2 What is the rate of change of $(5x^{-1} + 6)$ as x varies?

Ex. A.3 Consider the behavior of the derivative of $\cos(t)$ at $t = 0$ and $t = \pi/2$. Do you get values that make sense, given how the cosine is changing near those times?

Ex. A.4 Calculate $\dfrac{d[\sin(3t)]}{dt}$.

A.3 Answers to self-test

A.1 (page 3) $8t$

A.2 (page 3) $-5x^{-2}$

A.3 (page 3) At $t = 0$, the cosine is at a maximum, so its derivative should be zero; indeed, $-\sin(0) = 0$.

At $t = \pi/2$, the cosine is decreasing, so its rate of change should be negative; indeed, $-\sin(\pi/2) = -1$.

A.4 (page 3) $3\cos(3t)$

Greek alphabet, and its uses in this course

Alpha	A		α **alpha particle (helium-4 nucleus)**
Beta	B		β
Gamma	Γ		γ **relativistic factor, photon**
Delta	Δ	**change of; small quantity of**	δ
Epsilon	E		ε **a small quantity**
Zeta	Z		ζ
Eta	H		η
Theta	Θ		θ **angle**
Iota	I		ι
Kappa	K		κ
Lambda	Λ		λ **wavelength**
Mu	M	**mega** (10^6)	μ **micro** (10^{-6})**, muon, coefficient of friction**
Nu	N		ν **neutrino; frequency** ($=f$)
Xi	Ξ		ξ
Omicron	O		ο
Pi	Π		π **circle: circumference/diameter**
Rho	P		ρ **density, resistivity**
Sigma	Σ	**sum**	σ **conductivity**
Tau	T		τ **torque**
Upsilon	Y		υ
Phi	Φ	**flux**	φ **phase angle**
Chi	X		χ
Psi	Ψ		ψ
Omega	Ω	**# of microstates of a system, ohm, precession angular frequency**	ω **angular frequency, angular speed**

Units of measurement

kg = kilogram	m = meter	s = second	C = coulomb
N = newton (kg·m/s^2)	J = joule (N·m)	W = watt (J/s)	
A = ampere (C/s)	V = volt (J/C)	T = tesla (N/C)/(m/s)	eV = electron volt = 1.6×10^{-19} J

Important prefixes

pico (p) = 10^{-12}	nano (n) = 10^{-9}	micro (m) = 10^{-6}	milli (m) = 10^{-3}
centi (c) = 10^{-2}	kilo (k) = 10^3	mega (M) = 10^6	giga (G) = 10^9

Unit conversions

1 kg = 2.2 pounds	1 inch = 2.54 cm	1 foot = 30.5 cm	1 liter = 1.1 quart
1 cal = 4.2 J	1 kcal (1 food calorie, 1 "large calorie") = 4.2×10^3 J		$T_{Kelvin} = T_{Celsius} + 273$